CRYSTALLOGRAPHY IN NORTH AMERICA

EDITED BY DAN McLACHLAN, JR. & JENNY P. GLUSKER

Co-Editor:	Hugo Steinfink
Managing Editor:	Sidney C. Abrahams
Associate Managing Editor:	Alexander Tulinsky

Published by the American Crystallographic Association

First published 1983 by American Crystallographic Association.

Library of Congress Catalog Card Number: 81-71539 ISBN 0-937140-07-4.

Printed in the United States of America.

PREFACE

Dan MacLahlan, Jr.

During the year of 1976 when the United States was celebrating its two hundredth anniversary, people on this side of the Atlantic became unusually concerned with history, and books as well as journal articles about great men and great events were more numerous than ever before. The members of scientific societies, such as the American Chemical Society, the American Physical Society and the American Optical Society became aware that their organizations were then over fifty years old. Consequently, *Physics Today, Chemical and Engineering News* and the *Journal of the Optical Society of America* abounded in papers covering the origins of the Schroedinger equation in 1927, discovery of X-ray diffraction in 1912, relativity in 1905, electron diffraction in 1927 and the lives of men like Einstein, Debye, Compton, etc.

The same time, the crystallogrphers of North America began discussing at their semi-annual meetings the fact that the only material pertaining to the history of crystallography systematically collected since the publication in 1962 of Professor P. P. Ewald's *"Fifty Years of X-ray Diffraction,"* was the two-volume compilation on *"Early Papers on Diffraction of X-rays by Crystals"* edited by J. M. Bijvoet, W. G. Burgers and G. Hägg. The first official trend toward preserving the history of crystallographers was contained in an announcement in the *ACA Newsletter* of October 1977 by Professor Deane K. Smith entitled "Historical Material for ACA Records." This letter starts out by saying "Council is establishing a position of Historian for ACA." He goes on to say "Until a Historian has been named, the secretary will act as a collection center for items of interest." Evidently no one volunteered for a couple of years.

On May 15, 1978 Deane Smith wrote a circular letter to Brockway, Buerger, Burbank, Donnay, Gruner, Harker, Huggins, Jeffrey, McLachlan, Pepinsky and E. A. Wood asking for any documentary material on the CSA, ASXRED and ACA, especially pictures of people on special occasions. He went on to say "ACA Council has been quite concerned that the society may lose track of its early history." I responded with a couple dozen pictures taken at the 1946 meeting in London and at the Harvard Meeting of 1948. Others of us no doubt did likewise.

After several phone conversations in mid-1979 with Deane Smith and Jenny Glusker (The 1979 president of ACA) McLachlan agreed to be Editor of a history book for American Crystallographers. On October 19, 1979 Jenny wrote a letter making his Editorship official. It was agreed early in the project to have the book divided into two parts: (1) historical discourses by "Expert Authors" and (2) reports of the past-presidents of ACA on the events of importance during their tenure in office. This was announced in an *ACA Newsletter* for December 1979. An editorial committee was then set up.

Our greatest fear was that we might cause some bad feelings by failing to ask a worthy person to contribute accounts of the many really important subjects in crystallography to which they had made noteworthy contributions. I had written to and telephoned over two hundred potential authors. We also had notices inviting contributed chapters from anyone interested. These notices were published in the *ACA Newsletters* of February 1980, April 1980 and June 1980 as well as others since.

At a meeting in Calgary, Canada on August 19, 1980 the editorial committee for the book decided that the editor Dan McLachlan, Jr. should devote all his time to collecting chapters of the book from authors; and the responsibility of carrying out the publication procedures with the printers should be placed in the hands of the ACA publication manager Sidney Abrahams, assisted by Alexander Tulinsky. Also Jenny Glusker agreed to assist in the distribution of copies of manuscripts to the other co-editors.

A second meeting of the editorial committee was held at Texas A & M University, March 27, 1981 attended by the permanent members, Dan McLachlan (as chairman), Jenny Glusker, Hugo Steinfink, the ACA President Quintin Johnson, Charles Prewitt, W. O. Milligan and others. At this meeting the outline of the proposed book was approved. Rather than be redundant, the outline will not be given here in the preface; the reader is referred to the Table of Contents in the front of this volume. Sidney Abrahams and Al Tulinsky reported great progress with plans for the publication of the book. Jenny Glusker suggested that all final manuscripts be sent to her office where she would arrange to have them prepared in final form for the typesetters (through Abrahams and Tulinsky) and photocopies made for the authors. Jenny Glusker also offered at this meeting to handle all manuscripts in preparation for the typesetters. McLachlan realized Jenny's future burden was greater than his; so in order to give Jenny due credit for her contributions, he suggested that it should be a joint editorship between McLachlan and Glusker. The managing editor, Sidney Abrahams, was enthusiastic about the idea and its endorsement by Al Tulinsky and Hugo Steinfink followed.

Some extra problems were encountered in our desire to have the book contain portraits of all past-presidents, in addition to interesting group photographs and informal snapshots of crystallographers at national and international meetings. Sidney Abrahams solved these problems and we think we now have a fine collection of photographs that contribute greatly to the value of this book.

By this time it had been agreed that the book should be typeset by Aldinger's of East Lansing and printed by Book Crafters of Chelsea, Michigan which is located near enough to Michigan State University that Al Tulinsky could conveniently contact them in behalf of Sidney Abrahams and Jenny Glusker.

We wish to acknowledge the assistance of many people to whom we are greatly indebted. Not only do we owe thanks to all the authors who have worked so hard to produce such fine chapters but also those who have helped put the manuscripts in a form suitable for typesetting. These include, particularly, Mrs. Eileen Pytko of The Institute for Cancer Research, and also Mr. Leo Zwell of JCPDS, Swarthmore, PA. Ms. Sally Orr and Ms. Carolyn Gribbin of the Geology and Mineralogy Department of The Ohio State University, Mrs. Hilda Rosenberg, Ms. Kathy Glusker, Mr. Mark H. Carrell, and NIH grant CA-10925. We are grateful to ACA for financial support of McLachlan for the handling of manuscripts, to the Geology and Mineralogy Department of The Ohio State University for secretarial help, stationary and postage. We are especially indebted to The Institute for Cancer Research for the typing and photocopying of manuscripts to be sent to the co-editors, authors and the publication team.

We hope that this book will be a useful resource to many young crystallographers beginning their careers, that old crystallographers will read with pleasant nostalgia the happenings of the "old days," and that it will provide valuable insight to all who are interested in the history of science.

Columbus, Ohio
November 1982

DAN McLACHLAN
December 5, 1905 – December 3, 1982

Dan McLachlan, Jr., editor of this volume, died suddenly on December 3, 1982 during the final stages of the preparation of this book. He had worked with great dedication for the last three years on the history of "Crystallograhy in North America," and saw it in its final stages of proof correction. It is hoped that readers will appreciate the results of his labors.

Dan was born at Arcola, Saskatchewan, Canada on Dcember 5, 1905 of Scottish ancestry. He obtained a B.S. at Kansas State College in 1930 and a M.S. in Physical Chemistry at the Pennsylvania State College in 1933. While at Penn State he also worked at the Sinclair Refining Company. From 1933-1936 he was a graduate student at Penn State, obtaining a Ph.D. in the laboratory of Wheeler P. Davey. Then he moved to Corning Glass Works where he remained from 1936-1941. Following employment as a physicist at the American Cyanamid Company from 1941-1947, he was appointed Professor of Metallurgy, Mineralogy and Physics at the University of Utah, 1947-1953, after which he worked at the Stanford Research Institute from 1953-1961. In 1961 he moved to the University of Denver and was a Coordinator of Physics and then Professor of Metallurgy. From 1963 until 1982 he was at Ohio State University, first as a Battelle Professor, and from 1964 on as Professor of Mineralogy. He held this latter post in Emeritus status at the time of his death.

Dan was the ninth President of the American Crystallographic Association in 1958, a member of the U.S. National Committee for Crystallography and a delegate to the meetings and Congresses in London (1946), Harvard (1948), Stockholm (1951), and Montreal (1957). As a member of the U.S. delegation to the meeting in London in 1946 he took a significant part in the initiation of the International Union of Crystallography and Acta Crystallographica. As a scientist he made many contributions in a wide variety of areas of crystallography particularly in the interpretation of X-ray photographs, the rapid calculation and representation of electron density maps, the solution of the phase problem from the point of view of the Patterson function, the extension of the Donnay-Harker Law and representations of the results of crystal structure analyses. He was the author of numerous publications including two books on crystallography (*"X-ray Crystal Structure"*, published by McGraw-Hill in 1957, and *"Statistical Mechanical Analogies"* published in 1968), ten patents and a book of humor (*"Your Dog Died"* by Dok McMud). He was a member of many scientific organizations and also the Honorary Order of Kentucky Colonels. His family consisted of his wife Rachel, three sons (Edwin D. of Pullman, Washington, Dan H. of Olympia, Washington, and Wayne C. of Grass Valley, California) and four grandchildren.

There are several articles in this volume that are authored by Dan and in them his wit and modesty are evident. His talks at ACA meetings, while full of good scientific information, were also enjoyable and memorable. Dan, we will all miss you.

Jenny P. Glusker

DAN McLACHLAN, JR.
December 5, 1905 – December 3, 1982

I'm sitting in my father's office in my father's chair. About me are the shelves of books and the drawers of letters and the litter of research in the form of odd bits of machinery, lenses, rocks, glass, and blackboards of equations that seem to be present in all the offices of his friends, but particularly here. This is the office of a scientist who lived seventy-seven years with an insatiable curiosity about the world he walked through.

My father placed truth above all things, and in that regard, he was dedicated to determining those truths that reveal themselves so reluctantly to the world of science. What might appear to an outsider as drudgery was to him and his scientific friends a game of wits – an exciting contest with the unknown based on clues hidden in the obvious or in the terribly obscure, theoretical recesses of the world.

This book, I feel, says something else about him that its contributors and those people who assisted in its writing and publication know already, and that is that he considered himself as only a single part of a larger body of dedicated men and women that worked best and accomplished the most when they worked together on what he termed "the big problems." Consequently, he encouraged people to publish their ideas and to share them with the scientific community, and he would do everything in his power to assist them in their research and writing to make that possible. It distressed him when he saw problems that could be solved for the benefit of all mankind go unsolved because of personal, national or international jealousies.

Another thing that comes to mind as I sit here is where are the hundreds of oil paintings he did in his lifetime – they're nowhere about? The answer he would offer, of course, is that he gave them to his friends to punish them, for though he loved to paint, he really never felt he did it very well and enjoyed having his friends agree with him on at least one point.

And friends he had. His letters from them are here and in the archives of the Center for History of Physics in New York, and in The Ohio State University Archives. And if a man truly can be judged by the company he keeps, one can say of both him and his friends and colleagues that kindness, humor, mutual respect, caring, dedication, and dignity are virtues they valued.

My father came from the soil of a poor midwestern farm, and has now been returned there. In the intervening years he wanted to "contribute someting," and I believe that most of us that knew and loved him would agree that he contributed much more than he would ever admit to, but much less than he would have liked to.

his son,
Dan H. McLachlan
December 6, 1982

FOREWORD

Linus Pauling
Linus Pauling Institute of Science and Medicine
440 Page Mill Road, Palo Alto, California 94306

Crystallography and structural chemistry have made tremendous progress during the last seventy years, since the discovery of the phenomenon of the diffraction of X-rays by crystals. Our understanding of the nature of the physical world and of the biological world is now tremendously greater than it was seventy years ago. I feel fortunate to have lived through this period of astonishing discoveries.

When the first crystal-structure determinations were made, in 1913, I was a first-year student in high school, and in 1914 I learned about chemistry and decided to be a chemist (actually, to be a chemical engineer). At some time during the following years I learned about X-ray crystallography, and in the summer of 1922, before going to Pasadena to become a graduate student in chemistry, I read the book *"X-rays and Crystal Structure,"* by W. H. and W. L. Bragg. I can remember reading the book, but I do not remember having any special thoughts about it and about the field of the determination of crystals by the X-ray diffraction method, even though Professor Arthur A. Noyes had already suggested to me that I do my graduate work with Roscoe Dickinson in the X-ray diffraction field.

I remember very clearly, however, what the science of chemistry, especially structural chemistry, was like at that time, sixty years ago. It consisted of a multitude of facts, not very closely correlated with one another. In particular, inorganic qualitative analysis repelled me, because of its completely or nearly completely empirical character. Most of the methods of separation and detection of the different metals depended on the solubility of certain compounds, and there was essentially no theoretical basis for the difference in solubility of different substances. I disliked qualitative analysis, whereas the precision of quantitative analysis appealed to me.

I remember especially well the way in which my chemistry teachers and my textbooks discussed chemical bonding. Ionic bonding was presented for salts such as sodium chloride, and covalent bonding (although the word covalent had not yet been invented) for organic compounds. Sometimes the two concepts were not distinguished from one another. A molecule of sodium chloride was described as having a hook-and-eye bond. I put the eye on the sodium atom, and the hook on the chlorine atom. This permitted two chlorine atoms to form a chemical bond with one another by hooking their hooks together, whereas sodium atoms could not hook their eyes together (the existence of Na_2 in sodium vapor, in small concentrations, was not known to me at that time, and I do not remember wondering about the existence of the mercurous ion, Hg_2^{++}). So far as I was aware at that time, no chemists had any ideas about the nature of the forces that lead to the formation of stable aggregates of atoms, the forces involved in the formation of chemical bonds.

During the year 1919-1920 I read the early papers of Irving Langmuir on the electronic theory of valence, and went back to the 1916 issue of the *Journal of the American Chemical Society* to read G. N. Lewis's first paper on the electron-pair bond. I was much impressed by these papers. It seemed to me that they removed a good bit of the mystery from the chemical bond. To me, they made chemistry much more interesting. In those early years chemistry seminars were held only rarely, about one a year, in the Oregon Agricultural College. I spoke at the chemistry seminar that year, reporting on the electron theory of valence. In the fall of 1922, when I appeared in the California Institute of Technology, Dickinson immediately began instructing me in the Nishikawa-Wyckoff-Dickinson techniques of determining the structures of crystals from the X-ray photographs. The logic of his presentation of the procedure impressed me immensely, and I was also impressed by the power of the technique and the possibility that a tremendous number of questions in the field of chemistry and in other fields could be answered by its application. Dickinson and Raymond had just made the first precise determination of the structure of an organic compound, and Dickinson and Wyckoff, independently, had determined the structures of inorganic complexes of the Werner coordination type. Wyckoff, in his first structure determination, described in his doctoral thesis in Cornell, had shown that in cesium dichloroiodide the two chlorine atoms are quite close to the iodine atom, which accordingly seems to be bivalent, rather than univalent. A substantial body of information about interatomic distances in inorganic compounds had already been gathered, and W. L. Bragg at about this time published a paper in which he communicated the results of his efforts to set up a table of atomic radii that would reproduce the observed values of the interatomic distances in crystals. Thousands, even tens of thousands, of crystals presented themselves as candidates for X-ray examination, with the possibility of interesting new results to be obtained with any one of them. There was always the possibility that nature would provide a surprise to the investigator. The first crystal that, under the close supervision of Dickinson, I succeeded in analyzing was molybdenite, molybdenum disulfide, in which I expected the molybdenum atom to be surrounded by an octahedron of sulphur atoms. To my surprise the coordination polyhedron turned out to be different, a new one - the trigonal prism. Around ten years later, when I might have been getting bored with the usually expected results of a crystal-structure determination, I had another surprise. (I shall not mention the many intermediate ones.) My students and I had studied several sulfide minerals, and usually the structures were rather similar to those of simple sulfides, such as sphalerite, in which the zinc atom is surrounded by four sulfur atoms at the corners of a tetrahedron. For example, enargite, Cu_3AsS_4, has a structure closely related to that of wurtzite, the hexagonal form of zinc sulfide, with the three copper atoms and the arsenic atom all surrounded tetrahedrally by sulfur atoms. I expected a similar structure when I came into the possession of a few small crystals of the cubic mineral sulvanite, Cu_3VS_4. Ralph Hultgren prepared some X-ray photographs, and when we analyzed them we found that, as expected, the sulfur atoms in the unit cube are in essentially the same positions as in sphalerite and the copper atoms show tetrahedral coordination. The vanadium atoms, however, are not in the expected positions. Instead, each vanadium atom is surrounded by three sulfur atoms lying on the same side of it - three at three corners of a tetrahedron, and the fourth in the center of this group of three. This was an astonishing observation, and, even though I have discussed the reasons for this structure in several different papers, I am still somewhat puzzled by it.

As the number of X-ray crystallographers in many different countries increased and the number of crystals that had been analyzed became greater and greater, the information about the precise ways in which atoms interact with one another in crystals contributed more and more to the development of an extensive theory of structural chemistry. The technique of the determination of the structure of gas molecules by electron diffraction, originated by Hermann Mark in 1929, also led to the accumulation of a large amount of information about the structure of molecules.

The third development that was crucial to the progress of science was the discovery of quantum mechanics in 1925. The experimental data that led to the development of quantum mechanics were to a considerable part provided by spectroscopy, mainly atomic spectroscopy. Molecular spectroscopy began to be developed in the period around seventy years ago, but the band spectra could be interpreted only for quite simple molecules, and the technique of spectroscopy was far less important than the diffraction techniques in the development of structural chemistry.

Not only crystallography and structural chemistry, but also general inorganic chemistry, organic chemistry, molecular biology, solid-state physics, and the earth sciences benefitted greatly from application of X-ray crystallography to their problems. An example is provided by the work of Robert B. Corey and others in the California Institute of Technology on the structure of amino acids, peptides, and proteins. When they began their work, in 1937, the powerful methods of direct determination of complex crystal structures had not yet been developed. At that time there had not yet been correctly determined the structure of any amino acid, peptide, or protein. By making use of Patterson diagrams, E. W. Hughes' least-squares technique for refining structures, the Schomaker-Waser punched-card method of calculation, and other aids, Corey and his many collaborators succeeded during the next ten years in determining the structure of a dozen amino acids and several peptides, providing the structural information that permitted the discovery of the principal ways of folding of polypeptide chains in proteins. Information about many other advances in knowledge in various fields that were made possible through X-ray crystallography is provided in several chapters of this book.

I might ask what need there is for a history of X-ray crystallography in the United States. I think that the answer is that there has clearly been, over a long period of time, a difference between X-ray crystallography in North America and X-ray crystallography in most other countries, in particular in Britain and Germany. X-ray crystallography was discovered by physicists in Germany and England, and its development was for a long time carried out by physicists in departments of physics. In the United States the early work was done by chemists - by R. W. G. Wyckoff in Cornell and by C. Lalor Burdick and Roscoe G. Dickinson in the California Institute of Technology, in departments of chemistry. To many of the early X-ray crystallographers in the United States the motivation for carrying on the work was largely that of an endeavor to answer questions about chemistry rather than physics. Many X-ray crystallographers in North America have had a good background of training, experience, and knowledge in chemistry generally, which has not only influenced their selection of problems to be attacked but has also been of great help to them in the determination of the structures of the crystals, especially in the early days, when the power of the X-ray technique was still so limited that the help provided by some hint as to the nature of possible structures could be of great value in the structure determination. The result has been that X-ray crystallography in North America has developed a special flavor, which can be recognized in many of the accounts presented in the book.

I am sure that my fellow X-ray crystallographers join me in congratulating the officers of the American Crystallographic Association and especially the editor of this book, Dr. Dan McLachlan, Jr., on the success of their efforts.

TABLE OF CONTENTS

Page No.

Page No.

INTRODUCTION

AN OVERVIEW OF CRYSTALLOGRAPHY IN NORTH AMERICA

Clifford Frondel
Department of Geological Sciences, Harvard University
Cambridge, Massachusetts 02138

Introduction

This paper briefly reviews the development of crystallography in the United States from its beginnings before 1800 up to about World War II. It refers to chemical and physical crystallography as well as to the morphological aspects of the subject. Some connective material is given to the advent of X-ray crystallography in the 1920's and 1930's.

Among the colleges founded early in this country both crystallography and mineralogy have been taught continuously at Harvard since 1788, at Yale since 1807, at Amherst since 1832 and, with some interruptions, at Bowdoin since 1805, Rensselaer since 1824 and Columbia probably since 1792.

A brass contact goniometer purchased by Harvard University in London in 1797 may be the first crystallographic instrument used in the United States. This instrument, shown in Figure 1, is identical with that

Figure 1. Brass contact goniometer of the Carangeot type acquired by Harvard in 1797.

originally made by Arnoulf Carangeot in Paris in 1780 and supplied by him to the French crystallographers Rome Delisle and René J. Haüy. Delisle, first to make systematic measurements of crystal angles, illustrated Carangeot's goniometer and described its use in his *"Cristallographie"* of 1783.

The Harvard instrument was used by Benjamin Waterhouse for classroom instruction in mineralogy, then given in the Medical School. Waterhouse's handwritten lecture notes of 1809 and earlier, (still preserved) include reference to both Delisle's treatment of the primitive forms of crystals and to Haüy. With regard to Haüy's theory of crystal structure it ". . .seems in great measure founded on the measurement and the degree of the angles of each crystal, and may therefore be called a crystallonometrical system. It certainly is a very pretty French system, and so geometrically exact that it is enough to make a mathematician endure the name of mineralogy; but I am doubful if you can make every son of Euclid believe in this subterranean mathematics."

The existing histories of crystallography have reference almost exclusively to European work. They include the *"Geschichte der Kristallkunde"* of Carl M. Marx (1826), and the *"Geschichte der Mineralogie"* of Franz vonKobell (1864). Paul Groth's *"Entwicklungsgeschichte der mineralogischen Wissenchaften,"* published in 1926, and the review thereof by Arrien Johnsen (1927), include crystallographic developments up to about 1926 and give detailed accounts of work by Haüy and by Christian S. Weiss. Haüy's work is given an extended review in a series of papers published in the *American Mineralogist* in 1918, and some aspects of it are discussed by Hooeykas (1955). The book by John G. Burke, *"Origin of the Science of Crystals,"* published in 1966, outlines the major developments including their application to chemistry and optics. *"Das Geheimnis der Kristallwelt"* by H. Tertsch (1947) covers much the same ground.

Among shorter or more specialized contributions to the history of European crystallography may be mentioned articles by William Whewell (1833), A. Schoenflies (1905), Alfred Lacroix (1915), K. Mieleitner (1923), George Menzer (1945), Paul Niggli (1946a, b), Henri Longchambon (1954), P. Lacombe (1954), Paul Ramdohr (1959), A. J. Berry (1960), Jean Orcel (1938, 1962), K. C. Dunham (1977), and W. C. Smith (1978). Developments in classical Russian crystallography are outlined by S. I. Tomkeieff (1960) and S. H. Shafronovsky (1973). The development of crystallography and mineralogy and its scientific context in the United States around 1800 is described in publications by John C. Greene (1969), John C. Greene and John G. Burke (1978), E. H. Kraus (1921), I. Bernard Cohen (1950), Brooke Hindle (1956) and A. Oleson and S. C. Brown (1976). The early developments at Yale are described by E. S. Dana (1918), M. L. Jensen (1952), and Barbara L. Narendra (1979), and at Harvard by J. E. Graustein (1961). The history of the development of state and national geological surveys during the 19th century by Mary C. Rabbitt (1979, 1980) also is of value in this connection. The early history of chemistry in the United States, with which that of mineralogy is interwoven, as in Europe, is described by E. F. Smith (1914, 1919, 1926). The scientific environment in which X-ray crystallography developed in this country and abroad is seen in a work *"Fifty Years of X-Ray Diffraction"* edited by P. P. Ewald and published in 1962.

Crystallography relates to the solid state in general and its initial and long continued association with the particular field of mineralogy perhaps requires comment. Many minerals occur in well crystallized form. From the beginning days of modern mineralogy, in the latter decades of the 18th century, the study of crystal morphology and of other structure-dependent properties such as cleavage and optical properties has been an integral part of mineral science. The early pioneers of crystallography including Romé Delisle (1736-1790) and Réné J. Haüy (1743-1822) were mineralogists who specialized in crystal morphology and used minerals as the vehicle.

Inorganic chemistry in its developmental period around 1800 also was virtually synonymous with both mineralogy and crystallography. Here the early analytical work leading to the discovery of the laws of stoichiometry and to Dalton's atomic theory was based on crystals of minerals. These were visibly homogeneous, unlike the precipitates from reactive solutions, and also provided from their morphology and physical properties a means of characterizing chemical phases independently of chemical evidence. Much of the very early instruction in mineralogy in this country centered around the medical schools of Philadelphia, New York and Boston. Inorganic chemistry was taught therein as an aspect of materia medica, and mineralogy, then still closely allied to chemistry, was included. By the early 1800's both fields transferred to the colleges, growing both in number and their interest in science, with instruction in crystallography and the descriptive aspects of mineralogy continuing for some time in chemistry departments but ultimately falling entirely into association with geology.

Following the discovery of the diffraction of X-rays by crystals, crystallography extended from mineralogy into an independent field comprising all of the solid state. The early work on crystal structure analysis, and the early interpretive applications of the new structural knowledge, however, again involved mineralogy to a large extent. This came about because mineralogy, unlike chemistry and physics, contained almost the whole body of prior crystallographic description. Mineralogy also offered a challenging and already defined body of crystallographic problems, amenable to structure analysis and of general application. Among them were the constitution of the silicates - one of the first great rewards of structure analysis - together with the mechanisms of compositional variation, polymorphism, and especially the long-sought connections between internal structure and the morphology and physical properties of crystals.

Crystallographic Developments to 1850

Haüy was well known in the United States during his lifetime. A number of Americans went to Paris to study with him, especially from Philadelphia, where the practice of crystallography and mineralogy in this country initially centered. Several of the European mineralogists who immigrated to the United States in this period, bringing with them collections of minerals and crystals to illustrate the European textbooks, also had trained with Haüy. Among them was Gerhardt Troost (1776-1850), a pupil and companion of Haüy in 1807, who came to Philadelphia in 1810 with a one-circle reflecting goniometer of the type invented by William H. Wollaston in 1809.

In 1800 the Philadelphia chemist Thomas P. Smith contributed an article *"On Crystallization"* to the *Medical Repository*, published in New York City, in which he disagreed with Haüy's theory of crystal structure. He closed his article with the statement that Haüy ". . .appears to me to prove nothing more than that, if he had crystals to construct, that is the way that he would choose to make them." The opposition here and in Europe to Haüy's theory was based chiefly on chemical grounds. Greene and Burke (1978) note that Haüy had correspondence with Americans, especially Archibald Bruce of New York City, and sent mineral specimens to him, to a well-known mineral collector of the times, George Gibbs of Newport, Rhode Island, and to Charles Willson Peale's famous museum in Philadelphia. Haüy sent an inscribed copy of the 2nd edition of his *"Traité de Cristallographie"* to Gerhardt Troost in 1822. He also sent books on mineralogy to Peale in 1810 (Sellers, 1975). In an article translated and in part reprinted in the *Medical Repository* in 1810, Haüy himself acknowledged the gift of mineral specimens from Americans including Benjamin Barton, Archibald Bruce, Silvain Godon, William Maclure, S. L. Mitchell (the editor of the *Medical Repository*) and C. W. Peale.

The earliest known drawings of crystals to appear in the American literature are free hand sketches of pyrite cubes published in the *Columbian Magazine* in 1788 and 1789 by an author identified only as "B" (probably Benjamin Barton of Philadelphia). Descriptions of snow crystals were given by Jacob Green in 1820 and by Chester Dewey in 1821. Between 1821 and 1825 Gerhardt Troost contributed the first papers involving crystal measurements in this country. These dealt with the morphology of celestite, quartz, pyroxenes and other minerals and were published in the *Journal of the Philadelphia Academy of Natural Sciences*, founded in 1812. In the period 1813-1815 Troost lectured at the Academy on mineralogy and crystallography and demonstrated the use of his goniometer.

Troost's Wollaston reflecting goniometer was soon followed by others. Robert Hare, Professor of Chemistry at the University of Pennsylvania from 1818 to 1847, had an instrument made about 1815 by Dumotiez in Paris (Multhauf, 1961). A Wollaston reflecting goniometer, still preserved, was purchased by Harvard University in 1815. It was obtained for John Gorham (1784-1829), who taught chemistry and mineralogy in both the Medical School and the College. Gorham's *"Elements of Chemical Science,"* published in Boston in 1819 and well reviewed by Benjamin Silliman in the *American Journal of Science* in 1821, was the first systematic treatise on chemistry to be written in the United States. In it Gorham drew attention to crystallography and noted that crystal optics were useful in increasing our knowledge of the structure of crystals. David Brewster's work on crystal optics in this general period was well known in the United States.

The American talent for improving the design and construction of crystallographic apparatus was early demonstrated by Amos Eaton, who designed an improved version of the Wollaston goniometer in 1827 that was used by students at Rensselaer Polytechnic Institute. A further improvement was contributed by R. Graves in 1832.

The early teaching of crystallography, as now, was accompanied by the use of crystal models. John W. Webster, Professor of Chemistry and Mineralogy at Harvard from 1824 to 1850, noted in 1825 the acquisition of "400 large crystal models made in Paris for Count Bournon" [Jacques Louis Bournon (1751-1825), French crystallographer, known especially for his work on the morphology of calcite]. Webster did not publish any crystallographic work, but his interest and competence in this field is indicated by his review in 1823 of H. J. Brooke's *"Familiar introduction to crystallography, including an explanation of the principle and use of the goniometer,"* published in London in that year. Webster may have acquired through Harvard the fine reflecting goniometer made by Oertling in Berlin to Mitscherlich's design, reading to 10 seconds of arc by vernier, that was exhibited by J. D. Whitney in 1848 at a meeting of the Boston Society of Natural History. An inventory in 1821 of the equipment in the Harvard Chemistry Department lists "12 dissected models of crystals." These doubtless were the familiar wooden constructions showing Haüy's "molecules integrantes" aligned to illustrate the development of crystal forms. They presumably had been acquired by John Gorham. A wooden model of this type showing the "decrements on the cube" was presented to the Philadelphia Academy of Natural Sciences in 1821 (anon., 1822). F. Bache (1838) records that the collection of the American Philosophical Society contained transparent crystal models made of thin cleavage sheets of mica. Parker Cleaveland of Bowdoin College had 390 porcelain and painted clay models of crystals that had been purchased in Europe and given to him by John Bowdoin in 1812. Also goniometers and crystal models were acquired very early by Yale but the acquisitions have not yet been documented.

The total number of papers and textbooks with a significant content or use of crystallography that had been published in the United States up to 1850 is about 50. Virtually all of this work was descriptive and done in a mineralogical context. In this same period the number of persons with known or inferred competence in crystallography, aside from those who used it as a casual aid in mineral identification, totals only about a dozen. James D. Dana of Yale was foremost by far and remained so for some years after 1850. Among the others were Parker Cleaveland (1780-1858), Lardner Vanuxem (1792-1848). William Keating (1799-1840), Gerhardt Troost (1776-1850), Benjamin Silliman, Jr. (1816-1885) of Yale, Charles U. Shepard (1804-1886) of Amherst, Francis Alger (1807-1863) and J. E. Teschemacher (1790-1853) of Boston, John W. Webster (1793-1850) and John Gorham (1784-1829) of Harvard, Lewis C. Beck (1798-1853) of Albany, and Archibald Bruce (1777-1818) of New York City. The crystallographic work of all of these people was subordinate to interests in geology, chemistry and other fields.

James D. Dana left Yale in 1833 somewhat in advance of his graduation to take advantage of an opportunity to serve as an instructor in mathematics to midshipmen of a U.S. Naval vessel during a Mediterranean cruise. Here he occupied his leisure hours by ". . .working out, by methods of his own, many of the more intricate problems of mathematical crystallography" (E. S. Dana, 1895). In 1835 he constructed a set of crystal models in glass and published his first paper, on a new system of crystallographic symbols. Two years later, at the age of 24, he compiled the first edition of his *"System of Mineralogy,"* described later. J. D. Dana's numerous and important contributions to crystallography and mineralogy were mostly made in early and middle life and ultimately were overshadowed by his work in geology.

Lewis C. Beck and Francis Alger illustrate the general circumstances under which much crystallographic work was done in this period. In 1836 Beck was appointed to the New York State Geological Survey and was charged with that part of the Survey work that related to an examination, scientific description and chemical analysis of the minerals, mineral deposits and soils of the State. His *"Minerals of New York,"* published in Albany in 1842, contains 533 original drawings of crystals with tables of interfacial angles. Additional crystallographical descriptions were given in a supplement to this work in 1850. The nature of his instrumentation and the source of his crystallographic training is not known; both may have derived from Amos Eaton at Rensselaer. Beck's book of very creditable for American mineral science of the time. The scope and quality of the crystallographic description in this and other topographical mineralogy works in the 19th century, however, was not comparable to the best of the European work such as the *"Materialen zur Mineralogie Russlands"* by the Russian mineralogist N. I. Koksharov, published in parts between 1852 and 1877.

Francis Alger published a number of crystallographic descriptions between 1843 and 1850, and brought out a new and greatly enlarged American edition of a standart English work, Phillips' *"Mineralogy,"* in 1844. This work contained much new crystallographic information on American minerals. Alger was self-taught in crystallography and mineralogy. He had interests in mining and metallurgy and in 1856 succeeded his father in the management of a large and long-established iron works in South Boston. This supplied armament for both the War of 1812 and the Civil War. Alger died of pneumonia in Washington during the latter war, where his friend C. T. Jackson (1864) said he was "engaged in perfecting shrapnel in restoring the union of our divided States." Alger was a founding member (1830) and Curator of Mineralogy of the Boston Society of Natural History, and was active in promoting scientific education in the United States. The Society then and for a long while after was a center of activity in the geological sciences in the Boston area. At nearby Harvard in the 1840's science was at low ebb with instruction in chemical theory, chemical analysis, mineralogy, crystallography and geology wholly in the hands of one person, John W. Webster. Better practical instruction was available from private scientists in Boston including Alger and in particular the chemist and geologist Charles T. Jackson.

Virtually all of the crystallographic work done in this period was illustrative in nature and was applied in a mineralogical context. Only a few studies used crystallographic arguments to solve mineralogical problems. Among them were Francis Alger's demonstration on morphological grounds that several supposedly distinct mineral species were identical with heulandite, and C. U. Shepard's (1858) use of angular measurements to establish the separate identity of kyanite and sillimanite. Much of the crystallographic work was below the level of contemporary European practice. Among the first crystallographic descriptions to reach European standards of the time were studies by E. S. Dana of Yale in 1872 of the crystallization of datolite and, a little later, of chondrodite and danburite.

The low level of interest in crystallography as an intellectual or abstract pursuit in itself in the United States in this period and for a considerable time afterward is not surprising. It derived from a complex of social and economic factors attending the growth of a country that in 1800 hardly had been explored west of the Mississippi River. The pragmatic attitude of Americans, favoring applied rather than pure science, was an important factor. Mineral chemistry, responding to demand from the growing chemical, metallurgical and mining industries and from agriculture, early became the main theme in American mineralogy.

The main period of growth of crystallography and mineralogy began in the 1850's and 1860's. It was a consequence not of increasing pure science interests in American society but of the rapid growth of the mining industry and of associated technologies, especially in post-Civil War times in western states. The accompanying demand for persons trained in the geological and mineralogical sciences was met both in the universities and especially in the technical and scientific schools formed in this period in broad response to the increasing industrialization of the country from what was earlier an agrarian and rural society. Among the latter were the Lawrence Scientific School at Harvard, formed in 1858 and followed by the creation therein of a School of Mines and School of Practical Geology, the Columbia School of Mines, founded in 1864, and the Sheffield Scientific School at Yale.

George J. Brush (1831-1912) was the key figure in creating the scientific attitude at Yale in the latter 1800's in which crystallography and other sciences flourished. He obtained a B.S. degree at Yale in 1852, worked with a leading American chemist of the time, J. Lawrence Smith, at the University of Virginia, and spent two years abroad at Munich and Freiberg. He returned to Yale in 1856 as Professor of Metallurgy, later as Professor of Mineralogy in the Sheffield Scientific School, of which he was one of the organizers and Director from its inception in 1861 up to his retirement in 1898. A memorial to him in 1912 said ". . .it should be remembered that public sentiment toward science was very different then from what it is today. At that time men like Brush were needed who could strive against the general disfavor with which the American public then regarded scientific education, and especially against the disregard in the universities of instructional activities in pure science." At Harvard the analogues of Brush were the chemist Josiah P. Cooke, appointed in 1850, and Josiah D. Whitney (1819-1896), a mining geologist, appointed in 1865. The Lawrence Scientific School formally terminated in 1906 and its activities either were taken up locally by the Massachusetts Institute of Technology or, with increasing emphasis on pure science, were gradually absorbed into the larger University structure.

In the latter 1800's, in this general environment, mineralogy grew rapidly and maintained a strongly practical and descriptive character. Emphasis was placed on mineral chemistry and on the occurrence, association and use of minerals. Crystallography was taught chiefly as an adjunct to mineral description and identification. Blowpipe analysis received more attention than crystallography. Brush's book *"Determinative Mineralogy and Blowpipe Analysis"* passed through 14 editions between 1874 and 1898, with later editions written jointly with S. L. Penfield, and is said to have had the largest sale of any scientific book in the United States in this period. For some time in the 1800's the only instruction in crystallography at Harvard was a course titled *Chemistry 3: Mineralogy, including use of the Blowpipe and Crystallography.*

American Books on Crystallography, to about 1900

The first textbook by an American to deal with crystallography in detail was published in Boston in 1816 by Parker Cleaveland. His *"Elementary Treatise on Mineralogy and Geology"* had a 37 page section on crystallography with 39 drawn figures of crystals. The treatment was based on Brongniart's *"Traité de Minéralogie"* of 1807 and used a modification of Haüy system of familiar names for crystal forms that had been introduced by Frederick Accum in 1813. Cleaveland was a student under Aaron Dexter in the Harvard Medical School and graduated in 1799. He was appointed Professor of natural philosophy and mathematics at Bowdoin College in 1805. Cleaveland had both a contact goniometer and, it appears, a Wollaston reflecting goniometer. His book, with a 2nd edition in 1822, was well received in Europe but some American colleagues, more inclined toward chemistry and the Wernerian school, did not much like the emphasis on crystallography. The lengthy review by W. Channing (1817) indicates the status of crystallography in the United States at the time - he trembled for the fate of the book for lack of understanding readers - and adds a good discussion of the opposing French and German schools of thought on the role of crystallography in mineral classification and investigation. Cleaveland's book was said in 1842 to have been long out of print and a partly completed 3rd edition never materialized.

A major factor in the development of crystallography in the United States, and long the main avenue of entrance of Americans into the European crystallographic literature, were the successive editions of the *"System of Mineralogy"* by James D. Dana (1813-1895) and Edward S. Dana (1859-1935) of Yale University. The first edition in 1837 was titled *"A System of Mineralogy, including an extended treatise on crystallography with an appendix containing the application of mathematics in crystallographic instruction."* The crystallographic material totalled 145 pages, about 40 percent of the book, and included an 80 page abstract of C. F. Naumann's *Lehrbuch der Kristallographie* of 1830 with reference primarily to the use of analytical geometry in crystallographic computations. The formal treatment of crystallography diminished in later editions, and the title was shortened, because of the growing bulk of mineral descriptions. The 4th edition in 1854 for the first time used Naumann's notation for crystal planes. This edition is notable for the introduction of a chemical system of classification, much earlier introduced in Europe, in place of the irrelevant Linnean system of the preceding editions. In the 5th edition of 1868 the crystallographic material was squeezed down to 10 pages. The 6th edition by Edward S. Dana in 1892, a classic, and as free from errors as a book of this kind can be, rather belatedly used the Miller indices but still cited only interfacial angles for the more common forms. The crystallographic descriptions and angle tables in the incomplete 7th edition, by C. Palache, H. Berman and C. Frondel, are based on the Goldschmidt method of two-circle optical goniometry, described beyond. The treatment of morphological crystallography in the first six editions of the *"System"* was taken up in more detail, with added sections on physical and optical crystallography, in the successive editions of the *"Textbook of Mineralogy."* The first edition of this intermediate level textbook by Edward S. Dana appeared in 1877 and a fourth edition by William E. Ford in 1932. A third work, the *"Manual of Mineralogy,"* designed for the beginning course in mineralogy and crystallography, had a first edition by J. D. Dana in 1850 and a 19th edition by C. S. Hurlbut, Jr., and C. Klein in 1977. The *"Treatise on Mineralogy"* by Charles U. Shepard of Amherst College, with three editions between 1832 and 1857, included crystal drawings and figures of the Carangeot and Wollaston goniometers but was of little use to crystallographers.

Among the foreign works treating of crystallography, the best known in the United States in the early days were the successive publications of Robert Kirwan (1733-1812) of Dublin, Robert Jameson (1774-1854) of Edinburgh and especially of William Phillips (1775-1828). Phillips was a tradesman and private scientist in London. Kirwan's *"Elements of Mineralogy"* had three editions (1784, 1794, and 1810), as did Jameson's *"System of Mineralogy"* (1804, 1816, and 1820). Jameson was a leading proponent of Werner's ideas and numerous Americans went to Edinburgh to study mineralogy, crystallography, and geology with him, not always with success. In 1816 the Harvard administration proposed a Professorship in Mineralogy and Geology, the first by that title in the University, and offered it to Joseph G. Cogswell. Cogswell, who described himself as a disciple of Haüy, went off to Europe for training and in 1818 reported by letter from Edinburgh that "Nothing ever cooled my ardor for mineralogy so much as hearing it taught by such a cobbler as Professor Jameson" (Ticknor, 1874). Another Harvard visitor, Asa Gray, described Jameson in 1839 as a "stiff, ungainly, forbidding looking gentleman, who gave us the most desperately dull, doleful lecture I ever heard." On his return from Europe, Cogswell declined the post and entered a career as Librarian of Harvard and then of the Astor Library in New York City. Other Americans who went to Edinburgh to study mineralogy were Benjamin Silliman of Yale and his friend John Gorham of Harvard, in 1805-06.

The 1823 and earlier editions of Phillips' *"Introduction to Mineralogy"* were said by Benjamin Silliman in 1845 to have been the most popular mineralogy works in any language. The 1st edition was reprinted in New York City in 1818 with notes and additions by S. L. Mitchell. Phillips died in 1828 and later editions of his work were brought out by others including, as noted,

Francis Alger of Boston. H. J. Brooke and W. H. Miller brought out a final London edition, extensively modernized with regard to the treatment of crystallography, in 1852. It was widely used. Other early European works well known in the United States were William Haidinger's English translation of the *"Grundriss der Mineralogie"* by Friedrich Mohs (1825), Naumann's *"Lehrbuch"* already mentioned, and the *"Treatise on Crystallography"* by William H. Miller (1839). Thomas Thompson's *"Outlines of Mineralogy, Geology and Mineral Analysis"* (1802, 1836) and, especially with regard to the early development of American mineral chemistry, his *"System of Chemistry"* with a 1st edition in 1802 and a 5th edition in 1817 (reprinted in Philadelphia in 1818), were influential. Thompson, at the University of Edinburgh, had extensive contacts with Americans and analyzed numerous specimens of American minerals. He appears to be mainly responsible for the introduction of Dalton's atomic theory into the United States.

Outside of private possession, in the early 1800's the main library holdings in mineralogy and crystallography were associated with natural history societies and learned societies and are itemized in library catalogues of the time. In 1830 Harvard, still classical and theological in its leaning, had less than a dozen works in these fields, not including Delisle's *"Crystallographie"* of 1783 or Bournon's *"Traité"* of 1808; Professor Waterhouse, however, had a good personal library, mostly obtained from his friend J. C. Lettsom in London. Parker Cleaveland of Bowdoin also had a good personal library that included Haüy's works.

The first American book devoted wholly to crystallography as a field in itself, was the *"Elements of crystallography for students of chemistry, physics, and mineralogy"* published in 1890 by George H. Williams (1856-1894) of Johns Hopkins. Williams, primarily a petrographer, got his Ph.D. under Harry Rosenbusch at Heidelberg. A paper by Williams in 1883 in the *American Chemical Journal* drew attention to the usefulness of crystallography to chemists. One of his major crystallographic studies (1889) described morphological pseudosymmetry in the monoclinic pyroxenes, indicated that the point-symmetry was class m and not $2/m$. Williams' book, said to have had international use, was followed by a number of similar crystallographic works. Among them were books by Alfred J. Moses (1899) of Columbia University, Edward H. Kraus (1906) of the University of Michigan, M. E. Wadsworth (1909) of the University of Pittsburgh, W. S. Bayley (1910) of the University of Illinois, and G. M. Butler (1918) of the University of Arizona. Thomas L. Walker (1867-1942) of the University of Toronto published his *"Crystallography, an Outline of the Geometrical Properties of Crystals"* in 1914. Most of these books, as with the treatments of crystallography that continued to appear in general textbooks of mineralogy of the time, were elementary descriptions of the subject. Some of them also contained misstatements that brought specific comment in a paper by George Tunell and George M. Morey in 1932. The first American work to treat crystallography as an abstract science, rigorously based on symmetry theory, is the *"Elementary Crystallography"* published in 1956 by Martin J. Buerger of the Massachusetts Institute of Technology. It had forerunners in Europe.

European works influential in the United States around 1900 included N. Story-Maskelyn's *"Crystallography, a Treatise on the Morphology of Crystals"* of 1895, A. E. H. Tutton's *"Crystalline Structure and Chemical Constitution"* of 1910 and the editions of his *"Crystallography and Practical Crystal Measurement"* in 1911 and 1922. Numerous fine German texts on crystallography were available. Among them were Paul Groth's *"Physikalische Kristallographie"* of 1895, Theodor Liebisch's *"Grundriss der Physikalischen Kristallographie"* of 1896, and especially the 1920 and 1926 editions of the *"Lehrbuch der Mineralogie"* by Paul Niggli. The 1911 and 1926 editions of the *"Lecons de Cristallographie"* of George Friedel also were important. Some works, such as Woldemar Voight's *"Lehrbuch der Kristallphysik"* of 1910, were a little difficult for the ordinary reader.

Elementary and brief discussions of X-ray crystallography appeared in some American textbooks on mineralogy in the 1930's. Among them was the 1932 Dana-Ford edition of the *"Textbook on Mineralogy."* The earliest American books wholly devoted to X-ray crystallography were R. W. G. Wyckoffs *"The Structure of Crystals,"* with editions in 1924 and 1931, and M. J. Buerger's *"X-Ray Crystallography,"* treating primarily of the Weissenberg method, published in 1942. W. H. Zachariasen's more general *"Theory of X-ray Diffraction in Crystals"* appeared in 1945. The American X-ray work of this period in reference to group theory drew largely on European works including Harold Hilton's *"Mathematical Crystallography and the Theory of Groups of Movements"* (1903), Paul Niggli's *"Geometrische Kristallographie des Diskontinuums"* (1919), Arthur Schoenflies' *"Theorie der Kristallstruktur"* (1923) and George Friedel's *"Lecons"* of 1926.

Morphological Crystallography

For most of the 19th century American colleges provided little opportunity for study at the professional or graduate level in the sciences. Then and for some time beyond Americans who sought advanced training in crystallography generally went to Europe. This sometimes involved a formal degree, but the usual practice was to take a year or so abroad and visit several centers of learning. The financial support for this seems in most or all instances to have been wholly private. Edinburgh and Paris were early centers but in later years Paul Groth at Munich and Victor Goldschmidt at Heidelberg were the main attractions.

The first Ph.D. awarded in an American University on the basis of a thesis done wholly in crystallography appears to be that of Frank A. Gooch (1852-1919). His thesis, *"A Treatise on crystallography on the basis of Miller's system,"* was completed in 1877 under Josiah P. Cooke in the Chemistry Department at Harvard. After graduation Gooch visited England and met ". . .the elderly, weak and scholarly gentleman whose brief and concise 'Treatise' [1839] and 'Tract' [1863] upon crystallography I had tried to simplify as well as extend." The thesis was never published; Gooch was later appointed Professor of Chemistry at Yale and gained fame as an analytical chemist. The first Ph.D. to be attained by a woman in this country on the basis of a thesis on a crystallographic topic appears to be that of Gabrielle Hamburger Donnay in 1949, from the Massachusetts Institute of Technology.

The need for specialized European training decreased as American colleges began to offer advanced training in mineralogy and crystallography. Columbia University, for example, was one of the first and generated a succession of crystallographers. Thomas Egleston (1832-1900), founder of the Columbia School of Mines in 1864, was a metallurgist and mining engineer with a flair for crystallography. His *"Diagrams to illustrate the lectures on crystallography,"* published privately in 1866 with three later editions, attracted a student, Alfred J. Moses (1859-1920), with a mathematical bent. Moses got his Ph.D. in 1890 and, following the retirement of Egleston in 1897, became Professor of Mineralogy at Columbia. He published a number of papers on mathematical crystallography in the *School of Mines Quarterly,* including several that were reprinted in the *Zeitschrift für Kristallographie,* together with two textbooks. These were *"The characters of crystals; an introduction to physical crystallography,"* published in 1899, and a widely used textbook, written jointly with C. L. Parsons in 1897, *"Elements of mineralogy, crystallography and blowpipe analysis."* Moses was the first American author to recognize the rhombohedral and hexagonal subsystems in place of the single hexagonal system as used in the 1892 and earlier editions of Dana's *"System of Mineralogy."* One of his students was Austin F. Rogers (1877-1957), who obtained his Ph.D. in 1902. Rogers was appointed Professor of Mineralogy at Stanford in 1902 and became one of the leading crystallographers and mineralogists of his time. Among his students were Paul F. Kerr (1897-1981) and J. D. H. Donnay. Donnay went on to a career at Johns Hopkins, Laval University and McGill University in which he instructed both students and his peers. Kerr returned to Columbia as Professor of Mineralogy between 1924 and 1965 and generated several professional crystallographers. Kerr followed Alexis A. Julien (1849-1920), another Columbia graduate, who specialized in crystal optics and petrography. Julien together with C. E. Wright published in 1873 the first American study in optical petrography.

Two-Circle Optical Goniometry

Historically, the measurement of crystals initially involved only that of the interfacial angles by means of a contact goniometer. The first instrument of this type was that of Carangeot and it or slight modifications thereof later became known as the Haüy goniometer. The best known model in this country is the cardboard Penfield contact goniometer designed by Samuel L. Penfield in 1900, with a description in the *Zeitschrift für Kristallographie.* It sold widely for use in elementary mineralogy courses.

From 1809 onward, following the development of the one-circle optical goniometer by W. H. Wollaston, the interfacial angles could be measured in zones, with the crystal being reset for each zonal series. The early employment of Wollaston's goniometer is discussed by W. C. Smith (1978). In Wollaston's instrument the graduated circle was set vertical but, as in the design of Friedrich Mohs, the more common construction, especially later in the century, was to place the circle in the horizontal position. Some of the research instruments of this type could measure to a precision of 6 seconds of arc or better. This was much beyond the quality of any ordinary crystal.

The next development, following the plan of W. H. Miller in 1874 but of earlier origin (see Lewis, 1882), was the determination of the angular relationship by measurement of the polar coordinate angles on a two-circle optical goniometer. This method also extended to contact goniometers. A two-circle contact goniometer devised in 1896 by Victor Goldschmidt is shown in Figure 2. The use of this instrument was described in

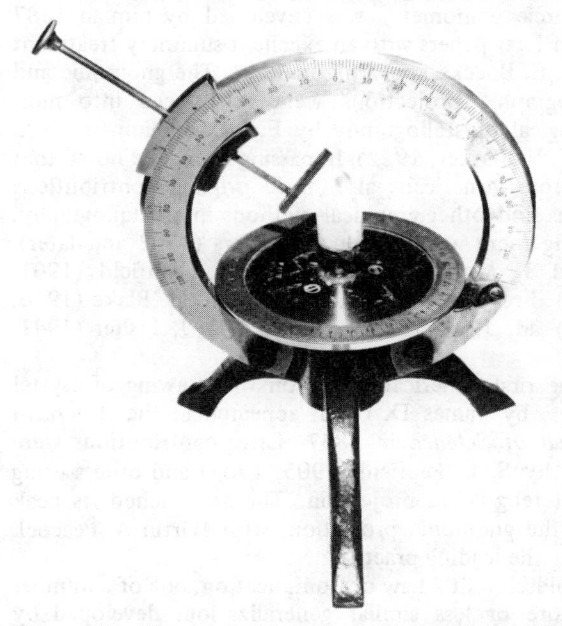

Figure 2. Two-circle contact goniometer of Goldschmidt's design. Acquired by Harvard about 1895.

1920 by Florence Bascomb (1862-1945) of Bryn Mawr College, who thereby acquired the distinction of being the first woman to publish in crystallography in this country in addition to being the first woman Ph.D. in geology (from Johns Hopkins in 1893). Two-circle goniometry was followed by three-circle optical goniometry. This permitted measurements of the polar coordinates and zonal relations at a single setting of the crystal (G. F. H. Smith, 1899). Only a few instruments of this type, one in the laboratories of the British Museum (Natural History), appear to have been constructed.

Two-circle optical goniometry based on instruments designed by E. S. Fedorov (1893), E. Czapski (1893), and especially by Victor Goldschmidt (1892) soon came into general use. The first two-circle optical goniometer was brought to the United States by Charles Palache of Harvard University in 1896. The second two-circle goniometer to be imported apparently was that of William Nicol of Queen's University in Kingston, Ontario, in 1897. By 1920 probably several dozen were in use here. The development of two-circle optical goniometry in hand with the gnomonic projection into a general system of descriptive morphological crystallography, often called the Goldschmidt method, was greatly facilitated in this country by a coordinated series of papers written by Palache and others that were published in the *American Mineralogist* in 1920.

Victor Goldschmidt (1853-1933), not to be confused with the geochemist Victor Moritz Goldschmidt, was an independent teacher at Heidelberg and the recognized grand master of crystal morphology. The use of the long known gnomonic projection as an adjunct to two-circle goniometry was developed by him in 1887 and in later papers with an excellent summary treatment by H. E. Boecke appearing in 1913. The gnomonic and stereographic projections were introduced into morphological crystallography by F. E. Neumann in 1823 (cf. T. V. Barker, 1922). In passing it may be noted that numerous Americans also made original contributions to this and other graphical methods in crystallography. Among them were Austin F. Rogers (1902 and later), Alfred J. Moses (1902), Samuel L. Penfield (1901, 1902), Frederick E. Wright (1913), J. M. Blake (1916, 1918), M. J. Buerger (1934), and D. J. Fisher (1941, 1952).

The first American paper on the drawing of crystal figures, by James D. Dana, appeared in the *American Journal of Science* in 1837. Later contributions were made by S. L. Penfield (1905, 1906) and others using the stereographic projection. The art reached its peak with the gnomonic projection, with Martin A. Peacock among the leading practitioners.

Goldschmidt's Law of Complication, one of a number of more or less similar generalizations developed by crystal morphologists over the years, was based on the nodal distribution and relative importance of forms as seen in the zone lines of the gnomonic projection. It was extended by Goldschmidt to a general theory of harmony involving musical scales, color spectrums, the spacing of the planets, and X-ray Laue photographs (1933). Contrasting discussions of these matters are given by M. A. Peacock (1932) and M. J. Buerger (1936). Another aspect of early 20th century crystallographic mysticism were the views of Rudolph Steiner (cf. E. Pfeiffer, 1936) on the influence of etheric formative forces and radiations from living organisms on crystal morphology.

The three persons mainly responsible for the widespread development of two-circle optical goniometry in the United States after about 1900 were Charles Palache of Harvard University - the acknowledged leader - Arthur S. Eakle (1862-1931) of the University of California (Berkeley), and Samuel G. Gordon (1897-1952) of the Philadelphia Academy of Natural Sciences. Eakle, Professor of Mineralogy at Berkeley from 1901 to 1930, obtained his Ph.D. from Paul Groth at Munich. His 1902 paper on the morphology of colemanite is an outstanding example of the Goldschmidt technique. Both Waldemar T. Schaller (1882-1967) of the U.S. Geological Survey and William F. Foshag (1894-1956) of the U. S. National Museum were Eakle's students. They were among the few mineralogists of their day who had the personal competence to do a complete phase description including the crystal morphology, crystal optics, chemical analysis (by non-spectrochemical methods) and a treatment of the mineral occurrence and association. Samuel G. Gordon's joint paper with Victor Goldschmidt in the *American Mineralogist* in 1928 helped make the Goldschmidt method useful in this country.

Morphological Crystallography at Harvard

Instruction in the bare rudiments of mineralogy and crystallography was contained in a course on natural history offered from 1788 to 1812 by Benjamin Waterhouse (1754-1846). Waterhouse, born in Rhode Island and educated in medicine at Leyden, was appointed to the Harvard Medical School at its founding in 1782. He became interested in mineralogy through friendship with J. C. Lettsom of London, an influential doctor and collector of minerals, and in 1784 established the Mineralogical Museum. The mineral collection immediately became a focal spot of scientific interest at Harvard and in the community generally. Recorded early visitors included George Washington and Timothy Dwight. Dwight, then president of Yale and responsible for the appointment of Benjamin Silliman to the Yale faculty in 1802, studied the collection in 1797. Silliman started teaching in 1807. Following this early start crystallography grew slowly at Harvard over much of the 19th century. In 1815 Parker Cleaveland, then the leading practictioner in the country, refused the offer of a post and in 1850 James D. Dana turned down an offer just before his professorial appointment at Yale. After the retirement of Waterhouse in 1812 instruction in mineralogy and crystallography and the custody of the mineral collection was the responsibility of the successive Erving Professors of Chemistry and Mineralogy, in the Chemistry Department: John Gorham (1816 to 1824), John W. Webster (1824 to 1850), and Josiah P. Cooke (1851 to 1894). Cooke, an able chemist and administrator who did much to advance the cause of instruction in science at Harvard, was among the leading American crystallographers of the second half of the 19th century. The first formal course in pure crystallography at Harvard, titled *Crystallography and the Physics of Crystals*, was offered by him in the Chemistry Department in 1868-69. Aside from other crystallographic work Cooke developed the idea, in studies extending from 1855, that the "chemical energy" associated with a fixed mass of an element varied slightly and had crystallographic consequences. Following Cooke crystallography was taught at Harvard by Charles Palache until 1941.

Charles Palache (1869-1954) was trained as a geologist under Joseph Le Conte and A. C. Lawson at the University of California (Berkeley) and was awarded his Ph.D., with a thesis in field geology, in 1894. In 1893 he travelled to Europe where he studied crystallography with Paul Groth and especially Victor Goldschmidt; here he found "my real goal which I have followed ever since." He entered the newly formed Department of Mineralogy and Petrography at Harvard in 1895. In the years to his retirement in 1941 almost every known crystallized mineral came to his attention. A number of his studies, among them his work on azurite, calcite and calaverite, stand as models of morphological description. The morphology of the gold telluride, calaverite, violates the law of form rationality and was the reigning crystallographic problem of the day. It successfully withstood several investigations including a joint study in 1931 by Palache, Victor Goldschmidt, and Martin A. Peacock, a Supreme Court of morphologists before which any crystal should tremble, but yielded to X-ray structural studies by George Tunell and C. J. Ksanda (1935, 1936) and Tunell and L. Pauling (1952). The full impact of X-ray diffraction techniques came midway in Palache's career but the resolution of crystals in terms of atoms and forces did not attract him. As a side-light, a new edition of Goldschmidt's *"Kristallographische Winkeltabellen,"* an invaluable aid to the morphological crystallographers of the day, was nearly completed by him but was put aside when it became apparent from X-ray work that the traditional morphological cell of many minerals did not jibe with the structure cell. Present-day computer-based crystallographers should reflect that the original 1897 German edition of the *"Winkeltabellen"* took Goldschmidt and one assistant, using logarithms, about two years to calculate the roughly 70,000 angles and coordinates.

Palache's work and most American work in morphological crystallography was an adjunct to descriptive mineralogy rather than a discipline or investigative method in itself. It is exemplified in Palache's 1935 study of the geology, mineralogy, and origin of the zinc deposit at Franklin and Sterling Hill, New Jersey. The group of five papers by Edward S. Dana and George J. Brush published between 1878 and 1890 on the Branchville pegmatite provides a much earlier example.

Among Palache's students who entered professional careers in crystallography and mineralogy were C. Wroe Wolfe (1908-1980) of Boston University, Cornelius S. Hurlbut, Jr., of Harvard, Horace Winchell of Yale, and Martin A. Peacock (1892-1952). Harry Berman (1902-1944) came to Harvard as Museum assistant to Palache and obtained his Ph.D. in 1936. Peacock obtained a Ph.D. from the University of Glasgow in 1925 and with the aid of a Commonwealth Fellowship began the study of crystallography with Palache in 1926. He published his first paper in crystallography, on coordinate angles in two- and three-circle goniometry, in 1929. Peacock, a brilliant crystallographer, with a sharp mind and tongue, went to the University of Toronto in 1937. Canadian crystallography owes much to him.

The lack of generality in the approach to crystallography at Harvard and in the United States in the early 1900's reflects social and economic factors earlier mentioned, and also educational administrative policies that developed in American universities during and after 1870-1880. The growth of the elective system and of a rigid departmental system of organization of the sciences were major factors. The development of graduate level instruction at Harvard is described by S. L. Morison (1946); the first Ph.D. awarded in the sciences was in 1873. At Harvard, typical of the general situation, instruction in crystallography was transferred in 1895 from the Division of Physics and Chemistry to the Division of Natural History. Here the new Department of Mineralogy and Petrography was allied to and, as in other universities, later merged with the Department of Geology. These actions greatly benefited mineralogy and geology, by putting them into their natural context, but restrained the much more general science of crystallography. It virtually isolated instruction in crystallographic theory and methods of phase investigation from other relevant sciences, and it specialized the thematic material.

The few prerequisites in the sciences and mathematics then existing for entrance into mineralogy and geology departments also was restrictive to both students and faculty. Palache's graduate course in *Advanced Crystallography* from 1905 onward required for entrance only a half course in general chemistry and a knowledge of trigonometry and plane and spherical geometry. The course dealt with the morphology of minerals. Students from the chemistry, physics and metallurgy departments who wanted entrance to this and an associated course in physical and optical crystallography were required to first take a course in elementary mineralogy. This was a deterrent and not much of a help anyway since the traditional elementary course involved only a few weeks of lectures on the point-symmetry and vectorial characters of crystals, accompanied by laboratory exercises on collections of wooden models and natural crystals, for the purpose primarily of using crystallographic characters in mineral identification. Palache's love of crystals and his deep sense of enquiry provided other requisites to students. The first Harvard course in crystal chemistry from a structural point of view was offered by the writer in the Geology Department in 1946. The entrance of X-ray crystallography to Harvard is noted in a following section.

In Europe, crystallography from the beginning remained much more closely integrated with chemistry and the physical sciences. At Paul Groth's institute at the University of Munich, as in his editorial conduct of the *Zeitschrift für Kristallographie,* no distinction was made between minerals, synthetic inorganic crystals, and organic crystals. The instruction and research included the physical, chemical and morphological aspects of crystallography with generality. It had no university analogue in this country, although there were some individuals who had the broad competence. In the 19th century the fundamental groundwork of crystallographic theory, to which Americans made little or no contribution, was laid in Europe in this general context.

Morphological Studies

Morphological crystallography culminated in the 1890's and the first few decades of the 20th century. The amount of morphological information that has accumulated over the past century and a half is enormous. In the case of calcite alone about 700 different crystal forms were reported in the period extending from 1781-1801, when R. J. Haüy developed a theory of crystal structure from the crystal form and cleavage of calcite, up to 1914, when W. L. Bragg determined the actual crystal structure. Aside from innumerable papers there have been 10 definitive monographic surveys of the morphology of calcite: Bournon in 1808 with 704 crystal drawings; Lévy in 1837 with 346 drawings; Zippe in 1851 with 700 drawings; Irby in 1878 with 156 drawings; a complete review of this and all other published work by Victor Goldschmidt in 1913 in his monumental *"Atlas der Kristallformen,"* with 2542 drawings; and five more surveys by Americans including Rogers (1901), Whitlock (1910, 1915), and Palache (1898, 1943). Most of these morphological surveys of calcite (as of almost equally extensive surveys of quartz, pyrite and some other minerals) included critical evaluations of the earlier angular measurements, for which Palache's work of 1943 is definitive. All are useful in matters such as the genetic dependence of crystal habit although the morphological literature rarely touched on the environmental parameters. They also provide a statistical base for theories of crystal growth and habit based on lattice theory.

Some of the specialized memoirs on calcite, such as that of the Belgian crystallographer G. Cesàro in 1889 on the morphology of calcite crystals from a locality at Rhisnes, Belgium, totalling 227 pages and 65 figures, are forgotten classics of crystallographic measurement and computation. Herbert P. Whitlock's memoir of 1910, with 136 figures and a review of the literature, was titled *"The Calcites of New York"* and simply described all of the interesting calcite crystals that came to hand from the State. Whitlock (1868-1948), a self-taught crystallographer, later deplored what was then called "goniometer pushing," putting the time as the 1890's and the place as Europe. (In 1920 Whitlock described a method for making models of crystal structures and he is responsible for what probably was the first public display of crystal structure models, in the American Museum of Natural History in the 1920's.)

Morphological studies of the general nature described ceased with the oncoming of X-ray crystallography. The problem of what all the crystal faces on minerals mean genetically still remains. Morphological crystallography was a valid approach and long one of the few approaches possible to the question of the internal structure of crystals and, regardless of its being displaced by later direct methods, a fate shared by many defined fields in the physical sciences, it must be recognized historically as a major organized contribution to science. Actually a morphological description still is necessary to the characterization of a crystalline phase, although the obligation is now often neglected. There is also the obligation in mineralogy, now less frequently neglected, for structural descriptions that are coordinated with descriptions of the chemistry, physical properties and morphology of minerals. A model description of this kind is the work by Gregori Aminoff (1885-1947) of the Museum of Natural History in Stockholm on the new mineral bromellite, BeO, done almost 60 years ago.

Crystal Determinative Tables.

The use of crystal angles as a systematic means for the identification of crystallized substances derives from the Russian theoretical and morphological crystallographer E. S. Fedorov (1853-1919) of the Mining Institute in Leningrad. The method culminated after decades of intensive effort by T. V. Barker of Oxford (who had worked with Fedorov in 1908) and an associated group of crystallographers with the publication in 1951 of the *"Barker Index of Crystals"* by M. W. Porter and R. C. Spiller. It is based on conventions for selecting a small number of classificatory angles for each substance that arranged into determinative tables. The *"Barker Index,"* aside from intrinsic limitations, faced strong competition here from optical methods and especially from the rapidly growing use of X-ray diffraction data in phase identification. Fedorov's universal microscope stage for the measurement of crystal optics found general use in the United States, but his work on morphology and its connections with crystal structure and the 230 space groups - derived by him in 1890 - attracted little attention in this country.

Goniometric Instrumentation

Most of the 18th and early 19th century scientific instrumentation, including telescopes and physical apparatus, used in this country was obtained in England. The early Harvard instrumentation has been described by I. B. Cohen (1950) and D. P. Wheatland (1968), and the early Yale equipment by H. M. Fuller (1938). The Harvard contact goniometer of 1797 and the Wollaston reflecting goniometer of 1815 were bought in London from W. and S. Smith and Accum and Garden, respectively. The first Yale mineralogical equipment including a contact goniometer and books on mineralogy were obtained by Benjamin Silliman in London in 1805 through Frederick Accum and Robert Banks. Carangeot of Paris also was a dealer in instrumentation and crystal models.

In the latter 1800's the main source of crystallographic apparatus, including reflecting goniometers and instrumentation for crystal optics and crystal physics, was Germany, with suppliers also in France and England but not in the United States. Illustrations of the crystallographic instrumentation can be found in many of the textbooks of the time such as Groth's works in crystallography. The German suppliers in some instances also provided detailed manuals and catalogues describing the use and adjustment of their instruments. The Fuess equipment was described by C. Leiss (1899). In 1895 a fine Fuess reflecting goniometer cost $340 in equivalent U. S. currency. Two-circle optical goniometers were first made in the United States by the Eichner Instrument Company after World War II.

A. Krantz of Bonn has long been the chief supplier of wooden crystal models for teaching purposes in the United States. Their comprehensive collection of 743 different models illustrating the morphology of the crystal classes sold for $300 in 1895 (Groth, 1895). Victor Goldschmidt described a bench saw adapted to cut blocks of plaster or pear wood at appropriate angles and similar equipment was in use at both Yale and Harvard in the 1920's (Blake, 1917; Palache and Lewis, 1927).

Mineral and Crystal Collections

It should be noted that the traditional teaching and research in crystallography was very largely supported by mineral collections kept in university museums. The earlier X-ray structural studies on minerals also were largely based on described or analyzed material kept therein. Mineral collections play a necessary role in mineral science. Several commentators have noted that the acquisition of the private mineral collection of George Gibbs by Yale in 1820 was the greatest single factor in the early development of mineralogy (and the associated work in crystallography) at Yale and in this country. The Harvard collection over the almost 200 years of its history has been the focal spot of crystallographic and mineralogical research in the University and an important source of investigative material to scientists over the world. The great collections of the U. S. National Museum in Washington and of the British Museum (Natural History) in London play a similar role.

The nature and role of American mineral collections have changed with time. The early Colonial collections such as that of John Winthrop (1681-1747) of Connecticut, catalogued and given to the Royal Society of London in 1734, were primarily to find and make known the identity and use of the mineral resources of a newly settled land. The collections around 1800 were based largely on correctly labeled material brought from Europe. There were not many Americans then who could tell one mineral from another, much less practice crystallography. These collections were primarily utilitarian with emphasis on materials such as refractories, abrasives, source materials for chemical manufactories, and the ores of the important metals.

Many very large and important private collections chiefly of aesthetic value were built up by purchase in the period roughly from the latter 1870's to about 1900. These were not a consequence so much of growing public interest in science as of growing private fortunes in the Gilded Age of American enterprise and industry. Collections of objects of art, paintings by European masters, and of books also flourished. Virtually all of these private collections ended in the hands of public and university museums. Comprehensive scientific mineral collections with emphasis on species representation, type material, crystallographic characters and kinds of mineral occurrence were less common. These were built chiefly by scientific curators in universities. Field collecting and research activities by faculty and students here were important sources of acquisition of specimen material. Some private collections also emphasized species and varietal representation. Among them were those of George J. Brush of Yale, Washington A. Roebling, whose collection went to the Smithsonian Institution (Natural History) in 1926, and Albert F. Holden, whose collection went to Harvard in 1913.

The most important ultimate source of supply to museum collections has been field collecting by commercial dealers in mineral specimens. Many X-ray crystallographers who do a structural study on a specimen borrowed from a museum have an unrecognized debt to the sharp eye and acquisitiveness of a mineral dealer. For example, our knowledge of the cuprous-cupric oxide called paramelaconite resulted from the activities of Albert E. Foote of Philadelphia. Foote, the first major dealer in mineral specimens in the United States, beginning in 1876, personally obtained the two known specimens from a miner in the booming copper camp of Bisbee, Arizona, in 1890. He sold them for $100 to Clarence S. Bement, a wealthy manufacturer and discriminating mineral collector of Pittsburgh. Bement's collection was purchased by John Pierpont Morgan and presented to the American Museum of Natural History. This Museum supplied the crystals for an X-ray study in 1941 that first drew modern attention to this unusual and not yet synthesized substance.

Special collections of large and perfect natural crystals arranged to illustrate the morphology of the crystal classes were a feature of many museum exhibits. J. W. Webster stated in 1824 that the Harvard Museum had a suite of wooden models and of natural crystals on exhibit to illustrate Haüy's primary forms. In the 1900's the Harvard crystal collection grew almost to the point of ostentation. Charles Palache, with justifiable pride, for he assembled much of it, would point to a huge complexly developed apatite crystal from the Zillerthal and remark to visitors that Paul Groth saw that and said it was better than the one in Munich. By the 1970's the monetary value of the 400 or so crystals became so high that the collection had to be removed from public exhibit for safekeeping.

Chemical Crystallography

William Whewell said "Now the knowledge which we principally seek concerning minerals is a knowledge of their chemical composition; the general propositions to which we hope to be led are such as assert relations between their intimate constitution and their external attributes . . . We cannot get rid of the fundamental conviction, that the elementary composition of bodies, since it fixes their essence, must determine their properties. . .We may begin with the outside, but it is only in order to reach the internal structure." This sounds valid today even though written in 1837.

The subject of chemical crystallography, later called crystal chemistry, was intensively developed in Europe in the 19th century with J. J. Berzelius (1779-1848), Eilhard Mitscherlich (1794-1863), Paul Groth (1843-1927), Gustav Tammann (1861-1938), and A. E. H. Tutton (1864-1938) among the leaders. Very little work was done in this field in the first half or so of the 19th century in the United States, even though mineral chemistry early became and remained the great strength of American mineralogy. The low level of crystallographic activity in the United States in this period, earlier noted, was the main factor.

The first significant American contributions to chemical crystallography appeared in the 1840's and 1850's. In this period J. D. Dana published a notable series of papers on the relation between the crystalline form and the chemical composition of crystals, and C. Gerhardt contributed a paper on the atomic volume of some isomorphous oxides. M. Scheerer's lengthy study of the supposed isomorphous substitution of H_2O for cations in various silicates, originally published in the *Annalen der Physik,* was translated and republished in the *American Journal of Science* in 1848. A paper on atomic constitution and crystalline form was published by the Canadian mineralogist E. J. Chapman (1821-1904) in 1857. In this same period, in 1853, T. Sterry Hunt (1826-1892) contributed a paper that is notable for anticipating the view of Gustav Tschermak, taken a decade later, that the plagioclase feldspars were an isomorphous mixture of albite and anorthite. Hunt obtained an M.A. from Yale in 1852. He was one of the 19th century pioneers in geochemistry, and in theoretical inorganic and organic chemistry, with wide geological experience in the United States and Canada (E. F. Smith, 1926; Boyle, 1971). In 1848 J. H. Alexander, a charter member of the National Academy of Sciences (Hilgard, 1877), announced a new law of crystallography and physics. Pending further confirmation, never published, the law was stated in a note in the *American Journal of Science* as an anagram with 118 identified plus 37 unidentified letters.

Most of the later 19th and early 20th century work on chemical crystallography centered at Yale with George J. Brush, Edward S. Dana, Samuel L. Penfield, H. W. Foote, H. L. Wells, and H. L. Wheeler among the major contributors. Edward S. Dana followed his father, James D. Dana, to become a leading morphological crystallographer and mineralogist of the later 19th century. A very careful and critical worker, he is said to have entirely recalculated the published crystallographic measurements of other workers before their inclusion in his 6th edition of the *"System of Mineralogy."*

Penfield (1856-1906), a graduate of Yale in 1877, was an outstanding chemical analyst and a skilled crystallographer. He was first to establish the structural equivalence of (OH) and F. This was based on analyses of triploidite, amblygonite, herderite, and topaz made in the period 1878 to 1894. This analytical finding was not immediately accepted in Europe, but in topaz he showed that the hitherto anomalous optical properties of this mineral were determined by the (OH)/F ratio with accompanying serial variations in the density, crystal angles and optic angle. He further showed that the mutual substitution of divalent Mn and Fe in the lithiophilite-triphylite series was accompanied by a systematic variation in the optical characters. Penfield also was first to establish that the chondrodite group constituted a morphotropic crystallographic series and predicted a fourth member, later found. The analysis of chondrodite by William Langstaff of New York in 1811 (published in 1823) and the analyses by Henry Seybert of Philadelphia in the 1820's of chondrodite and chrysoberyl were among the first notable contributions to American analytical chemistry. Outside of the Yale work the most important early extended study involving coordinated crystallographic, optical, and chemical data was that of Josiah P. Cooke (1827-1894) of Harvard, in the period 1867 to 1875, on the chlorite and vermiculite groups.

In the period from 1890 to 1920 Horace L. Wells (1855-1924) of Yale made important contributions to the crystal chemistry of water-soluble synthetic compounds, hitherto largely a European field of investigation. His work, with crystallographic descriptions by Penfield, involved thiocyanates, halides, polyhalides, and iodates of Cs, Rb, K, Tl, Ag, Au, and Hg. One of the first American contributors to this field was Josiah P. Cooke of Harvard, who in the 1850's and 1860's, published crystallographic studies of Rb and Cs tartrates. Another early American contribution to this general field was made in 1851-1856 by Frederick A. Genth (1820-1893) of Philadelphia who, in part with Wolcott Gibbs and with crystallographic measurements by James D. Dana, prepared numerous cobaltamines and related compounds. It was said in 1901 that this extended and elaborate research ranked among the highest chemical investigations ever made in this country. Genth was born in Germany and trained under Bunsen, Gmelin, and Liebig. Most of his work was in mineral chemistry with specialization in the then little known elements V and Te. Genth was one of the distinguished group of American mineral chemists that in the 19th century included Henry Seybert, J. Lawrence Smith, T. Sterry Hunt, Samuel L. Penfield, George A. Koenig, W. F. Hillebrand, and F. W. Clarke. Five of these men (Genth, Smith, Hunt, Hillebrand and Clarke) were elected to the presidency of the American Chemical Society; all were equally or better known for their work in geochemistry, crystal chemistry, mineral and rock analysis, descriptive mineralogy and meteoritics than for their contributions to pure inorganic or organic chemistry.

The Yale work was continued by William E. Ford (1878-1939), who took his Ph.D. in 1903 under Penfield. His work included studies of the crystal chemistry of the garnet group, calcite group and the amphiboles. Another notable series of Yale studies on solid solution in nepheline, albite and other minerals was contributed by H. W. Foote and W. M. Bradley in the period 1911 to 1914. Studies of this kind, using the optical characters and the density as the physical parameters, became quite general in the United States by the time of World War I. X-ray cell parameters soon were added as well, with C. H. Stockwell in 1927, for instance, contributing measurements of the unit cell edge to Ford's study of the garnet group. The compositional data base to which the measurements related was critically reviewed by Michael Fleischer (1937).

A massive American contribution to the crystal chemistry of organic compounds was made in 1909 by the physiologist E. T. Reichard and the mineralogist A. P. Brown, both of the University of Pennsylvania. It dealt with the crystallized hemoglobin derivatives from the blood of a wide range of animals. The isomorphous and crystallographic relations extending between the phases as determined under the polarizing microscope (with 411 crystal drawings and 600 photomicrographs) were found to reflect the phylogenetic and zoological relations of the animals. American contributions to descriptive organic crystallography were very scant prior to 1910, as shown by the summary of the world work in this field in Paul Groth's *"Chemische Krystallographie."* Much of it was done by chemists with mineralogical training in phase description or involved joint studies with mineralogists.

Geochemistry

In Europe chemical geology, later called geochemistry, had its beginnings before 1850 chiefly through the work of C. F. Rammelsberg (1813-1899), Gustav Bischoff (1792-1870) and Justus Roth (1818-1892). In this country interest in the analytical aspects of the field began early and by the later 1800's the field of mineral chemistry and geochemistry was the best in the world. The work centered in the then-called Division of Chemistry and Physics in the Washington laboratories of the U. S. Geological Survey beginning in 1883 under the direction of Frank W. Clarke (1847-1931). His best known work is the *Data of Geochemistry* with a 1st edition in 1908. Clarke's extensive development (1895, 1914) of the silicic acid theory of the constitution of the silicates was an early casualty of the new structural knowledge of crystals. Clarke graduated from the Lawrence Scientific School at Harvard in 1867. It may be noted here that the significant American contributions to crystallography during the 19th century were almost wholly in the general area of crystal chemistry and geochemistry and stemmed from observation rather than theory.

The crystallographic aspects of geochemistry, such as the structural factors controlling the distribution of minor and trace elements in mineral hosts, developed rapidly through the work of Victor Moritz Goldschmidt (1888-1947) and co-workers in Europe in the period from roughly 1920 to 1937. Trace element geochemistry was extensively taken up here during the 1930's and although spreading rapidly through the universities, the U. S. Geological Survey was the focal point. One of the largest programs undertaken in this field was the coordinated study of the crystal chemistry, geochemistry and mineralogy of U and Th carried on chiefly during the 1950's by the U. S. Geological Survey, the Atomic Energy Commission and their research contractors. Accompanying structural studies on uranium minerals were made in the Survey laboratories. Structural studies directed primarily toward geochemical and crystallochemical problems also were made later on V and B minerals. In the great program of research on lunar mineralogy and petrography supported by the National Aeronautics and Space Agency from 1969 on the main yield of information came through the interpretation of voluminous analytical data using crystallochemical arguments. For instance, the seemingly small matter of the substitutional relation between divalent Eu and Ca, established analytically by V. M. Goldschmidt in the 1930's and by subsequent structural studies of Eu compounds, was a key in tracing lunar petrologic history.

The use of the very early crystal structure determinations in the interpretation of geochemical problems was immediately taken up by mineralogists and others over the world. In 1923 the American mineralogist Edgar T. Wherry, independently of and almost simultaneously with the Italian mineralogist Ferruccio Zambonini, recognized that the relative size, or volume as then expressed, of atoms was the controlling factor in substitutional solid solutions. Zambonini's prior contribution, translated and republished by Henry S. Washington in the *American Mineralogist* (1923), considered the necessity for valence compensation in the substitutional mechanism, which Wherry did not suggest. Ralph W. G. Wyckoff (1923) further discussed the matter. Alexander N. Winchell of the University of Wisconsin also early contributed to chemical crystallography on an atomistic basis and indicated the coupled substitution of NaCa/AlSi in the zeolites. The definitive advance, however, was made by V. M. Goldschmidt in 1926 with the publication of his first paper on the laws of crystal chemistry. This was translated and republished in large part by Wherry in the *American Mineralogist* (1927). Goldschmidt's laws, later refined and extended by others are of general application to the solid state. Their first major application, however, was in geochemistry and was exploited by Goldschmidt and his co-workers in his classical *"Geochemische Verteilungsgesetze der Elemente"* published between 1923 and 1937.

Physical Crystallography

Crystal Optics

Americans early became informed about the optical characters of crystals since many of the more important European contributions were noted or abstracted in the *American Journal of Science* from 1818 onward. David Brewster's work, including his important contribution on the connection between crystallography and axes of double refraction received extended notice between 1824 and 1832. In 1829, J. W. Webster gave a good description of the use of the polariscope invented by H. J. Brooke.

The first American paper describing optical measurements on crystals is a study by Benjamin Silliman, Jr., of the optical axial angle (2V) of various types of mica. It was published in the *American Journal of Science* in 1850. Silliman's work was extended by a paper in the same journal by William P. Blake in 1851. Strain birefringence in rubber was recognized by C. G. Page in the same year. A very early effort to interpret optical characters from structural assumptions appears in J. P. Cooke's 1874 paper on vermiculite. A contribution notable for extremely careful experimental controls was the measurement in 1888, by Charles S. Hastings of Yale, of the indices of refraction of calcite to 7 significant figures to test Huyghens' law of double refraction. Hastings also measured on the same specimen what is probably the most precise value of the *rr′* cleavage angle (yielding the axial ratio of the morphological unit cell) yet reported. His work on calcite, however, and the later use by Siegbahn in 1925 of the $(20\bar{2}2)$ interplanar spacing of calcite from the same locality, Iceland, to establish his scale of X-ray wave-lengths, suffered from lack of a precise chemical analysis of the material employed. A Ph.D. thesis titled *Measurement of refractive indices* was completed at Yale in 1876 by Edward Alexander Bouchet (1852-1918). He is thought to be the first black to get a Ph.D. from an American University.

Crystal optics was the first field in physical crystallography to be extensively developed in this country. The impetus very largely came from two factors: the use of the polarizing microscope for the study of rocks in thin sections, and the identification of minerals through their optical characters as measured in crushed grains in liquids of known index of refraction. The latter technique, initiated by Friedrich Becke in Vienna but mainly deriving from J. C. Schroeder van der Kolk in Germany in 1900, was developed extensively in the early 1900's by Esper S. Larsen of the U. S. Geological Survey (later of Harvard) and by Alexander N. Winchell and Richard C. Emmons of the University of Wisconsin. Optical petrography began in the United States in the 1870's and grew very rapidly in the next few decades. It was supported chiefly then in both optical techniques and descriptive matter by the works of Harry Rosenbusch (1836-1914) and Ferdinand Zirkel (1838-1912). Many American petrographers of that time went to Germany to study with them.

Treatments of crystal optics soon appeared in American works on mineralogy and petrography. The first edition of a widely used textbook, *"Elements of Optical Mineralogy"* by Alexander N. Winchell and Newton H. Winchell of the University of Wisconsin, appeared in 1909. A feature of this work and of its later editions was the development of graphical representation of the variation in optical characters with composition. Similar works on mineral optics, especially as seen in thin sections of rocks, were published by Albert Johannson (1908, 1914), Joseph P. Iddings (1906) and later by Austin F. Rogers (1933), Ernest E. Wahlstrom (1943), and others. The definitive work on crystal optics at the time was F. Pockels' *"Lehrbuch der Kristalloptic,"* published in Germany in 1906.

Systematic determinative tables for the identication of transparent minerals by optical methods were devised by J. C. Schroeder van der Kolk in 1900, Austin F. Rogers in 1906, and by A. N. and N. H. Winchell in 1909. The main development of the method was done by Esper S. Larsen (1879-1961) in his *"The Microscopic Determination of the Non-Opaque Minerals."* The 1st edition appeared in 1921 as *Bulletin 679 of the U.S. Geological Survey* and a 2nd edition by E. S. Larsen and Harry Berman, both of Harvard, as *Bulletin 848* in 1934. The method found wide use in the United States but not so extensively abroad. It was extended to the identification of organic compounds and of synthetic inorganic compounds by A. N. Winchell. The 1st edition of his *"The Optical Properties of Organic Compounds"* appeared in 1943 and a 2nd edition, including about 2500 substances, in 1954. Here the response was relatively small. The 1st edition of Winchell's *"The Microscopic Characters of Artificial Inorganic Solid Substances or Artificial Minerals"* was published in 1927. The use of optical methods dropped sharply following the coming of X-ray powder determinative schemes. They remain useful, nevertheless, especially in dealing with mixtures, and are more broadly informative to a trained observer.

The identification of opaque minerals by polarized light reflected from polished surfaces, supplemented by physical and chemical tests made under the microscope, is a field initially developed in very large part in the United States. It began in 1906 with the work of William Campbell, a metallurgist at Columbia University, and was rapidly expanded up through the 1930's especially by L. C. Graton and his students at Harvard (Graton, 1937) and by M. N. Short of the U. S. Geological Survey and the University of Arizona. Joseph Murdoch's *"Microscopical Determination of the Opaque Minerals,"* published in 1916 and based on a Ph.D. thesis at Harvard, was the first book in the field. Ellis Thomson of the University of Toronto also was an early contributor. Hans Schneiderhöhn, H. van der Veen, Paul Ramdohr and, with regard to the theory of reflection of polarized light from absorbing crystals, Max Berek (1937) were leading European investigators with Ramdohr ultimately producing the definitive work (1980).

The work of the French crystallographer A. Des Cloizeaux in 1864-1867 and of other late 19th century French and German investigators in the field of physical optics did not evoke much response here. Studies made from 1912 to 1917 of the thermal variation of the optical properties of crystals by Edward H. Kraus (1875-1973) of the University of Michigan were among the earlier small American contributions. Kraus took his Ph.D. with Paul Groth in Munich in 1901 and became an influential figure in American mineralogy and crystallography.

Frederick E. Wright (1877-1953) was associated with the Geophysical Laboratory of the Carnegie Institution of Washington from its founding in 1906 to his retirement in 1941. Wright was a pioneer in this country in theoretical crystal optics and in the design and application of the polarizing microscope. He received his Ph.D. magna cum laude from Heidelberg in 1900, where part of his skill in instrumentation was gained from working in Peter Stöe's optical shop. His notable work *"The methods of petrographic-microscopic research, their relative accuracy and range of application"* was published by the Carnegie Institution in 1911.

The invention of the polarizing microscope was one of the most consequential advances in crystallographic instrumentation. The major part of the vast observational data base on compositional-variation in minerals, toward which modern crystallochemical and structural studies are directed, was gained through the recognition under the polarizing microscope of the variation in the optical characters of minerals. Direct chemical analysis was contributory to this in much smaller part until the advent of modern spectrochemical techniques. In the study of rocks, the polarizing microscope led first to the development of the descriptive field of petrography and through it to the petrologic study of the origin and interrelations of rock types. This in turn stimulated the study of the phase relations between rock minerals in relevant synthetic systems. The experimental study published by the Carnegie Institution in 1905 by A. L. Day, E. T. Allen, and J. P. Iddings on isomorphism in the feldspars was one of the first American responses in a rapidly growing field.

Crystal Physics

Crystal physics also developed slowly during the 19th century. An outstanding American contribution was made by Percy W. Bridgman (1882-1961) of Harvard. Bridgman was a Ph.D. from Harvard in 1908. He pioneered the study of phase transitions, elastic constants, electrical resistivity and thermo-electric behavior of single crystals at very high pressures. His first work on the pressure polymorphism of ice was done in 1911, with 35 later contributions to this general field. Bridgman's work on methods for the growth of large single crystals, for use in his experimentation, foreshadowed both zone refining and the use of necked-down tubes and crucibles to control nucleation. His designs for the construction and sealing of reaction vessels greatly aided the field of phase synthesis at extremely high pressures and temperatures.

The earlier American work in crystal physics dealt mostly with polycrystalline aggregates. George F. Becker (1847-1919) contributed a leading paper on *Finite homogeneous strain, flow and rupture of rocks* in 1893. Becker, a Harvard graduate of 1868 with an advanced degree from Heidelberg, was influential in the development of both geochemistry and geophysics in the U. S. Geological Survey and the Geophysical Laboratory. The Canadian geologist Frank D. Adams (1859-1942) of McGill University also did pioneer work on the elastic constants, compressibility and deformation of rocks. Many other aspects of physical crystallography did not receive much more than formal reference in American textbooks or an occasional paper until into the 20th century. Much of the early American work centered at the Geophysical Laboratory in Washington with Robert B. Sosman, Arthur L. Day, R. S. Shepard, E. T. Allen, H. E. Merwin, and F. E. Wright among the contributors. Among the notable early crystallographic studies made here were those of F. E. Wright and E. S. Larsen (1909) on the high-low thermal inversion in quartz near 573°C and its use as a geologic thermometer, and the work by C. N. Fenner (1913) on the stability relations among the silica polymorphs. F. E. Wright (1913) investigated the change in the crystal angles of quartz as a function of temperature. Sosman's book *"The Properties of Silica,"* published in 1927, is a model for the critical evaluation of experimental data on the physical properties of solids.

The relevance of crystal structure analysis to fields of interest in the Geophysical Laboratory was recognized early (Wright, 1920). X-ray diffraction equipment was installed following the appointment of R. W. G. Wyckoff in 1919. Among his early contributions were the crystal structure of some carbonates of the calcite group (1920) and the structures of high-cristobalite and high-quartz (1925, 1926). Wyckoff's *"The analytical expression of the results of the theory of space groups"* was published by the Carnegie Institution in 1922 with a 2nd edition in 1930. Maurice L. Huggins, later to go to the Eastman Kodak Laboratories, was appointed to the Geophysical Laboratory in 1926 and George Tunell in 1925. Tom F. W. Barth (1899-1971), whose interests were mainly in geochemistry and petrology, worked at the Geophysical Laboratory in 1930-36 and again in 1939. In 1932 he and E. Posnjak contributed an important early paper on order-disorder in the spinels. Sterling B. Hendricks worked cooperatively with the crystallographic group at the Geophysical Laboratory during the 1930's.

One field in crystal physics that has been extensively developed by Americans is the piezoelectric properties of single crystals. Among the leading contributors from the 1910's onwards were Walter G. Cady and Karl S. Van Dyke of Wesleyan University together with W. P. Mason and other scientists of the Bell Telephone Laboratories. The manufacture of quartz oscillator plates for frequency control in military radios during World War II and the earlier use of Rochelle salt and other synthetic crystals for similar applications in telephony constitute by far the largest practical application of crystallography to date. Up to 1945 over 80 million oscillator plates were cut from quartz crystals using X-ray, morphological, optical and etching techniques to control the crystallographic orientation (cf. Frondel, *et al,* 1945).

Among the earliest American contributions to the interpretation of the physical characters of crystals in terms of crystal structure was the paper by Maurice L. Huggins in 1923 on the structural control of cleavage. This work expanded rapidly to include the crystallographic mechanism of plastic deformation in single crystals (Buerger, 1928, 1930), problems of annealing and kindred subjects in metallurgy, polymorphism and other topics.

Crystal Growth and Solution

The study of etch pits and other symmetry-dependent solution phenomena of single crystals was almost wholly developed in Europe from the work of Widmanstätten in 1808 and of Daniell in 1816 onward with an important contribution by H. A. Baumhauer appearing in 1894. The earliest American work was the study by Meyer and Penfield in 1890 on the dissolution of a single-crystal sphere of quartz. The fine early study by the Harvard geologist and petrologist Reginald A. Daly (1899) on the etch figures of the amphiboles and pyroxenes was published in an American journal but the experimental work was done with Harry Rosenbusch at Heidelberg and Alfred Lacroix in Paris. The solution-bodies obtained from calcite spheres and the light-figures yielded by them were described by F. E. Wright and Victor Goldschmidt (1903, 1904) during Wright's sojourn in Heidelberg. The first major contribution by an American to this field was the book by Arthur P. Honess (1887-1942) of Pennsylvania State University, *"The Nature, Origin, and Interpretation of the Etch Figures on Crystals,"* published in 1927 but based on experimental work done in the period 1916-1923 at Princeton. The Canadian mineralogist William H. McNairn (1874-1953) of McMaster University also worked early in this field.

The interpretation of crystal habit in terms of lattice theory had European roots with a major domestic contribution coming from J. D. H. Donnay and David Harker in 1937. This work extended the law of Bravais to include the effect of space group symmetry in calculating the reticular density of lattice planes. A fundamental American contribution to crystal growth and habit was the treatment of crystal interfacial energies by Willard Gibbs of Yale in 1876 in his development of the phase rule. The discussion of crystal growth by James D. Dana in the 1850 edition of his *"System of Mineralogy"* is of interest in anticipating aspects of modern layer-spreading growth mechanisms. Edgar T. Wherry's paper of 1924 *At the surface of a crystal* was a factor in stimulating American crystallographic interest in crystal/solution interface phenomena. Early contributions to the modification of crystal habit by adsorption during growth and to sectoral compositional variation were made by A. J. Walcott (1926), W. G. France (1930), and C. Frondel (1935). The suggestion by George F.

Becker of the U. S. Geological Survey and Arthur L. Day of the Geophysical Laboratory in 1905 that crystals can exert a pressure during their growth and its geological implications led to a considerable literature (cf. Buckley, 1951). The effect of pressure on crystal habit was first investigated experimentally by F. R. Fraprie (1906) at Harvard.

Epitaxial growths, currently of great technological importance, were the subject of a large German and French literature, with Otto Mügge and Frederick Wallerant as important contributors, but the subject long received scant attention here. E. S. Dana's paper of 1877 in volume 1 of the *Zeitschrift für Kristallographie* on quartz-calcite intergrowths seems to be the first notable American contribution. The related field of oriented precipitation from the solid state early attracted attention from students of the feldspars and from metallurgists. Olaf Anderson's experimental study in 1915 of aventurine feldspar, made in the Geophysical Laboratory, is noteworthy. The microscopic study of opaque ore minerals in light reflected from polished surfaces also yielded from about 1906 onward a large amount of qualitative information on such intergrowths. The interpretive aspects of this field in both mineralogy and metallurgy gained pace following X-ray structural studies, especially in connection with polymorphic inversions and order-disorder effects. The relation of the actual crystal structures in determining the orientation of specific mineral intergrowths was first discussed in the American literature by John W. Gruner in 1929 and later by M. J. Buerger (1934).

American interest in the synthetic growth of large single crystals for commercial and research purposes began during World War I with the development of crystals such as ammonium dihydrogen phosphate for use as piezoelectric transducers. The commercial synthesis from melts of optically clear single crystals of substances such as CaF_2 and LiF began with the work of D. C. Stockbarger at the Massachusetts Institute of Technology in the 1920's. The crystal synthesis of quartz, alkali halides and a variety of other substances reached large-scale development during and immediately after World War II. The Verneuil technique for the synthesis of single crystals of Al_2O_3 and other refractory substances was introduced from Switzerland at that time and put into large scale production by the Linde Air Products Company. The hydrothermal synthesis of large quartz single crystals, chiefly for piezoelectric purposes, began on a commercial scale in the middle 1950's by the (then) Brush Development Company. Production by the Western Electric Company began in 1962. The first demonstrably valid synthesis of diamond was effected by H. T. Hall of the General Electric Research Laboratories in 1954.

Structural Crystallography

The first structural crystallographer to appear in our country, rebuffed in his time and forgotten today, was Gustavus Detlev Hinrichs (1839-1923). Born in Schleswig-Holstein, then a part of Denmark, and educated at the University of Copenhagen, he came to the United States in 1861. Here he spent 25 years as Professor of Physics and Chemistry at Iowa State University followed by 20 years at Saint Louis University. Most of his voluminous scientific work, listed by Keyes (1924), was published in European journals for want of domestic reception. His crystallographic work, aside from contributions to physical chemistry, atomic theory and meteorology, includes three books, *"Principles of Pure Crystallography"* (1871), *"Introduction to Crystallographic Chemistry"* (1904), and *"Principles of Chemistry and Molecular Mechanics"* (1874) [none seen by the writer]. Among his crystallographic papers is *Ueber den Bau des Quartzes* published in the *Sitzungsberichte of the Vienna Academy* in 1870. In this he took the structural motif of quartz to be a polar equilateral triangle with one Si and two O at the corners. These were then linked together in such manner as to acquire the polarity of the *a*-axes, the enantiomorphism and the rotatory polarization. An accompanying appreciative comment on the paper and on Hinrichs by William Haidinger states that Iowa was located "im fernen Western zwischen Mississippi und Missouri." Another paper by Hinrichs in 1870 dealt with the symmetry reasons for the observed statistical distribution of some 2000 crystallized substances among the crystal systems. Some of his ideas were novel. A paper probably of this nature titled *Physico-chemical remarks on the reality of rhombotesseral forms,* submitted to the Vienna Academy in 1869, was referred to a committee from which his thoughts never emerged. Other intriguing titles of papers given or listed by him at meetings of the American Association of Arts and Sciences in 1868 and 1869, but never published, included *The calculation of the crystalline form of the anhydrous carbonates, nitrates, sulfates, perchlorates, permanganates and other salts of the composition* AB_3C *or* AB_4C (where CB_3 and CB_4 are the anionic complex).

A few papers in X-ray crystallography were contributed by Americans before 1920, notably on the structure of chalcopyrite by C. L. Burdick and J. H. Ellis in 1917 and the structure of various metals by A. W. Hull in the same year. Rapid expansion began in the 1920's. This structural work was done mostly in chemistry, metallurgy, and physics departments in the universities and in industrial and other non-academic organizations. Before 1919 the only structural work in the United States was done at the General Electric Company laboratories in Schenectady, the Chemistry Department at Cornell University, and the Massachusetts Institute of Technology, with the California Institute of Technology soon becoming a leading center. X-ray structure analysis at the Massachusetts Institute of Technology was stimulated by a short series of lectures given there by W. L. Bragg in 1927. Among those in the small audience who went on to professional careers in this field were Bertram E. Warren, George L. Clark, John Norton, and Martin J. Buerger. The beginning American structural work was accompanied by fundamental contributions by G. N. Lewis, Irving Langmuir and Linus Pauling to the electronic theory of valence and the nature of the chemical bond.

There was a mixed response to the coming of X-ray crystallography by mineralogy and geology departments. With regard to the actual determination of crystal structures very few mineralogists and morphological crystallographers were equipped to go into this new field and very few did so. The first course in X-ray crystallography in a mineralogy-geology department was given by John W. Gruner of the University of Minnesota in either the winter quarter of 1928-29 or in the preceding year. Gruner learned X-ray techniques while on sabbatical leave at the University of Leipzig in 1926-27. Here he worked with F. Rinne and E. Schiebold and determined the structures of analcite and boracite using the oscillation method. Gruner was largely responsible for the introduction of the Schiebold oscillation camera technique into the United States and contributed a paper on it to the *American Mineralogist* in 1928. Lewis S. Ramsdell (1895-1975) of the University of Michigan described the structure of some metallic sulfides in 1925. He went to the University of Manchester for further study and after his appointment at Michigan organized a course in X-ray crystallography, in 1933, that was attended by mineralogists, chemists, metallurgists, and physicists. Others who then or somewhat later effected the transition were Adolph Pabst of the University of California (Berkeley) and J. D. H. Donnay. Aside from mineralogical studies Pabst's early crystallographic work included structure determinations on plazolite, gillespite and other minerals, and a structural classification of the fluoaluminates. Arthur L. Parsons (1873-1957) of the University of Toronto, a geologist, mineralogist, and crystallographer (specializing in the hexagonal system), contributed the first Canadian X-ray structural study. This was done jointly with G. Aminoff in 1928 on the crystal structure of sperrylite. The mineralogist Harry Berman, whose untimely death in 1944 cut short a promising career, did the first single-crystal X-ray work at Harvard in the 1930's and in 1937 published the leading early paper on the structural classification and crystal chemistry of the natural silicates.

It was not until the late 1950's and the 1960's that structure determination became an activity in some larger geology departments. The work was done primarily by persons brought in for the purpose. X-ray work started in the Washington laboratories of the U. S. Geological Survey in 1940. It was mostly service work on mineral identification by the powder method. Crystal structure analysis began in the period 1949-1953 with the appointments of C. L. Christ, H. T. Evans, Jr., and Joan R. Clark. The first structure determination was that of montroseite by Evans in 1953. Among other Survey work, both Christ and Clark contributed extensively to the structure, structural classification and crystal chemistry of boron minerals. A major factor in the rapid development of X-ray crystallography and the structural aspects of crystal chemistry and crystal physics after World War II was the beginning of large scale funding by Governmental agencies in support of pure and especially applied research in the Universities.

X-Ray Powder Diffraction

The initial use of X-rays in mineralogy was for purposes of mineral identification using the powder technique. This soon became a major determinative technique here as in other solid state fields. The first mineralogical application was made in 1923 at Stanford by Paul F. Kerr. At the urging of Austin F. Rogers he constructed an X-ray unit and did a Ph.D. thesis on *The identification of opaque ore minerals by X-ray powder diffraction patterns.* This was published in *Economic Geology* in 1924. Alexander N. Winchell of the University of Wisconsin also drew early attention to the utility of what he called X-ray finger prints and by 1927 he had a file of some 170 standard powder patterns. The first X-ray unit for instruction and research in the Department of Mineralogy and Petrography at Harvard was installed by Harry Berman in 1933. It comprised a homemade gas tube and a second-hand transformer (earlier used for medical radiography) together with some homemade powder cameras and a camera for oscillation photographs. The high tension lead was a 14 foot bare copper rod suspended from the ceiling, the whole unit generating crackling noises and an odor of ozone, with the operator, hair on end, occasionally rapping the high tension lead with a broom handle to stimulate a sticky meter attached thereto; the vacuum pump, working constantly against a needle valve on the gas tube to keep the voltage reasonably constant, resounded through the building. The early work on this primitive unit resolved some long-standing problems in mineralogy. Among them were the identity of various microcrystalline or supposedly amorphous substances such as limonite, wad, bauxite, opal, and clays. By the 1960's, using modern equipment, over 12,000 powder photographs had been taken on film. Work on X-ray physics had begun much earlier in the Department of Physics at Harvard under William Duane. In the early 1920's this had extended to the use of Fourier methods to study the electron distribution in atoms. The then Department of Metallurgy also had an X-ray facility with a Coolidge tube. This was used in 1934 by an enterprising mineralogist, Allen W. Waldo, to do a thesis on the identification of Cu ores by the X-ray powder method.

The first definitive compilation of spacing data for the ore minerals was contributed by L. G. Berry of Queen's University and R. M. Thompson of the

University of British Columbia in 1962. It was published as *Memoir 85 of the Geological Society of America*. The data of this compilation, unlike much of the earlier work, were accompanied by a careful characterization of the X-rayed specimens. The program of work was initiated by M. A. Peacock of the University of Toronto. The paper by J. D. Hanawalt *et al.* in 1938 drew general attention to the utility of X-ray interplanar spacing data as a means of identification and resulted in the American Society of Testing Materials (ASTM) system.

The unit cell dimensions and space group are employed as systematic determinative criteria in *Crystal Data.* The 1st edition by J. D. H. Donnay and W. Nowacki appeared in 1954 as *Memoir 60 of the Geological Society of America*, and a 2nd edition, with J.D.H. Donnay as general editor, was published in 1963 as *ACA Monograph 5*.

Publications

The crystallographic contributions in America in the 19th and early 20th centuries are scattered through a wide and varied literature. It included national and state geological survey publications and the proceedings, transactions, and bulletins of a variety of city, state, and national scientific societies and museums. Serial publications by university departments also were a factor. The *American Journal of Science,* founded by Benjamin Silliman in 1818, long carried the leading papers. The bulk of the professional crystallographic contributions began to appear in the *American Mineralogist* after its beginning in 1916. The American Mineralogical Society was founded in 1919 and the *American Mineralogist* became its official publication.

The first volume of the *Zeitschrift für Kristallographie* was published in 1877 with Paul Groth as editor. It carried papers by Americans from the start, including two in the first issue, together with lengthy abstracts or complete reprints of important work published in the United States. In 1878 papers by E. S. Dana on the crystallography of two organic compounds and by J. P. Cooke on the crystallography of some halogen compounds of Sb received detailed notice. A large proportion of the publications in the early period of the *Zeitschrift,* including those by Americans, were mineralogical in nature with a growing emphasis on crystallography when Paul Niggli became editor in 1921 (Groth, 1927). It would be interesting to know the subscription list of the *Zeitschrift* in the United States in its early issues but unfortunately all of the publisher's records are lost.

European work in crystallography was not easily accessible to Americans in the 19th century. Important works were reviewed in the *American Journal of Science* or were discussed in the Scientific Intelligence section of this Journal. The long series of separately published appendices to the various editions of Dana's *"System of Mineralogy"* were more or less current sources of information from 1850 onward to 1915. *The Annual Reports of the Smithsonian Institution* beginning in 1847, and other publications of the Smithsonian also were important in describing foreign developments. In the 1880's, for example, Edward S. Dana contributed annual detailed accounts of the worldwide progress in all aspects of crystallography and mineralogy. *Chemical Abstracts* began in 1907 and *Mineralogical Abstracts,* published by the Mineralogical Society of Great Britain, began in 1920. The European literature was relatively well organized in terms of abstract and reference publications. Particular mention may be made of the *Neues Jahrbuch für Mineralogie,* beginning in 1830, and of the *Zeitschrift für Kristallographie.*

In the 1920's the beginning literature on X-ray crystallography took a different course, with almost all early contributions appearing in American or European journals devoted to physics, chemistry, or metallurgy or in the *Zeitschrift für Kristallographie.* Some of the first literature on crystal structure analysis by X-rays was listed in the *American Mineralogist,* and a complete bibliography of the structural work in the early years was given in the 1924 and 1931 editions of *"The Structure of Crystals"* by R. W. G. Wyckoff. In later years only a small part of the rapidly growing structural literature was included in the mineralogical reference journals.

Societal Organization

The societal affiliations of the early structure analysts were quite varied and, as may happen with a small and new field, their work often received limited opportunity for presentation and discussion at the respective national meetings. Efforts to form a crystallographic society and a crystallographic journal in the United States go back in time to 1919 and probably before. It is noted by Phair (1969) that Frederick E. Wright refused to join the fledgling Mineralogical Society of America in 1919 as a founding member unless the name of the Society was changed to the Crystallographic Society of America since " crystallography is a much broader subject than mineralogy and if we look upon crystallography as the science which has to deal with matter in the solid state then crystallography is on a par with physics and chemistry." He added that there was more chance for a crystallographic society in view of the complete lack of publication outlets for papers on that subject. These became popular topics in later years. After efforts to compromise with the title Crystallographical and Mineralogical Society of America failed, the present Mineralogical Society of America gained its name. The heart of the problem is that mineralogy is allied with chemistry, crystallography and physics but, unlike these sciences, has as its primary interests the description, occurrence, origin, and utilization of crystallized chemical compounds - minerals - as they occur in and relate to natural environments. Geology is its closest relative.

Societal affiliations were not a problem in the 1840's since there were only a handful of crystallographers in the country. Separation seems to have been more needed than association since the group engaged in some lively controversies. One writer, in urging the identity of two minerals on crystallographic grounds, stated that his opponent had "a very superficial knowledge of the whole subject" and indicated that added factors were "conceit or lack of industry." In the latter 1800's and early 1900's, interaction came through meetings of the Geological Society of America and the American Association for the Advancement of Science. More important were meetings of local organizations such as the Boston Society of Natural History, the New York Academy of Sciences, the Philadelphia Academy of Natural Sciences, the Washington Academy of Sciences, and various state scientific societies. Organizations such as the New York Mineral Club and the Philadelphia Mineral Club then had largely professional memberships and played a significant role. The Philadelphia Club, in particular, encouraged many neophytes into professional careers in crystallography and mineralogy.

The annual meetings of the Mineralogical Society of America for some decades after 1919 seem to have represented the broad societal type of organization at its best. The attendence at the annual meetings numbered only in the low hundreds for some years, with most participants knowledgeable of each other and understanding to a considerable degree all of the scientific interests of the Society both in content and relevance. The extremely large constituency of science and the narrow and deep specialization therein makes such broad interaction difficult today.

The first successful effort by Americans interested in crystallography to form a professional society inclusive of chemistry, physics, mineralogy, ceramics and other fields of application was the organization of the Crystallographic Society of America in the Fall of 1939. It was comprised of a small group of people chiefly in the Cambridge-Boston area. A formal notice was later published in *Science* (1942) and five meetings were held, including invited lectures by Percy Bridgman and Isidor Fankuchen, before activities were suspended by World War II. The initial officers were M. J. Buerger, President, Harry Berman as Vice-President, and C. Frondel as Secretary-Treasurer. After the war a large meeting of the CSA was held at Smith College in March of 1946, with William Parrish as Secretary, and the society assumed a national status. Twelve papers were presented at the meeting and abstracts of these, as well as of papers given in later meetings, were published in the *American Mineralogist.* A second society, stressing crystal structure analysis, arose a few years later than the CSA. Primarily through the interest of the Committee on X-Ray and Electron Diffraction of the National Research Council, conferences were held in 1941 in New York City and Gibson Island, Maryland, at which the American Society for X-Ray and Electron Diffraction was organized. The first officers of ASXRED were M. L. Huggins, President, B. E. Warren, Vice-President, and George Tunell, Secretary-Treasurer. A joint meeting of the CSA and ASXRED at Yale in April 1948 did much to coordinate the interests of the two groups and, following further discussion at the first meeting of the International Union of Crystallography at Harvard in August 1984 (Plate 9) [see page 187] the two societies merged to form the American Crystallographic Association. The first officers of the new organization were I. Fankuchen, President, R. W. G. Wyckoff, Vice-President, H. T. Evans, Jr., Secretary, and J. Karle, Treasurer. The charter membership numbered almost 500. The first meeting of the ACA was held at Pennsylvania State University in April 1950. Detailed accounts of the history of the ACA are given by Elizabeth A. Wood and William Parrish (1957) and by M. J. Buerger (1974).

By the 1940's the rapidly growing volume of crystallographic contributions together with the suspension of the *Zeitschrift für Kristallographie* during the war years made it necessary to form a new journal devoted to crystallography. The formal steps were taken in July 1946 when a delegation of American crystallographers including M. J. Buerger, J. D. H. Donnay, I. Fankuchen, L. H. Germer, David Harker, Dan McLachlan, and W. H. Zachariasen met with W. L. Bragg, H. Lipson and other members of the X-Ray Analysis Group at Cambridge, England. From this meeting arose both *Acta Crystallographica* and the International Union of Crystallography. Volume 1 of *Acta Crystallographica* appeared in 1948 with P. P. Ewald as editor. In the following 30 years or so the periodical literature in crystallography has greatly expanded and specialized. The problems of the instructional organization, societal organization, and subject matter of crystallography also continue to enlarge as the science extends with increasing interest in partial order and disorder into the technology of the solid state.

Acknowledgments

I thank the Countway Library of the Harvard Medical School for permission to quote from the Benjamin Waterhouse manuscripts kept in its archives. Other unpublished material on the early history of mineralogy and crystallography at Harvard has been drawn from the archives of the University in the Pusey Library. Mr. Julian Green, Librarian of the Department of Geological Sciences at Harvard, has been of aid throughout. My deep appreciation is expressed to a number of persons who have aided in this reconnaissance of a large and diffuse field. They hold no responsibility for views expressed or for the selection of illustrative contributions by individuals. In particular, Professor Martin J.

Buerger has contributed from his knowledge of the development of X-ray crystallography. Supplementary information and comments also have been given by Professor George Tunell of the University of California (Santa Barbara), Dr. Michael Fleischer and Dr. Howard T. Evans of the U. S. Geological Survey, Professor Horace Winchell and Dr. Barbara L. Narendra of Yale University, Professor Edward W. Nuffield of the University of Toronto, Professor Leonard G. Berry of Queen's University, Dr. H. R. Steacy of the Geological Survey of Canada, Dr. Robert M. Hazen of the Geophysical Laboratory, and my colleagues Professors Cornelius S. Hurlbut and James B. Thompson and Dr. Carl Francis of Harvard University. My wife Dr. Judith W. Frondel has been of assistance throughout the study.

References

Accum, F. (1813). Elements of crystallography, after the manner of Haüy. London, 2nd ed., 1804.

Alger, F. (1844). Beaumonite and lincolnite identical with heulandite. Boston Jour. Nat. Hist. **4**, 422-426.

Aminoff, G. (1925). On beryllium oxide as a mineral and its crystal structure. Z. Kristallogr. **62**, 113-122.

Aminoff, G. and Parsons, A. L. (1928). Crystal structure of sperrylite. Univ. Toronto Studies, Contrib. Can. Mineral. No. **27**, 5-10.

Anderson, O. (1915). On aventurine feldspar. Amer. J. Sci. **40**, 407-454.

Anon, (1822) [no title]. J. Acad. Nat. Sci. Philadelphia **2**, 397.

Bache, F. (1938). [no title]. Proc. Amer. Phil. Soc. 1, 97.

Barker, T. V. (1922). Graphical and tabular methods in crystallography. London, 152 pp.

Barth, T. F. W. and Posnjak, E. (1932). Spinel structures with and without variate atom equipoints. J. Washington Acad. Sci. **21**, 255-258.

Bayley, W. S. (1910). Elementary crystallography. New York, 241 pp.

Beck, L. C. (1850). Report on the mineralogy of New York comprising notices of the additions which have been made since 1842. New York State Cabinet, Ann. Rpt. **3**, 109-153.

Becker, G. F. and Day, A. L. (1905). The linear force of growing crystals. Proc. Wash. Acad. Sci. **7**, 283-288.

Berek, M. (1937). Optische Messmethoden im polarisierte Auflicht, mit einer Theorie der optik absorbierenden Kristalle. Fortschr. Min. **22**, 1-104.

Berman, H. (1937). The constitution and classification of the natural silicates. Am. Mineral. **22**, 342-408.

Berry, A. J. (1960). A sketch of the study of crystallography and mineralogy in Cambridge, 1808-1931. Int. Union Cryst., Cambridge Meeting, England, Aug. 1960. 16 pp.

Blake, J. M. (1916). Plotting crystal zones on paper. Amer. J. Sci. **42**, 486-492.

Blake, J. M. (1917). Crystal drawing and modeling. Amer. J. Sci. **43**, 397-401.

Blake, J. M. (1918). Means of solving crystal problems. Amer. J. Sci. **46**, 651-662.

Blake, W. P. (1851). On a method for distinguishing between uniaxial and biaxial crystals when in thin plates. Amer. J. Sci. **12**, 6-9.

Boyle, R. W. (1971). Thomas Sterry Hunt, Canada's first geochemist. Proc. Geol. Assoc. Canada **23**, 15-18.

Brewster, D. (1818). On the laws of polarization and double refraction in regularly crystallized bodies. Phil. Trans. Roy. Soc. **108**, 199-273.

Brush, G. J. and Dana, E. S. (1878). On a new and remarkable mineral locality at Branchville, in Fairfield County, Connecticut. Amer. J. Sci. **16**, 33-46.

Buckley, H. E. (1951). Crystal Growth. New York, 1951, 468-479.

Buerger, M. J. (1928). The plastic deformation of ore minerals. Am. Mineral. **13**, 1-17, 35-51.

Buerger, M. J. (1930). Translation gliding in crystals. Am. Mineral. **15**, 45-64, 174-187, 226-238.

Buerger, M. J. (1934). Lattice indices and transformations in the gnomonic projection. Am. Mineral. **19**, 360-369.

Buerger, M. J. (1936). The law of complication. Am. Mineral. **21**, 702-713.

Buerger, M. J. (1974). Background and early history of the American Crystallographic Association. (Unpubl. manuscript of lecture given Aug. 1974 at Penn. State Univ.).

Buerger, M. J. and Buerger, N. W. (1934). Crystallographic relations between cubanite segregation plates, chalcopyrite matrix and secondary chalcopyrite twins. Am. Mineral. **19**, 45-54.

Buerger, M. J. and Butler, R. D. (1936). A technique for the construction of models illustrating the arrangement and packing of atoms in crystals. Am. Mineral. **21**, 150-172; 23, 471-512 (1938).

Burdick, C. L. and Ellis, J. H. (1917). X-ray examination of the crystal structure of chalcopyrite. J. Am. Chem. Soc. **39**, 2518-2525.

Butler, G. M. (1918). A manual of geometrical crystallography. New York, 155 pp.

Campbell, W. (1906). The microscopic examination of opaque minerals. Econ. Geol. **1**, 751-766.

Carangeot, A. (1783). Goniometre ou measur-angle. J. de Phys. **22**, 193.

Cesáro, G. (1889). Les formes crystallines de la calcite de Rhisnes. Ann. Soc. Geol. Belge **16**, 165-392.

Channing, W. (1817). (review of Cleaveland's Treatise on Mineralogy). North American Review **5**, 409-429.

Chapman, E. J. (1857). On atomic constitution and crystalline form as classification characters in mineralogy. Canadian J. (Toronto) n.s., **2**, 435-439.

Clarke, F. W. (1895). The constitution of the silicates. U.S. Geol. Survey Bull. **125**, 109 pp.

Clarke, F. W. (1914). The constitution of the natural silicates. U.S. Geol. Survey Bull. **588**, 128 pp.

Cohen, I. B. (1950). Some early tools of American Science. Harvard Univ. Press, Cambridge, MA, 201 pp.

Cooke, J. P. (1867). On cryophyllite, a new mineral species of the mica family. Amer. J. Sci. **43**, 217-230.

Cooke, J. P. (1867). Crystallographic examination of some American chlorites. Amer. J. Sci. **44**, 201-206.

Cooke, J. P. (1874). The vermiculites, their crystallographic and chemical relations to the micas. Proc. Am. Acad. Arts Sci **9**, 35-67; Phil. Mag. **47**, 241-272.

Cooke, J. P. (1875). On two new varieties of vermiculites, with a revision of the other members of this group. Proc. Am. Acad. Arts Sci. **10**, 453-462.

Cooke, J. P. (1878). Kristallographisch-chemische Untersuchung einiger Haloidverbindungen des Antimon. Z. Kristallogr. **2**, 633-642.

Czapski, E. (1893). Ein neues Kristallgoniometer. Z. Instrumentenkunde **13**, 242.

Daly, R. A. (1899). A comparative study of etch-figures: the amphiboles and pyroxenes. Proc. Am. Acad. Arts Sci. **34**, 373-429.

Dana, E. S. (1870). On the association of crystals of quartz and calcite in parallel position, as observed on a specimen from the Yellowstone Park. Z. Kristallogr. **1**, 39-42.

Dana, E. S. (1872). On the datolite from Bergen Hill, N. J. Amer. J. Sci. **4**, 16-22.

Dana, E. S. (1876). On the optical character of the chondrodite from the Tilly Foster mine, Brewster, N.Y. Amer. J. Sci. **11**, 139-140.

Dana, E. S. (1878). Kristallform des Aethylidenargentamine-Aethylidenammonium-nitrat. Z. Kristallogr. **2**, 205-206.

Dana, E. S. (1895). Memorial to James Dwight Dana. Amer. J. Sci. **49**, 329-356.

Dana, E. S. (1918). A century of science in America, with special reference to the American Journal of Science, 1818-1918. New Haven, Conn. 458 pp. Also in the Centennial Number, July, 1918, of the Amer. J. Sci.

Dana, E. S. and Brush, G. J. (1878). (Branchville papers). Amer. J. Sci. **16**, 33-46, 114-123; **17**, 359-368 (1879); **18**, 45-50 (1879); **20**, 257-285 (1880); **39**, 201-216 (1890).

Dana, J. D. (1850). On isomorphism and atomic volume of some minerals. Amer. J. Sci. **9**, 220-245.

Dana, J. D. (1853). On the isomorphism of sphene and euclase. Amer. J. Sci. **16**, 96-97.

Dana, J. D. (1835). A new system of crystallographic symbols. Amer. J. Sci. **28**, 250-262.

Dana, J. D. (1854). Contributions to chemical mineralogy. Amer. J. Sci. **17**, 78-88, 210-221, 430-434.

Dana, J. D. (1854). Homeomorphism of mineral species of the trimetric system. Amer. J. Sci. **18**, 35-54.

Day, A. L., Allen, E. T., and Iddings, J. P. (1905). The isomorphism and thermal properties of the feldspars. Carnegie Inst. Washington, Pub. **31**, 13-75.

Donnay, J. D. H. and Harker, D. (1937). A new law of crystal morphology extending the law of Bravais. Am. Mineral. **22**, 446-467.

Dunham, K. C. (1977). Progress in mineralogy. Miner. Mag. **41**, 7-26.

Eaton, A. (1831). Improvements in the reflecting goniometer. Amer. J. Sci. **20**, 158-159.

Evans, H. T. (1953). The crystal structure of montroseite. Am. Mineral. **38**, 1242-1250.

Fedorov, E. S. (1893). Universal- (Theodolith-) Methode in der Mineralogie und Petrographie. Z. Kristallogr. **21**,574-714 (1893).

Fenner, C. N. (1913). The stability regions of the silica minerals. Amer. J. Sci. **36**, 331-384.

Fleischer, M. (1937). The relation between chemical composition and physical properties in the garnet group. Am. Mineral. **22**, 751-759.

Foote, H. W. and Bradley, W. M. (on solid solution in minerals). Amer. J. Sci. **31**, 25-32 (1911); **33**, 433-439, 439-441 (1912) **36**, 47-50, 180-184 (1913); **37**, 339-345 (1914).

France, W. G. (1930). Crystal structure and adsorption from solution. Colloid Sympos. Monograph **7**, 59-87.

Fraprie, F. R. (1906). On the chromates of cesium. Amer. J. Sci. **21**, 309-316.

Friedel, G. (1911). Leçons de cristallographie, 1st ed.; 2nd ed., 1926.

Frondel, C., editor (1945). Symposium on quartz oscillator plates. Am. Mineral. **30**, 205-468.

Frondel, C. (1935). Oriented intergrowth and overgrowth in relation to the modification of crystal habit by adsorption. Amer. J. Sci. **30**, 51-56.

Fuller, H. M. (1938). Philosophical apparatus of Yale College. Papers in honor of Andrew Keogh. Privately printed, New Haven.

Gerhardt, C. (1847). On the atomic volume of some isomorphous oxides of the regular system. Amer. J. Sci. **4**, 405-408.

Gibbs, W. and Genth, F. A. (1857). Researches on the ammonia-cobalt bases. Amer. J. Sci. **23**, 234-265, 319-341; **24**, 86-107.

Goldschmidt, V. (1887). Ueber Projektion und graphische Kristallberechnung. Berlin: J. Spring, 1887.

Goldschmidt, V. (1892). Goniometer mit zwei Kreisen. Z. Kristallogr. **21**, 210-232.

Goldschmidt, V. (1896). Anlegegoniometer mit zwei Kreisen. Z. Kristallogr. **25**, 321-327.

Goldschmidt, V. (1933). Die Strahlenbüchse für Strahlen-Punktbilder und Reflex-Photographie. Cb1. Min. **1933**, 49-52.

Goldschmidt, V., Palache, C., and Peacock, M. A. (1931). Über Calaverite. Jahrb. Min., Beil.-Bd. **63**, 1-58.

Goldschmidt, V. and Wright, F. E. (1903). Ueber Aetzfiguren, Lichtfiguren und Lösungskörper, mit Beobachtungen am Calcit. Neues Jahrb., Beil.-Bd. **17**, 355-390.

Goldschmidt, V. and Wright, F. E. (1904). Uber Lösungskörper and Lösungsgeschwindigkeiten von Calcit. Neues Jahrb., Beil.-Bd. **18**, 335-376.

Goldschmidt, V. M. (1926). Die Gesetze der Kristallochemie. Naturwiss. **14**, 477-485; Skrifter Norske Videnskaps-Akad. Oslo, Math.-Nat. Kl. **1926**, No. 2.

Gordon, S. G. (1919). History of mineralogy in Pennsylvania. Am. Mineral. **4**, 16-17.

Graton, L. C. (1937). Techniques in mineralography at Harvard. Am. Mineral. **22**, 491-516.

Graustein, J. E. (1961). Natural history at Harvard College. Proc. Cambridge Hist. Soc. **38**, 69-86.

Graves, R. (1832). A simplification of Dr. Wollaston's reflective goniometer. Amer. J. Sci. **23**, 75-78.

Greene, J. C. (1969). The development of mineralogy in Philadelphia, 1780-1820. Proc. Amer. Phil. Soc. **113**, 283-295.

Greene, J. C. and Burke, J. G. (1978). The science of minerals in the age of Jefferson. Trans. Amer. Phil. Soc. **68**, 5-113.

Groth, P. (1927). Vorgeschichte, Gründung und Entwicklung der Zeitschrift für Kristallographie in den ersten fünfzig Jahren. Z. Kristallogr. **66**, 1-21.

Gruner, J. W. (1928). The oscillation method of X-ray analysis of crystals. Am. Mineral. **13**, 123-141.

Gruner, J. W. (1929). Structural reasons for oriented intergrowths in some minerals. Am. Mineral. **14**, 227-237.

Hanawalt, J. D. H., Rinn, H. W., and Frevel, L. K. (1938). Chemical analysis by X-ray diffraction. Ind. Eng. Chem. Anal. Ed. **30**, 457-512.

Hastings, C. S. (1888). On the law of double refraction in Iceland spar. Amer. J. Sci. **35**, 60-73.

Haüy, R. J. (1815). [Haüy on American minerals]. Medical Repository, New Ser. **2**, 86.

Hilgard, J. E. (1877). Memorial to John Henry Alexander. Natl. Acad. Sci., Biog. Mem. **1**, 213-216.

Hinrichs, G. D. (1869). Chemisch-physikalische Bemerkungen über die Realität rhombotesseraler Formen. Anzeiger Akad. Wiss. Wien **6**, 1.

Hinrichs, G. D. (1870). Zur Statistic der Kristal Symmetrie. Sitzber. Akad. Wiss. Wien. Abt. II, **52**, 345-361.

Hinrichs, G. D. (1870). Ueber den Bau des Quarzes. Sitzber. Akad. Wiss. Wien, Abt. I, **51**, 83-88.

Hooeykas (1955). The species concept in Eighteenth-century mineralogy. Archiv. Let. d'Hist. Sci. **31**, 45-55.

Hunt, T. S. (1853). On the constitution and equivalent volume of some mineral species. Amer. J. Sci. **16**, 203-218.

Iddings, J. P. (1906). Rock minerals, their chemical and physical characters and their determination in thin section. New York, 548 pp.

Jackson, C. T. (1864). Notice of the death of Francis Alger of Boston. Proc. Boston Soc. Nat. Hist. **10**, 2-6.

Jensen, M. L. (1952). One hundred and fifty years of geology at Yale. Amer. J. Sci. **250**, 625-635.

Johannson, A. (1908). A key for the determination of rock-forming minerals in thin sections. New York, 542 pp.

Johannson, A. (1914). Manual of petrographic methods. New York, 649 pp.

Johnsen, A. (1927). Groth's Geschichte der Mineralogie. Archly. Geschichte Math., Naturwiss. und Technik **10**, 367-368.

Kerr, P. F. (1924). The determination of opaque ore minerals by X-ray diffraction patterns. Econ. Geol. **19**, 1-34.

Keyes, C. R. (1923). Gustavus Detlev Hinrichs, mineralogist, meteorologist, and physical chemist. Pan-American Geol. **39**, 337-352.

Keyes, C. R. (1924). The crystallographic work of Gustavus Hinrichs. Am. Mineral. **9**, 5-8.

Keyes, C. R. (1924). Contributions to knowledge by Gustavus Detlev Hinrichs. Iowa Acad. Sci. **31**, 79-94.

Kleber, W. (1961). Die Entwicklung der Mineralogie an der Berliner Universität während der ersten Hälfte des 20. Jahrhunderts. Wiss. Zeits. Humboldt-Univ. Berlin, Beiheft zum Jubilaumsjahrgang IX (1959/60).

Kraus, E. H. (1906). Essentials of crystallography. Ann Arbor, Mich., 162 pp.

Kraus, E. H. and Young, L. J. (1912). Variation of the optic angle of gypsum with temperature. Jahrb. Min. **1**, 123-146.

Kraus, E. H. (1913). Variation in the optical angle of glauberite with temperature. Z. Kristallogr. **52**, 321-326.

Kraus, E. H. (1921). The future of mineralogy in America. Am. Mineral. **6**, 23-34.

Lacombe, P. (1954). Cristallographie et metallographie. Bull. Soc. Franc. Mineral. **54**, 163-191.

Lacroix, A. (1915). La minéralogie. Paris, Librairie Larousse **1**, 169-200.

Langstaff, W. (1823). [Cited in letter by Thomas Nuttall]. Amer. J. Sci. **6**, 172.

Leiss, C. (1899). Die optischen Instrumente der Firma R. Fuess, deren Beschreibung, Justierung und Anwendung. Leipzig, 397 pp.

Lewis, W. J. (1882). On the measurements of a bead of platinum. Proc. Cambridge Phil. Soc. **4**, 236-239.

Longchambon, H. (1954). L'apport de la minéralogie et de la cristallographie a la science. Bull. Soc. Franc. Mineral. **77**, 23-44.

Menzer, G. (1945). Zur Geschichte des Kristallographischen Grundgesetzes. Z. Kristallogr. **106**, 1-4.

Meyer, O. and Penfield, S. L. (1890). Results obtained by etching a sphere and crystals of quartz with hydrofluoric acid. Trans. Conn. Acad. Sci. **3**, 158-165.

Mieleitner, K. (1923). Die Angänge der Theorien über die Struktur der Kristalle. Fortschr. Min., Krist., Petr. **8**, 199-234.

Morison, S. L. (1946). Three centuries of Harvard, 1636-1936. Cambridge, MA: Harvard University Press.

Moses, A. J. and Rogers, A. F. (1902). Formulae and graphic methods for determining crystals in terms of coordinate angles and Miller indices. School of Mines Quart. **24**, 1-36.

Multhauf, R. P. (1961). Catalogue of instruments and models in the possession of the American Philosophical Society. Philadelphia, p. 29.

Narendra, B. L. (1979). Benjamin Silliman and the Peabody Museum. Discovery **14**, 12-29.

Neumann, F. E. (1823). Beiträge zur Kristallonomie. Berlin and Bonn.

Niggli, P. (1946a). Mineralogie und Petrographie. Festschrift Naturforsch. Ges. Zurich 1946, pp. 190-206.

Niggli, P. (1946b). Die Krystallologia von Johann Heinrich Hottinger. Aarau, Switzerland: H. R. Sauerlander and Co. 110 pp.

Oleson, A. and Brown, S. C. (1976). The Pursuit of Knowledge in the Early American Republic: American Scientific and Learned Societies from Colonial Times to the Civil War. Baltimore: John Hopkins University Press.

Orcel, J. (1938). Histoire de la chaire de minéralogie du Museum National d'Histoire Naturelle. Bull. Mus. Nat. Hist. Nat. **10**, no. 4, 328-354.

Orcel, J. (1962). Les sciences minéralogiques au XIXe siecle. Les conferences du Palais de la Decorverte, ser. D, No. 92. Univ. Paris, Paris.

Page, C. G. (1851). A new figure in mica and other phenomena of polarized light. Amer. J. Sci. **11**, 89-92.

Palache, C. (1935). The minerals of Franklin and Sterling Hill, N. J. U. S. Geol. Surv. Prof. Paper **180**, 135 pp.

Palache, C. (1943). Calcite: an angle table and critical list. Priv. printed, Dept. of Geology, Harvard Univ. 27 pp.

Palache, C. and Lewis, L. W. (1927). A saw attachment adapting Goldschmidt's model cutting machine to the sawing of wooden models. Am. Mineral. **12**, 154-156.

Peacock, M. A. (1932). Calaverite and the law of complication. Am. Mineral. **17**, 317-337.

Penfield, S. L. (1879). On the chemical composition of amblygonite. Amer. J. Sci. **18**, 295-301.

Penfield, S. L. (1900). Kontact Goniometer und Transporteur von einfacher Konstruction. Z. Kristallogr. **33**, 548-554.

Penfield, S. L. (1901). The stereographic projection and its possibilities from a graphical standpoint. Amer. J. Sci. **11**, 1-24, 115-144.

Penfield, S. L. (1902). On the solution of problems in crystallography by use of graphical methods. Amer. J. Sci. **14**, 249-284.

Penfield, S. L. (1905). On crystal drawing. Amer. J. Sci. **19**, 39-75.

Penfield, S. L. (1906). On the drawing of crystals from the stereographic and gnomonic projections. Amer. J. Sci. **21**, 206-215.

Penfield, S. L. and Forbes, E. H. (1896). On the optical properties of the chrysolite-fayalite group and of monticellite. Amer. J. Sci. **1**, 129-135.

Penfield, S. L. and Harper, D. N. (1886). On the chemical composition of herderite and beryl. Amer. J. Sci. **32**, 107-117.

Penfield, S. L. and Minor, J. C. (1894). On the chemical composition and related physical properties of topaz. Amer. J. Sci. **47** 387-396.

Penfield, S. L. and Pratt, J. H. (1895). Effect of the mutual replacement of manganese and iron on the optical properties of lithiophilite and triphylite. Amer. J. Sci. **50**, 387-390.

Pfeiffer, E. (1936). Formative forces in crystallization. New York, 64 pp.

Phair, G. (1969). The American Mineralogist: its first four years. Am. Mineral. **54**, 1233-1243.

Rabbitt, M. C. (1979, 1980). Minerals, lands and geology for the common defense and general welfare. U.S. Govt. Print. Office, Washington, D.C., **1**, 331 pp. (before 1879); **2**, 407 pp. (1879-1904).

Ramdohr, P. (1959). Entwicklung der Mineralogie während der letzten 50 Jahre in Deutschland und in der Welt. Fortschr. Min. **37**, 27-35.

Ramdohr, P. (1980). The ore minerals and their intergrowths. New York: Pergamon Press, 4th ed., 2 vols., 1205 pp.

Ramsdell, L. S. (1925). The crystal structure of some metallic sulfides. Am. Mineral. **10**, 281-304.

Reichert, E. T. and Brown, A. P. (1909). The differentiation and specificity of corresponding proteins and other vital substances in relation to biological classification and organic evolution: the crystallography of hemoglobins. Carnegie Inst. Washington, Pub. **116**, 338 pp.

Rogers, A. F. (1901). A list of the crystal forms of calcite with their interfacial angles. School of Mines Quart. **22**, 429-448.

Rogers, A. F. (1902). New graphical methods in crystallography. School of Mines Quart. **23**, 67-72; **24**, 1-36.

Rogers, A. F. (1907). The gnomonic projection from a graphical standpoint. School of Mines Quart. **29**, 24-33.

Rogers, A. F. and Kerr, R. F. (1933). Thin-section mineralogy. New York, 311 pp.

Scheerer, M. (1848). Upon a peculiar kind of isomorphism that plays an important part in the mineral kingdom. Amer. J. Sci. **5**, 381-389; **6**, 57-63, 189-206.

Schoenflies, A. (1905). Enzyklopaedie der mathematischen Wissenschaften **5**, 437-492, publ. 1922.

Sellers, C. C. (1975). Private communication.

Seybert, H. (1822). Analysis of the maclureite or fluosilicate of magnesium. Amer. J. Sci. **5**, 336-344; **8**, 105-112 (1824).

Shafronovsky, I. I. (1973). Lectures on crystal morphology. English transl. from Russian ed. of 1968. Nat. Bur. Stds., 1973.

Silliman, Jr., B. (1850). Optical examination of several American micas. Amer. J. Sci. **10**, 372-383.

Smith, T. P. (1800). On crystallization. Medical Repository **3**, 253.

Smith, G. F. H. (1899). A three-circle goniometer. Miner. Mag. **12**, 175-182.

Smith, E. F. (1914). Chemistry in America: Chapters from the history of science in the United States. New York, 356 pp.

Smith, E. F. (1919). Chemistry in old Philadelphia. Philadelphia, 106 pp.

Smith, E. F. (1926). Mineral Chemistry. J. Am. Chem. Soc. **48**, no. 8A (Jubilee Volume, Chapt. 6).

Smith, W. C. (1978). Early mineralogy in Great Britain and Ireland. Bull. British Museum (Natural History) **6**, 49-74.

Stockwell, C. H. (1927). An X-ray study of the garnet group. Am. Mineral. **12**, 327-344.

Ticknor, E. T. (1874). Life of Joseph Green Cogswell as sketched in his letters. Privately printed, Cambridge, Mass. 377 pp.

Tomkeieff, S. I. (1960). The progress of geology in the USSR. Geotimes **5**, 9-11.

Troost, G. (1821). Description of some new crystalline forms of the minerals of the United States. J. Acad. Nat. Sci. Philadelphia **2**, 55-58.

Troost, G. (1822). Some crystals of sulphate of strontian from Lake Erie. J. Acad. Nat. Sci. Philadelphia **2**, 300-302.

Troost, G. (1823). Account of the pyroxene of the United States. J. Acad. Nat. Sci. Philadelphia **3**, 105-124.

Troost, G. (1825). Notice of a new crystalline form of the yenite of Rhode Island. Trans. Amer. Phil. Soc. **2**, 478-480.

Tunell, G. and Ksanda, C. J. (1935). The crystal structure of calaverite. J. Washington Acad. Sci. **25**, 32-33.

Tunell, G. and Ksanda, C. J. (1936). The strange morphology of calaverite in relation to its internal properties. J. Washington Acad. Sci. **26**, 509-528.

Tunell, G. and Morey, G. W. (1932). Some correct and some incorrect statements of elementary crystallographic theory and methods in current textbooks. Am. Mineral. **17**, 365-380.

Tunell, G. and Pauling, L. (1952). The atomic arrangements and bonds of the gold-silver tellurides. Acta Cryst. **5**, 375-381.

Wadsworth, M. E. (1909). Crystallography, an elementary manual. Philadelphia, 299 pp.

Wahlstrom, E. E. (1943). Optical crystallography. New York, 206 pp.

Walcott, A. J. (1926). Some factors influencing crystal habit. Am. Mineral. **11**, 221-239.

Waldo, A. W. (1935). Identification of the copper ore minerals by means of X-ray powder diffraction patterns. Am. Mineral. **20**, 575-597.

Webster, J. W. (1823). (review of Brooke's Crystallography). Boston J. Phil. and Arts **1**, 298-299.

Webster, J. W. (1824). Cabinet of minerals at Cambridge. Boston J. Phil. and Arts **2**, 201-203.

Webster, J. W. (1829). Description of the polariscope, an instrument for observing some of the most interesting phenomena of polarized light. Amer. J. Sci. **15**, 369-373.

Wheatland, D. P. (1968). The apparatus of science at Harvard, 1765-1800. Cambridge, MA: Harvard Univ. Press, 204 pp.

Wherry, E. T. (1923). Volume isomorphism in the silicates. Am. Mineral. **8**, 1-8.

Wherry, E. T. (1924). Further notes on atomic volume isomorphism. Am. Mineral. **9**, 165-169.
Wherry, E. T. (1924). At the surface of a crystal. Am. Mineral. **9**, 45-54.
Wherry, E. T. (1927). The laws of chemical crystallography. Am. Mineral. **7**, 28-31.
Whewell, W. (1833). Report on the recent progress and present state of mineralogy. Report British Assoc. Adv. Science **1832**, 322-365.
Whewell, W. (1837). History of the inductive sciences. London, 1st ed., **3**, 227.
Whitlock, H. P. (1915). A critical discussion of the crystal forms of calcite. Proc. Amer. Acad. Arts Sci. **50**, 289-352.
Whitlock, H. P. (1920). A model for demonstrating crystal structure. Amer. J. Sci. **49**, 259-265.
Williams, G. H. (1883). Relations of crystallography to chemistry. Am. Chem. J. **5**, 461-464.
Williams, G. H. (1889). On the possibility of hemihedrism in the monoclinic system with reference to pyroxene. Amer. J. Sci. **38**, 115-120.
Winchell, A. N. (1927). Finger prints of minerals. Am. Mineral. **12**, 261-262.
Winchell, A. N. (1931). The microscopic characters of artificial inorganic solid substances or artificial minerals. New York, 1st ed. 1927, 2nd ed. 1931.
Wollaston, W. H. (1809). The description of a reflective goniometer. Phil. Trans. Roy. Soc. **1809**, 253-258.
Wood, E. A. and Parrish, W. (1957). History of the American Crystallographic Association. Norelco Reporter **4**, 48-50.
Wright, F. E. and Larsen, E. S. (1909). Quartz as a geologic thermometer. Amer. J. Sci. **27**, 421-447.
Wright, F. E. (1913). The change in the crystal angles of quartz with rise in temperature. J. Washington Acad. Sci. **3**, 485-494.
Wright, F. E. (1913). Graphical methods in microscopical petrography. Amer. J. Sci. **36**, 509-539.
Wright, F. E. (1920). Annual Report of the Director of the Geophysical Laboratory, Yearbook no. 19.
Wyckoff, R. W. G. (1920). The crystal structure of some carbonates of the calcite group. Amer. J. Sci. **50**, 317-360.
Wyckoff, R. W. G. (1923). On structure and isomorphism in minerals. Am. Mineral. **8**, 85-92.
Wyckoff, R. W. G. (1925). The crystal structure of the high temperature form of cristobalite. Amer. J. Sci. **9**, 448-459.
Wyckoff, R. W. G. (1926). Criteria for hexagonal space groups and the crystal structure of beta-quartz. Amer. J. Sci. **11**, 101-112.
Zambonini, F. (1923). The isomorphism of albite and anorthite. Am. Mineral. **8**, 81-92, (abstr. by H. S. Washington).

SECTION A

CRYSTALLOGRAPHIC LABORATORIES

CHAPTER 1

X-RAY CRYSTALLOGRAPHY IN THE CALIFORNIA INSTITUTE OF TECHNOLOGY

Linus Pauling
Linus Pauling Institute of Science and Medicine,
440 Page Mill Road, Palo Alto, CA 94306

X-ray crystallography has been prosecuted in the California Institute of Technology for sixty-six years, longer than in any other educational institution in North America. I was fortunate to have come to the California Institute of Technology, as a graduate student, in September 1922, about six years after the construction of a Bragg X-ray ionization spectrometer in the Institute.

The California Institute of Technology became a significant institution of higher education in the decade beginning in 1913. In that year Arthur Amos Noyes (1866-1936), director of the Research Laboratory of Physical Chemistry in the Massachusetts Institute of Technology and for two years acting president of that institute, became associated on a part-time basis, at the request of George Ellery Hale, with Throop College of Technology, which changed its name to California Institute of Technology a few years later. In 1919 Noyes resigned his post in M.I.T. and moved to California. In the meantime he had been building up the chemistry department in Pasadena. It was he who was responsible for initiating and fostering the work in X-ray crystallography there.

The first work in X-ray crystallography

One of Noyes' students at M.I.T. was C. Lalor Burdick, an American who had been born in Denver, Colorado in 1892 and who is now (1982), after a distinguished career with the DuPont Company, executive director of the Lalor Foundation, Wilmington, Delaware. Burdick had obtained a B.S. degree at Drake University in 1911, a B.S. from M.I.T. in 1913, and then an M.S. from M.I.T. in 1914. He was encouraged by Noyes to carry on advanced study in Europe, and he sailed for Germany in July 1914 to work at the Kaiser Wilhelm Institute in Berlin. He then shifted to the University of Basel, Switzerland, where he received the doctor of science degree in 1915. Noyes suggested that he get some experience in the field of X-ray crystallography by working with W. H. Bragg in University College, London. The only other person then working in the X-ray laboratory there was E. A. Owen, a convalescent wounded Australian soldier. Burdick and Owen, using the Bragg spectrometer, determined the structure of silicon carbide.

On Burdick's return to M.I.T. he constructed a Bragg spectrometer, with some innovations. Before he had a chance to use this instrument, however, he was asked by Noyes to come to Pasadena. There he supervised the construction of another Bragg spectrometer, built by the chemistry department's instrument maker, Fred Hensen. Burdick then collaborated with James Hawes Ellis, who had received his Ph.D. at M.I.T. and come to Pasadena with Noyes, to determine the crystal structure of chalcopyrite. Their report on this structure was published in the *Proceedings of the National Academy of Sciences* in 1917 and also, in the same year, in the *Journal of the American Chemical Society*. They had measured the intensities of reflection in two or three orders from each of seven planes, and had used these intensities to locate the copper and iron atoms (not quite correctly) and to determine the value of the parameter fixing the site of the sulfur atoms in this tetragonal crystal.

This was the first X-ray crystallographic study of a single crystal carried out in North America. In the same year, 1917, A. W. Hull of the General Electric Company published his report on the use of the X-ray powder method to determine the structure of some metals - the first use of the powder method in this country.

Burdick was inducted into the American army in 1918, and did no further work in the X-ray field. Ellis remained in the California Institute of Technology for a few years, but then left. He also did no further X-ray work, although he served as an advisor on some phases of the work to me and other students.

The work of Roscoe Gilkey Dickinson

Roscoe Gilkey Dickinson (1894-1945) was brought to M.I.T. as a graduate student by Noyes in 1917, and he immediately began work with the X-ray spectrometer that had been built by Burdick. He received his degree of doctor of philosophy in June, 1920; this was the first Ph.D. degree given by the California Institute of Technology. His doctoral thesis described the crystal structure of wulfenite, $PbMoO_4$ and scheelite, $CaWO_4$, and also the crystal structure of sodium chlorate and sodium bromate (Dickinson, 1920; Dickinson and Goodhue, 1921). At about this time Dickinson gave up the use of the Bragg spectrometer and adopted the procedure that had been developed in Cornell University by Ralph W. G. Wyckoff, with the advice and assistance of a Japanese scientist, S. Nishikawa, who had made use of Laue photographs and the theory of space groups in an early study of magnetite and spinel (Nishikawa, 1915). Wyckoff spent the year 1921-22 in Pasadena, continuing his own X-ray work. In the year 1922 Dickinson published four papers on the crystal structure of inorganic complexes - potassium and ammonium chlorostannate, complex cyanides of potassium with zinc, cadmium, and mercury, phosphonium iodide, and potassium chloroplatinite and potassium and ammonium chloropalladites. In the following year he and a student, Albert L. Raymond, reported on the crystal structure of hexamethylenetetramine, the first organic compound to have its structure determined by the X-ray diffraction method. Later work with inorganic substances was carried out in collaboration with Linus Pauling, James B. Friauf, and Sterling B. Hendricks. His last crystal structure paper, with Constant Bilicke, was on the crystal structures of two organic compounds, beta-benzene hexabromide and hexachloride. This paper was remarkable in providing evidence that the cyclohexane ring has a staggered conformation.

The Nishikawa-Wyckoff technique

I received notice of my appointment as a graduate assistant in the California Institute of Technology in the spring of 1922. At that time I was a senior in the Oregon Agricultural College, Corvallis, Oregon, and I had already developed an interest in the question of the nature of the chemical bond. Noyes suggested to me, in a letter, that I work with Dickinson on the determination of the structure of crystals by the X-ray diffraction method, and suggested also that I read the book *"X-rays and Crystal Structure"* by W. H. and W. L. Bragg, which I did during the summer. When I arrived in Pasadena in September 1922, Dickinson began immediately to teach me the Nishikawa-Wyckoff technique. I was very greatly impressed by the logical arguments that were used in this procedure and by the power and rigor of the structure determinations. The procedure involved great use of Laue photographs, which provided a tremendous amount of information, extending to crystallographic planes with quite small spacings. Usually only a few rotation photographs were taken with monochromatic radiation, from a tube usually with molybdenum anticathode, in order to get possible values of the length of the edges of the unit cell. All of the further analysis was based upon the Laue photographs.

Very little information was available at that time about the dependence of the scattering power of atoms of different elements for X-rays as a function of the diffraction angle. One factor was believed to be certain: that the scattering power, the f-value, decreased with increase in the value of the diffraction angle. On the Laue photographs the spots corresponding to reflection from different crystallographic planes usually involved different wave lengths of X-rays in the continuum from a tube with a tungsten target. The voltage applied to the X-ray tube, usually 54 kV, was such that the K-lines of tungsten were not excited, and the L-lines had wave lengths longer than those in the useful part of the continuum. The first task was to determine what multiples of the smallest lengths of the edges of the unit cell given by the rotation photographs needed to be taken in order to account for the existence of all of the observed Laue spots. This was a powerful technique, and it prevented errors that were sometimes made by other investigators. For example, Dickinson found the cubic unit of tin tetraiodide to have edge 12.23 Å, with 32 tin atoms and 128 iodine atoms in the unit cube (Dickinson, 1923), whereas Mark and Weissenberg in the same year had reported a unit with edge only half so great, containing only four SnI_4 molecules. The method used by Mark and Weissenberg (1923) depended almost entirely on rotation photographs, and the weaker reflections requiring a larger unit were overlooked by them.

The next step was to search for systematic extinctions, or observed reflections in certain forms, in order to eliminate some of the space groups with the symmetry of the Laue photograph and to indicate the probable space group or space groups. Following that, the various structures compatible with the size of unit and the selected space group or space groups were examined one by one, by the comparison of pairs of X-ray reflections on Laue photographs. In order to make such a comparison in the absence of quantitative information about the f-values and of a reliable way of estimating quantitatively the intensities of the Laue spots, reflections from two planes were found such that they were produced by the same wave length of X-rays and such that the plane with the smaller spacing was observed to produce a stronger spot than the plane with the larger spacing. It could be concluded from such an inequality that the structure factor for the first plane is necessarily larger than the structure factor for the second plane. In this way it was often possible to eliminate all structures except one, and for that one to eliminate all values of the variable parameters except those in a narrow range. Both Dickinson and Wyckoff succeeeded in this way in locating atoms quite precisely, usually within about 0.03 Å. Wyckoff's first published crystal-structure determination was that of cesium dichloroiodide, in 1920. He made a great contribution to X-ray crystallography in the early years not only by his publication of a large number of structure determinations, about 50 in the decade of the 1920's, but also by his publication of a very useful book, *"The Analytical Expression of the Results of the Theory of Space Groups,"* Carnegie Institution Publications, 1922.

Later work in X-ray crystallography in the California Institute of Technology

My first structure determination, published in collaboration with Dickinson (Dickinson and Pauling, 1923), was on the crystal structure of molybdenite. I was much impressed by this structure. Dickinson had verified the prediction by Werner that complexes such as the hexachlorostannate ion have an octahedral structure and the tetracyanides of zinc and its congeners have a tetrahedral structure, whereas the complexes of palladium(II) and platinum(II) have a square planar structure. In molybdenite we found that each molybdenum atom is surrounded by six sulfur atoms at the corners of a trigonal prism—a new kind of coordination polyhedron. Also, the interatomic distances were interesting, in that there were layers of sulfur atoms in contact with one another, with the sulfur-sulfur distance 3.49Å, far greater than the sulfur-sulfur distance in pyrite, 2.10Å. The distinction between covalent radii and van der Waals radii had not yet been made. Also in 1923 Richard M. Bozorth, who had been instructed in X-ray diffraction by Dickinson, reported his structure determination of KHF_2, which provided the first X-ray evidence for the hydrogen bond. Bozorth, who received his Ph.D. degree from the California Institute of Technology in 1922, also made structure determinations of some other crystals, including As_4O_6 and Sb_4O_6, which were the first inorganic crystals shown to contain discrete molecules.

X-ray diffraction was the principal field of research in chemistry in the California Institute of Technology for several years. Of the twenty papers that had been published by the end of 1922, fifteen were on the determination of the structure of crystals. As other fields of research began to be prosecuted the papers on the determination of the structure of crystals by X-ray diffraction dropped to about 20% of the total, and then to about 10%. During the fifty years from 1916 to 1965 about 400 papers on X-ray diffraction and crystal structure were published from the Gates and Crellin Laboratories of Chemistry of the California Institute of Technology, representing the determination of the structure of about 400 crystals. Many American scientists received their training in X-ray crystallography in the California Institute of Technology. Among the earlier ones were E. A. Goodhue, Albert L. Raymond, Linus Pauling, James B. Friauf, Sterling B. Hendricks, Constant Bilicke, J. H. Sturdivant, L. Merle Kirkpatrick, Edwin MacMillan, Fred J. Ewing, Ralph Hultgren, Jack Sherman, J. L. Hoard, B. N. Dickinson, A. M. Soldate, Norman Elliott, David Harker, Franklin Offner, Kenneth S. Pitzer, Jurg Waser, Jerry Donohue, K. J. Palmer, Kenneth Trueblood, William N. Lipscomb, Paul H. Emmett, M. D. Shappell, L. O. Brockway, S. Weinbaum, E. W. Neuman, Gunnar Bergman, You-chi Tang, Walter J. Moore, Richard E. Marsh, James V. McCullough, John L. T. Waugh, David P. Shoemaker, Verner Schomaker, G. B. Carpenter, P. A. Shaffer, Jr., D. R. Davies, R. A. Pasternak, H. A. Levy, W. G. Sly, Philip A. Vaughan, H. L. Yakel, Jr., James Ibers, Walter Hamilton, Barclay Kamb, Joe Kraut, Alex Rich, C. Fritchie, Jr., D. Duchamp, Kenneth Hedberg, A. F. Berndt, N. Webb, D. Britton, A. Hybl, M. E. Jones, A. Miller, D. R. Peterson, Max Rogers, and E. Goldish.

In addition a considerable number of chemists from North America and other countries – England, Germany, Switzerland, Belgium, France, Sweden, Australia – received post-doctoral training in this field in Pasadena. Among them, mention may be made of M. L. Huggins, Harold P. Klug, Lindsay Helmholz, H. D. Springall, G. C. Hampson, Henri Brasseur, A. N. Winchell, D. C. Carpenter, Chia-Si Lu, Hans Freeman, Rose C. L. Slater, Lynne L. Merritt, Jr., Barbara Low, Jenny Pickworth Glusker, Karst Hoogsteen, and J. D. Dunitz.

A very important part in the X-ray diffraction work in Pasadena was also played, from 1937 on, by members of the staff who had received their introduction to X-ray crystallography elsewhere. Especially important with respect to the work on the determination of the structure of amino acids, simple peptides, and proteins, beginning in 1937, were Robert B. Corey, who had been introduced to X-ray crystallography by Wyckoff, and E. W. Hughes, who had begun his X-ray work in Cornell, partially in association with W. L. Bragg, when Bragg was George Fisher Baker lecturer there. Another staff member who has contributed greatly to the X-ray diffraction work in Pasadena is Sten Samson, who came from Sweden, and soon began to make important contributions to the determination of the structures of alloys. Especial mention should be made to the late James Holmes Sturdivant, who for many years designed and supervised the construction of the X-ray diffraction apparatus and who was largely responsible for the smooth running of the Gates and Crellin Laboratories.

In addition to structure determinations of many crystals (simpler inorganic and organic compounds, silicate minerals, sulfide minerals, intermetallic compounds, and peptides and proteins), some general contributions to X-ray crystallography were made in the California Institute of Technology. I may mention the formulation of a set of structural principles for the silicates and similar crystals (Pauling, 1929), the theoretical calculation of f-values for all atoms (Pauling and Sherman, 1932), the development of the theory of sections and projections of Patterson diagrams (David Harker, 1936), the introduction of the use of anisotropic temperature factors (Lindsay Helmholz, 1940), and the introduction of the least-squares procedure for refining crystal structure determinations (Hughes, 1941).

Much use was made in the California Institute of Technology of another valuable method of determining the structure of molecules, the electron-diffraction method. This method had been devised and first applied by Hermann Mark in Ludwigshafen in 1929. When I visited him in 1930 I was much impressed by the possibilities of the new technique, and I asked him if he objected to our building an electron-diffraction apparatus. He not only said that he had no objections, but also gave me the plans of the apparatus. Our first ED apparatus was built by a graduate student, Lawrence O. Brockway, with advice from Professor Richard Badger. It was soon applied in the study of scores of molecules, providing information that, together with application of quantum mechanical principles, led to the answers to many questions about chemical bonds and molecular structure. Many of the persons whose names are given above participated in the electron-diffraction work. I may mention especially, in addition to Brockway, Verner Schomaker, J. Y. Beach, S. H. Bauer, John H. Saylor, and Walter Gordy.

I left the California Institute of Technology in 1964. I do not have detailed information about what has been done there during the last eighteen years, but I have been pleased to note the continued publication of results of important investigations in the X-ray field by E. W. Hughes, Richard E. Marsh, R. E. Dickerson, Sten Samson, William P. Schaefer, and other X-ray crystallographers. I judge that the California Institute of Technology, the first institution in North America to enter this field, will continue to make important contributions to it.

References

Burdick, C. L. and Ellis, J. H. (1917). Proc. Nat. Acad. Sci. U.S.A. **3**, 644-649.

Burdick, C. L. and Ellis, J. H. (1917). J. Am. Chem. Soc. **39**, 2518-2525.

Dickinson, R. G. (1920). Thesis for the degree of doctor of philosophy, California Institute of Technology: I. The crystal structures of wulfenite and scheelite. II. The crystal structure of sodium chlorate and sodium bromate.

Dickinson, R. G. (1920). J. Am. Chem. Soc. **42**, 85-93.

Dickinson, R. G. (1922). J. Am. Chem. Soc. **44**, 277-288.

Dickinson, R. G. (1922). J. Am. Chem. Soc. **44**, 774-784.

Dickinson, R. G. (1922). J. Am. Chem. Soc. **44**, 1489-1497.

Dickinson, R. G. (1923). J. Am. Chem. Soc. **45**, 958-962.

Dickinson, R. G. (1926). Z. Kristallogr. **64**, 400-404.

Dickinson, R. G., and Bilicke, C. (1928). J. Am. Chem. Soc. **50**, 764-770.

Dickinson, R. G., and Goodhue, E. A. (1921). J. Am. Chem. Soc. **43**, 2045-2055.

Dickinson, R. G., and Pauling, L. (1923). J. Am. Chem. Soc., **45**, 1466-1471.

Dickinson, R. G., and Raymond, A. L. (1923). J. Am. Chem. Soc. **45**, 22-29.

Harker, D. (1936). J. Chem. Phys. **4**, 381-390.

Helmholz, L. (1936). J. Chem. Phys. **4**, 316-322.

Hughes, E. W. (1941). J. Am. Chem. Soc. **63**, 1737-1752.

Hull, A. W. (1917). Phys. Rev. **10**, 661-696.

Hull, A. W. (1917). Proc. Natl. Acad. Sci. U.S.A. **3**, 470-473.

Mark, H., and Weissenberg, K. (1923). Z. Phys. **16**, 1-22.

Nishikawa, S. (1915). Proc. Math. Phys. Soc. Tokyo **8**, 199.

Pauling, L. (1929). J. Am. Chem. Soc. **51**, 1010-1026.

Pauling, L. and Sherman, J. (1932). Z. Kristallogr. **81**, 1-29.

Wyckoff, R. W. G. (1920). J. Am. Chem. Soc. **42**, 1100-1116.

Wyckoff, R. W. G. (1922). The Analytical Representation of the Results of the Theory of Space Groups, Washington, D.C.: Carnegie Institution of Washington.

CHAPTER 2a
X-RAY CRYSTALLOGRAPHY*

Edward W. Hughes
California Institute of Technology
Pasadena, CA 91125

The electromagnetic wave nature of X-rays was first firmly established in 1912 by the observation of diffraction of X-rays by crystals in the famous investigation of Friedrich, Knipping, and von Laue (Friedrich, Knipping and Laue, 1912). The application of X-ray diffraction methods to the determination of the detailed structure of crystals was demonstrated within less than a year by W. L. Bragg (Bragg, 1913) and since that time literally thousands of papers have appeared describing either improvements in these methods or giving the results of their application to crystals, the natures of which have become increasingly complex as newer methods of increased power have been developed.

Perhaps because of the war which began in Europe in 1914, it was nearly four years later before any results based on the new discovery were published in the United States. The first experimental work began in 1916 and the first papers were published in 1917.

In the Research Laboratory of the General Electric Company in Schenectady, Dr. A. W. Hull independently discovered the powder method of crystal-structure investigation and used it to work out the atomic arrangements in many metals. His results were first presented orally to the American Physical Society in 1916 but the first published abstracts (Hull, 1917*a*) appeared in 1917 and the first full paper (Hull, 1917*b*) somewhat later in the same year.

In the meantime, stimulated by the encouragement and support of A. A. Noyes, experimental work employing the single-crystal methods of the Braggs was started late in 1916 in the new Gates Laboratory of Chemistry at the California Institute of Technology in Pasadena by C. L. Burdick and J. H. Ellis. They soon determined the structure of the mineral chalcopyrite, the result appearing in two papers (Burdick and Ellis, 1917a, 1917b) in 1917.

To commemorate the fortieth anniversary of these pioneering works a one-day meeting of X-ray crystallographers was held at Pasadena on December 16, 1957, by the California Institute of Technology, using funds provided by the Lalor Foundation. About fifty scientists attended. There were eleven contributed papers dealing with various aspects of crystal structure research, representing achievements in eleven different laboratories in the United States, England, and the Netherlands. In a longer invited paper Dorothy Hodgkin of the Chemical Crystallography Laboratory, Oxford University, described the work of the group under her supervision which resulted recently in the full determination of the very complicated structure of the molecule of vitamin B_{12}. Thus the work of the American pioneers was contrasted with the most difficult and complicated application of these methods successfully completed to date.

In the evening there was a dinner after which Professor Pauling read a communication from Dr. Hull, who was unable to attend, and introduced Dr. Burdick (the only survivor of the Pasadena workers) who gave an account of the early days of X-ray crystallography both abroad and in the Gates Laboratory. The stories by Doctors Hull and Burdick have been written down by them and appear herewith.

References

Bragg, W. L. (1913). Proc. Roy. Soc. (London) **A89**, 248.
Burdick C. L., and Ellis, J. H. (1917a). Proc. Natl. Acad. Sci. **3**, 644.
Burdick, C. L., and Ellis, J. H. (1917b). J. Am. Chem. Soc. **39**, 2518.
Friedrich, W., Knipping, P., and Laue, M. von, (1912). Ber. Bayer, Akad. Wiss. (Math-Phys. Kl.) 303.
Hull, A. W. (1917a). Phys. Rev. **9**, 84 and 564.
Hull, A. W. (1917b). Phys. Rev. **10**, 661.

*Editorial note: The following three articles by E. W. Hughes, A. W. Hull and C. L. Burdick are reprinted with permission from *Physics Today*, Vol. 11, No. 10, 18-20, Oct., 1958.

CHAPTER 2b

AN ACCOUNT OF EARLY STUDIES AT SCHENECTADY

Albert W. Hull

My work in electronics was interrupted for several years by a lecture by Sir William Bragg. Sir William was in this country early in 1915, and we invited him to speak at our colloquium. He told us about his new work on the study of X-rays. At the close of the lecture, I asked him if he had been able to find the structure of iron. He had told us about some structures, sodium chloride and copper and a few other materials that he had successfully analyzed. His answer to my question was typical of him. He might have said, "No, but we think we'll have it very soon."

Instead of that he answered very simply, "No, we've tried, but haven't succeeded."

Well, that was a challenge–such a challenge as a young man needs–and I decided that I'd like to try to find the crystal structure of iron, reasoning that it might throw some light on the fact that iron is a magnetic material. I made this decision in spite of the fact that I was almost totally unfamiliar with either X-rays or crystallography.

Almost immediately, therefore, I began getting apparatus together to do X-ray crystal analysis research. I decided to use direct current for this study in order to make more precise measurements. We had available the new Kenotron rectifiers, which our laboratory had just developed, to rectify the alternating voltage.

It was necessary to smooth out the fluctuations of this rectified voltage, which I did in the way that any physicist would do, by using a pair of condensers across the line with an inducting choke between them. I had no idea of making an invention and when a young patent attorney, Mr. W. G. Gartner, who was following the work of the Laboratory for the General Electric Patent Department, came to see me to see if there was anything patentable in my apparatus, I told him, "No." But in looking it over, he noticed this filter circuit and patented it for me. Ten years later I was surprised to learn that this filter circuit was used in all broadcast receiver circuits, and that every receiver manufacturer in the country was licensed under my patent.

Bragg's X-ray crystal analysis had been done with single crystals, mounted so that they could be rotated to expose one face or set of crystal planes at a time at the reflecting angle for the single X-ray wavelength used. Since single crystals of iron were not available, I decided to try using a crystalline powder, reasoning that I would obtain reflections from all planes at once, and might be able to unscramble and identify them.

An interesting incident occurred in the early experiments with this method. I got some good diffraction patterns of iron powder quite early. But being still busy with dynatrons, I turned them over to an assistant to calculate on the basis of Bragg's formulae, to see whether they were consistent with any of the simple cubic structures. This young lady was quite able, but she made a mistake and reported that they were not consistent with any cubic structures. I proceeded to work with some other materials for the time being. Eventually, with the help of Dr. Frederick E. Wright, of the Geophysical Laboratory in Washington, who visited me for a couple of weeks, I got some data with a single crystal of silicon steel about three and a half percent silicon, which we guessed would have a structure similar to iron. We found that it had a body-centered cubic crystal structure.

Immediately I wondered why my original data for iron were not consistent with this structure. While riding home on my bicycle for lunch, I went over the data in my head and made the calculation, and I discovered that these original data were correct, and were consistent with a body-centered cubic structure for iron.

I published this work on X-ray powder diffraction in 1916. In the meanwhile, Debye and Scherrer had been working on this field at the University of Göttingen in Germany. This was during the war, and we got no journals from Germany. After the war was over, I discovered that they had developed just the same method and had published it earlier than I did, by several months, so this method is quite properly called the Debye-Scherrer method of crystal analysis.

As a philosophical afterthought, I must admit that the reason I was so late was that I had continued to be interested in the application of the dynatron. I was trying to develop the dynatron as an amplifier, with some success. That is why, instead of making the calculations about the iron crystals myself, I turned them over to an assistant. The philosophy behind that incident, which is typical of all the experience I've had, is that whenever I've been diverted into spending time on development, it turned out to be a poor investment of time. My advice to any true scientist who is interested in science and not in engineering is not to worry about the applications of what he discovers but to go on to discover something more.

Research at General Electric was interrupted by the war, our laboratories devoting all their energies to a study of the submarine detection problem.

After the war, I returned to X-ray crystal analysis and soon had analyzed all the easily available metals and other common elements. I might have gone on indefinitely analyzing more and more substances, but I got a little feeling from Dr. Whitney's attitude–he wouldn't advise me–that I had gone far enough. Dr. Langmuir advised me that there was no end to it and that I could do something more interesting. So I dropped the crystal-analysis work and went back to electronics.

CHAPTER 2c

THE GENESIS AND BEGINNINGS OF X-RAY CRYSTALLOGRAPHY AT CALTECH

C. L. Burdick

That Caltech was one of the first two centers in America where X-ray crystallography was started and continued was due entirely to the imagination, vision, and conviction of Dr. Arthur A. Noyes. In 1916, which was very early in Caltech's beginnings, Dr. Noyes was nominally visiting professor of chemistry, but actually he, with Drs. Millikan and Hale, was planning and shaping the future of this whole great institution. This particular specialty was only one of the many in which he saw great promise and forthcoming utility.

Prior to 1915 Dr. Noyes was occupied wholly as director of the Research Laboratory of Physical Chemistry at MIT in Boston, and it was there in 1913-14 that I came to know Dr. Noyes and to be a graduate student under him at the laboratory in MIT's old "Engineering C" building. This was near the end of the halçyon days of classical physical chemistry when on the staff could be noted among the younger men such names as G. N. Lewis, C. A. Kraus, and F. G. Keyes.

Dr. Noyes encouraged me to go abroad for doctorate work and postdoctoral study, and it turned out that I sailed from New York in July 1914 on the last regular German liner to get through to Hamburg. Because of war conditions my doctorate period was divided between Professors Fichter and Rupe in Basel and Professor Willstätter at the Kaiser Wilhelm Institute for Chemistry in Berlin-Dahlem. It might be noted that there were then two young, comparatively unknowns by the names of Hahn and Meitner in this same Institute. Following this, the writer had a brief period of postdoctoral work at the Kaiser Wilhelm Institute of Physical Chemistry under Professor Fritz Haber and then a later opportunity to work on "Reststrahlen" with Rubens at the University of Berlin and to sit in on the lecture courses of Nernst, Planck, and Einstein. What untold recognition and appreciation these names call forth today.

Early in 1916 it was only with some circumlocution, occasioned by the separate British and German censorship of all mail from America, that advice came to me from Dr. Noyes to try to find a way to spend my last months abroad working in the X-ray laboratory of Professor William H. Bragg at University College in London. In his letter Dr. Noyes expressed his strong belief in the importance of X-ray atomic structure analysis for the future of theoretical chemistry, and his wish to get something of the kind started at MIT.

It was not so simple for an impecunious, young PhD of American neutrality without connections to get from Berlin to London during the critical period of the Zeppelin raids and unrestricted submarine warfare. But due to the volunteered help and influence of Professor Kamerlingh Onnes of the Cryogenic Laboratory in Leiden, it was accomplished.

The six months in the laboratory of Professor Bragg were of inestimable value, even though its leader was largely occupied with war work. With what time Dr. Bragg had to spend with us in guidance, Dr. E. A. Owen and the writer were able to work out a tentatively satisfactory X-ray structure for carborundum, which was later published.

An interesting presentation could be made of the primitiveness of the equipment then available, the old-fashioned induction coil with Leyden-jar condensers and a mercury interruptor, the gas-filled X-ray tubes of unpredictable and uncertain output and "hardness," and gold-leaf electroscopes with the strangest static aberrations when it came to measuring ionization intensities.

On returning to MIT, which was in process of moving from Boston to Cambridge, my assignment from Dr. Noyes was to build a Bragg X-ray spectrometer with any improvements which the state of the art would permit. This meant, first, having at hand a really good X-ray transformer with rotary-disc rectifier and secondly, and of most importance, an only just then developed, new Coolidge X-ray tube fitted with a palladium target. Dr. James A. Beatty, then a student, was principal collaborator in building, putting this machine into initial operation, and testing.

In the latter part of 1916 when Dr. Noyes left for his annual tour of duty as visiting professor at Caltech he asked me to go with him and build another spectrometer embodying the refinements which our tests had shown could be made. The Caltech team on this was James H. Ellis, Fred Hensen (instrument maker), and myself. The things which made the original Caltech spectrometer probably the best of its day were its high-power input and relative constancy of measured electrical energy to the tube. This gave possibilities for narrower spectrometer slits, precise angle measurements, sharp reflection peaks, and better measurement of relative reflection intensities of the spectral orders than had probably ever been made before.

The measurements carried out in these months permitted the determination of the crystal structure of chalcopyrite and resulted in two papers by Ellis and Burdick. These were submitted to the *Journal of the American Chemical Society* and the *Proceedings of the National Academy of Sciences* early in 1917 and appeared as publication Number 3 of the Gates Laboratory of Chemistry.

Then the United States entered the war. Dr. Noyes' time was spent in large measure in Washington. Burdick was called from the US Ordnance Reserve into the nitrogen fixation program and for a time the X-ray crystal work at Caltech lagged. After the war, with the arrival of Dr. Roscoe G. Dickinson, the wheels hummed again. High-voltage sparks flew and research results by many collaborators came out apace. Today the successorship of Dickinson is in the notable hands of Pauling. Dr. Noyes' vision of the early days of 1916 has paid off and will continue to do so in generous measure.

CHAPTER 3

EARLY DAYS

Ralph W. G. Wyckoff
Duval Corporation, 4715 E. Fort Lowell Rd., Tucson, AZ 85712

In this country we were late in learning of von Laue's discovery of X-ray diffraction and the Braggs' use of it to determine the positions of the atoms in a crystal. These things happened just before the onset of the first world war and communications with Europe were slow and uncertain. I know that I became aware of them in 1916 when W. L. Bragg's *Proceedings of the Royal Society* paper (Bragg, 1914) was reviewed, shortly after its receipt, in a colloquium of Cornell's physics department. This was a year before I began working with Dr. Nishikawa.

I know of only two studies of crystal structure in the USA that antedate our work at Cornell. Both appear to have arisen out of contacts with W. H. Bragg either shortly before or shortly after the outbreak of the war. Burdick (Burdick, 1962) recounts that while studying abroad he was advised by A. A. Noyes to visit Bragg's laboratory before returning to MIT; there he built a Bragg spectrometer and together with J. H. Ellis determined the structure (Burdick and Ellis, 1917) of chalcopyrite, $CuFeS_2$ A. W. Hull (Hull, 1962) says that the use of X-rays resulting in his invention of the powder method (Hull, 1917), was begun after a discussion when Bragg was passing through the General Electric Laboratory.

My concern with X-ray diffraction at Cornell was almost an accidental consequence of a search for a doctoral thesis. Beginning graduate work in 1916, I had been given the general topic of the chemistry of cesium. The chemistry department had what was then a large supply of the mineral pollucite and my first job was to extract its cesium and see if by chance it contained any of the missing heavier alkali of atomic number 87. An extensive fractionation of the cesium alum prepared from this pollucite was more an exercise for the muscles than for the brain but I could show by spectroscopic examination that its end product did not contain a detectable trace of ekacesium (Wyckoff and Dennis, 1920). Such a result is now self-evident since Mlle. Perey discovered that element 87 is radioactive and short-lived. It was a pleasure to come to know her 50 years later but at the time I had to find a less negative thesis subject.

The nature of the polyhalides was then poorly understood and my committee chairman, L. M. Dennis, suggested I take advantage of the presence of a visitor, Dr. S. Nishikawa, in the physics department to see if more could be learned about them by determining the crystal structure of one of the cesium trihalides. Dr. Nishikawa had already determined a structure (Nishikawa, 1915), that of the spinel $MgAl_2O_4$, and he was hoping to learn more about Hull's recently invented powder method. He and Dr. G. Asahara were in the USA and Europe before becoming involved in the Japanese government's plans for establishing the equivalent of our Bureau of Standards. The war had made it impossible for them to work in Schenectady but Cornell welcomed them and put at their disposal the laboratory of a professor who had left for war service and whose X-ray installation had previously served him as radiographer for the Ithaca hospitals.

Discussing my projected thesis topic with Dr. Nishikawa, he expressed himself as very willing to teach me his X-ray diffraction techniques. After looking about for a chemically stable cesium compound that would yield good single crystals, I chose cesium dichloroiodide, $CsCl_2I$. We constructed a primitive arrangement for making oscillation photographs and a Laue camera from an old wooden box. With these the necessary photographs were taken and interpreted, and a structure finally obtained (Wyckoff, 1920a). To supplement this structure for my thesis I also determined the structure (Wyckoff, 1920b) of sodium nitrate, $NaNO_3$, which proved to be isostructural with calcite, as already established by the Braggs.

During my third year of graduate study I had, owing to the departure for war work of the professor of analytical chemistry, been assigned the job of teaching quantitative analysis, giving the lectures and supervising the laboratories. It was generally expected that after receiving my degree I would continue this teaching as an assistant professor, perhaps doing X-ray research in my spare time. My future changed when Dr. A. L. Day, Director of the Geophysical Laboratory in Washington, saw what I was doing during a visit to Ithaca. He wanted to start diffraction studies and offered me a position carrying the opportunity for full-time research. After a trip to Washington I took the offer.

The procedures which Dr. Nishikawa had been employing and which I would adopt were all photographic, in contrast to the spectrometry the Bragg school was using. In the Bragg spectrometer the detector was a simple ionization chamber and a gold leaf electroscope, the output of which was so weak that reflections from only a few planes of a large crystal could be registered. Many more reflections were recorded on Laue and oscillation photographs and this increase in the amount of data would be needed in studies we contemplated of more complicated crystals.

The atomic arrangements that had been found had been more or less intuitively guessed and then confirmed by reference to the measured reflections - they were thus probable structures compatible with these few data. Dr. Nishikawa had used space group theory to describe and illustrate his spinel structure and he introduced me to it. He had learned it in Japan from a professor who had been in Germany a generation earlier when Schoenflies was deducing the theory (Schoenflies, 1891). I felt that through its use one should be able to define all possible arrangements and, with enough X-ray information, select the correct one. This would require a more complete analytical expression of the theory than had yet been given and I began working it out soon after going to the Geophysical Laboratory. The Carnegie Institution of Washington published the finished results under the title *"An Analytical Expression of the Results of the Theory of Space Groups"* (Wyckoff, 1922). This and the working out of several relatively simple structures intended to illustrate how space group theory could be applied occupied most of my first years in Washington.

About a year after moving to Washington, J. H. Ellis paid me a visit en route to Pasadena. He was transferring from MIT to the recently established California Institute of Technology where Burdick had built a Bragg spectrometer and R. Dickinson was beginning X-ray studies. Impressed by the use that could be made of Laue photographs and the value of the theory of space groups, he arranged that Dickinson be for a brief time in my laboratory becoming familiar with my techniques. As a consequence of this visit I went to Pasadena for the following year and while there worked out several more structures. After returning to Washington, the years till 1927 were spent in writing *"The Structure of Crystals"* (Wyckoff, 1924) and carrying out X-ray studies of a number of minerals.

I went then to the Rockefeller Institute to start the examination of organic crystals and gradually devoted myself to problems of a more biophysical nature. Corey came from Cornell a couple of years later to begin X-ray work. We determined the structures of a variety of substances containing simple organic radicals, had developed equipment for quantitative spectrometry and were applying it to urea derivatives (Wyckoff and Corey, 1932) and amino acids. At this time we were also obtaining powder photographs of several crystalline proteins (Wyckoff and Corey, 1935; Wyckoff, 1936), when a change in the Institute's administration made it necessary to cease all X-ray work and leave the Institute. I have continued to work with X-ray diffraction but never have returned to the determination of crystal structures.

References

Bragg, W. L. (1914). The analysis of crystals by the X-ray spectrometer. Proc. Roy. Soc. (London) **A89A**, 468.

Burdick, C. L. (1962). The genesis and beginnings of X-ray crystallography at Caltech, in "Fifty Years of X-ray Diffraction." edited by P. P. Ewald. p. 556. Utrecht; Oosthoek.

Burdick, C. L. and Ellis, J. H. (1917). X-ray examination of the crystal structure of chalcopyrite. J. Am. Chem. Soc. **39**, 2518.

Dennis, L. M. and Wyckoff, R. W. G. A search for an alkali element of higher atomic weight than cesium. J. American Chem. Soc. **42**, 985.

Hull, A. W. (1962). Autobiography, in "Fifty Years of X-ray Diffraction," edited by P. P. Ewald. p. 582. Utrecht: Oosthoek.

Hull, A. W. (1917). A new method of X-ray crystal analysis. Phys. Rev. **10**, 661.

Nishikawa, S. (1915). The structure of some crystals of the spinel group. Proc. Math. Phys. Soc. Tokyo **8**, 199.

Schoenflies, A. (1891). Krystallsysteme u. Krystallstruktur. Leipzig.

Wyckoff, R. W. G. (1920a). The crystal structure of cesium dichloriodide. J. Am. Chem. Soc. **42**, 1100.

Wyckoff, R. W. G. (1920b). The crystal structure of sodium nitrate. Phys. Rev. **16**, 149.

Wyckoff, R. W. G. (1922). The analytical expression of the results of the theory of space groups. Washington, D.C.; Carnegie Institution of Washington.

Wyckoff, R. W. G. (1924). The structure of crystals. New York: Chemical Catalog Co. Also 1931, 1935. These volumes contained summaries of all structures determined up to that time, including results in ca 100 papers by the writer.

Wyckoff, R. W. G. and Corey, R. B. (1932). The crystal structure of thiourea. Z. Kristallogr. **81**, 386; (1934). Spectrometric measurements on hexamethylene tetramine and urea, ibid. **89**, 462; and others.

Wyckoff, R. W. G. and Corey, R. B. (1935). X-ray diffraction from hemoglobin and other crystalline proteins. Science **81**, 365.

Wyckoff, R. W. G. (1936). Long spacings in macromolecular solids. J. Biol. Chem. **114**, 407; X-ray diffraction patterns of crystalline tobacco mosaic proteins, ibid. **116**; and others.

CHAPTER 4a

CRYSTALLOGRAPHY
FIFTY YEARS OF X-RAY CRYSTALLOGRAPHY AT THE GEOPHYSICAL LABORATORY, 1919-1969*

Gabrielle Donnay
McGill University, 3450 University Street, Montreal, Quebec H3A 2A7
(With the assistance of R. W. G. Wyckoff, T. F. W. Barth, and George Tunell)

It is difficult for us today to realize the rudimentary state of the art of structural crystallography in 1919, only seven years after Laue's discovery of X-ray diffraction. A letter of December 22, 1920, from Dr. Arthur L. Day, the first Director of the Geophysical Laboratory, to Dr. Ralph W. G. Wyckoff, its first structural crystallographer (1919-1927), telling Wyckoff of his first raise, conveys the vision of this Director, whose birth dates back one hundred years, to 1869:

"The field of activity in which your work falls is entirely new to this Laboratory, and for the most part new in this country. It is, I think, rare that a man at the outset of his career has such an unusual opportunity to enter and exploit a field of such vital importance to our knowledge of the structure of matter, a field, by the way, which will probably take you far beyond the immediate application which this Laboratory will wish to make of it. I heartily commend both your choice and your opportunity, and wish you every success in your work."

Dr. Day's good wishes were fulfilled: of the many honors and recognitions later bestowed on Dr. Wyckoff, we need only mention that he became President of the International Union of Crystallography, which was founded in 1948. The pioneering nature of the studies of Dr. Wyckoff and his colleagues is evident from a perusal of Table 1.

The only laboratories in the United States where crystal-structure work was being attempted prior to 1919 were found at Massachusetts Institute of Technology, California Institute of Technology, the General Electric Company in Schenectady, and the Chemistry Department of Cornell University, where graduate student Wyckoff, under the guidance of Professor Shoji Nishikawa, became acquainted with space-group theory and its usefulness to X-ray diffraction. The first American crystal-structure determination based on single-crystal data, that of chalcopyrite, was published early in 1917 by C. L. Burdick and J. H. Ellis, as publication No. 3 of the California Institute of Technology. In 1919 the M.I.T. group moved to the California Institute of Technology, and Dr. Wyckoff joined the staff of the Geophysical Laboratory of the Carnegie Institution in Washington, D.C.; only in these two places has X-ray crystallography been carried on continuously ever since.

The first determinations of atomic positions in crystals were inspired guesses supported by the few X-ray diffraction intensities that could be obtained with the primitive spectrometers then available. Additional data were needed if many more structures were to be analyzed, and some orderly procedure was required for finding possible atomic arrangements from which the correct choice could be made. Wyckoff grasped the fact that space-group theory offered the means of enumerating possible structures, and following Nishikawa, he saw in the original patterns of Laue a promising source of additional data. These ideas were developed in his early publications from the Geophysical Laboratory. R. G. Dickinson, moving from M.I.T. to Pasadena, spent a short time at the Laboratory, familiarizing himself with these methods of approach, and Wyckoff supplemented this visit by working during the following year at the newly established California Institute of Technology. The work in Pasadena expanded rapidly, first under Dickinson and then under Linus Pauling; under Pauling it has played a dominant role in the development of structural chemistry throughout the world. In contrast, crystallography at the Geophysical Laboratory has remained in the hands of a very few people. Nevertheless, their contributions have dealt with many aspects of their science, such as theoretical crystallography, crystal-structure determinations and their interpretations and refinements, solid solutions, computing techniques, and biocrystallography. Compilations of crystallographic data, beginning with Wyckoff's *"Survey of Existing Crystal Structure Data,"* filling 68 pages in 1923, his first edition of *"The Structure of Crystals"* in 1927, through Donnay and co-workers' *"Determinative Tables of Crystal Data,"* comprising 1300 pages in 1963, show how aware of the retrieval problem were the "isolated" crystallographers at the Laboratory.

Let us look back at some of the highlights of the contributions, as we judge them now, for in our present-day group approach to scientific research, such a miniature crystallographic laboratory as the one at the Geophysical Laboratory might be called upon to justify its existence.

The value of space-group theory was not quickly recognized by the crystallographic fraternity, and Wyckoff's *"Analytical Expression of the Results of the Theory of Space Groups"* could not find a publisher. It finally appeared in 1922, as a publication of the

---

* Reprinted with additions and corrections from the *Carnegie Institute of Washington Year Book 68,* pp. 278-283, 1970. With permission of the Director of the Geophysical Laboratory, Carnegie Institution of Washington.

Carnegie Institution of Washington, through the personal interest and intervention of Dr. Robert S. Woodward, originally a mathematician, who was President of the Institution. This book was the forerunner of the *"International Tables of X-ray Crystallography"* in which the atomic positions in the 230 space groups are designated by their Wyckoff letters. Of the sixty-four crystallographic publications that came out of the Geophysical Laboratory during the nine-year period when Wyckoff was here, we need mention only the first crystal-structure determination to illustrate how much new information could then be packed into one structure paper. This one dealt with the "Crystal structures of some carbonates of the calcite group" and covered 44 pages in the *American Journal of Science.* In addition to determining the correct unit cell of calcite and the structures of magnesite, rhodochrosite, smithsonite, and siderite, and recognizing that dolomite would have a different structure, the author proved the existence of planar $CO_3^=$ groups, which Bragg's determination of the calcite structure with the X-ray spectrometer had failed to do, since the oxygen atoms could not be located. By making $CO_3^=$ the anions, the close relation of the calcite structure to the NaCl structure became evident. It was stressed that $NaNO_3$ and $CaCO_3$ are even more closely isostructural than $CaCO_3$ and $MnCO_3$ The faces of calcite were explained by recognizing that the nodes of a nonstructural lattice, based on the cleavage rhombohedron, are occupied by morphologically equivalent Ca^{2+} and $CO_3^=$ ions. The extinction criterion for a rhombohedral lattice ($-h + k + l = 3n$) was derived: the relation of the gnomonic projection to the Laue photograph was elucidated, and a ruler to construct the projection directly from the photograph was described. All this, in 1920!

Experimental work with gas X-ray tubes was very difficult and time-consuming in the early days, and the Laboratory was fortunate in having a Swiss-trained instrument maker, C. J. Ksanda, employed at the Laboratory from 1914 to 1940, who carried out much of the laboratory work and in 1932 designed the Ksanda twin gas tubes, which were built by a local firm.

A Laboratory colleague who was introduced to X-ray crystallography by Wyckoff was Dr. Eugene Posnjak (Staff Member, 1913-1947). Russian by birth, he was trained in physical chemistry at Leipzig, Germany, and had joined the Laboratory after a year with A. A. Noyes at M.I.T. Posnjak published "The crystal structure of ammonium chloroplatinate" jointly with Wyckoff in 1922. The authors proved once and for all the validity of Werner's Coordination Theory, which until then had been only a hypothesis: platinum is octahedrally surrounded by six chlorine ions. The structure is of the fluorite type with $PtCl_6^=$ replacing Ca^{2+} and NH_4^+ replacing F^-. Also jointly, they described the alkali halides and cuprous halide. Posnjak determined the crystal structure of the alkali metal potassium and together with Dr. Sosman discovered the naturally occurring ferromagnetic iron-deficient magnetite of composition Fe_2O_3, to which the name maghemite was later given. In 1928 he published the cell dimensions of spinel ($MgAl_2O_4$) and other compounds of the spinel group. He had exceptional skill in preparative and experimental work, and in order to study the magnetic properties of crystals, he produced various compounds and solid solutions containing both ferrous and ferric iron. He had begun a study of the powder diagrams of such samples exhibiting spinel structure when Dr. Tom F. W. Barth joined the Laboratory in 1929.

Barth (Staff Member, 1929-1936) had received his introduction to the field from the greatest geochemist of the time, V. M. Goldschmidt, then a professor at the University of Oslo, Norway. Barth let his active interest in petrology guide his choice of crystallographic problems and concentrated his efforts on important rock-forming minerals. He soon joined forces with Posnjak on the spinel problem, and this happy collaboration resulted in 1931 in their classical paper "The spinel structure: an example of variate atom equipoints," in which it was proved that crystallographically equivalent sites can be occupied by chemically different atoms. Five months later they showed that the structure of lithium ferrite, $Li_2Fe_2O_4$, belongs to the NaCl type, with univalent lithium and trivalent iron substituting randomly for each other on the sites of one and the same crystallographic position. The new concept of "variate atom equipoints" proved to be essential for understanding and solving most mineral crystal structures. Aluminum, for example, to some extent replaces silicon in all aluminosilicates, and studies of gehlenite, sodalite, related minerals, and the feldspars clarified the principle involved. Barth also studied the cristobalite structure, showed that nonsilicates can exist with the same structure type, and recognized considerable solid solution in the system $SiO_2 - Na_2Al_2O_4$, also with the cristobalite structure.

Dr. Sterling B. Hendricks, who had been a Fellow at the time of Wyckoff, was back in Washington, at the Fixed Nitrogen Laboratory, and cooperated throughout the thirties with the crystallographic group at the Laboratory. In 1931 and 1932 Hendricks, Kracek, and Posnjak verified Pauling's hypothesis that in sodium nitrate and ammonium nitrate molecular rotation takes place in the solid state. This was another new phenomenon first described at the Geophysical Laboratory. Brandenberger had pointed out that in the case of sodium nitrate there exists a particular stationary position for the NO_3 group which would be difficult to distinguish from the rotating group. It was important, therefore, that Barth could prove the existence of two polymorphic forms of KNO_3 one in which the nitrate group was rotating and another in which it had the location predicted by Brandenberger.

Dr. George Tunell (Staff Member, 1925-1947), a Harvard-trained economic geologist, mineralogist, and crystallographer, specialized in ore minerals. He promptly became the friend of Tom Barth and learned from him the techniques of crystal-structure determination. With his wife Ruth helping with the computations, he determined the structures of tenorite, calaverite, sylvanite, and krennerite, and later published a paper with Linus Pauling on "The atomic arrangement and bonds of the gold-silver ditellurides." Tunell also derived the Lorentz correction factor for equi-inclination Weissenberg films, without which the photographic intensities of diffracted X-rays could not be used. He had taken great interest in the computing of Fourier synthesis, as is evidenced by the "Patterson-Tunell stencils and strips," which rapidly became popular and helped crystallographers in their laborious calculations of the precomputer era. (They are, to this day, used as a teaching aid to make the student appreciate what really goes on in a Fourier summation.)

By 1933 the X-ray laboratory was so well equipped—it had a Weissenberg camera, the first to be converted to equi-inclination outside of M.I.T., an oscillation camera, several powder cameras, including one in which the sample could be studied at high temperature—that Dr. Barth could report to Dr. Day: "for a considerable length of time we have had the best X-ray goniometer [Weissenberg camera] in the U.S. and probably in the world" and "we have concrete problems already well under way which we believe are of more than ordinary interest, as is evidenced, for example, by the fact that Dr. [J.D.H.] Donnay, Professor of Mineralogy at the Johns Hopkins University, desires to study with us in the fall and that Professor Palache has sent us his assistant, Mr. Berman, for advice and instruction." The association with the morphological crystallographers of Harvard and of Johns Hopkins was both a fruitful and a happy one, as witness the formation of the delightful "Calaverite Club," composed of Palache, Peacock, Donnay, Tunell, and Barth. Calaverite is a mineral whose morphology apparently violates the Law of Rationality and which, even now, is not fully understood. In those days all the papers on calaverite were written by members of the club! The enthusiastic cooperation also led to a joint publication in 1934 on "Various modes of attack in crystallographic investigations."

The association with Professor Donnay continues to the present time. His wife, Dr. Gabrielle Donnay, joined the Laboratory in 1950 as a Fellow and in 1955 as a Staff Member. Her undergraduate training in chemistry at U.C.L.A. was followed by a Ph.D. in crystallography under M. J. Buerger at M.I.T. Her first research problem at the Laboratory, carried out with the help of the Director, Dr. L. H. Adams, was a test of the precision of the then-new powder diffractometer and led to the cell dimensions and cell volumes of a complete solid-solution series of well-documented alkali feldspars. It showed the existence of a high-order transition. A study of chalcopyrite, performed jointly with Drs. L. M. Corliss, J. D. H. Donnay, N. Elliott, and J. M. Hastings at Brookhaven National Laboratory, led to the first application of "generalized symmetry" to magnetic-structure determinations. Complex crystalline edifices resulting from twinning (in digenite), epitaxy, and syntaxy (in the bastnaesite-vaterite series) have been studied. Solid solutions, omission as well as substitution and organic as well as inorganic, have been investigated, and the theory of limited solid solutions between end members of different structure types has been considered. In addition to describing five new minerals (among them ewaldite, a simple new carbonate structure), a novel building block composed of three corner-linked SiO_4 groups was found in the structure of the rare silicate mineral ardennite. Together the Donnays have investigated the relation of morphology to structure, which led to further generalizations of the Law of Bravais. Most recently an X-ray survey of the orientation relations between crystallographic axes and morphological features of calcite "biocrystals" in Echinodermata was carried out with Dr. David L. Pawson of the Smithsonian Institution.

The tremendous help that modern computers can give crystallographers has not been overlooked at the Laboratory. Computing and its use for structure refinement to obtain meaningful temperature factors was of special interest to another student of Professor Buerger, Dr. Charles W. Burnham, who became a Fellow in 1961 and was a Staff Member from 1963 to 1966. He refined the structures of sillimanite and kyanite and wrote several computer programs that are now widely used.

This account would not be complete without mentioning the following postdoctoral Fellows who contributed to the crystallographic output: J. V. Smith (1951-1954), N. Morimoto (1957-1960, seven months in 1962, three months in 1963, and four months in 1966-1967), E. W. Radoslovich (1962-1963), N. Güven (1965-1967), and L. W. Finger (1967-1969).*

Looking back at the historical facts, it seems to us that the Geophysical Laboratory was different from most other places in its approach to X-ray crystallography. Most early workers in this field were classical physicists rather than crystallographers. After the initial discovery, the methods employed in working out actual crystal structures were slow and cumbersome, and lacked the elegance and ease that only crystallographic theory could provide. It was fortunate for the Geophysical Laboratory that Wyckoff, who demonstrated the use of the theory of space-groups and always

*A complete list of publications of crystallographers at the Geophysical Laboratory is included in the "Indices of the Annual Reports of the Director, 1905-1980," Paper No. 1860, 1981.

insisted on applying it in his work, became the first X-ray crystallographer of the Laboratory. The same approach was used by Tunell and Barth, both educated in classical crystallography; this trend was strengthened by the inspiring cooperation of morphological and optical crystallographers, in and out of the Laboratory, who brought in the best traditions of the European school. Together with this mode of approach, the freedom of research that has been the rule at the Geophysical Laboratory has served its crystallographers well. It is a pleasure, in conclusion, to quote Professor Barth's enthusiastic reply to our call for help when we started writing this short history: "I shall be happy to do so; I spent the most pleasant time of my life at the Geophysical Laboratory."

Table I

Crystal structures studied with X-radiation at the Geophysical Laboratory, 1920-1930 (compiled by Hatton S. Yoder, Jr.)

Authors	Date	Compound	Name	Reference
Wyckoff	1920	$CaCO_3$ $MnCO_3$ $FeCO_3$ $MgCO_3$	calcite rhodochrosite siderite magnesite	Amer. J. Sci. **50**, 317-360
Wyckoff	1921	MgO	periclase	Amer. J. Sci. **1**, 138-152
Wyckoff	1921	MnS	alabandite	Amer. J. Sci. **2**, 239-249
Wyckoff and Posnjak	1921	$(NH_4)_2PtCl_6$	ammonium chloroplatinate	J. Am. Chem. Soc. **43**, 2292-2309
Wyckoff and Posnjak	1922	CuCl CuBr CuI	cuprous halides	J. Am. Chem. Soc. **44**, 30-36
Wyckoff	1922	$(NH_4)Cl$	ammonium chloride	Amer. J. Sci. **3**, 177-183
Wyckoff	1922	Ag_2O	silver oxide	Amer. J. Sci. **3**, 184-188
Wyckoff	1921	NaBr, NaI KBr, KI, RbCl, RbI, CsBr, CsI	alkali halides	J. Washington Acad. Sci. **11**, 429-434
Posnjak and Wyckoff	1922	LiCl, LiBr, LiI NaF, KF, CsF		J. Washington Acad. Sci. **12**, 248-251
Wyckoff	1923	$N_2H_6Cl_2$	hydrazine dihydrochloride	Amer. J. Sci. **5**, 15-22
Wyckoff	1923	$KAl(SO_4)_2.12H_2O$ $NH_4Al(SO_4)_2.12H_2O$	alums	Z. Kristallogr. **57**, 595-609
Wyckoff	1923	Ir	metallic iridium	Z. Kristallogr. **59**, 55-61
Wyckoff	1923	LiI RbF	alkali halides	J. Washington Acad. Sci. **13**, 393-397
Wyckoff and Merwin	1924	$CaMg(CO_3)_2$	dolomite	Amer. J. Sci. **8**, 447-461
Wyckoff	1925	$CaCO_3$	aragonite	Amer. J. Sci. **9**, 145-175

Geophysical Laboratory

Authors	Date	Compound	Name	Reference
Wyckoff and Merwin	1925	$BaSO_4$*	barite	Amer. J. Sci. **9**, 286-295
Wyckoff and Merwin	1925	$CaMgSi_2O_6$*	diopside	Amer. J. Sci. **9**, 379-394
Wyckoff	1925	SiO_2**	high-cristobalite	Amer. J. Sci. **9**, 448-459
Wyckoff	1925	Ag_3PO_4 Ag_3AsO_4	silver phosphate silver arsenate	Amer. J. Sci. **10**, 107-118
Wyckoff	1925	SiO_2	α-quartz	Science **62**, 496-497
	1926			Z. Kristallogr. **63**, 507
Wyckoff	1926	SiO_2	β-quartz	Amer. J. Sci. **11**, 101-112
Wyckoff and Crittenden	1925	FeO	wüstite	J. Am. Chem. Soc. **47**, 2876-2882
Wyckoff and McCutcheon	1927	$Co(NH_3)_6I_3$	hexamine-cobalti-iodide	Amer. J. Sci. **13**, 223-233
Wyckoff and Müller	1927	Cs_2GeF_6	cesium fluogermanate	Amer. J. Sci. **13**, 347-352
Wyckoff, Hendricks and McCutcheon	1927	$[Co.6NH_3]\ (ClO_4)_3$	hexamine-cobalti-perchlorate	Amer. J. Sci. **13**, 388-398
Hendricks and				
Hendricks and Wyckoff	1927	$Al(PO_3)_3$*	aluminum metaphosphate	Amer. J. Sci. **13**, 491-496
Wyckoff and Hendricks	1927	$ZrSiO_4$	zircon	Z. Kristallogr. **66**, 73-102
Posnjak	1928	K	potassium	J. Phys. Chem. **32**, 354-359
Hendricks and Merwin	1928	$(NH_4)_2Pt(SCN)_6$ $K_2Pt(SCN)_6$ $Rb_2Pt(SCN)_6$	alkali platinithiocyanates	Amer. J. Sci. **15**, 487-494
Posnjak	1928	$MgAl_2O_4$***	spinel	Amer. J. Sci. **16**, 528-530
Roberts and Ksanda	1929	CuS	covellite	Amer. J. Sci. **17**, 489-503
Posnjak	1930	$MgFe_2O_4$ $ZnFe_2O_4$ $CdFe_2O_4$	ferrites	Amer. J. Sci. **19**, 67-70

*Space group only.
**Structure determined at high temperature from powder data.
***Lattice dimensions only.

CHAPTER 4b

CRYSTALLOGRAPHIC RESEARCH AT THE GEOPHYSICAL LABORATORY, 1970-1980

H. S. Yoder, Jr.
Director, Geophysical Laboratory
Washington, D C 20008

The substantial advancements in crystallography during the 1919-1969 period at the Geophysical Laboratory (Donnay *et al.,* 1970; Donnay, 1971) aided in the solution of many petrological problems. Crystallography continues to be an essential part of almost all observations involving the stability and behavior of minerals and is fundamental to the interpretation of their properties measured by other techniques. As the petrological questions become focused, the need grows for more precise determinations of crystal structures, with particular emphasis on site occupancies. A major goal is to relate mineral structures, determined under the same conditions under which they were observed to form and react, to their properties and behavior.

Initially, emphasis was given to the crystallographic details of rock-forming silicates in order to deduce geological history from such properties as twinning, transformations, cation ordering, and exsolution. The chain silicates have received the greatest attention, perhaps because their complexity results in the capacity to record more information about their conditions of formation. Investigation of the complex pyroxene and amphibole groups required new procedures for the collection and analysis of X-ray intensity data. Major advances were contributed in both automated data collection (Finger, Hadidiacos, and Ohashi, 1973) and structure refinement (Finger and Prince, 1975). Procedures for the refinement of partial occupancies, cation ordering, and complex thermal vibration were incorporated into the program RFINE, now in widespread use for X-ray and neutron diffraction studies. In his refinement of the structure of anthophyllite, Finger (1970) related the occupancies of the four octahedral sites to the temperature of formation, or annealing, in a quantitative way using those new techniques. Clinopyroxenes from a diamond-bearing pipe were annealed by McCallister, Finger, and Ohashi (1976), and a cation-ordering versus temperature calibration curve was determined. The clinopyroxene crystals themselves indicated an intracrystalline closure temperature of 530°C below which the rate of cation ordering was negligible, whereas the temperature below which the intercrystalline partitioning of Ca and Mg between clinopyroxene and orthopyroxene was negligible was 1375°C (Nixon and Boyd, 1973). The difference between these closure temperatures, together with the occurrence of nongraphitized diamonds, places important constraints on the rate of cooling of the rock. Other refinements of the common rock-forming minerals (e.g., olivine, pyroxenes, amphiboles) led to the development of time-temperature-transformation plots of special application to other petrological problems (Seifert and Virgo, 1975). Of special interest was the unraveling by Ohashi and Finger (1976) of the structural relations of the $(Mn,Mg)SiO_3$ pyroxenoids, which occur in several complex but related chain structures. The differences between bustamite and wollastonite, for example, could be related to the stepwise cation ordering in the octahedral strip.

By the mid-1970's the interests of the crystallographic investigators at the Geophysical Laboratory shifted to the study of crystals at elevated temperatures and pressures. The structure of sanidine was determined up to 800°C (Ohashi and Finger, 1975) and that of diopside, up to 700°C (Finger and Ohashi, 1976). Both studies illustrated the principle that relatively simple external changes in unit-cell dimensions with temperature change are the result of complex internal structural changes in bond lengths and angles. The investigation of the effects of high pressure on single crystals was initiated in 1976. Both Dr. Robert Hazen, a postdoctoral Fellow from Harvard University via Cambridge University, and Dr. Larry Finger, who had just spent his sabbatical leave at the Department of Earth and Space Sciences, SUNY, Stony Brook, were experienced in the use of a successful diamond-anvil, high-pressure cell modified from the design of Merrill and Bassett (1974). Additional improvements in the cell greatly increased the quality of results, and in the period 1977 to 1979 over 70 crystal structures were determined at high pressures jointly by Hazen and Finger! The crystals included the common rock-forming minerals: olivines, pyroxenes, garnets, spinels, layer minerals, zircon, and corundum- and rutile-type oxides. High-pressure phase transitions were also studied in analcite and manganese difluoride

Single-crystal studies at high pressures were limited to about 100 kbar for lack of a suitable hydrostatic medium. Fortunately, Finger *et al.* (1981) discovered that rare gases, including argon and neon, remain hydrostatic to much higher pressures. Furthermore, these gases crystallize in the diamond-anvil, high-pressure cell to *single* crystals that are probably the most compressible substances known. As a result, a single crystal of FeO was pressurized, for example, to 200 kbar in a neon-filled cell. It is possible, therefore, to study the structures of other single crystals immersed in such gases to pressures well above 100 kbar.

Concurrently with the single-crystal studies, the compressibility of magnesiowüstite, iron, and periclase, phases presumed to be of considerable importance in the earth's mantle, was determined with powder X-ray diffraction methods up to 865 kbar in an internally calibrated, diamond-anvil, high-pressure cell (Mao and Bell, 1976; Rosenhauer, Mao, and Woermann, 1976). The degree of nonhydrostatic stress was reduced by use of a deformable gasket that contained the sample. The compressibilities of those substances are being redetermined hydrostatically in a gas, and the results should be useful for interlaboratory pressure calibration. Of particular note is the demonstration by Mao *et al.* (1980) that an abrupt change in electronic structure takes place at about 200 kbar in metallic praseodymium. The collapse of the 4*f* electron shell results in a 19% volume change that is too large to be accounted for by a simple change in close-packed distortions.

Structures determined at either high temperature or high pressure are special cases of usually opposing effects. In 1980, a second automated diffractometer system was added, therefore, to study the combined effects. A high-temperature and high-pressure device was designed by Hazen and Finger (1981a) to collect intensity data and measure unit-cell dimensions under conditions up to 30 kbar and 450°C. The first structure measured in the device was fluorite, CaF_2, which now serves as an internal pressure reference in other determinations (Hazen and Finger, 1981b). The important pyroxene end-member diopside, $CaMgSi_2O_6$, was also studied under the same limits of pressure and temperature. Although the effects of compression and thermal expansion are compensatory for the most part, the net changes at high pressure and temperature do not exactly offset one another because of differences in the relative expansion and compression of different cation polyhedra. The characteristic expansion and compression coefficients of cation polyhedra may be used to predict many structural properties (Hazen and Finger, 1979). The bond angle deformations have yet to be evaluated.

Extension of the limits of temperature and pressure beyond 30 kbar and 450°C is essential so that structures can be determined under the same conditions at which relevant mineral reactions and melting take place in the earth. The structure of a crystal immediately below its melting temperature and the structure of the liquid itself, currently being studied with Raman, Mössbauer, and infrared techniques at temperature and after quenching to a glass, are of special interest. The determination of structures and the characterization of phase transitions in the common rock-forming minerals at combined high pressures and temperatures will no doubt remain a major program of research at the Geophysical Laboratory in the coming years. The Carnegie concept of a few talented investigators working within the framework of major significant problems continues to be a productive role for broad-based crystallographers.

References

Donnay, G. (1971). Corrigendum for "Fifty years of X-ray crystallography at the Geophysical Laboratory, 1919-1969," Carnegie Inst. Washington Year Book **69**, 314-315.

Donnay, G., R. W. G. Wyckoff, T. F. W. Barth, and G. Tunell (1970) Fifty years of X-ray crystallography at the Geophysical Laboratory, 1919-1969. Carnegie Inst. Washington Year Book **68**, 278-283.

Finger, L. W. (1970). Refinement of the crystal structure of an anthophyllite. Carnegie Inst. Washington Year Book **68**, 283-288.

Finger, L. W., C. G. Hadidiacos, and Y. Ohashi (1973). A computer-automated, single-crystal X-ray diffractometer. Carnegie Inst. Washington Year Book **72**, 694-699.

Finger, L. W., R. M. Hazen, G. Zou, H. K. Mao, and P.M. Bell (1981). Structure and compression of crystalline argon and neon at high pressure and room temperature. Appl. Phys. Letters **39**, 892-894.

Finger, L. W., and Y. Ohashi (1976). The thermal expansion of diopside to 800°C and a refinement of the crystal structure at 700°C. Am. Mineral. **61**, 303-310.

Finger, L. W., and E. Prince (1975). A system of Fortran IV computer programs for crystal structure computations. Natl. Bur. Stand. (U.S.) Tech. Note **854**, 133 pp.

Hazen, R. M., and L. W. Finger (1979). Bulk modulus-volume relationship for cation-anion polyhedra. J. Geophys. Res. **84**, 6723-6728.

Hazen, R. M., and L. W. Finger (1981a). High-temperature, diamond-anvil pressure cell for single-crystal studies. Rev. Sci. Instrum. **52**, 75-79.

Hazen, R. M., and L. W. Finger (1981b). Calcium fluoride as an internal pressure standard in high-pressure/high-temperature crystallography. J. Appl. Cryst. **14**, 234-236.

Mao, H. K., and P. M. Bell (1976). Compressibility and X-ray diffraction of the epsilon phase of metallic iron (ϵ-Fe) and periclase (MgO) to 0.9 Mbar pressure, with bearing on the earth's mantle-core boundary. Carnegie Inst. Washington Year Book **75**, 509-513.

Mao, H. K., R. M. Hazen, P. M. Bell, and J. Wittig (1980). Evidence for a localized 4*f*-shell breakdown in praseodymium under pressure. Carnegie Inst. Washington Year Book **79**, 380-384.

McCallister, R. H., L. W. Finger, and Y. Ohashi (1976). Intercrystalline Fe^{2+}-Mg equilibria in three natural Ca-rich clinopyroxenes. Am. Mineral. **61**, 671-676.

Merrill, L., and W. Bassett (1974). Miniature diamond anvil pressure cell for single-crystal X-ray diffraction studies. Rev. Sci. Instrum. **45**, 290-294.

Nixon, P. H., and F. R. Boyd (1973). Petrogenesis of the granular and sheared ultrabasic nodule suite in kimberlites, in "Lesotho Kimberlites," edited by P. H. Nixon. Lesotho: Lesotho National Development Corp., Maseru. pp. 48-56.

Ohashi, Y., and L. W. Finger (1975). An effect of temperature on the feldspar structure: crystal structure of sanidine at 800°C. Carnegie Inst. Washington Year Book **74**, 569-572.

Ohashi, Y., and L. W. Finger (1976). Stepwise cation ordering in bustamite and disordering in wollastonite. Carnegie Inst. Washington Year Book **75**, 746-752.

Rosenhauer, M., H. K. Mao, and E. Woermann (1976). Compressibility of magnesiowüstite ($Fe_{0.4}Mg_{0.6}O$) to 264 kbar. Carnegie Inst. Washington Year Book **75**, 513-515.

Seifert, F., and D. Virgo (1975). Kinetics of the Fe^{2+}-Mg, order-disorder reaction in anthophyllites: quantitative cooling rates. Science **188**, 1107-1109.

CHAPTER 5

THE WARREN SCHOOL OF X-RAY DIFFRACTION AT M.I.T.

Leonard Muldawer
Physics Department, Temple University, Philadelphia, PA 19122

Bertram Eugene Warren was born in Waltham (Massachusetts) in the year 1902 and went on to attend the Massachusetts Institute of Technology for the periods 1919-1925 and 1927-1929. In 1926 he was at the Technische Hochschule Stuttgart learning X-ray diffraction from R. Glocker and crystal physics theory from P. P. Ewald. Back at M.I.T. in 1927 Warren was assigned to build crystal structure models for Sir Lawrence Bragg during the latter's four month visit. (Interestingly, the crystal models to which most of us were exposed in Physics 8.27, Introductory X-ray Diffraction, were built by Richard P. Feynman when he was an M.I.T. student). During this period Warren worked with Bragg on the structure of diopside (Warren and Bragg, 1928). This resulted in a wholly unexpected chain structure, the first such silicate found. B. E. W. received his Sc.D. in 1929; his dissertation "X-ray Determination of the Structure of the Metasilicates" was under the supervision of Bragg. Then Warren spent a year at Bragg's home institution, the University of Manchester, where there was a very active group including R. W. James, from whom Warren learned a great deal. In 1930, Warren returned to M.I.T. as Assistant Professor of Physics after having cranked out a phenomenal number of silicate papers in the *Zeitschrift für Kristallographie.*

The Warren School, then, began in 1930 and was climaxed in 1969 with the appearance of the textbook by B. E. W. (Warren, 1969). During this forty year period, Warren carried out pioneering research, supervised 60 to 70 dissertation students and hosted many visitors and post-doctoral fellows for varying periods of time. He assisted many students who were not working in his own laboratory and, in many ways, aided researchers here and abroad. Charles Barrett has said, "it certainly would not be correct to say that I have not been a student of his, for I and countless others like me have studied his papers, listened to his talks and comments at crystallographic meetings, and even studied scraps of his unpublished (at that time) book." (Barrett, 1967). His introductory course in X-ray diffraction was given to an estimated total of close to a thousand students and its impact must be large.

I was part of the Warren School for most of the period from 1945 to 1951 – just about the middle of its existence. The names of the 1930's members were known to me and, as was usual for Warren's ex-students, I maintained close contact after leaving M.I.T. In particular, I followed the work of the group very closely during the 1950's since I was carrying out related research. In putting this narrative together, I have made use of reminiscences from *"Fifty Years of X-ray Diffraction"* (Ewald, 1962), a list of Warren's published papers, Barrett's unpublished manuscript on the Warren School, and correspondence which I had during preparation for the celebration of Warren's retirement from full-time teaching in 1967. I hope to give a balanced and accurate picture of the School.

Warren The Teacher

If Warren had not been a great teacher, it is doubtful that there would have been anything like the Warren School that did come about. Warren's teaching was on a par with his research; there were beautifully organized lectures, slides, models and examples from his own research. He began with the fundamentals and developed the solution to the problem within the problem with inexorable logic. There was never any fuzziness; only after the argument had been carried as far as possible were assumptions inserted. The working relations were then developed so that theory could be related to the results of experiment. When we worked on problems for Warren, we were experimental and theoretical physicists.

Once, for homework in our advanced course, Warren gave us a problem for which Bragg had asked him to provide a solution. George Kuczinski, Lester Siegel and I worked out a result which Warren said looked good. (But of course we used Warren's methods). Several years later I saw an article in *Acta Crystallographica* on this same subject and it contained essentially our results. Kuczinski has said that Warren was the best teacher he ever had.

Warren corrected every homework problem, every Laboratory report and the lengthy home examinations with meticulous care. We all complained about our 30 hour take-home finals and it must have taken Warren many hours to grade each one. We were concerned about his health during the 1950's when he had a great many students in the introductory course and in his Laboratory carrying out dissertations. With the time spent on each one, he must have been terribly overworked.

Warren's door at MIT (room 6 - 413) was open six days a week with the Metropolitan Opera a feature of Saturday afternoon. Students (and ex-students) could enter, ask questions, discuss results and learn physics. Often, graduate students who were working for other professors expressed envy of those who were in Warren's research group.

Clearly, the teaching made a great contribution to the success of Warren's group. His first course in X-ray diffraction was so good that it brought many capable students into his laboratory. In addition, his research reputation brought students from many foreign countries to share in the stimulating atmosphere among the members of the X-ray group. His help to those who were not members of his own Laboratory was also remarkably generous. From three of these evidence to that effect is clear.

J. B. Cohen: he allowed me to work in his Laboratory. This meant of course that he taught me the ropes and guided me to the answers.

B. D. Cullity: it was not so simple for me. . .it would have never been accomplished without his help. [Warren also allowed use of his problems in the text by the late Bernard Cullity (1956)].

G. H. Donnay: he told me that my first solution of my thesis problem was wrong and how to straighten it out.

I doubt that there would have been a school if Warren did not have the character that he did. While he set a high standard for himself and demanded the same from his students, he established a rapport with us which was uncommon. One of us has said that among his memories are two: (1) sometimes bloody but very inspiring series of informal X-ray seminars, (2) the Laboratory parties at Crane's Beach. C. S. Barrett noted that Warren has an exceptionally loyal group of former students and that there is an extremely delightful and helpful relation between them.

In my correspondence with graduates from Warren's group in preparation for the Warren retirement celebration, I found comments such as "fondest memories," "deep appreciation," "esteem and affection," "great help and inspiration," "often reflected about the good start I obtained in your lab," "really wonderful person he is," "my debt to him," etc. Another wrote "The key highlight that I recall was the calmness and kindness with which you (BEW) treated me when I informed you that I had burned out the main X-ray tube. . .probably represented a large portion of your research budget."

It is clear that Warren's example in the classroom and in the Laboratory has had a profound impact upon his students. We are very happy to share our research with him and we are certain that he is proud of our work.

Warren And His Research

Warren continued his early interest in silicate structures for a brief period at M.I.T. but soon moved away from pure crystallography (with some important exceptions). In 1933, he confirmed Zachariasen's hypothesis concerning the structure of glass (Warren, 1933) and continued these studies until 1942 with co-workers C. F. Hill, A. D. Loring, C. W. Morey, H. Krutter, O. Morningstar, J. Biscoe, C. S. Robinson, Jr., A. G. Pincus, M. A. Drussne and C. S. Smith. With his last student, R. L. Mozzi, he returned to a study of glasses in 1966 and obtained more definitive results with the aid of improved corrections and techniques (Warren and Mozzi, 1969, 1970).

His main efforts were not so much in expanding and improving methods of structure determination, although his contributions here were significant, but in developing a newer application of diffraction: the investigation of nonperiodic and nearly periodic structures through quantitative measurement of X-ray intensities. His greatest contributions to science lie in his creation of mathematical and experimental techniques for deriving information on the structure of matter from X-ray scattering not necessarily restricted to Bragg reflections. The glass studies were only the beginning of these. Warren himself has said, in this connection, "As with humans, it is the deviations from regularity that are more interesting."

The X-ray group in the early 1930's included N. S. Gingrich, G. G. Harvey, J. T. Serduke, R. R. Hultgren, A. L. Patterson, and graduate and undergraduate students. (Patterson was an unpaid associate from 1933 to 1936). With Gingrich, Warren was soon investigating liquids and the Fourier integral analysis of X-ray powder patterns (Warren and Gingrich, 1934). This important paper stimulated Patterson to produce his famous $|F^2|$ series paper (Patterson, 1934). In it Patterson stated, "The result obtained here is an extension of the application to crystals of the theory of scattering of X-rays. . .reported by Gingrich and Warren at the Washington meeting of the American Physical Society and arose in a discussion of that work, *Phys. Rev.* **46**, 368 (1934)." With this series, fundamental information regarding atomic separations in crystals could be obtained directly from measured intensities without making any assumptions about phase. This tool became very widely used.

Warren and Gingrich applied the Fourier inversion techniques to liquids and glasses and utilized crystal monochromatized radiation and accurate intensity measurements to give radial distribution functions. They gave the first accurate quantitative studies of scattering by liquids. Subsequent scattering studies were carried out with L. P. Tarasov on liquid sodium (1936) and with J. Morgan on water (1938). Each of these was an important contribution.

There were crystal structure determinations also being carried out but these were not of silicates. They were of simple substances which had not been previously done because of various difficulties. These included rhombic sulphur (Warren and Burwell, 1934), black phosphorus (Hultgren, Gingrich and Warren, 1935), crystalline bromine (Vonnegut and Warren, 1936) and uranium (Jacob and Warren, 1937).

While he was studying glass, Warren became interested in other amorphous materials as well. He studied carbon black (Warren, 1934) and showed that it was a two dimensional random layered structure. There was small angle scattering which had been observed by others but Warren gave the first correct explanation of the cause: large scale inhomogenities. Essentially, this was the beginning of the field of small angle scattering. Warren studied the mathematics of the scattering from such substances again (Warren, 1941) and the application to carbon black (Biscoe and Warren, 1942). As with glass, Warren returned to the carbon blacks years later. (Houska and Warren, 1954 and Warren and Bodenstein, 1965, 1966). There was also the study of amorphous rubber (Simard and Warren, 1936).

Warren then had the idea that short range order in binary alloys would be liquid-like and should produce a modulated intensity from which order parameters could be obtained by Fourier analysis. This work was begun by Z. Wilchinsky (1944) but this research interest was interrupted by the war. During the war, Warren carried out classified diffraction studies of neutron irradiated graphite (Bacon and Warren, 1956) as part of the Manhattan project.

The war produced a hiatus in Warren's X-ray Laboratory research but a group started up again at the end of the forties. Throughout the 1950's there was again great activity with studies in three main areas: order-disorder, temperature diffuse scattering and cold-work imperfections. There were studies of long range order by J. M. Cowley, B. W. Roberts, L. Muldawer, D. R. Chipman, D. T. Keating, R. T. Jones, C. B. Walker, and B. W. Batterman and studies of short range order by J. M. Cowley, N. Norman, B. Borie, E. Suoninen, and S. C. Moss. Warren loved to say that even when long range order was gone, there still was short range order indicating that, though atoms may be dumb, they are not fickle. The short range order measurements required measurement of diffuse intensities between Bragg peaks and necessitated many corrections, one of which was temperature diffuse scattering (TDS).

For a while TDS became a study interesting in itself (Cole and Warren, 1952). Other studies were carried out by H. Cole, R. E. Jacobsen and C. B. Walker. These studies were important in solid state physics since TDS resulted from the thermal vibration in crystals and thus were able to given phonon-dispersion curves and vibrational frequency distributions.

The last of the three studies listed above, cold work imperfections in metals, was started simultaneously with the order-disorder studies. Warren developed mathematical methods by which Fourier analyses of line profiles were able to give crystallite sizes, strain distributions, and stacking fault distributions. A large number of applications of these methods have been further developed at M.I.T. and elsewhere as a result of these pioneering investigations. Warren's co-workers in this field included B. L. Averbach, M. McKeehan, E. P. Warekois, O. J. Guentert, J. Despujols and C. N. J. Wagner. An overall picture of this field may be obtained from Warren's review articles (Warren, 1959 and Warren, 1963).

Thus the early interest in deviations from regularity has been continued by Warren into his most recent publications (Warren 1976, 1978) which were concerned with temperature effects on powder patterns and with powder pattern intensity distributions from very small crystals. He did not remain static in his interests, one subject leading him into the next. Then, when techniques and corrections were improved, he was right back into the old game and making order of magnitude improvements. This always produced excitement around the laboratory and this was especially obvious to the visitor.

The Continuation

The Warren School will continue in the work of the graduates who have gone into universities, industry and government laboratories. Those in the universities are now turning out their own students. Then there is Warren's textbook, a highly polished work which covers most of Warren's research areas. The final chapter on perfect crystal theory, the only chapter in which Warren has not worked, was written with the assistance of several former students. If one could not have Warren's course, one could have Warren's book.

On the occasion of Warren's retirement in 1967, we organized a symposium in his honor. Eight scientific papers were given, seven by former students. The eighth owed its basis to an idea suggested by Warren; the others were mainly concerned with order, thermal vibrations and amorphous materials. We also founded an award, the Bertram Eugene Warren Diffraction Physics Award, to be given to persons doing physics of solids research using diffraction techniques.

The lesson of the Warren School as stated by C. S. Barrett is as follows: learn everything that *can* be learned about atom arrangements by a rigorous mathematical analysis of the distribution of intensity of X-ray scattering, using the lines and their profiles, the intensities between the lines, or both, and making the absolute minimum of arbitrary assumptions – preferably no assumptions at all.

References

Bacon, G. E. and Warren, B. E. (1956). Acta Cryst. **9**, 1029.

Barrett, C. S. (1967). Unpublished manuscript, delivered at Symposium in honor of B. E. Warren. May 13, 1967.

Biscoe, J. and Warren, B. E. (1942). J. Appl. Phys. **13**, 364.

Cole, H. and Warren, B. E. (1952). J. Appl. Phys. **23**, 335.

Cullity, B. D. (1956). Elements of X-ray Diffraction. Reading: Addison-Wesley.

Ewald, P. P. (1962). Fifty Years of X-ray Diffraction. Utrecht: Oosthoek.

Houska, C. R. and Warren, B. E. (1954). J. Appl. Phys. **25**, 1503.

Hultgren, R. R., Gingrich, N. S. and Warren, B. E. (1935). J. Chem. Phys. **3**, 351.

Jacob, C. W. and Warren, B. E. (1937). J. Am. Chem. Soc. **59**, 2588.

Morgan, J. and Warren, B. E. (1938). J. Chem. Phys. **6**, 666.

Patterson, A. L. (1934). Phys. Rev. **46**, 372.

Simard, G. L. and Warren, B. E. (1936). J. Am. Chem. Soc. **58**, 507.

Tarasov, L. P. and Warren, B. E. (1936). J. Chem. Phys. **4**, 236.

Vonnegut, B. and Warren, B. E. (1936). J. Am. Chem. Soc. **58**, 2459.

Warren, B. E. (1933). Z. Kristallogr. **86**, 349.

Warren, B. E. (1934). J. Chem. Phys. **2**, 551.

Warren, B. E. (1941). Phys. Rev. **59**, 693.

Warren, B. E. (1959). Prog. Metal Phys. Vol. 8, Chapter III.

Warren, B. E. (1963). Acta Met. **11**, 995.

Warren, B. E. (1969). X-ray Diffraction. Reading: Addison-Wesley.

Warren, B. E. (1976). Acta Cryst. **A32**, 897.

Warren, B. E. (1978). J. Appl. Cryst. **11**, 695.

Warren, B. E. and Bodenstein, P. (1965). Acta Cryst. **18**, 282.

Warren, B. E. and Bodenstein, P. (1966). Acta Cryst. **20**, 602.

Warren, B. E. and Bragg, W. L. (1928). Z. Kristallogr. **69**, 168.

Warren, B. E. and Burwell, J. T. (1934). J. Chem. Phys. **3**, 6.

Warren, B. E. and Gingrich, N. S. (1934). Phys. Rev. **46**, 368.

Warren, B. E. and Mozzi, R. L. (1969). J. Appl. Cryst. **2**, 164.

Warren, B. E. and Mozzi, R. L. (1970). J. Appl. Cryst. **3**, 59.

Wilchinsky, Z. (1944). J. Appl. Phys. **15**, 806.

CHAPTER 6

CRYSTALLOGRAPHY AT M.I.T. – M. J. Buerger's Laboratory

A. J. Frueh, L. V. Azaroff, University of Connecticut, Storrs, CT, 06268
C. T. Prewitt, SUNY, Stony Brook, NY 11794 and
B. J. Wuensch, M.I.T., Cambridge, MA 02139

In 1927 W. L. Bragg came to the Massachusetts Institute of Technology as a visiting lecturer. He had, in the previous decade, solved the crystal structures of a number of minerals, and just the previous year he had published the structure of the magnesium-rich silicate, olivine. Among those attending his lectures were two graduate students—Bertram Warren of the Physics Department and Martin Buerger, a Geology Department student in the fields of Mineralogy and Ore Deposition. Both these graduate students were soon to become faculty members at the Institute, and for the next two score years under their leadership M.I.T. was to become an important center for teaching and research in crystallography.

Before Bertram Warren became a full-time faculty member of the Physics Department at M.I.T., his studies took him to Manchester to work with W. L. Bragg. Together they published their elucidation of the crystal structure of diopside in 1929. Still at Manchester, Warren published his work on the structures of tremolite and other common amphiboles. Then he returned to the United States and set up his own X-ray Laboratory at M.I.T. where, with two co-workers (J. Biscoe and D. T. Modell), he solved the crystal structures of enstatite and the monoclinic pyroxenes, vesuvianite, euclase, melilite and anthophyllite.

Meanwhile, Martin Buerger had turned his attention to the study of the physical and chemical properties of crystals, especially ore minerals. He was particularly interested in problems of plastic deformation, polymorphism and twinning. He soon found a need to determine the crystal structures of those minerals whose physical properties he was studying. In the days before the National Science Foundation, little commercially manufactured crystallographic equipment was produced and, when available, was not within the budgets of most Universities. Buerger had previously persuaded his department chairman to purchase an optical goniometer, but after that he was on his own. At first he had to use borrowed equipment, and the data for his first crystal structure, that of marcasite, were gathered in the Laboratory of his colleague, Bertram Warren.

In 1934, as the result of a grant-in-aid from the Rockefeller Foundation, Martin Buerger was able to set up an X-ray Laboratory in the Geology Department. In order to stretch the dollars as far as possible, X-ray generators were fashioned from discarded medical equipment. With protective import duties and restrictive patent rights, it was impossible to find the money for more than one sealed X-ray tube (Coolidge tube), and thus the homemade cold cathode tube was the order of the day. The current in the cold cathode tube was controlled by the quality of the vacuum, so that each tube had its own vacuum pump "chucking" away. A large vacuum chamber with a controlled leak was used to try to maintain a constant vacuum and, hence, a constant electrical current. It was an art to keep these tubes running, and one day, after spending some hours trying to get the vacuum under control, Buerger remarked to his assistant that he viewed these tubes as he did some women: "It takes a long time to get to understand them, and when you do, you still can't get them to do what you want."

As it turned out, the entire crystallographic community eventually reaped the benefits of Buerger's response to the lack of availability of crystallographic equipment. Using a portion of the above-mentioned Rockefeller grant, Buerger set up a machine shop in the Geology Department. This machine shop was staffed from the very beginning by excellent instrument makers (Otto Von der Hyde, 1934-1942; Charles Supper, 1942-1946; John Solo, 1946-1954) thus enabling Buerger to have equipment made of his own design. From this shop came many improved models of X-ray diffraction cameras that have withstood the test of time and are still in use today. In addition, of course, came the newly invented Buerger precession camera that so greatly simplified the interpretation of reciprocal lattice geometry.

By the end of the second world war, Buerger's reputation in crystallographic research had become worldwide and was enhanced by the publication in 1942 of *"X-ray Crystallography,"* the first of the twelve books he was to author. In 1946 he was one of a small group of American scientists that travelled to England to join with other crystallographers from around the world in organizing the International Union of Crystallography.

The last half of the decade of the "40's" was a period of great intellectual stimulation for the students of crystallography at M.I.T. A new doctorate program had been set up in crystallography through the cooperation of Martin Buerger, Bertram Warren, John T. Norton in Metallurgy and R. S. Bear in Biology. This arrangement facilitated the holding of seminars whenever distinguished guests visited any of M.I.T.'s crystallographic laboratories. Participating in these seminars were the graduate students of this era (John Cowley, Christopher Walker, Gabrielle Hamburger Donnay, Alfred Frueh, Benjamin Averbach and Gilbert Klein), as well as visiting senior scientists (Elysiario Tavora of Brazil and William H. Barnes of Canada).

In 1948-49, Buerger's interests in crystal-structure analysis focussed on the development of direct methods for solving the phase problem of X-ray crystallography. Leaving his earlier efforts in implication theory, which had foreshadowed the utility of inequalities, he started to consider vector sets and their interrelations in crystal space and Patterson space (Buerger, 1950a, 1950b). To assist him in these analyses, before the advent of high-speed computers, he also constructed an optical Fourier synthesizer (Buerger, 1950c, 1950d).

During a sabbatical in South America, Buerger continued his work on vector sets which ultimately led to the development of "A new approach to crystal-structure analysis" (Buerger, 1951) which he called the image-seeking function. Back at M.I.T. one of his graduate students, Leonid V. Azároff, was testing a single-crystal diffractometer that Buerger had designed on his ocean trip to Spain. Working with him at that time were three other graduate students, Leroy Jensen, Morse Klubock, and Virginia F. Ross. By 1951, the group of collaborators had grown to include one more graduate student, Nobukazu Niizeki, and Drs. Theo Hahn, Yoshio Takeuchi, and Anna and Joseph Zemann, whose marriage took place shortly after their arrival in Boston.

Buerger's broad interests in crystallography and mineralogy during these years included further work on unravelling vector sets in Patterson space (Buerger, 1953a, 1953b), an analysis of twinning (Buerger, 1954a), substructures (Buerger, 1954b), derivative structures (Buerger, 1954c), and determinations and refinement of the structures of nepheline (Buerger, Klein, and Donnay, 1954), cubanite (Buerger and Azároff, 1955), berthierite (Buerger and Hahn, 1955), Co_2S_3 (Buerger and Robinson, 1955), nepheline (Buerger and Hahn, 1955), wollastonite (Buerger, 1956), pentlandite (Buerger and Pearson, 1956), diglycine hydrobromide (Buerger, Barney, and Hahn, 1956), diglycine hydrochloride (Buerger and Hahn, 1957), livingstonite (Buerger and Niizeki, 1957a), and jamesonite (Buerger and Niizeki, 1957b). In recognition of these prodigious efforts, M. J. Buerger was awarded the Day Medal of the Geological Society of America in 1951 and the Roebling Medal of the Mineralogical Society of America in 1958, and was elected to the National Academy of Sciences in 1953.

The second half of the 1950's saw the advent of digital computers. One of the first large machines was M.I.T.'s Whirlwind, established in virtually an entire building on Massachusetts Avenue, adjacent to the main campus. Buerger's Laboratory was quick to incorporate the potential of high-speed computation in its crystallographic research with the calculation of two-dimensional and eventually three-dimensional Patterson functions and electron density maps. At the end of the 1950's Whirlwind was replaced by early versions of IBM 704 and 709 computers on which, in those early days, computation time was free of charge. For these machines Buerger's group developed data reduction programs for the Weissenberg counter diffractometer mentioned above, and early versions of programs for absorption corrections and full-matrix least-squares refinement that were widely distributed and remain in use today. The excitement of this period in the development of crystallography attracted a large number of students to Buerger's group. Students during the late 1950's and early 1960's were Tibor Zoltai, Charles W. Burnham, Donald R. Peacor, Charles T. Prewitt, Bernhardt J. Wuensch, Hilda Cid-Dresdner and Wayne Dollase. Present as post-doctoral fellows were Karl Fischer, Roberto Poljak, and Panos Rentzeperis. Using three-dimensional diffractometer data and least-squares refinement, the structures of a number of minerals and other materials were examined. Some represented solution of unknown structures, but an increasing number of studies in this period turned towards detailed refinements of poorly known or early structure determinations of minerals toward the end of providing insight into the geochemistry of these minerals. Several studies were conducted on families of polymorphic minerals, including SiO_2 (coesite, tridymite), the alumino-silicates (andalusite, kyanite, mullite), and pyroxenes (wollastonite, parawollastonite). Older silicate structures that were re-examined and refined included tourmaline, pectolite, rhodonite, and bustamite.

In 1963 Buerger's X-ray Laboratory was moved from its old quarters (previously the home of the M.I.T. Radiation Laboratory during W.W. II) to the eighth floor of the newly-constructed high-rise Green tower for the Earth Sciences. Although Buerger was to approach retirement age in 1968, the reputation of his Laboratory continued to attract a number of post-doctoral workers. Present in this capacity in the middle 1960's were Isabel Garochochea-Wittke, Hajo Onken, Jurgen Felsche, Peter Süsse, and Herbert Thurn. Felix Trojer, Richard Berger, Karlheinz Taxer, Martha Redden, and Lyneve Waldrop constituted the last of the succession of graduate students to study under Buerger's guidance at M.I.T.

In 1968 Buerger became Institute Professor Emeritus at M.I.T. As a measure of their respect and affection for Buerger, his former students began clandestine preparations for a banquet to mark the occasion. A network of seminar tours, carefully concealed from Buerger, was established to help finance the presence of even those former workers who resided abroad. Supposedly enroute to dinner with the Frondels, the Buergers were diverted under pretext to the M.I.T. Student Center to be greeted by a room filled with virtually all of his former students and post-doctoral workers. A festschrift of the *Zeitschrift für Kristallographie* (Vol. 127), composed of papers contributed by former students and colleagues, was also assembled on this occasion.

For the next five years Buerger divided his time between M.I.T. and the University of Connecticut. During these years, Buerger published 21 papers (seven with co-authors) and three books (one with two co-authors). Upon his seventieth birthday he retired from both institutions and joined the Hoffman Laboratory at Harvard University as an Honorary Research Associate during 1973-78. Two years later he returned to an office at M.I.T. where, in "retirement" he contributes to the revision of the *"International Tables for X-Ray Crystallography,"* pursues interests in vector sets and homometric structures, and continues as an Editor-in Chief of the *Zeitschrift für Kristallographie*.

Although this account has attempted to present the flavor of activities in Buerger's Laboratory at M.I.T., his influence extended far beyond M.I.T. For example, he played a major role in the resurrection of *Zeitschrift für Kristallographie* at the end of W.W. II, in the production of the second edition of *"International Tables,"* and the founding of The American Society for X-Ray and Electron Diffraction (ASXRED), the Crystallographic Society of America (CSA), and their successor, the American Crystallographic Association. An excellent reference for those interested in a more detailed account of these and other contributions is Robert R. Shrock's (1977) *"Geology at M.I.T. 1865-1965."*

References

Buerger, M. J. (1950a). Acta Cryst. **3**, 87-97.
Buerger, M. J. (1950b). Acta Cryst. **3**, 243.
Buerger, M. J. (1950c). Am. Mineral. **35**, 278-279.
Buerger, M. J. (1950d). Proc. Natl. Acad. Sci. U.S.A. **36**, 324-329.
Buerger, M. J. (1951). Acta Cryst. **4**, 531-544.
Buerger, M. J. (1953a). Proc. Natl Acad. Sci. U.S.A. **39**, 674-678.
Buerger, M. J. (1953b). Proc. Natl. Acad. Sci. U.S.A. **39**, 678-680.
Buerger, M. J. (1954a). Ac. Brazil de Ciencias Anais. **26**, 111-121.
Buerger, M. J. (1954b). Proc. Natl. Acad. Sci. U.S.A. **40**, 125-128.
Buerger, M. J. (1954c). Am. Mineral. **39**, 600-614.
Buerger, M. J. (1956). Proc. Natl. Acad. Sci. U.S.A. **42**, 113-116.
Buerger, M. J. and Azaroff, L. (1955). Am. Mineral. **39**, 213-225.
Buerger, M. J., Barney, E., and Hahn, T. (1956). Z. Kristallogr. **108**, 130-144.
Buerger, M. J. and Hahn, T. (1955). Am. Mineral. **40**, 226-238.
Buerger, M. J. and Hahn, T. (1957). Z. Kristallogr. **108**, 419-453.
Buerger, M. J., Klein, G. E., and Donnay, G. (1954). Am. Mineral. **39**, 805-818.
Buerger, M. J. and Niizeki, N. (1957a). Z. Kristallogr. **109**, 129-157.
Buerger, M. J. and Niizeki, N. (1957b). Z. Kristallogr. **109**, 161-183.
Buerger, M. J. and Pearson, A. D. (1956). Am. Mineral. **41**, 804-805.
Buerger, M. J. and Robinson, D. W. (1955). Proc. Natl. Acad. Sci. U.S.A. **41**, 199-203.
Shrock, R. R. (1977). Geology at M.I.T. 1865-1965. Vol. I. Cambridge: The M.I.T. Press, Cambridge.

CHAPTER 7

THE SHOEMAKER RESEARCH GROUP AT MIT, 1951-1970

David P. Shoemaker
Oregon State University, Corvallis, OR 97331

Shortly after arriving at MIT from Caltech in 1951 as a new Assistant Professor of Chemistry I set up a small laboratory with a Unicam Weissenberg camera, a Laue camera, and a Phillips powder camera on a General Electric XRD-3 generator and a continuously pumped cold cathode tube scrounged from someone else's garbage, and was in business. These sufficed for some years before I accumulated other generators and the normal complement of Charles Supper Weissenberg and precession cameras, and eventually spectrogoniometers.

The first structure to come out resulted from the summer 1952 visit of Dr. Hubert Kindler from Germany under the auspices of MIT's Foreign Student Summer Project. This was cyclooctatetraene (COT) monocarboxylic acid, provided by the late Dr. Arthur C. Cope. This was of interest then because of controversy about the shape of the COT ring skeleton, variously reported from gas phase electron diffraction and infrared spectroscopy to have symmetry D_{2d} (tub), D_4, or D_{4d} (crown). The structure was solved with the aid of a 3-D Patterson function, the first dimension of which was hand calculated with Beevers-Lipson strips providing coefficients for 2-D sections. These were produced by the famous analog computer XRAC at Penn State through Ray Pepinsky's characteristic kindness and unfailing generosity, of which we were to avail ourselves several more times before 3-D functions could be calculated on modern computers. The symmetry of the COT ring skeleton proved to be D_{2d} (Shoemaker *et al.*, 1965a), as it has been in other investigations since then, including a study of the dihydrated calcium salt of 1,2-COT-dicarboxylic acid (Wright, Seff and Shoemaker, 1972) by the New Zealander Dr. Don Wright who came to my Laboratory from Caltech in the sixties for this work (later to join C.S.I.R.O. in Australia).

In 1952 I began to be funded by the Office of Ordnance Research, and was supported by that agency and its successor, Army Research Office (Durham), continuously to 1969, to continue a program of research begun at Caltech under Linus Pauling: the structures of intermetallic compounds of transition elements, particularly phases related to the tetragonal *sigma phase*, the structure of which Gunnar Bergman and I determined at Caltech in 1950 (Bergman and Shoemaker, 1954). This program at MIT was aided immeasurably over the years by collaboration with Professor Paul Beck of the University of Illinois who did phase equilibrium studies on a wide range of binary and ternary transition alloy phases, of which he generously gave us specimens.

The decisive occurrence in the formation of my research group and in getting the new program going was the arrival in 1953 of Clara Brink, Ph.D. Leiden 1950, who had worked in the Laboratories of A. E. van Arkel, J. M. Bijvoet, Dorothy Hodgkin (on vitamin B_{12}), and Caroline MacGillavry. Jack Dunitz, after meeting Clara in Holland and learning of her plans to come to my laboratory, is reported to have said to mutual acquaintances: "I have just met Dave Shoemaker's future wife!" Well, it took two years for Dave Shoemaker to get the picture, but Jack's perceptivity was vindicated when the wedding took place in 1955.

Clara took over a problem that had been started by Alwyn Fox, an Englishman who came to MIT from Lynn Hoard's Laboratory at Cornell: the structure of the delta phase, MoNi, obtained from Paul Beck. Alwyn had collected Weissenberg data on a single crystal, employing the technique used in the sigma phase work and on all similar phases since, namely to crush the brittle material, mount tiny fragments (never more than 100 micrometers in any dimension) screen them with long exposure Laue photographs to identify ones that were single crystals, and orient them with Laue photographs by looking for mirror planes. It sometimes took months to find and orient a suitable crystal, and months more to photographically record the feeble intensities. The delta phase appeared to be tetragonal; we chose to ignore a few very small but definite violations of tetragonal symmetry in the intensities. From another 3-D Patterson produced by XRAC at Penn State we derived – and published – an incomplete (and wrong) structure in space group $P4_22_12$. Clara, however, would not let this rather unsatisfactory state of affairs rest, and eventually found an elegant structure in $P2_12_12_1$! (Shoemaker and Shoemaker, 1963).

That was not, however, before we had completed another structure, that of the P phase of Mo-Ni-Cr, also from Paul Beck. We found it to be very similar in structure to the sigma phase but with one basal axis nearly doubled, making it orthorhombic. This structure was not solved with XRAC, but with the first electronic digital computer of the modern stored program type. This was MIT's Whirlwind, constituting many dozens of huge relay racks of vacuum tubes (including the famous Williams electrostatic storage tubes), occupying fully a two-story Gothic brick building on Mass. Avenue just across the tracks, and having possibly all of the computing power and speed of today's $500 Radio Shack TRS-80. This structure (Shoemaker, Shoemaker and Wilson, 1957) opened our eyes to the fact (independently recognized by John Kasper at General Electric) that in the sigma and P phases we were beginning to deal with a unique family of structures having up to four (and only four) well-defined coordination types: CN 12 (icosahedral), CN 14, CN 15, and CN 16. Interstices are exclusively tetrahedra, generally somewhat distorted, leading to the designation "tetrahedrally close-packed" (tcp). When the correct delta phase structure was found, it too was tcp. Clara and I were to make something of a career of determing tcp and near-tcp structures in the years that followed (Shoemaker and Shoemaker, 1964, 1968).

Dr. Yukitomo Komura came to us from Toku Watanabe's laboratory in Japan, and undertook the study of another phase from Paul Beck, the R Phase of Mo-Co-Cr. This was very difficult in the beginning because Laue photographs failed to show any mirror lines of symmetry. We eventually found the space group to be $R\bar{3}$ ($\bar{3}$ is the only one of the 11 Laue symmetries except $\bar{1}$ to have no mirror planes). The structure, like σ, P, and δ, turned out to be ideal tcp (Komura, Sly and Shoemaker, 1960). Yuki later returned to Japan where he established in Hiroshima a very active research group working on polytypes of Laves phases (also tcp). Later tcp structures, done principally by Clara, were M phase (Nb-Ni-Al) (Shoemaker and Shoemaker, 1967) and υ phase (Mn-Si) (Shoemaker and Shoemaker, 1971). A graduate student in the middle to late sixties, Philip Manor, applied manually a direct method (Isabella Karle's symbolic addition method) to the X phase (Mn-Co-Si); much to our surprise, it worked! (We had suspected that the characteristic sigma-phase-type intensity distribution would disastrously affect the validity of the statistics.) That phase also proved to be tcp (Manor, Shoemaker and Shoemaker, 1971). Work on three others that were started at MIT – D(Mn_5Si_2, not ideally tcp), K(Mn-Fe-Si) and I(V-Ni-Si) – were completed not at MIT but rather many years later in our new laboratory at Oregon State University. The I phase, completed only in 1979, had low symmetry (*Cc*) and 57 different kinds of atoms! (Shoemaker and Shoemaker, 1981). Not all transition alloy specimens examined turned out to be tcp. The E phase, Ti-Ni-Si, was found to have the $PbCl_2$ structure (Shoemaker and Shoemaker, 1965). A young Russian, Yu. V. Voroshilov, came to us from the Laboratory of P. I. Kripyakevich and Y. B. Kuzma in Lvov, and determined the structure of NbCoB (Kripyakevich *et al.*, 1971).

Much went on beside the alloy work. My first graduate student, Frank C. Wilson undertook study of a compound obtained from Geoffrey Wilkinson and his then graduate student F. Albert Cotton at Harvard, containing, per molecule, two Mo atoms, two cyclopentadiene rings, and 6 CO. Could this be a triple decker sandwich with a planar ring of 6 CO's between the two Mo? That turned out not to be the case. The molecule is bis-[molybdenum cyclopentadienyl tricarbonyl], with a single bond connecting the two Mo atoms – the first example of a metal-metal bond in an organometallic complex. Later, Al Cotton, at MIT and later at Texas A & M, found many other examples of metal-metal bonds, including mutliple bonds. This structure was one of the first to be refined by least squares on a high speed computer; Frank and I took it to New York where David Sayre helped us put it on the IBM 704 with his program NYXR1. In a half hour we did 5 or 6 cycles (isotropic, of course) and got the R index down to a then acceptable level of 0.12 (Wilson and Shoemaker, 1957). Frank got his Ph.D. and went on to DuPont in Wilmington where he became very active in diffraction work on synthetic fibers.

In 1956 I began a consulting relationship with the Esso (later EXXON) Company in Baton Rouge (a fact which led Bill Lipscomb to remark "honey bees make honey; Esso bees make money"). This relationship, which has continued ever since, got me into a new research field – synthetic zeolites – and led to Esso providing some financial support for the work. We were fascinated by the possibility of determining the sites of molecules absorbed into zeolite cages. The work was necessarily by powder methods since the crystallites were only a few micrometers in size, but fortunately the zeolites studied – 4A, 5A, and 13Y ("synthetic faujasite") – were cubic. Dr. Walter Meier came to us from Caltech to spend a year, and determined the arrangement of Br_2 molecules in zeolite 4A (Meier and Shoemaker, 1966) before returning to his native Switzerland and an eventual professorship at the ETH, where he has achieved the status of foremost world authority on zeolite structure. A graduate student, Karl Seff, determined structures of 4A and 5A containing I_2 (Seff and Shoemaker, 1967), H_2O, Kr, and Xe. Later at the University of Hawaii, where he is professor, he succeeded in determining the structure of a number of inclusion complexes with single crystals ($\sim$50 μ) of zeolite A. In the late sixties, graduate students Philip Manor and Roberta Ogilvie verified the structure of a new zeolite, RHO, discovered by Dr. Harry Robson of EXXON in a synthesis involving cesium cations; the framework structure was a hypothetical one that had been proposed by Walter Meier (Robson *et al.*, 1973). Paul Manor is now Science Editor for Springer Verlag in New York; Roberta Ogilvie Day is at University of Massachusetts, Amherst.

In the middle fifties MIT got its own IBM 704. Bill Sly came to us from Holmes Sturdivant at Caltech and, among other things, programmed the 704 to do 3-D Fourier and Patterson syntheses (using procedures based on Verner Schomaker's "M-card" system for doing Fourier syntheses on an IBM 402 tabulator and sorter). I added some goodies to take care of symmetry and MIFR1 was born (Shoemaker and Sly, 1961), to remain one of the standard crystallographic computing programs for a decade or more (later as ERFR2, an adaptation of MIFR1 by Jan van den Hende of Esso for the IBM 709, 7090, and 7094). Bill is now professor at Harvey Mudd College, Claremont, California.

Although most of our work was with X-rays, we took brief excursions into neutron diffraction (courtesy of

Cliff Shull, whose equipment we used) and Low Energy Electron Diffraction or "LEED" (first with a machine "home built" by Leo Feinstein and improved by Max Taylor with NSF support, later with a Varian machine acquired by MIT Materials Center). Ron Chandross and Leo Feinstein did polarized neutron work on FeV to measure the spin of the iron atom (Feinstein and Shoemaker, 1968); John Mellor studied orientational disorder of the cyanide ion in zinc and cadmium cyanides, and studied the ordering of transition atoms in the P and R tcp phases (Shoemaker, Shoemaker and Mellor, 1965). With LEED, Leo Feinstein and John Keil studied (110) cleavage surfaces of II-VI semiconductors such as CdTe (Feinstein and Shoemaker, 1965), and Max Taylor studied low temperature adsorption of xenon on (0001) cleaved zinc. The LEED results were difficult to interpret, and our excursion into this field cannot be said to have been very successful. Ron Chandross is with Dick Bear at UNC, Chapel Hill; Leo Feinstein went to AT & T in Pennsylvania; John Keil went to Motorola; Max Taylor is at Bradley University.

Other activities and members of the group, with their present affiliations in parentheses where known, are as follows. Norman Weston (DuPont), my second graduate student, worked with Dr. Roberto Poljak from Argentina (John Hopkins) and Dr. Jørgen Rathlev of Denmark on Na-Pb alloys and ammonia solutions of them; Dr. Kutoputhur S. Chandrasekaran (Madurai, India) and Dr. Gunther Eulenberger (Hohenheim, BRD) worked on aluminas and on cation positions in zeolite Y (Eulenberger, Shoemaker, and Keil, 1967); Richard Nacciarone (now a Jesuit missionary in Africa) on metal COT complexes; Paul Metzger and John Hopps (MIT, physics) on heteropolymolybdates involving manganese and nickel; Dr. Ramesh Srinivasan (professor, Madras) on COT-monocarboxylic acid refinement and on the absolute stereochemical configuration of (–) *trans* cyclooöctene with the aid of a diastereomeric platinum complex, a project proposed by the late Professor A. C. Cope and successfully completed by Phil Manor and Dr. Alan Parkes. They found that the absolute configuration is the opposite of that predicted theoretically by Moscowitz and Mislow (Manor, Shoemaker and Parkes, 1970). Alan Parkes (Geology, MIT) was brought in to operate a card-programmed XRD 6 spectrogoniometer system acquired by Al Cotton and myself on an NSF instrumentation grant; he also worked with Roberta Ogilvie on anomalies in barium dicadmium hexachloride pentahydrate. William Moomaw (Williams College) took a non-crystallographic problem, the electrical conductivity of benzene and of solutions of iodine in benzene. Of a number of undergraduates who did baccalaureate theses in our group, two are worthy of mention: Earle Ryba (Professor, Penn State), heteropolymolybdates; Tom Margulis (Professor, University of Massachusetts, Boston), cadmium moncyclopentadienyl.

In August 1969, after a Washington meeting of the U.S.A. National Committee for Crystallography, Kenneth Hedberg of Oregon State University asked me (in the Men's Room of the National Academy of Sciences!) if I could consider accepting the Chairmanship of the Chemistry Department at Oregon State. Thus beginneth the end of our group at MIT. A year later only Roberta Ogilvie and Phil Manor were left, to finish writing their thesis. It was a sad time for crystallography at MIT; Bert Warren and Martin Buerger had already shut up shop, and Al Cotton was soon to leave for Texas A & M. Thanks however to Alex Rich and Bernie Wuensch, the crystallographic tradition at MIT continues on.

References

Bergman, C. and Shoemaker, D. P. (1954). Acta Cryst. **7**, 857-865.

Eulenberger, G. R., Shoemaker, D. P., and Keil, J. G. (1967). J. Phys. Chem. **71**, 1812-1819.

Feinstein, L. G., and Shoemaker, D. P. (1965). Surface Science, **3**, 294-297.

Feinstein, L., and Shoemaker, D. P. (1968). J. Phys. Chem. Solids. **29**, 184-188.

Komura, Y., Sly, W. G., and Shoemaker, D. P. (1960). Acta Cryst. **13**, 575-585.

Kripyakevich, P. I., Kuzma, Yu. V., Voroshilov, Yu. V., Shoemaker, C. B., and Shoemaker, D. P. (1971). Acta Cryst. **B27**, 257-261.

Manor, P. C., Shoemaker, C. B., and Shoemaker, D. P. (1971). Acta Cryst. **B28**, 1211-1218.

Manor, P. C., Shoemaker, D. P., and Parkes, A. S. (1970). J. Am. Chem. Soc. **92**, 5260-5262.

Meier, W. M. and Shoemaker, D. P. (1966). Z. Kristallogr. **123**, 357-363.

Robson, H., Shoemaker, D. P., Ogilvie, R. A., and Manor, P. C. (1973). in "Molecular Sieves", edited by W. M. Meier and J. B. Uytterhoeven. Advances in Chemistry Series 121. Washington, D.C.: American Chemical Society. pp. 106-115.

Seff, K., and Shoemaker, D. P. (1967). Acta Cryst. **22**, 162-170.

Shoemaker, D. P., and Sly, W. G. (1961). Acta Cryst. **14**, 552.

Shoemaker, C. B., and Shoemaker, D. P. (1963). Acta Cryst. **16**, 997-1009.

Shoemaker, C. B., and Shoemaker, D. P. (1964). Trans. Metall. Soc. AIME, **230**, 486-490.

Shoemaker, C. B., Shoemaker, D. P., and Mellor, J. (1965a). **18**, 900-905.

Shoemaker, C. B., and Shoemaker, D. P. (1967). Acta Cryst. **23**, 231-238.

Shoemaker, D. P. and Shoemaker, C. B. (1968) in "Structural Chemistry and Molecular Biology: A Volume Dedicated to Linus Pauling by his Students, Colleagues, and Friends." Edited by A. Rich and N. Davidson. San Francisco: W. H. Freeman and Co., pp. 718-730.

Shoemaker, C. B., and Shoemaker, D. P. (1971). Acta Cryst. **B27**, 227-235.

Shoemaker, C. B., and Shoemaker, D. P. (1981). Acta Cryst. **B37**, 1-8.

Shoemaker, C. B., Shoemaker, D. P., and Mellor, J. (1965a). Acta Cryst. **18**, 37-44.

Shoemaker, D. P., Kindler, H., Sly, W. G., and Srivastava, R. C. (1965b). J. Am. Chem. Soc. **87**, 482-487.

Shoemaker, D. P., Shoemaker, C. B., and Wilson, F. C. (1957). Acta Cryst. **10**, 1-14.

Wilson, F. C. and Shoemaker, D. P. (1957). J. Chem. Phys. **27**, 809-810.

Wright, D. A., Seff, K., and Shoemaker, D. P. (1972). J. Cryst. Mol. Struct. **2**, 41-51.

CHAPTER 8

CRYSTALLOGRAPHY IN BROOKLYN, 1942 - 1981

Ben Post
Polytechnic Institute of New York
Brooklyn, NY 11201

The story of crystallography in Brooklyn is very largely the story of Fankuchen at the Polytechnic Institute of Brooklyn. Isidor Fankuchen, or "Fan" as he was known to everyone, was awarded the Ph.D. degree in Physics by Cornell University in 1933. His doctoral research involved the determination of the crystal structure of sodium uranyl acetate, a formidable task in those computerless days. In the course of his work Fan built a new and improved type of cold-cathode X-ray tube as well as the X-ray generator and much of the accessory equipment needed for his research. He would later proudly recall to his students how he, working alone in the laboratory, had patiently wound the wiring in place for his high voltage transformer. This author recalls vividly when, in the informal and irreverent atmosphere which Fan generated, a student, pretending to be trying to downgrade Fan's achievement, commented "But you *bought* all the copper wire."

The award of a Fellowship enabled Fan to continue his studies with W. L. Bragg at Manchester (1934-36). There, he was one of the first to make use of the newly developed "Beevers-Lipson strips" method for summing Fourier series for crystal structure analysis, a procedure which, though more tedious and slower than modern electronic computer methods, was far superior to the latter in terms of the building of character.

Fan then spent two highly productive years (1936-8) at Cambridge. Together with Dorothy Crowfoot, J. D. Bernal and others he carried out detailed X-ray and optical studies of steroids and other molecules of biological interest. For this work he was later (1942) awarded his second Ph.D. degree by Cambridge University. Late in 1938 he moved on to Birkbeck College in London where, working in close collaboration with Bernal, he developed improved methods for X-ray investigations of proteins and conducted a number of fundamental studies of virus preparations.

He returned to the United States in 1939. After interludes as a National Research Fellow in Protein Chemistry at the Massachusetts Institute of Technology and as Associate Director of the Anderson Institute for Biological Research in Red Wing, Minnesota, he came to the Polytechnic Institute of Brooklyn in 1942, as Adjunct Professor of Crystal Chemistry, a designation changed a few years later to Professor of Physics.

Fan soon set about building a center of crystallography at the Polytechnic. His enthusiasm and persuasive powers played important parts in inducing Paul Ewald and David Harker to join him in Brooklyn in 1950, the former as Head of the Physics Department and the latter as Director of the newly organized Protein Research Project. Fan soon gathered about him a large number of graduate students, postdoctoral Fellows and faculty members of European and Japanese universities, all engaged in crystallographic research. The latter had been sent by their respective institutions to familiarize themselves with current techniques in crystallography and thereby make up, in part at least, for wartime deprivations.

Fan's research interests covered a wide range. He was deeply involved in conventional crystal structure determination, small-angle scattering, X-ray instrumentation, powder diffraction, low temperature X-ray studies and the optics of visible light and X-rays.

It was as a teacher, however, that he had the greatest impact. He offered courses in X-ray diffraction theory and in X-ray laboratory methods. His classrooms were consistently filled with students from all departments in the Polytechnic and from industrial and research laboratories throughout metropolitan New York. His lectures and lab sessions were scheduled for late evening hours to make them available to students who could not attend daytime courses. Fan's lectures were always clear and concise; they were carefully prepared and delivered with enthusiasm, patience and humor. He managed to convey to his students a sense of scientific excitement and a feeling of participation in the research he discussed. His courses were comprehensive; they included weekly two-hour lectures and four-hour laboratory sessions both of which extended over a full academic year.

In addition, early each summer he taught the rudiments of crystallography to large numbers of students in his intensive "X-ray Clinic" course, designed for engineers, scientists and technicians who could not spare the time to take the standard academic course. Fan taught the Clinic course for 23 years, each time to from 30 to as many as 50 students. Several of those who completed the course, i.e., Fan's "two-week wonders," returned to the Polytechnic later to work for advanced degrees in the X-ray Diffraction Laboratory. Fan performed a similar service at the annual schools in X-ray diffraction which he organized and conducted for the Philips Electronic Instrument Co. It is probably no exaggeration to state that as many as one half of the total membership of the American Crystallographic Association in the 1950's and early 1960's had, at one time or other, attended at least one of Fan's courses.

He also devoted much time and effort to organizational matters. In the early 1950's Fan recognized the need for a service which could make available to crystallographers books, charts and similar items at minimal cost. He therefore set up the Polycrystal Book Service and ran it – on a non-profit basis – out of his office at the Polytechnic with the help of his very efficient secretary, Doris Cattell. Fan also took an active part in the setting up of the International Union of Crystallography in 1948. He was the first President of the American Crystallographic Association, formed in 1950. He was also the first American Editor of *Acta Crystallographica.* More than 1500 manuscripts passed through his hands during his 17 years as Editor. He had an uncanny eye for errors of fact or of analysis. Crystallographers everywhere are indebted to him for perceptive comments, for the detection and corrections of errors, and for the rejection of manuscripts which did not merit publication.

Fan was extraordinarily skillful at arranging unusual and exciting seminars and colloquia. Within a relatively short period in the late 1940's, the principal speakers at four such meetings were W. L. Bragg, Kathleen Lonsdale, Max von Laue and C. V. Raman. Later, Fan organized, and for more than 15 years was the driving force behind the famous bi-weekly colloquia which he aptly named "The Point Group." The audiences at those colloquia usually included many of the leading figures in crystallography in the United States and large numbers of graduate students from the Polytechnic and other schools in the New York Metropolitan area. Fan usually chaired the Point Group sessions and used the prerogatives of that position as well as his keen crystallographic insights to provoke lively discussions that went on long after the formal terminations of the talks.

It is difficult at this late date to reconstruct, from scattered notes and faulty memory, the exciting flavor of those meetings. They clearly constituted major factors in the education of an entire generation of crystallographers.

A rough measure of the scope and technical level of the colloquia is furnished by the following very incomplete sampling of dates, speakers and topics:

Date	Speaker	Topic
October 5, 1950	Yvette Cauchois, Univ. of Paris	Reflection of X-rays from Bent Crystals
December 6, 1951	E. W. Muller, Max-Planck Instit.	The Point Projection Microscope
February 21, 1952	William Parrish	Instrumentation for X-ray Diffraction
March 6, 1952	David Sayre	Some Possibilities in Phase Determination
April 3, 1953	A. L. Patterson	Impossibilities in the Determination of Phases
February 25, 1954	Jerome Karle	Statistical Methods in Structure Work
October 28, 1954	David Harker	Phase Determination by Isomorphous Replacement
May 18, 1955	P. B. Hirsch, Cavendish Lab.	Problems of Small Angle Scattering
February 23, 1956	Clifford Shull	Diffraction With Polarized Neutrons
May 17, 1956	R. Uyeda, Nagoya Univ.	Studies in Electron Diffraction
January 30, 1958	J. C. Kendrew, Cavendish Lab.	3-Dimensional Fourier Synthesis of Myoglobin at Low Resolution
November 12, 1959	J. M. Hastings	Magnetic Symmetry

CRYSTALLOGRAPHIC LABORATORIES

Date	Speaker	Topic
December 10, 1959	L. H. Germer	Adsorbed Surface Films Studied by Electron Diffraction
February 19, 1960	N. V. Belov	Color Symmetry
November 17, 1960	Earl Ubell	A Science Writer Tangles with $P2_12_12_1$
February 23, 1961	J. D. H. Donnay	What Can Crystal Morphology Tell Us About Crystal Structure
March 9, 1961	G. Borrmann, Fritz Haber Institute Berlin	X-ray Wavefields in Crystals
October 31, 1963	Caroline MacGillavry, Univ. of Amsterdam	Absorption Coefficients in X-ray Crystallography
November 12, 1963	J. Monteath Robertson, Univ. of Glasgow	Atomic Bonds and Electron Density
November 21, 1963	A. Zajac and E. Saccocio	The Borrmann Effect and Simultaneous Diffraction
March 19, 1964	P. P. Ewald	Crystal Optics in the Visible and X-ray Regions

In addition to Fan's basic course in the theory and experimental methods of crystallography, more advanced and specialized courses were available to graduate students throughout the 1950's and 1960's.

Prof. Ewald offered a course in the Advanced Theory of Diffraction which dealt in detail with dynamical diffraction effects. Prof. Rudolph Brill regularly offered courses in Fourier Analysis of Diffraction Data, Particle Size Determination by X-ray Diffraction, and Crystal Structure and the Chemical Bond. Courses in Crystal Dynamics and Solid Phases in Metals were taught by Professor Juretschke, David Harker taught a course in the Physics and Structures of Alloys, and Ben Post taught Methods of Crystal Structure Determination.

Prof. Ewald's presence at the Polytechnic was an important factor in the School's important position in American crystallography in the 1950's. He was always available to faculty and students for help in matters related to the basic physics of diffraction. The apparently simple questions which he often raised at X-ray seminars often revealed implications of the seminar talk which were not clear to the audience and sometimes to the speaker as well. During his tenure at the Polytechnic Professor Ewald was the Editor of *Acta Crystallographica.* He read carefully each manuscript which passed through his hands and his comments on matters ranging from the technical content of the papers to details of presentation and even of grammar were always to the point and of great value to the respective authors.

Throughout the 1950's experimental research in crystallography and diffraction at the Polytechnic was under the direction of Professor Harker, of the Protein Research Project*, and Professors Fankuchen and Post of the Physics Department. In 1949 a program to investigate the crystal structures of substances which were liquids or gases at room temperature was begun in the Physics Department. A simple device was designed and built which made possible the convenient growth of single crystals from liquids or gases sealed in thin-walled glass capillary tubes (Post *et al.*, 1951b). A controlled stream of cold nitrogen gas was blown across the specimen to freeze it. Repeated cycles of melting and freezing usually resulted in the production of a single crystal suitable for X-ray analysis. The growth of the single crystals from minute nuclei was observed continuously with the aid of a crossed polarizer/analyzer arrangement mounted on a simple telescope. When a single crystal of suitable size and quality was obtained, the X-ray window was opened and data were collected in the usual way.

Initially, X-ray data were recorded using what are now considered to be primitive rotation and oscillation film methods. The precession camera, developed by

*See next chapter by Harker on the protein structure project at Brooklyn.

Martin Buerger of the Massachusetts Institute of Technology, became commercially available in the early 1950's. Its open construction, the readily accessible film holder, and the ease with which the diffraction pattern could be interpreted, made it almost ideally suited for low temperature studies. A minor modification of the original arrangement was used for low temperature equi-inclination Weissenberg photography, (Steinfink *et al.*, 1953).

Among the crystal structure of gases or liquids which were determined at low temperatures were those of: formic acid (Holtzberg *et al.*, 1953), sulfur dioxide (Post *et al.*, 1952); nickel carbonyl (Ladell *et al.*, 1952); formamide (Ladell and Post, 1954); octamethyl cyclotetrasiloxane (Steinfink, Post and Fankuchen, 1955); dimethyl acetylene (Pignataro and Post, 1955); N-methylacetamide, (Katz and Post, 1960); and dimethylfulvene (Norman and Post, 1961). Studies of ice (La Placa and Post, 1960) and of the D_2O/H_2O system (Katz and Handler, 1968) over wide ranges of temperature were also carried out.

The low temperature technique lent itself to the related studies of solid state polymorphic transformations. These included investigations of: t-butyl chloride and bromide (Schwartz *et al.*, 1951); cyclopentane and neohexane (Post *et al.*, 1951); carbon tetrachloride (Post, 1959, Rudman and Post, 1966); cyclooctanone and cyclononanone (Rudman and Post, 1968); dimethyl acetylene (Miksic *et al.*, 1959); and methyl chloromethyl compounds (Rudman and Post, 1970).

In some cases substances which were solid at room temperature were also investigated at lower temperatures either to increase the quality of the experimental data or to study thermal effects in crystals as functions of temperature rather than of $\sin^2\theta/\lambda^2$. In a study of urea, reflections were readily recorded at −140°C. out to a minimum d value of 0.35Å. A least squares analysis of the data for 2-dimensions yielded results comparable in precision with results obtained at room temperature with full data in 3-dimensions. Sklar, Senko, and Post (1961) worked on the thermal effects on urea.

A detailed study of atomic position parameters and amplitudes of thermal vibrations in alpha quartz, at room temperature and at 155°K and 233°K, was published by Young and Post in 1962. Studies of the corresponding behavior of the oxygen atoms in sodium nitrate (Cherin *et al.*, 1967) and calcite (Chessin *et al.*, 1965) were completed shortly thereafter.

Considerable attention has been paid above to low temperature X-ray investigations, largely because for more than a decade the group of graduate students working in that area constituted the largest research group in the Physics Department. The above should not be taken to imply that research interest in other areas languished. Fan continued to publish papers dealing with powder diffraction and micro camera techniques as well as studies of X-ray fluorescence and small angle scattering.

Post also was involved with a wide range of crystallographic problems. In the period from 1950 to 1970 he published several determinations of crystal structures based on powder diffraction data as well as studies of polymer configurations and epitaxial growths on single crystals. He supervised work on the structures and properties of metallic binary and ternary boride and silicide compounds and published the results of approximately 20 such studies. Many conventional crystal structure determinations of varying difficulty were also completed and published.

A series of events led to a significant decrease in the level of crystallographic activity at the Polytechnic. The Protein Research Project together with its Director, David Harker and his Associates moved to Roswell Park in the period 1957-9. This was followed by the retirement of Professor Ewald in 1959 and Fan's death in 1964, only a few days after he had delivered the final lecture of what turned out to be his last X-ray Clinic course. The financial "crunch" in the late 1960's also contributed to the decline of crystallographic activity. Nevertheless, though the decline was marked, a respectable level of research activity has continued to this day.

Since 1970 research emphasis has shifted to investigations of dynamical diffraction effects in single crystals. Detailed studies of simultaneous Bragg diffraction in germanium (Post, 1975a), gallium arsenide (Chang and Post, 1975), and diamond (Post, 1975b), were published. These demonstrated that such investigations can provide useful tools for the study of interference effects which take place within coherent domains in a crystal when Bragg's Law is obeyed. X-ray interferometry and n-beam Laue diffraction have also received attention: see for example work by Balter *et al.*, (1971), Feldman and Post (1972), Huang and Post (1973), Huang and Tillinger (1973), Post, Chang and Huang (1976) and Post (1975). Seven doctoral dissertations and 2 master's theses dealing with those topics have been completed during the past decade. Recently, attention has turned to the utilization of three-beam diffraction effects for the experimental determination of the phases of X-ray reflections, Post (1979). Work in that area is continuing.

References

Balter, S., Feldman, R. and Post, B. (1971). Multiple Borrmann diffraction. Phys. Rev. Lett. **27**, 307-311.

Chang, S. L. and Post, B. (1975). High order multiple diffraction in Ga As. Acta Cryst. **A31**, 832-835.

Cherin, P., Hamilton, W. C. and Post, B. (1967). Position and thermal parameters of oxygen atoms in sodium nitrate. Acta Cryst. **23**, 455.

Chessin, H., Hamilton, W. C. and Post, B. (1965). Position and thermal parameters of oxygen atoms in calcite. Acta Cryst. **18**, 689.

Feldman, R. and Post, B. (1972). Absorption and excitation in Borrmann diffraction. Phys. Stat. Sol. **12**, 273-276.

Handler, E. (1968). X-ray Diffraction Study of H_2O,D_2O, $H_2O^{18}D_2O^{18}$. Thesis, Ph.D. (Chem.) Polytechnic Inst. of Brooklyn.

Holtzberg, F., Post, B. and Fankuchen, I. (1953). The crystal structure of formic acid. Acta Cryst. **6**, 127-130.

Huang, T. C. and Post, B. (1973). Experimental methods for the study of Borrmann diffraction. Acta Cryst. **A29**, 35-37.

Huang, T. C., Tillinger, M. H. and Post, B. (1973). 6-beam Borrmann diffraction. Z. Naturforsch. **28**, 600-603.

Katz, J. L. and Post, B. (1960). The crystal structure and polymorphism of N-methyl acetamide. Acta Cryst. **13**, 624-628.

Ladell, J. and Post, B. (1954). The crystal structure of formamide. Acta Cryst. **7**, 559-564.

Ladell, J., Post, B. and Fankuchen, I. (1952). The crystal structure of nickel carbonyl, $Ni(CO)_4$. Acta Cryst. **5**, 795-800.

LaPlaca, S. and Post, B. (1960). Thermal expansion of ice. Acta Cryst. **13**, 503-505.

Miksic, M. G., Segerman, E. and Post, B. (1959). The solid phase transformation in dimethylacetylene at -119°C. Acta Cryst. **12**, 390-393.

Norman, N. and Post, B. (1961). Crystal structure of dimethylfulvene. Acta Cryst. **14**, 503-507.

Pignataro, E. and Post. B. (1955). The crystal structure of dimethyl acetylene at -50°C. Acta Cryst. **8**, 672-674.

Post, B. (1959). The cubic form of carbon tetrachloride. Acta Cryst. **12**, 349.

Post, B. (1975a). Accurate lattice parameters from multiple diffraction measurements. J. Appl. Cryst. **8**, 452-457.

Post, B. (1975b). Multiple diffraction in diamond. Acta Cryst. **A31**, 832-835.

Post, B. (1979). A solution of the 'phase problem'. Acta Cryst. **A35**, 17-21.

Post, B., Chang, S. L. and Huang, T. C. (1976). Simultaneous four-beam diffraction. Acta Cryst. **A33**, 90-97.

Post, B., Schwartz, R. S. and Fankuchen, I. (1951a). X-ray investigation of crystalline cyclopentane and neohexane. J. Am. Chem. Soc. **73**, 5113-5114.

Post, B., Schwartz, R. S. and Fankuchen, I. (1951b). An improved device for X-ray diffraction studies at low temperatures. Rev. Sci. Instrum. **22**, 218-219.

Post, B., Schwartz, R. S. and Fankuchen, I. (1952). The crystal structure of sulfur dioxide. Acta Cryst. **5**, 372-374.

Rudman, R. and Post, B. (1966). Carbon tetrachloride: A new crystalline modification. Science **154**, 1009-1012.

Rudman, R. and Post, B. (1968). Polymorphism of crystalline cyclooctanone and cyclononanone. Molecular Crystals **3**, 325-337.

Rudman, R. and Post, B. (1970). Polymorphism of crystalline methylchloromethane compounds. Molecular Crystals **5**, 95-110.

Schwartz, R. S., Post, B. and Fankuchen, I. (1951). X-ray investigation of t-butyl chloride and t-butyl bromide. J. Am. Chem. Soc. **73**, 4490.

Sklar, N., Senko, M. D. and Post, B. (1961). Thermal effects in urea: the crystal structure at -140°C and at room temperature. Acta Cryst. **14**, 716.

Steinfink, H., Ladell, J., Post, B. and Fankuchen, I. (1953). A low-temperature Weissenberg X-ray camera. Rev. Sci. Instrum. **24**, 882-883.

Steinfink, H., Post, B. and Fankuchen, I. (1955). The crystal structure of octamethyl cyclotetrasiloxane. Acta Cryst. **8**, 420-424.

Young, R. A. and Post, B. (1962). Electron density and thermal effects in alpha quartz. Acta Cryst. **15**, 337.

CHAPTER 9

THE PROTEIN STRUCTURE PROJECT AT THE POLYTECHNIC INSTITUTE OF BROOKLYN 1950 - 1959

. *David Harker*
Medical Foundation of Buffalo Inc., Buffalo, NY 14222

In late 1949, I accepted the responsibility for establishing a research group which would attack the problem of finding the atomic arrangement in a protein molecule. I had been granted enough money to start such a group, and to negotiate with various institutions for laboratory space, administrative assistance, and so forth.

Soon the United States was involved in the Korean War, and most of the large computing devices in the country were reserved by the Government for military use. The Watson Laboratory of the International Business Machines Corporation, at 116th Street and Broadway in New York City, was an exception and I was able to arrange a very favorable relationship with its management, whereby computations connected with the work on protein structure of my research group could be done on their computers by our people. This arrangement required us to locate our Laboratory in or near New York City.

My next effort was to find an institution which would provide me with laboratory space, which was sympathetic with our plans to find the structure of a protein, and where there was enthusiasm for X-ray crystallography. I visited several famous institutions in New York: New York Unversity, The Rockefeller Institute, Columbia University, and others, but my reception was, at best, lukewarm. At last I called on my friend Professor I. Fankuchen at the Polytechnic Institute of Brooklyn. He instantly telephoned the President, Dr. Rogers, who was immediately very enthusiastic about the idea of establishing what we decided to call "The Protein Structure Project" at Brooklyn Poly, and so it was decided.

At that time Brooklyn Poly was a very important center for X-ray crystallography. Professor P. P. Ewald had just been made Chairman of Physics, Professor I. Fankuchen was actively teaching crystal structure determination, and a group of junior faculty together with most of the graduate students in the Physics Department were all interested and working in one or another aspect of X-ray diffraction. This group was very helpful to us as we organized the Protein Structure Project (PSP).

Brooklyn Poly had just rented most of the building at 55 Johnson Street and the PSP was assigned about half of the 4th floor. We set up a row of three offices, a chemical laboratory, a window-less X-ray diffraction room with an adjoining dark room, and a machine shop.

Our initial staff consisted of myself as Director of the PSP, my wife Katherine as Secretary, Dr. Thomas C. Furnas, Jr. as X-ray diffraction Physicist, Dr. Beatrice S. Magdoff as X-ray Crystallographer, Dr. Murray Vernon King as Chemist, and Mr. William G. Weber as Instrument Maker.

The General Electric X-Ray Corporation (GE X-ray) in Milwaukee, Wisconsin was manufacturing and selling a very good X-ray powder diffractometer (XRD) which I had helped to design in my role as Consultant to GE X-ray. The PSP bought one of these. This XRD was well suited to conversion into a device for measuring the diffraction pattern of a single crystal. This conversion was carried out by Dr. Furnas and Mr. Weber. The result was the first so called Four Circle X-ray Diffractometer for Single Crystal Structure Determination (Furnas and Harker, 1955). The design of this single crystal XRD has been used throughout the world by many laboratories and manufacturers. The majority of crystal structures determined in the period 1960 until the present (1980) have been computed from X-ray intensity data obtained on such four-circle XRD's.

Concurrently with the design and construction of the four-circle XRD, Dr. M. V. King began experiments on the crystallization of globular proteins. We had decided to use the bovine pancreatic enzyme ribonuclease (RNAse), because it had been crystallized, by Dr. Moses Kunitz of the Rockefeller Institute, and because it was available in pure form from the research laboratory of Armour and Co. in Chicago, Illinois. For many months after the construction of the laboratories of the PSP in June 1950, no RNAse crystals were obtained. Late in 1950 several of us visited Dr. Kunitz in his laboratory, and conferred with him concerning his successful crystallization of RNAse. We could not find anything to criticize in Dr. King's methods, so we went back to Brooklyn quite discouraged. Quite soon thereafter, however, Dr. King found crystals of RNAse in several of his preparations. Perhaps we had picked up seed crystals in Dr. Kunitz's laboratory on our clothes, and carried them back to the PSP in Brooklyn! After that many different crystal forms of RNAse were prepared in the PSP and their X-ray diffraction patterns measured on the XRD (King *et al.*, 1956; Bello *et al.*, 1961; King *et al.*, 1962).

The only practical method of finding the positions of atoms in a crystal composed of molecules as large as those of RNAse (molecular weight 13,700) was to compare the intensities of the X-rays diffracted by crystals of the native protein with those diffracted by isomorphous crystals of the same protein to the molecules of which a few very heavy atoms had been attached. In principle, the structure of the protein molecule could be elucidated by using the intensities of the diffracted X-rays from three isomorphous crystals: one of the native protein, and two others with different arrangements of heavy atoms on their constituent protein molecules. The mathematics of this was first described in 1956 (Harker, 1955, 1956*a*, *b*, 1957).

Dr. M. V. King set about preparing RNAse entities to which side chains containing heavy atoms were attached. Most of these preparations failed to produce crystals, or, if crystals did indeed form, they were not isomorphous with crystals of native RNAse. Thus, this attack on the formation of isomorphous crystals of RNAse failed. Another way of achieving isomorphous crystals of RNAse containing heavy atoms was devised by Dr. King. Heavy atom compounds with small molecules were dissolved in the medium from which the RNAse crystals formed, and in which they were suspended. The crystals were then allowed to soak in this "heavy-atom solution" for several weeks or months, after which they were tested to discover whether their X-ray diffraction intensities had changed. This set of experiments succeeded, and several isomorphous series of heavy-atom-containing crystals were prepared. The intensity data so obtained eventually allowed the elucidation of the molecular structure of RNAse in 1967 (Kartha, Bello, and Harker, 1967), seven years after the PSP had been moved to the Roswell Park Memorial Institute (RPMI) in Buffalo, New York. The successful production of a series of isomorphous protein crystals by soaking in heavy atom solutions definitely belongs to the Brooklyn Poly era of the PSP.

During the almost ten years of the PSP at Brooklyn Poly, several distinguished scientists spent more or less time in the laboratory of the PSP. Dr. Erik von Sydow of the University of Uppsala in Sweden worked with us on the structure of sialic acid 1952-53. Dr. Alexander Tulinsky worked on one of the crystalline forms of RNAse at the PSP from 1952 to 1956, and greatly refined the structure of basic beryllium acetate. Dr. Vittorio Luzzati spent a year and a half in 1952-1953 working on the relations between the agreement of the observed and calculated diffracted X-ray intensities and the probable errors in the atomic coordinates in a crystal structure. Dr. F. H. C. Crick spent a year and a half in 1953-1954 working with Dr. Beatrice Magdoff on the alpha helix content of RNAse molecules in one of the RNAse crystalline modifications. Dr. Jake Bello joined the staff of the PSP in 1955 and worked with Dr. M. V. King on the crystallization of RNAse and their treatment with heavy atom compounds. Dr. Gopinath Kartha joined the PSP in 1966 and undertook to find the structure of the RNAse molecule. He accomplished this in 1967 at RPMI. Dr. C. Worthington, then a postdoctoral fellow, collected X-ray data from crystals of ribonuclease in the period of 1955-1957.

In summary, the PSP, as it existed at Brooklyn Poly during the 1950's, produced most of the experimental and theoretical techniques by means of which, up to the present time, the three-dimensional structures of protein molecules have been elucidated using X-ray crystallographic methods. The same group moved to RPMI at the end of 1959, and, in 1967, produced the structure of RNAse.

References

Some of the papers from the PSP at Brooklyn Poly:

Bello, J., Harker, D. and DeJarnette, E. (1961). X-Ray Investigation of Reduced-Reoxidized Ribonuclease. J. Biol. Chem. **236**, 1358.

Furnas, Jr., T. C. and Harker, D. (1955). Apparatus for Measuring Complete Single-Crystal X-Ray Diffraction Data by Means of a Geiger Counter Diffractometer. Rev. Sci. Instrum. **26**, 449-453.

Harker, D. (1953). The Meaning of the average of $|F|^2$ for Large Values of the Interplanar Spacing. Acta Cryst. **6**, 731-736.

Harker, D. (1955). Our Increasing Knowledge of the Structures of Crystalline Proteins. Trans. N.Y. Acad. Sci. **17**, 445-449, Series II.

Harker, D. (1956a). X-Ray Diffraction Applied to Crystalline Proteins. Adv. Biol. Med. Phys. **4**, 1-22.

Harker, D. (1956b). The Determination of the Phases of the Structure Factors of Non-Centrosymmetric Crystals by the Method of Double Isomorphous Replacement. Acta Cryst. **9**, 1-9.

Harker, D. (1957). The Structure of Crystalline Proteins. Ann. N.Y. Acad. Sci. **69**, 321-327.

Harker, D. (1962). Article in appreciation of the award of the Nobel Prize in Chemistry to Dr. John C. Kendrew and Dr. Max F. Perutz for their work on atomic arrangement and proteins. Science **138**, 668-669.

Kartha, G., Bello, J. and Harker, D. (1967). Tertiary Structure of Bovine Pancreatic Ribonuclease at 2Å Resolution. Nature **213**, 862-865.

King, M. V., Bello, J., Pignataro, E. M. and Harker, D. (1962). Crystalline Forms of Bovine Pancreatic Ribonuclease. Some New Modifications. Acta Cryst. **15**, 144-147.

King, M. V., Magdoff, B. S., Adelman, M. B. and Harker, D. (1956). Crystalline Forms of Bovine Pancreatic Ribonuclease: Techniques of Preparation, Unit Cells, and Space Groups. Acta Cryst. **9**, 460-465.

CHAPTER 10

CRYSTALLOGRAPHY AT BELL LABORATORIES

Elizabeth A. Wood
Red Bank, NJ 07701

Where should we draw the boundary of the field of crystallography in a research laboratory like Bell Laboratories where physicists, chemists, and engineers work together, some to try to understand the solid state; some to understand liquids and glasses; some to understand magnetism, the semi-conducting state, ferroelectricity, superconductivity, laser action, and the relation of these to composition and structure; others to try to put this understanding to use in materials and practical devices? That's a long sentence because it's a big question.

Bell Laboratories was created in 1925 out of the research section of the Engineering Department of the Western Electric Company, where, as early as 1917, a physicist named Sandy Nicolson had persuaded some of the chemists to grow Rochelle Salt crystals for his experiments with its piezoelectric effectiveness in a crystal-controlled oscillator. Nicolson (1919) wrote a paper in which he illustrated the piezoelectric polarization of tourmaline, boracite, and quartz.

A young man named Clinton Davisson had joined the Engineering Department of Western Electric in 1917 to help in the development of vacuum tubes. Ten years later he was studying the angular distribution of electrons scattered from a nickel target in an evacuated tube when the tube got broken and the nice clean surface of the nickel was ruined by oxidation. To get rid of the oxide coating, the target was heated very hot in an evacuated tube. During the heating large crystals grew at the expense of small ones, as they will when unwanted and won't when wanted. When Davisson and his colleague, Lester Germer, again studied the angular distribution of electrons scattered from the nickel target, there were wild peaks and troughs that hadn't been there before.

Now Davisson knew about X-rays. He probably even knew the Bragg equation by heart. But this had nothing to do with particles bouncing off a target, he thought, until he went to a meeting in Europe where the crazy ideas of de Broglie (who thought particles were waves and vice versa) were being incredulously discussed. When he and Germer (Davisson and Germer, 1927*a*) tried the Bragg equation on their electron scattering angles with the nickel-crystal spacings plugged in, it fitted so well that they wrote a paper about it. At first they spoke of it as "scattering," but a little later they called it "diffraction" (1927*b*). Ten years later Davisson shared the Nobel Prize for the discovery of electron diffraction with G. P. Thomson of England.

About ten years after that Davisson told me that he had regretfully decided not to attend scientific meetings anymore. "I get so excited when I listen to the papers that I invariably get sick," he said. "It happens every time." I first knew him when I came to Bell Laboratories in 1943. He had been urged to follow his own interests in research and not to feel the need to work on the practical problems of communications. However, he found that he wanted contact with the problems of production which were, at that time, related to the war effort. One of these problems was the frequency change ("aging") of quartz crystal oscillator plates as they sat on the shelf. It was finally found that this resulted from the release of quartz particles (resulting from surface grinding) that had adhered to the surface of the plate. Davisson devised a simple technique for measuring the angular distribution of these particles. An X-ray beam, collimated through slits, met the plate at the Bragg angle for its surface. The reflected beam was recorded at the center of a strip of film. Then the plate was turned one degree away from this reflecting position and another exposure was taken. In this exposure a centered line appeared from reflection of non-characteristic radiation from the crystal plate and, displaced from the center, a line from the characteristic radiation reflected by those particles misoriented by one degree. Successive exposures with increasing angles were recorded on a strip of film in a black paper envelope, held in a metal holder so that only a small area of it was exposed for each setting (Figure 1). The finished film looked rather like Figure 2. From this Davisson was able

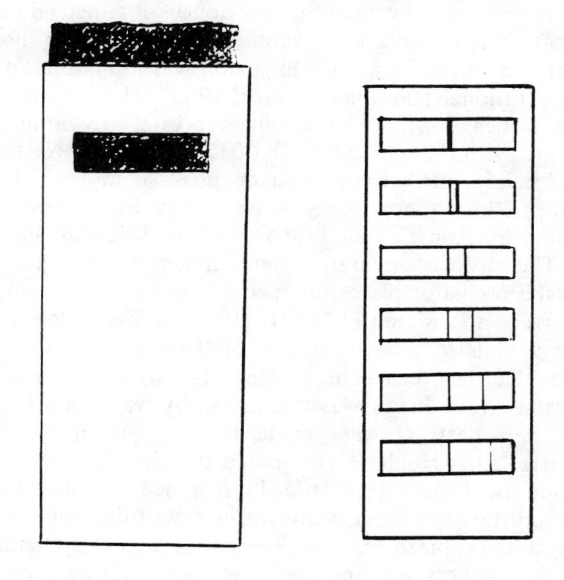

Figure 1 (left). The Davisson device for studying the angular distribution of ground particles on a quartz plate.

Figure 2 (right). The appearance of a film from the Davisson device.

to draw a distribution curve of the angular misorientation of ground quartz particles adhering to the quartz plate. When the "aging" was cured by an etch finishing process, the Davisson technique was used as a test of the cure.

"Davey" used to carry the paper-covered film strips around in the breast pocket of his sport jacket, along with his corn-cob pipe. One day he showed me several films that had large black streaks on them, so heavily exposed that they looked metallic. He was much amused by this evidence that he had inadvertently stood in the direct X-ray beam!

Walter Bond, who was introduced to crystallography in a Mineralogy course at Columbia University, ingeniously devised ways, during World War II, in which technicians with less than a high-school education could determine the orientation of thin small crystal plates within a few minutes of arc. His efforts to design foolproof jigs and fixtures were fraught with frustration. One time, when he and I were visiting a Western Electric plant where these crystal oscillator plates were being X-ray oriented, a supervisor demonstrated the safety device—a microswitch that allowed X-rays to be on only when a crystal was fastened against a pointer in the X-ray beam. To demonstrate it, the supervisor, with some difficulty, pushed her little finger against the pointer, thus turning on the beam which exposed her finger until we protested.

Walter Bond was an imaginative and skillful designer and builder of specialized crystallographic equipment. A symmetrical diffractometer he designed made possible the determination of lattice constants to a higher degree of accuracy than that previously obtained (Bond, 1960). Such accurate measurements were used to show that the oxygen impurity in silicon was interstitial, not substitutional (Bond and Kaiser, 1960). The increase in the lattice constant of a silicon crystal grown in an oxygen atmosphere was 0.000037. The high precision of Bond's diffractometer made possible more precise information about X-ray wavelengths from measurements by Ann Cooper (1965) who worked with Bond.

The proper orientation and dimensions of quartz-crystal oscillator plates for special uses were determined by Warren P. Mason (1942, 1950) from a knowledge of the symmetry and magnitude of the crystal's physical and electrical properties. Since the supply of quartz crystals from Brazil was threatened by wartime attacks on ships, efforts were made to find substitutes for quartz. Alan Holden, who joined the chemical research group in 1936, systematically searched the literature for crystals lacking a center of inversion that might, as a result, be piezoelectric. He used his skill as a chemist to find others and invented the rotary crystallizer to grow single crystals of the more promising substances (Holden, 1949). (See Figures 3 and 4.)

Meanwhile Ernie Buehler and Albert Walker (1949) were trying to grow quartz crystals on seeds in an alkaline aqueous solution under high pressure in sealed containers. After a couple of them blew up they were moved to a shack in a far corner of the laboratories' property. The process developed at that time is now being used, with minor modifications, to produce large quantities of quartz crystals, not only in the Western Electric plants, but also in other places all over the world.

Figure 3. The Holden Reciprocating Radial Crystallizer with ammonium dihydrogen phosphate seed crystals which have been grown from basal plates to pyramid faces, beyond which growth is clear. Note thermostatic control and metal lid from which condensed water drips onto surface of solution to dissolve unwanted seeds formed there. A central, low-wattage, heater in the base dissolves seeds swirling to the bottom.

Richard Bozorth was perhaps the first member of Bell Laboratories to do a crystal-structure determination. He determined the structure of barium platinocyanide before coming to Bell from Cal. Tech. in 1933. It was he who took a Laue photograph of the Davisson and Germer nickel target and showed that a large area of it was the surface of a single nickel crystal.

Lester Germer (who, in 1944 was President of the American Society of X-ray and Electron Diffraction, the precursor of the American Crystallographic Association) early developed an interest in the potential of low-energy electron diffraction for giving information about the top few layers of a crystal surface (Germer,

Figure 4. "How big an ammonium dihydrogen phosphate crystal can you grow, Dr. Holden, in your Reciprocating Radial Crystallizer?" Here clear growth is occurring beyond the pyramid faces, achieved by cloudy growth on the basal seed plate. Note distilled water dripping from the metal lid to dissolve unwanted surface seeds.

1929). The translation of his paper in the *Zeitschrift für Physik* was printed as *Monograph B-407* by Bell Laboratories. It has the following prophetic abstract: "Under appropriate experimental conditions, electron scattering by a single crystal of nickel can give rise to diffraction patterns of four quite distinct types. We attribute one of these patterns to the space lattice of the nickel crystal, one to the topmost layer of nickel atoms, one to a monatomic layer of adsorbed gas atoms, and one to a thick layer of gas atoms. From these phenomena some conclusions concerning gas adsorption have been drawn. We have at hand a new and important method of crystal analysis."

In spite of the attractive prospects, Germer dropped this work in favor of a prolonged study of the physical and chemical nature of the corrosion effects of the repeated making and breaking of electrical contacts. Only because of urgent importuning by Homer Hagstrum did he return, many years later, to the study of low-energy electron diffraction (LEED), first together with C. Hartman (Germer *et al.*, 1960) and then with A. U. MacRae (Germer and MacRae, 1962a,b and c), J. J. Lander, and J. Morrison. The relative ease of obtaining higher vacuum in later years made possible exciting new discoveries about crystal surfaces. Lander persuaded Elizabeth Wood (who was President of the American Crystallographic Association in 1957) to suggest a systematic notation and vocabulary for surface structures (Wood, 1964a), since they were being studied in several laboratories, each of which was using different terms to describe them. Her publication of the *"80 Diperiodic Groups in Three Dimensions"* (Wood, 1964b) was to aid those determining the surface structures revealed by LEED techniques.

In the course of an effort to understand the semiconducting state, the transistor effect was discovered in 1948 and Bardeen, Brattain, and Shockley received the Nobel Prize for this in 1956. The need for high-purity semiconducting crystals resulted in new crystal-growing techniques in the Metallurgy group. Jack Scaff (1949) and Henry Theuerer (1955) learned to produce large single crystals of silicon and germanium and Bill Pfann (1952, 1957, 1974, 1978) invented the zone refining process to achieve clean-slate crystals into which only the desired amounts of impurities could be introduced.

While we are on the subject of crystal growing, we should mention Kurt Nassau (1964, 1966) a crystal chemist who grew such crystals as calcium tungstate, lithium niobate, gadolinium molybdate, and many rare-earth iodates, some for ferroelectric and optical studies, others for investigation of laser action. In addition to a variety of solution techniques, he made use of Czochralski pulling.

We should also speak of Joe Remeika (1954, 1956, 1964, 1980a and b) who, through the imaginative use of fluxes and an aggressive attitude toward the periodic table, produced large crystals of many new substances with planned special properties. Among the crystals he grew were the so-called garnets which resemble natural garnets in structure, though not in composition. Their structures were exhaustively studied by Seymour Geller (1957, 1959) and their magnetic domains investigated by Joe Dillon (1958, 1968, 1978). Subsequently Van Uitert (1965) was able to grow very large garnet crystals (Figure 5) with highly mobile domain walls, useful in memory and logic devices. These mobile domains have been dubbed "magnetic bubbles."

The domain structure of magnetic substances had long been a subject of interest at Bell Laboratories. The story of its investigation is detailed in Richard Bozorth's monumental work, *"Ferromagnetism"* (1951). There is a classic paper in which a picture-frame-shaped specimen, cut from a single crystal of iron, is used to demonstrate the controlled movement of a domain boundary and its correlation with the hysteresis loop of magnetization

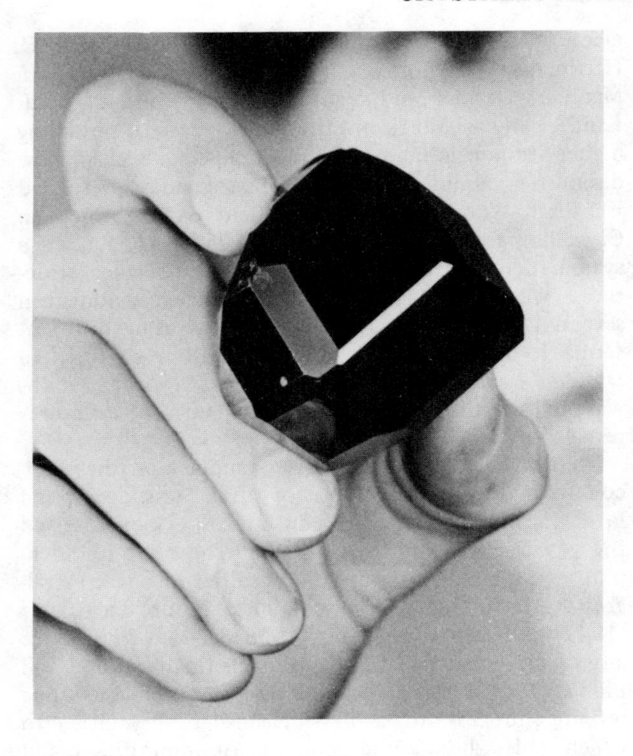

Figure 5. Yttrium iron garnet crystal, weighing approximately 200 grams, grown by Van Uitert.

(Williams and Shockley, 1949).

The study of magnetic materials by neutron diffraction was vigorously pursued by Sidney Abrahams (who was President of the American Crystallographic Association in 1968), working with the neutron source at Brookhaven National Laboratory (Abrahams and Prince, 1962; Abrahams, 1962; Abrahams, Guttman, and Kasper, 1962). He, being an impatient sort of person, not willing to baby-sit the diffraction equipment for long periods of time, and hungry for more data than could be collected in a working day, invented SCAND, the Single Crystal Automatic Neutron Diffractometer (Prince and Abrahams, 1959) with which he analyzed the arrangement of magnet dipoles in ferromagnetic, antiferromagnetic, ferrimagnetic (etc.) substances. SCAND was the forerunner of the various commercial automatic diffractometers that subsequently revolutionized structural crystallography.

When I went to school the inert gases were inert. Nobody tried to combine them with anything because they were, after all, the inert gases. Now they seem to have thrown off their inertia and developed a willingness to form compounds with other elements. The low-temperature structure of one such compound, xenon hexafluoride, was solved by Rob Burbank (1974) (who was President of the American Crystallographic Association in 1975). He found that it consisted of XeF_5^+ and F^- ions arranged in tetrameric and hexameric "rings" (nearly spheres) in the 1008-atom cell.

But what of the softer substances: the organics, the polymers, the stiff liquids called glasses? If we include those, we may as well break down and consider the runny liquids too. (Remember that one of the names suggested for the American Crystallographic Association at the time of its inception was "The American Society for the Study of the Orderly Arrangement of Matter in the Solid, Liquid, or Gaseous State.")

Work on the crystallography of polymers began in the late thirties with X-ray studies of homologous series of linear polyesters and polyamides by W. O. Baker (1943) (who later became President of Bell Laboratories), C. L. Erickson, C. S. Fuller (1937, 1940), C. J. Frosch, and N. R. Pape. About the same time Keith Storks (1938) interpreted electron diffraction evidence obtained from crystals of gutta percha as indicating chain folding. He was not taken seriously at the time, but nineteen years later he was proven to have been right when Keller at Bristol established chain folding in polyethylene crystals. Chain folding, I am told by H. D. Keith (of whom more anon), is common in polymer crystallization and recognition of this fact is one of the most important discoveries in the history of the subject.

X-ray diffraction work by Bill Slichter (1958, 1959) resulted in the determination of the structure of some of the amino-acid nylons.

Although all high polymers crystallize from the melt or from concentrated solution with a spherulitic habit (Figure 6), spherulites can hardly be said to belong to the polymer people. Geologists have known them for years, finding them in volcanic rocks and minerals associated with end-stage volcanic activity (e.g. pecto-

Figure 6. Spherulite of polypropylene blend, crystallized at 135°C., oil-bath quenched. Approximately a quarter of a millimeter in diameter.

lite). Doug Keith (1963, 1964) and his associates have critically considered the necessary and sufficient conditions for their growth. The mineralogists, Doug tells me, have confirmed their basic assertion concerning segregation and phase separation, using the electron microprobe on spherulites in andesite and rhyolite.

Subsequently Keith and his colleagues have shown that many organic polymers crystallize by chain folding. These include, for example, native (double stranded) DNA helices, many synthetic polypeptides (Keith 1969a and b; Giannoni, 1969), and many natural proteins in the beta form. Keith Storks' 1938 suggestion of chain folding had more widespread application, apparently, than he could possibly have guessed at the time.

In the midst of organics we return, rather surprisingly, to ferroelectrics and piezoelectrics. Andrew Lovinger (1979, 1981) in Doug Keith's group, has used electron diffraction to refine the unit cells and conditions for growth of the various polymorphs of poly(vinylidene fluoride), two of which are ferroelectric and piezoelectric.

It was George Brady, in the middle sixties and early seventies, who had the temerity to work with those diffracted beams that are so close to the direct beam that sorting them out is extremely difficult. But it is precisely those beams, as shown by Bernal, Warren, and Guinier and their colleagues in the late thirties and early forties, that give information about orderly arrangements in liquids. George Brady (1964, 1965, 1967, 1974) investigated the nature of what orderliness is left when large molecules go into solution, as well as the way in which a substance changes from the liquid state to the nematic ("liquid-crystal") state and then to the truly solid state (Gravatt and Brady, 1969).

In the late seventies, Jim Phillips (1980) and his colleagues are finding ways in which the medium-range order in some glasses can be understandably related to their properties, and in particular can be correlated with the composition dependence of the glass-forming tendency.

The writing of this account has been like a visit to an extensive smorgasbord. A bit of raw fish attracts me here, a special slice of cheese there, a ripe tomato quarter, crisp lettuce, some artichoke hearts, and my plate is full though more than half the interesting delicacies offered remain untouched. There was the whole story of ferroelectric crystals studied by Bernd Matthias, left untouched; the perovskite-type crystals grown by Remeika and studied by Geller and Wood – untouched. There was the fascinating investigation of whisker growth and the classic paper by Herring and Galt (1952) that showed that a single crystal metal whisker, being too skinny to include a dislocation, behaved elastically. There was the whole story of dislocations, including work by Thornton Read in elucidating the Frank-Read dislocation source of crystal growth; the study of the relation between etch pits and dislocations (Vogel, 1953); and Bob Heidenreich's use (1949) of the electron microscope to study slip and dislocations in thin metal specimens.

The interest in dislocations continued, with work by Patel (1976, 1977) and his colleagues showing that the addition of charged impurities ("p-type and n-type doping") to germanium and silicon affects the mobility of their dislocations. Do you know what goes on in certain types of lasers while they are lasing? Dislocations are developing and moving around! (Petroff and Hartman, 1974).

To have attempted the inclusion of all the crystallographic work at Bell Labs, where crystals have become central to communications technology, would have caused both the writer and the reader, if you will excuse the smorgasbord analogy, acute mental indigestion.

References*

Abrahams, S. C. and Prince, E. (1962). J. Chem. Phys. **36**, 50-55.

Abrahams, S. C. (1962). J. Chem. Phys. **36**, 56-61.

Abrahams, S. C., Guttman, L., and Kasper, J. S. (1962). Phys. Rev. **127**, 2052-2055.

Baker, W. O. and Fuller, C. S. (1943). J. Am. Chem. Soc. **65**, 1120-1130.

Bond, W. L. and Kaiser, W. (1960). J. Phys. Chem. Solids **16**, 44.

Bond, W. L. (1960). Acta Cryst. **13**, 814.

Bozorth, R. M. (1951). Ferromagnetism. New York: Van Nostrand.

Brady, G. W. and Salovey, R. (1964). J. Am. Chem. Soc. **86**, 3499-3503.

Brady, G. W. and Salovey, R. (1965). Biopolymers **3**, 573-583.

Brady, G. W. and Salovey, R. (1967). Biopolymers **5**, 331-336.

Brady, G. W. (1974). Acc. Chem. Res. **7**, 174-180.

Buehler, E. and Walker, A. C. (1949). Sci. Monthly **69**, 148-155.

Burbank, R. D. and Jones, G. R. (1974). J. Am. Chem. Soc. **96**, 43-48.

Cooper, A. S. (1965), Acta Cryst. **18**, 1079.

Davisson, C. and Germer, L. H. (1927a). Nature (London) **119**, 558-560.

Davisson, C. and Germer, L. H. (1927b). Phys. Rev. **30**, 705-740.

Dillon, J. F. Jr. (1958). J. Appl. Phys. **29**, 539.

Dillon, J. F. Jr. (1968). J. Appl. Phys. **39**, 922.

Dillon, J. F. Jr. (1978). In "Physics of Magnetic Garnets," edited by A. Paoletti, North Holland. p. 379.

Fuller, C. S. and Erickson, C. L. (1937). J. Am. Chem. Soc. **59**, 344-351.

Fuller, C. S. Baker, W. O., and Pape, N. R. (1940). J. Am. Chem. Soc. **62**, 3275.

Geller, S. and Gilleo, M. A. (1957). J. Phys. Chem. Solids **3**, 30-36.

Geller, S. and Gilleo, M. A. (1959). Acta Cryst. **10**, 239.

Germer, L. H. (1929). Z. Phys. **54**, 408-421.

Germer, L. H. Scheibner, E. J., and Hartman, C. D. (1960). Phil. Mag. **5**, 222-236.

Germer, L. H. and MacRae, A. U. (1962a). J. Chem. Phys. **36**, 1555-1556.

Germer, L. H. and MacRae, A. U. (1962b). J. Appl. Phys. **33**, 2923-2932.

Germer, L. H. and MacRae, A. U. (1962c). Proc. Natl. Acad. Sci. U.S.A. **48**, 997-1000.

Giannoni, G., Padden, F. J. Jr., and Keith, H. D. (1969). Proc. Natl. Acad. Sci. U.S.A. **62**, 964-971.

*In most cases the references are examples, chosen from many papers by the same author.

Gravatt, C. C. and Brady, G. W. (1969). Mol. Cryst. Liqu. Cryst. **7**, 355-369.
Heidenreich, R. D. (1949). J. Appl. Phys. **20**, 993.
Herring, C. and Galt, J. K. (1952). Phys. Rev. **85**, 1060-1061.
Holden, A. N. (1949). Discuss. Faraday Soc. **5**, 312-315.
Keith, H. D. and Padden, F. J. Jr. (1963). J. Appl. Phys. **34**, 2409-2421.
Keith, H. D. and Padden, F. J. Jr. (1964). J. Appl. Phys. **35**, 1270-1296.
Keith, H. D., Giannoni, G. and Padden, F. J. Jr. (1969a). Biopolymers **7**, 775-792.
Keith, H. D., Padden, F. J. Jr. and Giannoni, G. (1969b). J. Mol. Biol. **43**, 423-438.
Lovinger, Andrew J. and Keith, H. D. (1979). Macromolecules **12**, 919-924.
Lovinger, Andrew J. (1981). Macromolecules **14**, 322-325.
Mason, W. P., Editor (1942). Quartz Crystal Applications. Murray Hill, New Jersey: Bell Telephone Laboratories.
Mason, W. P. (1950). Piezoelectric Crystals and Their Application to Ultrasonics, New York: Van Nostrand.
Nassau, K. (1964). Quantum Electronics III, 833-839, Columbia University, Univ. Press.
Nassau, K., Levinstein, H. J., and Loiacono, G. M. (1966). J. Phys. Chem. Solids **27**, 983 and 989.
Nicholson, A. M. (1919) Trans. AIEE **38**, 1467-1493.
Patel, J. R., Testardi, L. R., and Freeland, P. E. (1976). Phys. Rev. **B13**, 3548-3557.
Patel, J. R. and Testardi, L. R. (1977). Appl. Phys. Letters **30**, 3-5.
Petroff, P. and Hartman, R. L. (1974). J. Appl. Phys. **45**, 3899-3903.
Pfann, W. G. (1952). J. Metals **4**, 747-753.
Pfann, W. G. (1957). Met. Rev. **2**, 29-76.
Pfann, W. G. (1974). Encyc. Brit. **19**, 1158-1160.
Pfann, W. G. (1978). Zone Melting, 2nd ed. Huntington, N.Y.: Robert E. Krieger Publishing Company.
Phillips, J. C. (1980). Phys. Stat. Sol. (b) **101**, 473-479.
Prince, E. and Abrahams, S. C. (1959). Rev. Sci. Instrum. **30**, 581-585.
Remeika, J. P. (1954). J. Am. Chem. Soc. **76**, 940.
Remeika, J. P. (1956). J. Am. Chem. Soc. **78**, 4259.
Remeika, J. P. and Comstock, R. L. (1964). J. Appl. Phys. **35**, 3320.
Remeika, J. P., Espinosa, G. P., Cooper, A. S., Barz, H., Rowell, J. M., McWhan, D. B., Vandenberg, J. M., Moncton, D. E., Fisk, Z., Woolf, L. D., Hamaker, H. C., Maple, M. B., Shirane, G., and Thomlinson, W. (1980a). Solid State Commun. **34**, 923.
Remeika, J. P. and Batlogg, B. (1980b). Mater, Res. Bull. **15**, 1179.
Scaff, J. H., Theuerer, H. C. and Schumacher, E. E. (1949). J. Metals **1**, 383-388.
Slichter, W. P. (1958). J. Polym. Sci. **35**, 77-92.
Slichter, W. P. (1959). J. Polym. Sci. **36**, 259-266.
Storks, K. H. (1938). J. Am. Chem. Soc. **60**, 1753.
Theuerer, H. C. (1955). Bell Laboratories Record **33**, 327-330.
Van Uitert, L. G., Grodkiewicz, W. H. and Dearborn, E. F. (1965). J. Am. Cer. Soc. **48**, 105-108.
Vogel, F. L., Pfann, W. G., Corey, H. E. and Thomas, E. E. (1953). Phys. Rev. **90**, 489.
Williams, H. J. and Shockley, W. (1949). Phys. Rev. **75**, 178-183.
Wood, E. A. (1964a). J. Appl. Phys. **35**, 1306-1312.
Wood, E. A. (1946b). Bell Sys. Tech. Jour. **43**, 541-559; Bell Tel. Laboratories Monograph No. 4680.

CHAPTER 11

CRYSTALLOGRAPHY AT PHILIPS LABORATORIES

Joshua Ladell
Computer Systems Research Group
Philips Laboratories, Briarcliff Manor, NY 10510

Philips Laboratories (PL) was established in 1944 during World War II. The parent organization, North American Philips Corporation, NAPC, is a family of diverse companies engaged in the fields of consumer products and services: electrical and electronic components and professional equipment and chemical products. NAPC ranks among the 150 largest industrial companies in the United States with sales in excess of $2.7 billion (1980). Some of the more familiar of the NAPC companies are Magnavox, Dialite, Ohmite and Philips Medical Systems. In the crystallographic community, the best known of the NAPC companies is probably Philips Electronics Instruments (PEI) since their product line includes X-ray analytic instruments which have found their way into most of the crystallographic laboratories. I shall say more about PEI below.

NAPC and Philips Laboratories (PL) were spawned from the giant N.V. Philips Gloeilampen Fabrieken, a multi-national corporation with headquarters at Eindhoven, The Netherlands. In an effort to thwart the acquisition of Philips property in North America by the Nazis, measures were taken which eventually led to the establishment of the publicly owned NAPC. When PL was established it was expected to play a role in the post war development of science, technology and new industrial processes. The mandate of PL was given to its first director, (the late) Dr. O. S. Duffendack who served from 1944 until his retirement in 1958. Dr. Duffendack concentrated the research efforts of PL in solid state physics. Specifically, experimental research was conducted in thermionics, fluorescence, ferrite magnetic materials, photoconductive materials for infrared radiation detectors, and crystallography. These activities were supported by chemical, optical, spectrographic, electronic circuitry, and X-ray diffraction laboratories.

The first site of the laboratories was the former manor house of the tobacco magnate, George Washington Hill, in Irvington on the Hudson. The X-ray and Crystallography section was housed in the former laundry room which was part of the garage structure. Despite the problems of transforming a private residence into a research laboratory, notable contributions were made in the fields of science, engineering, and new product development. As the staff grew, and the needs became increasingly complex, PL moved to its present facilities on a one hundred acre site in Briarcliff Manor, New York, with a panoramic view of the Hudson River. From 1958 through 1967, PL was under the direction of Dr. John Hipple who effected the move to Briarcliff in 1964. In the new laboratories the scope of PL activities included cryogenics, electro-optics, magnetics, solid state physics, devices, thermionics, X-ray diffraction, crystallography, chemistry and metallurgy.

The most significant growth and maturation of PL was effected during the tenure of the current president, Dr. Donald D. King. Dr. King reorganized the laboratories and concentrated most of the efforts in a program aimed at establishing a solid position for NAPC in silicon vidicon detectors. In the reorganization, the laboratories were divided into four groups. The Mechanical Systems Group, The Exploratory Research Group, The Electronics and Optical Engineering Systems Group, and The Device and Components Group. Later (1977), a fifth division, The Computer Systems Research Group was organized. The scope and aims of PL were clarified during this era. "The Laboratories direct their efforts mainly to scientific and technological advancements which will find potential industrial use within the North American Philips' companies." Regarding more basic research, a PL brochure reads: "A significant product of an industrial research laboratory is the contribution its staff makes to ongoing scientific knowledge. Philips' activities in this regard are evidenced by invited lectures, active participation in professional groups, publications in the form of papers to learned societies and to the technical press: every year sees numerous such contributions being made."

As PL enters the 80's, it continues to grow. Shortly (May, 1981) ground will be broken to begin a building expansion which will enlarge the physical facilities by 50%. Notwithstanding the significant growth and development of PL, it must be emphasized that PL is comprised of a small staff of approximately sixty-five professionals at the Ph.D. level and a total complement of about 250 employees. Before going into the crystallographic accomplishments of PL, and the personnel who were involved, it should be pointed out that the number of professionals (having Ph.D. degrees) specifically trained in crystallography working at PL never exceeded three, and the average was less than two! The names of PL personnel who spent a significant proportion of their time in the crystallographic effort and their tenure at PL are recorded in Figure 1.

In the beginning, the disciplines of PL were organized into sections. The X-ray and Crystallography section was organized and supervised by Dr. William (Bill) Parrish who held this position from 1944 through 1968. During his tenure, literally hundreds of papers flowed out of the laboratories. In attempting to classify which areas of crystallography were involved in this massive search, I found that I could not classify the work into less than 50 topics. Most of the contributions reported original,

Figure 1

NAMES AND TENURE OF PL PERSONNEL ENGAGED IN CRYSTALLOGRAPHY

Years 40 50 60 70 80

Name	**Specialty**	Tenure
G. Abowitz Ph.D.	MT	xxxx
P. G. Cath Ph.D.	EL	xxxxxxx
H. T. Evans Ph.D.*	CH,C	xxx
P. Goldstein Ph.D.*	CH,C	xxxx
B. Greenberg Ph.D.*	P,C	x
E. Hamacher*	EE	xxxxxxxx
T. R. Kohler	EE	xxxxxxxxxxxxxxxxxxxxxxxxxxxxxxxxxxxx
J. Ladell Ph.D.*††	P,C	x xxxxxxxxxxxxxxxxxxxxxxxxx
A. R. Lang Ph.D.*	C	x
K. Lowitzsch*	MDE	xxxxxxxxxxxxxxxxx
M. Mack*	CH	xxxxxxxxxxxx
J. Nicolosi*	P,C	xxxxxxxx
W. Parrish Ph.D.* †	M,C	xxxxxxxxxxxxxxxxxxxxxxxxxx
J. Reid*	EL	xxxxxxxx
W.N.S. Schreiner Ph.D.*	P,SS	xxxxxx
N. Spielberg Ph.D.*	P	xxxxxxxxxxxxxxx
C. Surdukowski*	SS,A	xxx
J. Taylor*	CH	xxxxxxxxxxxxx
I. Vajda*	MDE	xxxxxxxxxxx
A. Zagofsky*	MA,SS	xxx
G. Ziedens	EE	xxxxxxxxxxx

Years 40 50 60 70 80

Code: A: Astronomy, C: Crystallography, CH: Chemistry; EL: Electronics Eng. EE: Electrical Eng.
MA: Mathematics; M: Minerology, MDE: Mechanical Design; MT: Metallurgy; P: Physics;
SS: Software Specialist

*Member of "X-ray and Crystallography Section," or "Instrument Automation Group."

†Chief of Section
††Senior Program Leader

definitive and comprehensive research. The topics which kept the section busy throughout its history and references to key articles are summarized in Table 1. While at PL, Bill Parrish was an indefatigable researcher. In his passion for understanding and innovation he was uncompromising with his quest for truth. To Bill research was spelled with a capital R and science with a capital S. In choosing subordinates he sought out talented people and encouraged the best from them. In the early days, he worked closely with Eddie Hamacher, a brilliant electronics engineer, and Kurt Lowitzsch, a gifted mechanical design engineer. Together they constituted a formidable team; Bill was able to convey the scientific basis and experimental experience to them. They, in turn, were able to quickly absorb the concepts and convert them into investigational instruments of high quality. The passing of Eddie Hamacher (while still a young man) in 1954 and Kurt Lowitzsch in 1961 were great tragedies keenly felt by the section.

As a boss, Bill Parrish was kind and compassionate. The morale of the section was always high. The pace and intensity of work at this time were primarily set by his personal example. He insisted upon personally performing critical experiments, and aligning instruments. Since PEI was involved in the commercialization of X-ray analytic instrumentation a close liaison was always maintained with PEI. This liaison was a source of great opportunity on the one hand and frustration on the other. The opportunity was available to develop creative instrumentation and to carry out basic research. The section thereby maintained expertise somewhat forward of the state of the art in many technologies relevant to X-ray crystallographic instruments. The frustration arose from the restricted size of the section. From the point of view of the proliferation of new fields and disciplines which were branching out of X-ray diffraction (e.g., microdiffraction, small angle scattering, microprobe, crystal growth, molecular biology, X-ray topography, neutron diffraction etc.), the section was too small to

TABLE I

POWDER DIFFRACTOMETRY

Topic	Reference
1. Diffractometer Instrumentation	"The "Norelco" X-ray Diffractometer." W. Parrish, E. A. Hamacher and K. Lowitzsch (1954). *Philips Tech. Rev.* **16**, 123-133.
2. Extinction in Powders	"Extinction in X-Ray Diffraction Patterns of Powders." A. R. Lang (1953). *Proc. Phys. Soc.* **66**, 1003-1008.
3. Systematic Errors in Powder Diffractometry	"X-ray Diffractometry Methods for Complex Powder Patterns." W. Parrish (1968). *X-ray and Electron Methods of Analysis.* New York: Plenum Press.
4. Radioactive Samples	"X-ray Diffractometry of Radioactive Samples." T. R. Kohler and W. Parrish (1955). *Rev. Sci. Instrum.* **26**, 374-379.
5. Alignment Apparatus	"Geometry, Alignment, and Angular Calibration of X-Ray Diffractometers." W. Parrish and K. Lowitzsch (1959). *Am. Mineral.* **44**, 765-787.
6. Precision Lattice Parameters	"Center-of Gravity Method of Precision Lattice Parameter Determination." J. Ladell, W. Parrish and J. Taylor (1959). *Acta Cryst.* **12**, 253-254. "Precision Measurement of Lattice Parameters of Polycrystalline Specimens," W. Parrish and A. J. C. Wilson (1959). International Tables for X-ray Crystallography Vol. II, 216-234.
7. Particle Statistics	"Experimental Study of the Effect of Crystallite Size Statistics on X-ray Diffractometer Intensities." P. M. deWolff, J. Taylor and W. Parrish (1959). *J. Appl. Phys.* **30**, 63-69.

TABLE I (CONTINUED)

8. Intensity Statistics	"Statistical Factors in X-ray Intensity Measurements." M. Mack and N. Spielberg (1958). *Spectrochimica Acta* **12**, 169-178.
9. Fluorescent Sources	"Fluorescent Sources for X-ray Diffractometry." W. Parrish, K. Lowitzsch and N. Spielberg (1958). *Acta Cryst.* **11**, 400-405.
10. Rotating Specimen Device	"Rotating Flat Specimen Device for the Geiger Counter X-ray Spectrometer." K. Lowitzsch and W. Parrish (1958). U.S. Patent No. 2,829,261, April 1, 1958.
11. Profiles Line Interpretation of Diffractometer	"Interpretation of Diffractometer Line Profiles Distortion Due to the Diffraction Process." J. Ladell (1961). *Acta Cryst.* **14**, 47-54.
12. Transmission Powder Diffractometry	"X-ray Diffraction Method and Apparatus." P. M. deWolff and W. Parrish (1959). U.S. Patent No. 2,887,585, May 19, 1959.
13. IUCr Lattice Parameter Project	"The IUCr Lattice Parameter Project." W. Parrish (1959). IUCr Conferences, Karolinska Institute, Stockholm.
14. Minimal Detectable Limits in XRD	"Factors in the Detection of Low Concentrations in X-ray Diffractometry." W. Parrish and J. Taylor (1960). Proceedings of the 2nd International Symposium on X-ray Microscopy and X-ray Microanalysis, Stockholm, 1960.
15. Recommended Practices	"Outline of Recommended Practice for X-ray Diffractometry of Polycrystalline Substances." W. Parrish (1960). Proceedings of the 5th International Instruments and Measurement Conference, Stockholm, 1960.
16. K-alpha Satellite Interference in XRD	"K-alpha Satellite Interference in X-ray Diffractometer Line Profiles." W. Parrish, M. Mack and J. Taylor (1963). *J. Appl. Phys.* **34**, 2544-48.
17. Seeman-Bohlin Diffractometry	"Seeman-Bohlin X-ray Diffractometry II. Comparison of Aberrations and Intensity with Conventional Diffractometer." M. Mack and W. Parrish (1967). *Acta Cryst.* **23**, 693-700.
18. Cu K-alpha2 Elimination Algorithm	"Cu K-alpha2 Elimination Algorithm." J. Ladell, A. Zagofsky and S. Pearlman (1975). *J. Appl. Cryst.* **8**, 449-506.
19. Studies of Phase Transitions	"Phase Transitions in Cs $(D_xH_{1-x})_2AsO_4$." G. M. Loiacono, J. Ladell, W. N. Osborne and J. N. Nicolosi (1976). *Ferroelectrics* **14**, 761-765.
20. Texture Sensitive Diffractometry	"Asymmetric Texture Sensitive X-ray Powder Diffractometer." J. Ladell (1980). U.S. Patent No. 4,199,678, April 22, 1980.
21. Computer Controlled Diffractometry	"The APD-3600, A New Dimension in Qualitative and Quantitative X-ray Powder Diffractometry." R. Jenkins, Y. Hahn, S. Pearlman, and W. Schreiner (1979). *Norelco Reporter* **26**, 1.

TABLE I (CONTINUED)

22. Peak Hunting in Statistical Data	"A Second Derivative Algorithm for the Identification of Peaks in Powder Diffraction Patterns." W. N. Schreiner and R. Jenkins (1979). *Adv. X-Ray Anal.* **23**, 287.
23. Automated Search/Match Procedures	"A New Minicomputer Search/Match/Identify Program for Qualitative Phase Analysis with the Powder Diffractometer." W. N. Schreiner, C. Surdukowski, and R. Jenkins (1981). *J. Appl. Cryst.* (in press).
24. Identification of Solid Solutions	"An Approach to Isostructural and Solid Solution Problems in Multiphase X-ray Analysis." W. N. Schreiner, C. Surdukowski and R. Jenkins (1981). *J. Appl. Cryst.* (in press).
X-RAY PHENOMENA	
1. X-ray Analysis	"Modern X-ray Chemical Analysis." W. Parrish and A. Engstrom (1956). *Svensk Kemisk Tidskrift* **68**, 437-454.
2. X-ray Optics	"Method of Generating An X-ray Beam Composed of a Plurality of Wavelengths." N. Spielberg and J. Ladell (1968). U.S. Patent No. 3,418,467, Dec. 24, 1968.
3. X-ray Interference	"Method of Obtaining X-ray Interference Patterns." N. Spielberg and J. Ladell (1969). U.S. Patent No. 3,439,164, April 15, 1969.
4. Multiple Diffraction	"An Interactive Computer Controlled System for the Accurate Measurement of Lattice Parameters by Post's Method of Multiple Diffraction." J. Ladell and J. Nicolosi (1975). Abstract - A.C.A. Meeting, Charlottesville, VA, March 12-14, 1975.
X-RAY TOPOGRAPHY	
1. Texture of Large Crystals	"A New Photographic Method for Studying the Texture of Large Single Crystals." A. R. Lang (1954). *Acta Cryst.* **7**, 583-587.
2. Topography of DTGFB	"X-Ray Topographic Analysis of Dislocation Line Defects in Solution Grown DTGFB." J. Nicolosi and J. Ladell (1980). *J. Cryst. Growth* **49**, 120-124.
DETECTORS	
1. Single Crystal Detection	"Counters for X-Ray Analysis." P. H. Dowling, C. F. Hendee, T. R. Kohler, and W. Parrish (1956). *Philips Tech. Rev.* **18**, 262-275.
	"Use of a Geiger Counter for the Measurement of X-Ray Intensities from Small Single Crystals." H. T. Evans (1953). *Rev. Sci. Instrum.* **24**, 156-161.
2. Counting Electronics	"Effect of Anode Material on Intensity Dependent Shifts in Proportional Counter Pulse Height Distributions." N. Spielberg (1967). *Rev. Sci. Instrum.* **38**, 291.

TABLE I (CONTINUED)

3. Electric Devices	"Circular Slide Wire Resistance Element." E. A. Hamacher and K. Lowitzsch (1953). U.S. Patent No. 2,658,131, Nov. 3, 1953.
4. Electronic Intensification	"A Proposed Method for Electronic Intensification of Single-Crystal X-ray Diffraction Images." A. R. Lang (1954). *Rev. Sci. Instrum.* **25**, 1032-3.
5. Conversion of Quantum Counting to Roentgens	"Conversion of Quantum Counting Rate to Roentgens." T. R. Kohler and W. Parrish (1956). *Rev. Sci. Instrum.* **27**, 705-706.
6. Electronic Apparatus for Structure Determination	"Apparatus for Determining Crystal Structure." E. A. Hamacher (1957). U.S. Patent No. 2,802,947, Aug. 13, 1957.
FLUORESCENT SPECTROMETRY	
1. Methodology	"X-Ray Spectrochemical Analysis." W. Parrish (1956). *Philips Tech. Rev.* **17**, 269-286.
2. Particle Absorption	"Measurement of Particle Absorption by X-ray Fluorescence." P. M. deWolff (1956). *Acta Cryst.* **9**, 682-683.
3. Calibration Techniques	"Calibration Techniques for X-Ray Fluorescence Analysis of Thin Nickel-Chromium Films." N. Spielberg and G. Abowitz (1966). *Anal. Chem.* **38**, 200.
4. Simultaneous Spectrography	"Laue Spectrometer for Multichannel X-ray Spectrochemical Analysis." J. Ladell and N. Spielberg (1960). *Rev. Sci. Instrum.* **31**, 23-29.
5. Instrumentation	"Scanning Single Crystal Multi-Channel X-Ray Spectrometer." J. Ladell and N. Spielberg (1963). *Rev. Sci. Instrum.* **34**, 1208-1212.
6. Intensity Theory	"Intensities of Radiation from X-Ray Tubes and the Excitation of Fluorescence X-Rays." N. Spielberg (1959). *Philips Research Reports* **14**, 215-236.
7. Non-focussing Geometry	"Geometry of the Non-Focussing X-Ray Fluorescence Spectrograph." N. Spielberg, W. Parrish and K. Lowitzsch (1959). *Spectrochimica Acta* **8**, 564-583.
8. Minimal Detectable Limits in XRF	"Instrumental Factors and Figure of Merit in the Detection of Low Concentrations by X-ray Spectrochemical Analysis." N. Spielberg and M. Bradenstein (1963). *Appl. Spect.* **17**, 6-9.
9. Regression Analysis	"The Non-Linear Least Squares Fitting Routine for Optimizing Empirical XRF Matrix Correction Models." W. N. Schreiner and R. Jenkins (1979). *X-Ray Spectrometry* **8**, 33.
SINGLE CRYSTAL INVESTIGATION	
1. Linear Tracking	"Automatic Single Crystal Diffractometry I. The Kinematic Problem." J. Ladell and K. Lowitzsch (1960). *Acta Cryst.* **13**, 205-215.

TABLE I (CONTINUED)

2. Intensity Measurement Technique	"Theory of the Measurement of Integrated Intensities Obtained with Single-Crystal Counter Diffractometers." J. Ladell and N. Spielberg (1966). *Acta Cryst.* **21**, 103-118.
3. Crystal Structures [Under a special contract with the Philips Research Laboratories (The Netherlands), a series of organic phosphate pesticides were studied. The complete structures were refined and internally reported. Data were obtained with the PL PAILRED diffractometer]	"The Crystal and Molecular Structures of: A & B: 3-methyl-5-phenylamino-1 and 2-bis(dimethylamido)phosphoryl-1,2,4-triazole. C: 3-phenyl-5-methylamino-1-bis(dimethylamido)phosphoryl-1,2,4-triazole. D & E: 5-phenylamino-3-ethyl-1 and 2-bis(dimethylamido)phosphoryl-1,2,4-triazole. F & G: 5-phenylamino-3-isopropyl-1 and 2-bis(dimethylamido)phosphoryl-1,2,4-triazole". P. Goldstein and J. Ladell (1966). Internal PL Reports 1966. A.C.A. Meeting, Atlanta (Abstracts) 1967. "Refinement of the Crystal and Molecular Structure of 1,2,4-triazole ($C_2H_3N_3$) at Low Temperature." P. Goldstein, J. Ladell, and G. Abowitz (1969). *Acta Cryst.* **B25**, 135-143.
4. Absorption for Spheres	"Tables of Absorption Factors for Spherical Crystals." H. T. Evans and M. G. Ekstein (1952) *Acta Cryst.* **5**, 540-2.
5. Crystal Orientation	"Symmetry of Interface Charge Distribution in Thermally Oxidized Silicon." G. Abowitz, E. Arnold and J. Ladell (1967). *Phys. Rev. Lett.* **18**, 543-6.
6. PAILRED Diffractometer	"Principles and Design of the Automatic Single-Crystal Diffractometer System PAILRED." P. G. Cath and J. Ladell (1968). *Philips Tech. Rev.* **29**, 165-85.
7. Low Temperature Diffractometer	"A Low Temperature System for Automatic Single Crystal Diffractometry." G. Abowitz and J. Ladell (1968). *J. Sci. Instrum. Ser. 2,* **1**, 113-117.
8. Biaxial Diffractometer	"An Interactive Computer Controlled Biaxial Single Crystal Diffractometer." J. Ladell, J. Nicolosi and S. Pearlman (1977). A.C.A. Meeting, Asilomar Calif. (Abstract) Feb. 1977.
9. Alpha-Keratin	"An X-ray Study of Alpha-Keratin. I. A General Diffraction Theory for Convoluted Chain Structures and an Approximate Theory for Coiled-Coils." A. R. Lang (1956). *Acta Cryst.* **9**, 436-445.

effectively 'cover' the general field. From the point of view of NAPC, the size of the section was large when considered in terms of R. & D. effort for the anticipated X-ray analytic instrument market. The PL X-ray and Crystallography section was therefore always keenly interested in the success of PEI. Hopefully, if their share of the market increased, some expansion of the section might be possible. Other means were used to enhance the research effort of the section. Invitations for extended visits were made to such distinguished scientists as Prof. A.J.C. Wilson of Cardiff (1950's) and P.M. deWolff of Delft (1954), both of whom spent some time at PL. In recent years the needs for materials characterization in the Exploratory Research and other groups at PL have widened the horizons of the Crystallographic group, more emphasis is being directed towards materials research and analysis.

Space does not permit a thorough account of the crystallographic accomplishments at PL. I will, therefore, restrict the discussion to four topics 1) powder diffractometry, 2) fluorescent spectrometry, 3) lunar effort, and 4) single-crystal diffractometry.

The "focussing circle" which is the geometric basis for the modern powder diffractometer had been known for some time. Retrospectively, the concept was associated with Bragg, Brentano, Seeman and Bohlin. It was Parrish, Hamacher and Lowitzsch, however, who developed the concept to produce a practical, high intensity, high resolution, direct photon counting instrument. The feasibiltiy of the concept was enhanced by the synchronization of the specimen post to rotate at half the angular speed of the counter detector and the effective use of Soller collimators which allowed a larger surface area of sample to diffract from the full width of the line focus. Not satisfied with the innovation of a unique multipurpose powder diffractometer which remains virtually unaltered in basic design over a thirty year period (about 10,000 X-ray powder diffractometers have been sold to date), Parrish carried out a life-long research effort to develop the potential of this sensitive device for crystallographic research. From the experimental side he studied all the factors which defined the powder diffractometric experiment: the X-ray source and methods for improving source stability; the detector, its sensitivity, spectral response, escape-peak phenomena, quantum counting efficiency, dead time, advantages of Geiger, proportional and scintillation detectors; the physical and geometric aberrations of the diffractometer, the intensity vs. angle of view, the flat specimen error, the vertical and axial divergences, the specimen displacement and the effects of Lorentz, polarization, dispersion and absorption factors, the effect of crystal monochromatization, balanced filters, screening, and other conceivable factors, such as optimum wavelength, alignment and calibration, specimen preparation and counting statistics which were in any way related to the understanding and improvement of the X-ray powder diffractometric technology. Having developed the techniques for properly using the instrument, he went on (with the help of his colleagues A.J.C. Wilson and his co-workers Josh Ladell, Jeanne Taylor and Marian Mack) to develop the theory of measuring diffractometer line profiles, the treatment of aberrations and the use of these theories for the precision measurement of lattice parameters. The path was treacherous. The best of our efforts only succeeded in revealing the complexity of the processes which were involved in the development of diffractometer line-profiles. Nevertheless the basic approach was sound, and with the facility of modern computational power, the approach is becoming practical. The story of powder diffractometry at PL did not end with Bill Parrish's leaving PL in 1969. Instead, powder diffractometry at PL took off in an entirely new direction, that of computerized powder diffractometry. This development led to the first commercially available computer-controlled powder diffractometer, the APD-3500 (Automated Powder Diffractometer). In the wake of the minicomputer revolution [when the price of the Digital Equipment Corporation (DEC) PDP8 dropped to less than $10,000 in 1964], it was apparent that computers with their extensive and versatile software capabilities would become the "intelligent" controllers of sophisticated analytic instrumentation. Just as the "electronic age" replaced the "mechanical age," the age of computers would replace the age of electronics. A program designated "Automated Instrumentation" was organized in the Electronics and Optical Systems Engineering Group, under my leadership. Its objective was to develop computer-linked analytic techniques which were unfeasible in the days of costly computers. In a symbolic link to the past, a gold-plated powder diffractometer which was presented to Parrish by PEI on the occasion of the sale of the 2000th diffractometer in the U.S., was retrofitted with a computer interfaced stepper motor to become the PL prototype APD minicomputer-controlled powder diffractometer. Our initial success in computer-controlled powder diffractometry was due to the concentrated cost-effective software package we developed for the (Ferroxcube Digital Controller) FDC 302, a 4096 word minicomputer (controller). The FDC 302, developed by Ferroxcube Corp., one of the NAPC companies, was an interesting and unusual computer. Its instruction set consisted of 11 basic instructions, none of which could add or make a logical comparison! It's 18-bit word length organized into six 3-bit bytes was very convenient for text manipulation, as was its unique scheme for implementing input/output processes which did not require the standard priority interrupt logic. Starting with a virtually "naked" computer, a complete operating system, assemblers, editors, floating-point software, extended memory in the form of digital cassette transports were developed by the group to enable the implementation of control and on-line analytic capability. The hardware design and implementation of the cassette transports, scalers, timers, analog to digital, digital to analog interfaces, stepper motor drivers and literally dozens of peripheral interfaces were designed and constructed by G. Ziedens and Joe Reid. Milestones in the powder diffractometry effort included the development of an interactive Spectral Stripping powder diffractometer which used the Cu K-alpha2 elimination algorithm on-line and the development of a plotter package which provided a versatile graphics capability.

In 1975, to improve the coordination between the PL Automated Instrument Group and the PEI Engineering Group, a joint software effort was initiated with Ron Jenkins of PEI serving as liaison. Similar software stations equipped with modern minicomputers and supporting peripherals were installed at PEI and PL. A software task of mutual interest to PL and PEI was then partitioned among the groups, and the development carried out in close cooperation. The first closely coordinated project was the development of AXS (Automated X-ray Spectrometer) software (which will be discussed below). Two years later the team was ready for bigger and better assignments. In late 1977, the groundwork was prepared for the development of the APD 3600 - a modern minicomputer-controlled powder diffractometer - having sufficient computational power and graphics display capability to carry out all known powder analytic procedures. At PL the focus of this effort was the program SANDMAN designed and implemented by W.N. Schreiner with the help of Carol Surdukowski. SANDMAN (Search and Match Nova) is a multiphase identification program which searches the entire JCPDS file of diffraction patterns. Early experience with the program indicates that it is highly effective in solving identification problems and performs as well or better than search/match programs designed for very large computers. Because of its potential for multiphase identification SANDMAN may prove to be one of the most significant developments in X-ray powder diffractometry. Another milestone is our recent development of the Texture Sensitive Diffractometer (TSD). The computer-controlled interactive TSD was designed to establish experimentally whether a powder sample consisted of randomly distributed or preferentially oriented crystallites. It has the capability of measuring diffraction with the diffraction vector varying from the specimen surface normal. Focussing is maintained using an asymmetric focussing geometry. For texture studies, pole-figures are generated.

With the same zeal manifest in the powder diffractometric effort, Parrish attacked the problems of X-ray Fluorescence Analysis (XRF) with the view of categorizing the bases for the development of effective spectrometers. The major effort of the XRF program was carried out by N. Spielberg, an X-ray physicist. Research in XRF had been directed primarily at a quantitative understanding and analysis of factors of significance for wavelength dispersive instruments, i.e., geometrical ray optics, analyzing crystals, excitation of X-ray fluorescence in analytical specimens, and detection of X-rays. The last two factors are also significant for energy dispersive instruments. Criteria and figures of merit for evaluating some of these factors, and their relationship to instrumental resolution and analytical sensitivity of the total instrumental system, were also studied. Geometric optics studies considered the effect of parameters of Soller collimators on the resolution and intensity of the standard instrument using large flat crystals. Of particular crystallographic interest is the interplay of the geometric optics and the selection and orientation of the analyzing crystal in suppressing spurious upper level reflections, and in the design of multichannel spectrometers using simultaneous reflections from different sets of crystallographic planes within the same analyzing crystal. Excitation studies revealed that a ten-fold increase in signal intensity for excitation of light elements could be obtained using properly designed thin-window X-ray tubes with suitably chosen targets.

A recent accomplishment of PL in X-ray spectrometry was regression analysis software developed by W.N. Schreiner in conjunction with the AXS project. A unique feature of the regression analysis program is that it allows the fitting of a generalized concentration calibration function which is non-linear in form. This permits separation of absorption dependent parameters from instrument dependent parameters, thereby greatly simplifying instrumental calibration once the interelement effects have been determined by fitting a set of reference standards.

Parrish was keenly interested in the space program and felt that extraterrestrial investigations by XRD (X-Ray Diffraction) and XRF would be very worthwhile. In an unsolicited proposal to the National Aeronautics and Space Administration (NASA), he proposed building a miniature diffractometer to go on the Surveyor spacecraft. The proposed lunar diffractometer consisted of a miniature X-ray tube and detector each mounted on plates of a hinge. A motor driven leadscrew would open or close the hinge (closing or opening the scattering angle 2 theta). The device had a flat bottom with a rectangular opening. When deployed, whatever was beneath it was the sample. Parrish planned that the device be placed upon the lunar surface and then activated. The data obtained would be telemetered back to Earth. At first NASA decided that the proposed time for implementing the study was too short to obtain effective results. In the end, the contract was let for less time than Parrish estimated was needed to get the job done. Stringent constraints were imposed - only 50 watts total power was allowed. Also, NASA insisted that the diffractometer be designed along more conventional lines. An elaborate specimen preparation was planned including grinding, sifting and conveying the sample to the diffractometer using belt driven hoppers. The original specimen would come from a core dug out by a drill. The miniature X-ray tubes were designed by the X-ray tube group at the Philips Research Laboratories in Eindhoven, The Netherlands. Half the power allocation was allowed for producing X-rays. Using his expert knowledge, Parrish redetermined the parameters which hold the balance between high resolution and high intensity in the diffractometric set-up. He made the necessary trade-offs, thereby recovering a forty-fold increase in intensity without too much loss in resolution. Working with care, diligence and speed, the lunar diffractometer was constructed in time for the last of the Surveyor probes. We all shared Bill Parrish's pride when the lunar diffractometer survived the pre-launch test and his disappointment when the lunar diffractometer was scrubbed - because they couldn't get the drill to work. Under another NASA contract, as part of the post-Apollo program a minicomputer powder-diffractometer and X-ray spectrometer were designed, fabricated and installed at the Lunar Receiving Laboratory. The system operated within a biological barrier and was remotely controlled. The system was used to quickly study returned lunar samples. The design of the diffractometer was based upon specifications laid down by Parrish; specific features were the use of Cr K-alpha radiation for improved resolution, a shorter radius and a 15° angle-of-view to enhance intensity. Appropriate entrance, receiving and antiscatter slits were mounted on indexable handwheels so that slit changes could be effected by remote control. The X-ray Spectrometer was designed by modifying a standard Norelco instrument. Most of the modifications introduced by Spielberg were made to incorporate technical advances not yet commercially available or to make the various components of the system more readily accessible either through glove ports or by remote control. I succeeded N. Spielberg as program leader in 1969, and formulated the software specifications. The software for the control of both instruments was written by Y. Okaya, who was engaged as a consultant for this program. As Okaya developed and tested his software using the DEC PDP8/L, I developed FDC 302 software which controlled another diffractometer. (The choice of the DEC PDP8/L was made at the insistence of NASA.) This activity led to the development of the APD.

A market survey conducted by PEI in the late 50's indicated that they were losing sales of X-ray generators to a competitor who was supplying a large table top which could accommodate single-crystal apparatus. PEI approached the X-ray and Crystallography group for advice concerning what single-crystal apparatus could be put into their product line to offset the loss of X-ray generator sales. We accepted this request as a mandate to actively go into single-crystal instrumentation. Parrish and I visited various laboratories in the U.S. to evaluate prototype single-crystal diffractometers which were then coming into use. After completing the survey, we felt that none of the existing configurations was sufficiently general for single crystal structure analysis to recommend to PEI for commercial exploitation. Some pioneering work in single-crystal diffractometry had been done earlier (1950) at PL by H. T. Evans. We decided to follow his lead in adapting the Weissenberg method. While studying the various approaches I conceived a mechanical linkage which would replicate a reciprocal lattice construction and thus permit the analog calculation of the crystal and detector angles. I sketched out my thoughts on one page and showed them to Parrish, whose response was, "It will never work!" I then showed the sketch to Kurt Lowitzsch who nodded and came back a half hour later with a working cardboard model of the linkage. We went back to Parrish whose revised response was, "Write a patent disclosure." This was the beginning of the PAILRED (Philips Automatic Indexing Linear Reciprocal-Space Exploring Diffractometer) story. While I experimented with some improvised mock-up equipment (and wrote a 42 page disclosure) Lowitzsch proceeded with the design and mechanical detailing. To expedite the work he submitted drawings to the shop as soon as they were ready. He rarely asked questions after the initial communication we had which defined the motions the diffractometer was to perform. When he had to quit working because of terminal illness, I was called down to the shop and confronted with a maze of components for which there were no assembly drawings! Armed only with the knowledge of the principles, I had to literally solve the puzzle of how all the pieces fit together! The "feasibility" prototype was completed with the help of engineer Imre Vajda. The electronics and electromechanical control devices were designed and executed by T. R. Kohler (who later became the Director of the Electronics and Optical Systems Engineering Group). For this model (1960), we used brakes and clutches, optical encoders, electrical readout mechanical counters and an 102 conductor slip ring assembly which was originally designed for the Polaris missile. The control "program" consisted of a bank of stepping switches. In the early 60's we were awarded a contract by the Philips Research Laboratories (Eindhoven) to build a PAILRED diffractometer to be used in Paul Braun's laboratory in Eindhoven. This task was assigned to P. G. Cath, a brilliant electronics engineer who greatly simplified the mechanical structure and redesigned the control circuitry using state-of-the-art digital electronics. Three identical PAILREDs were constructed. The first PAILRED was delivered to Paul Braun after its odyssey to Rome. The second PAILRED remained at PL and the third eventually ended up in Sweden. Three months prior to the scheduled delivery to Paul Braun's Laboratory, his PAILRED was airlifted to Rome to be displayed at the IUCr meeting. At Rome, PAILRED was the only automatic single-crystal diffractometer which was operating and actually collecting

3-dimensional data from a real crystal. The PAILRED was a great technical success. Special features included: linear tracking capability, crystal monochromatized incident radiation, a 65° Eulerian cradle for orienting the crystal, full zone or level automation, and means for electronically changing scan and slew rates. To evaluate PAILRED, the crystal and molecular structures of a series of phosphoryl triazole compounds were determined. About 30,000 independent reflections were measured from eight different crystals. The molecules on the average contained 40 to 45 atoms. The data were so good that Paul Goldstein was able to interpret the 3-dimensional Patterson maps within twenty minutes of obtaining them. A low-temperature extension for the PAILRED was designed and implemented with the help of G. Abowitz. With the low temperature facility crystals were maintained at -150°C for as long as six weeks. The implications of the process control computer on single crystal diffractometry were investigated. The equi-inclination Weissenberg type approach used for PAILRED was generalized by the Biaxial diffractometer in which the feasibility of generating integrated intensities by rotation about the μ axis was exploited. To test the new ideas, PAILRED was updated. The instrument was put under the control of an FDC mini-computer. The equi-inclination crystal and detector motions were automated. The function of the linkage was altered to enable the system to be rigid at low 2θ angles. The improvement anticipated by the biaxial concept were evaluated with the updated PAILRED. Recently, a completely new Biaxial Diffractometer has been designed and constructed. Controlled by a Data General Nova minicomputer, most of the necessary software has been written. The single-crystal diffractometry effort at PL continues.

Contributions to the crystallographic effort at PL have been made by many people. In some instances, the nature of their effort is identified by authorship in one or more of the references, in other cases the involvement, though significant, has been more subtle. The instrumentation program over the years could hardly have succeeded without the support of an efficient machine-shop. PL was lucky to have superb foremen over the years, R. Johnson, R. Hamilton and presently R. Wassman. Dotty Barrett was secretary for the X-ray Crystallographic Section at its most prolific era - in the days before word processors! From the early days when the PL mainframe computer was an LGP30 (4096 word drum memory), the crystallographic effort received the support of Therese Gendron who is presently the manager of the PL Computer Center. Mechanical engineers G. Pitcher and Aharon Sereny designed the updated PAILRED and the Biaxial Diffractometer respectively. To complete the story, it should be mentioned that the distinguished crystallographers, Howard T. Evans and Andrew Lang, launched their careers at PL.

CHAPTER 12

X-RAYS AND CRYSTALLOGRAPHY AT GENERAL ELECTRIC

J. S. Kasper and W. L. Roth
General Electric Research and Development Center
P.O. Box 8, Schenectady, NY 12301

From 1914 to the present, there have been parallel streams of effort at General Electric that have made major contributions both to X-ray technology and crystallography. Although the scientific and technological work occasionally coalesced, for the most part crystallographic research proceeded separately from instrument developments concerned with manufacturing X-ray apparatus for medical, dental, and industrial applications. There are two distinct periods during which outstanding crystallographic advances took place, roughly spanning the periods from 1914 to 1921 following World War I and 1945 to the present following World War II.

In 1914, W. D. Coolidge, following Langmuir's discovery of electron space-charge, utilized electron emission from tungsten to invent the hot-cathode X-ray tube. The sealed X-ray tube was rapidly applied throughout the world for medical, dental, and industrial X-radiography, as well as for structure analysis by X-ray diffraction. General Electric today is a major manufacturer of X-ray equipment for medical, dental, and industrial X-ray radiography, including high voltage X-ray tubes and CAT scanners, although it has discontinued manufacturing spectrometers and diffractometers for X-ray diffraction.

Structure analysis at G.E. began when Sir William Bragg visited the General Electric Research Laboratory and spoke about X-ray crystal analysis work then underway at Cambridge. A. W. Hull became interested in the subject and decided to determine the crystal structure of iron to find a clue to its magnetism. Single crystals were not available and to solve the structure Hull developed in 1917 the method for solving structures from powder patterns. After World War I, when scientific exchange of information was resumed, it was discovered that Debye and Sherrer had independently discovered the X-ray powder method and published it nearly a year before Hull's paper. By 1917, Hull had used the method to solve the structures of Al, Mg, Na, Li, and Ni. In 1921, in collaboration with W. P. Davey, the logarithmic strip technique for solving cubic, tetragonal, and hexagonal structures was developed. The structures of the following common metals were solved, Cr, Mo, W, αCo, Ni, Rh, Pd, Ir, Pt, βCo, Zn, Cd, and also that of graphite.

Following World War II, there was a substantial growth in basic research at General Electric; crystallography shared in a stimulating scientific atmosphere that resulted in many major advances in solid state and materials science. From 1940 to the present, a large number of people in the General Electric Research Laboratory (now the Research and Development Center) have made substantial contributions to crystallography in diverse fields ranging from theory to practical applications. The list of people includes David Harker, John S. Kasper, and Walter L. Roth, who devoted most of their careers to crystallography, and many others who used crystallography to solve important problems in metallurgy, surface physics, and solid state chemistry: E. Alessandrini, S. Bartram, K. Browall, B. F. Decker, E. F. Fullam, A. Geisler, R. Goehner, M. L. Kronberg, C. M. Lucht (Kasper), J. Lukesh, J. B. Newkirk, L. Osika, S. M. Richards, B. W. Roberts, C. Tucker, L. Vogt, and R. Waterstrat.

During the 1940's and early 1950's some notable pioneering crystallographic research was accomplished. A fundamental contribution to the very basic phase problem was made by Harker and Kasper, with the derivation of inequalities relating modified structure factors of certain reflections to the intensities of other reflections, a discovery that has inspired the development of modern direct methods. The first elucidation of structure for a boron hydride (decaborane) was made, which was a major contribution to understanding boron bonding chemistry; similarly, the first structure of a silicone was determined (hexamethyl disiloxane) which for the first time explained the origin of low intermolecular interactions in this family of compounds. On a more practical note, the single crystal goniometer and the XRD-3 diffractometer for X-ray structure analysis were designed and evaluated in the Research Laboratory and an Automatic Pole Figure Goniometer was developed and extensively applied to solve fundamental problems of metal deformation and recrystallization.

With the advent of neutron diffraction, the General Electric Research Laboratory was the first industrial laboratory to undertake a full-time research program with its own spectrometer at the Brookhaven National Laboratory Reactor. Most of the effort was with magnetic structures: quite noteworthy was the elucidation of the nature of antiferromagnetic domains, as well as studies of the magnetic structures and magnetic defects in transition metal monoxides, spinels, and perovskites. Also, the magnetic structures of α- and β-manganese were investigated. The nature of ordering was established for various transition element alloys, a point of considerable uncertainty at the time.

The leadership of the Research Laboratory in high pressure research following the first successful production of man-made diamonds quite naturally resulted in efforts in the crystallography of high pressure phases. In addition to characterization of diamonds and subsequently boron nitride, there was the elucidation of "hexagonal diamond," and the structures of the new forms of elemental silicon and germanium. Structures at high pressure were studied as well as those that could be retained after release of pressure, for a large variety of materials.

Concern for the structural basis of physical properties has been a continuing effort. Noteworthy examples are the nature of faulting in ZnS and its relation to luminescence, the crystallographic basis of magnetic exchange anisotropy in structures with mixed ferromagnetic-antiferromagnetic order, dependence of electrical and magnetic properties of graphite on long-range order, and the nature of several modulated structures. The concept of superionic conductivity originated in the Research Laboratory as an outgrowth of structure studies of calcia-stabilized zirconia and beta alumina.

In the realm of metallurgy there have been basic studies of recrystallization phenomena, the nature of phase transformations and of deformation. On the other hand, there have been structure determinations for transition metal alloy phases, such as sigma phase, and the development of a theory of such alloy phases based on packing principles.

The structures of many diverse materials have been determined over the years, but one class of compounds that has received continuing attention is that of elemental boron and higher borides, such as MB_{66}.

More recent activities have been the structural studies of superionic conductors, particularly members of the β- and β''-alumina family which have been investigated by X-ray and neutron diffraction to establish the structural basis for high conductivity and stability, and those conductors of lower dimensionality such as intercalated graphites (two-dimensional) and charge transfer compounds that result in stacks of flat molecules that exhibit one-dimensional behavior.

CHAPTER 13

A HISTORY OF CRYSTALLOGRAPHIC STUDIES AT SANDIA

Bruno Morosin
Sandia National Laboratories†
Albuquerque, New Mexico 87185

Sandia National Laboratories had its beginnings as a group within Los Alamos Scientific Laboratory that was situated in Albuquerque to be near the airport and railroad. As a branch of Los Alamos, the group had a part of the Los Alamos task of designing and building atomic weapons. In 1949 Sandia was split off from Los Alamos and formed into a separate laboratory responsible for the ordnance phases of weapons. Such ordnance engineering was subsequently also provided for Lawrence Livermore Laboratory by a branch of Sandia located adjacent to that facility. The growth of research functions within Sandia's engineering responsibilities involved activities which were specifically and directly related to the atomic weapon task and which were driven by the continuous introduction of highly sophisticated techniques in such ordnance engineering. Research into components used in such devices or the resulting weapons' effects upon materials provided the beginning of the solid state research at Sandia. More recently, energy related research and development activities have broadened the scope of studies at Sandia, culminating with the "national" laboratory designation in 1980.

From a crystallographic point of view, structural phase transformations and their resultant changes in physical properties, and more generally, the effects of structure on physical properties have played a central role in Sandia's research efforts. Effects under extreme conditions of high pressure, high temperature, shock wave loadings require fundamental knowledge of piezoelectric, ferroelectric, and ferro- and ferrimagnetic properties and phase transformations of materials. Sandia has played a leading role in practical application of such materials to transducers, light shutters, active and passive pressure gauges and shock-activated power supplies.

At Sandia, Neilson and co-workers (Neilson, 1957; Kulterman, Neilson and Benedick, 1958) carried out the first reported studies on shock demagnetization of a magnetic material. Shock demagnetization is a term which describes changes in ferromagnetic or ferrimagnetic states of shock-loaded samples due to either first-order or second-order phase transitions or to stress-induced magnetic anisotopy. Neilson's early studies were interpreted with the assumption that the pressure and temperature in the shock-compressed material was sufficient to induce a Curie-point transition. However, many subsequent observations on various different magnetic materials have required alternate interpretations on these different magnetic materials. A summary and extension of this area with many Sandia contributors can be found in the review by Duvall and Graham (1977) on phase transitions under shock wave loading. Neilson and his associates also developed nanosecond time-resolved stress gauges, employing piezoelectrics. Sandia's early studies were on quartz (Neilson, Benedick, Brooks, Graham, and Anderson, 1962). These led to our eventual development of the quartz gauge as the electrical waveforms produced were understood and the elastic properties were determined (Graham, Neilson, Benedick, 1965); more recent work has been carried out on lithium niobate by Graham (1972). A collection of selected reprints concerning piezoelectricity under shock loading and related piezoelectric gauges has been published by Graham and Reed (1978).

Our detailed understanding of the electrical properties of several piezoelectrics under shock-loading has been detailed to a larger extent than those for ferroelectrics. This arises from the highly nonlinear, irreversible and strongly-coupled nature of the mechanical and electrical states even though the crystal structure properties are known. On the other hand, structural properties of the perovskite class of materials are important in governing these properties. Sandia has had a continuing effort in this area beginning with the pioneering work by Neilson (1957). This experimental effort has been summarized in a recent review on shock compression of solids by Davison and Graham (1979). Chen and co-workers at Sandia have developed theories of plane-wave propagation and electromechanical coupling in ferroelectrics which include domain switching and dipole dynamics. Recently, Chen and Tucker (1981) obtained for the first time excellent agreement with experimental results in describing hysteresis and butterfly loops under the conditions of a slowly varying cyclic field loading ($>$1s) with dipole dynamics (μs time scale) at rest.

Shock induced phase transitions are important in another area, that of passive pressure gauges. Many times costs or convenience require pressure measurements which are not actively instrumented and in such circumstances passive gauges can be employed (Samara, 1967).

Complementing the shock-induced phase transition studies at Sandia are the many studies employing hydrostatic pressure. Besides employing conventional diffraction techniques, the dielectric behavior of solids together with the more recent ability to employ Raman scattering under pressure has allowed detailed structural

*This work sponsored by the U.S. Department of Energy (D.O.E.) under Contract DE-AC04-76-DP00789.
†A U. S. Department of Energy facility.

information on a large class of transitions based on the concept of soft modes, i.e., a normal mode of vibration whose frequency decreases and approaches zero as the transition is approached and thereby the crystal lattice becomes unstable. Samara and Peercy (1980) have summarized Sandia's and other contributions in this area. The dielectric behavior of other materials has led to various applications, i.e., values of temperature and pressure derivatives along very specific crystallographic directions have yielded quartz pressure gauges for geothermal well logging (E. P. EerNisse, 1978; 1979; also see Davison and Graham, 1979 for other examples).

The perovskite class of ferroelectric materials also has an important application as light valves or shutters for goggles used for eye protection. At Sandia, Land and Haertling were responsible for contributions on such electrooptic ceramics which also have important applications for image storage and display devices. This area has been reviewed by Land, Thacher and Haertling (1974). Essentially, based on the optical behavior of high density ceramics of lead lanthanum zirconate-titanate ferroelectric alloys, structural properties governed by compositional changes in rather complex phase diagram fields play an important role. The image storage process relies on optically induced changes in the switching properties of ferroelectric domains (photoferroelectric effect). Recently, enhanced photosensitivity in such materials has been produced by Peercy and Land (1980 a, b) by use of ion implantation to disorder the near-surface structure of these ceramic materials.

Another area of importance for materials related to these perovskite structures concerns Sandia's radioactive waste disposal process. The same alkoxide of Ti, Nb and Zr used as starting materials for the ferroelectric and electrooptic ceramics are employed in a process developed by Dosch which will incorporate radioisotopes in rather stable crystal structure forms (Peercy, Dosch and Morosin, 1977); some of these forms and the intermediate products have been shown to consist of unusual oxide complexes (Morosin, 1977; Graeber and Morosin, 1977; 1980a,b).

Sandia has had a strong effort in solid state physics; this has provided techniques other than conventional diffraction which can be powerful tools in understanding the perfection and symmetry of crystals or their surfaces. A few of these were developed in this laboratory and are, therefore, subsequently mentioned. Hydrostatic pressure-low temperature techniques for examination of the Fermi surface of metals were developed by Schirber (1978, 1979) and these have been applied to various electronic metal-semiconductor and more complex structural phase transitions. The requirement of understanding atomic displacement in some of the more complex metal systems led to the development of a low-temperature, high-pressure X-ray cell suitable for single-crystal studies (Morosin and Schirber, 1974). A continuing effort on material preparation and crystal growth has produced unique crystals for study, selected examples being Czochralski - prepared Na β-alumina (Lefever and Baughman, 1975), fused SiO_2 nutrient growth of α-quartz (Baughman, 1980), anomalous vapor phase growth of gold (Schwoebel,1966) and implantation-prepared equilibrium alloys (Myers, 1981). Dielectric measurements have been carried out as a function of pressure and temperature and these have provided understanding of the role which hydrogen bonds play in various materials (Samara, 1978). Many important physical properties of materials are controlled by the crystallographic location of impurities in solids. In addition, the interpretation of solid state experiments by theoretical calculations often requires a knowledge of the lattice location of impurities. Ion channeling has been used to directly determine the location of impurities within a crystal lattice and Picraux (1975) and co-workers were the first to do detailed analysis of such low solubility implanted ions. Of particular interest to crystallographers is the use of such methods as ion back scattering as well as ion-induced nuclear reaction analysis to determine hydrogen and other light atom positions in crystals (Picraux, 1975). Resonance techniques have been employed to elucidate the structure of defects in silicon (Brower, 1977) and in hydrides (Venturini, 1981). Emin (1979, 1980) has developed a theoretical description of defects in solids, initially for amorphous materials, based on the small-polaron concept. The structural aspects on a microscopic basis are of importance. This work has been extended to hydrogen in metals, an area of importance to Sandia. In other work, Switendick (1979, 1980) has been able to predict the structure of many hydrides using band calculations. Recently, Knotek and Feibelman (1979) have developed a core-hole Auger decay model which interprets electron- photon- and ion-stimulated desorption data, making this a useful atomic-specific, valence-sensitive surface probe for ionic solids. The departure of the model due to covalency effects may be understood in the future by considering structure-related information on bond lengths. Also recently Panitz has produced field ion images of large molecules by employing carefully prepared field-emitter tips. Such studies may eventually offer structural information on biological systems which conventional diffraction studies cannot provide (Panitz, 1978; Panitz and Giaever, 1980).

Complementing Sandia's efforts in solid state physics have been a large variety of diffraction studies. A few key topics allow a measure of the depth of subjects covered or of unusual experimental techniques employed. Occasionally NMR or other techniques proved the published crystal structure to be incorrect and this served as the motivation for the diffraction study (Morosin and Narath, 1964; Smith and Morosin, 1967). Resonance techniques have been coupled with diffraction studies to understand strucure-dependent magnetic interactions (Hughes, Soos, Morosin, 1975;

Hughes, Morosin, Richards, Duffy, 1975; Wada, Kashima, Haseda, Morosin, 1981; Azevedo, Schirber, Switendick, Baughman, Morosin, 1981). Hydrogen bonded crystalline materials, particularly those useful as ionic conductors, have been examined (Morosin, 1966b; 1967; 1975; 1975 or 1976; 1978a and 1978b). Organic hydrogen getters and the resulting products have been determined (Morosin and Harrah, 1977). Many materials which at low temperature exhibit one-dimensional properties were studied (Morosin, 1966a; 1967; 1969; 1975; 1976; Peercy, Samara and Morosin, 1973; Bartkowski and Morosin, 1972; Hughes, Soos and Morosin; 1975; Lowdnes, Finegold, Rogers and Morosin, 1969). Exchange striction in simple magnetic systems has been carefully determined by diffraction from the small changes in size and shape of the unit cell (Jones and Morosin, 1967; Morosin, 1970a; Bartel and Morosin, 1971). The unusual thermal expansion behavior of various materials has led to the development of a new high temperature goniometer furnace (Lynch and Morosin, 1971, 1972; Morosin and Lynch, 1972; Takeda and Morosin, 1975). The pressure dependence of various properties has required study of the compressibility of these materials (Morosin and Schirber, 1979; Schirber and Morosin, 1979; Jones, Shanks, Finnemore and Morosin, 1972), or accompanying pressure induced phase transitions (Johnson and Morosin, 1976; Morosin, 1970b; Morosin and Peercy, 1975). The stability of PdH alloys has required high-pressure synthesis and low-temperature handling and measurements (Schirber and Morosin, 1975). The influence of crystal structure on the response of photo-electrodes employed in the electrolysis of water has been determined (Morosin, Baughman, Ginley and Butler, 1978). Sandia has had a continuing interest in explosives which are employed in various components. In particular, hexanitroazobenzene has five crystalline polymorphs (Graeber and Morosin, 1974). Substituted tetrazolatopentaammine-cobalt perchlorates have been employed for their relative explosive stability and have been the subject of various structure studies (Graeber and Morosin, 1978, 1980; Ortega, Campana and Morosin, 1980). Very recently, studies on shock-induced chemistry have been pursued; this area has been examined in the past by Soviet scientists, with various points needing clarification and further amplification (Graham, 1980; Morosin, Graham, Richards and Stohl, 1980; Graham and Dodson, 1980). X-ray diffraction studies will be an important tool in this research area and will continue to impact on a broad spectrum of Sandia programs.

References

Azevedo, L. J., Schirber, J. E., Switendick, A. C., Baughman, R. J. and Morosin, B. (1981). J. Phys. F: Metal Phys. **11**, 1521-1530.

Bartel, L. C. and Morosin, B. (1971). Phys. Rev. **B3**, 1039-1043.

Bartkowski, R. R. and Morosin, B. (1972). Phys. Rev. **B6**, 4209-4212.

Baughman, R. J. (1980) Unpublished results.

Brower, K. L. (1977). Rev. Sci. Instrum. **48**, 135-143; references therein.

Chen, P. J. (1980). Int. J. Solids and Structures **16**, 1059-1067; also see references given in Davison and Graham, 1979.

Chen, P. J. and Tucker, T. J. (1981). Int. J. Eng. and Sci. **19**, 147-158.

Davison, L. and Graham, R. A. (1979). Physics Reports **55**, 255-379; references therein to Halpin, W. J., Lysne, P. C., and Steutzer, O. M.

Duval, G. E. and Graham, R. A. (1977). Rev. Mod. Phys. **49**, 523-579. References therein to Edwards, L. R. and Lysne, P. C.

EerNisse, E. P. (1978). Sandia Laboratory Report SAND 78-2264.

EerNisse, E. P. (1979). Proc. 33rd Annual Freq. Control Sym., Atlantic City.

Emin, D. (1979). Hall Effect and Its Applications, pp. 281-298. Edited by C. L. Chien and C. R. Westgate. New York: Plenum Press.

Emin, D. (1980). Polycrystalline and Amorphous Thin Films and Devices, pp. 17-57. Edited by L. L. Kazmerski. New York: Academic Press.

Feibelman, P. J. and Knotek, M. L. (1978). Phys. Rev. **B18**, 6531-6535. See also Knotek and Feibelman (1979).

Graeber, E. J. and Morosin, B. (1974). Acta Cryst. **B30**, 310-317.

Graeber, E. J. and Morosin, B. (1977). Acta Cryst. **B33**, 2137-2143.

Graeber, E. J. and Morosin, B. (1978). Amer. Crystallogr. Assoc. Program Abstr. Ser. **2**, Oklahoma Meeting, Paper PA8.

Graeber, E. J. and Morosin, B. (1980a). Amer. Crystallogr. Assoc. Program Abstr. Ser. 2, Eufaula Meeting, Paper PB4.

Graeber, E. J. and Morosin, B. (1980b). Amer. Crystallogr. Assoc. Program Abstr. Ser. 2, Calgary Meeting, Paper PA10. PA10.

Graham, R. A., Neilson, F. W., and Benedick, W. B. (1965). J. Appl. Phys. **36**, 1775-1783.

Graham, R. A. (1972). Phys. Rev. **B6**, 4779-4792.

Graham, R. A. and Reed, R. P. (1978). Sandia Laboratories Report SAND 78-1911.

Graham, R. A. (1980). Bull. Am. Phys. Soc. **25**, 495; also see contributions by Dodson, B. W., Graham, R. A., Keck, J. D., Morosin, B., and Venturini, E. L., at the March 1981 APS meeting and the June 1981 APS Topical Conference on Shock Waves in Condensed Matter.

Graham, R. A. and Dodson, B. W. (1980). Bibliography on Shock Induced Chemistry, Sandia Laboratories Report SAND80-1642.

Hughes, R. C., Soos, Z. G., and Morosin, B. (1975). Acta Cryst. **B31**, 762-770.

Hughes, R. C., Morosin, B., Richards, P. M., and Duffy, Jr., W. (1975). Phys. Rev. **11**, 1795-1803.

Johnson, R. T. and Morosin, B. (1976). High Temperatures-High Pressures. **8**, 31-34.

Jones, E. D. and Morosin, B. (1967). Phys. Rev. **160**, 451-454.

Jones, R. E., Shanks, H. R., Finnemore, D. K., and Morosin, B. (1972). Phys. Rev. **B6**, 835.

Knotek, M. L. and Feibelman, P. J. (1979). Surface Science **90**, 78-89.

Kulterman, R. W., Neilson, F. W., and Benedick, W. B. (1958). J. Appl. Phys. **29**, 500-501.

Land, C. E., Thacher, P. D., and Haertling, G. H. (1974). Applied Solid State Science **4**, 137-233.

Lefever, R. A. and Baughman, R. J. (1975). Mater. Res. Bull. **10**, 607-611.

Lowndes, D. H., Finegold, C., Rogers, R. N., Morosin, B. (1969). Phys. Rev. **186**, 515-521.

Lynch, R. W. and Morosin, B. (1971). J. Appl. Cryst. **4**, 352-356.

Lynch, R. W. and Morosin, B. (1972). J. Am. Cer. Soc. **55**, 409-413.

Morosin, B. and Narath, A. (1964). J. Chem. Phys. **40**, 1958-1967.

Morosin, B. and Schirber, J. E. (1965). J. Chem. Phys. **42**, 1389-1390.

Morosin, B. (1966a). J. Chem. Phys. **44**, 252-257.

Morosin, B. (1966b). Acta Cryst. **21**, 280-283.

Morosin, B. (1967). Acta Cryst. **23**, 630-634.
Morosin, B. (1969). Acta Cryst. **B25**, 19-30.
Morosin, B. (1970a). Phys. Rev. **B1**, 236-243.
Morosin, B. (1970b). Acta Cryst. **B26**, 1635-1637.
Morosin, B. and Lynch, R. W. (1972). Acta Cryst. **B38**, 1040-1046.
Morosin, B., Berman, J. G., and Crane, G. R. (1972). Acta Cryst. **B29**, 1067-1072.
Morosin, B. and Schirber, J. E. (1974). J. Appl. Cryst. **7**, 295-296.
Morosin, B. and Peercy, (1975). Phys. Rev. Lett. **53A**, 147-148.
Morosin, B. (1975). Acta Cryst. **B31**, 632-634.
Morosin, B. (1976). Acta Cryst. **B32**, 1237-1240.
Morosin, B. (1977). Acta Cryst. **B33**, 303-305.
Morosin, B. and Harrah, L. (1977), Acta Cryst. **B33**, 1760-1765.
Morosin, B., Baughman, R. J., Ginley, D. S., Butler, M. A. (1978). J. Appl. Cryst. **11**, 121-124.
Morosin, B. (1978a). Acta Cryst. **B34**, 3730-3732.
Morosin, B. (1978b). Acta Cryst. **B34**, 3732-3734.
Morosin, B. and Schirber, J. E. (1979). Phys. Rev. Lett. **73A**, 50-52, References therein.
Morosin, B., Graham, R. A., Richards, P. M. and Stohl, F. V. (1980). Amer. Crystalogr. Assoc. Program Abstr. Ser. 2, Calgary Meeting, Paper PA7.
Myers, S. M. (1981). Treatise on Material Science and Technology. Vol. 18. Ion Implantation. Ed., Hirvonen, J. K., Acad. Press. pp. 51-83; references therein.
Ortega, R., Campana, C. F., and Morosin, B. (1980). Amer. Crystallogr. Assoc. Program Abstr. Ser. 2, Calgary Meeting, Paper PA24.
Neilson, F. W. (1957). Bull. Am. Phys. Soc. **2**, 302.
Neilson, F. W., Benedick, W. B., Brooks, W. P., Graham, R. A., and Anderson, G. W. (1962). In "Les Ondes de Detonation" (Editions du Centre National de la Recherche Scientifique, Paris), pp. 391-419.
Panitz, J. A. (1978). Prog. Surface Sci. **8**, 219-228.
Panitz, J. A. and Giaever, I. (1980). Surface Science **97**, 25-33.
Peercy, P. S., Samara, G. A., and Morosin, B. (1973). Phys. Rev. **B8**, 3378-3388 and references therein.
Peercy, P. S., Dosch, R. G., and Morosin, B. (1977). Sandia Laboratory Report SAND 76-0556, and references therein.
Peercy, P. S. and Land, C. E. (1980a). Proc. Ion Beam Modification of Materials. Conf. Albany, NY.
Peercy, P. S. and Land, C. E. (1980b). IEEE - SID Biennial Display Research Conference Record.
Picraux, S. T., (1975). New Uses of Ion Accelerators. pp. 229-281. Edited by J. F. Zigler. Plenum Pub. Co., also see work of Vook, F. L. and Borders, J. A., listed therein.
Samara, G. A. (1967). Passive Pressure Gauges, Sc-TM-67-729 (A Sandia Laboratory Report).
Samara, G. A. (1978). Ferroelectrics **20**, 87-96; references therein.
Samara, G. A. and Peercy, P. S. (1980). In "Solid State Physics," Vol. 36, Edited by F. Seitz and D. Turnbull. New York: Academic Press. (in press); also references therein to Abel, W. R., Fritz, I. J., Gillis, N. S., Morosin, B., and Nettleton, R. E.
Schirber, J. E. and Morosin, B. (1975). Phys. Rev. **B12**, 117-118.
Schirber, J. E. (1978). High Press. and Low Temp. Physics, pp. 55-65. Edited by C. W. Chu and J. A. Wollam; also see references therein.
Schirber, J. E. (1979). High Pressure Science and Technology, **1**, 130-140; also see references therein.
Schirber, J. E. and Morosin, B. (1979). Phys. Rev. Lett. **42**, 1485-1487; references therein.
Schwoebel, R. L. (1966). J. Appl. Phys. **37**, 2515-2516.
Smith, D. L. and Morosin, B. (1967). Acta Cryst. **22**, 906-910.
Switendick, A. C. (1979). Z. Phys. Chem. N.F. **117**, 447-453.
Switendick, A. C. (1980). J. Less Common Met. **74**, 199-206.
Takeda, H. and Morosin, B. (1975). Acta Cryst. **B31**, 2444-2452.
Venturini, E. L. (1981). In "Nuclear and Electron Resonance Spectroscopies Applied to Material Science." Edited by E. N. Kaufman and G. K. Shenoy. New York: Elsevier North-Holland. pp. 321-326; references therein.
Wada, N., Kashima, Y., Haseda, T., Morosin, B. (1981). J. Phys. Soc. Japan, **50**, 3876-3881.

CHAPTER 14

HISTORY OF A COMMERCIAL CRYSTALLOGRAPHIC SERVICE COMPANY

Bertram A. Frenz
Molecular Structure Corporation, College Station, TX 77840

The idea of a commercial crystal structure determination service is not original; many crystallographers have toyed with the thought. But actually converting the thought into reality is original to Molecular Structure Corporation.

In the spring of 1973, C.M. Lukehart, a synthetic inorganic chemist, was developing research proposals that he planned to put into operation as soon as he obtained an academic position. Although he could lay out his plans for the synthetic work and the analytical analyses (ir, nmr, mass spec., etc.) his research area very much depended on definitive structural information such as that obtained by X-ray crystallography. Inquiries at analytical labs revealed no commercial service was then available. He discussed this with one of his colleagues, J.M. Troup, who was collaborating with him on a crystal structure. In the course of Troup's research, he had developed methods to reduce the time, and thereby the cost, in determining crystal structures. Lukehart and Troup immediately recognized the potential of forming a commercial service and they related their ideas to a third postdoc, B.A. Frenz. Based on some investigations that he had been making, Frenz suggested that the cost of structural studies could be substantially reduced in all calculations were performed on an in-house mini-computer.

Other ideas were incorporated into the plan after consultation with their research advisor, F.A. Cotton, and with other chemists as well as lawyers, accountants, and bankers. Convincing the banker of the feasibility of the proposed company proved to be the most difficult step. Bankers are not accustomed to loaning money to buy diffractometers and do not value them for collateral as highly as crystallographers do. In August of 1973 Molecular Structure Corporation was incorporated, but the decision to actually pursue the idea was not made until late in the year when its validity was more established and the banker was convinced to finance the undertaking.

The doors were opened for business on January 1, 1974, in a small 220-square foot office located across the street from Texas A & M University in College Station, Texas. Business was slow but steady during the first year, slower and less steady the second year, but stabilizing and expanding in the third and subsequent years. The company moved into a larger office after two years and in July, 1979, it expanded once more, this time into its own custom built 2750 sq. ft. office/laboratory building.

The original concept of offering complete crystal structure determinations primarily for academic chemists was expanded to include other services, such as data collection for crystallographers without modern diffractometers, special techniques such as high or low temperature data collection, handling air sensitive, heat sensitive, or explosive crystals, electron density studies, powder diffraction experiments, and construction of custom molecular models. Clientele shifted from primarily academic to primarily industrial. While academic chemists could occasionally obtain funding for one crystal structure determination, industrial chemists could budget a half dozen or more structures per year. For them, quality and turn-around were more important criteria than price.

One of the surprises to the original founders of MSC was the very important role that crystallography can and does serve in industrial research. Single crystal diffraction studies have elucidated the chemistry of homogeneous catalysts, one-dimensional conductors, cancer-related drugs, natural products, explosives and detonators, and numerous other chemicals that are only one step removed from practical application. To a large extent, Molecular Structure Corporation has been responsible for the expanded interest in crystallography in chemical companies, both through its crystallographic service and through its development of computer software now in use in many laboratories.

Looking back on the success of Molecular Structure Corporation, it appears that the company entered the field at a transition point in crystallography and to some extent it was responsible for enhancing that transition. Small molecule crystallography was in the process of moving from a Ph.D. level project to a routine chemical analysis. Highly automated diffractometers took the time and drudgery out of data collection; direct methods solution techniques made structures easier to solve routinely, and the mini-computer, coupled with commercial software packages, lowered the cost of computing and also accelerated the time in moving from data reduction to final table production.

The presence of a crystallographic service company by no means signaled the end of independent crystallographic work in academic and industrial laboratories. Instead it established even more firmly that crystallography is an essential tool to the scientist and the use of that tool is now more widespread than ever before.

CHAPTER 15

THE DENVER X-RAY CONFERENCES

Charles S. Barrett
Denver Research Institute, University of Denver, Denver, Colorado 80208

Every August for the past 29 years a large group of X-ray-oriented scientists and engineers has gathered in Denver, Colorado, to discuss their common technical interests. This annual conference has come to be widely known as the Denver Conference on Applications of X-Ray Analysis.

The first organized meeting was arranged in 1952 by Professor James Blackledge and Mr. Merlyn Salmon, then of the Denver Research Institute, and Mr. Stephen Knight of Technical Equipment Corporation. Dr. I. Fankuchen of Brooklyn Polytechnic Institute, well known to the world of X-ray crystallography, was the featured lecturer. At this first meeting there were three local speakers and a total of 78 people in attendance. That year and for the next two years the X-ray conferences were held on the University of Denver campus with Fankuchen and about six others as speakers. Attendance dwindled from the original 78 to 58 in 1953. In 1954 the meeting was moved to a downtown Denver hotel. Yet only nine papers were offered and 50 people attended. Undaunted by these figures and convinced that X-ray analysis had a growing importance in national technology, a concerted effort was mounted to put the meeting on a national scale. The efforts showed positive results the next year (1955) when 28 papers were heard by 78 people. Thereafter the value of the Denver X-ray conference was given increased national recognition and it grew steadily to the present five-day meeting with an attendance that has reached 365 practicing X-ray specialists plus numerous exhibitors. Speakers and attendees come from all parts of the globe and include chemists, physicists, metallurgists, mineralogists, microscopists, engineers, instrument designers, sales and maintenance personnel, and technicians. Also, hard cover volumes of the conference proceedings, entitled *Advances in X-Ray Analysis*, are to be found in technical libraries throughout the world. A local committee plans each annual conference after canvassing numerous people who had previously attended. W. M. Mueller headed the local committee while he was in residence at the University of Denver, and the late Mrs. Marie Fay handled much of the routine paper work. Later, J. B. Newkirk headed the local group for several years, assisted in 1963-1969 by G. R. Mallett as co-chairman and Mrs. Jeanne Cochran as secretary, to be followed by C. O. Ruud, who undertook the responsibility until he resigned 10 years later. He was aided by others at the Denver Research Institute, including C. S. Barrett, J. B. Newkirk, P. K. Predecki, and since his joining the Institute in 1976, by D. E. Leyden, who took responsibility for the portions of the program dealing with analysis by X-ray fluorescence. Mrs. Mildred Cain has served as conference secretary since 1975. Numerous other Institute employees and students also spent effort on the conference. In 1980 P. K. Predecki began as chairman of the conference, again with D. E. Leyden as cochairman on the X-ray fluorescence side, and with C. S. Barrett and J. B. Newkirk as honorary chairmen.

The Denver Conference has traditionally been self-supporting, though it has been officially sponsored and guaranteed by the University of Denver Research Institute, S. A. Johnson, Jr., Director. Efforts to keep down the cost to participants sometimes ran the conference into debt, resulting in the need for the University to bail it out. Some travel assistance for a few invited speakers from abroad was obtained in 1959 from ONR, in 1975 from NSF and in 1976 from JCPDS. During the last decade important support has come from X-ray related industries who recognized this as a good opportunity to make effective contact with potential customers. More recently the conference has been notably aided by the JCPDS-International Centre for Diffraction Data. Since 1978 this organization has been a co-sponsor of the conference on alternate years when the topic of emphasis is diffraction. Members of this organization together with many other experts in the X-ray field have also provided invaluable help as workshop organizers and instructors.

The technical policy by which each Denver X-ray conference is structured is set by the local committee, all of whom are employed by the University of Denver. As the conference title implies, any high quality paper whose subject falls within the broad field of Applications of X-Ray Analysis is appropriate for presentation. Papers involving other radiation, such as electron diffraction and gamma rays, are also within the scope of the conference. To bring focus to the meetings, about a year in advance of the meeting date the local committee identifies a technical subject which appears to be ready for organized emphasis at the conference. Through direct contacts with persons who are active in that subject the committee selects an acknowledged leader in the subject and invites him to serve with them as Invited Co-chairman of the Denver Conference. Together, he and the local committee assemble a panel of known leaders in the selected field. The panelists then are invited to present keynote papers at the opening session of the conference. The subjects they cover are carefully chosen and interrelated so as to present as complete and balanced a picture as possible of the subject picked for emphasis

Early in the calendar year, wide publicity is given to

the next August meeting, including the subject to be emphasized, the names of the invited speakers and the titles of their talks. At this same time a general call for contributed papers is issued. This is distributed to a mailing list of several thousand and can also be obtained by writing to the Metallurgy and Materials Science Division, Space Sciences Building, Denver Research Institute, Denver, Colorado 80208 (Tel. (303)-753-2141). Later the full schedule of the meeting is circulated. The effect of this procedure is to stimulate the contribution of papers which deal with the emphasized subject and to promote increased attendance of technical people who share the common specialized interest.

As a result of this technical policy and procedures, the proceedings volumes stand as state-of-the-art summaries of certain techniques, applications, etc., involving X-rays or related radiation. Since 1966 the emphasized subjects and the respective invited co-chairmen have been:

Volume in Adv. in X-Ray Analysis	Year	Emphasis	Invited Co-chairman
10	1966	X-ray Diffraction Topography	None
11	1967	Chemical Analysis by X-ray Fluorescence	H. G. Pfeiffer
12	1968	X-ray Diffraction – General	C. S. Barrett
13	1969	Low Energy X-ray Phenomena	B.L. Henke
14	1970	No subject emphasis	None
15	1971	X-ray Detection Methods	K. F. J. Heinrich
16	1972	X-rays in Environmental and Biomedical Applications	L. S. Birks
17	1973	Standards and Sampling Methods	C. L. Grant
18	1974	The Application of X-ray Technology to Current Problems in Energy and Resource Development	W. L. Pickles
19	1975	Mathematical Correction Procedures for X-ray Spectro-chemical Analysis	R. W. Gould
20	1976	X-ray Diffraction for Chemical Analysis	H. F. McMurdie
21	1977	New X-ray Techniques in Chemical Analysis	D. E. Leyden
22	1978	Special Techniques in Powder Diffraction	G. J. McCarthy
23	1979	Field Applications of X-ray Fluorescence Analysis	J. R. Rhodes
24	1980	Practical Applications of Automated Analysis of Diffraction Data	D. K. Smith

Contributed papers that do not fall within the announced subject for emphasis also are welcomed. In this way the Denver conference also serves as a forum at which new subjects may be discussed and where young or relatively unknown contributors can be heard. The only criteria for participation are an active interest in the subject matter and, if work is to be presented, that it be technically sound, original work of significance, and within the broad subject scope of the conference.

The subjects covered by the individual papers, which recently have averaged nearly 70 per year, include practically every known area of the science and application of fluorescent analysis, diffraction, absorption, scattering, detection, and data reduction. For example in the 1970's there were papers on the analysis of coal, uranium ores, natural waters, airborne particulates, cements, stainless and low alloy steels, steelmaking slags, and many mineral phases. There were also papers on stress analysis in pipes, in thin films, and in polymers by diffraction; automation in powder diffraction data collection and analysis, position sensitive detector design and use, applications of microbeam and Gandolfi techniques, proton induced emission analysis of minerals and biomedical samples, new designs of equipment for fluorescent analysis including energy dispersive types, advances in fluorescent data interpretation and of course new X-ray tubes, detectors, and computation schemes.

Almost all papers that are orally presented at the Denver conference are also submitted in manuscript. Soon after the conference is held in August, the local committee and the invited general co-chairman study the manuscripts. Most of the papers are ultimately accepted for inclusion in *Advances in X-Ray Analysis*. By spring of the following year the volume is available for distribution, free to the conference attendees and at moderate cost to others.

The detailed arrangements for the meeting are made entirely by the local committee. Booths are made available near the meeting rooms where the latest commercial X-ray equipment is exhibited throughout the conference period. Previously Denver area hotels were used as the meeting place, but beginning in 1976 this was moved to the Denver University campus.

For a few years in the 1970's an elementary short course was given by the local committee, just before the conference. Since 1978, however, the need for basic instruction was met by Workshops scheduled on the two days preceding the conference, taught by numerous experts from industry, universities and the government. Several evenings during the week are devoted to social mixers, a special dinner, and occasionally a conference tennis tournament.

In its 29 years of existence, the annual Denver conference on Applications of X-ray Analysis has apparently filled a need. In the foreseeable future, its continuing function will be to bridge the gap between pure X-ray science and the applied technologies which have their bases in X-ray and related phenomena.

CHAPTER 16

CRYSTALLOGRAPHY AT PENN STATE

R. E. Newnham
Materials Research Laboratory
Pennsylvania State University
University Park, Pa 16802

Penn State is in the middle of Pennsylvania, halfway between Philadelphia and Pittsburgh. Some people say it is in the middle of nowhere, and thirty years ago when I began graduate work here it was even more isolated.

My first memories of Penn State were like the opening scenes of a Frankenstein movie. My parents and I had driven most of the night through the forests of Northern Pennsylvania and at dawn we reached the fog-shrouded Nittany Valley. A large castle-like building surrounded by a high wall rose above the mist in front of us. Friends had told me Penn State was isolated, but I never expected anything like this: barred windows, guard turrets, and locked gates. And then we saw the sign – Rockview State Penitentiary. We had mistaken the state pen for Penn State! A few miles down the road we came to the town of State College and the beautiful Penn State campus.

A few days later, I was settled in a rooming house on South Allen Street, eating my meals at the Penn State Diner, and meeting my first crystallographers.

Wheeler P. Davey

One that made a big impression on me was Wheeler P. Davey, who worked in the office next to mine in the basement of Osmond Laboratory. Professor Davey was nearing the end of his career, having officially retired several years earlier, but he still came to work every day. He and Arthur Beward were busy collecting Debye-Scherrer patterns and recording powder data for the ASTM file. Dr. Davey spent a lot of time filling in *d*-spacings on 3 x 5 cards in his neat legible handwriting. Badly afflicted with arthritis, his hand would sometimes curl tight, lock in position and begin to shake, but he would bring it under control and keep working.

Professor Davey always had a smile for new students and stories about the early days of X-ray diffraction when he worked at General Electric with Hull, Whitney, and Coolidge. It was then he developed the Hull-Davey Charts for indexing powder patterns, did some pioneering radiographic studies of steel castings, and published a paper with the intriguing title "Effect of X-rays on the Length of Life of Triboleum Confusum." Unlike most life forms, Confusum seem to live longer when irradiated with X-rays; perhaps this explains the longevity of X-ray crystallographers.

Dr. Davey had joined the Penn State faculty in 1926, and for the next 33 years was Research Professor of Physics and Chemistry. Precision measurements of lattice constants were carried out in Pond Laboratory with Dan McLachlan, Charles Siller and Sid Smith among his first graduate students. In 1933 they moved to a large cinderblock building with a big room called the "Bull Pen", filled with seven X-ray machines, each with a Coolidge tube and a "Ashcan" powder camera. During this period, metallurgical investigations were carried out on preferred orientation, cold rolling, and thermal expansion. *"A Study of Crystal Structure and its Applications"* (McGraw-Hill), based on Professor Davey's lectures on crystal structure analysis, was published in 1934.

When the heavily insured "Bull Pen" burned down, Professor Davey helped design Osmond Laboratory, one of the most modern physics buildings of its time. The Physical Structure Laboratory (Room 6 – later to become Ray Pepinsky's Lab) contained twelve X-ray stations, each with a 42-kV transformer and kenotron housed in a wire cage. Specialty equipment included X-ray spectrometers for studying the structure of liquids and mechanically worked metals, and a McLachlan pole-figure machine.

By 1939 Davey had made Penn State the head-quarters of the X-Ray Powder Data File, having established close connections with Hanawalt, Rinn, Frevel, Fink and Fankuchen. In 1952 he received the Award of Merit of the American Society for Testing Materials in recognition of "extremely significant leader-ship in the work on chemical analysis by X-ray diffrac-tion methods and the preparation and publication of X-ray patterns in a card index universally used." Dr. Davey died in 1959 and a few years later his memory was honored with the naming of the Davey Laboratory which houses classrooms, offices and research facilities of the Departments of Physics and Chemistry.

Arthur P. Honess

One of Davey's collaborators in the early days was mineralogist Arthur P. Honess, who joined the faculty in 1924 after completing his doctorate at Princeton. An enthusiastic teacher, Dr. Honess inspired an extra-ordinary amount of love and loyalty among students in the School of Mineral Industries. "Doc's" office was a gathering place for old grads whenever they returned to campus.

Scientifically, he was a world authority on the etching of crystals. His book, *"The Nature, Origin, and Interpretation of the Etch Figures on Crystals"* (Wiley, 1927) was regarded as the outstanding reference on the subject. One of his special interests was the theory of etching and its significance in the symmetry classification of minerals. Professor Honess pioneered the use of optically active reagents in the study of etch figures, and his books and research papers are filled with photographs of remarkably beautiful etch patterns.

Ray Pepinsky and X-RAC

The golden days of crystallography at Penn State were the fifties when Ray Pepinsky succeeded Dr. Davey as Research Professor of Physics. For fourteen years he presided over a staff of twenty-thirty scientists in the basement of Osmond Laboratory, and continually provided the fresh ideas and hard work it took to keep it active and well financed.

When I arrived in 1952, I remember being very impressed with Ray's work habits. He came to the lab about ten in the morning to handle his correspondence and teaching duties. His graduate lecture course on X-ray structure analysis began at eleven and finished at noon. Later he went home to lunch and spent some time with his family. We seldom saw him in the afternoon but about ten in the evening Ray reappeared at the lab and worked late into the night, writing reports and proposals, and growing crystals. In those days he was searching for new ferroelectrics – mostly sulfates and tartrates – and grew many of the crystals himself. Some of the new ferroelectrics he and Bernd Matthias discovered were based on ideas obtained from Groth's *"Chemische Kristallographie"*. They knew how important it was to read the early literature and make good use of it. In later years Professor Pepinsky founded a crystallographic information center and named it the Groth Institute.

My first assignment in Pepinsky's lab was to grow crystals in the $NaNbO_3$ – $KNbO_3$ system for ferroelectric studies. Two young post-docs – Gen Shirane from Japan and Franco Jona from Switzerland – were my immediate supervisors. Jona and Shirane were later to collaborate in writing *"Ferroelectric Crystals"* (Pergamon, 1962), which is still one of the best books on the subject. They helped me build a furnace for flux growth, and taught me a lot about powder diffraction and the polarizing micróscope. A seminar series they organized on crystal physics awakened my interest in the tensor properties of crystals and their relation to crystal structure.

Three other participants in our crystal physics seminar were Roland Good (now chairman of the Penn State Physics Department), Jack Tessman, and Erwin Felix Bertaut, a young French crystallographer who taught us a lot about the relationship between ferroelectricity and ferromagnetism. It was at Penn State that Bertaut developed some of the ideas leading to the discovery of ferrimagnetism in the garnet family.

In the summer of 1953 I went to Brookhaven with Ray Pepinsky and Betty Jane Rock to carry out neutron diffraction experiments on Rochelle salt. I stayed in an old army barracks near the reactor building. We worked with Chalmers Frazer–one of Ray's former students – who taught me a lot about neutron single crystal methods and counting statistics. Chalmers, Gen Shirane, and the other Penn Staters at Brookhaven used neutrons to determine the polarization mechanisms in potassium dihydrogen phosphate, lead titanate and a number of other ferroelectrics. In later years they extended the studies to neutron inelastic scattering and soft modes.

But crystal physics and ferroelectricity were only a small part of what went on in Ray Pepinsky's lab. Many people were working on structure analysis, on techniques for solving the phase problem, and on crystallographic computing methods. There was a room full of Weissenberg and precession cameras, darkroom facilities and a large analog computer called X-RAC.

Dr. Pepinsky had been involved in radar oscilloscope studies at the M.I.T. Radiation Laboratory during World War II, and after the war, worked on oscilloscope systems in Alabama. He set about designing an analog computational system which presented electron density or Patterson maps on an oscilloscope screen. X-RAC (X-ray analog computer) was built by students at Auburn University and transported to Penn State in 1949. Using a room full of sine wave generators, X-RAC carried out the summation

$$\rho(x,y) = \sum_{h=-20}^{+20} \sum_{k=0}^{+20} A_{hk0} \cos 2\pi (hx+ky) + \sum_{h=-20}^{+20} \sum_{k=0}^{+20} B_{hk0} \sin 2\pi (hx+ky)$$

and the resulting contour maps were plotted and photographed from the screen. Amplitudes and phase factors were set on dials in front of each of the oscillators. The computer occupied a large room in the basement of Osmond Laboratory under the eagle eye of Paul Jarmotz, an electronics engineer who kept X-RAC in running operation and later designed and built much of the circuitry for S-FAC, an electronic analog computer for calculating structure factors.

Crystallographers came from all over the world to carry out their calculations at Penn State. Shortly after X-RAC was set up, a conference on "Computing Methods and the Phase Problem in X-ray Crystal Analysis" was held at Penn State in April, 1950. The list of participants reads like a crystallographic all-star team and included such famous names as:

"Fourier Transforms"
P.P. Ewald (Brooklyn Polytechnic)

"Patterson Syntheses"
C.A. Beevers (Edinburgh)

"Symmetry Maps Derived from $|F|^2$ – Series"
A.L. Patterson (Philadelphia)

"Application of Image Theory"
M.J. Buerger (M.I.T.)

"Relation between Harker-Kasper Inequalities and Buerger Equalities"
Caroline H. MacGillavry (Amsterdam)

"Phase-Determining Relationships between Structure Factors"
J.A. Goedkoop (Penn State)

"Mathematics of the Phase Problem"
Bernard Friedman (N.Y.U.)

"Isomorphous Substitution Method"
J.M. Bijvoet (Utrecht)

"Structure Determination from X-ray Scattering"
Max Born (Edinburgh)

"Phase Determination by Dynamical Theory and by Thermal Scattering"
W.H. Zachariasen (Chicago)

"Diffuse Scattering Regions in Reciprocal Space and Phase Relationships"
M.S. Ahmed (London)

"Physical Methods for Phase Determination"
William Lipscomb (Minnesota)

"Mechanical Methods for Investigating the Phase Problem"
J. Monteath Robertson (Glasgow)

"Optical Analogue Methods"
C.W. Bunn (I.C.I.)

"Improvements in the Bragg X-ray Microscope"
C.A. Taylor and Henry Lipson (Manchester)

"Punched Card Methods in Crystal Analysis"
E.G. Cox (Leeds)

"Calculations by High-Speed Digital Computers"
Fred Ordway (N.B.S.)

"X-RAC: Electronic Analogue Computer for X-ray Crystal Structure Analysis"
Ray Pepinsky (Penn State)

"X-ray Analysis of Proteins"
Max Perutz (Cambridge)

"Computing Techniques"
Rudolph Brill (Brooklyn Poly.)

This would make a great program for an American Crystallographic Association meeting.

Vladimir Vand

In addition to many distinguished visitors, there were some excellent "staff" crystallographers in Pepinsky's group. Some I remember were Yoshiharu Okaya, Frank Eiland, K. Vedam, Klaas Eriks, Arrigo Addamiano, Tommy Doyne, June Turley, and especially Dr. Vladimir Vand.

Professor Vand was a remarkable physicist with an unusually wide range of interests and achievements. He never had difficulty in changing from one field of research to another and in attaining results of importance and originality in an astonishing short period of time. He made important contributions to astronomy, metal physics, rheology, and polymer science as well as to crystallography. His paper on the diffraction transforms of helices preceded the work of Watson and Crick and was an important step in solving the structure of DNA. He also worked on methods for solving the phase problem, and on information retrieval systems for the rapid comparison of crystallographic data. Other interests included polytypism, the structure of water, and molecular orbital calculations.

I remember Dr. Vand best for his kind and gentle ways with students. Vladimir taught us about the intricacies of group theory and its applications to physics and crystallography. He had an excellent sense of humor and lectured with a charming Eastern European accent: all the grad students would smile and nudge one another when his pronunciation of "matrices" came out sounding like "mattresses." He would wink back, and we all laughed at the joke. It was a sad day for Penn State when he died in 1968.

George Brindley and MRL

In 1954 I left the Physics Department and went to work with George W. Brindley in the College of Earth and Mineral Sciences. Dr. Brindley had just been appointed Professor of Mineral Sciences after a distinguished career at Leeds University in England. Two of his graduate students, Howard Gillery and Fred Harrison, came with him. The three of them together with Herb McKinstry and myself occupied rooms on the top floor of the Mineral Science Building. Our X-ray group was part of the Mineral Constitution Laboratory which was (and still is) one of the best characterization labs in the country. In addition to X-ray diffraction, there were facilities for electron microscopy, and for atomic, infrared, and mass spectrometry, and wet chemical analyses.

Although his earlier work was on metals and scattering factors, George Brindley is best known for his X-ray studies of clay minerals. He began investigating kaolinite and other fine-grained silicates during World War II and is still at it, although officially retired. *"Crystal Structures of Clay Minerals and Their X-ray Identification"* (Mineralogical Society of London,

1951) now in its third edition, summarizes his many contributions to the field.

Professor Brindley was a fine lecturer and an excellent thesis advisor. He was always the first one to arrive in the morning and kept our noses to the grindstone. I don't believe there ever was an advisor who worked closer with his students and took a greater interest in their work.

One of our colleagues in the College of Earth and Mineral Sciences was a young Assistant Professor named Rustum Roy. Back in 1955 he taught a course in crystal chemistry which embodied most of the topics of a modern materials science course. It came as no surprise to me when in 1961 he and Dr. Brindley were instrumental in founding the Materials Research Laboratory. Vladimir Vand, Herb McKinstry, Bill White, K. Vedam, Eric Cross and I joined the lab a few years later. Penn State's MRL is perhaps the only Materials Research Lab in the country overrun with crystal chemists and crystal physicists!

In recent years there have been a succession of good crystallographers at Penn State, all of whom are still going strong: J. V. Smith, Jerry Gibbs, Deane Smith, Earle Ryba, Ian Harrison, Marguerite Bernheim, Gerry Johnson, Joe Stanko, and Richard Morgan. They have all made important contributions to crystallography, but we will leave their stories (and mine) to a later and hopefully superior historian.

CHAPTER 17

TWENTY-FIVE YEARS OF CRYSTALLOGRAPHY IN PITTSBURGH: A PERSONAL ACCOUNT

G. A. Jeffrey
University Professor of Crystallography; Chairman, Department of Crystallography; University of Pittsburgh, Pittsburgh, PA 15213

My association with the American Crystallographic Association began very soon after I arrived in the U.S. In 1955 I became the Treasurer. David Sayre had been re-elected to the office, but resigned when he received a fellowship to study in Oxford with Dorothy Hodgkin. I welcomed the invitation from the President, Eddie Hughes, to take his place. This gave me an excellent opportunity better to get to know, at least, the names of my future colleagues, the American crystallographers. The membership at that time was less than 500. Each year there was one mailing from the Treasurer which combined the dues notices and voting slips for the next year's officers. Very soon, the members were divided in my mind into three categories, those who paid and cared, those who paid but did not care, and those who neither paid nor cared. There was also a small minority of enthusiasts who regularly paid twice. This classification, which remained constant through my four years in office, was useful some years later when I had to select chairmen to organize sessions for the IUCr Congress in Stony Brook, and later avoid black-hole referees for *Acta Cryst.* The ACA didn't have many financial responsibilities in those days; there was one meeting a year which was small and self-supporting. I was quite relaxed about the penalty for not paying dues, but my successor, Tom Furnas, was more of a disciplinarian. I recall that he struck Lawrence Bragg from the rolls for delinquency. Sir Lawrence was rather offended by this. He took a convenient opportunity some years later to admonish me with a mile rebuke to the ACA. I am sure that his ghost will be pleased at this opportunity to deliver it.

The relative proximity of Pittsburgh to State College, Pennsylvania, the location of Pepinsky's X-RAC, provided a unique opportunity for me to meet, not only the American crystallographers, but many other crystallographers from all over the world who came to use Pepinksy's analogue Fourier synthesizer and work in his large research group. The X-RAC at Penn State was the mecca for crystallographers in the early 1950's. With Bill Kehl from Gulf Research, we made numerous pilgrimages there, both on my first visit to the U.S. in 1950, and on my return in 1953. However, this preoccupation with analogue computers, which had engaged crystallographers for two decades, was fast pre-empted by the arrival of the commercial digital computers. At Pittsburgh, we obtained our first University computer, an IBM 650, in 1956. Ryonosuke Shiono, on his way back from England to Japan, wrote our first crystallography programs. Serious computer users were rare in those days and we had no problem in reserving the University computer for our crystallographers two days a week. We operated it ourselves, of course. In 1957, we held a tutorial on crystallographic computing on the IBM 650, which was probably the first such specialized computing workshop in the U.S. A few years later, a meeting of an ad-hoc committee of crystallographic computing for the IUCr led to the formation of the IUCr Commission on Crystallographic Computing. I was chairman of the ad-hoc committee, and Theo Hahn was host. It took place in a beautiful castle on the Rhine and included a memorable wine-tasting party given by the mayor of Hochheim.

The exciting X-ray equipment of the early 1950's were the new Geiger-counter diffractometers. Very soon after my arrival at Pittsburgh, I obtained a General Electric XRD-3, with a grant from Pittsburgh Plate Glass, which was matched by the "General" through his notable representative Howard Pickett. This was quite a cause for celebration, since it was the first piece of GE equipment in Pittsburgh, which was regarded at that time as a Philips stronghold. Later, this instrument became our first manual single-crystal diffractometer, using the Furnas Eulerian cradle. Our first research was on simple inorganic materials; and we even attempted charge density studies on AlN and BeO. More rewarding, in retrospect, was an exploration of the polytypism of some aluminum carbonitrides synthesized by ALCOA in Pittsburgh. This research was supported by the Army Office of Ordnance Research in the days when they supported long term basic research. One of the fringe benefits of such support was access to MATS transportation across the Atlantic, and thence use of the Air Force logistic air arm throughout Europe. The service was cheap and extremely efficient. The prop planes had uncomfortable canvas seats and the trans-Atlantic flight with stops took at least fourteen hours. It made communication with European science very much more accessible to young scientists than it is these days. I used it to attend the Fourth IUCr Congress in Paris, and thereafter to Germany where I made some contacts which led to some research with Erich Wolfel and the formation of Stoe Instruments, Pittsburgh. Similar excursions took me to conferences in England, Spain and Greece, all of which were scientifically stimulating and rewarding. Contacts were made with young European crystallographers, many of whom came to work with my group in Pittsburgh as visiting post-docs or professors. It is ironical that when the U.S. was leading the world with its technology, it was easy for U.S. scientists to travel abroad and spread their knowledge. Now that the U.S. is lagging behind in many areas of science and technology, that inexpensive and very efficient method of international communication is no longer available to younger colleagues who could benefit so much.

Of the ACA meetings that took place in the 1950's, one of the more unusual, for me at least, was that at French Lick, Indiana in 1956. The meals were included, as were the free bottles of Pluto-water that became necessary to ensure bowel motion later in the week. I recall that a hotel employee told me that the crystallographers couldn't be very important conventioneers or big spenders, because the management had not deployed the usual jazz band and bevy of cheer-leaders to welcome them. I sensed that I was being told about vanishing mores. The excitement at ACA meetings in those days were the spirited arguments between Jerry Karle and Ray Pepinsky as to whether the new Karle & Hauptman phase determing formulae were a new and potentially more powerful way of solving the phase problem, or simply a restructuring of the Patterson method with the same limitations. Posterity has provided the answer to that controversy in the widespread use of such programs as MULTAN and SHELX, which, interestingly, originated in England rather than the U.S.

In 1957, there was a joint meeting between the ACA and the Pittsburgh Diffraction Conference. It was at this meeting that the ACA past-president, Dan McLachlan, presented his famous one-hour shaggy dog story entitled "Crystallography in the Geophysical Year". The Pittsburgh Diffraction Conference looms large in my experiences, since, like taxes, the problem comes up every year. The Conference started in 1943, with an informal meeting with presentations by Sidhu of the University of Pittsburgh, Gulbransen of Westinghouse Research Laboratories and Barrett of Carnegie Institute of Technology. The object clearly stated on the program was "to provide an atmosphere where new diffraction techniques, as well as new problems, may be put forth and intelligently discussed for the mutual benefit of all concerned." A similar meeting was held in the following November with an evening discourse by I. Fankuchen on "The X-ray Diffraction Studies of Biological Interest." The 1945 meeting was a two-day affair with Maurice Huggins talking on "The Use of Fourier Synthesis in Crystal Structure Analysis." The next two annual meetings were larger, since they were joint meetings of the Electron Microscope Society of America and the American Society for X-ray and Electron Diffraction. For two decades, the Conference was a primary forum for papers on neutron and gas-phase diffraction, low-angle scattering, and instrumentation for diffraction, with related fields such as proton scattering, image reconstruction, structure of polymers, nucleic acids, and viruses represented from time to time. Attendances at the Pittsburgh Diffraction Conferences were in the range of two to three hundred in the fifties and sixties, but dropped to less than one hundred in the seventies. Since the ACA meetings were very much oriented toward single crystal determinations prior to 1970, the Pittsburgh Diffraction Conference provided a complementary forum for many aspects of applied crystallography. The meetings were held at the Mellon Institute, where there was a large and active X-ray diffraction group under the direction of Harold Klug and Leroy Alexander, with a more applied emphasis than ours at the University. Invited guest speakers to the Diffraction Conference included R. W. G. Wyckoff, A. L. Patterson, Dorothy Hodgkin, R. Heidenreich, J. Turkevitch, R. Pepinsky, Francis Schmitt, Dorothy Wrinch, Caroline MacGillavry, W. Zachariasen, R. E. Rundle, H. L. Klug, Leroy Alexander, John Slater, Bert Warren, Charles Barrett, Peter Debye, Kathleen Lonsdale, and Betty Wood. In 1948, the guest speaker was Sir Lawrence Bragg, and in 1949 William Lipscomb described the low-temperature methods that he would later use to work on the structures of the boron hydrides, for which he was awarded a Nobel Prize in 1976. André Guinier, later to become a President of the International Union of Crystallography, was the guest speaker in 1950; Dave Harker talked about the X-ray crystallography of proteins in 1951. The sixteenth meeting coincided with the Pittsburgh Bicentennial and John Bardeen was the guest speaker. Professor Kato, the future President of the International Union of Crystallography, gave a paper at that meeting.

More recently, the Pittsburgh Diffraction Conference has featured symposia on special topics. It has played away in alternate years, and was held at the Medical Foundation in Buffalo in 1977 and at Purdue University in 1979. Since 1967, a fund collected at the death of one of the founders has been used to provide the Sidhu Award for outstanding research in X-ray diffraction by a young investigator. The Awardees have been A. I. Bienenstock (1967), R. M. Nicklow (1968), T. O. Baldwin (1969), S. H. Kim (1970), L. K. Walford (1971), Dale E. Sayers (1972), B. C. Larsen and N. C. Seeman (1974), P. Argos (1975), G. T. DeTitta and K. O. Hodgson (1978) and G. Petsko (1980).

I became President of the ACA in 1963. This was fortunate for me, because it was the centennial anniversary of the National Academy of Sciences and I was invited to represent the ACA at the celebrations. Two events stick in my mind. One was an address by J. F. Kennedy that left me, and possibly the other less politically sophisticated members of the audience, with the strong impression that JFK had an unusual comprehension of science and might have been an excellent scientist had he not been distracted elsewhere. The other was part of the scientific sessions, which spanned the whole of science. Linus Pauling was responsible for taking the Academy from the carbon atom to the protein molecule. Linus had a beautiful series of slides which were eventually published in his book,

"The Architecture of Molecules." The talks were given in the cavernous auditorium of the U.S. State Department. Unfortunately, the projectors had inadequate heat protection for Linus' colored slides which survived less than a minute. Linus rose to the occasion, as ever, but the impact of his science was lessened somewhat by the audience's fascination by the ensuing race between the speaker and the projector.

1963 was also the Sixth IUCr Congress in Rome, which was a very enjoyable affair. It made a bad start by garbling Leslie Orgel's slides, to his very obvious indignation, but compensated with the banquet at the newly opened Rome Hilton and the romantic cocktail party given by Philips at the Castel Saint Angelo. At the Rome meeting, I reported some of the first results of a program of research on the clathrate hydrates, which I had started with Dick McMullan, now at Brookhaven National Laboratory. Between 1958 and 1970, some thirty papers were published on these fascinating structures, mostly in the *Journal of Chemical Physics.* They were challenging problems both experimentally and interpretatively. The data were collected on Weissenbergs, lubricated and geared to operate at -30°C in refrigerated boxes and almost all the structures were solved by intuition from the Patterson syntheses. At that time, the U.S. dollar was strong and highly desirable, so we could recruit assistance from Visiting Scientists from other countries, especially Mario Bonamico from Italy, Dave Hall from New Zealand, Dirk Feil and Paul and Gezína Beurskens from Holland, Dieter Panke from Germany, Stan Nyburg from England, Kalman Sasvari from Hungary and Tom Mak from Hong Kong. This research was supported initially by the Air Force Office of Scientific Research and later by the Office of Saline Water of the Department of the Interior. A highlight of that period was a one-month Research Conference on Water Research, organized by the National Academy of Sciences and held at the Whitney estate at Woods Hole in the summer of 1961. The report from this meeting, which included contributions from many distinguished American scientists, served as the 'bible' for the really excellent research program developed by the Office of Saline Water during the next decade. This first-class basic research program perished, as did many others, during the Nixon debacle.

By 1957, my earlier interest in organic structures was revived with some cyclopropane fatty acids synthesized by Klaus Hofmann. Bryan Craven came from New Zealand to work with me on what was my first of several NIH-supported research projects. Craven's professor in New Zealand was the older graduate student from whom I had learned much of my crystallography at Birmingham in the late 1930's. We also initiated some structural research on the barbiturates which still persists with Bryan's charge density research. Early collaborators in this program from overseas were Yukio Kinoshita and Tosio Sakurai from Japan, Dieter and Edith Mootz from Germany, Sagrario Martinez-Carrera from Spain, Bryan Gatehouse from Australia, and Yvonne Mascarenhas from Brazil. Neville Stephenson from Australia also enlivened our environment by spending a sabbatical with us.

The graduate programs at the University of Pittsburgh in the 1950's were very much night-school oriented, long on course work and short on laboratory equipment and experimental research. The evening crystallography courses were very well attended by excellent students working full-time in the local industries. There were lectures and problems based on the methods and materials from the summer schools taught in the late forties at Leeds and Cambridge Universities in England. By 1960, I was beginning to attract some very exciting and gifted full-time students, such as Shirley Chu, now Professor of Electrical Engineering at Southern Methodist University in Dallas, Maynard Slaughter, now Professor of Chemistry at the Colorado School of Mines, Martin Sax, now Head of the Biocrystallography Laboratory and Chief of Staff of the Veterans Administration Medical Center in Pittsburgh, and Muttaiya Sundaralingam, Professor of Biochemistry at Wisconsin.

My long involvement with the USA National Committee for Crystallography started in 1955. I served as Secretary for the period which included the IUCr Congress in Montreal. The big political question at that time was whether the Germans should have one or two adhering bodies. This East-West question was presided over with great control and infinite patience by Betty Wood. I guess the State Department lost that one, but in retrospect, it hasn't seemed to matter very much. The Montreal Congress was small enough that all the attendees could be accommodated on a boat for a day cruise up the St. Lawrence. This was such a success that we talked about an IUCr Congress on a transatlantic liner. Unfortunately, plans for that didn't materialize before that particular nautical species became extinct.

I was Chairman of the USANCCr during the period which included the 1966 Moscow meeting and thereby became Chairman of the USA delegation. That meeting got off to a bad start since it coincided with a summer season TWA airline strike. I managed to get on the last TWA plane to London prior to the strike, but Jim Ibers, who was the Secretary of the USANCCr, chickened out and never did get to Moscow. That caused some dismay, since he had the records of all the careful deliberations of the USANCCr prior to the Congress. However, as often happens on these occasions, no great decisions were affected one way or the other. As I recall, Dave Shoemaker came to my rescue and between us we remembered most of our instructions. The odd feature about the Russian tourist system is that it never learns or improves. My scenario was that I was bumped off my British Airways flight from London on to a later Aeroflot, and in consequence when I arrived at Moscow airport, my name wasn't on the official list. Some eight hours later I entered my bedroom at the Ukrainian Hotel. I suspect it would have been much later if Sundaralingam had not banged his walking stick, Kruschev style, on the reception desk. The two highlights of that meeting for me were the opportunity to talk with the U.S. Ambassador for half an hour prior to an excellent reception at the U.S. Embassy, and Kitaigorodski's tour-de-force at the Russian reception. Each of the more than one hundred guests were toasted individually in turn, with vodka. Kitaigorodski introduced each guest with an appropriate comment in his native language. At least, I think so!

In 1967, I undertook a UNESCO mission on behalf of Crystallography to East Pakistan, now Bangladesh. It is questionable how much Crystallography or Pakistan benefited, but the visit added an important component to my understanding and respect for the problems facing scientists in poorly developed countries. A much more scientifically productive international experience was with the crystallographers in Sao Carlos, Brazil. Yvonne Mascarenhas came to work with us in 1963, and out of this developed a very enjoyable and rewarding collaborative program which involved a joint research project and a number of personnel exchanges between our respective laboratories in both directions during the next fifteen years. In 1977, I had the pleasure of participating in a meeting of the Society of Brazilian Crystallographers in Sao Carlos which was very reminiscent of an early ACA meeting.

In 1968, crystallography at the University of Pittsburgh received a significant stimulus from the National Science Foundation through a substantial Center of Excellence Award, which provided funds for new equipment and programs. The new programs started on the nuclear quadrupole resonance of crystals and diffraction physics were reasonably productive, but unfortunately did not survive the rigorous University tenure tests of the mid-seventies.

On the other hand, our other component, biochemical crystallography, which was supported by an NIH training grant in the late sixties, continued to flourish, in keeping with the national trend. This aspect of crystallography in Pittsburgh was greatly enhanced by the very successful development by Martin Sax of the Biocrystallography Laboratory at the nearby Veterans Administration Hospital. This large research group is closely affiliated with the Department of Crystallography by means of adjunct and joint appointments, to the mutual advantage of both. The opportunity to do their research in a biomedical hospital environment is particularly advantageous to the students who took advantage of this, one of whom, Phillip Pulsinelli, went on to post-doc with Max Perutz, and is now on the faculty of our School of Pharmacy.

We had another exciting group of students about this time, which included Helen Berman, now at the Institute for Cancer Research in Philadelphia, Sung Hou Kim, who is Professor of Chemistry at the University of California at Berkeley, and later Ned Seeman and George DeTitta. Another interesting student of this period was Edward Fasiska, who is President of the Materials Research Corporation in Pittsburgh.

The Crystallography program was now too large and too interdisciplinary to reside comfortably in any of the traditional University departments, and the Department of Crystallography was created in 1969.

In 1969, I was program chairman of the Eighth Congress in Stony Brook, New York. In retrospect, I often wonder why the use of the poster session was not implemented sooner at large international meetings. It would have made both the eighth and ninth Congresses so much easier to program and provided better results. As it was, at Stony Brook, we had over 700 oral presentations crammed into one week, with undigestible results.

For me, my colleague, B-C. Wang, and especially our departmental administrative secretary, Joan Klinger, our most notable contribution to the Ninth Congress in Kyoto was to get the U.S. crystallography charter flight off the ground. Organizing a charter flight in itself is a once-only experience, but the blow fell when Northwest Airlines went on strike. At the very last moment, China Airlines came to the rescue; we arrived behind schedule but in time for the meeting.

Our first automatic diffractometer, a Picker, was acquired in 1966, marking the end of the time-honored (30 year) method of eye-estimation of intensities by the students, sons, daughters and sometimes wives and mothers of crystallographers. It was also apparent that the direct methods of Karle and Hauptman, particularly the tangent formula, were going to solve the phase problem much more quickly and certainly for non-centric crystal structures of medium-sized organic molecules. Seeking an area of structure-relevant chemistry, I recalled some of the conformational problems associated with the carbohydrates which I had studied as a student in the department of W. N. Haworth in Birmingham, England. Providing a wide variety of good crystals and some interesting chemistry, this proved to be a very fertile area of crystallographic research. I was fortunate at that time in having some exceptionally able students and colleagues in the department, all of whom gave the carbohydrate project an excellent start. Sundaralingam, who had been a student in the department some years earlier, was also interested in carbohydrates. We divided the field between us, Sunda taking the furanoses, nucleosides and nucleotides and exploring that field with great personal success, while I concentrated on the pyranose sugars. By 1970, the ab-initio molecular orbital quantum mechanics practiced by John Pople across the street at Carnegie-Mellon University reached the stage of development such that it could be applied to some fundamental conformational and structural questions in pyranose sugars. We published our first joint paper with Leo Radom, now in Canberra, Australia, in 1972, which was very successful, as judged by its selection as a "citation classic" eight years later. This collaboration continued and developed into a most productive combination of crystallography and quantum mechanics as applied to small organic molecules, and has extended to the present.

I spent the first six months of 1974 on sabbatical leave in Lisbon, Portugal, as a Senior Fulbright Scholar teaching crystallography and structural biochemistry. This was a most enjoyable experience since prior to its revolutions, Lisbon was one of the most beautiful cultural centers in Europe.

On return to the U.S., I accepted an invitation to go the the Chemistry Department at Brookhaven National Laboratory. This was a stimulating experience work-wise. There is nothing quite like the thrill of the big science which takes place around a nuclear reactor and similar national facilities. However, I missed the social variety and complexity of life in the microcosm of society provided by an urban University of 30,000 persons. I didn't succeed in acquiring the necessary enthusiasm for fishing, swimming, sailing, or boating necessary for life on the Long Island Sound. The most profitable part of my two year stay at Brookhaven was the series of theoretical chemistry and experimental neutron diffraction studies on the hydrogen-bonding in carbohydrates carried out in collaboration with Marshall Newton, Dick McMullan, and an ex-graduate student of mine, Shozo Takagi.

On my return to the University of Pittsburgh in 1976, I continued the carbohydrate research, became interested in molecular mechanics, and re-established collaboration with the quantum mechanics group at Carnegie-Mellon University. The carbohydrate research, which has been funded by NIH since 1965, had been extremely productive, with more than one hundred papers published. This was recognized in 1980 by the Claude S. Hudson Award of the American Chemical Society.

In addition to his continuing research on charge densities in collaboration with Bob Stewart at Carnegie-Mellon University, Bryan Craven has initiated an exciting program of research in the very difficult field of cholesterol fatty acid structures. Protein crystallography has been added to our program in the form of Azotobacter ferredoxin and nitrogenase with Dave Stout, transferrin with Jaime Abola, and serum albumin with Dick McClure. In the associated Biocrystallography Laboratory of the VA Medical Center a very successful study of the immunoglobulin proteins is underway under the direction of Martin Sax and B-C. Wang.

In 1980, crystallography is doing what it has done since the time of Niels Stensen, alias Nicholaus Steno; responding aggressively to changing fashions and times, and as a consequence prospering at Pittsburgh and elsewhere.

CHAPTER 18

CRYSTALLOGRAPHIC ASPECTS OF AN AUTOBIOGRAPHY

William N. Lipscomb
Gibbs Chemical Laboratory, Harvard University,
12 Oxford St., Cambridge, MA 02138

Kentucky 1937-41

No X-ray diffraction existed where I studied at the University of Kentucky, but when I was an undergraduate there I tried to interest Hahn of Physics, who was studying X-ray absorption.

Caltech 1941-46

I had come to Physics to study quantum mechanics with Houston, but switched to Chemistry after the first half-year. Linus Pauling outlined a thesis in electron diffraction, X-ray diffraction and magnetic anisotropy, but the war intervened from mid-1942 until late 1945. Even so I studied the chloroethylenes, diketene, tetramethyldiketene (Lipscomb and Schomaker, 1946) and vanadium tetrachloride (Lipscomb and Whittaker, 1945) by electron diffraction, and in my spare time completed methylamine hydrochloride (Hughes and Lipscomb, 1946) so that Pauling would have a reliable C-N distance (1.465 ± 0.010 Å) for protein structures. I also took data on α-glycylglycine in 1946, but the structure was later solved by Hughes (Biswas, Hughes and Wilson, 1950; Biswas, Hughes, Sharma and Wilson, 1968) using the first algebraic method: that of Banerjee (1933) and Hughes (1949).

Sturdivant's course was incredibly good. I still have his clear derivation of how one could use anomalous scattering to establish absolute configuration, but the effect was thought to be too small to see in visual estimates of intensities in 1945. Before my Ph.D. in 1946, I worked closely with both Sturdivant and Corey, but on war projects, not on crystallography. Whenever Sturdivant ("Tex", because he was from Texas) wanted to express some qualifications he would address me as Dr. Lipscomb; otherwise, Willie. Sturdivant's gracious permission for me to make use of his notes on symmetry for Section III of my account of X-Ray Crystallography (Lipscomb, 1960) is an example of his unselfish generosity.

Hughes was in Berkeley for some of my Caltech years. My correspondence of that time is remarkable. It was then that I worked out the general theory of errors for both the Fourier and least squares methods at Hughes' direction, but only a tiny fraction of the work was accepted by the journal (Hughes and Lipscomb, 1946). Some of the letters begin, "My God, Willie!" This was not a promotion: rather, I had committed some highly original blunder. One letter of 1945 from Hughes refers to Schomaker's discovery in 1941 of fractional cell projections, which Verner tested on glycine in 1943, and which Hughes reported at the ASXRED in 1944. The occasion was my letter to Hughes noting Booth's publication (Booth, 1945).

Lindsay Helmholtz, when at Caltech, had taken a photograph of a single crystal to find a powder diffraction pattern. While Hughes and Sturdivant were pondering over this result, Helmholtz took more photographs of the same sample and obtained a single crystal pattern. The result was all the more remarkable in that the pattern was already indexed. Helmholz had become so familiar with the darkroom that he never turned on the safelight: these experiments had been done using a box of Corey's previously exposed films!

Minnesota 1946-59

I had thought of doing systematic studies of molecules in crystals at low temperatures, even before the invitation to Minnesota, because of my uncertainties about electron diffraction methods. Even while still at Caltech, I wanted to work on the $B_{10}H_{14}$ structure, but Pauling told me that Harker's group was already doing that structure (Fig. 1) (Kasper, Lucht and Harker, 1950). My students and I set out first to develop the X-ray diffraction techniques to the point of reliability, and we studied a number of hydrogen bonded structures and other structures in which residual entropy was thought to occur (Abrahams, Collin and Lipscomb, 1951; Collin and Lipscomb, 1951; Dulmage, Meyers and Lipscomb, 1953; Tauer and Lipscomb, 1952; Zaslow, Atoji and Lipscomb, 1952; Atoji and Lipscomb,

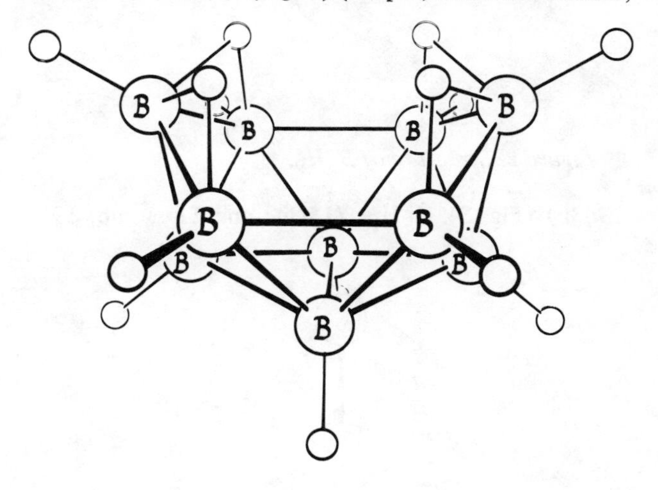

Figure 1. Structure of $B_{10}H_{14}$.

1953b). At first there was so little support for research that we obtained X-ray transformers which had been discarded by the medical and dental professions. Deliveries of liquid nitrogen were unreliable, so we took matters into our own hands, only to be detained by the police on suspicion of stealing large metal dewars from the University. Shortly before we solved $(NO)_2$ (Dulmage, Meyers and Lipscomb, 1953) m.p. – 165°C, we were ready to tackle the boranes. In order, they were B_5H_9 (Dulmage and Lipscomb, 1951; 1952) (Fig. 2), B_4H_{10} (Nordman and Lipscomb, 1953a;

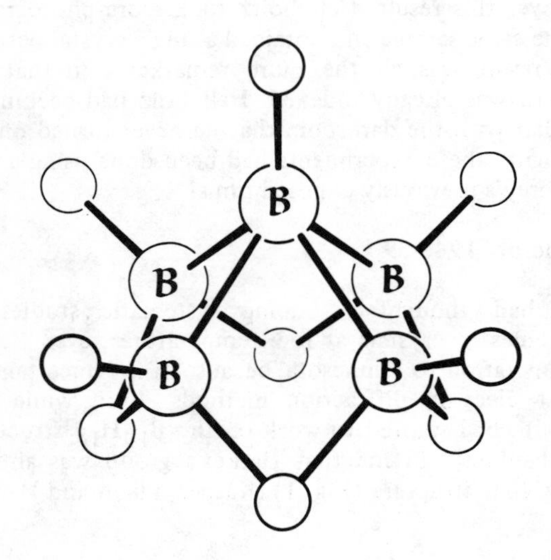

Figure 2. Structure of B_5H_9.

1953b) (Fig. 3), B_5H_{11} (Lavine and Lipscomb, 1953;

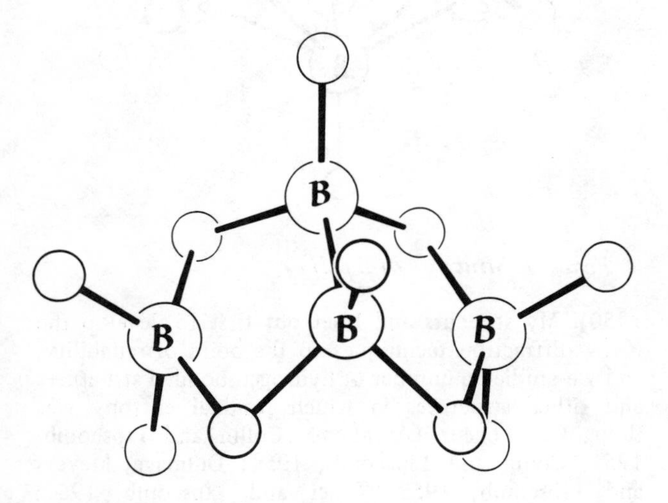

Figure 3. Structure of B_4H_{10}.

1954) (Fig. 4), B_6H_{10} (Eriks, Lipscomb and Schaeffer,

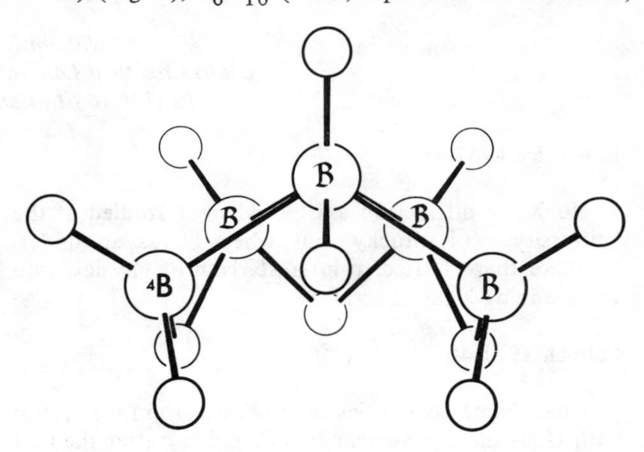

Figure 4. Structure of B_5H_{11}.

1954; Hirschfeld, Eriks, Dickerson, Lippert, Jr. and Lipscomb, 1958) (Fig. 5), B_9H_{15} (Dickerson, Wheatley,

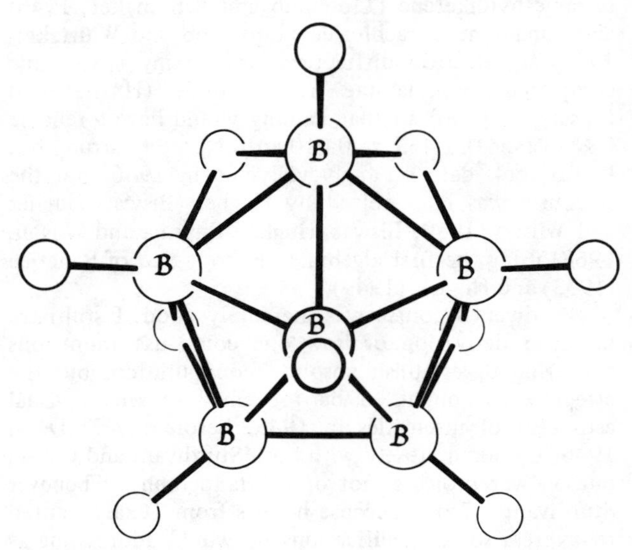

Figure 5. Structure of B_6H_{10}.

Howell, Lipscomb and Schaeffer, 1956; Dickerson, Wheatley, Howell and Lipscomb, 1957), and others. The first two were independently done by electron diffraction by Schomaker, et al. (Hedberg, Jones and Schomaker, 1951; 1952; Jones, Hedberg and Schomaker, 1953) at Caltech, almost unambiguously. The B_9H_{15} structure was solved by the method of S. Holmes [Exhausting all other possibilities but the correct one (Dickerson, Wheatley, Howell, Lipscomb and Schaeffer, 1956; Dickerson, Wheatley, Howell and Lipscomb, 1957) inasmuch as it had been given to us by R. Schaeffer as a B_8 hydride.] What is still unusual about these and many of our later studies based on visual estimates is that the only chemical analysis was the count of borons and hydrogens in the electron density maps. The hydrocarbon-like structures of the early electron diffraction work of Bauer were wrong, and the geometrical principles were clearly based upon polyhedra of boron atoms or fragments of those polyhedra, consistent with the indication of Kasper, Lucht and Harker (1950) on $B_{10}H_{14}$. Also the theory of diborane fragments by Pitzer (1946) was wrong. The double bridged diborane structure of Stitt (1940, 1941) and Price (1947, 1948) (Fig. 6) did not yield a

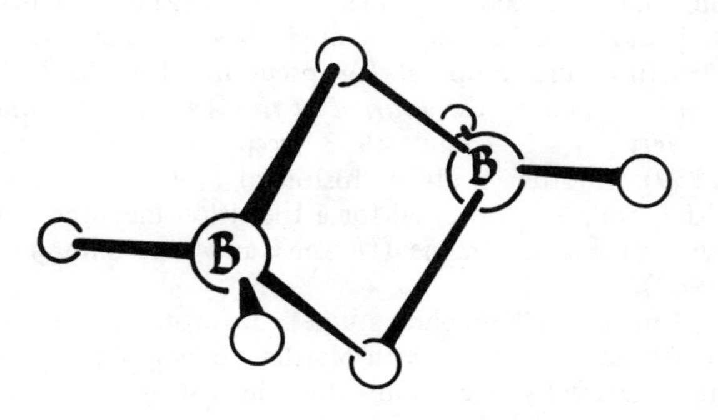

Figure 6. Structure of B_2H_6.

generalization to a valence theory, nor did $B_{10}H_{14}$ because the valence theory was too complex for this example. It was the simpler structures listed above that led Eberhardt, Crawford and me (1954) to the theory of three-center bonds in the boron framework. This was a difficult idea at that time, because hybrid orbitals were thought to be directed along the internuclear directions. In our structures, hybrids sometimes were directed toward the center of a B_3 triangle. Extensions of these ideas, and many further structures were accompanied by an enormous expansion of this area of chemistry (Lipscomb, 1977) including theoretical studies by Longuet-Higgins (1949), Longuet-Higgins and Bell (1943), Longuet-Higgins and Roberts (1954, 1955) and by us (Lipscomb, 1963) and then experimental studies on the closed polyhedra for which examples or analogues are now known for $B_nH_n^{-2}$ for n from 5 to 15, inclusive, along with many stable fragments of these polyhedra. The vertices of these polyhedra or their fragments now are represented by most of the elements of the periodic table, and examples are especially numerous when both boron and carbon are present. While the perhalides B_4Cl_4 (Atoji and Lipscomb, 1953a) and B_8Cl_8 (Jacobson and Lipscomb, 1959) are also polyhedral, these examples are neutral because of back coordination of electrons from chlorine. Subsequently, other polyhedral perhalides were studied by X-ray diffraction, including negatively charged species, or their isoelectronic carborane equivalents, e.g. $C_2B_{10}Cl_{12}$ (Potenza and Lipscomb, 1963).

The B_6H_{10} study was terminated twice, once when an electrician dropped a screwdriver on the power supply to the building and again when a technician tripped on the stairs in the very early hours of the morning as he went to fill the dewar with liquid nitrogen. Our accidents with the vacuum lines filled with explosive boranes were somewhat lucky ones, without severe injuries. After one of these explosions, I took Russell Grimes to a hospital in Cambridge only to hear the doctor say "Louis Fieser sends me much more interesting cases." I was reminded recently of an example of group enthusiasm: the study of the B_4H_{10} structure was facilitated when I brought a calculator to the bedside of Chris Nordman as he was recovering from mononucleosis. His wife, Barbara, thought we should have had our heads examined. Nordman and I (1953c) also explained the expansion of hydrogen bonds when deuterium is substituted, an experimental result of Robertson and Ubbelohde (1939). This was a nervous time: the title of our second paper was printed, "A One-Dimensional Equation for Hydrogen Bombs," on the cover of that issue of the *Journal of Chemical Physics.*

A Guggenheim Fellowship allowed me to go to Oxford in 1954-5 primarily to do theoretical chemistry with C. A. Coulson. Having previously established the hour of tea with my students in Minnesota, I found time nearly every day to have tea with Dorothy Hodgkin's group, and particularly to follow the work on vitamin B_{12}. (Kenneth Trueblood, a member of our Caltech Chemists baseball team in the 1940's was then doing some of the extensive calculations.) Dorothy's abilities at pattern recognition in Patterson functions and preliminary electron density maps were absolutely incredible. G. Zhdanov came to visit us then, as a member of the Program Committee for the Fourth International Congress in Crystallography (Montreal, 1957) and we assembled for a whole day at the Hodgkins (in 1955) ending with the dinner created by Thomas. Zhdanov and Sevast'yanov (1941) had found the icosahedron of boron atoms in the boron carbide structure (Fig. 7) (See next page). On another happy occasion I went to Cambridge to celebrate the announcement of Pauling's Nobel Prize in Chemistry at Francis Crick's flat, along with Peter Pauling, Linda Pauling and others. I had originally planned in this sabbatical to turn my research efforts to theoretical chemistry. However, I was unable to leave crystallography. Perhaps my rather speculative nature is at its best when it starts from the relatively secure results of new crystal structures. Certainly my interests in the chemical ideas are greatest in the molecules which are studied in my research group.

We then moved to a variety of new crystallographic studies, including some large organic molecules, and I remember encouraging Richard Dickerson to work with John Kendrew and Michael Rossmann to study with Max Perutz, after having become convinced by J. M.

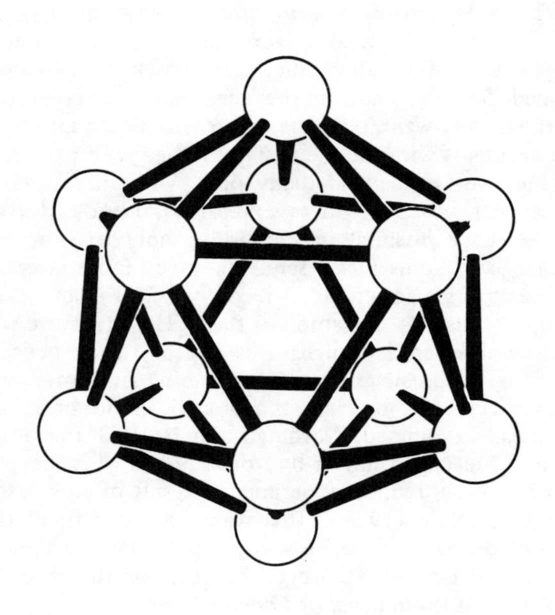

Figure 7. The B_{12} icosahedron, which occurs in boron hydride $B_{12}C_3$, in elementary boron, and in $B_{12}H_{12}^{2-}$. The three isomers of $C_2B_{10}H_{12}$ also have this icosahedral arrangement in which there is one externally bonded hydrogen on each boron or carbon atom.

Bijvoet's results (Bokhoven, Schoone and Bijvoet 1949; 1951) on organic molecules that protein structures could be solved.

Harvard (1959-present)

Actually, the crystallographic studies of boranes also continued as we moved to Harvard, yielding many complex boranes and derivatives solved by vector superposition and minimum function techniques. By then, a chemical program and a theoretical program were strong. Notable was the isolation of small quantities of new boranes (B_8H_{12}, Enrione, Boer and Lipscomb, 1964, $B_{10}H_{16}$, Grimes, Wang, Lewin and Lipscomb, 1961 and $B_{20}H_{16}$, Friedman, Dobrott and Lipscomb, 1963; 1964) from electrical discharge reactions, although one new "borane" turned out to be cyclo-$(Me_2Si)_4O_4$ (Boer and Lipscomb, 1964). from the stopcock grease. Unfortunately that structure had already been published.

Starting with mineral structures, one of which is now named lipscombite (Katz and Lipscomb, 1951), our research also included many organic structures, both natural products and synthetics. Perhaps notable is the first stereochemistry for the anti-leukemic agents of the vincristine type (Moncrief and Lipscomb, 1965; 1966), and our recent structure of the cyclic peptide β-amantin (Kostansek, Lipscomb, Yocum and Thiessen, 1978), the most common mushroom toxin. The gibberelic acid (Hartsuck and Lipscomb, 1963) and tetrodotoxin (Gougoutas, Paul and Lipscomb, 1964) structures were helpful in the synthetic efforts of E. J. Corey and R. B. Woodward, respectively. But stereochemistry can be ahead of its time. Our structure of the cyclic 1, 4-dithiadiene (Howell, Curtis and Lipscomb, 1954) was rejected by the *Journal of Organic Chemistry,* with the comments of the referee, "I fail to see any excuse at all for publishing Lipscomb's paper in our journal. It is already known to be a cyclic compound, which is all that is needed by the average organic chemist." Rejection can be a sign of originality; for example, my structure and proposal for pseudorotation in $B_3H_8^-$ were rejected by the *Journal of the American Chemical Society* in 1957 and 1958, respectively (Lipscomb, 1959). Another study of historical interest is the first "ring whizzer" proposal for a transition metal complex with cyclooçtatetraene (Dickens and Lipscomb, 1961; 1962).

Our crystallographic studies of protein structures, which began in 1962 with Martha Ludwig, were greatly accelerated by the termination in 1964 of my grant from National Science Foundation for research in boron chemistry (resumed in 1977). Support from the National Institutes of Health enabled us to obtain, by mid-1966, an excellent three-dimensional electron density map of carboxypeptidase at 2.8 Å resolution. (Lipscomb, 1968). Inasmuch as the chemical sequence was then unknown, we proceeded to 2.0 Å (Lipscomb, Hartsuck, Reeke, Jr., Quiocho, Bethge, Ludwig, Steitz, Muirhead and Coppola, 1968), and just now have refined the structure to 1.7 Å resolution. The structures of several complexes of the enzyme with inhibitors or poor substrates have illuminated the initial stages of the mechanism (Lipscomb, 1980) by which this enzyme cleaves off the C-terminal amino acid of a peptide substrate. And we have recently solved the structure of a complex of carboxy-peptidase A with a protein inhibitor which contains 39 amino acids (Rees and Lipscomb, 1980).

The three dimensional structure of concanavalin A was solved to 4 Å in early 1971 at Harvard (Quiocho, Reeke, Becker, Lipscomb and Edelman, 1971), and completed elsewhere. In the case of glucagon, solved to 6 Å, our derivatives were poor at high pH, and the structure was finally solved at a lower pH in Blundell's laboratory (Sasaki, Dockerill, Adamiak, Tickle and Blundell, 1975).

Our major work since 1967 has been a study of the allosteric mechanism in aspartate transcarbamylase from *E. coli.* We now have three structures: the enzyme without ligands at 3 Å (Monaco, Crawford and Lipscomb, 1978), the enzyme liganded to the allosteric inhibitor cytidine triphosphate (CTP) at 2.8 Å (Monaco, Crawford and Lipscomb, 1978) and the enzyme bound to the substrate analogue N-phosphonacetyl-L-aspartate to 4 Å resolution (Kuttner, 1980). This last structure was obtained without heavy atom derivatives, with the use of the solutions of the two other structures. The methods, most completely developed by Bricogne (1972, 1976), appear to have converged, but we are still testing various aspects of this result. This method involves non-crystallographic averaging in real space, and structure factors calculated directly from the averaged electron density. We had used this method earlier on non-crystallographic symmetry in the CTP-enzyme, and had incorporated data from the different crystal form of the unliganded enzyme.

Finally, other X-ray diffraction results included the study of bond distances from molecular wave-functions instead of spherical atoms in B_2H_6 (Jones and Lipscomb, 1970), and a series of studies of crystal structures near liquid helium temperatures including α-N_2 and β-N_2 (Jordan, Smith, Streib and Lipscomb, 1964), β-F_2 (Jordan, Streib, Smith and Lipscomb, 1964; Jordan, Streib and Lipscomb, 1964) γ-O_2 (Jordan, Streib, Smith and Lipscomb, 1964) and CH_4 (Streib and Lipscomb, 1964).

Before the Cambridge Electron Accelerator was closed for lack of support, we obtained a few X-ray diffraction photographs with the use of synchrotron radiation, very shortly after K. Holmes (Rosenbaum, Holmes and Witz, 1971) had obtained results at Hamburg. This, along with wire counters, or television tube recording of intensities, plus high speed computers, plus low temperature studies promises the next revolution in X-ray diffraction studies.

Two sabbaticals at the Medical Research Council, Cambridge, England and one in Martinsried near Munich developed a close relationship to the stimulating research groups there. Discussions with Max Perutz, Aaron Klug, David Blow, Robert Diamond, and more recently, with Robert Huber have provided the genial atmosphere of science in its best aspects. The free interchange of ideas is far more stimulating when it occurs in discussions than when it is read in the journals.

The Wolcott Gibbs Laboratory at Harvard, where our recent studies were carried out, has recently been the center of striking advances in the crystallography of large molecules independently of my own contributions. The first structure of a virus has been completed here by Stephen Harrison's group (Harrison, Olson, Schutt, Winkler and Bricogne, 1978), in particular bushy stunt virus. In addition, Don Wiley's group (Wilson, Skehel and Wiley, in press 1980) has solved the structure of influenza protein coat virus. Finally, I should like to call attention to the first t-RNA structures solved independently by Alexander Rich's group (Suddath, Quigley, McPherson, Sneden, Kim, Kim and Rich, 1974) at M.I.T. and Aaron Klug's group (Robertus, Ladner, Finch, Rhodes, Brown, Clark and Klug, 1974) in Cambridge, England.

Finally, I add a brief account of my presidency of the ACA. My term of office occurred during my first sabbatical. As I was returning to the USA by boat, I drove to the harbor at Le Havre to find no boat at all. There was a strike, and I missed the 1955 meeting in Pasadena, where my good friend and first teacher of crystallography, E. W. Hughes, filled in my duties.

About 60 Ph.D.'s from my research group have been in crystallography from 1946-1980. They, together with the others (theoreticians, inorganic and organic chemists, biochemists, mineralogists, chemical physicists and biophysicists) have joined to make a constantly creative research effort.

I express far more than an acknowledgement: rather, a debt of gratitude to these and other members of my research groups through these years.

References

Abrahams, S. C., Collin, R. L. and Lipscomb, W.N. (1951). Acta Cryst. **4**, 15-20.

Atoji, M. and Lipscomb, W.N. (1953a). Acta Cryst. **6**, 547-550.

Atoji, M. and Lipscomb, W. N. (1953b). Acta Cryst. **6**, 770-774.

Banerjee, K. (1933). Proc. Roy. Soc. (London). **A141**, 188-193.

Biswas, A.B., Hughes, E. W., Sharma, B. D. and Wilson, J. N. 1968). Acta Cryst. **B24**, 40-50.

Biswas, A. B., Hughes, E. W. and Wilson, J. N. (1950). Am. Crystallogr. Assoc. Program Abstr. Ser. **2**, Aug. 21-25, p. 8.

Boer, F. P. and Lipscomb, W. N. (1964). Study unpublished.

Bokhoven, C., Schoone, J. C. and Bijvoet, J. M. (1949). Proc. Acad. Sci. (Amsterdam) **52**, 120-121; (1951). Acta Cryst. **4**, 275-280.

Booth, A. D. (1945). Trans. Faraday Soc. **41**, 434-438.

Bricogne, G. (1974) Acta Cryst. **A30**, 395-405; (1976) **A32**, 832-837.

Collin, R. L. and Lipscomb, W. N. (1951) Acta Cryst. **4**, 10-14.

Dickens, B. and Lipscomb, W. N. (1961). J. Am. Chem. Soc. **83**, 4862; (1962). J. Chem. Phys. **27**, 2084-2093.

Dickerson, R. E., Wheatley, P. J., Howell, P. A. and Lipscomb, W. N. (1957). J. Chem. Phys. **27**, 200-209.

Dickerson, R. E., Wheatley, P. J., Howell, P.A., Lipscomb, W. N. and Schaeffer, R. (1956) J. Chem. Phys. **25**, 606-607.

Dulmage, W. J. and Lipscomb, W. N. (1951). J. Am. Chem. Soc. **73**, 3539; (1952). Acta Cryst. **5**, 260-264.

Dulmage, W. J., Meyers, E. A. and Lipscomb, W. N. (1953). Acta Cryst. **6**, 760-764.

Eberhardt, W. H., Crawford, B. L. Jr. and Lipscomb, W. N. (1954). J. Chem. Phys. **22**, 989-1001.

Enrione, R. E., Boer, F. P. and Lipscomb, W. N. (1964). Inorg. Chem. **3**, 1659-1666.

Eriks, K., Lipscomb, W. N. and Schaeffer, R. (1954) J. Chem. Phys. **22** 754-755.

Friedman, L. B., Dobrott, R. D. and Lipscomb, W. N. (1963). J. Am. Chem. Soc. **85**, 3505; (1964) J. Chem. Phys. **40**, 866-872.

Gougoutas, J. Z., Paul, I. C. and Lipscomb, W. N. (1964), unpublished study.

Grimes, R., Wang, F. E., Lewin, R. and Lipscomb, W. N. (1961). Proc. Natl. Acad. Sci. USA **47**, 996-999.

Harrison, S. C., Olson, A. J., Schutt, C. E., Winkler, F. K. and Bricogne, G. (1978). Nature (London) **276**, 368-374.

Hartsuck, J. A. and Lipscomb, W. N. (1963). J. Am. Chem. Soc. **85**, 3414-3419.

Hedberg, K., Jones, M. E. and Schomaker, V. (1951). J. Am. Chem. Soc. **73**, 3538-3539; (1952). Proc. Natl. Acad. Sci. USA **38**, 679-686.

Hirschfeld, F. L., Eriks, K., Dickerson, R. E., Lippert, E. L., Jr. and Lipscomb, W. N. (1958). J. Chem. Phys. **28**, 56-61.

Howell, P. A., Curtis, R. M. and Lipscomb, W. N. (1954). Acta Cryst. **7**, 498-503.

Hughes, E. W. (1949) Acta Cryst. **2**, 37-38.

Hughes, E. W. and Lipscomb, W. N. (1946) J. Am. Chem. Soc. **68**, 1970-1975.

Jacobson, R. A. and Lipscomb, W. N. (1959). J. Chem. Phys. **31**, 605-609.

Jones, M. E., Hedberg, K. and Schomaker, V. (1953). J. Am. Chem. Soc. **75**, 4116.
Jones, D. S. and Lipscomb, W. N. (1970). Acta Cryst. **A26**, 196-207.
Jordan, T. H., Smith, H. W., Streib, W. E. and Lipscomb, W. N. (1964). J. Chem. Phys. **41**, 756-759.
Jordan, T. H., Streib, W. E. and Lipscomb, W. N. (1964). J. Chem. Phys. **41**, 760-764.
Jordan, T. H., Streib, W. E., Smith, H. W. and Lipscomb, W. N. (1964). Acta Cryst. **17**, 777-778.
Kasper, J. W., Lucht, C. M. and Harker, D. (1950). Acta Cryst. **3**, 436-455.
Katz, L. and Lipscomb, W. N. (1951). Acta Cryst. **4**, 345-348.
Kostansek, E. C., Lipscomb, W. N., Yocum, R. R. and Thiessen, W. E. (1978). Biochemistry **17**, 3790-3795.
Kuttner, P. G. (1980). Ph.D. Thesis, Harvard University.
Lavine, L. R. and Lipscomb, W. N. (1953). J. Chem. Phys. **21**, 2087-2088; (1954). **22**, 614-620.
Lipscomb, W. N. (1959). J. Inorg. Nucl. Chem. **11**, 1-8.
Lipscomb, W. N. (1960). Techniques of Organic Chemistry. Vol. 1, Physical Methods, Part II, Third Edition, 1641-1738.
Lipscomb, W. N. (1963). Boron Hydrides, New York: W. A. Benjamin Publishing Company.
Lipscomb, W. N. (1977). Science **196**, 1047-1055.
Lipscomb, W. N. (1980). Proc. Natl. Acad. Sci. USA **77**, 3875-3878.
Lipscomb, W. N., Hartsuck, J. A. Reeke, G. N. Jr., Quiocho, F. A., Bethge, P. H., Ludwig, M. L., Steitz, T. A., Muirhead, H. and Coppola, J. C. (1968). Brookhaven Symposia in Biology, **21**, 24.
Lipscomb, W. N. and Schomaker, V. (1946). J. Chem. Phys. **14**, 475-479.
Lipscomb, W. N. and Whittaker, A. G. (1945). J. Am. Chem. Soc. **67**, 2019-2021.
Lipscomb, W. N. (1968). In "Structural Chemistry and Molecular Biology" edited by A. Rich and N. Davidson. (A volume dedicated to Linus Pauling) p. 38. This was the second enzyme, and the third protein, at high resolution.
Longuet-Higgins, H. C. (1959). J. Chem. Phys. **46**, 268-275.
Longuet-Higgins, H. C. and Bell, R. P. (1943). J. Chem. Soc. pp. 250-255.
Longuet-Higgins, H. C. and Roberts, M. deV. (1954). Proc. Roy. Soc. (London) **A224**, 336-347; (1955). **A230**, 110-119.
Monaco, H. L., Crawford, J. L. and Lipscomb, W. N. (1978). Proc. Natl. Acad. Sci. USA **75**, 5276-5280.
Moncrief, J. W. and Lipscomb, W. N. (1965). J. Am. Chem. Soc. **87**, 4963-4964; (1966). Acta Cryst. **21**, 322-331.
Nordman, C. E. and Lipscomb, W. N. (1953a) J. Am. Chem. Soc. **75**, 4116; (1953b) J. Chem. Phys. **21**, 1856-1864.
Nordman, C. E. and Lipscomb, W. N. (1953c). J. Chem. Phys. **21**, 2077.
Nordman, C. E. and Lipscomb, W. N. (1965). J. Am. Chem. Soc. **87**, 4963-4964; (1966). Acta Cryst. **21**, 322-331.
Pitzer, K. S. (1946). J. Am. Chem. Soc. **67**, 1126-1132.
Potenza, J. A. and Lipscomb, W. N. (1963). In "Boron Hydrides" edited by W. N. Lipscomb. New York: W. A. Benjamin Publishing Company p. 26.
Price, W. C. (1947). J. Chem. Phys. **15**, 614; (1948). **16**, 894-902
Quiocho, F. A., Reeke, G. N. Jr., Becker, J. W., Lipscomb, W. N. and Edelman, G. M. (1971). Proc. Natl. Acad. Sci. USA **68**, 1853-1857.
Rees, D. C. and Lipscomb, W. N. (1980). Proc. Natl. Acad. Sci. USA **77**, 4633-4637.
Robertson, J. M. and Ubbelohde, A. R. (1939). Proc. Roy. Soc. (London) **A170**, 222-240.
Robertus, J. D., Ladner, J. E., Finch, J. T., Rhodes, D., Brown, R. S., Clark, B. F. C. and Klug, A. (1974). Nature (London) **250**, 546-551.
Rosenbaum, G., Holmes, K. C. and Witz, J. (1971) Nature (London) **230**, 434-437.
Sasaki, K., Dockerill, S., Adamiak, D. A., Tickle, I. L. and Blundell, T. L. (1975). Nature (London) **257**, 751-757.
Stitt, F. (1940). J. Chem. Phys. **8**, 981-986; (1941). **9**, 780-785.
Streib, W. E. and Lipscomb, W. N. (1964), unpublished results.
Suddath, F. L., Quigley, G. J., McPherson, A., Sneden, D., Kim, J. J., Kim, S. H. and Rich, A. (1974) Nature (London) **248**, 20-24
Tauer, K. J. and Lipscomb, W. N. (1952). Acta Cryst. **5**, 606-612
Wilson, I. A., Skehel, J. J. and Wiley, D. C. (1980) Nature (London), **289**, 373.
Zaslow, B., Atoji, M. and Lipscomb, W. N. (1952). Acta Cryst. **5**, 833-837.
Zhdanov, G. A. and Sevast'yanov, N. G. (1941). C. R. Acad. Sci. USSR **32**, 432.

SECTION B

IN MEMORY OF SOME PAST-PRESIDENTS

CHAPTER 1

A. L. PATTERSON

Jenny P. Glusker
The Institute for Cancer Research
The Fox Chase Cancer Center, Philadelphia, PA 19111

Lindo Patterson, in articles published in 1934, showed how it was possible to determine interatomic distances in a crystal by use of a Fourier calculation involving values of $|F|^2$ (which are phaseless) as coefficients. Thus, structural information can be obtained without any knowledge of the relative phases of the diffracted beams. It was immediately recognized that a method had, at last, been proposed that could be used to solve larger structures, and the method was eventually to play an important role in the subsequent determinations of protein and virus structures. The technique for structure solution had been changed from a trial-and-error procedure to an analytical procedure.

There is a story that J. M. Robertson and Lindo shared a room during the April 1950 conference at State College on "Computing Methods and the Phase Problem" run by Ray Pepinsky. Lindo drove there late one evening when Robertson had already retired and locked the door. When Lindo knocked on the door and woke him, Robertson said: "I see before me a dimly resolved Patterson". There is space here for only a few aspects of his life, giving only a "dimly resolved" view of this great man.

Historically, the first big step toward crystal structure determination came when von Laue, as a result of Barkla's work on X-ray polarization, estimates of X-ray wavelengths and his own earlier work on scattering from periodic structures, realized that if X-rays were wave-like rather than particle-like, they could be diffracted by crystals. This was demonstrated by Friedrich and Knipping. The second step was taken by W. L. Bragg when he realized that the internal structure of the crystal could be determined from the X-ray diffraction pattern. Interatomic distances could also be measured if the X-ray wavelength were known. Young Bragg asked some chemists which were the interesting structures to do and was sent to see Pope and Barlow. They told him to look at the alkali halides and see if the ions were arranged as packed spheres as they had predicted. As a result the Braggs (father and son) determined the structures of sodium and potassium chloride. Other structure determinations followed in their footsteps and some interesting salts, alloys and organic compounds were studied.

But the crystals had to have simple structures if they were to be determined. The frustrations of the time were very clear. Structures could only be solved if the symmetry of the crystal meant that only one or two general parameters were needed to define the entire structure. For example, Pauling said that when he got to Pasadena, in 1922, Dickinson told him that the thing to do in order to get answers to questions was to determine crystal structures. But he also told him that it was earier to determine the structure of a cubic crystal than of a crystal of low symmetry. So Linus crystallized about a dozen cubic crystals in the first few months there. One (cesium mercuric bromide) had 32 molecules in the unit cell in general positions, another (potassium nickel sulfate) had 19 atoms in the asymmetric unit, and a third (the alloy, sodium dicadmium) had 1200 atoms in the unit cell. Then the next (dimagnesium stannide) had the fluorite structure so that all atomic positions were immediately known. Linus then got tired of not being able to solve the structures of crystals because there were too many parameters that needed to be evaluated and so he, like many others, determined structures by "trial-and-error" methods. A structure was derived by model building or other means, and the agreement between the observed and calculated diffraction intensities was used to test the validity of the proposed structure.

But how could crystal structures be determined more directly? The only hope at that time, as Cork realized when he studied the alums, was the use of isomorphous replacement methods. But such small structures are rarely truly isomorphous even if some metal replacement is possible.

The frustration in not being able to solve more complicated structures was well-recognized by Patterson, but he was convinced that there was some information in the diffraction pattern that could be better used to analyze the structure. His paper (on the use of the map described by him as the $|F|^2$ map, but known by us as a Patterson map) was an instant success, and was rapidly used by many for years until direct methods were introduced – but that is another story of other people. The Patterson function is, of course, still used extensively by the macromolecular crystallographers. But even for smaller structures it is often very instructive to study the Patterson map of a compound that cannot readily be solved.

Lindo was born in New Zealand on July 23, 1902 in Nelson (near Auckland) but his family left for Canada when he was eighteen months old. He went to school in Montreal, but when he was 14 he was sent to an English school (public to the English, private to Americans), Tonbridge (1916-1920). He had an older brother who died in World War I of gangrene from his wounds but Lindo was never old enough during the war to serve in the armed forces. However, later, he always hated peanut butter, a reminder of the poor diet at that time.

From Tonbridge he then went to McGill University (1920-1924) where he majored in physics, receiving a bachelor's degree in 1923 and a master's in 1924. The subject of the master's thesis was the production of hard X-rays by the interaction of radium β-rays with solids. He said that his professors decided that the bachelor's degree was not as good as it might be because he had too many friends in Montreal and was addicted to activities such as skiing, bridge, and dancing.

His graduate studies were done under the supervision of A. S. Eve, on "The application of X-rays to the study of organic substances". In 1924 McGill announced the endowment of two Moyse Travelling Fellowships, and one was awarded to Lindo. Lindo's parents were in England, and so he wrote to W. H. Bragg and was able to work with him at the Royal Institution (R. I.) in London for two years. Bragg was interested in naphthalene and anthracene, and Muller and Shearer were studying long chain compounds. Patterson worked on determinations of the unit cell dimensions and space groups (at that time no structure determinations were contemplated) of various phenylaliphatic acids, since these had characteristics of both sets of compounds being studied at the R.I. To do this he had to build his own X-ray equipment.

The Pulp and Paper Industry Research Laboratory at McGill University was interested in the work of Herzog and Jancke on the X-ray diffraction of cellulose and provided Lindo a fellowship for such studies. Therefore he spent a year with Hermann Mark at the Kaiser-Wilhelm Institute in Berlin, on the "particle size" of cellulose. Here he worked on the theory of particle-size line broadening and his contributions were substantial, as shown by a note in *Zeitschrift für Physik.*

While he was in Berlin in 1926-1927 he kept a diary which shows that music was his first love, but he was very stimulated by the atmosphere provided by all the scientists there and was rapidly able to contribute. At first he had to struggle with the problems of learning German. He wrote "Tonight my landladies have given me some light reading "Fürst Bismarcks Briefe an seine Braut und Gattin" (Fürst Bismarck's Letters to his Bride and Wife). I really cannot face it so I think that bed is indicated." But later he was able to write "I am very proud to say that yesterday I separated my first prefix successfully. I think that the sensation must be similar to that of a doctor removing his first appendix". His love of music resulted in a very interesting analogy. Lindo wrote: "This leads me rather to a discussion which I had with Mark, and a comparison, I think due to him the discussion, between music and crystallography. He is not so particularly keen on Bach, from the music point of view and although he acknowledges him as the founder of modern music, only from the mechanical point of view. The comparison is between the mathematical theory of crystallography as developed by Schönflies and von Fedorov, and Bach. The mathematical theory of music was provided by Bach, and just as Fedorov and Schönflies would not have been able to make the beautiful applications of their theories to the interpretations of crystal structure, so was Bach unable to write the Ninth Symphony. The Goldberg Variations fall very nicely into the analogy, as being the space groups of a certain symmetry class. The Well Tempered Clavichord is more or less the set of examples appended to show what can be done with the beautiful new mathematics. The St. Matthew Passion, and the Mass, and the Brandenburg Concertos are more difficult to fit in. The two, three, and four piano concertos, are more or less the examples of the cubic system in extension of W.T.C. [Well Tempered Clavicord]. Carrying the analogy further, von Laue, Friedrich and Knipping compare with Mozart and Haydn, who found out what could be done with the mathematics in applying it to the expression of these ideas. W. H. and W. L. Bragg acquire the cloak of Beethoven. W. L. [Bragg]'s *Cam. Phil. Soc.* paper and the structure determinations of salt and diamond compare with the first two symphonies of Beethoven, the first, which also might have been written by Mozart, and the second showing the hand of the master. The Eroica, when the artistic sense of the master begins to take hold, one can compare with the application of the work to the organic substances. Fedorov, Schönflies or even Laue would have wanted a little more to go on, but the application of the touch of the artist brought about this beautiful work, which may lack the preciseness of geometrical definiteness, but which is truth for all that. The work of W. L. Bragg on beryl, I should say, ranks very high in the string quartets of Beethoven when we carry the comparison on, but I am afraid, that the V and IX Symphonies have yet to be written." Remember that Lindo wrote this in Germany, long before the Patterson function was thought of!

Lindo enjoyed the scientific atmosphere in the Germany of the times (1926-1927) and wrote: "There really are an amazing collection of "wissenchaftliche" stars here in Dahlem. At the colloquium in Haber Institute there were: Haber, Freundlich, Weissenberg, Polanyi, Hahn, Meitner, and then they were not nearly all met as the session has not really got going. The Berlin Physics Colloquium is I believe absolutely magnificent. One name only need be mentioned, and that is Einstein, but there are many others. Today I assisted in the setting up of a new tube, and find that troubles are more or less the same the world over. Today I think for the first time I succeeded in cracking a joke in German without having to explain it about three times; I also succeeded in making a successful pun, so I am feeling pretty good." His further description of the scientific atmosphere is described as follows: "In the afternoon, I went to my first Physics Colloquium in the University. Einstein, Laue, Planck, Nernst, form the nucleus, with goodness knows how many other stars. It really was wonderful. I understood most of everybody but Laue, and the Germans themselves find him difficult to follow. Einstein is quite a cheery soul. I met Ewald who is a first rate sort, who speaks, as do most people, about 3 languages fluently. I don't think that even in the R.I.

or at the B.A. I have ever been in the same room with so many scientific bloods. They all seem quite ordinary people, but Einstein and he seems extremely human in character, but he has one of the most marvellous faces of any one I have ever seen. A perfectly marvellous head as well. There is a photograph of him in a shop in the Uhlandstrasse that I would like to buy, but I don't know if such is allowed. When you see the man you realize how it is possible that he has done such wonderful work."

On Good Friday in 1927 the Davisson and Germer paper on the diffraction of electrons was read by Lindo. He describes the events as follows:

"Sat 16th received copy of *Nature* with Davisson's paper and made everybody's life miserable with my enthusiasm. Sunday, went to Potsdam with Stark [whom he describes as S] and talked Davisson. Monday we worked Davisson's paper out completely. Nothing of note seems to have happened then until Friday 29th Ap. when someone rang me up, (later found to be S). Saturday W came and asked me for my copy of *Nature* with the Davisson paper in it (for S). Monday at the Wellenmechanik Colloq. I gave it up to him. Wednesday before the Colloquium I asked S if Laue was going to speak on the Davisson paper. He said no. I had invited the Rosses to the Colloquium to look at the Lions. As soon as the Colloquium started Laue started off very dramatically with the Davisson work. Announced that they had obtained interferences from single crystals with electrons. Sensation! Much discussion, in which someone asked Laue if anything was said about velocities, he said no, I put up my hand and said yes. Laue goes on with the discussion of the paper and gets it all wrong, bad English and general unclearness in the paper being the cause. He said they had used the wrong lattice constants, that the effect did not agree with a surface reflection. I tried to speak, but everyone was talking at once, and I did not get a chance to speak until Pringsheim, called Laue's attention to the fact that I wanted to speak. He then called on me, and I got up and told him that I had worked the thing through and it all agreed perfectly. He said that he and S had worked it out and it didn't agree. I am inclined to think that S lead him astray. It was lucky that I knew the paper by heart, as S had my copy the whole time. After my remarks, Laue said that as I read the paper so thoroughly I had better come and talk about it. Which I did, with the worst wind up which I have had for a long time. I talked for about 15 minutes."

He was so scared, Lindo related later, when he talked about the Davisson and Germer paper that "the first line he drew on the blackboard came out dotted".

Lindo returned to McGill in 1927 and finished his Ph.D. in 1928. He then was a demonstrator in the physics department for a year. He moved to the Rockefeller Institute in New York and Thomas White, his student from McGill, accompanied him. They worked in R.W.G. Wyckoff's group on cyclohexane derivatives, mainly hexols, from 1929-1931. But he wrote that he had an "obsession with the notion that something was to be learned about structural analysis from Fourier theory". So in 1930 he spent many hours looking through the tables of contents of all the mathematical journals in the New York Public Library, searching for some clues on how he should proceed.

After two years at the Johnson Foundation in Philadelphia studying biological systems he decided he had saved up enough money to be able to work independently on the problem of solving structures and so he went as an unpaid guest scientist to Bert Warren's laboratory at M.I.T. He was particularly interested in going there because he could talk to Norbert Wiener who, he said, "knew as much about Fourier integrals as anyone in the world." His main interest was in the Faltung or convolution and its Fourier transform. He discussed Fourier integrals with Weiner while singing Gilbert and Sullivan operas. In fact he put the words of the question he had into the music of the song they were singing together. Lindo asked Norbert Wiener "What do you know about a function representable by a Fourier series when you know only the *amplitudes* of the Fourier coefficients?" Wiener's laconic reply was "You know the Faltung."

Lindo wrote "The understanding of the Faltung came, of course, from the work on liquids and their radial distributions. Warren with Gingrich and others had perfected the techniques used by Debye and Menke in the study of the X-ray scattering from liquids. These were of course based on the original suggestions of Zernike and Prins. Warren and Gingrich had already had the idea that these methods applied to powders would give the radial distribution in a crystal. While trying to learn about their work I noticed that the mathematical form of the theory given by Debye and Menke would be identical with that of the Faltung if the integrations over random orientation were left out and the randomness of choice of origin was left in. What was immediately apparent was that the crystal contained atoms and that the Faltung of a set of atoms was very special in that it would consist of a set of atom-like peaks whose centers were specified by the distances between the atoms in the crystal." Thus the Patterson or $|F|^2$ series was proposed.

He continued, "All this happened on a Tuesday, and Friday was the deadline date for the Washington spring meeting (1934) of the American Physical Society. An abstract had to be prepared in a hurry to go in with that of Warren on the radial distribution in carbon black and that of Gingrich and Warren on the radial distribution in powders which was basic to my work. The only $|F(h)|^2$-- series which I was able to compute [with the primitive computing equipment of the time] in the month between the deadline and the meeting was the $(hk0)$ of KH_2PO_4 and a one dimensional series for a simple layer structure. All three papers were very well received and had very full discussion with A. H. Compton in the chair and W. L. Bragg in the audience to ask the right questions."

It was tedious and time-consuming to compute such Fourier maps and therefore Patterson, with George Tunell, devised the Patterson-Tunell strips, similar to the Beevers-Lipson strips of the English crystallographers. Then it was possible to compute Patterson maps (generally only in projection) for a whole series of compounds. However, Lindo was always sorry that he had not thought of the fact, noticed by Harker, that certain areas of the Patterson map contain a lot of structural information.

Due to the depression, Lindo had been 3 years without a job, until he joined the faculty of the physics department at Bryn Mawr College, near Philadelphia. His colleague was Michels, who was interested in solid state physics. The combination of names meant that they were affectionately called "Pat and Mike". Together they wrote a textbook. In this text book there are some out-of-the-ordinary diagrams, including a schmoo (drawn by Al Capp), to illustrate Gauss' law and a Charles Addams child (with no neck) illustrating the principle of the solitary see-saw. The year before going to Bryn Mawr, Lindo married Betty Knight, whom he had met at the Rockefeller Institute. She is still a lively and active scientist interested in the biochemistry of enzyme reactions.

During the war, the government was interested in the mining of harbors and the effects of waves on the triggering of these explosive devices. Lindo worked with Michels, who was a Commander in the Navy, on this, first at Bryn Mawr and later in Washington, (as a foreign national) for the U.S. Naval Ordinance Bureau. In 1945 he received the Meritorious Civilian Service Award for his work on submarine warfare. He became a U.S. citizen in the same year.

After the war he continued his main interests which were particle-size line broadening and the study of "homometric structures". These are different atomic arrangements with the same Patterson function. Pauling tells the story of this. Apparently about the time that the Patterson function was first published Linus had a graduate student who was studying bixbyite, originally studied by Zachariasen. This structure contains 32 metal atoms and 48 oxygen atoms in the unit cell. Linus discovered that in bixbyite the metal atoms could be arranged in two physically quite distinct arrangements that gave exactly the same calculated X-ray diffraction pattern. So, when he read, in a copy of *American Physical Society Bulletin*, about a meeting, he found that Lindo planned to present a proof that no two physically distinct structures can have the same Patterson map or diffraction pattern. So he wrote to Lindo saying that he had an example of two structures that were physically distinct but that had exactly the same Patterson diagram, and when Lindo gave the paper the following month it was on the conditions under which two structures can have the same Patterson map.

In 1949 Lindo came to the Institute for Cancer Research where he studied the structures of compounds of biological interest, such as citrates. But his interest in homometric structures continued, and he was working on the development of a theory on them when he died.

Lindo served the ACA with great dedication throughout its history and played an important role in its formation in 1950 when the American Society for X-ray and Electron Diffraction (ASXRED) and the Crystallographic Society of America (CSA) merged. He was President of ASXRED in 1949 and served on the ACA Publications Committee in 1955-1957 and 1959-1961. Another great service to the Crystallographic Community was his section on "Fundamental Mathematics" in Volume II of *"International Tables."* He contributed to international crystallography by his service on the U.S.A. National Committee for Crystallography in 1951-1955, 1957-1962 and 1964-1966. He was a major writer of the by-laws of the International Union of Crystallography.

Lindo died suddenly and unexpectedly of a cerebral hemorrhage in November 1966 at the age of 64. Unfortunately, although he was very goodlooking, not many good photographs of him remain. He tended to "freeze" in front of a camera and his dynamic character was lost in the resulting photograph.

Of course, the insight gained from his scientific discoveries tells us nothing of the kind of person Lindo Patterson was. How would he appear to each of you at a meeting? He was a tall, thin, freckled, red-head with a ready smile, and yet he could be firm. Even though crystallographers are particularly nice and friendly people, he ranked among the best for friendliness, thoughtfulness and integrity. He almost never had a bad word to say about anyone, and no one said a bad word about him. The older members of ACA knew Lindo well because he was an active member, attended all meetings, and made sure he had a word of encouragement for new young crystallographers giving their first papers. He also gave young scientists an opportunity to meet older scientists and to take part in discussions at a level that might otherwise not have been possible to them for many years. Yet he was shy and was always nervous in a meeting before he asked a question.

He thoroughly enjoyed good conversation, good food, and good music. He could sing most of the Gilbert and Sullivan arias, having acted in productions of these operettas in his earlier years. He had a delightful sense of humor and said that the best crystallographic information was obtained by informal discussion, preferably at the bar. That was in the days before poster sessions.

At one stage Lindo became tired of reading lots of acronyms and yet realized that jargon was taking over crystallography and therefore wrote an article, signed A. L. Pon, in which he tried to save space by condensing words. The article, which is very readable, was accepted by *Acta* since it was good science, but, since there were financial problems in publishing *Acta* at that time people had been requested to shorten submitted papers. As a result, Lindo withdrew this, but offprints marked "Not reprinted from *Acta Crystallographics*" were produced. A "translation" is in order.

"On the symmetry of white radiation streaks produced by the Buerger precession camera" by A. L. Patterson, Senior Member, Department of Physics, Institute for Cancer Research, Philadelphia, Pennsylvania, U.S.A.

Recent publications on the reciprocal lattice theory of diffracted reflections of X-rays have suggested the mathematical techniques used in the preliminary note.

It is clear that in all discussions of diffraction problems, the radius of the Ewald sphere can be set at unity (Verner Schomaker, unpublished). If then a given reciprocal lattice point is associated with monochromatic radiation of a given wavelength, then a reciprocal vector, joining this reciprocal lattice point with the reciprocal lattice origin corresponds to the white radiation which is necessarily produced with the monochromatic radiation in the X-ray tube (see any textbook on quantum theory). Thus any diffraction phenomena produced by the Buerger precession camera possesses plane symmetry about a symmetry plane through the reciprocal lattice point, the reciprocal lattice origin and the diffracted ray" (Not reprinted from *Acta Crystallographica).*

(Adapted from the Past-President's Address, Calgary, August 1980).

Not reprinted from *Acta Crystallographica*

PRINTED IN DENMARK

On the symmetry of the wheaks produced by the Bucessera. By A. L. Pon, *Senics, Incanearch, Philpa, U. A.*

(*Received* 5 *December* 1952)

Recent publications on the reory (some workers prefer the term rekhaning) of diffions of X-ray (Raster, 1951*a*, *b*; Helister, 1951; Hoester, 1952*a*, *b*) have suggested the mathiques used in the prote.

It is clear that in all discussions of diflems, the radius of the Ewere can be set at unity.* If then a given relp is associated with monion of a given wangth, then a relve joining this relp with the relor corresponds to the whion which is necessarily pruced with the monion in the X-rube (see any took on the quary). Thus any difromena pruced by the Bucessera possesses planretry about a sylane through the relp, the relor and the difray.

Rences

HELISTER, N. H. A. (1951). *Intergraphs.* London: Macan.
HOESTER, J. A. (1952*a*). *Experientia*, **8**, 297.
HOESTER, J. A. (1952*b*). *Acta Cryst.* **5**, 626.
RASTER, G. A. (1951*a*). *Acta Cryst.* **4**, 335.
RASTER, G. A. (1951*b*). *Acta Cryst.* **4**, 431.

* Vernished.

CHAPTER 2

WILLIAM H. ZACHARIASEN

Robert A. Penneman
Los Alamos Scientific Laboratory, Los Alamos, NM 87549

Professor (Fredrik) William H. Zachariasen was born February 5, 1906 in Langesund, Norway and died December 24, 1979, Santa Fe, New Mexico.

Professor Zachariasen, one of the outstanding scientists of the 20th century, is gone. "Willie" leaves to his countless friends the warm remembrance of his gracious manner, his great good humor, and his generous friendship. His rich and varied scientific contributions engaged him to the last. He was one of the giants in X-ray crystallography and was involved in every major advance in that field. Linus Pauling wrote of him:

> "I have known Professor Zachariasen for nearly fifty years. His principal field of work has been the determination of the structure of crystals of inorganic substances by use of the X-ray diffraction technique. This is a field in which I also have done a large amount of work, and I believe that I am in a position to form a sound opinion about his ability and his contributions. It is my opinion that he has been and is the world leader in this field.
>
> I feel that he is to be classed among the outstanding scientists of the twentieth century, and at the top in the field of inorganic crystal structures."

The Formative Years

Zachariasen began his career, unknowingly, with his youthful exploration of the islands in Langesundfjord near his home, islands rich in well-crystallized rare earth minerals. He was to return to those islands later as a student at the University of Oslo under the great geochemist Professor Goldschmidt, who had purchased one of the islands for his studies. "Willie" rowed him out on occasion, a short journey for the son of a sea captain. The trips to the islands combined work with pleasure. Mrs. Zachariasen supplied a photograph of a picnic on one of the islands; in it Professor Goldschmidt is seen talking to guest Albert Einstein. The vivid memory of those days never left Zachariasen: some dozens of years later he correctly identified (on sight) one of the crystal specimens from the Langesundfjord islands – a specimen that had been mislabeled by a commercial firm.

Zachariasen spent his student days (1923-1928) in the midst of that exciting period when Goldschmidt and his collaborators worked out general laws governing distribution of chemical elements in minerals; they were the first to apply X-ray diffraction to the study of geochemistry. During those years Zachariasen read hundreds of X-ray films, an activity that he continued throughout his life. Where the results were of particular significance, as in the recent controversy involving the *a′* and *a″* phases of cerium, he would read each film several times.

Zachariasen published his first paper at age 19, after presenting it before the Norwegian Academy of Science the year before. It was on X-ray diffraction studies of oxides. With its publication in 1925 he began a period of contributions to the scientific literature, most of them singly authored, which would span 55 years. At age 22, he received his Ph.D., the youngest person ever to receive it in Norway. In 1928-29, he was a Postdoctoral Fellow at the laboratory of Sir Lawrence Bragg, where he began his study of silicate structures, a study that would later culminate in the first real understanding of the structure of glass. He returned to the University of Oslo, but within a year accepted a call from Nobel Laureate Arthur Compton to join the faculty of Physics at the University of Chicago. He was 24.

Before leaving Norway in 1930, he married Ragni Durban-Hansen, the striking granddaughter of the pioneer Norwegian geochemist, W. C. Brögger, who discovered and first described the extensive mineral deposits of the Langesund area. Willie and his bride spent their honeymoon on the ship that brought them to the United States to fulfill his commitment to the University of Chicago. There he spent the next 44 years. The Zachariasens had two children, Fredrik Zachariasen (at the California Institute of Technology, Pasadena) and Ellen Z. Erickson of Santa Monica.

Many honors came to him during the 44 years at the University of Chicago. He advanced to full Professor, then to Department Head (during the critical postwar years when the department was rebuilt), and finally to Dean of the Physical Sciences. He was a kindly administrator and solicited advice, but always maintained a policy that one had to "select the few from the many." He was a member of the Norwegian Academy of Science, the U.S. National Academy of Science, the American Academy of Arts and Sciences, the American Physical Society, the American Crystallographic Association, and the Executive Committee of the International Union of Crystallography. He was presented with the honorary degree of Doctor of Science by the New York Polytechnic Institute.

The 1930's were years of financial hardship. Travel funds were nonexistent and Zach would sit up on a night train to New York, give his paper at a scientific meeting the next day, and return that night to save the cost of a hotel room. During that same period X-ray work was shut down for six months for lack of a $75 tube.

But those were also years of high scientific productivity. Following his work with silicates and other oxy-anions, Zachariasen began to think about how the glass structures were built. In 1932 he published his landmark paper on the structure of glass. Referring to this paper in 1961, Charles H. Green wrote: "The present day understanding of glass rests heavily on a singly lucid paper, only 12 pages long, written in 1932 by W. H. Zachariasen." Zachariasen continued through 1941 to study complex oxy-anions and to develop his work on the diffuse scattering of X-rays caused by thermal motion.

In 1941 he became an American citizen. He was then in his mid-30's and already a world figure, having published 80 experimental and theoretical papers, including major papers on diffuse scattering, oxide structures, and the structure of glass. A significant change was soon to occur following the onset of World War II.

The War Years

In parallel with his academic career, Zachariasen began in 1943 to immerse himself in a then-secret activity, doing all the X-ray identification work for a new project on the Chicago campus. As part of a major wartime effort, scientists had gathered at the Metallurgical Laboratory to work with new elements that had not yet been seen, new elements whose chemistry was largely a mystery.

Recall the situation – Enrico Fermi had just demonstrated the existence of the chain reaction at Chicago in December 1942, using uranium[1]. Plans were being rushed for the pilot plant at Oak Ridge, Tennessee, and the production reactors at Hanford, Washington. This meant that the new element, plutonium, would be made in large quantity using neutrons from a nuclear reactor. Before this, plutonium could be made only in microgram quantities by tedious cyclotron irradiation.

There still remained the formidable chemical separations problem. How could pounds of plutonium be isolated from tons of uranium containing radioactive fission products? Chemical processes had to be devised that would work efficiently on large scale and in high-radiation fields that required remote handling. Ultimately, plutonium metal had to be made. What would be its properties? Totally unexpected were the great complexities soon to be encountered with the plutonium metal phases, their unusual number (six), and the confusion caused by the ease of transfer among the four plutonium aqueous valence states.

Seaborg, the co-discoverer with Kennedy, Segré, and Wahl of the fissionable isotope ^{239} Pu, was responsible for directing the efforts of some 60 chemists attempting to elucidate its chemical properties and to develop reliable plutonium separation processes. This was a heavy responsibility and work was conducted with great urgency (July 4th was just another work day). Zachariasen's X-ray analyses provided the information essential to the understanding of plutonium chemistry, deciphering single-handedly the composition of countless samples that were prepared by the chemists. His work was indispensable in replacing mystery with fact and guesses with quantitative structure identification.

Zachariasen was uniquely equipped for this challenge, given his prodigious skill in elucidating structures from powder diffraction data, plus his long familiarity with the rare earths and their 4f-contraction [the decrease in size with increasing atomic number at constant oxidation state (valence)]. Zachariasen's application of crystallography to the elucidation of the nature and chemistry of the transuranic elements, their compounds and metals, is probably his most celebrated work, covering 37 years and several dozen papers.

In the introduction to a book, which did not (and now cannot) get beyond that preliminary stage, Zachariasen wrote:

> "In 1943 Pu metal and some compounds were prepared in microgram amounts; but the ultra-microchemical studies of plutonium presented great difficulties of interpretation. Because of the small amounts of material, it was difficult to prepare single phase samples and to establish chemical identities.
>
> In the autumn of 1943 I demonstrated that satisfactory X-ray diffraction patterns of Pu-preparations on the 10 microgram scale could be obtained and that the interpretation of the X-ray pattern often could provide positive identification of the phase or phases present in the preparation. Thus, in the period from November 1943 to June 1944 a number of plutonium compounds were identified and their crystal structures determined in full or in part."

A fledging chemist who was impressed by Zachariasen, and who later became an important crystallographer in his own right and Dean of Chemistry at Berkeley, David Templeton wrote:

> "Of those early days, I know from personal observation that Zachariasen's work strongly influenced the development of the separation processes for the Hanford plant. As a young chemist, quite unblemished with any understanding of crystallography, I heard him report to the Seaborg team the identity of this or that new compound as they tried to identify oxidation states. Zachariasen's identifications were almost the only reliable analyses available. He solved the structures of hundreds of substances. . ."

The 5f Series: Thorides vs. Actinides

In "History of the Met. Lab. Section C-1," Seaborg recounts that on June 21, 1944, a sample of what was thought to be NpO_2 was sent to Zachariasen. By 11:00

[1]During the Fall of 1942 I was in Massachusetts as part of the Chicago team aiding in the production of sufficiently pure uranium metal.

a.m. on June 22, 1944, his X-ray analysis had confirmed the existence of NpO_2, and Zachariasen had written his memo discussing the thoride series. It states in part:

"The radius of Np^{+4} is thus 0.015Å larger than that of Pu^{+4}, 0.016Å smaller than that of U^{+4}, and nearly identical with that of Ce^{+4}. I believe that a new set of "rare earth" elements has made its appearance. I believe that the persistent valence is four, so that thorium is to be regarded as the prototype just as lanthanum is the prototype of the regular rare earth elements."

W. H. Zachariasen

There had been earlier qualitative observations supporting formation of a new inner transition series of elements, in particular, the narrow absorption features in the plutonium spectra suggested it. However, Zachariasen's quantitative data showing the progressive 5f contraction provided the key confirmation.

On July 14, Seaborg dictated a memo containing the sentence, ". . . I suggest that the elements heavier than actinium be placed in the Periodic Table as an 'Actinide Series'." The actinide name prevailed for the 5f series, but Zachariasen wrote, "The name actinide is not acceptable because thorium is never actinium-like . . ." He called them the 5f-series and would point out that not until the elements 95 and 96 were the metals rare-earth-like, and that the dioxide structure persisted from ThO_2 to CfO_2, elements 90 through 98.

Twenty years earlier, Niels Bohr had predicted the occurrence of transuranium elements as a 5f series, with the series beginning at element 95. We now know conclusively that in the metals of this series, localized 5f electrons first appear with element 95 (americium) and in neighboring trivalent curium, element 96, the 5f shell is just half filled.

In an interview recounting those days and events pertaining to the history of Zachariasen's association with the University of Chicago, we find the following quotation with its significant ending:

". . . usually we found that the compounds which were present in the sample were not what they had intended to make. We had a very exciting time struggling with all these patterns over the various plutonium compounds, identifying what the chemists had made and, hence, getting information about the chemistry of plutonium that was essential. . . I remember working like hell on New Year's Day and all holidays; often I worked late for many, many hours to get the work done. I had a wonderful time. . ."

The Book

In 1945, Zachariasen published his classic and tightly written book, *"The Theory of X-Ray Diffraction in Crystals."* Professor Pepinsky, himself a noted crystallographer, wrote of it: "The physical chapters serve as a basis for most of the developments in scattering theory since their publication; and many a contemporary paper is no more than a direct expansion of one or another paragraph." In 1948-49, Zachariasen published 26 papers, an heroic effort.

Direct Methods

In 1952, Professor Zachariasen pioneered the use of "Direct Methods," for determining crystal structures. The extraction of detailed structural information from X-ray measurements requires the knowledge of two fundamental classes of information: The magnitudes and the phases of the structure factors. The values of the X-ray intensities are obtained directly, and are equal to the squared magnitude of the structure factors. For a complete solution of the structure, knowledge of the phases of the structure factors are necessary. Before 1952 the inability of the X-ray experiment to provide this latter information was referred to as the *phase problem.* This problem severely limited the application of the potentially powerful X-ray technique. The structure of solids containing small numbers of heavy atoms could be deduced more or less directly from the intensity pattern, but other substances were impossible to decipher except for the rare occasion when a model could be devised intuitively. The more powerful methods for deriving phase information depend on the statistical distribution of the intensities. This methodology is referred to as the *direct method* in crystallography.

In 1952 Zachariasen devised an algorithm for applications of this new method and successfully used it to determine the structure of boric acid. Although direct method techniques have been refined considerably in the ensuing years, Zachariasen's work remains pertinent to the modern methodology.

The Extinction Problem

In a series of brilliant papers published over the years 1963-1967 he described a practical solution to the problem of secondary extinction, one of the major unsolved problems in X-ray crystallography. Secondary extinction arises for reflections of such large intensity that an appreciable amount of the incident radiation is reflected by the first planes encountered by the beam. Since deeper planes are thus presented with less incident radiation, their reflections are reduced in intensity. This effect is strongly dependent on the types and degrees of imperfections of the scattering crystal. The stronger the scattered intensity the more serious is the extinction. Scattered rays closest to the incident beam are generally most intense and it is specifically these low-angle reflections that are important in the accurate determination of valence electron distributions and the positions of light atoms such as hydrogen. In severe cases, secondary extinction problems can lead to erroneous crystal structure determinations.

Some exceptionally accurate measurements on quartz caused Zachariasen to recognize a fundamental error made by diffractionist Darwin more than 50 years before in deriving the equations for the extinction effect. By the end of 1966 Zachariasen had circulated a preprint of a paper in which a practical solution to the extinction problem was described. It is a landmark of modern diffraction theory, ameliorating many of the problems that had held back precise structure determination.

Los Alamos Associations

In the early 1950's Zach began to consult at Los Alamos during the summers. During that period he accomplished a scientific *tour de force*. He succeeded in solving the complicated structures of plutonium metal, where others had labored fruitlessly even with computer assistance. In their book, *"The Metal Plutonium,"* Coffinberry and Miner state:

> "It is highly improbable that any scientist other than Zachariasen could have solved the three structures as complex as those of alpha, beta, and gamma plutonium from powder patterns alone."

He was simply unique in his ability to derive quantitative information from a powder diffraction pattern of a complex mixture. Often after what seemed like a glance, he would suggest cell size and composition. He had an uncanny knack for suggesting the correct atom positions. Only in recent years would he use a pocket calculator. He memorized trigonometric and log values to save time, and combined this with an incredible store of crystallographic information.

I was exposed to this ability and his extraordinary recall in 1953. We were then working with americium compounds. I showed him a film of one and he remarked, "Oh yes, I remember a similar pattern from an unknown plutonium compound, about six years ago." Our knowledge of the composition of the americium compound in combination with his X-ray data provided compositions and structures of both compounds. That was the beginning of a collaboration that continued, often including Larry Asprey, on the structures of the transplutonium metals. Zachariasen also continued his collaboration with Finley Ellinger and Fred Schonfeld. In such collaborations he never took offense in the rare instances when he was wrong and was able to offer guidance in such a tactful way that you were never offended.

With retirement from the University of Chicago and his move to Santa Fe in the 1970's, there were dinner parties given by Willie and Mossa (his pet name for her in their Santa Fe home and gracious entertaining of friends from Santa Fe and Los Alamos. On the Hill there were parties involving the Agnews, Argos, Bradburys, Cowans, Evanses, Halls, Hoerlins, Hoyts, Kings, Marks, Matthiases, Metropolises, Richardsons, Rosens, Spences, Steins, Suydams, Tucks, Turkeviches, and Ulams. There were also poker sessions with a precise dichotomy: the *Chamber Music Society* allowed poker only in the purest form, while the *Symphony Society* permitted a wider variety of play.

For several years a Thursday ritual was established: Zach would come to Los Alamos to consult, first with Finley Ellinger and Fred Schonfeld in the morning; then after their lunch he would come to visit Larry Asprey, Bob Ryan, and me. In mid afternoon he would then go to the labs of Al Giorgi and Gene Szklarz, where he was invaluable in identifying the complicated mixtures involved in their superconductivity work. He went yearly to Bernd T. Matthias's lab in La Jolla for identification of minor but important components in the compounds that resulted from their search for high-temperature superconductors. That particular collaboration produced a series of important papers.

Recently he returned to the theme of 5f element bond lengths and bond strengths, providing the definitive tabulation of their values and simple equations to reproduce them.

A World Figure

On the world scene, in 1975 I had occasion to introduce Professor Zachariasen at the Baden-Baden International Conference on Plutonium and The Other Actinides. I mentioned that he was then in his 50th year of publication, publishing his first paper before most in the audience were born. He received a standing ovation. In his self-effacing way he dismissed the tribute by saying, "I was born young." This year is his 55th year of contributing to the literature of science, and I shall always be proud of sharing in his final publication.

To an extent he was an anachronism, a scientific giant out of the times when science was funded from personal resources. For years, he paid his own way to meetings and was proud that he did not seek grants. Others took full advantage of the ready support for post-war science to build personal scientific empires. Zach did not; he felt it took the scientist-administrator too far from the science of the matter.

One measure of scientific impact is the extent that others make use of your work. It is scientific courtesy to acknowledge that debt by citing the former work. The system breaks down, because work of great importance often becomes just part of the lore and the more recent users are cited, not the original. This is true of Zach's work. Nonetheless, he was cited 3600 times in the period 1965-1975. This is an average of once for each day of that 10 years. It is clear that his name will remain bright in the literature of science.

To close, it is appropriate to repeat the quotation from Zachariasen's discussion of his arduous work during the war years, "I had a wonderful time."

Willie, so did those who know you!

CHAPTER 3

PETER JOSEPH WILHELM DEBYE

Dan McLachlan, Jr.
Ohio State University, Columbus, OH 43210

In 1950, the last year of the two societies CSA and ASXRED, our president of ASXRED was Peter Debye. We have never before or since had a fellow crystall-lographer whose career and achievements so thoroughly overwhelmed our imaginations. He had come from Berlin in 1940 to avoid becoming a citizen of Germany at the insistence of the Hitler regime. He had already become a Foreign Member of the Royal Society of London in 1933, a Foreign Associate Member of the National Academy of Sciences in 1931 (later to become a full member in 1947) and had earned the Nobel Prize in Chemistry in 1936. Dr. Debye has won several medals including the Rutherford Medal, the Lorenz Medal, the Franklin Medal, the Willard Gibbs Medal and the Max Planck Medal. He has also received honorary doctor degrees from the Universities of Liege, Brussels, Oxford, Sofia, Harvard, Brooklyn and St. Lawrence and has been elected to membership in fifteen prestigious academies.

When Debye was invited to give the Baker Lectures in Cornell in 1940, he decided to stay. He used this invitation as a means of getting out of Germany. World War II was becoming an extremely serious event in Europe, and the United States officials could see that tooling up for what was to follow would require contri-butions from the best minds available. Debye's response to these needs was quick, enthusiastic and effective. He immediately embarked on research and consulting activities at Oak Ridge, Bell Labs and many other organizations bent on catching up and surpassing the scientific and military progress of Germany and her allies. Even after the war was over he continued to help the United States promote science and education in this country. For example, after the death of Robert A. Welch of Texas in 1952, Debye was elected as one of the nine original Scientific Advisory Board Members of the Robert A. Welch Foundation, later to include such men as Henry Eyring, Arthur Cope, Glenn Seaborg, Wendell Stanley and Roger Adams. To show how beneficial to American science this organization has been under the Director of Research, W. O. Milligan, we can cite its achievements in 1975-76. The Foundation gave grants to "1500 persons including 347 principal investigators, 261 postdoctoral and 546 predoctoral fellows and 346 undergraduate scholars." The Foundation was so appreciative of Debye's achievements that they had Dr. W. O. Baker, President of Bell Labs give an address on "Peter Joseph Wilhelm Debye" at their *Proceedings of the Robert A. Welch Foundation, Conference on Chemical Research, XX, American Chemistry-Bicentennial*, 1976. The reader is advised to read the 45 pages of Dr. Baker's text in which he furnishes equations and figures referring to Debye's chief works.

From Dr. Baker's address one can make the following summary of Debye's movements through his life.

1884	Born in Maastreicht in the Netherlands
1901	At age 17 moved to Aachen, Germany
1905	Graduated from Aachen under Sommerfeld. He was also influenced by Max Wien
1906	He followed Sommerfeld to Munich when Sommerfeld was offered a head position there.
1908	Graduated from Munich at age 24
1911	Moved to Zurich to take the place of Einstein, who was moving to Prague
1912	After only one year at Zurich he went to Utrecht, the Netherlands
1913	Again, after only a year he moved from Utrecht to Göttingen, Germany, where he developed the powder method with Scherrer
1920	After staying seven years at Göttingen he went to Zurich where he stayed seven years and established a strong school of X-ray diffraction
1927	Moved to Leipzig and stayed seven years
1934	He went to Berlin to be the Professor at the University of Berlin and Director of the Kaiser Wilhelm Institute.
1940	Arrived at Cornell University to give the Baker Lectures and stayed
1946	American citizenship
1950	Professor Emeritus
1966	Died

The information contained in the above was taken partially from Dr. W. O. Baker's address and also from a biography written by Raymond M. Fuoss, published by Interscience Publishers, Inc., New York (1954) V-XXI. In the same issue is a list of Debye's publications selected by Debye himself. There are in this list 11 papers on X-ray scattering, seven papers on dipole moments, 11 papers on electrolytes, 12 papers on light scattering and 11 miscellaneous papers, making a total of 52 papers of which Debye was proud enough to mention. In this same issue Herman Mark discusses the papers on X-ray scattering, Charles P. Smythe discusses the papers on molecular dipole moment and polarizability, Raymond M. Fuoss discusses electrolytes, and Herman Mark discusses light scattering.

Other sources on the life and works of P. Debye are to be found in *Biographical Memoirs of Fellows of the Royal Society* (1970) Vol. 16, page 175, by M. Davis and *Biographical Memoirs of the National Academy of Sciences* (1975) Vol. 46, page 23 by J. W. Williams. Among his friends at Cornell published in *Science* (1967) Vol. *155*, page 979, F. A. Long wrote a praise-worthy account of Debye and his work.

To quote Raymond M. Fuoss, "Debye is known to his colleagues through his published work, but better by his active participation in scientific meetings, and by the lectures he delivers on various occasions. When it is known that Debye will be at the meeting, it is invariably the cue for everyone who can get there because we have learned that worthwhile things will be said. . ." It is in light of this remark by Fuoss that the writer of this chapter of history was always eager to go to meetings where Debye was likely to attend. I saw Debye at meetings of the American Physical Society, and at the informal meetings on solid state physics in New York during the forties. I had only one long conversation with him and that was during a train ride during the war; and when he was the ASXRED president in 1950 we all benefited from Debye.

An article that is well worth reading by crystallographers and physicists in general was written by Professor Felix Bloch of Stanford University in *Physics Today*, Vol. 20, No. 12, page 23 (1976) on the 50th anniversary of quantum mechanics entitled "Heisenberg and the Early Days of Quantum Mechanics." He has much to say about Debye. He heard Debye say something like, "Schrödinger, you are not working right now on very important problems anyway, why don't you tell us something about that thesis of de Broglie which seems to have attracted some attention." And Schrödinger, *did* later give just such a speech.. But Debye was still not satisfied. He remarked that as a student of Sommerfeld he had learned that, to deal with waves, one had to have a wave equation. At the next meeting, Schrödinger gave his famous Schrödinger wave equation. I tell this in this chapter to show how Debye influenced scientific gatherings and to remind us in North America how fortunate we were to have had the benefits of his presence for 26 years.

It is interesting that although Debye was with us during the years 1940 to 1966, while the phase problems and computers of various kinds were revolutionizing structure analysis, he never threw himself wholeheartedly into either field. I suppose crystallographers will remember him most for the Debye temperature factor, the atomic form factor and the introduction of the powder method.

CHAPTER 4

J. W. GRUNER

Tibor Zoltai
University of Minnesota, Minneapolis, MN 55455

John Walter Gruner was born in 1890 in Neurode, a small town in the Sudetenland which is now part of Poland. After finishing his high school education he worked for two years and saved enough money to come to the U.S. in 1912. Three years later he enrolled at University of New Mexico and received his B.A. in 1917. He continued his studies at the University of Minnesota and earned an M.S. in 1919. After that he accepted an assistant professorship at Oregon State University for an academic year in 1919 then returned to Minnesota as an instructor and completed the Ph.D. requirements in 1922. Soon after that he was promoted to assistant (1923), to associate (1926) and to full professorship (1944). With the exception of two sabbatical leaves (1926-27 and 1937-38) he taught continuously at the University of Minnesota until his retirement in 1959. He was an excellent and thorough teacher. He set very high standards of performance for himself in teaching and research. Only the best students could satisfy his expectations and in spite of that 12 have completed Master's and 28 Ph.D. programs under his direction. They all became successful members of the scientific or the industrial community. The number of undergraduate students who took his mineralogy and lithology courses (or physics and geography courses during the war) number over a thousand.

Dr. Gruner, J. W. to his friends, was an equally productive research scientist. He published over 100 scientific papers of which many became classics in mineralogy and crystallography. He was among the first professors of mineralogy in the US to teach X-ray diffraction (1927) in a geology department. He and his students determined a number of crystal structures, including several difficult layer silicate structures. All of his early crystal structure models have since been refined and almost all were found to be basically correct. For decades he was the leading authority in the mineralogy and geology of the Iron Formations of Minnesota and of the radioactive mineral deposits of Colorado. The value of his contributions to mineralogy and geology were recognized by his colleagues, who elected him as their president of the Crystallographic Society of America (1947-1948); Mineralogical Society of America (1949-1950). After his retirement his contributions to science were recognized by the award of the Roebling Medal (1962) of the Mineralogical Society of America, the highest scientific award in the field of mineralogy.

Other recognition of his achievements included: an Honorary Doctor of Science degree of the University of New Mexico (1963), and the Distinguished Service Award of the University of Minnesota Chapter of Sigma Xi (1960), in which chapter he served as president (1953-1954). In 1965 the University of Minnesota sponsored and NSF supported an "International Conference on Rock-Forming Minerals" which was dedicated to his 75th birthday. In 1972, a group of his former students wrote and dedicated in his honor GSA Memoir No. 135 entitled *"Studies in Mineralogy and Precambrian Geology"*. This publication is referred to as the *"Gruner Volume"*.

In 1919 he married an equally exceptional person, Opal Garrett. In the following years they had three children: Wayne (1921), Hazel (1924) and Garrett (1928), and maintained an ideal and happy marriage till her death in 1966.

All of us who were fortunate to know him as a scientist and as a man bend our heads with sorrow to the will of God. It appears to be appropriate to close our commemoration with the message of S. A. Underwood, the favorite poem of Opal and J.W.:

"We need it ev'ry hour – a purpose high,
To give us strength and pow'r to do or die,
We need it ev'ry hour – a firm, brave will,
That though hate's cloud may low'r, shall conquer still."

Biographical Date of John W. Gruner

B.A., University of New Mexico	1917
M.S., University of Minnesota	1919
Ph.D., University of Minnesota	1922
Assistant Professor, Oregon State University	1919-1920
Instructor, University of Minnesota	1920
Assistant Professor, University of Minnesota	1923
Associate Professor, University of Minnesota	1926
Professor, University of Minnesota	1944
Director, Rock Analysis Laboratory, University of Minnesota	1947-1949
Sabbatical leave, University of Leipzig, Germany	1926-1927
Sabbatical leave, consultant to H. A. Brassert Co., Germany	1936-1937

Member of the Geological Survey of Minnesota	1920-1948
Fellow of Geological Society of America	1923
Fellow of Mineralogical Society of America	1927
Fellow of Society of Economic Geologists	1928
Member of Advisory Board, American Gem Society	1939
Vice-president, Mineralogical Society of America	1943
Membership on committees of National Research Council	1940-1944
President of Crystallographic Society of America	1947-1948
President of Mineralogical Society of America	1948-1949
Vice-president of Geological Society of America	1949-1950
Committee Chairman of Mineralogical Society of America Award	1951
Committee Chairman of Roebling Medal Award of Mineralogical Society of America	1953
Nominating Committee of Mineralogical Society of America	1954
President of Sigma Xi Honorary Scientific Society (local chapter)	1953
Member of Fulbright Fellowship Awards Committee	1954
Recipient of Roebling Medal of Mineralogical Society of America	1962
Honorary Doctor of Science, University of New Mexico	1963

CHAPTER 5

LESTER H. GERMER

Alfred U. Mac Rae
Bell Laboratories
Murray Hill, NJ 07974

The December, 1927 issue of the *Physical Review* contained an article, "Diffraction of Electrons by a Crystal of Nickel", by Clinton J. Davisson and Lester H. Germer of Bell Telephone Laboratories. This publication is a classic in the history of science, since it established the wave nature of the properties of the electron. It confirmed the theoretical work of Louis de Broglie and stands today as one of the most significant events in the establishment of quantum mechanics as a viable description of matter. This work led to a Nobel Prize for Davisson, which was shared with G. P. Thompson. However, students know of this experiment as the Davisson-Germer experiment. It involved impressive experimental techniques, the handling of very low energy electrons ($\sim$ 100 eV), the establishment of very high vacuum and the measurement of low currents. The American Physical Society established the Davisson-Germer Prize for outstanding work in the field of electron and atomic physics as a tribute to this early work.

Following this experiment, Lester Germer continued to use high energy electron diffraction techniques to study a wide variety of materials and phenomena. In 1958, he returned to the field of low energy electron diffraction, a technique that was rarely used after the original work of 1927. He is recognized as the father of the present equipment, which is used all over the world for the studies of surfaces. This equipment provides for the display of the diffraction pattern directly on a fluorescent screen. He publicized the technique in stimulating lectures and papers and was directly responsible for urging the commercial availability of the equipment. His strong impact on the surface community is evidenced by the great number of papers that are published and presented at meetings each year on the subject of low energy electron diffraction.

Lester H. Germer's scientific achievements are well documented in the technical literature. However, his personal traits are indelibly recorded in the memories of the many people who interacted with him. There are numerous "Lester" stories, most of which are associated with his vigorous physical pursuits which were a very important part of his life. He had disdain for people who used elevators. One of my first memories of Lester, after I joined the Labs in 1960, was chasing after him as he ran up the four floors from our lab to the stockroom. He introduced me to rock climbing and we spent many days together on the cliffs of the Shawangunks in New York State, followed by sleeping out in the open with the bare minimum of camping gear. Although he was a latecomer to rock-climbing, he distinguished himself by leading difficult climbs at an age when most people would be content with fireside naps. He was a one man climbing school, introducing hundreds of people to this activity with his enthusiasm and patience with beginners. At the conclusion of my first climb with him, I can remember telling him that "I was scared." His response was, "Good, you'll make a good climber (not true), if you don't experience a thrill over this activity you might as well walk up a stairs." Lester tried to participate in "thrills" via new experiences in both science and his outdoor activities. His science was characterized by being alert to the occurrence and interpretation of new results. He was devoted to hiking and led many a weekend hike, stooping to identify mushrooms, ferns and ants. Shortly before his death, he made a very strenuous hike into the foothills of Mt. Everest.

He served in World War I as a fighter pilot in the U.S. Army. For his service on the Western Front, he received a citation from General Pershing. He was credited with downing four German warplanes. He enjoyed telling stories about flying those rickety planes, and hiking into Paris.

Following his retirement from Bell Labs in 1961, he joined the staff of Cornell University, where he continued his low energy electron diffraction studies and enjoyed intimidating younger staff members who were unable to beat him in his daily mile run.

Lester died of natural causes while rock climbing on the Shawangunk cliffs on Sunday, October 3, 1971. He died as he loved to live, in the pursuit of physical activity with the companionship of friends. Lester rarely made concessions to his chronological age; his death occurred while negotiating a series of strenuous moves to get past an overhang, halfway up the last climb of the weekend. This weekend of climbing followed a Friday lecture on surfaces and was to precede a week of research at Ithaca. Lester was an exciting personality who was continuously "high" on life, and enjoyed the companionship of young people. He introduced a countless number of people to the excitement of outdoor activities and scientific research.

CHAPTER 6

LAWRENCE OLIN BROCKWAY

Jerome Karle
Naval Research Laboratory, Washington, DC 20375
and Lawrence S. Bartell
University of Michigan, Ann Arbor, MI 48109

As a graduate student, Lawrence Brockway and his thesis adviser, Linus Pauling, introduced gas electron diffraction into the United States in 1932. He carried out a great number of structural investigations in the 1930's in which he established the characteristic configurations and dimensions of a variety of important structural groups. This work played a significant role in facilitating understanding of the nature of chemical bonding and in paving the way for subsequent advances in the fields of chemistry and molecular biology. For his work in structural chemistry, Lawrence Brockway was an early recipient of the American Chemical Society Award in Pure Chemistry in 1940.

Shortly before he died on 17 November 1979, Lawrence Brockway was preparing an article for a book commemorating *"Fifty Years of Electron Diffraction"* edited by P. Goodman (Dordrecht: Reidel) and it appeared in 1981. We present here the incompleted manuscript that describes briefly in his own words his early experiences.

"My own connection with electron diffraction began in the fall of 1930 when I appeared as a new graduate student at the California Institute of Technology. It was the custom there to start each new student on a research project very early in his first term, and when I expressed an interest in the structure of matter the department chairman, Professor A. A. Noyes, turned me over to Linus Pauling. Pauling's first suggestion was that I should embark upon the seas of crystal structure determination by X-ray diffraction. For some reason still unknown to me the various projects he suggested seemed unattractive and I kept refusing. Finally, in desperation he spoke of an experiment he had seen in the summer of 1930 while he was visiting the laboratories of the I. G. Farben Industrie carried out by Mark and Wierl. They had given Pauling some prints of diffraction patterns obtained from carbon tetrachloride using a stream of electrons crossing a jet of the vapor.

"Although there had been no publication describing either the equipment or the method of interpreting the recorded pattern, I felt I should agree to try the experiment before Pauling became completely disenchanted with his new graduate student. It was agreed that I should turn to Professor Richard M. Badger, an infrared spectroscopist, for guidance in the design and construction of a vacuum unit enclosing a source for an electron beam, a vapor jet and photographic plates for recording the scattered electrons. Our first try utilized a hydrogen discharge tube as a source of electrons with the hydrogen pressure adjusted by temperature control on a palladium filter. This was so unreliable that we next turned to a heated filament surrounded by an electron lens with adjustable bias potential. The vacuum pumps on the system were of the mercury-in-glass type necessarily constructed in our own shop but the pumping speed was too slow to cope adequately with the jet of vapor injection for each exposure. Nonetheless, after about three months we had obtained diffraction patterns from thin gold foils and also from carbon tetrachloride gas, showing barely detectable maxima and minima." (See Plate 10, page 188).

"My contact with Pauling in the meantime had been very scant, and he was surprised and delighted when I was able to show him diffraction patterns made in our laboratory. After a series of improvements in the apparatus we began to get much clearer patterns and proceeded to apply the method to as many substances as we could readily obtain. It was early in 1931 when the first publication from Wierl's experiments appeared, and we used his simplified expression relating the intensity of scattered electrons at various angles to the interatomic distances in the gaseous molecules.

"The calculation of theoretical intensity for various molecular models involved a double summation over terms of the type $(\sin sr_{ij})/sr_{ij}$. Our first calculations used a plot of $(\sin x)/x$, drawn on a very large scale, from which we tabulated values of the function with an appropriate adjustment of scale factor for each of the terms. The next improvement was a table of $(\sin x)/x$ constructed by Jack Sherman. After that Paul Cross assisted in the preparation of sets of strips containing $(\sin ax)/ax$ values with each strip having a fixed a value. To carry out the calculation for a given molecular model the procedure boiled down to selecting the set of strips with a values corresponding to the interatomic pairs in the model and then summing up at each value of the angular coordinate x the terms on the strips with an appropriate coefficient applied to each one. At first the strips were laid out on a drawing board with a straight edge to mark the particular x value being summed, but shortly we devised a cylinder around which the strips were wrapped, and later students and I spent many days with the cylinder-mounted strips adding up sums of products with the aid of the then standard desk calculator. Because of the extensive labor involved in the calculations, the selection of models had to be judged carefully or we found ourselves spending weeks on models which were unrelated to the structure at hand.

"The comparison between theoretical curves and the experimental data was based at first on a comparison of the positions of the maxima and minima in the curves. The diffraction negatives were observed visually and the diameters of the apparent maxima and minima were measured using fine pointers. Measurements of the negatives on a recording microphotometer failed to show any maxima and minima but only fluctuations about a very rapidly falling background. For this reason our observations of the first hundred or so substances we analyzed were all reported in terms of the diameters of the maxima and minima which appeared in the visual measurements.

"By 1933 when I received my doctoral degree, other laboratories had entered the field and major improvements in both the experimental technique and the methods of calculation were beginning to appear. I stayed on at the California Institute of Technology for four years, and then joined the faculty of the University of Michigan.

"The last electron diffraction unit I designed and constructed at Cal Tech was supposed to be greatly improved and included a cylindrical camera which could record electrons scattered up to about 165°. I am not sure what enthusiastic dream led to this, but it is no surprise now that our molecular patterns never even reached the edge of our flat plate camera and the cylindrical camera was relegated, untested, to a museum.

"Subsequent developments are well known. Experimentally the most important of these was the introduction of a rotating sector before the recording photographic emulsion; the sector opening could be cut in such a way as to compensate largely for the rapidly falling background and real maxima and minima then appeared in the microphotometer records. The fit between calculated and observed modified intensity curves was considered over the whole continuous range of scattering angles, an improvement over looking only at positions of maxima and minima. Major improvements in the calculating methods lay first in the use of better computing techniques, such as IBM punched cards, and later in the introduction of more complicated theoretical expressions made feasible by the adoption of high speed computers. The earliest refinement which led to great improvement in the comparison between experiment and theory was the use of proper atomic scattering amplitudes expressed as functions of the scattering angle.

"The developments after the later 1930s are fairly well known. My own recollections of the earliest days are centered around the sense of excitement and fun, and an appreciation of the opportunity to work in a major scientific development while still enrolled as a new graduate student."

Lawrence Brockway was born in Topeka, Kansas in 1907. He received his bachelor's and master's degrees from the University of Nebraska in 1929 and 1930 and his doctoral degree from Caltech in 1933. He held a four-year appointment as senior fellow in research at Caltech and spent a year during 1937-1938 as a Guggenheim Memorial Foundation Fellow at the Royal Institution and at Oxford University where he was associated with N. V. Sidgwick. He joined the faculty of the University of Michigan in 1938 where he completed his career as professor emeritus of physical chemistry. During the years of World War II, he applied his expertise to defense projects with the Naval Research Laboratory, the National Advisory Committee for Aeronautics and the General Electric Company. This consulting work involved the application of electron diffraction to solid surfaces and influenced the broadening of his interests to include surface phases, an area of major interest to him in the last two decades of his career. It also stimulated interest in the somewhat esoteric study of the structure of absorbed monolayers of hydrocarbon films that have the property of being oleophobic and are of value in applications to boundary lubrication (Karle and Brockway, 1946). His pioneering research correlating chemical reactivity at surfaces with the underlying structure is summarized in the literature citations of his final article on the subject (Fan and Brockway, 1974).

Lawrence Brockway helped to found the American Crystallographic Association in 1950. It was formed from a merging of the Crystallographic Society of America and the American Society for X-ray and Electron Diffraction. He was appointed to the executive committee, division of physical sciences, of the National Research Council and served as vice chairman and chairman of the NRC National Committee for Crystallography during the period 1951-1953. As an active participant in the affairs of the International Union of Crystallography, Lawrence Brockway twice served as chairman of the Commission on Electron Diffraction of the IUCr.

Lawrence Brockway's scientific career developed with unusual rapidity. His outstanding review article on gas electron diffraction that appeared in the *Reviews of Modern Physics* in 1936 illustrates how soon he mastered his subject and made major contributions to the field of structural research. His measurement of the distribution of atomic electrons a third of a century ago remains unsurpassed in resolving power and is still found in current textbooks (Bartell and Brockway, 1953). Having achieved considerable recognition as a scientist early in his career, he applied himself with enthusiasm and dedication to education and became a distinguished teacher. His graduate students went on to make significant contributions to the field of structural research. As the years passed, his interests evolved more and more toward the humane rather than the technical aspects of science. He became a valued counselor of students and devoted much time and effort to this activity. Many students in need of encouragment and program counseling benefited from these efforts. He was called back two years after his retirement in 1976 to teach a special seminar course for highly motivated undergraduates. The seminar was very successful and the administration asked him to continue despite his retirement. Lawrence Brockway was a man of great vitality who had a zest for all aspects of living, artistic and humanitarian as well as scientific and technical. He enriched the lives of students, colleagues, friends and acquaintances in many ways.

References

Brockway, L. O. (1936). Rev. Mod. Phys. **8**, 231-266.

Bartell, L. S. and Brockway, L. O. (1953). Phys. Rev. **90**, 833-838.

Fan, P. L. and Brockway, L. O. (1974). J. Electrochem. Soc. **121**, 1534-1537.

Karle, J. and Brockway, L. O. (1946). J. Chem. Phys. **15**, 213-225.

CHAPTER 7

ROBERT E. RUNDLE

R. A. Jacobson
Ames Laboratory, Iowa State University, Ames, IA 50011

Robert E. Rundle was Vice President of the American Crystallographic Association in 1958 and became its President in 1959. The choice of Bob Rundle as one of the early ACA Presidents was natural in light of his numerous accomplishments in crystallography, the impact of his work on the scientific community, and the enthusiasm he had for science in particular and life in general.

Bob Rundle obtained a B. S. degree with high distinction in 1937 from the University of Nebraska, and went on to receive his M.S. degree there in 1938, completing sufficient research in this period for four publications. He then proceeded to enter California Institute of Technology and obtained the Ph.D. degree in 1941. His research there involved both X-ray and electron diffraction and was under the direction of Drs. Don Yost, J. H. Sturdivant, Linus Pauling, and Verner Schomaker. He joined the staff of Iowa State University as an Assistant Professor of Chemistry in 1941, and then after a brief stay at Princeton University, returned to Iowa State as Professor of Chemistry in 1946. At the time of his death – October, 1963, he was Distinguished Professor of Chemistry and Physics in the College of Science and Humanities and Senior Chemist in the Ames Laboratory of the Atomic Energy Commission.

Robert Rundle was the author of more than one hundred publications in chemistry. His research career was largely devoted to the study of the nature of chemical bonding, and his contributions have been many and significant, including extensive investigations into the chemistry and structures involved in starch and starch complexes, uranium and thorium compounds, hydrogen bonding, charge transfer complexes and transition metal complexes, as well as some pioneering work in neutron diffraction. Bob Rundle's broad outline of the theory of bonding in electron deficient compounds is still widely held today. He used this theory not only to explain existing compounds but also to predict the structures of compounds then unknown. He correctly predicted the structures and relative bond lengths of xenon difluoride and xenon tetrafluoride in advance of measurements.

CHAPTER 8

WALTER CLARK HAMILTON

T. F. Koetzle
Brookhaven National Laboratory
Upton, New York 11973
and
S. C. Abrahams
Bell Laboratories
Murray Hill, New Jersey 07974

Walter Hamilton, the twentieth president of the American Crystallographic Association, was a man of extraordinary vitality and talent. His untimely death at age 41 deprived the crystallographic community of the anticipated flow of further insights from his brilliant research and exuberant personality. An indication of the impact of his research on current work is provided by the more than 300 citations made to his publications in 1979, six full years after his death.

Walter was born in Texas on February 16, 1931. He was brought up in Stillwater, Oklahoma and graduated in 1950 with a B.S. degree in chemistry from the Oklahoma Agricultural and Mechanical College, where his father was professor of mathematics. Following a year at the Eidgenössiche Technische Hochschule in Zürich, he entered the graduate school in chemistry at the California Institute of Technology with Verner Schomaker as his research adviser. Walter's thesis topic was mainly concerned with gaseous electron diffraction.

Following his graduation with a Ph.D. degree in 1954, Walter worked for a year as a National Science Foundation Postdoctoral Fellow with Charles Coulson in Oxford, England. He joined the scientific staff at Brookhaven National Laboratory as a Research Associate in September, 1955 where he remained until his death. He became a Senior Chemist at the Laboratory and was appointed Deputy Chairman of the Chemistry Department in 1968. He was an excellent administrator, and the quality of his judgment is remembered by all who were privileged to work with him.

During his career at Brookhaven, Walter created a group that was at the forefront of research on structure analysis using both X-ray and neutron diffraction techniques. Walter had outstanding mathematical abilities, and contributed greatly both to the theory and to the practice of crystallography. Perhaps most notable among his theoretical works are those on extinction, and the development of statistical tests applicable to alternative structural models.

Walter was a driving force in the development of the neutron diffraction facilities at Brookhaven, first at the Brookhaven Graphite Research Reactor and later at the High Flux Beam Reactor. He creatively applied these facilities to the investigation of a wide range of structural problems including, but certainly not limited to, magnetic systems, ferroelectrics, the various structural forms of ice and, toward the end of his career, the naturally-occurring amino acids. Walter's work contributed tremendously to our knowledge of hydrogen-bonded systems.

Walter's enthusiasm and energy were almost legendary. He was the author of *"Statistics in Physical Science"* (1964) and also co-authored *"Hydrogen Bonding in Solids"* (1968), and *"Symmetry"* (1972). He published over 125 research papers. However, the record of his research accomplishments alone cannot adequately convey the extent of his importance to the scientific community. He was always a central figure at ACA meetings, in the thick of many discussions and ready to offer his insights on difficult problems. His distinctive voice, which retained a good deal of the tones of his Oklahoma upbringing, always heralded his stimulating presence. He served with great panache on the U.S.A. National Committee for Crystallography from 1964, and was elected a U.S. delegate to the International Union of Crystallography General Assemblies in Moscow, in Stony Brook and in Kyoto. He was an active member of the National Research Council Committees on Chemical Crystallography and on Computers in Chemistry. He organized a symposium on the topic "Computational Needs and Resources in Crystallography" in the spring of 1972, the proceedings of which were published as an influential report by the National Academy of Sciences after his death.

He was a Co-editor of *Acta Crystallographica* for five years and served on the I.U.Cr. Commission on International Tables from 1966 to 1972. He was Co-editor of Volume 4 of the *"International Tables for X-Ray Crystallography."* As a member of the I.U.Cr. Commission on Crystallographic Apparatus, he co-directed the I.U.Cr. Single Crystal Intensity Measurement Project which showed that the normal least-squares derived estimates of the standard deviations in atomic coordinates are not infrequently too small by a factor of about two. He also served on the I.U.Cr. Commission on Crystallographic Computing.

The year he was president of the American Crystallographic Association coincided with the Eighth International Congress of Crystallography which was held in Stony Brook, New York. Walter was Chairman of the Local Committee for the Congress and, in the eyes of many overseas visitors who attended, he personified

U.S. crystallography.

Walter had numerous qualities that made him loved and admired by his colleagues. He combined the ability to grasp the essence of a problem quickly with willingness to assist others, and was an inspiration to many younger scientists. He was scrupulously fair, always crediting an idea to its rightful originator. Walter also was a delightful and charming man who radiated enthusiasm in every action. Even today, more than eight years after his death, his memory remains vivid for many of us.

Following Walter's death, a W. C. Hamilton Memorial Scholarship Fund was established, with contributions received from friends and colleagues, to support research in neutron diffraction at Brookhaven by promising young scientists. Through early 1981, seven Hamilton Scholars have participated in this program, which perpetuates his concern for a field that interested him so profoundly.

CHAPTER 9

MAURICE LOYAL HUGGINS

Dan McLachlan, Jr.
Department of Geology and Mineralogy
Ohio State University, Columbus OH 43210

When our book "Crystallography in North America" was started in November 1979, several of our past presidents were already dead and their lives and the lives of those who have died since have been discussed in the previous chapters of Section B. Now we are confronted with the sad fact that our ninth past-president passed away in December 1981. He is no other than Dr. Maurice Loyal Huggins, the first president of ASXRED in 1940.

In discussing Dr. Huggins as one of our past-presidents, the task is made easier by the fact that his name is prominent in Section C 1 of this book where we discuss "The Organizations of American Crystallographers" and also in Section E 6 where Maurice himself presents a chapter on "The Introduction of Some Important Concepts and Principles."

Now it is fitting to enumerate the highlights of his career.

–Born in Berkeley, California, September 1897
–A.B. and B.S. University of California, Berkeley, 1919 after serving in World War I.
–Ph.D., University of California, Berkeley, 1922
–California Institute of Technology, 1922-25
–Stanford University, Instructor 1925 to 1926 Assistant Professor 1926-1933, Associate Professor 1933 to 1938
–Eastman Kodak Company, Research Associate 1936-1958
–Stanford Research Institute 1959-1962

Dr. Huggins has been awarded the Morrison prize from the N. Y. Academy of Sciences, 1941, the American Ceramic Society prize in 1956 and others. He has done research on intermolecular forces, hydrogen bonding, high polymers, solutions, photographic theory and crystal structure, all of which have attracted much attention and resulted in extensive travel throughout the world.

Dr. Huggins will be remembered for his researches and his instant cooperation in any matters dealing with the operation and organization of people in the field of crystallography.

SECTION C

ORGANIZATIONS

CHAPTER 1

THE ORGANIZATIONS OF AMERICAN CRYSTALLOGRAPHERS

Compiled by Dan McLachlan
Department of Geology and Mineralogy
Ohio State University, Columbus, OH 43210

(Taken from letters and notes supplied by Don Hanawalt, Maurice Huggins, Adolph Pabst, and others)

The historical progress of any branch of science is not a continuous process, but may be halted to wait for a crucial discovery, postponed because of events such as war, epidemic or depression, or speeded up by the needs of other branches of science. A good example of such effects is illustrated by the very origin of the discovery of X-rays. As far back as 1705 researchers in Europe had been interested in the phenomena produced by high potential electrical discharges through vessels from which air had been pumped out as completely as possible. Men who worked in this field were Hauksbee in 1705, Abbe' Nollet in 1753, Morgan in 1785, Faraday in 1825, Plücker in 1859, and Hittorf in 1869. (See G. L. Clark†). Nearly two centuries of research thus preceded Roentgen's discovery of X-rays in 1895. Seventeen years later, von Laue demonstrated X-ray diffraction and less than a year later the Braggs started determining structures. But not until 1916 did the first Americans, C. L. Burdick and J. H. Ellis, publish a structure.

The technique of X-ray diffraction from crystalline powders was developed in 1916 almost simultaneously by Debye and Scherrer in Germany and by Hull in the United States. These developments had a tremendous influence along four lines:

(a) the need for standardizing powder data
(b) the need for a formal committee to assemble and distribute the information
(c) the need for a society to represent and arrange for meetings of American crystallographers
(d) the possibility of starting a new journal just for crystallographers

The reasons for (a) and (b) are that powder patterns are different for different substances and therefore the powder camera is an analytical tool. In those days, chemists did not yet have the infrared spectrographs, mass spectrographs, electromagnetic resonance equipment, chromatographic techniques and other modern developments at their disposal. Chemists needed the results of powder data and cooperated wholeheartedly. There are on file at least fifty-five pages of correspondence between Maurice Huggins and others from 1936 to 1940 regarding this matter. Not only the American Chemical Society but also the American Society of Testing Materials, the National Bureau of Standards, the National Research Council and several other prestigious bodies lent encouragement.

We do not have space in this book for all the details of committee meetings and letters exchanged between the concerned people on the subject of organizing the crystallographers but one circular letter from Maurice Huggins indicates that there was, in 1940, a National Research Council Committee on X-ray and Electron Diffraction with members:

*Brockway	Hicks	*Richmond
*Buerger	*Huggins	Sisson
*Davey	*Jette	Tunell
Debye	Kerr	Warren
Fink	*Mark	Wyckoff
*Hanawalt	*Milligan	*Wyman
Hendricks	Nelson	Zachariasen

The committee met at Detroit on September 9, 1940. Those starred in the above list were present.

Copies of this 1940 circular were sent to the following:

Barton	Germer	Pauling
Crane	Magos	Proskauer
Fajans	Mayer	Urey
Fankuchen	Moody	

This circular is of historical importance because it refers to the first joint N.R.C. and A.S.T.M. subcommittee on powder diffraction, consisting of:

Davey	Hicks	Nelsŏn
Fink	Kerr	Richmond
Hanawalt	Magos	Wyman

This subcommittee stemmed from a previous cooperative committee, which existed in 1938, between chemists and crystallographers called the Committee on Application of X-rays to Chemistry and Chemical Technology.

More about the Powder File is included in a chapter in this book by Professor J. D. Hanawalt.

The year 1940 was important for crystallographers, with at least four letters written to Dr. M. L. Huggins by Professor Dr. P. Niggli of Zurich, Switzerland expressing his concern that the *Zeitschrift für Kristallographie* might have to be discontinued as well as the *Struktur Berichte.* These concerns added to the already established desire of the Americans to have a society for crystallographers and also a journal of their own. But Professor von Laue rescued the *Zeitschrift* and the question of a journal was postposned and eventually became unnecessary in 1946 when a meeting was held in London to establish the international journal, *Acta Crystallographica,* and also the International Union of Crystallography. This important meeting is discussed in separate chapters in this book by Dr. David Harker and Dr. Paul Ewald. Two years after this organizational meeting in London, the first meeting of the International Union of

†G. L. Clark, Applied X-rays. 4th edn. New York: McGraw-Hill, 1955, pp. 3-15.

Crystallography met at Harvard University. Dr. Robert C. Evans of Cambridge, England discusses this important meeting in a later chapter. The Union is a permanent body and meets triennially.

Prior to the establishment of the International Union, the Americans and Canadians continued their efforts to organize. Even before we had a name or a constitution there was a meeting at Gibson Island, Maryland in 1940. The subjects dealt with can be judged from the program which follows.

APPLICATIONS OF X-RAY AND ELECTRON DIFFRACTION

*****Maurice L. Huggins, Chairman***

July 29. J. L. Hoard, "X-ray Studies of Complex Inorganic Compounds"
H. Mark, "X-ray Studies of Long-Chain Compounds"

July 30 **I. Fankuchen, "X-ray Studies of Proteins"
D. Harker, "The Interpretation of X-ray Data on Proteins"

July 31 Wm. L. Fink and D. W. Smith, "Preferred Orientations in Metals"
Charles S. Barrett, "X-ray Diffraction from Strained Metals and Alloys"

August 1 **P. Debye, "Theory of X-ray and Electron Diffraction"
J. Y. Beach, "Electron Diffraction Studies of Gaseous Molecules"

August 2 H. R. Nelson, "Electron Diffraction Studies of Thin Films"

Not counting wives and children, thirty-eight people participated in the conference. Several are still active in crystallography today.

This first meeting was a very important one, not only because of the fine papers presented but also because of the work of the organizing committees and the plans for a 1941 meeting at the same place.

By 1941 we had a powerful committee comprised of the following:

Prof. L. O. Brockway, Prof. Peter Debye, Dr. Wm. L. Fink, Dr. J. D. Hanawalt, Dr. Sterling B. Hendricks, Dr. Victor Hicks, Prof. Eric R. Jette, Prof. Paul F. Kerr, Prof. H. Mark, Prof. W. O. Milligan, Dr. H. R. Nelson, Dr. Wayne A. Sisson, Prof. B. E. Warren, Dr. Ralph W. G. Wyckoff, Dr. L. L. Wyman, Prof. W. H. Zachariasen, Dr. Herbert R. Moody, Prof. M. J. Buerger, Prof. Wheeler P. Davey, Dr. W. E. Richmond.

This committee sent out a questionnaire to all known American and Canadian workers asking them to choose a name for the society. The American Society for X-ray and Electron Diffraction (ASXRED) was adopted, and 135 charter members were enrolled. Maurice Huggins was elected President for the remainder of the year with Bert Warren as Vice-President and George Tunell as Secretary-Treasurer.

**Deceased

The first official meeting of the new organization was held at Gibson Island, July 28-Aug. 1, 1941. About 75 were present . . . most of these, but not all, were members of the Diffraction Society.

PROGRAM

GIBSON ISLAND MEETING (1941)

AMERICAN SOCIETY FOR X-RAY AND ELECTRON DIFFRACTION IN

Cooperation with

Section C, AMERICAN ASSOCIATION FOR THE ADVANCEMENT OF SCIENCE

Monday, July 28
10:00 a.m. B. E. Warren, "Theory and Practice of Particle Size Determination."
7:30 p.m. M. J. Buerger, "New Experimental Techniques in Crystal Structure Analysis."

Tuesday, July 29
10:00 a.m. A. L. Patterson, "Fourier Series as Applied to Crystal Structure Determination."
7:30 p.m. C. S. Fuller, "X-ray Studies of Linear Polymers of Known Chemical Constitution."

Wednesday, July 30
9:00 a.m. N.R.C. Committee on X-ray and Electron Diffraction
10:00 a.m. Sir Lawrence Bragg, "Long Range Regularities in Crystals."
2:00 p.m. Business meeting
7:30 p.m. E. W. Hughes, "Recent X-ray and Electron Diffraction Work at the California Institute of Technology."

Thursday, July 31
10:00 a.m. L. O. Brockway, "Recent Electron Diffraction Work on Gas Molecules."
2:00 p.m. M. L. Fuller, "Applications of Electron Diffraction to Industrial Problems."

Friday, August 1
10:00 a.m. E. F. Burton, "Applications of the Electron Microscope."
2:00 p.m. V. K. Zworykin, J. Hillier, and A. W. Vance, "Preliminary Report on 300,000 Volt Electron Microscope."

This was the year that a constitution for ASXRED was written and adopted.

The second meeting of ASXRED was also held at Gibson Island, Maryland, with B. E. Warren as President.

The places and dates of ASXRED meetings are shown in the following table furnished by E. A. Wood and William Parrish.

MEETINGS OF ASXRED

1941 July 28-August 4	Gibson Island, Maryland
December 30-31	Joint meeting with Mineralogical Society of America, Boston, Mass.
1942 July 27-August 3	Gibson Island, Maryland
1943 January 23	Joint meeting with American Physical Society, Columbia Univ., New York, N.Y.
June 7-11	Univ. of Michigan, Ann Arbor, Mich.
1944 August 21-25	Gibson Island, Maryland
1945 October 1-4	Lake Geneva, Wisconsin
1946 June 10-14	Lake George, N.Y.
December 5-7	Joint meeting with Electron Microscope Society, Pittsburgh, Pa.
1947 June 23-26	Ste. Marguerite, Quebec, Canada
1948 March 21-April 3	Joint meeting with C.S.A. at Yale Univ., New Haven, CT.
December 16-18	Battelle Memorial Institute, Columbus, Ohio
1949 June 23-25	Cornell Univ., Ithaca, N.Y.
December 1-3	Franklin Institute, Philadelphia, Pa.

During these same years a group in New England guided by Martin J. Buerger and his colleagues organized the Crystallographic Society of America. The dates and places of their meetings were:

MEETINGS OF CSA

1946 March 21-23	Smith College, Northampton, Mass. See *Am. Mineral.*, 31, 508 (1946)
1947 March 19-21	U.S. Naval Academy Postgraduate School, Annapolis, Md. See *Am. Mineral.*, 32, 684 (1947)
1948 March 31-April 3	Joint Meeting with ASXRED at Yale Univ., New Haven, Conn. See *Am. Mineral.*, 33, 749 (1948).
1949 April 7-9	Univ. of Michigan, Ann Arbor Mich. See *Am. Mineral,* 35, 122 (1950).

In 1949 a new idea began to ferment in the minds of the leaders in Crystallography, that is the amalgamation of the two existing societies. An account of the proceedings leading to amalgamation is furnished by Dr. Pabst in the following news item entitled "Amalgamation of ASXRED-CSA" and this news item is followed by a letter from Dr. Howard T. Evans, Secretary of the new society, American Crystallographic Association (ACA) welcoming all new members and referring to the first (and very interesting) meeting at Penn State, January 1, 1950.

U.S.A. National Committee for Crystallography

The U.S.A. National Committee for Crystallography (USANCCr) is a committee of the National Academy of Sciences - National Research Council (NAS-NRC), appointed by its President. Members of ACA Council are *ex-officio* voting members of USANCCr. The purposes of the USANCCr are to promote the advancement of the science of crystallography in the U.S.A. and throughout the world and to effect appropriate U.S.A. participation in the International Union of Crystallography (I.U.Cr.) through the NAS-NRC, which adheres to the I.U.Cr. on behalf of the crystallographers of the U.S.A.

American Crystallographic Association

Inside the ACA, to bring people closer together whose interests are related, we have special interest groups as follows:

Applied Crystallography Group
Biological Macromolecule Group
Small Angle Scattering Group
Small Molecule Group

It is very difficult to list all the committee members and representatives over the years in this volume and even more difficult would be documenting all the communications and other diligent efforts they have made. An example of a huge task is the work of our Managing Editor, Sidney Abrahams, and his Associate Editor (now Editor of the most recent edition) Allan L. Bednowitz, in compiling a "Subject Index" correlating all fields of interest in crystallography with the names of the researchers listed in the fifth and sixth editions of the *"World Directory of Crystallographers"* for 1977 and 1981 put together by the same authors under the auspices of the International Union of Crystallography.

Aside from papers published in the leading scientific journals (including the *Zeitschrift für Kristallographie, Acta Crystallographica* and the *Journal of Applied Crystallography)* ACA has other publications as follows:

Publications of the American Crystallographic Association

Transactions

Volume 1
"Accuracy in X-ray Intensity Measurement"
S. C. Abrahams, Ed., 1965

Volume 2
"Machine Interpretations of Patterson Functions and Alternative Direct Approaches" and "The Austin Symposium on Gas Phase Molecular Structure"
W. F. Bradley and H. P. Hanson, Eds., 1966.

Volume 3
"Thermal Neutron Scattering Applied to Chemical and Solid State Physics" H. G. Smith, Ed., 1967.

Volume 4
"Low Energy Electron Diffraction"
D. H. Templeton and G. A. Somorjai, Eds., 1968.

Volume 5
"Crystal Structure at High Pressure"
D. B. McWhan, Ed., 1969.

Volume 6
"Intermolecular Forces and Packing in Crystals"
W. R. Busing, Ed., 1970

Volume 7
"Mechanisms of Phase Transitions"
S. Block, Ed., 1971

Volume 8
"Experimental and Theoretical Studies of Electron Densities" P. Coppens, Ed., 1972

Volume 9
"Biophysical Applications of Crystallographic Techniques" W. Love and E. Lattman, Eds., 1973

Volume 10
"Liquids and Amorphous Materials"
A. Bienenstock, Ed., 1974

Volume 11
"Applied Crystal Chemistry and Physics"
R. E. Newnham, Ed., 1975

Volume 12
"Instruments for Tomorrow's Crystallography"
H. Cole, Ed., 1976

Volume 13
"Fifty Years of Electron Diffraction"
L. O. Brockway, Ed., 1977

Volume 14
"Structural Aspects of Homogeneous, Heterogeneous, and Biological Catalysis" S. D. Christian, J. J. Zuckerman and L. J. Guggenberger, Eds., 1978

Volume 15
"Chemistry and Physics of Minerals"
G. E. Brown, Ed., J. R. Clark, Sup. Ed., 1979

Volume 16
"Structure and Bonding: Relationships between Quantum Chemistry and Crystallography"
T. F. Koetzle, Ed., 1980

Volume 17
"Diffraction Aspects of Orientationally Disordered (Plastic) Crystals"
R. Rudman, Ed., 1981

Volume 18
"New Crystallographic Detectors"
R. C. Hamlin, Ed., 1982

Monographs

Number 1
"The Photography of the Reciprocal Lattice"
M. J. Buerger, 1944

Number 2
"Fourier Transforms and Structure Factors"
Dorothy Wrinch, 1946

Number 3
"Solution of the Phase Problem. I. The Centrosymmetric Crystal"
Herbert Hauptman and Jerome Karle, 1953

Number 4
"On the Systems Formed by Points Regularly Distributed on a Plane or in Space"
M. A. Bravais, translated by A. J. Shaler, 1949

Number 5
"Crystal Data, Determinative Tables," 2nd Edition
J.D.H. Donnay, G. Donnay, E. G. Cox, O. Kennard and M. V. King, 1963

Number 6
"Crystal Data, Systematic Tables, 2nd Edition
Werner Nowacki, 1967

Number 7
"Symmetry of Crystals"
E. S. Fedorov, translated by David and Katherine Harker, 1971

Number 8
"Further Studies of Three-Dimensional Nets"
A. F. Wells, 1979

National Research Council Committee for Chemical Crystallography

This Committee was first organized in 1961 within the Division of Chemistry and Chemical Technology, to serve as a national focal point in chemical crystallography. At that time, the USA National Committee for Crystallography was a committee of the NRC Division of Physical Sciences, and it was felt that representation in both divisions would be beneficial for US crystallography. The first chairman was G. A. Jeffrey, 1961-1964 who was followed by D. H. Templeton, 1964-1968 and finally by S. C. Abrahams, 1968-1976. The Committee merged with the USA National Committee for Crystallography in 1977, becoming the Standing Subcommittee on Interdisciplinary Activities of the USANCCr.

Outstanding among its various activities were the formulation of a checklist for authors, referees and editors of crystal structure papers, the organization of a conference in Dartmouth College on the *"Critical Evaluation of Chemical and Physical Structural Information"* and another in Clemson on the application of current advances in computer science to crystallography.

The checklist was widely distributed and has now largely been adopted by many major journals. The proceedings of the Dartmouth meeting were published by the National Academy of Sciences in 1974, with title identical to that of the conference. The major papers presented at the Clemson conference have been published in *Acta Cryst.* (1977) **A33**, 4-24.

National Academy of Sciences of the U.S.A.

A list of crystallographers who are members of the National Academy of Science as compiled by Isabella Karle follows.

Physics
Bernd Matthias — U. of California, La Jolla (dec.)
William Zachariasen — New Mexico (dec.)

Chemistry
F. A. Cotton — Texas A and M
David Harker — Buffalo
J. L. Hoard — Cornell
I. L. Karle — NRL
J. Karle — NRL
W. N. Lipscomb — Harvard
L. Pauling — California
Gabor Somorjai — California
R. W. G. Wyckoff (emeritus) — Arizona

Geology
M. J. Buerger — MIT

Biochemistry
David Davies — NIH
James Watson — Cold Spring Harbor

Engineering
Charles Barrett — Denver

Applied Physics and Mathematical Sciences
Clifford Shull — MIT

Foreign Associates
D. Hodgkin — Oxford
J. C. Kendrew — Oxford
M. F. Perutz — Cambridge
Francis Crick — Salk Institute

Amalgamation ASXRED – CSA (as of Aug. 19, 1949)

This section has been prepared to give all members a general knowledge of developments leading to the formation of ACA.

1. At the Harvard Congress of the International Union of Crystallography several members of C.S.A. and A.S.X.R.E.D. were invited to a luncheon arranged by Prof. A. Pabst (then President of C.S.A.) to explore the possibility of forming a new Society which would combine the functions and have the objectives of the present two Societies. It was decided that, in principle, the idea was good and some working plans were considered. A committee consisting of M. J. Buerger, I. Fankuchen, A. Pabst, W. Parrish and E. A. Wood met at M.I.T., August 4, 1948, and discussed the problem in greater detail. Preliminary plans were drawn up and distributed to the Council members of both Societies.
2. In early October, 1948, J. D. H. Donnay, I. Fankuchen, W. Parrish, A. L. Patterson and E. A. Wood, representing the Councils of both Societies, met in New York City to consider furthering the preliminary plans. The results of these discussions were mimeographed and distributed to the members of both Societies November 30, 1948. The scope, name, meetings, publications, membership and procedure were briefly described. The scope is now stated as follows:
The object of this Society is given in Article II of the

proposed Constitution which states in part, "to promote the study of the arrangement of the atoms in matter, its causes, its nature and its consequences, and of the tools and methods used in such studies." Thus the new Society would have as its object the promotion of crystallography in its widest sense. It would be interested in all aspects of structural crystallography and in the physical manifestations of that structure. The experience of the two Societies has shown that in any crystallographic problem all branches of the science have application and that no single line of approach can stand alone. Thus in addition to the basic approaches of X-ray, electron, and neutron diffraction to structural problems the society would be interested in all the classical techniques of morphological, optical and physical crystallography. Opportunity would be provided for the separate discussion of scientific and industrial problems and for their joint discussion by groups with common interests.

3. The procedure proposed was outlined below:
   (a) Discussion of proposals and plans at business meetings of both Societies (see 4, below).
   (b) Preparation by an organizing committee of a draft of Statutes, By-Laws, Plans, etc., which should be approved by the National Research Council Committee on Crystallography.
   (c) Voting by members of both Societies as listed below.
4. The proposals were discussed at the business meeting of ASXRED at Columbus, Ohio, December 7, 1948, and of CSA at Ann Arbor, Michigan, April 8, 1949. Stenotypist transcription of the ASXRED meeting and wire recording and stenographic records of the CSA meeting were made and studied. The principal results of these discussions are indicated below:

PROPOSAL	ASXRED		CSA	
	For	Against	For	Against
Approval in principle of there being one Society instead of two (omitting questions of name and procedure).	44	4	36	3
The name should include American and Crystallography in some form.			38	0
Would be satisfied with a name which included American, Crystallography and Diffraction in some form.	51	0		
The name should include American and Crystallography, but not Diffraction	25	30	36	0

It can be seen that there was a strong preference for the plan. The principal point of non-agreement was in the choice of the name. A group in ASXRED wanted to have the word "Diffraction" in the name while the CSA was strongly opposed to this. Therefore two names covering these two possibilities were proposed and prospective members were able to choose the one they preferred.

5. A working sub-committee of the N.R.C. Crystallographic Committee (I. Fankuchen, W. Parrish, A. L. Patterson and E. A. Wood) meet in May and in June, 1949, to draft the Constitution and By-Laws and to discuss various problems relating to the formation of the new Society. The former were prepared in written form and distributed to the Councils of both Societies and members of the N.R.C. Committees.
6. A meeting at which the enclosed proposals were discussed and conditionally approved was held at Cornell University, Ithaca, N.Y., June 23, 1949. It was attended by:

C. S. Barrett	D. Harker	A. L. Patterson
M. J. Buerger	M. L. Huggins	R. Pepinsky
P. Debye	C. C. Murdock	B. E. Warren
I. Fankuchen	W. Parrish	E. A. Wood

who represented the Councils of both Societies and the N.R.C. Committee. Final approval by the whole N.R.C. Committee was given by letter ballot.

7. A letter sent to the membership on August 19, 1949, stated "If these plans are approved by the membership, both Societies will go out of existence December 31, 1949. Officers of the new Society will be elected in the early Fall of 1949. The funds of both Societies, after a certified audit has been made, and records of both Societies will be turned over to the new Society which will come into existence January 1, 1950. If the plans are defeated, both existing Societies will continue as before."

The following flyer was sent out once the plan was approved.

Letter mailed to membership, January 30, 1950

Successor to:
American Society for X-ray and Electron Diffraction
Crystallographic Society of America

TO ALL MEMBERS:

1. **Welcome to the new Society!** On January 1, 1950, the activities of the two societies, the American Society for X-ray and Electron Diffraction and the Crystallographic Society of America were officially ended, and the new society, the American Crystallographic Association, was created. The activities of the ACA, it is hoped, will carry on more effectively

and expand the former activities of the CSA and the ASXRED. The aims of the ACA are stated in its constitution: "The object of this Society shall be to promote the study of the arrangement of atoms in matter, its causes, its nature and its consequences, and of tools and methods used in such studies." The principal events leading up to the formation of the new society are briefly outlined in a leaflet enclosed with this letter.

2. **Election of Officers.** The officers of the ACA have been duly elected by the members.

   The names and addresses of the members of the Council are as follows:

President:	**Vice-President:**
Prof. I. Fankuchen Polytechnic Institute of Brooklyn 85 Livingston St. Brooklyn 2, N.Y.	Dr. R. W. G. Wyckoff National Institutes of Health Bethesda 14, Maryland
Secretary:	**Treasurer:**
Dr. Howard T. Evans, Jr. Phillips Laboratories, Inc. Irvington-on-Hudson, N.Y.	Dr. Jerome Karle Naval Research Laboratory Washington, D.C.
Last President of CSA:	**Last President of ASXRED:**
Prof. J.D.H. Donnay Dept. of Chemistry John Hopkins Univ. Baltimore, Maryland	Prof. Peter Debye Dept. of Chemistry Cornell University Ithaca, N.Y.

3. **Committee Nominations.** The constitution provides for five standing committees to be elected as follows: "Each Standing Committee shall have three members, elected by the membership of the Society by letter ballot from a list of candidates submitted by the Council which may be supplemented by nominations from the general membership, signed by five members. Plurality vote shall constitute election. Members of Standing Committees shall each serve for a period of three years and shall be eligible for reelection but shall not serve for more than six consecutive years. The chairman shall be chosen by the Council from among the elected members. In the case of the formation of a new standing committee, the initial members shall be elected for 2, 3 and 4 years." These committees are listed below, with the names of candidates which have been proposed by the council. Further nominations by the members must be sent to the Secretary before February 22. A ballot will be sent out about March 1.

Publications Committee:	C. S. Barrett, L. O. Brockway, J. D. H. Donnay, P. P. Ewald, D. Harker
Nomenclature Committee:	W. L. Bond, M. J. Buerger, J. D. H. Donnay
Apparatus and Standards Committee:	R. Bailey, M. J. Buerger, F. G. Chesley, W. Parrish, R. Pepinsky, C. G. Shull
Crystallographic Data Committee:	J. D. H. Donnay, C. Frondel E. W. Hughes, R. L. Mooney
Powder Diffraction Data Committee:	F. W. Matthews, W. Parrish, H. W. Rinn, E. A. Wood

4. **Spring Meeting.** The American Crystallographic Association has been invited to hold its first meeting at Pennsylvania State College, State College, Penn. Plans have been made to hold this meeting on Monday, Tuesday, and Wednesday, April 10-12, 1950. Details concerning titles and abstracts of papers, hotel reservations, travel routes, etc., are given in an accompanying folder. *Please note:* (1) Titles of papers should be sent to Prof. Pepinsky as soon as possible before February 21, to assist him in making up the program; and (2) the reservation blank should be sent in before February 21.

5. **Membership:** This notice has been sent to all 1949 members of ASXRED and CSA as well as new members of ACA. Hereafter, *only ACA members will receive notices.* So far, 496 applications have been received. If you have not already done so, be sure to fill in the enclosed membership form and send it together with dues to the Secretary. Endorsements are not required of former ASXRED and CSA members. The charter membership deadline will be extended to March 1.

6. **Dues:** The first annual dues of the ACA are now payable. Regular dues are \$5.00; dues for students during the period of their studies are \$3.00. Application for student membership should be endorsed by the Professor in charge. Receipts will be sent out to those few who have at the date of this mailing already sent their dues. If you do not receive such acknowledgement within a day or so, please send a check or money order payable to the American Crystallographic Association to the Treasurer, Dr. Jerome Karle, Naval Research Laboratory, Washington, D.C. No further receipts will be issued.

7. **Notice to Bibliographers.** It is planned to expand and improve the semi-annual bibliography published until now by the ASXRED. Until the Publications Committee is elected and can send out instructions to those who have so generously helped collect entries for the bibliography in the past, these people should continue their work if they are willing, and send their cards before April 1 to Elizabeth A. Wood, Bell Telephone Laboratories, Murray Hill, N. J.

Howard T. Evans, Jr., Secretary
American Crystallographic Association

ORGANIZATIONS

ACA Meetings

The meeting places and dates of the new society ACA are as follows:

ACA MEETINGS

Dates	**Location**	**President**
April 10-12, 1950	Pennsylvania State College, State College, Pa.	I. Fankuchen
August 21-25, 1950	New Hampton, N. H.	
February 15-17, 1951	National Bureau of Standards, Washington, D.C.	R. W. G. Wyckoff
June 27-July 5, 1951	Second I.U.Cr. Congress, Stockholm, Sweden	
October 24-27, 1951	Chicago, Ill. (Joint Meeting with A.I.P.)	
June 16-20, 1952	Tamiment, Pa.	P. P. Ewald
June 22-26, 1953	University of Michigan, Ann Arbor, Mich.	L. O. Brockway
April 5-9, 1954	Harvard University, Cambridge, Mass.	E. W. Hughes
July 21-30, 1954	Third I.U.Cr. Congress, Paris, France	
November 3-5, 1954	Mellon Institute, Pittsburgh, Pa. (Joint Meeting with Pittsburgh Diffraction Conference)	
April 11-13, 1955	Polytechnic Institute of Brooklyn, Brooklyn, N.Y.	W. N. Lipscomb
June 27-July 2, 1955	California Institute of Technology, Pasadena, Calif.	
June 11-15, 1956	French Lick, Ind.	J. D. H. Donnay
July 10-14, 1957	Fourth I.U.Cr. Congress, Montreal, Canada	E. A. Wood
November 6-8, 1957	Mellon Institute, Pittsburgh, Pa. (Joint Meeting with Pittsburgh Diffraction Conference)	
June 23-27, 1958	Marquette University, Milwaukee, Wis.	D. McLachlan
July 19-24, 1959	Cornell University, Ithaca, N.Y.	R. E. Rundle
January 25-27, 1960	Washington, D.C.	J. Waser
August 15-24, 1960	Fifth I.U.Cr. Congress, Cambridge, England	
July 30-August 3, 1961	University of Colorado, Boulder, Colo.	K. N. Trueblood
June 18-22, 1962	Villanova University, Philadelphia, Pa.	V. Schomaker
March 28-30, 1963	Massachusetts Inst. of Technology, Cambridge, Mass.	G. A. Jeffrey
September 9-19, 1963	Sixth I.U.Cr. Congress, Rome, Italy	
July 26-31, 1964	Montana State College, Bozeman, Mont. (Joint Meeting with Min. Soc. Amer.)	H. T. Evans
February 24-26, 1965	Motel-on-the-Mountain, Suffern, N.Y.	H. A. Levy
June 27-July 2, 1965	Gatlinburg, Tenn. (Joint Meeting with Min. Soc. Amer.)	
February 28-March 2, 1966	University of Texas, Austin, Tex.	B. Post
July 12-21, 1966	Seventh I.U.Cr. Congress, Moscow, USSR	
January 25-27, 1967	Georgia Institute of Technology, Atlanta, Ga.	J. S. Kasper
August 21-25, 1967	University of Minnesota, Minneapolis, Minn.	
February 4-7, 1968	University of Arizona, Tucson, Ariz.	S. C. Abrahams
August 12-16, 1968	State University of New York at Buffalo	
March 23-29, 1969	University of Washington, Seattle, Wash.	W. C. Hamilton
August 7-27, 1969	Eighth I.U.Cr. Congress, State University of New York at Stony Brook	
March 1-5, 1970	Tulane University, New Orleans, La.	D. P. Shoemaker
August 16-21, 1970	Carleton University, Ottawa, Canada	
February 1-5, 1971	University of South Carolina, Columbia, S.C.	W. R. Busing
August 15-20, 1971	Iowa State University, Ames, Iowa	
April 4-7, 1972	University of New Mexico, Albuquerque, N.M.	J. Karle
August 27-September 7, 1972	Ninth I.U.Cr. Congress, Kyoto, Japan	
January 15-18, 1973	University of Florida, Gainesville, Fl.	R. A. Young
June 17-22, 1973	University of Connecticut, Storrs, Conn.	
March 24-28, 1974	University of California, Berkeley, Calif.	E. C. Lingafelter
August 19-23, 1974	Pennsylvania State University, University Park, Pa.	

Organizations

Dates	Location	President
March 9-13, 1975	University of Virginia, Charlottesville, Va.	R. D. Burbank
August 7-15, 1975	Tenth I.U.Cr. Congress, Amsterdam, Netherlands	
January 18-22, 1976	Clemson University, Clemson, S.C.	I. L. Karle
August 8-13, 1976	Northwestern University, Evanston, Ill.	
February 21-25, 1977	Asilomar, Pacific Grove California	C. K. Johnson
August 7-12, 1977	Michigan State University, East Lansing, Mi.	
March 19-24, 1978	Oklahoma University, Norman, OK.	P. Coppens
August 3-12, 1978	Eleventh I.U.Cr. Congress, Warsaw, Poland	
March 26-30, 1979	University of Hawaii, Honolulu, HI	J. P. Glusker
August 12-17, 1979	University of Boston, Boston, MA	
March 17-21, 1980	Lakepoint Convention Center, Eufaula, AL	H. W. Wyckoff
August 17-22, 1980	University of Calgary, Calgary, AB Canada	
March 22-26, 1981	Texas A & M University, College Station, TX	Q. C. Johnson
August 15-25, 1981	Twelfth I.U.Cr. Congress, Ottawa, Canada	
March 29-April 2, 1982	National Bureau of Standards, Gaithersburg, Md.	J. B. Cohen
August 15-20, 1982	University of California, La Jolla, CA	

Over the years the American Crystallographic Association has diligently kept all members informed as to what is going on within the organization and with outside organizations. We maintain an executive office consisting of:

Membership Secretary
Newsletter Editor
Managing Editor

and have ACA representatives to:

AMERICAN INSTITUTE OF PHYSICS
Board of Govenors and other AIP committees

AMERICAN NATIONAL STANDARDS INSTITUTE

AMERICAN SOCIETY FOR TESTING MATERIALS
Joint Committee on Chemical Analysis by Powder Diffraction Methods
Now separately incorporated as:

JCPDS – INTERNATIONAL CENTRE FOR DIFFRACTION DATA

CRYSTAL DATA DETERMINATIVE TABLES

RADIATION SAFETY

CHAPTER 2

THE BEGINNINGS OF THE UNION OF CRYSTALLOGRAPHY

P. P. Ewald
Ithaca, New York 14850

During the war, I was living in Belfast, Northern Ireland, separated from England by the Irish Sea. I needed a special permit to cross over to England when I received an invitation to give a lecture at the yearly meeting of the X-Ray Analysis Group at Oxford in March 1944. This meeting took place in the newly built lecture theatre of physical chemistry. The lecture was to be a review of former dealings in crystallography on an international scale, but I extended the topic into the post-war future. As far as I am aware, this was the occasion on which the foundation of an International Union of Crystallography was suggested, and of the creation of a journal for international publication of modern crystallographic papers. A survey of my talk was published in *Nature* Volume 154 of November 18, 1944.

The idea for creating new journals for the publication of papers on crystal structure had arisen elsewhere. In the USA two journals were planned once the war was concluded, one under the leadership of Martin Buerger, and the other by a group among whom I. Fankuchen was most active. The crystallographers in England had discussed creating a journal in their country, and it seemed likely that after a period of recovery from the war there would be further national crystallographic journals.

The proposal in my Oxford talk was

(1) to establish an International Union of Crystallography, similar to the greater Unions of Physics and of Chemistry;
(2) that this Union act as a publishing center for crystallographic research;
(3) that the Union establish a collection of tables of experimental details of structural papers which would interest only a limited group of readers; copies of the full data would be obtainable from the Union;
(4) that the Union act as a centre for the collection of all structural work, including phase diagrams;
(5) that a new edition be prepared of the *"International Tables for the Determination of Crystal Structures"* as one of the Union's activities;
(6) that the Union keep in touch with the Unions of Physics and Chemistry and with mineralogical and biological societies and with those for testing structures and materials.

The idea of an International Union was taken up by W. L. Bragg and the British crystallographers in general. Correspondence took place with the leading crystallographers in USA, in France, the Netherlands, Sweden and other countries. The English group had several meetings to arrange the program of the first international meeting after the war. Finally this meeting took place in London in July 1946 under the chairmanship of W. L. Bragg. The group from the USA was particularly strong with Buerger, Donnay, Fankuchen, Germer, Zachariasen, Warren, Harker, McLachlan, Brockway, and others; from France there came Wyart, Mathieu, Guinier, and Yvette Cauchois; from Holland W. G. Burgers, from Sweden G. Hägg. The Russians had been invited but came late when the London meeting was over; they took part in special discussions in Cambridge and Shubnikov proposed the name *Acta Crystallographica* for the journal that was to be created.

The result of the meetings was threefold: the creation of an International Union of Crystallography, corresponding to the much larger Unions of Astronomy, of Physics, of Chemistry and Biological Sciences; the creation of an International Journal of Crystallography to replace and extend the *Zeitschrift für Kristallographie* which seemed dead at least for several years to come; and thirdly the continuation of the collection of crystal structures as they were being determined - a task that had been fulfilled by the *Strukturbericht* of the *Zeitschrift.*

These three topics required some time for their preparation. Ewald in Belfast and R. C. Evans in Cambridge looked after the first two, while the responsibility for the *"Structure Reports"* rested with A. J. C. Wilson, who eventually edited the first 10 volumes of *"Structure Reports."*

Correspondence and discussion with the General Secretary of the Council of Scientific Unions, F. J. M. Stratton, showed that he was very much in favour of the creation of international unions for more limited fields than general physics or chemistry. In fact, he carried our proposal through the next meetings of the Council and we had the Statutes of our new Union readily acknowledged.

Regarding *Acta Crystallographica,* Evans and I obtained from seven firms prices and specimen settings. That of a French firm was the cheapest but the least appealing; the Swedish offer was the most impressive but also the most expensive. Finally negotiations were begun by Evans with the Cambridge University Press and this led to entrusting them with the production of the first volumes. Evans was not only the English co-editor, but, following his genuine interest in all aspects of print production, took over also the task of Technical Editor. The first numbers of the new journal, *Acta Crystallographica,* appeared early in 1948 and by the time of the Foundation Assembly of the Union of Crystallography, which took place in Harvard University in August 1948, several numbers of the new journal were on show.

In the next years it turned out that so many good papers were handed in for publication that the number of sheets, grudgingly granted by the Cambridge University Press was insufficient. So, after the second Assembly of the Union of Crystallography in Sweden, in 1951, Ewald stopped in Copenhagen to find out whether Ejnar Munksgaard would be in a position and willing to take over the production of the journal. Ewald found the firm headed by a very enterprising and energetic lady, Mrs. Thorning-Christensen, who saw no obstacle in fulfilling all desires for more space.

So a transfer of the publishing was arranged, starting with Volume 4 of *Acta*. The Munksgaard firm is still publishing *Acta* for the Union, now in a general part for Theory (A) and a second part for Crystal Structure Determination (B). The yearly volume is about 10-times the volume of the 1948, an indication that *Acta Crystallographica* is still favoured by crystallographers the world over for publication of their research. Ewald remained Editor in Chief for the first 12 years with, initially, R. C. Evans as English Co-editor, I. I. Fankuchen as American, and J. Wyart as French co-editors. At present there are no less than 14 co-editors in 8 countries to help the main Editor with the much increased work; besides, a separate *Journal of Applied Crystallography* is being published by the Union.

The collection of *"Structure Reports"* has grown to two volumes each year and has changed the chief editor several times. The *"International Tables for the Determination of Crystal Structures"* have also developed under the editorship of Dame Kathleen Lonsdale and her successors from the original two volume to a four volume work. The adherence to the Union has grown with its activities, and the membership list of 1977 lists a total of 30 countries adhering.

CHAPTER 3

THE U.S.A. NATIONAL COMMITTEE FOR CRYSTALLOGRAPHY

David P. Shoemaker
Department of Chemistry, Oregon State University
Corvallis, Oregon 97331

During the last thirty years the U.S.A. National Committee for Crystallography has been one of the most important institutions of American crystallography. Its prime *raison d'être* was and is to represent American crystallography internationally – primarily in its interactions with the International Union of Crystallography. In this role it communicates extensively with officials of the Union, sends delegates to General Assemblies of the Union, organizes Congresses and inter-Congress international meetings in the United States, and exercises surveillance of the work of the Union (especially Union journals) on behalf of American Crystallographers. The Committee also looks after the interests of American crystallography in relations with the U.S. government and in other contexts. Details follow.

In 1948 (the year of Volume 1 of *Acta Crystallographica)* the newly formed International Union of Crystallography (IUCr) held its first Congress and General Assembly at Harvard University in Cambridge, Massachusetts, U.S.A. As is usual for recognized international scientific unions, the "members" are not individual scientists but rather scientific societies, national academies, or scientific agencies of national governments. There is generally one such member or "adhering body" per country, which in the case of the United States is generally the National Academy of Sciences – National Research Council. When the NAS-NRC adhered to IUCr it needed to form a body of advisory scientists to deal on a day-to-day basis with its relationships with the Union and to provide representation of it in the General Assemblies of the Union. This body, the U.S.A. National Committee for Crystallography (USANCCr, more recently USNCCr, here for brevity often simply NC) was formed in 1951, the year of the second IUCr Congress and General Assembly in Stockholm, Sweden. Its Constitution was written by David Harker.

In principle the NC members are appointed by the National Academy of Sciences after receiving appropriate recommendations, which after the early days were made by the NC itself in accordance with its own Constitution and By-Laws. I do not know precisely how the initial committee was formed, but do know that it was constituted of key members of the American Society for X-ray and Electron Diffraction (ASXRED), Crystallographic Society of America (CSA), and Mineralogical Society of America (MSA) who since 1945 had been active in making contacts with other crystallographic leaders in the post-WW II world to form the Union and launch *Acta Crystallographica*. Its membership surely represented a very appropriate cross section of the scientific leadership of American crystallography of that day: Ralph Wyckoff (Chairman), Lawrence Brockway (Vice Chairman), Ray Pepinsky (Secretary-Treasurer), Peter Debye, Ludo Frevel, Rose C. L. Mooney (later Mooney-Slater), C. C. Murdock, Bill Parrish, Willie Zachariasen, Charlie Barrett, Martin Buerger, Maurice Huggins, Lindo Patterson, Betty Wood, José Donnay, Isidor ("Fan") Fankuchen, David Harker, and Bert Warren.

The relationships between the NC and the NAS-NRC staff have generally been very good through the years, and Academy approval of recommendations for appointment of members has been, in actuality, automatic. The NC is a continuing body in that its voting membership "elects" (i.e., chooses by poll to recommend for appointment) all except certain *ex officio* voting members. The latter includes all five members of the Council of the American Crystallographic Association (since 1969; before that time ACA *ex officio* members of the NC were elected explicitly by the ACA membership), all officers of the NC (who need not be members of the NC when nominated or elected), and any members of the IUCr Executive Committee who are resident in the United States. The Chairman of the cognizant body of the NAS-NRC and the Foreign Secretary of the NAS-NRC are *ex officio* non-voting members. Four voting members are elected each year to a three-year term, and are eligible for re-election unless they are in their fifth or later year of continuous NC membership (regular or *ex officio*), in which case they must spend one year off the Committee before again becoming eligible. This feature was intended to prevent entrenchment and provide new blood. Well, it has allowed some new blood but it has not altogether prevented entrenchment; by managing to hop from NC officerships (3-year term) to ACA Council (up to 3 years) to IUCr Executive Committee (6 years, more or less) a few individuals have managed to stay on almost continuously for the better part of 20 years. (Recently a limitation has been instituted against further election of those who in consequence would serve more than 12 years on the NC as an officer or voting member.) The NC has three officers elected by its voting membership, serving three year terms: Chairman, Vice Chairman, and Secretary-Treasurer; since 1958 these terms have been staggered so that one officer is elected each year.

As has been the case generally throughout its history this Committee of about 20 individuals meets twice per year on the occasion of ACA meetings (usually all day Sunday, the day before the ACA meetings), except in a General Assembly (GA) year, when it meets at the winter ACA meeting, then (usually) in Washington to "instruct" the delegates to the GA, and then several times at the GA site while the GA is going on. The delegation to the GA consists of the NC Chairman as the Chairman of the delegation, and four individual delegates elected by the NC and appointed by the NAS-NRC. At the Washington meetings the Committee is often briefed by the NAS Foreign Secretary and others (e.g., the Department of State) on international political realities as they might affect U.S. interests within the purview of the NAS-NRC. My own perceptions as a delegate in the years following is that the briefings in the Washington meetings were basically only informational, and I found them very interesting. In the earlier days there was at times a strong perception that the State Department was trying to instruct the delegates how they should vote on certain matters in the GA: I remember Betty Wood, who was Chairman of the delegation at the 1957 Montreal Congress, telling us how she vigorously resisted this. The NC similarly succeeded in convincing the State Department in 1960 that delegates should vote their own conscience at the Cambridge, England General Assembly of that year.

Concerns of the NC and of the delegates to GA's covered a very wide range of matters relating to IUCr and its operations. These included admission of new adhering bodies and changes in adherence. Most important in this category in the 1960's was the relationship between the two Germany's. In the late sixties the new Germanys agreed to merge their adherance, forming a single adhering body and national committee. This arrangement fell apart three years later, but relations between the delegations have seemed amicable. In recent years the two Chinas threatened to be a sticky problem, but when the People's Republic of China joined the Union in 1978 (Warsaw) the U.S. delegation gave its approval on the stipulation that admission of the PRC would not prejudice the future admission of any other country. Choice of candidates for the Union officers, members of the Executive Committees, and Chairmen and members of Commissions occupied considerable attention of the NC, as did the efficacy (or lack of it) of the work of the Union Commissions. The publication program of the Union, particularly *Acta Crystallographica,* was constantly discussed. With U.S. authors supplying a third of the authorship and a third of the readership, the Committee found difficult to accept the rather minor share of influence that it found itself able to exert through its members on the Union Executive Committee and on the Journals Commission, both of which seemed to be largely a UK hegemony. The NC fought against splitting *Acta* into A and B, and lost (perhaps, in retrospect, fortunately). Several times it proposed US-style page charges to solve *Acta's* financial problems, again without success.

The NC spent much of its time and energy attempting to play the roles of constitutional lawyers and legislators, with regard both to its own constitution and By-Laws and to the Statutes and By-Laws of the Union. As to the first, I remember much polishing-up activity in the early sixties, partly in my role as Secretary-Treasurer. During one meeting in Boulder, Colorado, I remember our exasperated Ken Trueblood demanding "What does this Committee exist for, anyway?" The most wheel-spinning, however, arose later out of USANCCr frustration over the politics of the Union, mainly the tendency of the Executive Committee to be even more self-perpetuating than the USANCCr itself. The remedy seemed to lie in constructing amendments to Statutes and By-Laws, to be submitted for a vote at the GA. These were often quite elaborate and much argument was spent on the turning of words and phrases which never did eventually see the light of day. I plead guilty to having been as much a part of this as anyone, even at times a ringleader, and yet today I can barely remember the substance of the proposals (having to do, for example, with more democratic generation of nominees for Executive Committee members and members of Commissions). The Executive Committee generally regarded the formal proposals of these upstart Americans with bewildered amazement. Although in my own experience the U.S. proposals were virtually never accepted in the form submitted, they did lead to some awareness of U.S. concerns, and some measures (far fewer than we wanted) were undertaken to redress the problems pointed out. For example, the Executive Committee undertook to solicit from the various National Committees nominations for Executive Committee and Commissions well ahead of General Assembly time. Years earlier in response to USANCCr anger over the decision by split postal ballot – without opportunity for adequate discussion – to divide *Acta Crystallographica* into Sections A and B, the General Assembly in Stony Brook in 1969 accepted a compromise suggestion that in any postal ballot two adverse votes could require that the question at hand be held over to a meeting of the Executive Committee in person. On another matter GA appointed a special Commission having an American as one of its members to formulate recommendations for Statute changes to the next GA. In the course of all this the Committee members became more aware of international sensitivities and of the advisability of subtler approaches to problems of Union governance. These were lessons worth learning.

After six Congresses and General Assemblies abroad it seemed manifest that it was USA's turn to host a second Congress and General Assembly. A delegation to the Moscow General Assembly in 1966, chaired by George Jeffrey, conveyed the invitation of the U.S. NAS-NRC to the Union to hold the Eighth International Congress of Crystallography in the United States in Summer 1969. The following year as newly elected Chairman of the NC (and thereby Chairman-to-be of the Congress Organizing Committee) I chaired a NC meeting in Atlanta, at which it was decided that the Eighth Congress would be in three parts: (1) a topical meeting on Biologically Important Structures at the State University of New York in Buffalo, N.Y., Aug. 7-12; (2) the main Congress (General Meeting) and General Assembly of the Union at the State University of New York at Stony Brook, N.Y., Aug. 13-21; and (3) a windup at Washington, D.C. with laboratory visits and special events, ending Aug. 27. Later was added a symposium on Chemical and Physical Aspects of Neutron Diffraction at Brookhaven National Laboratory (near Stony Brook), Aug. 22. This multi-meeting feature may have set a strong precedent for "satellite" meetings connected with later Congresses of the Union. Program Chairmen were David Harker* for Buffalo, George Jeffrey* for Stony Brook, and Lester Corliss for Brookhaven. Local Chairmen were J. Berger for Buffalo, Walter Hamilton* (then ACA President) for Stony Brook, B. C. Frazer for Brookhaven, and Jerry Karle* for Washington. All of these except Corliss were members of the Organizing Committee, which also at the time of the Congress included Ben Post*, Congress Treasurer and NC Vice Chairman; John Kasper, earlier Vice Chairman; Bill Parrish*, NC Secretary-Treasurer; Ray Young*, Chairman for Laboratory Visits; Sidney Abrahams* and Betty Wood*, Liaison, American Institute of Physics (which handled the Congress publications); and Natalie Fiess, our paid and hard working Executive Secretary. Those whose names are starred (*) were NC members at the time of the Congress; other NC members at the time were Robinson Burbank, Gabrielle Donnay, Ken Hedberg, Lyle Jensen, Bill Kehl, Dick Marsh, Dave Templeton, Bert Warren, and Willie Zachariasen, plus Harrison Brown and R. Smoluchowski *ex officio* non-voting members.

The NC and most particularly the Organizing Committee worked very hard during the latter half of the inter-Congress period. The Organizing Committee met frequently at the AIP headquarters in New York and at Stony Brook where the Lecture Hall Building to be used for the Congress was still under construction. Much effort was devoted to fund raising. At its 1967 Summer Meeting in Minneapolis, the ACA voted a substantial levy from its membership to support the Congress. U.S. Government agencies – Army Research Office (Durham), National Institutes of Health, National Science Foundation, and Office of Naval Research – granted funds for the meeting, as did the New York State Science and Technology Foundation. Contributions and other support were received also from American Society for Testing Materials, Carnegie Institution of Washington, Alfred P. Sloan Foundation, 27 industrial companies, and 240 individual American crystallographers, in response to our solicitations. In addition to concerns about money there were concerns about access to the meetings for individuals from Eastern bloc countries, especially for the meeting at the Brookhaven site which had historically been very security conscious. Painstaking negotiations with government agencies forestalled major problems. The Congress logo, replicating the famous Patterson projection of KH_2PO_4, was executed in gold plated metal (rhodium plated for NC members) for tie or lapel pins, and on plastic circular diffraction grating for stickers to be attached to books, luggage, or other possessions. Much effort went into planning innovations, such as ad-hoc sessions and, courtesy of Yoshi Okaya and the IBM Company, a network of computer terminals to provide program update information and message services. We took a giant gamble with the weather and planned, for the major social event of the Congress, a clambake picnic on Fire Island. Fortunately the weather was splendid – no rain.

The Congress took place much as planned, in very hot weather. There were at least 1200 active members. There was much concern about finances during the Congress, owing in part to a breakdown of the arrangement with the Stony Brook caterer, requiring painful renegotiation on the spot. (We had to renege on our promise to provide wine with meals.) Scientifically, perhaps the most exciting event was Dorothy Hodgkin's report of the newly determined structure of rhombohedral insulin. The Congress opened in the backwash of excitement over the Apollo 11 landing on the moon, which gave Bill Parrish, then of the National Aeronautics and Space Administration, occasion to report at the opening ceremonies some preliminary crystallographic results on lunar materials. The opening address was given by Linus Pauling, the country's only crystallographer at that time who was a Nobel laureate.

It took most of a year for the finances of the Congress to get sorted out, under the NC Chairmanship of Sidney Abrahams. When that was done it was discovered that, far from being broke, we had an embarrassingly large surplus! This occasioned much discussion in the NC and also in the Council of the ACA, of which I had in the meantime become President and was beginning to have acute feelings of conflict of interest since a large part of that surplus was the ACA contribution. It however became clear as a practical matter that contributions could not be selectively returned, and technically were not the NC's to return since all solicitations had been made on behalf of the NAS-NRC, the Congress' official host. Thus it was that the "Stony Brook Fund" in the NAS-NRC coffers was created, which has been used by the NC since that time for purposes reasonably consonant with the understandings on which it was raised. For example, some of the monies were used to enable young crystallographers to travel to the International Union Meeting in Ottawa in 1981.

Sidney Abrahams took the Chairmanship of the NC in 1970, receiving most of the Stony Brook windup headaches. He organized the NC into a number of subcommittees which have been very active ever since.

In addition to its activities relative to the IUCr, the

NC has looked out for the welfare of crystallography in the U.S. in many ways. After statements somewhat damaging to the image of crystallography had appeared here and there, the NC, under Jerry Karle's chairmanship, organized a conference on Feb. 10-11, 1975, in Washington, on "Status and Future Potential of Crystallography," and published an excellent report with that name for fairly broad distribution. In more recent years the NC exerted effective influence for the formation and NSF funding of the National Resource for Computation in Chemistry – now, unhappily, moribund due to the discontinuation of NSF funding for this project.

This piece has been a difficult one to write because I have unfortunately disposed of most of my records of the NC and Congress. Much here is based on my memory, which I hope is correct. Much more that might have been worth telling is glossed over or left out because I could not trust my memory enough and had no time to consult others adequately. I thank Ray Young (present NC chairman), Sidney Abrahams, Mel Mueller and Jerry Karle (past NC chairmen), Hugo Steinfink (present NC Secretary-Treasurer), and Jenny Glusker (past NC Secretary-Treasurer) for reading the draft and in most cases offering corrections or suggesting changes or additions. I thank Dave Harker for some information about the early days. Those many past and present members and officers of the NC whose names I have not mentioned are asked to forgive me for not attempting completeness, on account of space.

This article can be ended on a happy note because, as it goes to press, it has been announced that Jerry Karle has been elected the President of the International Union of Crystallography, 1981 - 1984.

CHAPTER 4

THE 1946 CONFERENCE IN LONDON

Dan McLachlan, Jr.
The Ohio State University, Columbus, OH 43210

In 1946, in the wake of World War II, an almost miraculous event occurred: that is, the assemblage of delegates from a dozen or so nations (regardless of their wartime affiliations) to enhance the progress of crystallography the world over. This meeting was arranged by the X-ray Analysis Group of the Institute of Physics with Sir Lawrence Bragg presiding. The item on page 142 lists the delegates from the various nations who had accepted the British invitation as of July 1, 1946.

The reason for thinking of this meeting as a miraculous event is not only that the results included establishment of an international journal *Acta Crystallographica,* and the organization of the International Crystallographic Union, but also the difficulties overcome by various nations in sending delegates. The countrysides and cities of Europe were war-torn, the industries of the world were trying to make the transition from the manufacture of war products to peacetime essentials, much of the scientific research was militarily oriented and facilities for tourist travel were strained by the conditions of aeroplanes and ships and also the question of visas and passports.

Regarding passports, I remember being questioned by the U.S. authorities: "What is this meeting in London for?" "It is for the gathering of crystallographers." "What are the crystallographers trying to do?" "We are at present interested in locating the positions of atoms in crystals. The word "atom" caused the questioner to flinch. The atomic bomb was a horrible memory to everyone. He questioned, "The position of atoms?" I said, "Yes we are pretty good at it too. It is beginning to appear that soon we will be able to locate hydrogen atoms." The word "hydrogen" had disastrous effects for me. It took three days to get from under the shadow of that word and reschedule my flight on BOAC to London.

Upon arriving in London all the foreign delegates were handed the following welcoming letter. I will use mine as an example (item 2, page 143). The letter shows the generosity of the British in spite of their war losses and the urgencies of rebuilding.

One of the first things that the British had planned for us was a conference held at the Royal Institution on July 9, 10 and 11 to get reports from the various nations on the work in crystallography that had been accomplished during the war years. The printed program as handed to the delegates is shown on page 143 as item 3. This program had to be altered in two places. First, the Russians were late in arriving and the Russian paper which was to be given between 2:30 and 3:15 Wednesday, July 10 was replaced by a paper by Dr. R. C. Evans on X-ray Crystallography in Germany. Also another paper was added and it was hoped that D. McLachlan would speak on "The Sand Machine." However, Dr. W. T. Astbury used extra time for his paper and left only three minutes for McLachlan's paper, which was probably as much time as the paper deserved. Also there was no time for discussion of the papers which was a pity. In general, I can say that the content of all these papers were very exciting to me and I was thrilled to hear and see the great scientists living at that time. These papers will not be reviewed in this book since abstracts of them have been published elsewhere. The reader is advised to see "Summarized Proceedings of Conference on X-ray Analysis-London, 1946" by G. W. Brindley, G. A. Jeffrey and I. McArthur, in the *Journal of Scientific Instruments* (British) Vol. 24, No. 1, January 1947, and also "Wartime Progress in X-ray Analysis" by Audrey M. D. Douglas (Parker), H. S. Peiser and Barbara W. Rogers (Low) in *Nature,* Vol. 158, page 260, August 24, 1946.

During the meetings, three of the Americans who were employed by industry (and presumed to be well supported) entertained some of the British and Russian delegates to luncheons at the Brown Hotel. The hosts were: Dave Harker of General Electric, Lester Germer of Bell Labs and Dan McLachlan of American Cyanamid. To my luncheon, Dr. Bernal brought a guest named Iona Canavisher, an artist, who was painting a portrait of Zachariasen. Fankuchen, Zach and I went to Iona's apartment one evening to see the portrait, but it wasn't finished and we did not get to see it. Bernal played a prominent role in all the events associated with the conferences. I had learned a great deal about Bernal when he visited the United States in 1945 on a mission to look into the matter of temporary housing in the U.S. for the British government.

On Friday and Saturday, July 12 and 13 the conferences were held pertaining to the proposed new international journal and the organization of an International Union of Crystallography. Preparatory to these meetings, our hosts, Bragg and his team which included Bannister had already done a lot of work, so that prior to the meetings all delegates were handed a bundle of pages containing the proposed agenda of the meeting. The matters to be discussed regarding the journal were:

1. Scope of Journal
2. Title of Journal
3. Location of center of publications
4. Financial responsibility
5. Editorial arrangements, etc.,

and finally tables of classification of subjects which the journal should include. In the American delegation were:

Dr. David Harker, president of the six year old ASXRED and Dr. M. J. Buerger, president of the new C.S.A. Harker was second in position to Sir Lawrence Bragg as chairman of the first day of meeting and took the chair on the second day.

Professor P. P. Ewald was persuaded to be the first editor of *Acta Crystallographica.* Hopes ran high as we went to the organization of the International Union, which Dr. R. C. Evans of the Cavendish Laboratory, secretary of the Provisional International Crystallographic Committee carried through till March 1947. In 1948 the first issue of *Acta Crystallographica* began to appear and issues have continued monthly since.

In preparation for the discussions of the possible formation of an International Union, the British came through again with foresight and we were handed another bundle of pages discussing the items to be resolved. The items were:

1. Objects of the Union
2. Membership
3. Administration
4. Finance
5. Status
6. Duration of the Union
7. Authoritative text.

Then there were a few pages under the title of By-Laws which discussed the seven items and really was equivalent to a constitution for the Union.

The first meeting of the International Union was held in Harvard in 1948 and is reported on by Dr. Robert Evans in the next chapter of this book. Further meetings of the International Union have been held every third year since then at the larger institutions of all the great nations of the world. These meeting places are listed in the chapter "Organizations of North American Crystallographers" page 132.

Now that the main purposes of the London meeting were completed, we were free for sightseeing, being educated and entertained by local people. For example, I spent an evening in each of the homes of Kathleen Lonsdale and Dorothy Crowfoot Hodgkin and at a banquet I heard a fine speech by David Harker which drew loud applause. One evening Bernal took a group of us to his exclusive club where we heard a fine program of music, dancing and humor on the stage directed by a very versatile and talented master of ceremonies. This was an evening of pure fun. Also, on an evening whose date I have forgotten, the Americans were invited to a party at the home of a friend of Fankuchen and Bernal, Mrs. Geraldine Godfry. This was one of the most enjoyable parties I ever attended. Some of the guests thought they had seen some evidence of sloppy housekeeping when they spied a dead rat under the davenport in front of the fireplace. It later turned out that Mr. Godfry was a taxidermist and his masterpiece had been knocked off of the library table and had rolled toward the davenport.

In one of our sightseeing ramblings over London, Lester Germer and I got kicked out of the Ritz. In my mind I partly blamed Lester for it. When we entered the large lobby with its twelve foot ceiling, that mountain-climber spied a stairway on the opposite wall, leading to the upper floor and insisted on making ascent. When we got about halfway up, the man at the desk called us back. He said, "How would you feel if two strangers were to enter your home and make their way upstairs to your bedrooms?" He accepted my apology with all the warmth of a half melted snow-man. As we departed through the front door, Lester said, "Dan, you shouldn't let these hotel clerks push you around like that."

On Monday, July 15 we went to Oxford and had the wonderful experience of spending the night in the famed Christchurch. Dr. Hodgkin and Barbara Rogers (Low) and I spent the following morning going over the work of my team at the American Cyanamid Co., on the structure of penicillin. We agreed on the cell dimensions, intensities, space groups and Patterson maps but my phases were non-existent. Dr. Hodgkin and her group were eventually successful in completing the structure and went on to the structure of Vitamin B_{12} for which she won the Nobel Prize.

On Wednesday, July 17 we were in Cambridge all day and enjoyed the beautiful town and visited many laboratories, and Thursday we were in Leeds visiting Astbury's Laboratory and I got acquainted with an enthusiastic young graduate student who was D. W. J. Cruickshank, destined to go on to do great things.

It is difficult to recall all the exciting programs and fine lectures we attended but I do remember that during the week of July 22-27 a series of programs were scheduled to take place in the big lecture rooms of the Arts School, Examination Hall and the Maxwell Lecture Room of the old Cavendish Laboratory on the subjects of Fundamental Particles and Low Temperatures. I attended several of the sessions and heard lectures by N. Bohr, W. Pauli, P. A. M. Dirac, Max Born, F. London, L. Onsager, W. H. Taylor, F. Simon, and Sir K. S. Krishnan. Years later L. Onsager came to the United States to stay at Brown University. Adding this list of good scientists to those I had met at the crystallographic conference produced a roster of people about whom I had learned and whose work I had tried to read and understand in journals and books throughout my college and research careers. I am still thankful for all these opportunities furnished by the British. I had never dreamed that I would ever get so close to so many people who had done so much to better the world.

Toward the end of July, the crystallographers started to disperse. Dr. Harker, for example, embarked on a four week tour of the mainland of Europe and has since reported that he was well treated, saw old friends and made new ones. I went to Glasgow on July 28 and visited a man whom I had admired for many years for his work on the structure of organic compounds. Dr. J. Monteath Robertson. He showed me his laboratory and introduced me to his graduate students. His researches had been curtailed during the war and he was trying to get reorganized and re-established so he could continue his fine work.

From Glasgow I went to nearby Kirkintilloch, a small town on the Glasgow-Edinburgh Canal from which my grand parents, Peter and Mary McLachlan, had migrated to Lanacona, Maryland about 1870. I found the Alt Isle Church where they were married and also an old abandoned graveyard containing a tower (still standing) for the guards to oversee the cemetery to keep the students from stealing the fresh bodies for sale to the Medical College in Glasgow. I also learned that my great-great-grandfather owned a boat which he used to smuggle Scotch whiskey from the Highlands to the docks in Glasgow without paying the excise duty. Being chased by the red-coats he went underground, literally, to work in the coal mines of Kirkintilloch and Tintuck. I have read about a gang of McLachlan cut-throats whose main occupation was relieving cattle drivers of their stock while on their way to market with their herds. Also, I was asked by a woman to help her find out what had happened to the estate of her bachelor kin, President William McKinley when he was shot. But my most interesting ancestor was a McLachlan who was a fine gadgeteer. He made it his life ambition to build a boat that would surpass all boats. He put all his daily income into brass screws, fine mahogany and ebony planking and sail cloth. And he achieved his goal with a magnificent structure. But the problem of moving the huge boat from his back yard to the canal was never solved. After he died, the neighbors borrowed the fine brass screws one by one and finally the beautiful planks; and all that marks the spot now is a pen containing 47 chickens. I wonder if the McLachlan characteristics are inheritable.

Finding no crystallography in Kirkintilloch, I drifted back to London where I visited Westminster Abbey, the Saint Paul Cathedral, a fine old castle and grounds on the Drinkwater Estate where I learned a great deal about how the royalty operated many years ago to make England a great nation and then to the House of Lords. I asked why the members of parliament were meeting in the House of Lords. An M.P. (member of Parliament) told me that the House of Commons had been bombed out and they had no other place to go. So I asked about the Lords and he said, "They don't matter; they don't have streetcar fare to go to a meeting anyhow." Later while sitting in a chair in a waiting space to the back of the auditorium an M.P. said to me, "You are now sitting in the very chair where Sir Winston Churchill sat when he delivered the most profound speech of his entire career." I said, "I feel honored. It must have been an arousing oration; what was his subject?" "He was giving his son a lecture on the evils of strong drink."

I visited Bernal at Birkbeck and then home to Old Greenwich, Connecticut on August 7, 1946.

Item 1. Visitors from abroad coming to the July Conference as of 1st July, 1946.

U.S.A.
Brockway, L. O.
Buerger, M. J.
Donnay, J. D. H.
Fankuchen, I.
Germer, L. H.
Harker, D.
McLachlan, D.
Wyckoff, R. W. G.
Zachariasen, W. H.

France
Bernard, P.
Cauchois, Mlle, Y.
Chaudron, G.
Grison, E.
Guinier, A.
Hocard, R.
Laval, J.
Moiround, Mlle. J.
Morel-Klopstein, L.
Paic, Dr. M. et Mme.
Petitpas, Mlle. G.
Petitpas, Mlle. T.
Stora, Mlle. C.
Tertian, R.
Trillat, J. J.
Walter-Levy, Mme. L.
Weill, Mme. A. R.
Wyart, J.

Czechoslovakia
Kochanovska, Mrs. A.
Novak, J.
Petrzilka, V.

Sweden
Hambraeus, G.
Linden, Miss E.
Sillen, L. G.
Waller, I.

Denmark
Jensen, A. T.

Holland
Bijvoit, J. M.
Burgers, W. G. et Mme.
Favejee, J. C. L.
Gusters, J. F. H.
Hermans, J. J.
Jong, W. F. de
Ketelaar, J. A. A.
Lange, J. J. de
MacGillavry, Miss C. H.
Perdok, W. G.
Reyen, L. L. van
Stryk, Miss B.
Terpstra, P.
Wiebenga, E. H.
Woldringh, J. S.
Kolkmeijer, N. H.

Finland
Wasastjerna, J.

Germany
Hermann, C.
Laue, M. von
Laves, F.

Switzerland
Meyer, K. H.
Nowacki, W.
Scherrer, P. et Mme.
Weigle, J.

Norway
Finbak, Ch.
Hassel, O.
Oftedal, I.

Belgium
Brasseur, H.
Homes, G. A.

Russia
Borovsky
Konobeyevsky
Kurdyumov, G. V.
Shubnikoff

England
Too numerous to list here.

Item 2. Welcoming Letter

THE INSTITUTE OF PHYSICS

19, Albemarle Street,
London, W. I.

Dr. D. MacLachlan

Dear Dr. MacLachlan,

The Chairman, Sir Lawrence Bragg, and the Committee of the X-ray Analysis Group of the Institute of Physics welcome you to London and hope that your visit will be profitable and, as far as the present conditions permit, your stay will be a pleasant one.

The Conference Committee hopes that you will be among its guests at the following functions, tickets for which are enclosed. Should you be unable to accept, please return the tickets to the Institute of Physics, as soon as possible.

Wednesday, 19th July
7 p.m. for 7:30 p.m.
-Dinner at Frascati's Restaurant, Oxford Street, W.1. (Informal dress)

Thursday, 11th July
6:45 p.m.
-Performance by New York Ballet Theatre, The Royal Opera House, Covent Gardens. (Informal dress)

Tickets for tea on each of the three days of the Conference are also enclosed.

Yours truly,

F. A. BANNISTER
Honorary Secretary
X-ray Analysis Group

P.S.

As a contribution to your expenses during your stay here I am desired by my Committee to hand you the enclosed sum (fifteen pounds). Would you kindly sign the enclosed receipt and hand it to the official on duty in the Institute's offices.

Item 3. Program of the meeting

THE INSTITUTE OF PHYSICS X-RAY ANALYSIS GROUP

1946 CONFERENCE
X-RAY ANALYSIS DURING THE WAR YEARS

To be held on 9th, 10th and 11th July, at the Royal Institution,
21, Albemarle Street, London, W.1
(By kind permission of the Managers)

SCIENTIFIC PROGRAMME
Chairman: Sir Lawrence Bragg, F.R.S.

July 9th.

	Time	Speaker	Title
Morning	10:00-10:45	R. W. G. Wyckoff	Electron-microscopy in radiography
	10:45-11:00	K. Lonsdale	Review of work done in Great Britain,

	Time	Speaker	Title
			1939-1946: 1. Thermal and other perturbations of crystal structure
	11:00-12:00		Discussion
Afternoon	2:30-3:15	J. Wyart	French work during hostilities: A. X-ray work at the Sorbonne.
	3:15-4:00	W. H. Zachariasen	Crystal chemistry of plutonium and neptunium
	4:00-4:30		Discussion

July 9th.

Evening Lecture:	8:00	Sir Lawrence Bragg	Metals

July 10th

	Time	Speaker	Title
Morning	10:00-10:45	J. M. Bijvoet	X-ray research in Holland
	10:45-11:30	J. D. Bernal	Review of work done in Great Britain, 1939-1946: II. Organic structures.
	11:30-12:00		Discussion
Afternoon	2:30-3:15		Russian paper
	3:15-3:35	J. J. Trillat	French work during hostilities: B. Electron micro-radiography.
	3:35-4:00	A. Guinier	C. Various X-ray studies
	4:00-4:30		Discussion

July 11th

Morning	10:00-10:20	D. Harker	Crystallography of Metals.
	10:20-10:45	L. O. Brockway	Electron diffraction.
	10:45-11:30	W. T. Astbury	Review of work done in Great Britain, 1939-1946: III. X-rays and biology.
	11:30-12:00		Discussion
Afternoon	2:30-3:15	P. Scherrer	Swiss review.
	3:15-4:00	I. Waller	Swedish review.
	4:00-4:30		Discussion

At some suitable time during the proceedings Sir K. S. Krishnan, F. R. S. (Allahabad University) will talk on the "Diffuse Scattering of Electron in Metals and Alloys in relation to their Resistivities," and Dr. R. C. Evans will give an account of X-ray crystallography in Germany.

CHAPTER 5

THE FIRST GENERAL ASSEMBLY AND INTERNATIONAL CONGRESS OF THE INTERNATIONAL UNION OF CRYSTALLOGRAPHY

R. C. Evans
Elsworth, Cambridge, CB3 8JQ, U.K.

On Monday 7 April 1947 F. J. M. Stratton, General Secretary of the International Council of Scientific Unions, wrote me to say that the International Union of Crystallography had that day been admitted as a constituent member of the Council. Some of the circumstances that led to this event are related elsewhere in this volume in the articles by P. P. Ewald and D. Harker.

Once the Union had been formally admitted to ICSU it became the duty of its provisional Executive Committee (of which Ewald was Chairman and I was Secretary) to make arrangements for its inaugural General Assembly. It was therefore with much pleasure that I received a letter, dated 15 July 1947 and signed by Elizabeth Armstrong Wood and William Parrish on behalf, respectively, of the American Society for X-ray and Electron Diffraction and the Crystallographic Society of America, inviting the Union to hold its first international conference in the United States in the summer of 1948. This invitation was all the more welcome because it confirmed the dedication to our international endeavour of the one country that could so easily have stood aside on account of its own strength in crystallographic research and its isolation from the crystallographers of other continents. The Executive Committee gladly accepted the invitation.

It so happened that I had already arranged to visit the United States in August 1947 and my trip provided a convenient opportunity for some discussions about the proposed congress. This was before the days of regular air services, and ocean passages at the time were difficult to secure. With the backing of the Royal Society, however, I managed to obtain a berth (in an eight-berth cabin) on board the *Aquitania,* in which floating nursery - for my fellow passengers were almost all "G.I. brides" and their infants - I crossed the Atlantic to Halifax.

My discussions in the United States could only be of a preliminary nature for at the time neither the venue nor the date of the congress had been fixed. It was, nevertheless, useful for me to exchange views with many of the American crystallographers who had been at the Royal Institution conference in London the previous year and also with A. L. Patterson, who had not been in London but who later (in February 1948) became Chairman of the United States National Committee, the body set up by the National Research Council as the official organization through which the United States "adhered" to the Union. Patterson showed himself from the beginning to be dedicated to the cause of the Union and in a long letter to me, dated 2 March 1948, urged that as many countries as possible should be encouraged to join and that if necessary a revolving fund should be set up by private subscription in the U.S.A. and the U.K. to meet the dues of those countries unable to pay owing to the then prevailing exchange controls and other financial difficulties. I also took the opportunity during my trip to meet H. A. Barton, Director of the American Institute of Physics, who gave me an insight into the editorial procedures involved in the production of the Institute's many journals and arranged for me to visit the Lancaster Press, where they were printed. All this was invaluable to me in my work as technical editor then engaged in launching *Acta Crystallographica.*

I returned to the U.K. in October, and shortly thereafter an invitation was received from J. B. Conant, President of Harvard University, inviting the Union to hold its meeting at that university from Wednesday 28 July to Tuesday 3 August 1948. Once this invitation had been accepted it naturally became a local responsibility to organize the programme of scientific sessions and to arrange the necessary domestic details. These responsibilities were undertaken by a programme committee under B. E. Warren (of M.I.T.) and a local committee under C. Frondel. Neither of these committees had any precedents to guide them and it is greatly to their credit that the pattern of administrative, scientific and social events adopted at Harvard has been repeated with little change at successive triennial congresses of the Union for thirty years.

My own responsibilities in the period prior to the conference were limited to the arrangement of the formal business of the Union to be transacted at the sessions of the General Assembly. The provisional Statutes had to be ratified, Officers and members of the Executive Committee had to be elected, the publishing activities of the Union had to be discussed, and Commissions to oversee these and other activities had to be appointed. Here I was in difficulties for although the scientific sessions of the conference would, of course, be open to all interested crystallographers the formal Union business would be the concern only of the nominated representatives of those countries adhering to the Union; but at the time when preparations had to be made the number of such countries could be counted on the thumb of one hand: only the U.K., which joined on 9 October 1947, was formally a member. I therefore felt compelled to extend my consultations to all countries in which crystallographic research was believed to be in progress in the hope that some of these would have completed formalities for joining the Union by the time of the congress. In fact the U.S.A. joined on 7 April 1948 (a welcome birthday present for the Union), followed shortly thereafter by Canada

(28 April) and Norway (31 May). By the time we assembled in July four countries had thus been formally admitted to the Union. Now there are 32.

I travelled to Harvard with Patterson from his home in Bryn Mawr. My first impression on reaching Cambridge was of intense heat and high humidity. This was no surprise to me, for it was not my first summer visit to the Atlantic seaboard, but some European delegates visiting the United States for the first time and inclined to dress more formally than their American colleagues may well have found the climate overwhelming; certainly those who sunbathed intemperately during the Sunday picnic on Ipswich Beach suffered sorely in every sense of the word.

My second impression was of the austere accommodation provided for the students in the Harvard dormitories. As an undergraduate in the other Cambridge I had had a comfortable "set" consisting of a bedroom, a separate sitting room with easy chairs, table, desk and coal fire, and a small kitchen or "gyp room" as it is called. In Harvard I found myself in a room, obviously occupied normally by two students, with a two-tier bunk bed, two desks, two hard chairs, and nothing else!

The total number of active participants who registered for the congress was 310. Understandably, the great majority of these (265) were from the United States but there were sizable delegations from the U.K. (20) and from Canada (14). Other countries represented by smaller numbers were Belgium, France, Germany, Italy, the Netherlands, Spain, Sweden and Switzerland. There were no representatives of the U.S.S.R. although A. V. Shubnikov (who had been at the London conference) had written to me in October 1947 implying that as many as seven might attend.

One notable absentee was J. Wyart of Paris. He had attended the Royal Institution meeting in 1946, was a member of the provisional Executive Committee of the Union and a co-editor of *Acta Crystallographica.* The reason for his absence was that he had been denied an entry visa by the American authorities even though he presented a letter of recommendation from the French Government. What lay behind this was never made known to the Union but it was a matter of extreme embarrassment to the United States National Committee and precipitated lengthy correspondence between that committee, the National Research Council and the President's Advisory Commission on Educational Exchange; it was referred to obliquely in an editorial in the *Washington Post* of 21 October 1948. The embarrassment lingered, and when nine years later it seemed appropriate that the Union should again meet in the American continent the National Committee preferred not to issue an invitation but instead promised financial support and other assistance for a meeting in Canada. Thus it was that the Fourth General Assembly was held in Montreal. It was not until 1969 that the Union again met in the United States.

Most of those who enrolled for the congress may be identified from the group photograph (pp. 150 and 186). A number, however, are not in the picture, among them C. Hermann and M. von Laue from Germany; G. N. Ramachandran from India; E. Onorato from Italy; C. H. MacGillavry from the Netherlands; I. Waller from Sweden; A. J. Bradley and E. G. Cox from the U.K.; and P. Debye, R. C. Gibbs and A. Pabst from the U.S.A. (On the other hand a number of individuals, whom I shall not name, contrived to appear in the picture although apparently not enrolled for the congress.)

The expenses of the conference were largely met by generous donations from American industrial and other organizations, as were the expenses of some foreign delegates who might have otherwise been unable to attend. This American generosity was not, however, matched by corresponding support for other activities of the Union, such as its programme of publications. At the London conference in 1946 American and British crystallographers had pledged themselves to attempt to raise funds to underwrite the launching of the Union, and more particularly of *Acta Crystallographica.* In this endeavour the British were conspicuously the more successful, and this, too, was an embarrassment to American crystallographers which prompted Patterson to write to me on 21 June 1948: "X-ray crystallography is less recognized by American industrialists than it is by British. In this country we do not have publicizers of the caliber of Bragg, Bernal, Astbury and Lonsdale; in fact we have had a number who have gotten us into considerable difficulties."

Three sessions of the General Assembly were held during the week. These were open to all, and all were invited to participate in discussion although, if any contentious issues had arisen, voting would have been restricted to the nominated representatives of the three adhering bodies present. At the first session C. Palache welcomed the visitors on behalf of the United States National Committee and a motion that M. von Laue be elected Honorary President of the Union was carried with acclamation, as was a second motion that W. L. Bragg be elected President for the period until the Second General Assembly. (Bragg was not able to attend the Harvard meeting but before the close of the congress a message was received from him accepting this office.) Also at the first session the provisional statutes and by-laws were confirmed and reports were received from the provisional commissions on *Acta Crystallographica, International Tables* and *Structure Reports* set up the London meeting. These commissions, and also newly established commissions on crystallographic data, crystallographic apparatus and crystallographic nomenclature, met throughout the week and presented reports on their deliberations to the closing session of the General Assembly on 3 August.* Arising from the recommendations in these reports permanent Commissions of the Union were established to coordinate activity in each of these various fields. The Executive Committee of the Union was also elected at this final session. There were three American members: R. W. G. Wyckoff as a Vice-President, and M. J. Buerger and A. L. Patterson as ordinary members. W. H. Zachariasen proposed that the Second General Assembly should be held in Norway (his homeland) but in the absence of an invitation from that country no decision could be reached. In any event, the 1951 Assembly was held in Sweden.

I can recall little of the scientific sessions of the meeting because I was too much preoccupied with Union business to attend many of the papers. (The titles of the papers presented are on record in *Acta Cryst.* (1948) **1**, 342-343.) I do, however, vividly recall the brilliant evening discourse by C. G. Shull on neutron diffraction. At the time this was a field in which few crystallographers had had an opportunity to work and it was a subject of which I was totally ignorant. Yet so masterful was Shull's presentation that I can recall his lecture almost word for word to this day and felt qualified on my return to Cambridge to lecture on the subject myself, as indeed I did!

A number of social events were included in the programme. I have already referred to the picnic in searing sunshine on Ipswich Beach. There was a party generously offered by Messrs. Philips which has since become an established feature of the triennial General Assemblies and which, in 1975, in the company's homeland of the Netherlands, took the memorable form of a concert by the Philips Orchestra in the Concertgebow in Amsterdam. On the penultimate day of the meeting the congress banquet (now another established feature) was held at the Commander Hotel. I have never known a banquet to be served at such breakneck speed: the meal was all over in twenty minutes or thereabouts. But the occasion was none the less pleasurable for its informality and was memorable for delightful reminiscences of the early days of X-ray diffraction from P. P. Ewald and M. von Laue. Ewald spoke of his life as a student in Munich in the first decade of the century (this century) and von Laue vividly described his excitement when the success of Friedrich and Knipping's experiment was revealed to him at the Café Lutz, where the physicists of Munich assembled regularly for afternoon coffee. X-ray equipment at the time was primitive and that in von Laue's laboratory was operated by a Rühmkorff coil excited by a Wehnelt interrupter. The effect of this was to inject into the electricity supply of the university a modulation at 1000 hertz and this in its turn, caused the then common arc lights to emit a high-pitched tone of this frequency. It was convenient for von Laue thus to be able to monitor activity in his laboratory from any point on the campus but less so for his colleagues in other faculties trying to lecture against the musical accompaniment; so much so in fact that von Laue was ordered by the Rector to discontinue his experiments. But for the friendly intervention of Röntgen, who provided an independent electricity supply from his own institute, the diffraction of X-rays might never have been discovered.

W. L. Bragg often recalled with pride his period as President of the Union and would frequently remark that no matter how distinguished his successors in office he alone could enjoy the distinction of having been its *first* President. In the same way American crystallographers may take pride in the fact that the *first* General Assembly of our infant Union was held at their invitation and in their country. Later Assemblies have been bigger, more and more papers have been presented, social events have been more numerous and sometimes grander, but throughout all these Assemblies the imprint of the original meeting has been clearly discernible. And no subsequent Assembly has quite succeeded in recapturing the friendly informality, the comradeship and the intimacy of that first small pioneering gathering at Harvard.

*The meetings of the commissions were held formally and informally, at Harvard and on Ipswich Beach, by day and by night. That on *International Tables*, of which Kathleen Lonsdale was Chairman, often deliberated until long after midnight. A few hours later members would find on their breakfast tables the minutes of the meeting, each copy transcribed in her neat and distinctive hand and individually marked in red to indicate action for which the recipient was responsible.

ADDRESS BEFORE THE FIRST CONGRESS OF THE INTERNATIONAL UNION OF CRYSTALLOGRAPHY AT HARVARD UNIVERSITY - AUGUST 1948

Professor Max von Laue

Ladies and Gentlemen:

Mr. Ewald has told you in general about the conditions at the University of Munich which led in 1912 to the discovery of the X-ray interference in crystals and of the activities of the various physicists there, especially of their regular afternoon meeting at the Cafe Lutz in the Odeonsplatz. I should like to continue his report by recalling events in which he did not share, for the reason that he left Munich before Friedrich and Knipping had commenced their experiments.

In their first experiment the latter had placed the photographic plate so that it could catch rays which were deflected at an angle of about ninety degrees. This experiment produced no results. We have never fully succeeded in explaining the reasons for this. Probably the spectral range was not great enough to include the wave lengths which such rays would have had. I then effected a change so that the plate was struck by the direct ray, thus providing the existence of less deflected rays. As with the first experiment, the exposure was to last for several days. The X-ray tubes of that period could not stand up well under such extended use. Knipping had therefore constructed an automatic device which switched on the current from the University's electric system for about five seconds and then off again for twice that length of time. This experiment had commenced some time before the day about which I will now speak, and I naturally knew of it. It must have been in the beginning of March; the exact date can no longer be ascertained.

On this beautiful, warm spring day I joined the usual group for coffee in the garden of the Café Lutz. I remember that the physicists P. P. Koch and P. Epstein were there, also the mathematician Rosenthal, and, I believe, the physicists E. Wagner and W. Lenz. Friedrich and Knipping were missing. But an unusual atmosphere prevailed at the table reserved for the physicists. Instead of conversing as usual, each one silently read a newspaper. I sat down with them, ordered my coffee, and probably took up a newspaper, waiting until a conversation should begin. But none did. One of the company merely made an obscure remark, incomprehensible to me. Shortly thereafter another man made a similarly incomprehensible remark, and so it went, right around the table. I grew more and more mystified. But finally it dawned on me what had happened, and I said, "Well, gentlemen, I assume from your remarks that the interference experiment had a positive result and that each one of you has been told this confidentially. I knew nothing about it until now." And this was indeed what had occurred.

I soon went to the Institute for Theoretical Physics, whose rooms were in the University building and where the experiments were taking place. There I was shown the first interference photograph with copper sulphate, which was later published.

Lost in thought, I then went home. I took no notice of my surroundings. But after some time I found myself standing before the house at number ten Siegfriedstrasse, from which, a year and a half earlier, I had taken my bride. There, on the street, the idea came to me for that theory of space-lattice interference with which you are familiar. It consists essentially of three interference conditions, each of which implies a group of cones. I at once suspected that the circles into which the interference points had grouped themselves on that photograph must have something to do with these ones. And so I returned to my home at twenty-two Bismarckstrasse, full of hope.

This occurred in the Easter holidays of 1912, and it was still during this vacation that the first diagrams with symmetry were obtained. Sommerfeld was in Munich and naturally knew about the progress of the experiments and theory. But Röntgen was away and did not return until the beginning of the summer semester, early in May. Although the experiments were discussed only in a very small circle, word leaked through to him. He immediately hastened from his institute over to Sommerfeld's which was diagonally opposite, and had Friedrich show him the apparatus and the photographs. He arrived in the highly critical mood which was usual with him on such occasions. But when he saw those symmetrical photographs, he was forced to abandon all his experimental objections. He congratulated Friedrich on his splendid success cordially and sincerely, but added, turning to me, "But those are not interference phenomena. They look different to me."

Despite his doubts, however, Röntgen entered helpfully into the project. I must tell you a rather humorous story about this.

At that time X-ray tubes were still operated in physical laboratories with a Rühmkorff induction apparatus and with interrupted direct current. We used an electrolytic Wehnelt interrupter which interrupted the current about a thousand times per second and transmitted the vibrations of this frequency into the air but also into the University's electric system. In the evenings as we walked through the wonderful courtyard of the University, which unfortunately lies in ruins today, we could tell by the sound of the arc-lights hanging there whether our experiment was in progress. For these lights gave out a sound corresponding to those vibrations, lasting for five seconds, and then were silent for ten seconds. The connection was clear to us. We did not give it a second thought, and during the vacation nobody indeed objected to it.

But then the summer semester started in the beginning of May, and with it the lectures on history of art which played a most prominent part in the art-minded city of Munich. Many slides were naturally shown during these lectures, and the projector contained an arc-light, as was usual at that time. That this lamp should have made that same interrupted sound when our experiments were going on, would be clear to physicists. It must be due to some psychological law that this primitive music was contagious to the students. They thought it a great joke to hum along with it. The merriment grew greater and greater until finally the whole lecture was ruined.

The rector of the University naturally ordered a strict investigation into the cause of the disturbance. All of the many committees which are part of a university were set into motion. But in vain. We physicists, who could have explained the whole thing, knew nothing about it.

At the end of the first three weeks of the new semester the matter accidentally came to light. A mechanic, who had also been ordered to look for the source of the disturbance came into the cellar where the Wehnelt interrupter stood, listened, and at once reported it to the higher authorities. Then the waves of general indignation broke over all our heads. All the various committees came and certainly did not show us the most agreeable side of their natures. They demanded an immediate remedy or else suspension of the experiments. We should gladly have kept the vibrations away from the University current by means of filter coils, if we had only had such coils.

Faced with this need, we turned to Röntgen to ask whether we might draw our current from his institute. Technically this was very simple. We need only to carry a conducting wire across the university court from the window of one institute to that of the other. And as soon as it was established that the University would thus no longer be disturbed, Röntgen gladly gave his consent.

Just as matters had reached this point, the building committee walked into Sommerfeld's institute. They were the most powerful of all the university committees and apparently the least popular with the professors. They too wanted to let us feel the force of their anger, but we did not give them a chance to speak. Instead we at once told them of the arrangement which we had made with Röntgen. They were nevertheless suspicious. They went to Röntgen themselves to have this confirmed. And they returned a few minutes later in a state of indignation. We had deceived them. Röntgen was absolutely opposed to supplying current from his institute. We must therefore discontinue our experiments at once.

So the four of us sat there, Sommerfeld, Friedrich, Knipping, and I, and actually did not know what we should do next. Luckily our quandry did not last long. The solution came a few minutes later in the person of a mechanic from the Röntgen institute, a fat, affable Bavarian. In his deep bass and local dialect, which considerably increased the humor of the situation, he said, "The Geheimrat," (meaning Röntgen), "told me to tell you that you can go ahead and put up the wire. He is keeping to his agreement. It is just that whenever the building commission comes to him, the Geheimrat always says NO to them!"

Thus we could carry out our experiments to the end. In the course of the year Röntgen was persuaded that the new phenomena were of an interference nature after all. What convinced him most was the work of Sir William and Sir Lawrence Bragg and an experiment carried on in the Röntgen institute by Ernst Wagner and Johannes Brentano, with two crystals, which demonstrated the monochromatization of the rays in the first crystal.

FIRST CONGRESS OF THE INTERNATIONAL UNION OF CRYSTALLOGRAPHY

Harvard University, July 28 - August 3, 1948

A photograph of those attending the conference is shown on page 186 as Plate 8. Names of attendees follow:

FIRST ROW, *left to right: C. C. Murdock, A. D. Booth, C. R. Berry, D. Wrinch, T. Richards, J. D. Bernal, N. W. Buerger, I. Fankuchen, B. Warren, H. Mark, C. V. Raman, M. J. Buerger, C. Frondel, A. L. Patterson, P. P. Ewald, R. C. Evans, W. H. Taylor, Mrs. R. J. Bailly, R. J. Bailly, G. Hamburger, D. McLachlan, Jr., D. C. Hodgkin, S. C. Abrahams, H. M. Powell, K. Lonsdale.*

SECOND ROW, *left to right: E. A. Wood, W. L. Roth, E. W. Hughes, R. L. Griffith, M. L. Huggins, B. W. Low, H. P. Rooksby, R. Brill, W. O. Statton, R. D. Burbank, H. T. Evans, L. W. Strock, G. Switzer, H. S. Kaufman, G. Goldschmidt, H. Lipson, A. N. Winchell, S. B. Levin, R. O. Jackel, E. Grison, A. Frueh, W. Parrish.*

THIRD ROW, *left to right: J. D. McCullough, H. P. Klug, W. H. Zachariasen, M. C. Bloom, R. Pepinsky, L. Pepinsky, D. Fankuchen, R. C. L. Mooney, L. E. Lynd, H. Sigurdson, A. de Brettville, Jr., O. R. Trautz, C. B. Slawson, D. J. Fisher, D. McConnell, I. Weil, W. J. McCaughey, A. Van Valkenberg, Jr., H. F. McMurdie, H. W. Rinn, A. W. Kenney, S. S. Sidhu, C. E. Black, K. N. Trueblood.*

FOURTH ROW, *left to right: A. G. Brook, D. Ellis, D. A. Vaughan, O. N. Frey, L. Thomassen, L. H. Germer, A. M. Dowse, W. P. Mason, J. L. Abbott, C. G. Whinfrey, H. W. Dunn, H. W. Keilholtz, W. L. Kehl, M. E. Straumanis, R. A. Van Nordstrand, M. L. Fuller, W. F. Bradley, L. Alexander, E. J. Ritchie, L. Schulz, C. S. Barrett.*

FIFTH ROW, *left to right: J. G. White, P. Raesz, J. T. Edsall, M. Semchyshen, E. S. Greiner, C. D. West, H. Ekstein, W. L. Bond, M. Collins, H. F. Beeghly, D. Sayre, B. C. Frazer, C. W. Wolfe, R. B. Ferguson, R. J. Arnott, L. G. Berry, D. F. Clifton, J. M. Cowley, H. S. Yoder, N. F. M. Henry, George Vaux.*

SIXTH ROW, *left to right: S. Beatty, E. Kummer, M. E. Merchant, A. McIntosh, F. W. Matthews, M. A. Peacock, C. S. Hurlbut, Jr., C. Palache, W. H. Barnes, B. H. Billings, H. N. Campbell, A. H. Ehrhardt, A. E. Smith, H. M. Long, Jr., W. J. Roberts, P. F. Eiland, Jr., R. V. Gaines, A. F. Wells, F. A. Bannister.*

SEVENTH ROW, *left to right: G. L. Clark, R. W. G. Wyckoff, R. C. Taylor, H. Hughes, Z. W. Wilchinsky, J. W. Fitzwilliam, M. A. Bredig, C. G. Shull, A. C. Eckert, M. L. Jolley, M. L. Kronberg, G. Pish, J. H. Wilson, G. I. Faust, P. A. Bergmann, T. H. von Laue, M. E. Bergmann, B. M. Siegel, F. C. Brenner, S. H. Blank, R. S. Schwartz, R. Ward, A. J. C. Wilson.*

CHAPTER 6

LAKE GEORGE: 10 JUNE 1946

Andrew D. Booth
Lakehead University, Thunder Bay, Ontario, Canada P7B 5E1

In 1946 I was actively engaged in the design of a digital computer for structure analysis. This project was supported by the late Desmond Bernal and, in even larger measure, by the late John Wilson, Director of Research for the British Rubber Producers' Research Association. During the early months of the year, a visit to the U.S.A. to look at the state of the electronic computer art was under discussion. Matters were brought to a head by the arrival of an invitation to talk at the A.S.X.R.E.D. meeting at Lake George in June.

At that time a trip to America was no minor undertaking. Foreign currency was unobtainable, transport unavailable, and passports and visas uncertain. Bernal undertook transport and some funds, Wilson considerably more finance, and I set about the passport and visa business. The former went smoothly; however when I visited the U.S. Consulate, in the upper regions of Selfridges London Store, I found myself buried under a mass of squalling infants and distraught war-brides. After waiting for over 6 hours I eventually met the Consul. He was anything but friendly. In fact, after for the 7th time denying that I wanted to come to the U.S.A. to marry an American girl and thus obtain resident status, I was ready to wish the Consul and the U.S.A. to the devil! Fortunately he relented in time and I was led off to another room where a Hollywood detective type took my fingerprints. I thought this rather funny and laughed, only to be reprimanded and told that this was a serious business.

The next obstacle was Dollars. In my innocence I penetrated the inner offices of the Bank of England. The bureaucrats were horrified. They told me that the public was not allowed in and inquired how I had effected entry. I declined to answer, or to leave, until I had the necessary Dollars, and eventually, to get rid of me I suppose, I was given $200 in notes. This was, I hasten to say, much more valuable than it would be now, as will appear later.

Still there was no boat. On the 20th of May, however, I heard from Bernal that I was to sail on the Ile de France on May 27th, for New York. Great excitement! But, on the 24th I was notified that the boat had developed mechanical trouble and was delayed. By the 26th it transpired that the scow had lost a propeller so that the trip, at least, was cancelled - horror!

By now I was not prepared for failure. I tried airlines - a great adventure in those days - I haunted shipping offices and eventually, with some string pulling by Bernal got passage on the Cunard Liner Aquitania sailing on 3rd June, for Halifax. Late, but not too late, thank goodness.

The trip itself was fairly uneventful. It was a "dry" ship carrying war brides and returning G.I.'s. Despite this Anthony Eden (later Earl of Avon of Suez fame), arrived on board at the last minute, drunk, a condition which he maintained to the end of the voyage and thus started my lack of appreciation for *all* politicians. On the 5th we were off the Bay of Biscay and my diary simply says "sick." On the 6th, however, things were looking up. I had several good and large meals and apparently saw the film "Laura," although I have no recollection of it now.

On the 9th June, Whit Sunday, we picked up the pilot at Halifax in brilliant sunshine and docked at 4 p.m. There followed my first experience of North American trains, steam of course at that time. It really did stop at little villages but seems to have made good time since we arrived in Montreal at 5:10 a.m. on the 10th. It is a commentary on the dismal conditions in England at that time that I noted in my diary "Bought BANANAS!!" They were unobtainable at home then. At 9:10 I caught the train for Fort Ticonderoga, the nearest stop to Lake George. My note says that I arrived at 1:10 p.m. and took a taxi to the Hotel Uncas at Lake George. The cost was $5. I wonder what it would be now? The evening was spent pleasantly at a session and later with Dorothy Wrinch and Fankuchen (Fan to us all), both, sadly no longer with us. This was the start for me of friendships which lasted until both of these fine scientists, and most lovable of human beings, died.

On the 12th of June I note that I presented my paper if I recollect right, on the state of Crystallography in England. I well remember the long discussion which followed. It was my first exposure in the delightful informality of North American Scientific meetings, quite different from the stodgy formality of British conferences then. Next day I saw my first chipmunk and later spent the evening with Dorothy and Fan at the Mohican House Pub.

The 14th saw the business meeting and the end of a delightful experience. Dorothy had taken me under her wing and drove me over the Taconic and Mohican trails to her summer home at Woods Hole, driving in half a day, over superb roads, what would have been a summer vacation expedition in Britain or Europe at that time. At Woods Hole I met Dorothy's charming daughter, Pam, tragically burned to death in a fire a few years ago, and Otto Glaser, Dorothy's quiet husband. Although we had a great deal of pure fun, all was not idleness. To celebrate, as I remember, we decided to write a joint paper which, being theoretical, was finished next day and was published the same year.* On each of my subsequent visits to the U.S.A. we repeated this process so that it became almost an annual event, but the 16th June 1946 marks the first of the series.

All good things must come to an end and, on June 17th I took the train to Boston (fare $1.80) and on the same day visited with Martin Buerger at M.I.T., and later with Cutler West and Edwin Land at Polaroid. I still have two mounted Polaroid discs presented to me by Land, and an interesting optical gunsight from Cutler. This was the end of my crystallographic visiting in Boston although I spent some time with Vannevar Bush and Sam Caldwell. The real end of Crystallography, for the time being, came on June 18th with a visit to Fan at Brooklyn Poly, where I deposited a suitcase until my return from California to see Pauling and Corey. Alas, I had laid in a stock of chocolate, unobtainable in the U.K., to take back for friends. Upon my return to Poly, I found that mice had not only demolished the chocolate but also the suitcase!

The trip was the start of a long and fruitful connection with colleagues in the U.S.A. and I look back on it with pleasure and affection. One last reminiscence: Anita Rimel, Bernal's secretary, had asked me to give a note to Dina Fankuchen and to bring back for her (Anita) the parcel which Dina would produce. I did this and forgot about the matter until, having made my customs declaration (forgetting the parcel) at Southampton, the Inspector unearthed it and asked me what it was. As I said that I did not know, he opened it, and to my horror and embarrassment, but to his amusement, waved in the air for all to see, a pair of corsets!

*Booth, A. D. and Wrinch, D. (1946). Synthetic Patterson Maps. J. Chem. Phys. **14**, 503.

CHAPTER 7

BACKGROUND AND EARLY HISTORY OF THE AMERICAN CRYSTALLOGRAPHIC ASSOCIATION*

Martin Buerger
Institute of Material Science, University of Connecticut, Storrs, Connecticut 06268

The American Crystallographic Association began life on January 1, 1950. It held its first meeting on April 10-12 of that year at Pennsylvania State College in State College, Pennsylvania. Today*, a little over a quarter of a century later the society holds its 25th meeting in the same locality, to find that the place is called University Park, the home of Pennsylvania State University.

Six years after its first meeting the origin of the ACA was outlined by William Parrish and Betty Wood in Volume IV of the *Norelco Reporter.* There it is recorded that the ACA is the successor of the American Society of X-Ray and Electron Diffraction and the Crystallographic Society of America. These two societies, in turn, had begun their lives as the second world war was beginning to disturb the world. Two separate groups had begun to raise the question of the usefulness of establishing a journal for crystallographic research and this eventually stimulated the formation of the International Union of Crystallography which then not only took over the *"International Tables for X-Ray Crystallography,"* but a short time later established *Acta Crystallographica.* Few of you here recall the birth of the ACA, and even fewer remember the origins and acts of the two societies which eventually merged to become the ACA. It seemed useful, then, while there are a few left who took part in the acts which led to our beginning, to briefly outline the history on this occasion.

One pre-existing branch of the ACA, the Crystallographic Society of America, had its beginning as a local organization of crystallographers in Cambridge, Massachusetts and vicinity. As I recall it, at first most of its members were a few mineralogists from the neighborhood who felt that the future of mineralogy lay in understanding the roles of crystal structures in the properties, genesis and relations between minerals. The membership included Harry Berman, Newton Buerger, Martin Buerger, William Dennen, Clifford Frondel, Cornelius Hurlbut, Joseph Lukesh as well as the chemists Cutler West, I. Fankuchen and others. According to a remark in a letter from M. J. Buerger to Paul Kerr dated May 1, 1941, this group began to meet in the winter of 1939 for the purpose of discussing their research results with each other. The meager records which remain show that I. Fankuchen addressed the second meeting, held May 1, 1941 at Harvard University, on the subject "Preparation and Handling of Small Crystals" in which, among other things, the prototype of all twiddlers was described. Volume 95 of *Science,* which appeared on January 2, 1942, contains a note from the secretary of the society, Clifford Frondel, to the effect that after the business meeting of November 17, 1941, held at M.I.T., Joseph Lukesh addressed the assembled crystallographers on "The Tridymite Problem." On the occasion of the fifth meeting on April 24, 1942, at M.I.T., Percy Bridgman gave an invited lecture on a crystallographic aspect of his high-pressure work, which he entitled "Polymorphism at High Pressures." After this there appears to have been a hiatus in the activities of the society due to preoccupation with war work. During this period the society lost its vice-president, Harry Berman, as a casualty of the war. When the war finally ended the society prepared for a large meeting, held at Smith College on March 21-23, 1946. On this occasion the small local society of crystallographers assumed the status of a national society. At the Smith College meeting, papers were presented by Richard S. Bear, J. D. H. Donnay, Howard Evans, I. Fankuchen, Samuel Gordon, Joseph Lukesh, Dan McLachlan, Benjamin Schaub, Newman Thibault, Edward Washken, Cutler West, and Dorothy Wrinch. The abstracts of their papers were published in the *American Mineralogist,* as were the abstracts of papers given at later meetings.

From the time of the earliest meetings of the Cambridge crystallographers, the possibility of launching a journal of crystallography was considered; indeed an estimate of possible subscribers was made and the financing of the undertaking discussed. Unfortunately it was found that a subsidy would be necessary to begin publication, and the society had no success in arranging for one.

The crystallographers of the American Crystallographic Society, which included biologists, ceramists, chemists, mineralogists and physicists, had banded together to present the results of their crystallographic research to one another because their own professional societies offered them little encouragement and limited discussion. In such meetings their results were overwhelmed by the routine and classical results, and their own results were little appreciated.

Meanwhile certain chemists had found some relief from this situation. Many interested in the new results coming from the study of the atomic arrangement in solid matter were members of the National Research Council's Division of Chemistry, Committee on X-Ray and Electron Diffraction, then chaired by Maurice L. Huggins. Stimulated by the efforts of the ACS to establish a journal devoted to crystallography, Huggins called a conference under the auspices of the N.R.C. Committee on X-Ray and Electron Diffraction to be held at the American Museum of Natural History in New York, June 10-11, 1941. Although no action was taken to go ahead with the journal, the attendants did agree to form a society of structure researchers. The NRC Committee on X-ray and Electron Diffraction (to which I had been appointed in 1940) took the initiative in organizing the society, and its constitution was adopted at the Gibson Island meeting of July 30, 1941. Huggins was elected its first president and Warren its first vice-president, to succeed to the presidency the next year. The second meeting was held in Cambridge, Massachusetts, December 31, 1941. Meanwhile, many names had been considered for the new society; the one finally chosen by written ballot from the many submitted was 'The American Society for X-Ray and Electron Diffraction," which was abbreviated ASXRED. Thus this lineage of the ACA took its name from the name of the committee which brought it into existence.

There were now two societies concerned with different aspects of crystallography. Some believed that the societies were in competition and felt it would be to the advantage of both to join forces. Others pointed out that the ASXRED was named after a tool which, though important to crystallographic testing and research, did not begin to represent the whole science of crystallography, and could, indeed, pass out of existence, as did the optical goniometer. In 1948 a joint committee of the ASXRED and the CSA was formed to consider the consolidation of the two societies. Their report was sent to both societies. The membership of the ASXRED discussed the proposal for joining on December 19, 1948 and the CSA on April 8, 1949. Both memberships voted for consolidation. The two societies merged into the American Crystallographic Association on January 1, 1950. The new society held its first meeting on April 10-12 at Pennsylvania State College, as a continuation of the Conference on Computing Methods and the Phase Problem, which had been organized by Ray Pepinsky.

From all this it is apparent that, while the ACA is formally 25 years old, its roots range farther back. The ASXRED and the CSA are not the parents of the ACA, but rather they are earlier divisions of the ACA, so that our beginnings extend back at least to 1938 and 1941.

From this broader point of view, what has been accomplished until now? Perhaps most important is that the interests of the root societies in a journal devoted to crystallography was the stimulus which eventually launched the International Union of Crystallography and *Acta Crystallographica.* This came about in the following way: As president of the ASXRED in 1943, I realized the need of a vehicle to publish articles too long for the ordinary journal. Accordingly, I appointed a committee consisting of J. D. H. Donnay, George Tunell, and myself to see what could be done. We circulated a memorandum to the membership of the ASXRED sounding them out on their interest in a possible monograph series. They authorized us to go ahead. To finance the undertaking, I solicited contributions from companies who were in the business of making X-ray diffraction equipment and received support from The General Electric X-Ray Corp., Machlett Laboratories, North American Philips, and Picker X-Ray Corp. These contributions were placed in a revolving fund. The monographs were distributed free to all members of the ASXRED but sold to all others. Two monographs were published by the ASXRED before the advent of the ACA. The CSA published one monograph during the same period.

On instruction from the membership of the ASXRED, and as chairman of the Monograph Committee, I wrote on October 16, 1944 to H. Lipson, Secretary of the X-Ray Analysis Group of Cambridge, England, telling them of our Monograph Series and informing them of our interest in establishing a journal. I expressed the hope that the X-Ray Analysis Group would join with us in this project. Lipson called a meeting of that group on November 18, 1944 to consider this and other business. Eventually Sir Lawrence Bragg, chairman of the X-Ray Analysis Group, suggested that the Americans send a delegation to England to discuss the possiblity of publishing a journal and to consider the formation of an International Union of Crystallography. The meeting was eventually set for July 12 and 13, 1946. The American delegation included Fankuchen, Donnay, Germer, Harker, Wyckoff, McLachlan, Zachariasen and myself. From this meeting arose the International Union of Crystallography, *Acta Crystallographica* and plans for a new edition of the *"International Tables."*

Among the several noteworthy accomplishments of the ACA are the solution of numerous crystal structures whose reports have so filled the pages of *Acta Crystallographica* that the journal had to be split into two parts per volume and other material separated into another journal. It would be interesting to see an analysis of these contributions, but I make no attempt to do that here. I would, however, like to note that in one field of crystallography, the main channel has been outlined by members of the ASXRED, although contributions have come from many countries, namely, the direct methods of crystal-structure analysis.

I believe I had the honor of presenting the break which led to the hope of developing direct methods. At the Lake George meeting of the ASXRED in 1946 I presented the *implication diagram.* This results from a simple characteristic rotation and shrinking of the Harker section of the Patterson function. It has the property, in favorable space groups, of mapping the locations of atoms in a projection of the crystal structure. In less favorable space groups this desired result is accompanied by certain ambiguities and satellites. I demonstrated the use of the theory by solving the structure of the mineral nepheline, space group $P6_3$. That a structure could be solved by use of reflection magnitudes alone was a surprising result and it was greeted with no little incredulity. But Fankuchen immediately pointed out that since this could be done it implied that there must exist phase information hidden among the collection of reflection magnitudes. This was a stunning conclusion, for up to this time it had been believed that since the phases were experimentally unobservable, they were hopelessly missing and there was not any direct route to the solution of a crystal structure. By the time of the next ASXRED meeting at Ste. Marguerite in 1947, the Harker-Kasper inequalities were presented. These provided phase information under certain conditions, so there could no longer be any doubt that phase information was contained in the set of reflection magnitudes.

In subsequent meetings of the ASXRED many kinds of inequalities were reported. It is interesting that at the Conference on Computing Methods and the Phase Problem just preceding the first meeting of the ACA here in 1950, David Sayre showed that the comparison of the electron density with its square revealed that there existed some quite simple relations between the signs of certain structure factors. Sayre's conclusions inspired a spate of theoretical investigations of sign relations between structure factors. These culminated in the statistical work by Herbert Hauptman and Jerome Karle entitled *"The Solution of the Phase Problem,"* published as Monograph No. 3 of the ACA. Later this was followed by a long series of papers, mostly in *Acta Crystallographica* by Isabella Karle and J. Karle, which taught crystallographers a routine called the "Symbolic-addition method" which led to a direct determination of the crystal structure.

Although the "direct" route from diffraction intensities to the crystal structure is commonly thought to lead only through Fourier space, it may also be followed through Patterson space. In the same Conference on Computing Methods and the Phase Problem 25 years ago, I showed that a Patterson function can be decomposed into images of the crystal structure. In another paper in Acta Crystallographica entitled "A New Approach to Crystal-Structure Analysis" I presented image-seeking functions by which a Patterson function can be transformed into an approximation to the electron density function. Although many crystallographers have used this "direct" route through Patterson space, the route is less popular than the symbolic-addition method, probably because it has not been reduced to a computing routine. It is, however, a powerful "direct" method which has had no difficulty in solving an inorganic structure which has 15 non-heavy atoms and 20 oxygen atoms per asymmetric unit.

I have presented an outline of how the ACA arose from its two root societies, and some of the achievements of that period. Having solved some of the problems of the past so thoroughly that the solutions are available to other sciences as computer routines, let us not make the mistake of believing that crystallography has reached its zenith and is coasting downhill. Like all other sciences crystallography is open-ended, and the solution of crystal structures by diffraction is not its sole objective. I need only mention another phase problem as an example. It should be possible to theoretically predict the structure of a phase in any phase field, and not have to discover it experimentally. At present we cannot even do this for one-component systems. Understanding phase fields and their structures is one of the many problems which should engage our attention during the next quarter century. Ladies and gentlemen of the American Crystallographic Association, please present your report before the end of nineteen hundred ninety nine.

*Address at 1974 Summer ACA Meeting, Pennsylvania State University.

CHAPTER 8

ACCOUNTS OF ACA PRESIDENTS

ACA in 1951 - 1953. P. P. Ewald

I had come to the United States from Belfast, Northern Ireland, in the fall of 1949 as the Head of the Physics Department of the Polytechnic Institute of Brooklyn, largely through the endeavours of Herman Mark and I. Fankuchen. Rudolf Brill was already settled at Poly, and Dave Harker came shortly after me with his co-workers. This made Poly one of the strongest institutions for X-ray crystallography in the USA.

As a member of the Executive Committee of the International Union of Crystallography I was invited to attend the meetings of the National Committee of Crystallography in Washington, and took part in their deliberations for 4 or 5 years, for instance in those concerning the International Union of Crystallography meetings in Stockholm and Uppsala in Sweden in 1951.

At the committee meeting in Chicago in October 1951 I was nominated as the next President of ACA and instructed to prepare next year's general assembly. The ACA had received an invitation from Dr. Schwarz to hold the next meeting at Tallahassee. Correspondence with Schwarz indicated however that any coloured members of ACA would not be permitted to use the same entrance to the lecture halls as the other members and that they would be excluded from attending the banquet. Under these restrictions the meeting in Tallahassee was not possible and a different location had to be found. Luckily the United Garment workers offered to accommodate us at their Tamiment Lodge in the Pocono Hills. After several visits there (one with Fankuchen, another with him and Dave Harker) the Assembly was held there from 16th to 20th June 1952.

The 1953 meeting took place in Ann Arbor from 22nd to 26th June. On Wednesday 24th June the banquet was held with Dean Kraus and myself as speakers. Several of the members of ACA, among them myself, stayed on to give lectures or to help in other ways.

Thus my year as President of ACA came to an end.

ACA in 1959. J. Waser

The president of the American Crystallographic Association during 1959, its tenth president, was Robert E. Rundle. Due to his untimely death on October 9, 1963, at the age of 47, this report is being written by the Association's vice president for 1959.

The 1959 annual meeting took place in Ithaca, New York, on the campus of Cornell University, July 19-24. President Rundle was in England at the time, being on leave of absence at Clarendon Laboratory, Oxford. His place was taken by the vice president.

Among the highlights of the meeting was an invited lecture by Peter Debye on the investigation of random structures by scattered radiation, in which the details of Professor Debye's pioneering approach and the customary lucidity of his presentation kept the audience spellbound for an hour. Spellbinding in a different vein was a movie presented by Elizabeth A. Wood and S. Dworkin, its main theme being the relation between the physical properties of a crystal and the symmetry of its structure. Spellbinding in yet another vein was the witty after-dinner address by the ACA past president Dan McLachlan, Jr., on the theme "Lost in the Reciprocal Lattice." Also memorable was a well-attended barbecue picnic at Taughanock Falls.

The meeting was dominated by papers on the X-ray determination of crystal structures. A smaller number of presentations concerned neutron diffraction, and a smaller one yet electron diffraction. On the theoretical side there were papers on lattice vibrations, Pendellösung fringes, anomalous dispersion, and atomic structure factors. Further papers dealt with automated diffractometers that were beginning to be taken seriously and on computer programs. Of special interest were papers on phase determination, including reports on two crystal structures solved by I. L. Karle, J. Karle, and H. Hauptman by the use of their probabilistic methods.

The meeting was attended by a number of foreign scientists, among them P. S. Aggarwal and A. Goswami of India, James Trotter, H. E. Petch, F. Holuj, A. D. B. Woods and B. N. Brockhouse of Canada, C. H. MacGillavry and A. H. Gomes de Mesquita of the Netherlands, and W. Cochran of the United Kingdom.

A congratulatory message to Professor Max von Laue was signed by more than two hundred crystallographers in anticipation of the celebration of his eightieth birthday on October 9. The text of the scroll was:

To Professor Max von Laue:

Your friends assembled at Cornell University for the Annual Meeting of the American Crystallographic Association join together in extending to you their congratulations and warmest wishes on your eightieth birthday. We wish to honor the discoverer of X-ray diffraction by crystals and to pay our respects to the colleague who has contributed so much to the development of the field in which we all work. In addition we want to express our admiration to the very great citizen of Germany and of the world for the many contributions which he has made to human understanding and freedom.

Ithaca, New York
July 20-24, 1959

Professor von Laue was touched by these birthday greetings, as indicated by his reply:

An die
American Crystallographic Association

Unter den vielen Glückwünschen, die ich zu meinem 80. Geburtstag empfing, finde ich besonders rührend, dass Ihre Mitglieder -soweit sie in Ithaca versammelt waren- anscheinend alle die Adresse unterzeichnet haben.

Ich erinnere mich noch mit grosser Freude der Kristallographentagung vom August 1948 in Cambridge/Mass., an der sicherlich sehr viele der Unterzeichneten auch schon teilgenommen haben, so dass also eine persönliche Bekanntschaft mit Ihnen auf diese Weise hergestellt worden war.

Mit herzlichem, kollegialem Gruss
verbleibe ich,
Ihr ganz ergebener
(signed) M. v. Laue

The year 1959 marked the death of Wheeler P. Davey, a great American pioneer in the field of X-ray diffraction by crystals. He died on October 12 and President Rundle sent a message of sympathy to Mrs. Davey on behalf of the ACA membership.

ACA in 1960. J. Waser

In 1960 the Association became affiliated with the American Institute of Physics. The annual ACA meeting took place early that year, January 24-27, in Washington, D.C. The reason for this early date was the scheduling of the Fifth International Congress of the International Union of Crystallography for August, 1960 in Cambridge, England.

A high point of the Washington meeting was the invited paper entitled "Electrons in Crystallography" by L. L. Marton, a pioneer in the physics of electron microscopy and electron diffraction. Aside from sessions on structure determinations there were sessions on ferroelectrics, low temperature and neutron diffraction, precision refinement, and experimental and computing techniques. Among the foreign scientists that presented papers were Academician N. V. Belov of the USSR who discussed the crystal chemistry of silicates, and Alfred Niggli of Switzerland and Hans Wondratschek of Germany who talked about "Cryptosymmetry: A Generalization of the Point-Symmetry Concept." Among other scientists from abroad who presented papers were R. Brill of Germany, P. B. Braun and J. A. Goedkoop of the Netherlands, and D. Hall of New Zealand.

The speaker at the meeting's banquet was Wallace R. Brode, Science Advisor to the Secretary of State, whose topic was "National and International Science." At the ACA business meeting there was prolonged and stormy discussion of Dr. Brode's statements, such as the one to the effect that scientists whose attendance at meetings abroad was financed, at least in part, by the government were in a sense official representatives of the government and should therefore express only views that were supportive of government policy. It was then voted by a large majority of those present that a letter, the text of which follows, be sent to Dr. Brode, explaining the ACA viewpoints:

"We of the American Crystallographic Association, assembled in Washington for our annual meeting, wish to express our thanks to you for addressing us at our banquet on the evening of January 25, 1960. In your address you presented to us what we take to be the official State Department viewpoint on certain aspects of the relation of the American government to international scientific activities and to participation of American scientists therein.

"It is entirely fitting that we should be fully informed of the State Department view. However, it is also fitting that the State Department be fully informed regarding the views of American scientists on these matters.

"We feel it is our duty to inform you frankly not only that most of us are in strong disagreement with many of the views you expressed, but also that considerable apprehension was generated concerning the possibility that the State Department is not well enough informed regarding the attitudes of American scientists toward these matters, at or somewhere near the grass roots level.

"We cannot, of course, speak for the entire American scientific community, but only for ourselves; however, in view of the spontaneity of the reaction generated, it would be surprising if an altogether dissimilar attitude were found to exist in the community at large. We therefore feel compelled to state emphatically some views held by most of us that seem to differ importantly from those expressed by you.

"1) We maintain that the prime responsibility of American delegates of non-governmental American scientific bodies, in regard to international activities, is to represent the scientific bodies concerned, and not the government or anyone else.

"2) While such delegates should always welcome the opportunity to receive information from the government that will aid them in protecting the best interests of the United States, attempts to instruct or otherwise exert pressure to determine or influence the votes of such delegates, against their conscientious judgments, will generally be regarded as deplorable.

"3) All of these considerations should properly be entirely independent of any questions of financial support.

"4) We believe that State Department policies restricting the travel of foreign scientists - as for example that of Academician N. V. Belov - as well as that of American scientists abroad are not in the best interests of national and international scientific progress.

"We affirm that these views are largely matters of principle and accordingly are strongly held. We hope that they will come to the attention of the Department of State. A copy of this letter is enclosed which we respectfully suggest be transmitted to the Secretary of State. We also hope that the attitudes of other American scientific bodies that participate in international activities will be sought at the grass roots level. Because of our great interest in these matters we would very much appreciate it if you would keep us informed.

"Adopted by a large majority of those present and voting at a business meeting of this Association, open to the general membership, held on this day, January 27, 1960."

The chairman of the USA National Committee for Crystallography, Dr. B. E. Warren, also brought the matter to the attention of the President of the National Academy of Sciences, Dr. Detlev Bronk.

Since no answer to the foregoing letter had been received by the end of May the ACA Council decided to bring the matter to the attention of the Secretary of State, in accordance with instruction received at the business meeting. Later a letter of Dr. Brode's was received by the President of ACA, together with reprints of an article in *Chemical and Engineering News* (Volume 38, page 140, April 18, 1960). Dr. Brode implied that his failure to answer the original letter was due to a misunderstanding. He stated that many scientists agreed with his point of view and requested "specific comments of a substantive nature." In answer, a letter to Dr. Brode expressed regret that the situation had arisen but pointed out that the original letter to him was a concise statement of the point of view of a large majority of those present at the business meeting. The fear was again expressed that that point of view was not represented in the Department of State.

No further action was taken, in part because Dr. Brode resigned his advisory position in September. Many of the crucial statements made by Dr. Brode in his ACA address were not in the *Chem. and Eng. News* article, while his other statements were much weakened. At about this time the Governing Board of the National Academy of Sciences - National Research Council came out with its own policy statement regarding delegates to meetings of International Unions, a statement that was much in line with the views expressed in the ACA letter quoted earlier.

Many American crystallographers attended the Congress and Symposia of the International Union of Crystallography in Cambridge, England, August 15-24, 1960. As a result of the then prevailing political climate the eminent American scientist, Linus Pauling, failed to receive a passport, as had happened to him before. On an earlier occasion it had prevented Dr. Pauling from seeing the famous and crucial X-ray diffraction photographs of one form of DNA taken by Rosalind Franklin, pictures that were crucial to the work of James D. Watson and F. H. C. Crick. On the present occasion it precluded Dr. Pauling from presenting an invited paper on his theory of metallic bonding and from participating in a discussion of these ideas and those of J. C. Slater and N. F. Mott that many had looked forward to.

ACA in 1968. S. C. Abrahams

The American Crystallographic Association met this year at the Ramada Inn in Tucson, Arizona on February 4-7 and at the State University of New York at Buffalo on August 11-16. Both meetings were well attended, with over 330 registrants in Tucson and more than 480 in Buffalo, in addition to the accompanying wives and children.

The Tucson meeting began with a symposium on low energy electron diffraction, the proceedings of which have been published as Volume 4 of the *Transactions of the American Crystallographic Association.* Slightly more than half the program consisted of crystal structure determinations, with 18 papers on organic or biological molecules, 18 on inorganic and 13 on organometallic materials. The remaining papers were distributed with 7 on the symposium topic, 7 on thermal motion, 11 on the physics of solids and surfaces, 5 on apparatus and techniques, and 5 on diffraction theory. About 10 percent of the authors were from overseas. The program chairman was David Templeton, with Gabor Somorjai as symposium chairman and John Schaefer as chairman of the local committee.

Several organic structure determinations on molecules containing over 20 carbon and oxygen atoms were reported: these generally contained a heavy atom as in $C_{27}H_{26}O_3Br$ (J. Bordner and R. E. Dickerson), and were solved by standard heavy atom methods. Organometallic structures were not presented as such but were included either in the inorganic sessions, e.g., $[Cu_2\{(CH_3)_2NCH_2CH_2N(CH_3)_2\}_2(OH)_2]Br_2$ (T. P. Mitchell, W. H. Bernard and J. R. Wasson) or in the organic session, e.g., Cu_2(gly-L-leu-L-tyr)$_2$ $8H_2O.4C_2H_5OC_2H_5$ (W.A. Franks and D. van der Helm). The latter structure, with a unit cell volume of 5052Å^3, was one of the largest discussed. Several determinations gave R-factors comparable to values still regarded as entirely acceptable e.g., cubic $[Cr(NH_3)_6]$ $[CuCl_5]$ with a = 22.24Å had R = 0.037 (K. N. Raymond, D. W. Meek and J. A. Ibers). The majority of the organic and biological structure determinations were based on diffractometer measurements, whereas film methods were more commonly used for organometallic molecules.

Values of the effective ionic radii in oxides and fluorides were presented by R. D. Shannon and C. T. Prewitt: the paper resulting from this work has subsequently been very heavily cited. Among the interesting structures reported in the inorganic sessions were a pair of isomorphous triclinic crystals that differed only by the replacement of iridium in one structure by rhodium in the other. The ratio of electrical conductivity along the needle-axis in the two crystals was 10^6: both metals have similar square-coplanar coordination (D. Ülkü, G. C. Oldham and J. C. Morrow). Nearly equal use was made of diffractometer and of film methods in the inorganic studies.

Two papers on thermal motion in crystals attracted considerable attention, one by E. N. Maslen from Australia, the other by P. Coppens. The former showed that separating the rigid-body motion in molecular crystals from the internal modes reduces the number of variables, gives better atomic positions and more reliable difference densities. The latter discussed the origin of differences observed between anisotropic thermal parameters obtained by accurate X-ray and by neutron diffraction analysis in terms of aspherical vibrational corrections. Several contributions to the session on the physics of solids and surfaces were related to the symposium topic: the latter papers will not be discussed here since they have all been published in the *Transactions of the American Crystallographic Association.* R. L. Park presented a matrix formalism for LEED crystallography, based on the 1946 textbook by L. Brillouin.

The session on diffraction theory and experiment was unusually interesting, with a discussion by Andrew Lang of Bristol on the use of Moiré fringes produced by the superposition of separate crystals of silicon or quartz, followed by two papers on proton scattering by C. S. Barrett. Proton scattering in the 100 keV range gives information concerning zones that lie deeper than those producing LEED patterns but shallower than those causing normal X-ray scattering. With wavelengths on the order of 10^{-3} Å, Kikuchi-and Kossel-line patterns can be obtained that have intensities proportional to the structure amplitudes of the scattering regions. W. H. Zachariasen extended the theory of X-ray diffraction to highly absorbing specimens in which extinction was also taken into account. He showed that measurement of a CaF_2 sphere with both Mo and Cu radiation, after correction for absorption and extinction, gave an R-factor less than 0.02 and a mean domain radius of $3x10^{-4}$cm.

R. W. G. Wyckoff, the second ACA President (1951), presented a delightful address following the banquet on the topic "Crystal Analysis, Yesterday and Today" (see next chapter). It was a particular pleasure for many of the registrants to renew their friendship with Ralph Wyckoff, one of America's truly great men of crystallography.

ACA Council met all day Sunday preceding the Tucson meeting. Dick Marsh was appointed Program Chairman for the 1969 winter meeting in Seattle and Charles Fritchie was appointed Local Chairman for the 1970 winter meeting at Tulane University. A number of changes in the ACA Constitution were considered for action at the next business meeting. A beginning was made on drafting a meetings manual to contain guidelines for Local and Program Chairmen. In view of the forthcoming IUCr Congress in Stony Brook, a joint meeting was held by the USANCCr, ACA Council and the Stony Brook Steering Committee.

A high point of this winter meeting for most out-of-state registrants was the combination of bright sunny weather in a beautiful setting and the convenience of the meeting site in which most activities took place at the convention hotel. The midweek barbecue in Old Tucson added a special Western flavor to the meeting.

The Buffalo meeting of the ACA, with about 150 papers distributed over 19 sessions, had a strongly biological emphasis. In the opening session David Harker, who was also program chairman, presented a moving tribute to A. Lindo Patterson in the second Patterson Memorial Lecture, entitled "Appreciative reminiscences of A. L. Patterson." Structural papers generally dominated the meeting, with 55 on biological, 24 on organic and 12 on organometallic molecules, 22 on inorganic and 10 on mineral structures. In addition there was a symposium with 8 papers on small angle X-ray scattering, also sessions with 9 papers on methods of structure determination, 4 on the physics of X-ray diffraction, and 5 on dislocations and defect structures. The very able local chairman was Dorita Norton, whose untimely death we were to mourn four years later.

Six papers dealt with protein structures: that on subtilisin BPN′ by C. S. Wright, R. Alden and J. Kraut, for example, revealed the entire polypeptide chain of 275 amino acid residues, based on a difference Fourier series calculated with 2.5Å resolution data. Among the steroids and alkaloids discussed was wightionolide (K. T. Go and G. Kartha), with 72 non-hydrogen atoms in the asymmetric unit of two $C_{26}H_{28}O_8Br_2$ molecules, solved by the heavy-atom method on the basis of diffractometer measurements. The *t*-RNA nucleotide Na^+-5′-inosine monophosphate was handled similarly by S. T. Rao and M. Sundaralingam, who showed that the Na^+ ion in this highly hydrated salt is coordinated by water molecules and the two hydroxyl groups of the ribose unit, with the phosphate group linked through a water molecule to Na^+. Among the organic structures reported was the absolute configuration of (+)methyl p-tolysulfoxide, by H. Hope and U. de la Camp, based on the anomalous scattering by the sulphur and oxygen atoms. The molecular parameters of thiophene and its bromo- and chloro-substituents were determined by electron diffraction studies on the vapor (W. Harshbarger and S. H. Bauer).

The structure of the xenon compound $[Xe_2F_3][AsF_6]$ was solved by N. Bartlett, F. O. Sladky, B. G. de Boer and A. Zalkin: the $Xe_2F_3^+$ ion forms a V-shaped chain of F-Xe-F-Xe-F atoms with an angle of about 151° at the central F atom and 178° at the xenon atoms, both with an e.s.d. of 2°. A neutron diffraction study by J. M. Williams and S. W. Peterson on $HAuCl_4 . 4H_2O$ showed that the structure contains the $[H_2O . H.H_2O]^+$ ion, which has a *trans*-configuration and a symmetric central O-H-O bond.

André Guinier opened the small angle X-ray scattering (SAXS) symposium with a review of the progress made in the field since it began about thirty years ago, and continued with a preview of the advantages of SANS (neutron scattering). W. W. Beeman, followed by D. F. Parsons and C. K. Akers, discussed the application of SAXS to biological molecules, R. W. Hendricks did similarly for metals: C. C. Gravatt and G. W. Brady and also H. Brumberger considered SAXS by liquids. P. W. Schmidt reviewed current methods for correcting experimental data, and D. Harker commented on the ambiguity inherent in recognizing specific particle shapes that are compatible with given sets of distances between identical subunits having spherical symmetry.

The session on methods included a report by J. M. Stewart of his progress in producing the integrated set of crystallographic programs later known as XRAY 67. Dan McLachlan investigated methods of resolving overlapped peaks in a Patterson function. The physics of X-ray diffraction was advanced by the introduction of a formalism for treating anisotropy in the extinction correction, using Zachariasen's approximation (P. Coppens and W. C. Hamilton). A new neutron diffraction technique in which a pulsed electron linear accelerator was used with a time-of-flight detection system was described by M. J. Moore, J. S. Kasper and J. H. Menzel. In the session on defects, A. Taylor and N. J. Doyle discussed the production of vacancies in TiO caused by subjecting the material simultaneously to pressures as high as 78kb and temperatures ranging to 1760°C. They showed that the superconductive phase transition temperature increases from 0.7 to 2.3K as the vacancy density progressively decreases.

ACA Council again met throughout the Sunday preceding the Buffalo meeting. The amendments to the ACA constitution to be presented at the Annual Business Meeting included the provision that members of Council shall also serve on the USA National Committee for Crystallography, and that a temporary committee shall cease to exist immediately following its final report to an ACA business meeting. Draft guidelines for the Nominating Committee were discussed. It was decided that appointed representatives to other bodies should not be routinely reappointed. Contributed papers at forthcoming ACA meetings are to be presented only by the first author, who must be an ACA member: each member may be first author only once per meeting. The requirements for ACA membership may be waived in the case of foreign scientists.

The traditional past-presidential banquet address was given by John Kasper, who presented a fascinating account of his experiences with five-fold symmetry in crystallography. The midweek picnic held on Goat Island just above Niagara Falls and the tour of the Canadian side of the Falls were both very successful.

ACA in 1975. R. D. Burbank

As 1975 commenced our first concern was the precarious financial status of the ACA. The problem had been worried over the previous year and corrective measures had been initiated. However it was clear that none of the needed adjustments could take effect until 1975. Some of them involved delicate contractual negotiations with the American Institute of Physics and others required the approval of the membership at the next annual business meeting.

The most important change made in 1975 was a reorganization of the secretaryship. The Office of Administrative Secretary which had grown to three salaried days per week at the AIP was modified to the Office of Membership Secretary with one salaried day per week. The appointive post of Newsletter Editor was created and we were fortunate to have Jenny Glusker accept the initial appointment. At the annual business meeting a temporary committee on the Employment Clearing House was organized and Helen Berman was appointed Chairwoman. In addition an increase in the annual dues from $15 to $20 for regular members was approved.

In the flurry of financial problems we realized only belatedly that 1975 was our 25th anniversary year and arrangements for a suitable celebration at the Charlottesville meeting were improvised rather hastily. However it turned out to be a grand and glorious occasion. Of the thirty individuals living at that time who had been presidents of ACA or the predecessor ASXRED and CSA Societies twenty were able to attend the meeting including the founding presidents of both ASXRED and CSA. Our most venerable past president, Paul Ewald, wisely decided not to undertake a winter journey to a mountainous region. However he sent greetings to the Society which were read at the opening ceremony:

To ACA on its twenty-fifth anniversary:

The twenty-fifth anniversary of ACA presents me with a fitting occasion for thanking the U.S. crystallographers for the friendly reception they gave me (and my wife) when we first came to this country for good in the fall of 1949. To my astonishment and great pleasure the new immigrant was elected President of ACA for 1952, at a time when he had scarcely found his bearings in the new country. I appreciated this friendly gesture all the more when the ravages of the McCarthy era set in with the witchhunting and distrust of "unAmericanism." Of course I was no stranger to many of my U.S. colleagues; since the mid-twenties I had had personal contact with R. W. G. Wyckoff, Linus Pauling and Sterling

Hendricks, and had, on my first visit to the States in 1936, met many others who at that time were interested in X-ray crystallography and optics. When at the end of WWII the isolation of European from American crystallographers came to an end at the Bragg Conference of 1946, it was a joyous renewal of old friendships and the beginning of new ones. Yet, even with all these personal links it was an unexpected and gracious gesture of the American colleagues to entrust me with the presidency of their but recently united societies. They also allowed me, by a suitable formulation of the statutes of the USA National Committee of Crystallography to sit for the next 10 or 12 years on that body as a full-fledged member as belonging to the Executive Committee of the International Union of Crystallography.

Now that crystallographers have experienced, so to say, a population explosion, ACA meetings are much more formidable affairs than in the 1950's. Nevertheless, ACA performs the noble task of bringing together those who are curious to discover the internal fine-structure of matter—any kind of matter—and its meetings present the opportunity of helpful discussion and personal acquaintance.

I am sorry that circumstances prevent me from joining you at this 25th anniversary, but my main message of thanks and good wishes can be brought across in a more economical way by this letter.

May ACA flourish as long as there remain crystals to study and friendships to form between those trying to solve the same or kindred problems.

Paul P. Ewald

At the banquet, following the introduction of all past presidents in attendance, we were treated to a memorable address by Betty Wood (see Chap. 11). She was in superlative form as she gave an evocative picture of the early days of the Society. A prolonged and rousing standing ovation followed her brilliant presentation.

With the benefit of hindsight it is probably safe to suggest that the most significant scientific event of the meeting was a special evening session on uses of synchrotron radiation for diffraction which was organized and chaired by George Jeffrey.

There was no ACA meeting in the summer of 1975 since it was the occasion of the Tenth I.U.Cr. Congress at Amsterdam. The official American delegation was led by Jerry Karle, Chairman of the USANCCr. At a ceremonial session commemorating the Tenth Congress I.U.Cr. President Dorothy Hodgkin asked all those present who had attended the first congress at Harvard in 1948 to stand up. It appeared to this observer that there were less than a dozen Americans present in this category and no more than 25 or 30 counting all nationalities. It was a vivid reminder of the growth and changes that had occurred in crystallography in the preceding 27 years.

In mid-1975 we submitted a nomination for the National Medal of Science that had been prepared by a committee consisting of Ken Trueblood, Chairman, John Kasper, and Betty Wood. In September it was front page news when Linus Pauling received the medal from President Jerry Ford.

At the White House ceremony President Ford also expressed his intention of reinstituting a Presidential Office of Science and Technology. The Committee of Scientific Society Presidents with which ACA was affiliated at that time was one of the organizations which had urged President Ford to take this action.

The administration of the Fankuchen Award had been a problem to the ACA from its inception because the award fund is in the possession of the Polytechnic Institute of New York and no set procedures had ever been established between PINY and ACA for its use. Late in 1975 after extended negotiations a formal agreement was finally executed between the presidents of PINY and ACA which eliminated the earlier ambiguities. We received substantial assistance from Ray Young and Betty Wood in working out the details of the agreement.

With each passing year the membership of the council is altered by one or two individuals. Three years following the actions of a given council there is normally not a single person on council who participated in the actions. In a few instances this period has been extended to five years by reason of a secretary or treasurer serving two successive terms. In either event the time soon comes when no one on council has first hand knowledge of commitments of a contractual nature which were entered into by their predecessors. This lack of knowledge was brought home forcibly to us in the course of negotiations with both the AIP and PINY. Consequently it was decided to prepare a definitive set of copies of all documents of a contractual or business nature to be presented to each new member of council when they assume office. At the end of 1975 these documents included besides the incorporation papers various agreements and communications with the Internal Revenue Service, AIP, Polycrystal Book Service, the founders of the Warren Award Fund, PINY, and the National Science Foundation. It was hoped that this procedure would help provide a needed continuity to our successors in the governance of the Society.

ACA in 1979. J. P. Glusker

The winter meeting of ACA was held in Honolulu, Hawaii and was a joint meeting with the Crystallographic Society of Japan. This was the "brain-child" of Philip Coppens. In addition, two crystallographers from the Chinese mainland attended. The invitation to the Chinese took much effort in 1978 on the part of ACA Council, and, finally, an invitation was sent in Chinese since none of our letters were answered and we were afraid that the letters were sitting on someone's desk waiting to be translated. The letter in Chinese resulted in an acceptance in English. We then had to contact the Liaison Office in Washington to make sure the Chinese would get visas (there was no Embassy because formal diplomatic relations had not been established at that time between the U.S. and China). Dr. You-chi Tang gave a talk on the state of crystallography in China.

A new format for meetings, including invited speakers, was introduced at the Honolulu Meeting. The theme of the Meeting Symposium, organized by Gordon Brown, was geological crystallography. Many ACA members enjoyed their interactions with the Japanese and Chinese crystallographers. Dr. Olga Kennard, from Cambridge, England, discussed the Cambridge Crystallographic Data File and her structural work on the tetranucleotide dpApTpApT. Of course, the surroundings in Hawaii were very pleasant. The ACA meeting banquet, attended by Professor Linus Pauling and his wife, featured an outstanding display of Hawaiian dancing. Unfortunately a strike by United Airlines greatly complicated the return of many to the United States mainland.

The ACA meeting was preceded by a meeting on "Modulated Structures" held at the King Kamehameha Hotel at Kailua Kona, Hawaii. The meeting included a detailed discussion of the meaning of the term "modulated structure" which was defined, for the present, as a general term covering a wide variety of long period, more-or-less regular structural modifications. Immediately following the ACA meeting was an ACS meeting, also in Honolulu. At this meeting there was a joint ACA-ACS symposium on one-dimensional conducting solids. There are a large number of these because there are many possibilities for chemical substitutions which affect the crystal disorder which in turn affects physical properties such as electrical and thermal conductivity.

The summer ACA meeting was held in Boston, Mass. Unfortunately the weather did not favor us but the scientific program was most interesting, particularly accenting macromolecular crystallography, thanks to some hard work by Martha Ludwig. The first report of Z DNA was given at this meeting. Many structures of proteins and viruses were also presented in sessions organized by the macromolecular special interest group. The Warren Award was presented to Drs. F. W. Lytle, D. E. Sayers and E. A. Stern for their pioneer work on EXAFS and Professor Warren attended the Award Ceremony.

At the business meeting the initiation of a Patterson Award was announced, and a by-law change to merge the Crystallographic Computing and the Crystallographic Data Committees and to make Continuing Education a standing committee were approved by proxy ballot in November. A Sagamore Meeting, covering charge density studies and momentum and energy analyses of radiation and particles impinging on solids, followed the ACA meeting. It was held at the Mont Tremblant Lodge in Canada.

A monograph by A. F. Wells on *"Three-dimensional Nets"* was published by AIP. This year also saw the initiation of this volume on reminiscences of American crystallographers, edited by Dan McLachlan, Jr.

Concern about the discontinuation of Ilford G film was expressed and alternatives were sought. Also concern for the continuation of the lease payment for the Cambridge Crystallographic Files by NIH was alleviated by a letter from Dr. William Raub of NIH, who promised to inform ACA well ahead of time if the lease were ever discontinued.

Finally, all did not go smoothly. The projected meeting site for the winter ACA meeting in 1980 on the Gulf Coast in Alabama was literally blown away by the hurricane Frederick. However, by hard work on the part of Jerry Atwood, a new site was found at Eufaula, Alabama.

CHAPTER 9

ADDRESS AT ACA MEETING, 1968

Ralph W. G. Wyckoff
Tucson, Arizona 85712

After Dr. Templeton had suggested that I talk to you tonight, I felt some uncertainty as to my subject, for I have not been able in recent years to take an active part in the determination of structures. We all like nothing better than telling others of what we are doing and I toyed with the idea of describing our current work with fossils aimed at gaining evidence of chemical, as contrasted with anatomical, evolution. In it we make extensive use of X-rays and electron microscopy and it is thus a chapter in our study of solids; but I have resisted this impulse because of a feeling that occasions like the present ought to be devoted to a discussion of matters which concern us all but are not appropriate to our scientific sessions. One of these is the general way in which our subject is developing.

Sometimes it is still hard for me to remember that I am now among the oldest generation of crystallographers still active. For I have been concerned with crystal structures longer than any other native American, and longer than anybody now living in this country, except of course Ewald. In 1917, four years after Bragg's publication of the structure of potassium chloride, Burdick and Ellis published from MIT the structure of chalcopyrite and in that same year Hull at the G. E. developed the powder method and gave the structures of several metals. None of these persons continued to pursue crystal structure though Ellis, largely through Roscoe Dickinson, initiated what has become the Cal. Tech. school. As a graduate student at Cornell I began in that same year to learn about X-ray diffraction and space group theory under the tutelage of Dr. Shoji Nishikawa, who later became Professor of Physics in the University of Tokyo, and I published the structures of sodium nitrate and cesium dichloroiodide as part of my doctoral thesis.

It is not within my powers of description to give you a lively impression of the very different world of science that prevailed at that time and of the trivial place that crystals held in it. Graduate training still harked back to the 19th century. Chemistry was taught as a highly empirical subject with theory limited to electrochemistry, a few applications of the phase rule and the beginnings of thermo-chemistry. There were rules, but no theory, of valence. Ideas about the role of electrons in chemical combination were yet to come and my teachers thought and talked about valence bonds as if they were little hooks attached to solid atoms; and there still were a few eminent men who seriously debated the reality of atoms anyway. Synthetic organic chemistry was just emerging out of the very practical search for dyes that would not fade in the sunlight and most of the rest of chemistry devoted itself to the properties of aqueous solutions. Nobody thought much about solids except as the objects used for analyses of gross composition. Crystallography was rarely taught, and then only as an aid in the identification of minerals or the recognition of pure organic compounds. Neither the geometrical crystallography that expressed itself in the theory of space groups nor the physical crystallography developed in the little-known book of Voigt had claimed the attention of scientists - and this was quite right because by themselves neither led to something new. It was against this background of complete disinterest that W. L. Bragg suddenly demonstrated how X-ray diffraction, just discovered by von Laue, Friedrich and Knipping, could show where the individual atoms are in a crystal. The impact of this was immediate, but for the next five or six years the first World War was to occupy nearly everybody's attention. With the renewal of academic research at its end the first phase of our present preoccupation with solids began.

Those of you doing crystal structure now, when practically any crystalline substance can be tackled successfully, should not imagine that Bragg's description of the atomic arrangement in potassium chloride made everything clear sailing. In using the powerful experimental and computational techniques available today, you should not forget that it has taken the concentrated efforts of many individuals over two scientific generations to bring them into being. I do not intend to subject you to a historical review of the development of crystal structure in this country and abroad - it can be read, as seen through various eyes, in Ewald's *"Fifty Years of X-Ray Diffraction."*

But I do want to spend a few minutes on certain aspects of this development. In coming back to a University after a scientific lifetime in research institutes, I am struck by the lack of a sense of history, not only in the outlook of students but in the way science is being taught. It is easy to see how the rapid expansion of all science in this century has provided an almost unmanageable number of new things the student must learn. Nevertheless the very rapidity with which our ideas alter, almost from day to day, should be a sufficient demonstration of how transient is the point of view that we tend at any moment to accept as final. We all need very badly to realize that in the work we do, as in all other aspects of our lives, we live in a moment of time whose important characteristics have been set by all that has gone before; we really understand what we are doing only if we appreciate how we have come to be doing it, and it is only through intelligent participation in this broad flux of events that we can perceive the real meaning of our efforts and get a glimpse of where we are all going. It is easier to do this as one gets older but a certain preparation is needed and the educative process should provide this; graduate training ought to lead to technical competence but at the same time it should initiate the development of this broader outlook.

We who are interested in the solid state have had in this respect the advantage of being a rather heterogeneous lot. In England physicists, following the Braggs, dominated the earlier work in crystal structures; in this country chemists played a larger role. In Germany mineralogists soon became involved, and now biologists and organic as well as physical chemists have become actively concerned with crystal structures. This diversity of interests can, if we choose, be our insurance against becoming trapped in a sterilizing specialization.

We who started crystal analysis saw before us problems very different from those you have today; our immediate, all-consuming task was the development of methods. Now these methods are so powerful that atomic positions can be determined in practically any crystal. You can choose those substances which can contribute most directly to the general questions, in valence or solid state theory for instance, that concern you, and you can have full confidence in the results obtained. We were limited to the simplest of structures and always foremost in our minds was the problem of how to deal with more complicated compounds and how to make our results more sure. You must remember that the first structures were mere guesses, supported by a few qualitative observations. It was clear that in order to do more than suggest the atomic arrangements for a few very simple crystals, we would have to be able to get many more data, be able to interpret them quantitatively in terms of a more complete theory of X-ray diffraction and find ways to deduce from these data the only structure that was possible. To a few of us at that time it seemed that the almost forgotten theory of space groups provided the requisite cornerstone for such a deductive procedure. We take its use for granted now but that was not initially the case. It encountered much opposition, and to take a personal example, it was only a most unexpected chance that permitted the publication of my *"Analytical Expression of the Theory of Space Groups"* after it had been completed.

In order to appreciate the present state and future prospects of crystal analysis we should keep in mind the three distinct stages through which it has grown. The first, reaching from 1913 to the late 1920's, involved the development of ways to establish with certainty the structures of simple crystals (those with no more than a handful of variable parameters) and the successful investigation thereby of a few hundred such compounds. The second period, reaching from the late 1920's to the second World War, saw the analysis of compounds of intermediate complexity. This was made possible by the introduction of Fourier methods, by the development of experimental techniques giving hundreds instead of tens of quantitative reflections per crystal and by a mounting confidence in the quantitative interpretation of these reflections. By the end of this period we knew the structures of two or three thousand compounds but were still very severely limited in those that could be successfully analyzed. The barrier to further expansion was of course computational. It was breached and the present, third, period began when computers became available and automatic spectrometers began to accumulate quantities of data that it would have been impractical to amass by hand.

To me as one who has followed this growth of our subject from its earliest beginnings, the question of overweening importance is: what next? In spite of all that has been learned in the last 50 years, our knowledge of the solid state is still fragmentary. It is up to your generation to formulate out of these beginnings a comprehensive science of solids. How this will come about I don't know, though we can see at least some of the things that should be done.

Personally I think we should begin by amalgamating into a whole all the diverse interests there now are in microstructure. We know the atomic arrangements in thousands of crystals - inorganic, mineral, organic and biological - and this knowledge is increasing faster than we are properly assimilating it. Quantum mechanics and valence theory can now use what we find to further in fundamental fashion our understanding of the combining forces between atoms and the physical properties of the solids they form; and this understanding is applicable in all branches of natural science. Furthermore there exists a wide spectrum of techniques to supplement and vastly extend our X-ray diffraction - other modes of diffraction, spectroscopy of various sorts and, not least of all, the direct visualization typified by electron microscopy. The future student of solids should be conversant with the potentialities of each of these and he needs to look at solids in terms of the total knowledge they have contributed. This is a teaching problem that it will not be easy to solve. In the universities where this synthesis should take place we seem too specialized in our interests to feel an urgent need to work together; and besides that, the rapidity with which each of the established sciences is growing puts such a load on the adequate teaching of them that there is little time for innovation. Nevertheless I believe that we must continue to make the effort to create, in our own minds and if possible in the minds of our students, a lively sense of the unity of our knowledge of solids.

There are always two aspects to the teaching of a growing branch of science. One involves familiarizing younger workers with its problems and training them in its special techniques. This is an affair of graduate and post graduate instruction; and it seems to me that it is being very satisfactorily handled by our many active centers of crystal analysis. The other phase of teaching involves the presentation of a subject in a form that will introduce it to future specialists and at the same time familiarize a wider group of coming scientists with what has been learned. It is such a presentation of the overall science of solids which seems to me woefully lacking. I would urge that this is necessary if our efforts are not to fragment into supplying useful but subsidiary techniques to physicists, chemists and biologists acting as our patrons.

In the past the idea of such an eventual presentation to a more general audience underlay my crystal structure compilations. For a number of years I thought of them as becoming a sourcebook for such a future course. Things have not worked out to make this possible but neither has my *"Crystal Structures"* developed as I expected it would. I am convinced that some kind of consolidated, uniform description of all structural data is essential and in their different ways both my compilations and the *Structure Reports* were intended to provide this; but it now seems obvious neither will be able to cope with the great mass of new data made possible by the computers. The *Structure Reports* are falling farther and farther behind and they are already as clumsy to use as was my loose-leaf edition. Simply for reasons of space, and cost, my latest volumes are degenerating into little more than incomplete abstracts of finished structures - incomplete because if I were to include even the thermal and oscillatory motions that are an essential part of an up-to-date determination, not to mention a critical discussion of accuracy, I would never finish a volume and you could not afford to buy one if printed. Let me illustrate in terms of the volume now under preparation: Volume VI dealing with benzene derivatives. Almost as many structures have been worked out in the last two or three years as in the preceding forty and still the rate of output of new structures continues to mount. I am not yet prepared to leave the task unfinished but my will is weakening and of one thing I am sure: there would be no more editions - even if I were a young man. Some other way must be found to coordinate our structural results. Often I hear it said that we only need to put everything on IBM cards which can then be sorted to supply whatever information may be desired. This I am sure will have to be done, but it is not enough. Undoubtedly the future will bring excellent reviews that will gather together all the evidence bearing on specific questions of topical concern. But neither they nor the cards can be a substitute for that broadened outlook and understanding which comes only from having oneself survey, if in no more than cursory fashion, all aspects of a subject. Somehow we must find a practical way for each of us to do this and to make accessible to the students who will follow us a total picture of what has been learned about solids. Research, in contrast to engineering, flourishes in an environment more interested in the unknown than in the already-known. But we need a thorough grounding in the known in order to venture fruitfully beyond it. Perhaps none of us has an inkling of what the next large step forward in our knowledge of solids will be, but judging by the past we may be pretty sure that it will be something we may not now reasonably expect. We cannot prophesy the future but we can prepare ourselves to accept and exploit it and to mold our efforts to it as it emerges.

CHAPTER 10

CRYSTALLOGRAPHY AND THE GEOPHYSICAL YEAR

Dan McLachlan, Jr.
Stanford Research Institute
Menlo Park, California 94305

(An evening lecture before the Joint Meeting of Pittsburgh Diffraction Conference and the American Crystallographic Association, November 7, 1957, at the Mellon Institute, Pittsburgh, Penn.)

Mr. Chairman, Members of the American Crystallographic Association, Members of the Pittsburgh Conference, Ladies and Gentlemen:

I am very glad to be here, especially after the friendly introduction by your Chairman. This introduction made me eager to hear the speech myself. I could not help thinking, while Dr. Taylor was speaking, of one of the ancient and distant cousins in the McLachlan clan who said that there was nothing in this world more disgusting than a woman who takes the hard-earned family cash and spends it for such trifling things as oatmeal, soap, and potatoes when there is not a drop of liquor in the house. To me, there is nothing more discouraging than an after-dinner chairman who spends the entire introduction time telling what a nice fellow the speaker is and fails to tell how much he has contributed to the discomfort, trials, and tribulations of future graduate students. But Abe Taylor did not let us down; he told you that I had written a book. As you can judge from the title *"X-ray Crystal Structure"* it is chuck full of tender feelings, sweet sentiments, and good will. After it is circulated, no doubt the sparrows will sit closer together on the telephone wires of at least two continents.

When I received the telephone call from Dr. Taylor eight months ago asking me to speak tonight I was so flattered that I said yes before I knew what I was getting into. The subject, "Crystallography and the Geophysical Year," has kept me sitting up every night since that fateful telephone call. I sat up at nights for two reasons. First, it is very difficult to prepare a talk on a subject of which one knows nothing. The second reason for sitting up at *night* is because I have learned from long years of experience that the man who sits up in the *daytime* doesn't get any sympathy from *anyone.*

The Geophysical Year is a very unusual year from several standpoints. In the first place it is supposed to last eighteen months and if it holds out, it will win the record of being the longest year in the memory of man. In the second place, all the nations of the world are to cooperate in collecting data about the earth, its atmosphere and its environment among the stars and, if this plan comes to fruition, this year will mark the first occasion upon which all mankind has worked together for any single purpose other than war.

As you know, the earth is a sphere 8,000 miles in diameter. The outside crust, about 300 miles thick, is solid crystalline material. The next 2,000 miles in depth is thought to be somewhat plastic but so thoroughly crystalline that crystallographers cannot ignore it. The inside core is suspected to be fluid, as judged by the fact that seismographers can pass compression waves through it but cannot pass transverse waves through it. It is under such tremendous pressure however, that some semblance of ordering of the atoms cannot fail to exist to delight crystallographers. On the surface of this earth, the winter suburban resident sweeps the crystalline snow off of the doorstep before bringing in his milk bottles. The polar wanderer finds himself treading plateaus of snow thousands of feet thick and during storms finds himself in an endless whiteness of which every minute speck is a crystal of jewel-like charm. When the storm clears, he sees sundogs as evidence of crystals in the upper atmosphere. In the mountains the volcanoes belch forth vapors that crystallize in midair and great rivers of lava congeal to polycrystalline masses as they flow down the colder slopes. The growth of plants generates organic fibers of which a great fraction are crystalline and Eskimos eat frozen fish with crystalline teeth. During all this, crystalline star dust and meteorites are coming in to us from outer space at the rate of several tons per day. This crystalline world is to be examined this year.

The data-collecting team of this Geophysical Year is a body comprising the world's best professional talent: explorers, meteorologists, astronomers, geologists, mineralogists, chemists, physicists, politicians, mathematicians, biologists, the army, navy, and marines, cooks, boot-blacks and scattered members of the medical profession. All professions are represented except one, crystallography. Therefore those of you who have your programs handy may turn to the section where the title of this speech is given as "Crystallography in the Geophysical Year" and draw a pencil line through the word "crystallography." You may do that now. Or, if you wish to save time, you may wait until we get to that portion of the lecture when you will become convinced that a line should be drawn through the remainder of the title also.

The earth is large compared to everyday things. Suppose that a 10 foot circle were drawn with a piece of chalk on one of the plain wall areas of this Mellon Institute to represent our globe and that the artist were requested to add to his design a profile of Mt. Whitney drawn to scale. It may surprise him to find that his chalk has already made a mark wide enough to cover Mt. Whitney. The earth is not really very rough. This can be visualized by assuming, for a moment, that the earth is shrunken to three inches in diameter. Then it is about the size of an orange and about as smooth as an orange, with an atmosphere hanging about it like an aroma.

But if you did have such an earthly orange in your hand, you would have to handle it carefully. Large things are like that. A sick elephant goes to the river for the comforting buoyancy of water and a whale cast ashore dies from the torture of its own weight. The dinosaurs probably reached the greatest size possible for a self-supporting body made of the materials, flesh and bone. Even the earth, made as it is largely of stone, is a delicate object. If you open your hand to let the orange rest on it without support, it would deform like under-cooled (or too dilute) jello, cracks would form and grow until the sphere is transformed to a shapeless mass. This orange, when suspended in space, would be relatively undistorted. However, it has an uneven distribution of radioactive material in it, hot spots form from time to time, local swelling occurs, and subsequent strains are formed. These strains may be relieved by spontaneous formation of cracks, or the starting of the cracks may be induced by the gravitational pull of the moon in its cycles. When the frequency of earthquakes resulting from cracks (geological faults) is plotted in lunar time instead of solar time, an interesting correlation exists. Thus, we see that the earth is a very tender body of matter. The men who live on it are even more tender; even in the best of athletic condition, a man is really a colloidal suspension reinforced with bones. When the cracks form, the orange trembles and the dwellings which man has built for shelter fall down upon him and crush him. In very active geologic eras, the hot spots cause volcanic action. Thus while earthquakes shake cities to the ground, volcanoes pour lava down upon villages.

Looking at the orange again, we cannot see anything as small as a human being. In special places like New York or London we might see spots which look like fungi but even those are not as pronounced as our forested areas. History, as it is written by mere mortals, is the story of war, of how these fungi have won out in their contests of throwing their manufactured goods at one another.

In this Geophysical Year, more men will go into the snowy north where they may see a human principle illustrated. Inside those lonely cabins of the northern homesteader, the children quit quarreling when they hear the wolves howling outside, and dogs that fought in the summer huddle together for warmth in the winter. Apparently the best environment for maintaining internal harmony between people is the knowledge of mutual dangers on the exterior. It has been said that our bitter civil war could not have been waged if, in 1860, the entire nation had been under threat from a foreign power. If a complete awareness could be brought to the peoples of the world of the vast natural and little understood cosmic powers both within the earth and surrounding it, this awareness would perhaps bring about some unity. In long future planning, the dangers that harbor on political boundaries are feeble in comparison to the natural dangers. Let us look at some of these big ones that we in the Geophysical Year may aid in understanding.

Let us say that the two powers already mentioned, earthquakes and volcanoes, do not need, forever in the future, to remain, as they have in the past, unpredictable calamities from which we are destined helplessly to suffer. With the aid of modern instruments, the murmurings inside the earth become voices with meaning, voices based on the past and predicting the future. The cosmic data are the many and mysterious tongues of nature and it is the task of scientists to interpret them.

Since long before Galileo and Kepler, men have charted the heavens and plotted the orbits of planets and stars. Although the interest is largely academic, some practical results are obtained. For example, the second appearance of Halley's comet was predicted and did not cause the fright that its first startling appearance caused. In general a study of the composition of comets has resulted in the descriptive phrase "big bags of nothing" and fear of them has been largely arrested. However, on the other side of the fright-ledger, the investigation of asteroids, beginning with the discovery of the first one in 1801, has justified serious doubts about the earth's security in the heavens. The asteroids are large, irregular shaped, tumbling mountains 500 miles in diameter, "pell-melling" through space in elliptical orbits at about seven miles per second. Eight of these objects cross the earth's orbit with an estimated probability of one in 30,000 of colliding; and more of these obstacles are discovered every year.

Also we must contend with meteorites, objects which are large enough to penetrate the earth's atmosphere, create a shock wave with a luminous flash and strike the earth with varying degrees of damage. They vary in size from ounces to tons, but do not vary greatly in initial velocities in the neighborhood of seven miles per second. Simple computations give estimates of the energies in one of these fragments. Using $E = \frac{1}{2} mV^2$, one finds that the energy of a meteoric body is equivalent to about 400 pounds of TNT per pound of meteor, so that a three ton meteorite is equivalent to an atomic bomb. The occurrence of these large ones, although infrequent in the life of the individual man, is not rare in the history of the earth, as attested by numerous samples found, or their craters such as the one in Arizona or the one in Quebec. The smaller objects about the size of pinheads flying through space strike the earth's atmosphere and burn out before reaching the ground, their energy consumed in a flash of light equivalent to that of a giant searchlight. These are called shooting stars and occasionally the earth experiences a shower of them. Just as the asteroids or "flying mountains" can be charted as to orbit, so can the "flying sandbars" and the flying "gravel beds" of space. They tend to have orbits which mark the hazardous course of other heavenly bodies. We know only in part when to expect a shower of stars or a close brush with an asteroid and there is much more to be done.

Also flying through space are the electrons and protons which are boiled out of the surface of the sun, which, after traveling ninety two million miles under the gentle but tireless push of the sun's radiation, reach a velocity of 6,000 miles per second. Because of the earth's magnetic field, they are deflected into the atmosphere at the poles giving the well known and beautiful auroras. Although the electrons produce 100 kilovolt X-rays and ionize the oxygen, nitrogen, and moisture of the atmosphere, they cause no harm.

The interstellar materials, especially those in the form of dust, influence the amount of sunlight reaching the earth and have been thought to cause the glacial periods, although this theory cannot as yet be proved. We believe that a variation of five or seven degrees in annual average temperature of the earth is enough to cause glaciation to the extent of causing the massive ice caps to creep ruthlessly toward the equator, destroying all life in their paths. But we don't know whether the variation should be an *increase* or a *decrease* of five to seven degrees. These things and their predictions would be good to know. When the weight of the ice becomes so great, the orange is deformed and the swelling at the equator accompanying the loss of water from the oceans causes the land masses to rise above sea level, exposing the continental shelves. England is no longer an island but a European peninsula and Australia is a part of the continent of Asia. This gaining of land masses has been used to account for part of the past world wide migrations of animal species and mankind. The other extreme, that is the melting of the ice in Antartica and Greenland, would cause the levels of the oceans to rise 100 feet and inundate most of the major cities of the world, including New York and London.

Looking again at our orange with its molten core, we have experienced in our own lifetimes a shifting of the magnetic poles. This is troublesome in navigation and we wonder how great such variations might become. Attemps are being made in this Geophysical Year to investigate the variations of the distant past. When molten lava from volcanoes melted magnetite thousands or millions of years ago, the magnetite became permanently polarized upon cooling with its field parallel with the magnetic lines of the earth. As the poles migrated, the magnetite was left in the wrong orientation. The variations between the directions of magnetization of the ancient lava flows and the direction of the compass are now being measured.

Also there is evidence of the migrations of the earth's crust. The migration of continents is a theory which creeps into the geological back door every time it gets thrown out of the window. But the mineralogical and geological similarity between the west coast of Africa and the east coast of South America is embarrassing to those who have difficulty in believing that the two continents were once joined and with time split and migrated apart. Troublesome too is the fact there are thick veins of coal under the ice of Antarctica and the ferns and other plant life embedded therein suggest that Antarctica was once a tropical continent. All these things are of interest to a mankind which wishes to know its future climate.

Inside the earth and outside, there are mighty things and mighty forces which, if understood, could be applied to our gain and, if ignored, can be disastrous.

But the world is at best a mysterious place to man. We do not know how to measure absolute position, only relative position. We have chronometers which can bite off chunks of time in equal lengths but we have no zero reference. So we do not know when or where we are. The theory of relativity only helped to remind us that we know nothing about absolute velocities either. So man, using a meridian on a spinning earth, referring to the sun which moves in our galaxy about a greater star which is also moving, is like a very small fish deep in the middle of the Pacific Ocean who gets smug contentment from the knowledge of its whereabouts by using as a landmark another, larger fish.

And in all this blue space of danger and mystery we have to face the terrible problem of overpopulation. Of course, I for many years now, have been convinced that the earth has *always* been overpopulated by definition; so I am no more afraid of it than I am of *death* – which I think has done, in the past, so much to prevent overpopulation. But fortunately for the maintenance of the sad tone of this speech, I know how to make overpopulation sound *terrible.* Eight thousand years ago there lived in Mesopotamia a civilization comprising thousands of people who lived in pairs in a husband-wife relationship and each pair had a new baby every year or so from the age of 18 to 40. Like all civilized people, they no doubt felt that from all of these children a goodly number, say four or six, should reach maturity. Suppose the number of survivors were four per marriage. What would happen? This means that the population would double every 33 years from 8000 years ago till now; and if there were only two Mesopotamians to start with, there would be 10^{80} people now. This is enough people to cover the surface of the earth, fill the interior of the earth if it were hollowed out, fill all the heavenly bodies similarly, and populate interstellar space with people one yard apart. But we would not have material to make the people out of because there are only 10^{78} particles in the universe. Wouldn't it be nice to develop interstellar, or space, flight so that some of us could escape the first stages of this ensuing mess? At present we believe that the capacity of the earth for people is one person per 2.5 acres of cultivatable land and we expect to reach this limit in about 70 years. The last earthly continent has been discovered. We need a new home.

Space flight has its hazards too. The aluminum gondola touring space without a blanket of air, such as the earth has for protection, would experience dangers even from the pinhead-sized particles going at seven miles per second. Ballistics experts have not yet figured out what kind of a wound the gondola would suffer. The results of collisions with larger particles up to the size of gravel and asteroids is gruesomely obvious. We should know more about their paths and the numbers of these bodies. Also there is the problem of temperature during flight. Other items are the cosmic rays, gamma and X-rays of interstellar space which are known to cause sterilization and mutation among plants and animals. Dr. Friedman has informed us that the chance of mutation from gamma rays is small enough to be neglected. I know that this will come as a disappointing revelation to those young people who would like to look forward to the birth of every one of their children as an unpredictable surprise.

The problem of finding a heavenly body to which to migrate is not a difficult one. There are estimated to be hundreds of bodies in our own galaxy with the right temperatures for human habitation, the correct atmospheric composition of oxygen, carbon dioxide and nitrogen, and where the body itself is not so large that its gravitational pull is strong enough to flatten us out like tar on the garage floor. In the entire universe there may be millions of habitable bodies. But the distance gets to be a problem. Some of the people who homestead the distant stars will not get back to eat Thanksgiving turkey with Grandma and Grandpa. With the optical barrier of 186,000 miles per second facing us we cannot hope to reach a habitation 1000 light years away in less than 1000 years.

The greatest contribution to space flight is perhaps the artificial satellites program. These satellites are to be loaded with equipment, some of which is to send data by radio to the earth regarding the temperatures, pressure and radiations of various kinds as a function of time and altitude. The data which are retained inside of it, such as photographs and material damage, must be brought down with the satellite. It is apparently a difficult feat to bring down a satellite. I have toyed with the idea of telling you that a Canadian wild goose hunter was going to bring it down with a fowling piece. But knowing that this scientific audience would not accept this, I have decided to tell you that the Monsanto Chemical Company is working on an insecticide which will make the satellite roll over on its back and flutter, disparaged, to the ground.

There are moments when the conscientious observer of the events of this Geophysical Year may feel disheartened. There are times when one feels convinced that all the noble plans of the endeavor are being more effective in sponsoring animosity and fear than good will between the nations. But I think we should take heart; because there has been at least one achievement that could not have been accomplished in any other way. I refer to the discovery of the Sputnik cocktail. I think that I am not giving away any bartender's secret when I give you the recipe: Take two jiggers of vodka and add a dash of bitters. Unlike martinis, where you use olives, the Sputnik cocktail calls for sour grapes.

I would like to thank you all for your patience during this talk and I close, hoping with you, that the Geophysical Year will give us much knowledge of our cosmic environment. I thank you.

CHAPTER 11

ADDRESS AT ACA MEETING, 1975

Elizabeth A. Wood
Red Bank, New Jersey 07701

Introduction by R. Burbank

And so the curtain falls on the first part of our program. We have displayed the family tree, but there is still the merger or union of ASXRED and CSA to consider. Back in 1957, at the time of the 4th Congress, of the International Union of Crystallography in Montreal, the *Norelco Reporter* came out with a special commemorative issue. It included an article on the history of the American Crystallographic Association which seems to have been forgotten by most people over the intervening years. It was co-authored by two people who played very active roles in the committee work, discussions and negotiations that led to the formation of ACA. One of those people was the secretary of CSA, Bill Parrish, who unfortunately could not be with us tonight. The other person was the secretary of ASXRED who later became the 8th president of ACA. I first met her at a joint meeting of ASXRED and CSA in 1948. I have never ceased to admire her outstanding attributes as a stage woman. She has graced our speaker's podium at a banquet on more than one previous occasion and it is a pleasure to welcome her back this evening. She is going to speak to us tonight on some of the history of the ACA and possibly will convey a little bit of the flavor of our earlier meetings. Ladies and gentlemen, I am pleased to introduce Elizabeth Armstrong Wood.

Address by Betty Wood

You know, when Rob first asked me if I would come here tonight, he said, "Would you be willing to say a few words, you know, 15 or 20 minutes that would give the flavor of those early meetings?" I said sure, I could do that. Then I saw the title of the talk listed as the history of the American Crystallographic Association. Well, that was not just what I contracted for, and finally it came out in the program as "a historical address." I have seen a news release in which it was called "The Historical Address." This progression towards formality or formidableness I think deserves to be called some kind of a law, and since I assume that it is the work of our esteemed program chairman, I suggest that we call it Stewart's Law. Well, when the title of a talk threatens a speaker, with its inhibiting influence, the speaker disregards it . . .that's Wood's Law. So I shall proceed as I originally intended with giving you the flavor of some of those early meetings.

History is often written in a desiccated, lifeless sort of way. For example, one reads, "On January 1, 1950, the ASXRED and CSA ceased to exist and the ACA began." Well, that doesn't begin to give you the idea of the turmoil that surrounded the Amalgamation, with a capital A. I remember a meeting chaired by Adolph Pabst. That was the CSA meeting just before the amalgamation. That was in Ann Arbor, Michigan. Remember the Crystal Bell in Ann Arbor, where they used to have beer in pitchers on the tables? That was why we enjoyed going to Ann Arbor, Michigan. Adolph Pabst was chairing this meeting. Now, he is a gentleman of the old school. In his mind, no one speaks without being acknowledged by the chair. Well, the heated argument about the amalgamation got so violent, that people leapt to their feet and argued with each other across the room and Adolph Pabst said, "Gentlemen, gentlemen." This didn't stop them, so he took the very violent means of hitting the table with the gavel. This didn't stop them and he hid his head in his hands in despair. Then, taking heart, he picked up his gavel and held it out, saying, "Please, somebody, *somebody,* take the chair and control this meeting, I can't." You see, that doesn't come out in history.

Then there was Martin Buerger. Like many of us who came into crystallography from mineralogy, he knew that there was a great deal of crystallography before X-rays were ever shone upon a crystal surface. He also knew that there was still a lot of crystallographic research that was not related to solving crystal structures and he felt that this was what the Crystallographic Society of America had made possible, had given a forum for. So he said, "If there is an amalgamation of these two societies, I'll just have to start another one." He didn't.

Then there was a lot of talk about the name of the society. You couldn't call it the Crystallographic Society of America because obviously, that was a direct snub to the ASXRED. They were just being absorbed by the CSA, and that wouldn't do. You couldn't call it the American Crystallographic Society because then its initials would be ACS and that was the American Chemical Society. There was a good deal of feeling for calling it the American Society for the Study of the Orderly Arrangement of Matter in the Solid, Liquid or Gaseous State. Well, some people thought that was too long and the acronym was not pronounceable. So quite a few people were in favor of calling it the American Crystallographic Association. But then there was Bert Warren. Bert Warren said, "I'm a physicist and I'll be (word deleted) if I'll be called a crystallographer; if you call this society the American Crystallographic Association, I simply can't belong to it." Well, you may remember that he subsequently became a member of the Executive Committee of the International Union of you know what. It's very fortunate for us that he did stay with us and shared with us his beautiful work on the non-Bragg diffraction effects which have in them information that, until he showed us how to get it out, was hidden from most of us.

Of course, the real history of an association such as this lies in the interchange of ideas, the stimulation its members get, the papers in *Acta Crystallographica,* in *Journal of Applied Physics,* and in the Metallurgical Journals that were first given at an ACA meeting and that benefitted from the exchange of ideas and the discussions in ACA meetings.

In the very early days, the meetings were very small . . .I mean really small: 35, 40, 45 people. In the pre-history entry in this historical document that Rob has referred to, of the various meetings and where they were held, there is one that reads "October 1-4, 1945, ASXRED Meeting at Lake Geneva, Wisconsin." Now there's the bare fact. But you should have seen where we met. You went to the bar in the place where we stayed and beside the bar there was a panel that didn't look like a door, but it was a door. And if you knew the way, you could go through that door into a room behind the bar; and that was quite a big room. It was about 30 x 20 feet and held all of us and there was a screen in it and a projector and it was air-conditioned, I seem to remember; at least it was ventilated because we didn't suffocate. We didn't know what that room was doing behind the bar. We suspected it was a bookie joint. Anyway, that's where we held our meeting. That was the meeting at which Willie Zachariasen abolished any committee that didn't have a report ready for him. Now he pointed out that that was a perfectly reasonable thing to do because if the work of the committee was really necessary it would report; if it wasn't necessary, it should be abolished. That's the kind of logic that Willie is good for. . .you know he wrote a book that was so comprehensive that it covered just about everything in X-ray diffraction, but it was so elegant in the mathematical sense that many people were unable to get out of it the information that was in it and if you asked Zach a question, he'd simply say, "you haven't read my book; it's in there, it's in the equations."

This reminds me of Paul Ewald's doctoral dissertation. You know, that dissertation had to do with the interaction of electromagnetic radiation and solids and you may know the story of his going to talk to Max von Laue about it. Max von Laue had just been writing an encyclopedia article. You know instructors have to do things on the side to earn a little extra money and he'd been writing an encyclopedia article on electromagnetic radiation: in particular, light and the associated effects such as diffraction and other effects of light. When Paul Ewald talked to him about his dissertation, according to Paul, that was the first time that Max von Laue had ever heard that a crystal, according to the mineralogists, was made up of very small units; they didn't know their size, but extremely small units of matter that were arranged in a geometric way. Max von Laue had heard from Roentgen, who was also in Munich, that maybe his X-rays were electromagnetic radiation of very short wavelengths.

When Paul Ewald talked to Max von Laue about his dissertation, von Laue asked Ewald, "what would happen if the radiation falling upon your crystal were of such a wavelength that it was close to the spacings of those pieces that you think the crystal is made up of" and Ewald said that his answer was, "Well, that should come right out of the equations of my dissertation because they are perfectly general. There are no boundary conditions on the equations, so you just have to read the dissertation." And then he went on trying to discuss the dissertation with von Laue, but von Laue was paying no attention to him and finally he gave up trying to talk about his dissertation with him because he was obviously very excited about something and was not listening.

I have here with me something that Bill Parrish sent to me, written by Ewald, which is in his handwriting. It had to do with a talk that he was going to give and I think it's worth quoting. He writes, "The discovery of the diffraction of X-rays by crystals by Max von Laue in 1912 is an example of a discovery originating entirely in the mind of one man and coming as a surprise to even the best experts in the field. It was, in the accepted sense of the word, a 'stroke of genius'." And then he goes on to speak about the surroundings in Munich at the time that this occurs, mention Roentgen, Sommerfeld and Groth, who was a mineralogist; and farther on he speaks of the influence of the work done by these scientists on the conception of Laue's brilliant idea. "Apart from these famous professors, there was a great crowd of eager pupils and assistants who met at their table in the Hofgarten Cafe after their weekly colloquium at the Institute and after playing 9-pins. Work and pleasure were blended in one and the contact between the younger men created an atmosphere of friendship and mutual help and stimulation." It is quite likely that without this background, not all of the threads would have come together in the one brain that possessed the originality to combine them in the prediction of a new phenomenon and the tenacity to see the experiment carried out. In that case, not only crystal structure analysis, but also X-ray spectroscopy, and indeed the rapid and safe development of atomic physics following Niels Bohr's first papers, would have been held back for an indefinite period. So you see, the Pretzel-Bell aspect is important.

Actually, our early meetings were ones that combined playing and work. We always kept the afternoon free for sports and sitting around on the grass. We met in the springtime where we could be outdoors, where we could swim together or play tennis together and enjoy the out-of-doors together, so that we came together as a group of friends. There was a great deal of the kind of interchange that Paul Ewald just described. Everyone knew everyone else, everyone did everything together; there were no multiple sessions, no parallel sessions, so there was nobody milling around in the hall. When it came time for the meeting to be held, everybody went to the meeting. When the meeting was over everybody went to dinner or went outdoors to play or whatever it was we were going to do. We did things together as a group. I remember in one of these early meetings—and now this is going back to the third year of the ACA when Paul Ewald was president—we had the meeting at Camp Tamiment in Pennsylvania. . . . Somehow that was a very nice meeting. Do you know how Paul Ewald found that place? Instead of asking around at universities for an invitation, he asked the conference bureau, the visitor's bureau in New York City, to find him a place where we'd be far away from a city, where people wouldn't be distracted, where we could swim and have outdoor sports together and where we could have a little meeting room where we could meet, and they came up with Tamiment.

So that's where we met and it was there that a young man named Clifford Shull gave a paper on the new neutron diffraction work. He had maxima that were very, very broad and spread out and then he applied some rather sophisticated mathematics and came out with rather precise figures at the end. I remember that Peter Debye, speaking in that typical German-professor kind of way he had (genial, but clearly with a fatherly sort of attitude towards the younger students) said to Clifford Shull, "Young man, your data are fuzzy, your mathematics should be fuzzy!"

Well, Debye was not the only one who criticized speakers. We all remember Fankuchen and the way he would get up again and again after a paper and say, "Now there was just a point I didn't understand. Would you put your third slide on again please . . . now you claim . . ." and then he would simply tear apart the student's evidence. As Bill Parrish put it, he would comment in an abrasive way to stimulate discussion. We all know that Fan was a great teacher and he was determined that the younger generation of crystallographers should grow up with very self-criticizing habits of work . . . that they should not put on work that was shoddy, that they should not claim they had a result unless they had very, very good evidence for it, and he was acting always as a teacher and a missionary, I think, when he did this. He liked to keep the discussion hot and going in a lively way. This was what made him the terror for all young students who were giving their first papers.

A much less abrasive teacher, also one of our great teachers who was determined to keep us honest, is sitting right down here in the front row; this is José Donnay who would insist on people using terms correctly. This is not a trivial matter because confusion of ideas can so easily result from misuse of terms. We have all heard endless discussions which finally became resolved when it developed that the two people who were arguing were just putting entirely different interpretations on a particular term that was used. So your service to these students was an extremely important one. It is always stimulating for the younger people in the Society to have the founders of the Crystallographic discipline present not only questioning them, criticizing their work, but arguing with each other.

I remember one time when Paul Ewald and Peter Debye were on their feet in a meeting room arguing back and forth violently about the dynamical theory of diffraction and Paul Ewald quoted an equation, whether out of his original dissertation or one of his later papers on the subject—I don't know—and Debye in his German professorial way . . . (I know he isn't German, but he talks like a German professor) said, "Well, if you want to get mathematical, I can get mathematical too." He was always the completely self-confident person, a genial, completely relaxed individual. The last time I saw Peter Debye, he was in Newark Airport waiting for an airplane. He was sitting there with his feet out in front of him, crossed, reading a paperback 'whodunit.'

One of the things that Lindo Patterson used to do which I think is a really great thing was to seek out a student or anybody who had given his first paper or a very early paper, especially one who seemed a little uncertain as he gave his paper, and comment on the paper, leading him into a discussion about the paper. If it had been well given, he commented on this particularly. It was very stimulating to a person giving his first paper to have Patterson come and discuss his work with him.

There was a lot of teaching and learning going on, not only for the young, but for everybody. X-ray crystallography was a new discipline to everybody. That genius in geometry, Martin Buerger, was always a lucid teacher, and he took part in one of the very exciting periods of our Association, the development of direct methods of structure analysis. In 1946, the ASXRED met at Lake George. This was a summer resort just before it had opened for the summer, and the idea was that we would use it before it was filled up with other guests. Well that was fine, but it wasn't filled up with hotel staff either. As a result, the wives of the crystallographers (fortunately I wasn't a wife . . . I was a crystallographer under those circumstances) just moved into the kitchen and were the kitchen staff. The head of the hotel claimed that he was constantly trying to get a kitchen staff, but since he had one for free, we had our suspicions. Well, it was at this meeting that Martin Buerger presented the implication diagram and he credited Fankuchen for pointing out that since he was able to proceed with his analyses using the implication diagram, there must be phase information lying hidden in the assemblage of diffraction data. And you all know the story of what happened after this. Soon after this came the Harker-Kasper inequalities and

David Sayre's paper on the relation of signs and certain structure factors and then the Hauptman and Karle paper on solution of the phase problem, *ACA Monograph No. 3*.

Another exciting sequence has been the development of computer methods, beginning of course with Ray Pepinsky's marvelous machine. People came from all over the world to twiddle the dials and push the buttons that would change the magnitude and signs of the terms in the Fourier summation and instantly show on the oscilloscope the effect on the electron-density projection. Perhaps this meeting gave a special send off to our interaction with friends from abroad. ACA has always had a very close relationship with many crystallographers in Europe, people who come to see us often like Kathleen Lonsdale, (*Are* there any people *like* Kathleen Lonsdale?) Dorothy Hodgkin, Caroline MacGillavry, André Guinier, Beevers, Lipson, Robertson, sometimes Sir Lawrence Bragg.

I remember Robertson at that meeting at Ray Pepinsky's place in Penn State. You know Penn State was placed in the Middle of the state so that it would be accessible to everybody easily. Well, I can see by the laughter that many of you have tried to get there. Poor Monteath Robertson had been travelling for a very, very long time when he arrived in Penn State for that famous meeting which was the first meeting of the American Crystallographic Association. The following day he described to us what finally happened when he arrived at his room in the College at about 2:00 o'clock in the morning. He was more dead than alive and he knocked on the door of the room which he was to share with Lindo Patterson. There was no sound at all. So he knocked again, very much harder and finally he heard a shuffling sound inside and, as Robertson told us the next day, "The door opened and there before me was a dimly resolved Patterson in pajamas."

Well, there are many other exciting sequences of discovery and some of you I know are waiting for me to mention the particular sequence of discovery in which you were involved, but you remember that I was supposed to speak for 15 or 20 minues, and I'm afraid I have already exceeded my time.

You know, as we meet again these past presidents that we just had standing in the room and other senior citizens of the Society, they seem now as malleable and ductile as pure copper. They stood up when they were told to stand up and they sat down again when they were told to sit down. It is hard to believe that they were once as fractious as an ionic or covalent crystal. But you know, old crystallographers don't melt away, they just get a few more dislocations.

CHAPTER 12

ADDRESS AT ACA MEETING, 1976

Robinson Burbank
Bell Laboratories, Murray Hill, New Jersey 07701

One of the predictable consequences of being an ACA president is that after a year you become the past president. At that point you might think your troubles are over, but that isn't exactly so. Among other things, by custom and tradition, you are expected to engage in the exercise known as the past president's address.

While mulling over how best to discharge this obligation I thought of all the subjects which have been discussed through the years by my many distinguished predecessors. It seemed to me that previous addresses have tended to cluster into three broad categories. One type has treated serious social or political issues of concern either to the future of our organization, the further development of our science, or the welfare of society at large. Other speakers have given a summation and integration of a life time's research in a particular area of crystallography and a revealing picture of the understanding and satisfaction that have come from such an endeavor. Then we have had those who attempt to penetrate the misty veils of the future, who venture into the realm of prediction, with lots of imagination, a dash of science fiction, and a generous mixture of good, clean fun.

What I shall try to do doesn't quite fit into any of these categories. There is something in the make-up of the human species that gives a special emphasis to the beginnings and the endings of things, the alpha and the omega. The workings of our memories are such that the start and the termination of important phases of our lives are sharply etched in the mind's eye while all that passed between is a vague, grey blur that we cannot grasp.

A cherished part of my own life has been my associations with crystallography and therefore I retain a very vivid memory of how it all began. What I want to do is present some reminiscences of the first meeting I attended. In doing so I have three ulterior motives in mind. First, I will try to convey some aspects of the atmosphere and color of our unique little society as it existed a generation ago as perceived by an inexperienced and very impressionable neophyte. Second, I will try to describe how incredibly exciting and stimulating the scientific developments appeared to me. Finally, I will express the hope that enough of those qualities may still be present in our meetings of today to stir up and inspire our beginning crystallographers into achievements of which we have not yet dreamt.

A little background may now be in order. Consider a callow small town boy whose interest in science was aroused by reading about the Lewis octet theory of valence in a high school chemistry text. Add four undergraduate years in a small, isolated liberal arts college, then proceed to the World War II frenzy of the intellectual factory known as MIT. Spend most of the war years at chemical analysis in a spectroscopy laboratory as a small cog in an obscure corner of the vast enterprise known as the Manhattan project. Then transfer to the laboratory of a man with a panoramic vision, who proclaimed that the mission of his laboratory was to understand and measure the dielectric properties and behavior of materials over a frequency range extending from direct current up to cosmic rays.

In a few months the war ended and this professor, Arthur von Hippel, announced that of course I would resume work for a doctorate degree and that henceforth I would be his X-ray man, because, after all, the frequency range of X-rays was only a little higher than those in emission spectroscopy which I was already familiar with. Besides, it was all electromagnetic radiation and there were no new principles involved, only a few minor extrapolations. I was pretty well struck dumb by this unified perspective and the conclusions it led to.

I frantically thumbed through the MIT catalog and found that there were men who dealt with X-rays in four departments: metallurgy, biology, physics, and geology. But in that fall of 1945 the best bet looked like the courses in the physics dept. given by Professor B. E. Warren. I soon discovered that I had stumbled onto one of the world's great teachers with a lucidity in his methods that could be compared with those of his mentor, Sir Lawrence Bragg. Bert Warren's interests had already progressed beyond single crystal analysis to studies of imperfections, order-disorder, liquids and so on. After the fundamentals had been covered in the first semester I had to look elsewhere concerning more conventional structure analysis.

I was stymied in the practical laboratory details of how to handle single crystals until I was joined by an experienced colleague two years my senior, Howard Evans. We soon were equipped with a precession camera with the original hack-saw arm linkage that is illustrated in Martin Buerger's ASXRED *Monograph No. 1.* Under Howard's skilled tutelage I was soon into the happy adventure of exploring in reciprocal space.

Early in 1948 Howard announced that it was time for me to become a member of ASXRED. Furthermore, there was to be a joint meeting of ASXRED and CSA at Yale where I should present a paper on our joint endeavors.

And so in the early spring of 1948 a rather apprehensive young crystallographer arrived in New Haven. The meeting attracted the largest group of crystallographers that had ever been assembled in America up to that time with 124 ASXRED members and 81 CSA members, a total of 205. The number of papers to be presented was also a record with 57 titles distributed over four days. You can better appreciate the magnitudes of these numbers when I tell you that four months later that same year, the first Congress of the International Union of Crystallography was held at Harvard with 310 crystallographers assembled from eleven countries and 86 papers presented during the course of seven days.

This vast throng at Yale was housed and fed in Silliman College, one of the small self-contained residential units for upperclassmen. The technical sessions were held in a science building about 100 feet away, directly across the street. One of the most startling experiences for me was to see so many names that were enshrined in my mind from the crystallographic literature suddenly become living flesh and blood. They were real people who laughed and joked and conversed as though they had known me all my life. The business of sitting down at a different table for each meal in the dining hall greatly facilitated this delightful phenomenon. In addition there was a large, well appointed lounging hall, much like a clubroom or library which the crowd could retire to directly following meals to continue the process.

The central attraction in the lounging hall for the duration of the meeting was a display that Ray Pepinsky had brought up from Alabama comprising a couple of the sine wave generators and the oscilloscope viewing screen that would soon go into his great analog computer. We all marvelled, young and old alike. I first met Lindo Patterson and experienced the magic of his smile as we stood side by side nodding our heads in admiration as Ray explained to a gathering of electronic illiterates how simple it was to create contour lines and make the negative contours come out dotted instead of continuous on the oscilloscope screen.

I also have a recollection of a large basement-level room furnished with many tables and chairs where the crowd could sit around and take refreshment. Considerable mirth and merriment transpired in that room between technical sessions and that brings the subject of wit to mind. Given the right setting, circumstances, and mix of personalities wit will spontaneously flourish. And of course what I am referring to is wit in the sense of swift perception of the incongruous which depends for its effect chiefly on ingenuity or unexpectedness of turn. There was something about that meeting and many more to follow in which wit thrived. Unfortunately, it is virtually impossible to recapture those evanescent bubbles which appear out of nowhere to surprise and delight the mind for a moment and then float off and vanish into vapor. However, I can roughly reconstruct two examples which were perpetrated by Lindo Patterson on other occasions.

Verner Schomaker in his Cal Tech days was notorious for his reluctance to interrupt his prolific activities long enough to write things up for publication. As a consequence there are many instances in the literature where a reference says only unpublished- or unpublished material- or unpublished results- Verner Schomaker. Patterson observed that the situation could be summed up much more concisely by introducing a new word into the literature which would simply state that the research or result or conclusion had been "Vernished."

The other instance had to do with a paper from J. Monteath Robertson's laboratory. In those pre-electronic computer, pre-automatic diffractometer days Robertsons's people made a monumental, laborious, meticulous effort in the realm of aromatic hydrocarbons. They photographed thousands of reflections out to the vanishing point of visual intensity estimation, and hand-summed hundreds of Fourier sections with Beevers-Lipson strips ultimately to obtain the electron density in the plane of the molecule. They progressed from naphthalene to anthracene through molecules with more and more fused benzene rings until at the stage of this story they were up to some impressive number like 6 or 7 or 8 rings. Always there was a beautiful electron density map with remarkably smooth, even contours, a truly esthetic summation of all the labors that had preceded it. At the end of the paper Patterson got up, tongue in cheek, devilry in his eye, slowly and carefully got the full attention of the audience, commended the speaker on his efforts, then went on to say that he found this excessive preoccupation with what could only be described as flat chemistry to be absolutely deplorable and wasn't it time that they got on to something more interesting.

The scientific content of the Yale meeting influenced me for many years to come. There were several topics that carried significant omens for the future. I shall allude to three or four of them soon but first I want to describe something of the atmosphere of those technical sessions. All except the very youngest of us once had to make an oral defense of a doctoral dissertation. In general it is an event in which the outcome is fairly certain. You face a small group of your institution's professors. At least one of them really knows what you are talking about, in rare instances there may even be two or three people in this category. The remainder are experts in other fields and will only ask you peripheral questions that bear on their own specialties. You are knowledgeable and prepared for this event as no one else in the room and usually you experience one of the triumphal days of your life.

Imagine, however, another setting in which you are confronted by a dozen, maybe two dozen people, who know your specialty inside out, who have their antennae attuned for the subtlest nuances in your presentation. They have built the edifice on which you presume to add your increment of knowledge. They will praise and encourage you if your work rings true, but if it rings false they will defend their domain unto the death and let the tattered corpses, including your own, fall where they may. Thus you entered an intellectual arena in which Patterson, Harker, Huggins, Fankuchen, Zachariasen, Buerger, Donnay, Warren, Parrish, McLachlan, Pepinsky and a host of other eager participants were lying in wait to enter the fray. The defense of a paper was often protracted for 25 or 30 minutes. Sometimes the point at issue would be generalized and suddenly a couple of these senior members would be up on their feet or at the blackboard having a go at each other with all the wit and elegance and sarcasm at their command while the younger members retreated to the sidelines to take in the jousting and sparring and were educated, enlightened, and entertained all in one glorious melange. Often there was no clear cut victor in the battle but it was a marvelous and unique form of education for the young crystallographers.

One of the triumphs of the Yale meeting was a report by John Kasper, Charlys Lucht, soon to be married to John, and David Harker on the structure of decaborane. As luck would have it they had unknowingly chosen the most intractable problem in the entire field of boron hydrides. They had been bedevilled by micro twinning, high specimen volatility, and a host of other technical difficulties. More importantly, the entire structural concept of the boron hydrides as embodied in Pauling's *"Nature of the Chemical Bond"* was dead wrong! This meant that every conceivable model that might be postulated was doomed to failure. Finally, in desperation John turned back to the phase problem and in a moment of inspiration discovered that it was possible to deduce something about the signs of the structure factors using the Schwartz inequality. David pounced on this breakthrough and it was rapidly developed into the tool that solved the decaborane structure. The open clam shell configuration defined by an incompleted icosahedron is now in the freshman chemistry books but until that paper, it had never penetrated the mind of man. The Harker-Kasper inequalities proved to be a life-saver for me the following year as I struggled with my doctoral thesis. Two other young men, Herb Hauptman and Jerry Karle, immediately took up the inequalities and for a full generation developed and expanded that beginning into an imposing body of theory on direct methods.

As was the custom of the day, Fankuchen read a paper for Guinier and Fournier who sent a contribution from Paris but did not attend the meeting. Since the authors were not present for discussion Fan took the remaining time and launched into an impromptu paper of his own on low temperature techniques with a cold gas stream. Fan predicted that there was a gold mine waiting for anyone who wanted to study small molecules that normally formed gases or liquids by growing them as single crystals as he had done for SO_2. Another young man at that meeting fulfilled Fan's prophecy. He took the low temperature technique and the newly gained knowledge on decaborane and completely cleaned up the field of boron hydrides in a few years. He was, of course, Bill Lipscomb.

The wartime Manhattan Project research was just becoming declassified at the time of the meeting. Willie Zachariasen was thus able to give a summary of the structures of many of the oxides and halides of the transuranium elements. He had carried out a remarkable tour de force at a time when the chemistry of these materials was completely unknown and only microgram quantities were available of most of them. Time after time he was able to take a nearly invisible speck of powder, obtain a Debye-Scherrer photograph, and from it deduce the crystal structure, stoichiometry, and oxidation state. He was the salvation of the project chemists at several critical junctures. How he did it I will never know since Zach is not much given to low level explanations. At any rate it was a product of nearly pure mental effort, and even today all our computers and automated diffractometers would hardly be a fair exchange for whatever it was that went on in Zach's brain.

There was a session on computing, then as now, but with what a difference! Nearly everyone was restricted to two dimensional Fourier projections, hand calculated with Beevers-Lipson strips or Patterson-Tunell strips and an adding machine. Structure factors were especially nettlesome. One had to draw form factor curves, pick off values by eye, look up tables of trig functions, hope to God that he had not made any mistakes in reducing the general formulas from the tabulations in the *"Internationale Tabellen"* of 1935 or Kathleen Lonsdale's handwritten tables—she had feared typesetters would make mistakes so her handwriting was reproduced directly. One was especially prone to make mistakes in the rhombohedral, trigonal, and hexagonal space groups. Then all the multiplications had to be done, one by one, on a desk calculator. The least squares method had been demonstrated seven years earlier by Eddie Hughes but no one used it because of the prohibitive labor involved. It had recently been appreciated that punched cards could be adapted to crystallography if one had a key punch, a sorter, and a tabulator. It was already clear that Fourier summations were made to order for this approach by punching the Beevers-Lipson strips on cards. Jerry Donohue and Verner Schomaker gave a paper showing that if one could also get access to a reproducing punch and a multiplying punch then structure factors would become a snap. I still recall Donohue glibly running through the routine of how easy it was to punch this deck and that, sort them and combine them thus and so, a little more punching and sorting and presto one had results. In fact they gave an example where they had been able to calculate 600 structure factors involving eight nonequivalent atoms in a noncentrosymmetric cell in only 24 hours of actual calculation as contrasted with 240 hours required by traditional methods.

Ray Pepinsky's electronic Fourier synthesizer as it was then called undoubtedly attracted the most attention because it was clear that it would be cranking out results within the next year. However the true portent of the future came from the final speaker of the session on computing, a very business-like looking man in a grey flannel suit, not a crystallographer but a mathematician, Herman Goldstine from John von Neumann's group at the Princeton Institute for Advanced Studies. In the phrases of the late entertainer Al Jolson his message was "you ain't seen nothing yet, folks." He spewed forth a bewildering array of facts about something called the electronic digital computer in a language replete with so and so many microseconds or milliseconds for this, that, or the other kind of arithmetic operation. He spoke of a revolution in the making but I think that very few people in the room grasped the implications of what was in store for all of us.

There were many other speakers at that meeting that still register in my memory, especially David Harker with a variety of the theoretical developments that arose from the decaborane problem. But also, Martin Buerger and implication theory, Charles Barrett with a new modification of sodium, Rudolph Brill seeking details of the chemical bond in electron densities, José Donnay with a complex classical twinning problem, Bill Parrish and the new powder diffractometer, Berndt Matthias on ferroelectricity. But I've said enough to indicate some of the flavor and variety that was on the menu.

Now I want to direct my closing remarks to the young people in the audience who are the future of our science and our association. You have heard me enthuse over how great it was when the world of crystallography was bright and shiny and new for me. That was more than 28 years ago. I suspect that for many of the senior members of that day the world didn't appear quite as remarkable as it did to my naive mind. This is to be expected and would be in the natural order of things. As we accumulate knowledge and experience we pay a certain price. In exchange for a modicum of the answers we inevitably lose some openness of mind as well as the energy and flexibility to pursue every new idea that comes along. We become far too prone to muster up a long list of reasons why a novel concept won't work instead of jumping on to an experiment to find out whether it will work. There may be significant advances presented at this meeting that will not penetrate my opaque vision, yet will be obvious and exciting to you. I hope it will always be so that an ACA meeting reflects a vital and growing science. I hope that among you there is someone who about 25 years from now will be up on the podium, getting ready to commence the past president's address, who will be able to say that back in 1976 at the Northwestern Meeting there was an intellectual ferment and a flow of ideas that influenced crystallography for years to come.

PORTRAITS AND SNAPSHOTS

S. C. Abrahams
Bell Laboratories, Murray Hill, New Jersey 07974

Among the actions leading to the present history of crystallography in North America was a call for "photographs of crystallographers at work and at play" that was published in the American Crystallographic Association Newsletter of October 1977. The response to this call formed the nucleus of a collection that is presented in the following twenty-two plates. A considerably larger number of publishable-quality photographs of crystallographers is known to exist in the form of many private collections. It is hoped that the following collection will encourage readers to submit copies of historically interesting photographs from their own collection to the American Institute of Physics Center for History of Physics, where they can be catalogued and archived for future use. The copies in the present collection have been presented to this Center.

Photographic portraits of the presidents of the American Crystallographic Association and of its predecessor societies, the American Society for X-ray and Electron Diffraction and the Crystallographic Society of America are presented, in chronological order, on plates 1 to 5. The originals of most of these photographs are stored in the Niels Bohr Library of the American Institute of Physics (including the ACA Presidential Gallery). That of Peter Debye is in the W. F. Meggers Collection at the AIP. The majority of the photographs were taken during the incumbent's presidency, although some are of later date. The names of presidents now deceased are accompanied by a dagger-sign. Amy Weiner of the AIP Center for History of Physics provided generous assistance in arranging for prints to be made from the original portraits for use in this book. The following photocredits are due: B. E. Warren, Paone, M.I.T.; J. D. H. Donnay, William C. Hamilton; R. W. G. Wyckoff, National Institutes of Health; W. N. Lipscomb, Harvard University News Office; E. A. Wood, Bradford Bachrach; S. C. Abrahams, Fabian Bachrach; H. W. Wyckoff, J. D. Levine/Yale University.

The portraits of the officers for 1948 of the American Society for X-Ray and Electron Diffraction and of the Crystallographic Society of America on plates 6 and 7 are from W. Parrish's collection, that on plate 8 of the members of the First Congress of the International Union of Crystallography meeting in Cambridge, Massachusetts was provided by R. C. Evans: a list that identifies members is given on page 150. A distinguished subgroup of these members is shown on plate 9, and a 1935 photograph of Lawrence Brockway on plate 10, both from the collection of C. Frondel.

All photographs on plate 11 are reproduced by courtesy of D. McLachlan Jr., except for that of B. S. Magdoff, L. O. Brockway and J. Karle which was taken by I. Karle. The photographs on plate 11 date from the 1948 Harvard or 1951 Stockholm Congresses of the International Union of Crystallography except for two that were taken at a Gibson Island meeting of the American Society for X-Ray and Electron Diffraction.

The photograph taken at the dedication of Ray Pepinsky's X-Ray Analog Computer and shown on plate 12 is from W. Parrish's collection, that on plate 13 is from S. C. Abrahams' and that on plate 14 from E. K. Patterson's. The crystallographers on plate 13 are J. M. Bijvoet, H. Lipson, C. H. MacGillavry, C. W. Bunn and J. M. Robertson from left to right, front row: J. F. Schouten, E. G. Cox, M. F. Perutz, C. A. Beevers, R. Pepinsky and E. Grison from left to right, back row. The Conference on Computing Methods and the Phase Problem in X-Ray Crystal Analysis immediately preceded the first meeting of the American Crystallographic Association: both were held at the Pennsylvania State College. An identification chart for the group attending the second meeting of the American Crystallographic Association in New Hampton, New Hampshire which is shown on plate 14 was not available.

Four of the photographs on plate 15 are from the Fankuchen Collection in the American Institute of Physics Niels Bohr Library, through the cooperation of Amy Weiner; the three at the left side, and that at the top right which was taken at Gibson Island in 1944. The remainder are from W. Parrish's collection, as are those on plates 16 (1948-1949) and 17 (1957-1959), except for the photograph of J. Karle which was provided by E. K. Patterson.

The photograph taken during a lecture at the Fourth International Congress of Crystallography in Montreal on plate 18 is also from E. K. Patterson's collection. The signatures on the scroll sent to Max von Laue on his eightieth birthday are reproduced on plate 19 from a photograph in D. McLachlan's collection. The portraits on plate 20 (1969-1970), plate 21 (1970-1973) and plate 22 (1973-1977) were taken by S. C. Abrahams.

The total collection of photographs from which the plates that follow were chosen was not large, resulting in a selection that is necessarily quite arbitrary with many conspicuous gaps. A potentially rich and previously untapped source for a future pictorial compilation of active crystallographers would be the photographs used for identifying the authors of papers presented at poster sessions of American Crystallographic Association meetings.

M.L. Huggins †
1941
ASXRED

B.E. Warren
1942
ASXRED

M.J. Buerger
1943 ASXRED
1945-6 CSA

L.H. Germer †
1944
ASXRED

W.H. Zachariasen †
1945
ASXRED

D. Harker
1946
ASXRED

C.S. Barrett
1947
ASXRED

J.W. Gruner †
1947
CSA

A.L. Patterson †
1948
ASXRED

A. Pabst †
1948
CSA

P. Debye †
1949
ASXRED

J.D.H. Donnay
1949 CSA
1956 ACA

I. Fankuchen †
1950
ACA

R.W.G. Wyckoff
1951
ACA

P.P. Ewald
1952
ACA

L.O. Brockway †
1953
ACA

E.W. Hughes
1954
ACA

W.N. Lipscomb
1955
ACA

Plate 2

E.A. Wood
1957
ACA

D. McLachlan †
1958
ACA

R.E. Rundle †
1959
ACA

J. Waser
1960
ACA

K.N. Trueblood
1961
ACA

V. Schomaker
1962
ACA

G.A. Jeffrey
1963
ACA

H.T. Evans
1964
ACA

H.A. Levy
1965
ACA

B. Post
1966
ACA

J.S. Kasper
1967
ACA

S.C. Abrahams
1968
ACA

W.C. Hamilton †
1969
ACA

D.P. Shoemaker
1970
ACA

W.R. Busing
1971
ACA

J. Karle
1972
ACA

R.A. Young
1973
ACA

E.C. Lingafelter
1974
ACA

Plate 4

R.D. Burbank
1975
ACA

I.L. Karle
1976
ACA

C.K. Johnson
1977
ACA

P. Coppens
1978
ACA

J.P. Glusker
1979
ACA

H.W. Wyckoff
1980
ACA

Q.C. Johnson
1981
ACA

J.B. Cohen
1982
ACA

D. Sayre
1983
ACA

Plate 6

OFFICERS OF THE AMERICAN SOCIETY FOR X-RAY AND ELECTRON DIFFRACTION, 1948

C.S. BARRETT A.L. PATTERSON E.A. WOOD C.C. MURDOCK

OFFICERS OF THE CRYSTALLOGRAPHIC SOCIETY OF AMERICA, 1948

S.G. GORDON J.W. GRUNER A. PABST W. PARRISH

FIRST CONGRESS OF THE INTERNATIONAL UNION OF CRYSTALLOGRAPHY
HARVARD UNIVERSITY, JULY 28 - AUGUST 3, 1948

Plate 8

J.D. BERNAL, C.V. RAMAN, C. PALACHE, P.P. EWALD, A.L. PATTERSON

L.O. BROCKWAY, CALIFORNIA INSTITUTE OF TECHNOLOGY, 1935

A.F. Wells

R.C.L. Mooney

B.S. Magdoff (left)
L.O. Brockway (center)
J. Karle (right)

B.W. Low

D. Harker (left)
D. McLachlan (right)

L.H. Germer

A.L. Patterson (left)
E. Patterson (right)

F.W. Matthews

H.E. Buckley

C.W. Bunn

E.W. Hughes (left)
D.F. Clifton (right)

I. Fankuchen

M. von Laue

D.C. Hodgkin

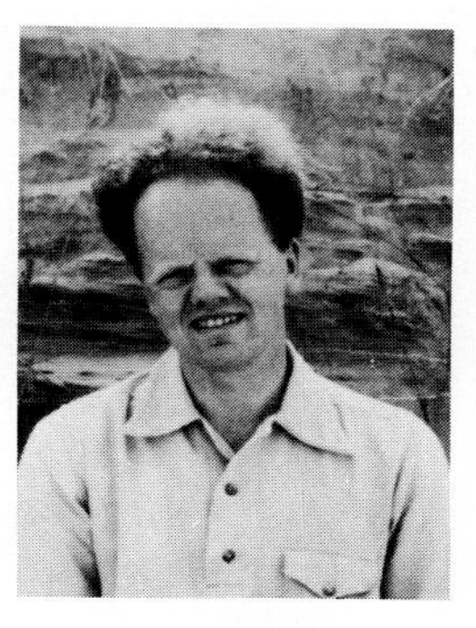

A.J.C. Wilson

R.C. Evans

DEDICATION OF THE PEPINSKY SYNTHESIZER, ALABAMA POLYTECHNIC INSTITUTE
APRIL 21, 1949

M.J. BUERGER C.H. MAC GILLAVRY D. HARKER R. MOONEY A.L. PATTERSON

CONFERENCE ON COMPUTING METHODS AND THE PHASE PROBLEM IN X-RAY CRYSTAL ANALYSIS, PENNSYLVANIA STATE COLLEGE, APRIL 1950

AMERICAN CRYSTALLOGRAPHIC ASSOCIATION, NEW HAMPTON, N.H.
AUGUST 21-25, 1950

M. von Laue (left)
W.L. Bragg (center)
I. Fankuchen (right)

J.D.H. Donnay

B.E. Warren, J.D.H. Donnay,
L.H. Germer, W.H. Zachariasen,
M.J. Buerger, M.L. Huggins

M.J. Buerger (left)
H.W. Wyckoff (center)
K. Lonsdale (right)

M.J. Buerger (left)
C.S. Barrett (center)
J.T. Norton (right)

V. Vand

I. Fankuchen

J. Donohue

W. Parrish

A.L. Patterson

D. Wrinch

W.H. Zachariasen

C. Frondel

G.L. Clark

M.L. Huggins

R.W.G. Wyckoff

J. Karle

P.J.W. Debye (left)
J.C.M. Brentano (center)
W. Parrish (right)

Plate 16

W.L. Bond

L.O. Brockway

R.E. Rundle

E.W. Hughes

R. Pepinsky

J.A. Bearden

H.T. Evans (left)
R.D. Burbank (center)
B. Post (right)

R.C.L. Slater (left)
J.C. Slater (right)

D. HARKER, B.E. WARREN, E.A. WOOD, A.L. PATTERSON, E.W. HUGHES,
W.H. TAYLOR, K. LONSDALE, UNIDENTIFIED, J.M. ROBERTSON

FOURTH INTERNATIONAL CONGRESS OF CRYSTALLOGRAPHY
MONTREAL, JULY 10-14, 1957

To Professor Max von Laue:

Your friends assembled at Cornell University for the Annual Meeting of the American Crystallographic Association join together in extending to you their congratulations and warmest wishes on your eightieth birthday. We wish to honor the discoverer of x-ray diffraction by crystals and to pay our respects to the colleague who has contributed so much to the development of the field in which we all work. In addition we want to express our admiration to the very great citizen of Germany and of the world for the many contributions which he has made to human understanding and freedom.

ITHACA, NEW YORK
JULY 20-24, 1959

CONGRATULATORY SCROLL WITH 235 SIGNATURES, MAX von LAUE'S 80th BIRTHDAY

W.H. Zachariasen

W. Parrish

W.L. Kehl

M.E. Straumanis

C.B. Shoemaker

L.D. Calvert

B.N. Brockhouse

W.F. Bradley

C.M. Mitchell

Plate 20

C.K. Johnson

L.H. Jensen

J.S. Kasper

T.C. Furnas

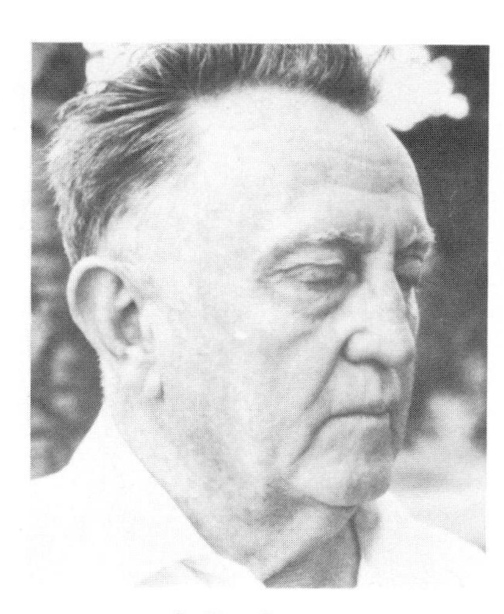

A.E. Smith

W.C. Hamilton

J.R. Clark

H.T. Evans

R.D. Burbank

J. Karle

K. Knox

I. Karle

J.M. Reddy

L.O. Brockway

M.J. Buerger

M. Przbylska

V. Schomaker

Plate 22

SECTION D

APPARATUS AND METHODS

CHAPTER 1

HISTORY OF THE X-RAY POWDER METHOD IN THE U.S.A.

William Parrish
IBM Research Laboratory
San Jose, California 95193

Introduction

The powder method has played a very important role in the development of science and technology. Virtually the entire industrial complex has benefitted from it. In fact, it is difficult to envisage what it would be like without the method.

This brief history of the X-ray powder method in the United States is based on my intimate involvement with the method for 40 years. A detailed history would require a separate volume to evaluate the thousands of papers published on the developments and applications of this method. Space limitations preclude the inclusion of many contributions which have advanced the method and I apologize to those authors whose work had to be omitted. It was often difficult to limit the discussions to the work of Americans because many of the developments were growing in parallel here and abroad.

The history can be divided into three time frames:

(1) The first period from Hull's discovery of the powder method in 1916 to the end of World War II.

(2) The 1945-1975 period which encompassed the development of the diffractometer, the JCPDS file and many new methods and applications in a renaissance of the powder method.

(3) The present period beginning around 1975 in which there has been a large increase in the use of computers, particularly powerful mini-computers and microprocessors. They have made possible expanded use of automation, high speed data reduction, profile fitting, comprehensive search/match, crystal structure refinement using powder data and other developments which have brought us to a second renaissance.

Applications of the Powder Method

The 65 years of development of the powder method have produced dozens of general category applications and hundreds of specific uses of powder diffraction in research, development, quality control, materials characterization and industry. The broad categories of materials analyzed are inorganics, minerals, ceramics, metals and alloys, organics and polymers. The specimens may be in any crystalline form such as powders, flat sections, dusts, residues, deposits, precipitates and thin films.

Most of the following types of analyses cannot be done (or done as well) by other methods: identification of crystalline phases, qualitative and quantitative analysis of mixtures and minor constituents, identity of the crystalline and amorphous states and devitrification, following solid state reactions occurring with heat treatment or other processing, distinction of substitutional and interstitial solid solutions, polymorphism and determination of phase diagrams, lattice parameter and thermal expansion, preferred orientation, physical properties such as crystallite quality, small crystallite sizes, strain, disorder stacking and lattice damage, in situ high pressure studies up to several hundred kilobars, high/low temperature studies from 2000°C to liquid helium, and many others.

Polycrystalline X-ray methods have been developed and refined for the determination of texture, preferred orientation, pole figures and localized elastic stresses. They are widely used in the development and process control of synthetic fibers, polymers, wire drawing, rolling metal sheets and various other metallurgical procedures and constitute a considerable fraction of the industrial applications. These methods are described in detail in the books of Alexander (1969), Barrett and Massalski (1980), Cullity (1978), Klug and Alexander (1974), and Taylor (1961).

In mineralogy, for example, the powder method is extensively used for identification and characterization, to analyse minerals from such diverse origins as the moon, meteorites and the ocean basins, the incorporation of radioactive materials in solid solutions as a means of solving the nuclear waste problem, the analysis of shales for oil recovery and the mineral phases in coals, for large scale geological explorations requiring the analyses of thousands of core samples, and many others.

The Early History, 1916-1945

The discovery of the powder method by A. W. Hull in the U.S.A. at virtually the same time as P. Debye and P. Scherrer (1916) made their discovery abroad, and the classification and use of powder patterns for identification by J. D. Hanawalt, H. W. Rinn and L. K. Frevel (1938) were the most important developments in the early history.

A. W. Hull/Discovery of the Powder Method. The powder method was introduced in the U.S.A. with two landmark papers by A. W. Hull: A New Method for Crystal Analysis (1917), and A New Method for Chemical Analysis (1919). He was unaware of the Debye and Scherrer paper presented at a meeting in Gottingen in December 1915 because World War I prevented many foreign journals from reaching the U.S. Hull's achievement is a classic example of the right person being at the right place at the right time. The fortunate environment was the General Electric Research Laboratory where W. D. Coolidge was developing the hot tungsten cathode X-ray tube and Saul Dushman the Kenotron rectifiers.

Hull was interested in determining the crystal structure of iron to learn if it would give a clue to its magnetism. He planned from the outset to use powders because single crystals had not yet been grown. He described the method as a modification of the Laue and Braggs single crystal methods in a paper presented to the American Physical Society in October 1916. He recognized from the beginning that a fine powder of randomly oriented crystallites should make it possible to obtain reflections from every possible plane in the crystal. He was probably the first to use filters to remove the $K\beta$ lines. The beam passed through the specimen and a flat photographic plate or curved film with intensifying screen recorded the patterns which were sufficiently clear to allow him to determine the structures of the simple metals he was interested in. Hull thus made the first crystal structure determinations in the U.S.A.

In the second paper, he wrote ". . . that every crystalline substance gives a pattern; that the same substance always gives the same pattern; and that in a mixture of substances each produces its pattern independently of the others . . . the method is capable of development as a quantitative analysis." The Debye and Scherrer paper did not explicitly mention the possibility of using powder patterns for identification. Like the first Hull paper, they were interested mainly in the crystal structure. Not long after Hull's papers were published, W. J. Mead began to use the method to identify minerals in the Geology Department at University of Wisconsin.

The Dow Tabulated Powder Data. in 1938, J. D. Hanawalt, H. W. Rinn and L. K. Frevel of Dow Chemical Company published their paper Chemical Analysis by X-Ray Diffraction, which contained tabulated data on the diffraction patterns of 1,000 substances and described a classification scheme based on the three most intense lines. They prepared a search manual by dividing the d-values from 0.8 to 20Å into 77 Δd size groups for the strongest line and the sub-group was determined by the d of the second strongest line. There was a large advantage in the group system using a pair of lines rather than a sequential listing of single d-values. This important paper opened the way to the use of the powder method in a systematic way for identification and provided the basis for starting the Joint Committee File. The editor wrote in the prologue "Industrial and Engineering Chemistry considers itself fortunate in being able to present herewith a complete, new, workable system of analysis, for it is not often that this is possible in a single issue of any journal."

They used the old GE X-ray unit and a number of 8-inch radius quadrant cassettes arranged around the Mo X-ray tube (Davey, 1921). An intensifying screen was placed behind the film and a direct comparison film intensity scale was used. All the powder patterns were prepared in their laboratory so that there was a unity of methodology. Many of the patterns were verified by comparison with published crystal structures and in general, their data were the best available for a number of years.

Early Instrumentation. Setting up an X-ray Laboratory in the early days was far more difficult than it is today. There was no X-ray diffraction equipment industry. Most of the X-ray generators had to be designed and assembled from medical or radiographic parts and the cameras had to be built. Demountable X-ray tubes were generally used and there was quite a ritual as to the best ways to keep them operating. In the late 1930's, General Electric X-ray Corporation and Philips Metallix Company developed self-contained X-ray units with sealed-off X-ray tubes and it became possible to purchase well-engineered push-button X-ray equipment for the first time in the U.S.A. This was very important to the development of X-ray crystallography.

M. J. Buerger (1936) developed a cylindrical powder camera which was later manufactured by Charles Supper and Otto von der Heyde. The Straumanis film mounting was used and the camera diameter, 57.3 or 114.6 mm, permitted reading the film with a millimeter scale. Later, collimators were designed using ray diagrams to eliminate unwanted scattering, and increase the intensity and resolution (Parrish and Cisney, 1948). In that era, the powder method did not require elaborate equipment.

The Recent History, 1945-1975

The bulk of this paper is devoted to this period. The dominant events were the development of the powder diffractometer and the establishment of the powder data file. A number of important developments in the instrumentation and methods took place and there was a huge growth in the use and applications of the powder method.

Development of the Powder Diffractometer. The development of the diffractometer began near the end of World War II. The manufacture and large sales of the new instruments revolutionized the powder method and greatly enlarged its use. The Geiger counter tube replaced film and made it possible to observe the X-rays instantly and to measure the intensities with precision. The pattern could be seen on a strip chart recorder instead of waiting to develop a film. Experimental parameters could be easily varied. The large space around the specimen could accommodate devices to vary the specimen temperature, apply stress and to perform many other types of experiments that could not be easily done within the confines of the powder camera. The shapes of the reflection profiles could be precisely determined and a large amount of work was done to learn the physics and geometry of the method, to increase the precision, to determine particle size and strain, and many other properties of polycrystalline specimens. The number of users and the applications of the method thus increased rapidly.

The development of the diffractometer took place in two overlapping stages. The first instrument assembled by H. Friedman (1945) at the Naval Research Laboratory was a direct outgrowth of the quartz orientation apparatus (Parrish, this volume, page 345). North American Philips Company (N.A.P.) engineered and sold several hundred. The development of the diffractometer used today began in 1945 and the first commercial instruments were sold by N.A.P. in 1950 (Parrish, 1965, p. 71). Philips Electronic Instruments celebrated the production of its 2,000th vertical diffractometer in 1964 by presenting me a gold plated instrument which is still in use at Philips Laboratories. There are now probably over 10,000 diffractometers which have been sold by a score of manufacturers and are in use throughout the world, making it the most widely used of all crystallographic instruments.

Early Diffractometer. The use of Geiger counters for X-rays was tried by LeGalley (1935) who made a simple instrument to rotate the counter around a powder slab irradiated by a collimated beam of MoKα X-rays. The poor results were caused by the inadequate geometry and the unfortunate choice of Mo radiation for which the Geiger counter had very low efficiency.

It was not until 1947 that I learned of the paper by Lindemann and Trost (1940) describing the development of a "Bragg" focusing instrument for powders. They used a Geiger counter and measured the current to determine the intensities. World War II had interrupted the delivery of German journals, the same situation as had occurred with Hull in World War I.

The success of the quartz orientation counter tube method made it clear that a similar technique could be developed for powders and I had many discussions with Friedman. The large single crystal quartz plates gave very high intensities and required only the rotation of the crystal with the detector in a fixed position to locate the peak. Powder speciments presented a more difficult problem because the intensities were orders of magnitude lower, and the system had to be stable to measure the relative intensities point by point. Friedman (1945) used a divergent beam focusing geometry with the specimen axis equidistant from the source and receiving slits, added a 2:1 counter:specimen gear coupling and motor drive to the quartz goniometer. He was an expert on detectors and developed a Geiger counter which operated in the plateau region to count the individual pulses with a scaling circuit and a rate meter strip chart to record the pattern. The first commercial instruments manufactured by N.A.P. looked similar to the quartz instruments and used the same low power horizontal X-ray tube (Bleeksma et al. 1948).

This early instrument was the first X-ray apparatus used in many industrial applications and greatly increased the scope of the powder method. However, it had many limitations for serious X-ray analysis. There were complaints of the "lunch effect," i.e., the different readings obtained before and after lunch. It had an inherent 0.2° to 0.5° angular error in the bent rack and the instrument was limited to the front-reflection region.

Present Diffractometer. After some experience with the first instrument, it became evident that the low intensity and limited resolution could be improved. N.A.P. was manufacturing vertical mounted water-cooled diffraction tubes, which at that time could be operated at six times higher power than the air-cooled tube. In order to utilize all four windows, I conceived the design of a diffractometer to scan in the vertical plane thereby allowing the remaining windows to be used for cameras or another diffractometer. The first diffractometer of this type with the geometry described below was built in a few weeks at Philips Laboratories from pieces of the horizontal instrument and gears taken from a recorder,

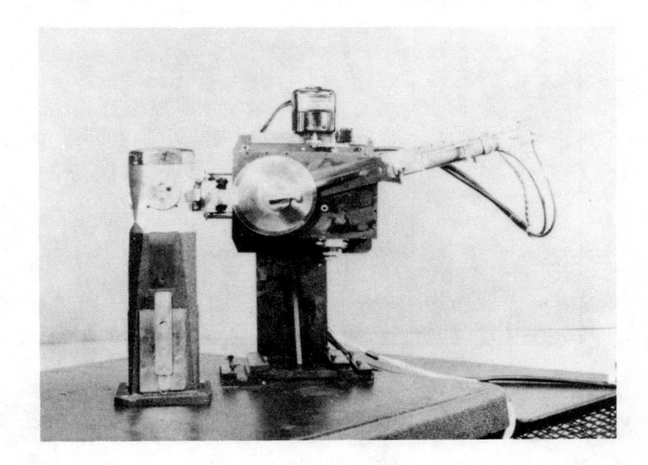

Fig. 1. The first vertical-scan powder diffractometer designed at Philips Laboratories, 1947.

(Fig. 1.) I was lucky because unlike most experimental research, the first runs were successful and the high intensities and resolution that had been calculated were immediately realized.

The greatly increased quality of the pattern was the result of the radically different X-ray optics used in the instrument geometry (Parrish, 1965, p. 76). The early instrument used the spot focus with an effective size

0.24 by 2.5 mm, and the line focus in the new geometry was effectively 0.06 by 10 mm. The large axial divergence would have caused severe asymmetric profile broadening with the wide specimen used; it was reduced by a set of long parallel equally-spaced thin molybdenum foils which prevented the beam from crossing channels. (Soller (1924) had used parallel metal sheets now called Soller slits to obtain narrow triangular peaks without focusing; these were opposite to the orientation used in the diffractometer.) The long X-ray source thus became effectively a number of short overlapping very narrow sources parallel to the specimen axis of rotation and had the double advantage of high intensity and high resolution. A second set of parallel foils in the same orientation was placed in the diffracted beam. The large increase in resolution decreased the FWHM of the peaks from about 0.35° to 0.10° (2θ).

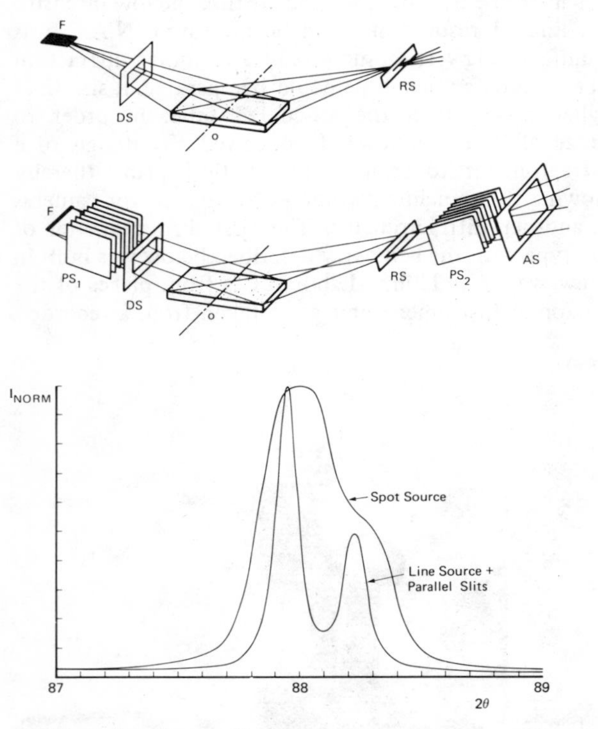

Fig. 2. X-ray Optics of the early and present diffractometers (above) and their profiles: Si(422), CuKα, 4° divergence slit (below).

Fig. 2 shows the optics and line profiles of the two instruments and Fig. 3, made in 1952, compares quartz patterns made with the 114.6 diameter powder camera and the vertical-scan diffractometer. The mechanical design of the goniometer was done by my late colleague Kurt Lowitzsch, (Parrish, 1965, p. 79). Most of these instruments are still in use throughout the world.

Other geometries were added to the diffractometer: Lang (1956) and deWolff (see Parrish, 1965, p. 122) independently developed the transmission specimen-

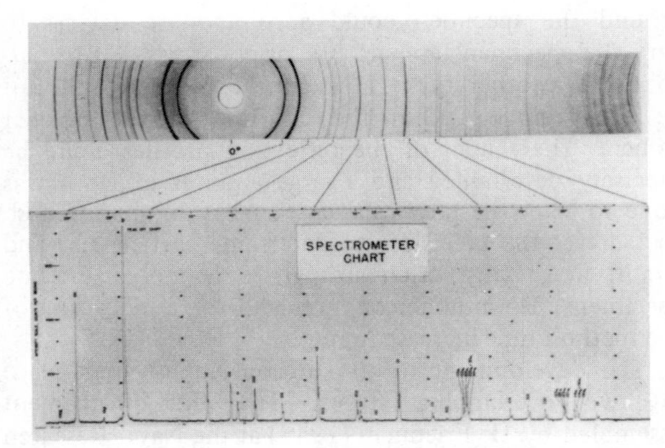

Fig. 3. Comparison of 114.6 mm diameter powder camera film and vertical scan diffractometer pattern of quartz with CuKα (Trans. Inst. Meas. Confs. Stockholm, 1952).

diffracted beam monochromator method and Parrish and Mack (1967) developed the Seemann-Bohlin geometry with source slit or incident beam monochromator on the Norelco diffractometer. The use of the curved diffracted beam monochromators with the reflection specimen geometry increased rapidly in the 1970's when graphite became available.

The name "Geiger Counter Spectrometer," often prefixed with "X-ray" was used in early publications. The name "Diffractometer" was substituted in 1952 and "Spectrometer" used only for the sister development of X-ray fluorescence analysis which used a similar instrument and actually measured wavelengths.

The literature often incorrectly refers to the "Bragg-Brentano parafocusing principle" in describing the powder diffractometer. W. H. Bragg (1921) published a short note showing patterns of Al, Si, and LiF made on his ionization chamber spectrometer but there was no focussing. Brentano (1946) summarized his work on the use of film cameras with large angular apertures and extended powder specimens to obtain higher intensities than Debye-Scherrer cameras. He distinguished real focusing as from a mirror or single crystal and the ray-collective properties of a powder which he called "parafocusing." He noted that the powder should have a double curvature – a toroidal surface which changes with diffraction angle to avoid aberrations. However, the instruments and geometries used by Bragg and by Brentano were totally different from the counter tube diffractometer geometry which was granted a patent (Parrish, 1949).

X-ray Sources. Sealed-off X-ray tubes with the common anodes rated at 50-60 kV and 1-2 kW for standard and fine focus are now used for virtually all powder work. The early tubes were made with Lindemann glass windows which were later replaced by vacuum tight thin mica backed by non-vacuum beryllium; pure beryllium windows are used today. The life has generally been increased and the contamination reduced. Rotating anode tubes with standard focus that can be operated at ten times higher power are available but are not yet widely used; rotating anode microfocus tubes with very high brilliance are used for small samples. The new synchrotrons at Stanford, Cornell, and Brookhaven are extraordinary sources of high intensity parallel continuous radiation. Their use for powder methods will require special instrumentation and we can expect many new developments.

The X-ray generators, originally half- or full-wave rectified, have been replaced by constant potential DC operation units with highly stabilized kV and mA circuits. The older equipment sometimes had radiation leaks from faulty rectifiers, X-ray tube shields and improperly enclosed collimating systems. State and national laws have been enacted governing safe practices and radiation protection has become an important factor in the design. Most present day equipments have safety shutters, enclosed beam paths and shielded couplings between the instrument and X-ray tube. The present-day cost of a 3 kW generator with X-ray tube, diffractometer, detector, and circuits is in the range of \$50-\$60K not including special accessories.

Detectors. Several types of detectors have been used in powder diffractometry. The end-window Ar-filled Geiger counters used in the early instruments had a quantum efficiency of about 15% for CuKα, which was later improved to 60% and the radial sensitivity enlarged for the new geometry. Their most serious disadvantage was the 50-300 μsec dead time which limited the linearity to low count rates.

A large improvement in detectors occurred in 1954 with the introduction of proportional and scintillation counters (Parrish 1962, 1965, p. 244). The side-window proportional counter had an effieiency of 60% for CuKα and the scintilliation counter nearly 100%. Both had dead times of the order of 0.25 μsec, and the pulse amplitudes were proportional to the energies of the X-ray quanta. The resolution expressed as the FWHM/average pulse amplitude of the amplitude distribution was 20% for proportional and 50% for scintillation counters using CuKα radiation. This made it possible to use a pulse height analyser to accept only pulses of amplitudes within the selected limits and thereby greatly decrease the background with only a few percent loss of characteristic line intensity. Fig. 4 shows the effect of pulse height discrimination with a scintillation counter on the spectrum of a Cu tube at 40 kV peak.

The combined action of the Ni filter and electronic discrimination increased peak-to-background ratio (P/B) well beyond the Geiger counter. The later introduction of graphite monochromators eliminated all the X-ray background except the scattered characteristic radiation, and discrimination was then needed only to remove electronic noise and the subharmonic wavelengths reflected by the specimen.

The solid state detector system has much higher energy resolution but limited resolving time. Because a liquid nitrogen Dewar is required it is not often used in scanning. However, it has opened the field of energy dispersive methods. Position sensitive proportional counters are still in a development stage and have been used to record an extended angular range without scanning; Gobel (1979) in Germany introduced scanning with this detector.

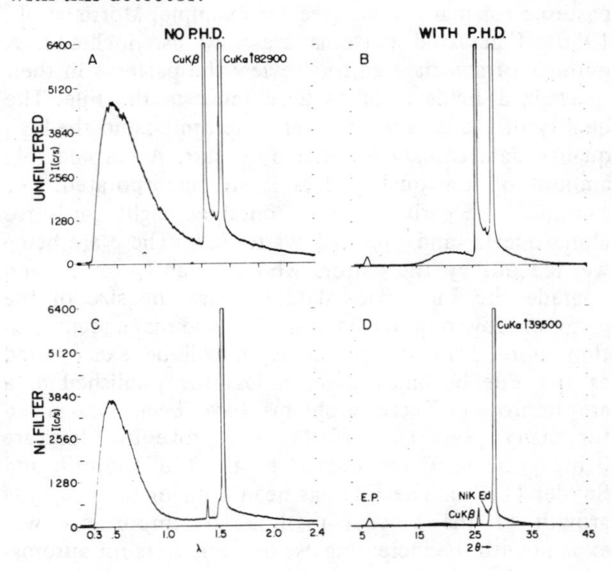

Fig. 4. Effect of pulse height discrimination (PHD) and Ni filter on spectrum of Cu X-ray tube operated at 40 kVp, scintillation counter (Rev. Sci. Instrum. **37***, 795, 1956).*

Standard Powder Data File. W. P. Davey and representatives of ASTM Committee E-4 arranged in 1941 to publish a powder data file using the 1,000 Dow patterns as a base. A joint committee was subsequently formed with representatives from ASXRED, British Institute of Physics, and later other Societies were added. In recent years, it was reorganized into a nonprofit group with a full time staff now called International Centre for Powder Diffraction, which publishes the JCPDS File (Joint Committee for Powder Diffraction Standards).

The File contained about 38,000 patterns in 1981 and each year a new set of 1,000-2,000 patterns is added; set 31 was issued in 1981. Each set is divided into inorganic (about 75%) and organic (25%) sections. It is issued in various forms: three by five inch data cards, microfiche, books which include several sets, and magnetic tape for computer use. Large Search Manuals (running to 1,000 pages) are updated each year. There are three Manuals for the inorganics – an Alphabetical index, Hanawalt, and Fink listings, and Manuals for organics and frequently encountered phases. Specialized data books and Search Manuals have been prepared for the NBS data, minerals, metals and alloys. Several subfiles are provided on magnetic tape, e.g., FEP, minerals, metals and alloys, etc., and various computer search programs such as Johnson/Vand are supported. The data base is usually rearranged by the commercial suppliers of Search/Match computer programs.

Most of the data are taken from the literature, and a number of groups produce patterns for the File. The National Bureau of Standards has provided high quality patterns for many years (see for example, Morris et al., 1980). Calculated patterns are now also included. A number of associate editors review the patterns in their specialized fields prior to acceptance in the File. The quality of the data varies over wide limits, and the high quality data cards are noted by a star. A considerable amount of low quality data is still incorporated. For example, the early editions contained eight cards for alpha quartz, and one had *d*'s to 30Å. These are being weeded out by the editors who constantly review and upgrade the File. Poor data increase the size of the error window required in searching and make identification more difficult; the situation will be exacerbated as the File becomes larger unless the published data are improved. These problems have been recognized for many years and efforts to improve the data are being publicized; see Calvert et al. (1980), Smith and Snyder (1979). The File has been a major factor in the growth of the powder method. Its importance will expand with the increasing use of computers for automation and Search/Match.

Some Problems in Phase Identification. Frevel (1944) noted early in the development of identification methods that there were a number of potential practical difficulties, and many more have been recognized since then. For example, different experimental methods may give different relative intensities even when they are properly used. This problem may arise from preferred orientation in the sample (reflection specimen on the diffractometer vs. transmission with Guinier or powder cameras), differences in resolution which affect the $K\alpha$ doublet separation and overlapping reflections, peak-to-background, counting statistics, specimen preparation, and many other factors. It is surprising that X-ray fluorescence spectrometry has not been more widely used for identification because it quickly provides chemical information to facilitate Search/Match.

Several authors have noted problems that occur in the identification of solid solutions which frequently occur in minerals and metals and alloys, and may shift peaks out of the error window range. Frevel and co-workers (1972) developed systematic procedures for identifying isomorphous and isostructural substances. They tabulated powder data for a large number of cubic, tetragonal, hexagonal, and orthorhombic phases. The earlier compilations beginning in 1942 need updating and computer programs developed to make it easier to use these procedures for identifying unknown phases with standard known structures. This could lead to a different type of searching procedure.

The use of the File for organic substances has been more limited and we can anticipate a large growth in this area. There have been a number of publications giving powder data for groups of related organic compounds. For example, Hofer et al. (1963) developed a file of 178 patterns of aromatic hydrocarbons and certain derivatives, phenols, and organic bases for identifying organic substances in coals, air pollution, and related fields. Parsons and co-workers (1968) published patterns of 523 steroids over a period of years. Both used powder cameras and today much higher quality organic data can be obtained with reflection and/or transmission specimen diffractometers using $CrK\alpha$ radiation and vacuum path (Parrish, 1968). The identification schemes may also have to be altered.

Calculated Powder Patterns. The computer generation of a theoretically correct powder pattern from good crystal structure data was done independently by D. K. Smith in 1963, T. Zoltai and L. C. Jahangabloo in 1963, and W. Jeitschko and E. Parthé in 1965. Borg and Smith (1969) have published a volume containing all the silicate mineral patterns. Smith (1968) has made available updated versions of the program to many users. The method computes the intensities and d's of all possible reflections, and by using the peak widths of a particular instrument, generates a pattern which looks like the experimental pattern, Fig. 5. This widely used

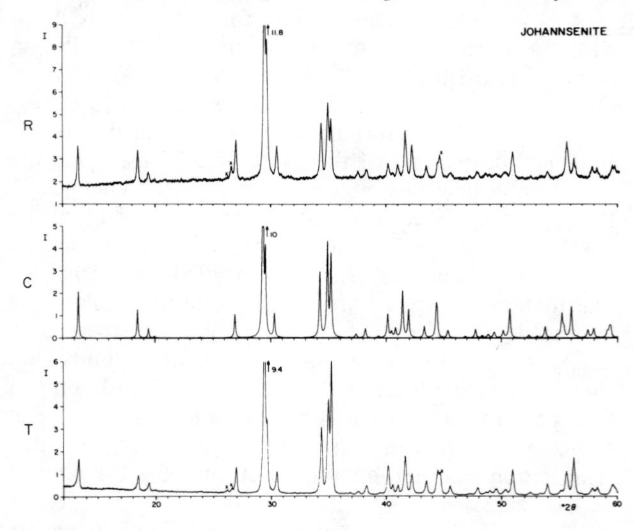

Fig. 5. Calculated powder pattern (C), reflection specimen (R, no monochromator), transmission specimen (T, diffracted beam monochromator). $CaMnSi_2O_6$, $CuK\alpha$. Peaks marked x are impurities.

development has many important applications such as checking experimental data, specimen preparation and preferred orientation, chemical differences and solid solution, indexing and trial crystal structures.

Indexing. The indexing of powder patterns is often required to make certain the material is single phase, for lattice parameter determination and other crystallographic studies. Direct indexing, i.e., without prior knowledge of the unit cell and symmetry is easy for high symmetry materials but becomes increasingly difficult with lower symmetry and increasing cell dimensions. Hull and Davey (1921) published charts for indexing tetragonal, hexagonal and rhombohedral crystals but they were difficult to use because of the large number of lines. Improved charts, analytical methods and Ito's method have been used (Azároff and Buerger, 1958). Zachariasen (1963) published a good method for monoclinic crystals.

Indexing usually begins with the lowest angle reflections where the systematic errors are largest. Corrections may be required because it is essential to use high quality data to avoid errors and ambiguous results. A number of programs have been written to do the extensive computations with computers but more development is needed for a completely satisfactory solution (Smith, 1980).

Precision Lattice Parameters. Accurate lattice parameter values are important in determining physical constants and in solid state theory, solid solution chemistry and the determination of substitutional and interstitial types, indexing, instrument calibration, calculating densities, crystallite perfection and a host of practical applications. There have been more papers written on this subject than any other topic on the powder method. The methods have been summarized by Klug and Alexander (1974), Parrish and Wilson (1959), and Straumanis (1959).

Bearden (1964) published a tabulated list of all X-ray wavelengths in which the previously published values were recomputed on a consistent basis, and new measurements were made in his laboratory. The data were converted to an absolute basis $\Lambda=\lambda_g/\lambda_s$ with Λ=1.002076. The wavelength standard used was $WK\alpha_1$= 0.29090100A* (The A* may differ from Å by ±5 ppm). He also established several secondary wavelength standards.

The question of how accurately lattice parameters can be measured by the powder method aroused considerable heated debate. In the 1950's, the highest precision claims varied from 0.002% to 0.00005%. An international effort to answer this question was made by the IUCr Commission on Crystallographic Apparatus. Sixteen laboratories, mainly using film methods, participated in the late 1950's using the same diamond, Si and W powders and wavelengths. The results gave surprisingly low agreement of around 0.01% for Si due largely to systematic errors as shown in Fig. 6 (Parrish,

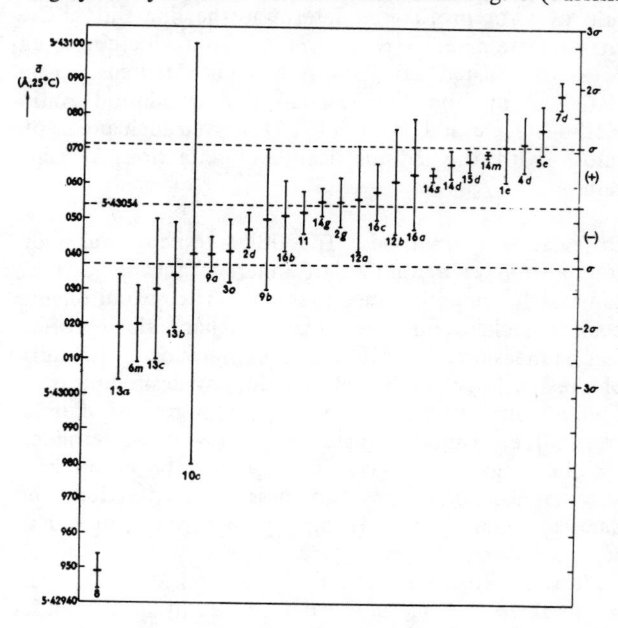

Fig. 6 Results of the IUCr lattice parameter project for Si. Distribution of individual mean values (short horizontal lines) and error limits (vertical lines) reported by various laboratories. Dotted lines show composite mean of the mean values and the standard deviation (Acta. Cryst. **13**, *838, 1960).*

1960). It is likely that the same type of round-robin conducted today with larger use of diffractometers would give better agreement, but I don't know how much better.

Jennings (1969) ran a round robin on the accuracy of intensity measurements in which eleven samples of nickel powder were measured in ten laboraotries. He concluded the integrated intensities could be relied on to no better than 5%.

Powder Camera Methods. The general strategy that evolved was to use the highest angle reflections (because $\Delta d/d=\cot\theta\Delta\theta$), and to correct for systematic errors which arise mainly from specimen absorption. Bradley and Jay plots of calculated a's vs $\cos\theta$ with extrapolation to $\cos^2\theta=0$ have been widely used as well as other functions derived by Nelson and Riley and by Taylor and Sinclair. Warren (1945) analysed the basis of extrapolation plots and showed there was a linear dependence of line displacement on $\phi=(\pi/2)-\theta$ for $\theta > 60°$ and this error vanishes at $\theta=90°$. Straumanis (1960) avoided plots by using thin specimens to reduce absorption.

Cohen (1935) developed a least squares method to correct the errors and Hess (1951) added weighting factors. (Such calculations can now be done with a computer in a few seconds.) Beu and co-workers (1967) developed a very elaborate method using the convolution of error profiles to determine the line shifts. The lattice parameter was computed from the corrected data (they listed calculations to eight decimals!) by a statistical method they named the Likelihood Ratio Method. Jette and Foote (1935) derived high precision values with the symmetrical back-reflection focusing camera.

Diffractometer Methods. The diffractometer was soon found to have a number of inherent advantages over cameras for precision measurements. The resolution and peak-to-background are much higher, the profiles can be measured directly with high precision, the quality of the specimen can be determined, any desired measure of 2θ can be selected (peak, midpoints of chords, centroid, etc.) and the angle scale accuracy is determined by the large worm gear. It may now be completely automated and various programs used to reduce the data, including profile fitting for high precision and to separate overlaps.

It was recognized early that the precision that could be obtained in practice was determined by many factors which were studied in great detail at Philips Laboratories and by A. J. C. Wilson and collaborators in England; Wilson published the rigorous mathematical theory (1963, and in Azároff et al. 1974 p. 439). His formulae for systematic errors had to be applied to the centroid rather than the peak and a method for measuring the centroids of the profiles was developed (Ladell et al., 1959; Taylor et al., 1964; Parrish, 1965, p. 161). The method also required the determination of the centroid of the X-ray spectral distribution (Mack et al., 1964); the Bearden data are peak values. The idea was to correct each profile to see if the systematic errors were eliminated, i.e., all the reflections should then give the same value of the lattice parameter. The major disadvantage of the centroid method was that the long profile tails limited it to very simple patterns.

The largest and most common source of systematic error is caused by displacement of the specimen from the axis of rotation. Materials with well-established lattice parameters are often used as standards to aid in correcting errors. High precision also requires temperature control of the specimens. A present practice is to use careful specimen preparation and experimental techniques, a computer-controlled diffractometer with step scanning, peak angles determined by profile fitting and a weighted least squares refinement. A precision of 0.01% to 0.002% is readily attainable using front reflection patterns (Parrish, Huang and Ayers, 1980); the use of back reflection patterns can increase the precision.

Quantitative Phase Analysis. Quantitative analysis is used in many applications, e.g., following solid state reactions, mineral mixtures, metallic phases, cements, and others. It is also widely used to analyze pollutants, industrial and mine dusts where the samples may be small and contaminated with other constituents and may require special specimen preparations prior to X-ray analysis.

Although the quantitative analysis of phases in a mixture was first mentioned by Hull (1919) it was 20 years before the first U.S. publication by Clark and Reynolds (1936). They analyzed the quartz content in mine dusts by using an internal standard and microphotometering the film. Alexander, Klug and Kummer (1948) developed the theoretical basis using integrated intensities and added a constant weight fraction of an internal standard. Leroux and co-workers (1953) developed a direct method by measuring the specimen absorption but it was limited to low absorbing specimens. Copeland and Bragg (1958) added a known weight fraction of the analyte instead of a separate standard. They analyzed N phases using m lines ($m > N$) to solve a set of simultaneous equations. Chung (1974) developed the "flushing agent" method which takes into account the absorption effects in multicomponent analyses and also permits determining the amorphous fraction.

Success with any of the quantitative methods is strongly dependent on specimen preparation and very careful experimental technique. The necessity of using small crystallite sizes and a rotating specimen device were studied in detail by Alexander et al. (1948) and by de Wolff et al. (Parrish, 1965, p. 130). Profile fitting can be used for cases where overlaps prevent direct measurement of the integrated intensities. Alexander (1977) has reviewed the present status of quantitative analysis.

Energy Dispersive Powder Diffractometry. This new powder method, a type of Laue powder diffraction, was developed by Giessen and Gordon (1968). It uses continuous radiation and a fixed selected scattering angle. All reflections occur simultaneously and are measured with a solid state detector (Li-drifted Si which must be kept at LN temperature, or intrinsic Ge) and multichannel pulse height analyser to separate the spectrum into small energy channels. The pattern contains both the spectral peaks of the element in the sample (and the X-ray tube) and the diffraction pattern (Sparks and Gedcke, 1972). Energy dispersive systems are used widely for X-ray elemental analysis with electron beam excitation in scanning and transmission electron microscopes and in X-ray fluorescence analysis.

In this method, the Bragg equation is used in the form $d=12.398$ (keV A)/E.2sinθ and each reflection arises from a different X-ray energy E. The present-day detectors have a Gaussian distribution with FWHM about 150 eV at 5.9 keV and the resolution and quantum efficiency vary with E. Although this is far better energy resolution than that of a proportional or scintillation counter, a conventional diffractometer pattern has much higher resolution and peak-to-background ratio. The energy dispersive peaks are an order-of-magnitude broader thereby limiting the use to relatively simple patterns. There is a considerable specimen transparency aberration which increases with increasing E. Conversion of the data to relative intensities is also considerably more difficult.

A recognizable pattern can be obtained in a minute or less, and in a second with a synchrotron, but it takes 10-20 hours to record patterns with good statistics (Mantler and Parrish, 1977). This apparent anomaly arises from limitations in present-day electronics which require the total detector input be less than 10,000 to 60,000 c/sec (divided into all the channels) to avoid peak shifts, changes in resolution and non-linearity. Fig. 7 compares conventional diffractometer and EDX

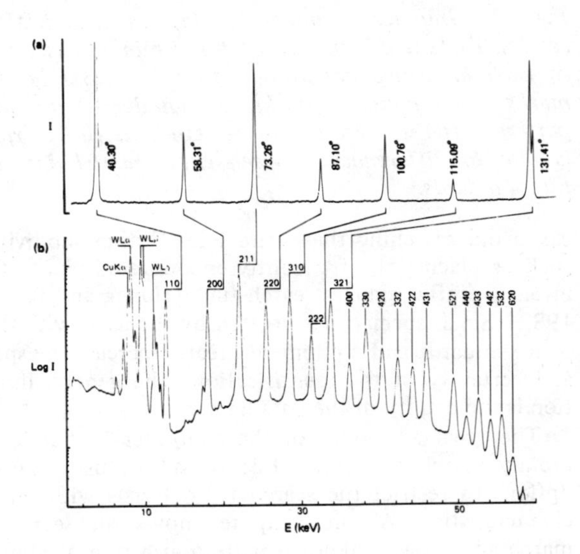

Figure 7. (a) Diffraction pattern of W powder with conventional diffractometer and diffracted beam monochromator, (b) EDX pattern with Cu tube operated at 60 kV, 2θ=25° (Adv. X-ray Anal. **20**, *171, 1977).*

recordings of the very simple tungsten pattern. The high EDX background corresponds to the continuous radiation of the Cu tube at 60 kV.

The method has been very useful in special applications where conventional scanning geometry is not possible such as in high pressures, high/low temperatures and phase transitions.

Graphite Monochromators. Monochromators have been used for many years particularly in certain research studies but they did not become popular for general work until the late 1960's when the Union Carbide Corporation succeeded in making flat and bent highly oriented pyrolytic graphite for X-ray use (Gould et al. 1968). The great advantage of graphite (first pointed out by Renninger) is that it reflects about 50% of CuKα compared to 10% for LiF (Sparks, 1981); hence by eliminating the nickel filter there is no loss of intensity. When placed in the diffracted beam of a diffractometer, the second set of parallel slits can be eliminated because the monochromator also acts as a collimator; in this case, there is actually a factor of two gain in intensity. The intensity and profile width are dependent on the receiving slit width of the diffractometer.

Many diffractometers are now equipped with diffracted beam graphite monochromators which give a large gain in peak-to-background ratio by eliminating all scattered and fluorescent X-rays except the wavelength they are set to reflect. This has been an important development because the intensity gain over other monochromators is almost as much as would be realized in switching from a stationary to a rotating anode X-ray tube.

The Guinier and deWolff powder cameras with incident beam focusing quartz monochromator and transmission specimen in Seemann-Bohlin geometry have been widely used in Europe. Their use in U.S.A. is increasing but the number is far below the conventional powder camera or diffractometer.

The Present Period, 1975-1980.

The most notable change has been the large increase in the use of computers and this stage in the development of the powder method may eventually be called the computer period. The development of the synchrotron X-ray sources has had little impact on the powder method so far, although they make it possible to obtain patterns with energy dispersive methods in a fraction of a second.

Small Computers. Computers were not widely used in powder work until recent years when powerful mini-computers and microprocessors were introduced. We can already see that their use will have a large impact as did the diffractometer 30 years ago. Their role appears similar to their use in the development of single crystal structure analysis. They make possible automation of the instruments and data reduction with a speed and precision that is not possible with manual methods. They are essential for sophisticated methods such as profile fitting, comprehensive search/match, structure refinement and many others which require large amounts of data and extensive calculations.

Although small (or mini) computers do not have the capacity of large host computers, they make it possible for individual laboratories to have their own facility. Their capabilities are constantly being improved with the extensive developments of large scale integrated circuits on small chips. They already far surpass the large computers used in the earlier stages of X-ray crystallography. Systems can now be purchased as a virtually complete package consisting of computer, interface and programs which are integrated with the new X-ray equipment or added to existing diffractometers. They are often designed so that computer experience is not essential and it is now not too difficult to convert to computer operation. Prices currently range from about $25K to $75K for the package, not including the X-ray equipment.

Microprocessors to control various instrument functions such as shutter operation, voltage and current settings, specimen changers, scanning and many other functions are being incorporated in the instrumentation to replace manual operations. The trend in instrumentation is towards electronics and computer operation in the entire field.

As an example, I will use the automated diffractometer package recently developed at the IBM San Jose Research Laboratory. For automation, a stepper motor is mounted on the diffractometer and controlled by the computer to scan the pattern. Several methods for scanning any angular range can be used such as step scan with selectable angular increments and count times, continuous scan with read-out on the fly, and point to point. The pattern can be seen on the CRT terminal screen or strip chart recorder, and the digital data are stored for analysis. Experimental parameters can be entered for automating a large number of runs and one computer can operate several instruments simultaneously.

A peak search program derives the 2θ's, d's, relative peak intensities and number of counts at each peak in less than a minute. The data can be transferred by the computer for Search/Match identification. It is now possible to collect data at high speed and make the identification of a not too complicated pattern in 5 to 15 minutes (Huang and Parrish, 1982).

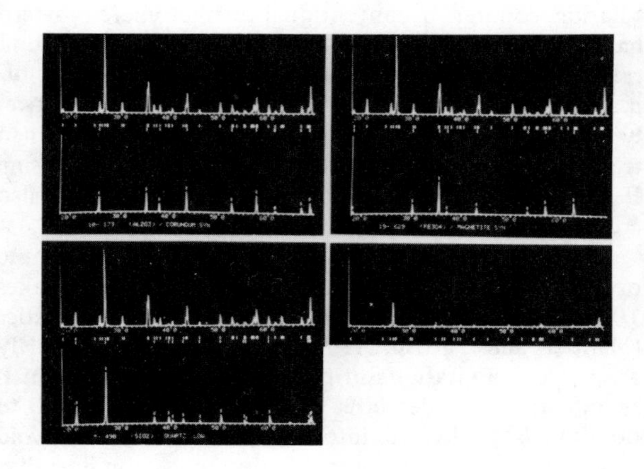

Fig. 8. Graphics terminal display of Search/Match results. Pattern of mixture of three minerals is on top of each identified constituent from JCPDS File. Tick marks show matched peaks. Remainder of unknown patterns after subtracting three standards (lower right) is due to differences between experimental data and File standards.

Computer Search/Match. The use of large computers to systematically search the JCPDS File for identification was introduced over 20 years ago by the pioneer work of the late V. Vand and G. G. Johnson, Jr. (Johnson, 1977). Johnson has been intimately involved in the development and has continually improved and updated the program which is now in version 20 and distributed by the ICDD. M. Nichols (1966) wrote a program which introduced a number of important features and it has been used as the base method by others to make further developments.

The recent availability of minicomputers with large internal disks and/or diskettes (floppy disks) to store the large File stimulated the writing of a number of new programs for Search/Match. These included rearrangement of the File for more efficient searching, the use of chemical composition restrictions, graphics for direct pattern comparison (Fig. 8) and other interactive options. When the File is rearranged into subfiles of inorganics, organics, minerals, metals, and alloys, many users can avoid searching the entire File; further subdivision such as placing the frequently encountered phases first in each subfile reduces search time (Huang and Parrish, 1982). Small special files created by the users with their own standards and equipment greatly increase the speed and certainty of the identification. The general limitation is the quality of the data base.

The great advantage of the computer is that it can rapidly search the entire File or subfile using various options to restrict the search to materials with certain characteristics. A minicomputer now can search to match an average unknown pattern at a rate of about a hundred cards in a few seconds; large computers are more than an order of magnitude faster. Mixtures with several components are handled in routine manner. Currently, there are several groups improving the programs and it is likely this will continue for several years. See also Frevel (1977), Huang and Parrish (1982), Jenkins et al., (1980), and Marquart et al., (1979).

Profile Fitting. Rietveld (1967) published a method for refining structures obtained by neutron diffraction by adjusting the various parameters to get the best fit with the complete experimental pattern. It has been widely used in several hundred refinements and its application is growing. Young (1981) and his collaborators and others have applied the method to X-ray powder diffractometer data to refine several dozen structures including hydroxyapatite and human tooth enamel. The X-ray results are not as good as the neutron results probably due in part to the asymmetric forms of the X-ray profiles.

A profile fitting method developed for automated X-ray powder diffractometry determines the integrated and peak intensities, and reflection angles of the individual reflections with high precision, and resolves overlapping reflections with much higher resolution than the original experimental data as shown in Fig. 9

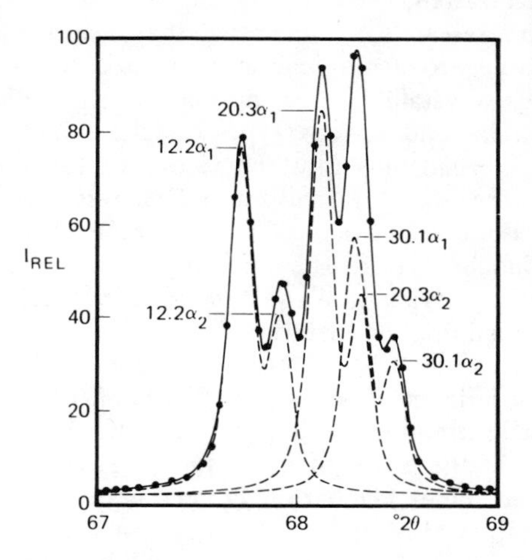

Fig. 9. Profile fitting of portion of quartz pattern. Dots: experimental points from step scanning; dashed lines; profile fitting of individual $K\alpha_{1,2}$ reflections which gives the correct peak and integrated intensities; solid line: sum of profile fitted reflections.

(Parrish et al. 1976). If the relative $K\alpha_1$ peak intensities in this illustration were taken from a ratemeter recording or computer peak search they would have been 79.3: 84.7:100. The correct values from profile fitting which takes the overlapping into account are 96:100:65.3. The instrument function W*G, which is a convolution of the spectral distribution W and the instrument aberrations G, is determined by careful measurements of the profiles of several standard specimens free of profile broadening. The computer adjusts a number of Lorentzian curves until their sum best matches the experimental profile and stores the data to compute the true diffraction of the specimen S from (W*G)*S. The method has been used in various applications requiring high precision such as in least squares refinement of crystal structures where low R-factors are obtained, direct indexing, lattice parameter and quantitative analysis. The method is relatively new and its full impact on crystallography is yet to be realized.

Learning X-ray Analysis.

It was not easy to learn powder methods and how to apply them to practical problems in the early days. There were few books in English, the literature was small and it was not taught in formal university courses. A few schools with teachers who had interest in the method included some instruction as part of a laboratory course in metallurgy or mineralogy but in the main the students had to dig it out for themselves.

Papers presented at meetings and related small group discussions were sources of information. The ACA and its predecessor societies ASXRED and CSA, the Denver and Pittsburgh Conferences, and the various discipline societies such as MSA and AIME held sessions on X-ray powder methods. (The establishment of the Applied Crystallography Group in the ACA five years ago may create a better forum for discussing advanced developments in the powder method.) Short intensive courses at industrial companies and universities and the growing tide of publications provided the basis for the expansion.

The Annual Conference on Applications of X-ray Analysis, sponsored by the Denver Research Institute and the University of Denver, has become the largest meeting for applied X-ray analysis. It began in the summer of 1951 encouraged by the presence of Professor I. Fankuchen with a one-day meeting attended by 35 people mainly interested in industrial applications. They have grown in size and scope to a three-day meeting in early August with 300-400 in attendance. The papers, divided roughly equally between powder diffraction and X-ray fluorescence analysis, are mainly of the application type. A hard cover volume of the papers presented has been published annually by Plenum Press since 1957. These volumes thus trace the development of applied X-rays over more than two decades although the quality of some of the papers is quite spotty. In recent years, workshops have been held in the two days preceding the meetings.

The Pittsburgh Diffraction Conferences were two day meetings held annually in the autumn at the Mellon Institute (now Carnegie-Mellon University) in Pittsburgh, Pennsylvania, beginning in 1943. Many papers on the applications of powder diffraction to industrial applications were presented in the 1950's and 1960's.

The X-ray division of North American Philips Company (now Philips Electronic Instruments) held its first "School" in 1946 in New York City. Its purpose was to promote sales which were limited by the small number of people trained to operate the equipment and do X-ray analyses, and the general lack of literature on industrial applications. The schools ran for a week, lectures were given in the mornings, and hands-on equipment training in the afternoons and evenings. A typical early "faculty" included full morning lectures by I. Fankuchen and later Ben Post on elements of crystallography and X-rays, W. Parrish on diffractometry, H. Friedman on fluorescence analysis and a number of speakers on various applications on the last morning. Occasionally, a well-known visitor such as Martin Buerger, Kathleen Lonsdale or Ralph Wyckoff would speak. Attendance grew from about 40 to over 200 and several schools were held annually in various sections of the country. Over 200 schools have now been given, attended by many thousands including a large number of practicing X-ray analysts. The *Norelco Reporter* began publication in September 1953 and is still published several times a year. It includes papers on X-ray diffraction, fluorescence analysis, electron microscopy and instrumentation, specializing in but not limited to Norelco equipment.

The most famous course was Professor Fankuchen's "Clinic" begun in 1943 at Polytechnic Institute of Brooklyn. This two-week course on powder and single crystal methods was usually given in June with attendance limited to 20 to permit intensive instruction by Fan and his assistants. Professor Ben Post was an active associate and took over when Fan died in 1964. Known as Fan's or Ben's "two-week wonders," many students became successful in industrial and government laboratories.

Training in powder diffraction was given by the General Electric X-ray Corporation in the 1950's and 1960's. Short courses are presently given at University of Texas (Hugo Steinfink), State University of New York at Albany (Henry Chessin), and others.

Some Major Accomplishments.

There have been a large number of people who have contributed to the development of the powder method in U.S.A. in addition to those mentioned above. It is regretted that space requirements limit the following additional descriptions to only a few individuals.

M. E. Straumanis/Powder Camera Technique. Professor Straumanis began his work in Riga, Latvia but World War II caused him to leave and in 1947 he took a position at the Missouri School of Mines and Technology, Rolla, where he continued until his death. He made an important contribution to powder camera technique with his asymmetric film mounting and the careful procedures he developed to determine lattice parameters accurately. His philosophy was to do the best possible experimental work, use the highest angle reflection and avoid extrapolation plots and other corrections. He used a small camera (usually 64 mm diameter) with temperature control, small diameter specimens to reduce absorption effects and give sharp lines, and a precision comparator for film measurements. Straumanis applied the technique to many materials to determine thermal coefficients of expansion, Avogadro's number, densities, and similar properties, and published a large number of papers (see for example, Straumanis, 1959). I was never able to convince him to try his hand on a diffractometer.

B. E. Warren/Diffraction Physics. Professor Warren developed the leading U.S. center in diffraction physics during his more than 40 years at MIT. After his collaboration with W. L. Bragg to solve silicate crystal structures he spent the rest of his career on the application of physics to non-crystal structure problems. He was fortunate in having a large group of outstanding graduate students, many of whom have also become important contributors to the field. His classic book rigorously derived the basis of temperature vibration effects, order-disorder, crystal imperfections and amorphous structures (Warren, 1969).

Warren's research demonstrated that accurate analysis of profile broadening can lead to the determination of average crystallite sizes normal to the reflecting planes, strain, and various types of lattice faults. The instrument broadening must be accounted for in carrying out these types of analyses and he was among the first to use the powder diffractometer with incident beam monochromator. The Fourier method of Warren and Averbach (1950), Warren (1959), is widely used for studying deformed metals.

W. H. Zachariasen/5f-Series Crystal Chemistry. One of the most extraordinary applications of the powder method was the work of Professor W. H. Zachariasen in the then secret Manhattan Project at the Metallurgical Laboratory, University of Chicago, in World War II. A large group of chemists was working on chemical properties of plutonium to develop methods of isolating it. Large numbers of samples were prepared but they were in too small quantities to make chemical identifications. Zacharaisen used powder cameras to provide positive identification of samples as small as 10 μgm. Often a partial structure determination was necessary to establish the chemical identity. The compounds in the samples were not usually what the chemists intended to make and he made virtually all the reliable analyses to support that huge program. He had to create his own reference file and build the chemical and structural information as he went along.

It is difficult to handle those highly dangerous materials, and once the floor of his laboratory had to be torn up because of an accidental Pu spill. "Willie" worked out the crystal structures of a large number of these compounds directly from the powder patterns, a skill in which he was an unsurpassed virtuoso. When the secret classification was removed, he published dozens of papers in *Acta Crystallographica* and other journals on the identity, crystal structures and crystal chemistry of the compounds of Th, U, transuranic elements, Ac and Pa (see for example, Zachariasen, 1948). I remember sitting with him when he was being questioned on this work by a number of Russian crystallographers at the 1959 Fedorov Commemoration in Leningrad. They found it hard to believe that there was no large staff and he did all the work himself with Anne Plettinger taking the photographs. His obituaries use the word miracles in describing this work - it was not an exaggeration.

References

(With Few Exceptions, the Publications are by U.S. Authors)

Books on Powder Method:

Azaroff, L. V. and Buerger, M. J. (1958). The Powder Method in X-Ray Crystallography. New York: McGraw Hill.

Cullity, B. D. (1978). Elements of X-Ray Diffraction, 2nd ed. Reading, Massachusetts: Addison-Wesley.

Klug, H. P. and Alexander, L. E. (1974). X-Ray Diffraction Procedures for Polycrystalline and Amorphous Materials, 2nd ed. New York: John Wiley.

Parrish, W. (1965). X-Ray Analysis Papers. Eindhoven: Centrex.

Books Containing Some Descriptions of the Powder Method:

Alexander, L. E. (1969). X-Ray Diffraction Methods in Polymer Science. New York: John Wiley.

Azaroff, L. V., Kaplow, R., Kato, N., Weiss, R. J., Wilson, A. J. C., and Young, R. A. (1974). X-Ray Diffraction. New York: McGraw-Hill.

Barrett, C. S. and Massalski, T. B. (1980). Structure of Metals, 3rd rev. ed. New York: McGraw-Hill.

Clark, G. L. (1955). Applied X-Rays, 4th ed. New York: McGraw-Hill.

Clark, G. L. (Editor). (1963). Encyclopedia of X-Rays and Gamma Rays, New York: Van Nostrand Reinhold.

Cohen, J. B. (1966). Diffraction Methods in Materials Science. New York: MacMillan.

Davey, W. P. (1934). A Study of Crystal Structure and its Applications. New York: McGraw-Hill.

Kable, E. (Editor). (1967). Handbook of X-Rays. New York: McGraw-Hill.

Schwartz, L. H. and Cohen, J. B. (1977). Diffraction from Materials. New York: Academic Press.

Sproull, W. T. (1946). X-Rays in Practice. New York: McGraw-Hill.

Taylor, A. (1961). X-Ray Metallography. New York: John Wiley.

Warren, B. E. (1969). X-Ray Diffraction. Reading, Massachusetts: Addison-Wesley.

Literature References

Alexander, L. E. (1977). Forty Years of Quantitative Diffraction Analysis, Adv. X-Ray Anal. **20**, 1-13.

Alexander, L. E., Klug, H. P. and Kummer, E. (1948). Statistical Factors Affecting the Intensity of X-Rays Diffracted by Crystalline Powders. J. Appl. Phys. **19**, 742-754.

Bearden, J. A. (1964). X-Ray Wavelengths, U.S. A.E.C. Report NYO-10586.

Beu, K. E. and Whitney, D. R. (1967). Further Developments in a Likelihood Ratio Method for the Precise and Accurate Determination of Lattice Parameters. Acta Cryst. **22**, 932-933.

Bleeksma, J., Kloos, G., DiGiovanni, H. (1948). X-Ray Spectrometer with Geiger Counter for Measuring Powder Diffraction Patterns. Philips Tech. Rev. **10**, 1-12.

Borg, I. and Smith, D. K. (1969). Calculated X-Ray Powder Patterns for Silicate Minerals, Geol. Soc. Am. Memoir 122.

Bragg, W. H. (1921). Applications of the Ionization Spectrometer to the Determination of the Structures of Minute Crystals. Proc. Phys. Soc. **33**, 222-224.

Brentano, J. C. M. (1946). Parafocusing Properties of the Microcrystalline Powder Layers in X-Ray Diffraction Applied to the Design of X-Ray Goniometers. J. Appl. Phys. **17**, 420-434.

Buerger, M. J. (1936). An X-Ray Powder Camera, Am. Miner. **21**, 11-17; (1945). The Design of X-Ray Powder Cameras. J. Appl. Phys. **16**, 501-510.

Calvert, L. D. et al. (1980). Standards for the Publication of Powder Patterns. Nat. Bur. Stand. Special Pub. **567**, 513-535.

Clark, G. L. and Reynolds, D. H. (1936). Quantitative Analysis of Mine Dusts. Ind. Eng. Chem. Anal. Ed. **8**, 36-40.

Chung, F. H. (1974). A New X-Ray Diffraction Method for Quantitative Multicomponent Analysis. Adv. X-Ray Anal. **17**, 106-115; (1974). Quantitative Interpretation of X-Ray Diffraction Patterns of Mixtures. I. Matrix Flushing Method for Quantitative Multicomponent Analysis. J. Appl. Cryst. **7**, 519-523.

Cohen, M. U. (1935). Precision Lattice Constants from X-Ray Powder Photographs. Rev. Sci. Instrum. Inst. **6**, 68-74; (1936) Z. Kristallogr. **94**, 288-298; 306-310.

Copeland, L. E. and Bragg, R. H. (1958). Quantitative X-Ray Diffraction Analysis. Anal. Chem. **30**, 196-208.

Davey, W. P. (1921). J. Opt. Soc. Amer. **5**, 479; Gen. Elect. Rev. **25**, 565.

Debye, P. and Scherrer, P. (1916). Interferenzen an Orientierten Teilchen Röntgenlicht. Physik. Z. **17**, 277-283.

Frevel, L. K. (1944). Chemical Analysis by Powder Diffraction. Ind. Eng. Chem. Analyt. Ed. **16**, 209-218.

Frevel, L. K. (1972). Prototype Charts for Identifying Biaxial Phases. Anal. Chem. **44**, 1840-1856 (contains earlier references).

Frevel, L. K. (1977). Quantitative Matching of Powder Diffraction Patterns. Adv. X-Ray Anal. **20**, 15-25.

Friedman, H. (1945). Geiger Counter Spectrometer for Industrial Research. Electronics (May issue) **18**, 132-137, (1945); U.S. Patent 2,286,785.

Giessen, B. C. and Gordon, G. E. (1968). X-Ray Diffraction: New High Speed Technique Based on X-Ray Spectrography. Science **159**, 973-975.

Gobel, H. E. (1979). A New Method for Fast XRPD Using a Position Sensitive Detector. Adv. X-Ray Anal. **22**, 255-265.

Gould, D. R. W., Bates, S. R. and Sparks, C. J. (1968). Application of the Graphite Monochromator to Light Element X-Ray Spectroscopy. Appl. Spect. **22**, 549-551.

Hanawalt, J. D., Rinn, H. W. and Frevel, L. K. (1938). Chemical Analysis by X-Ray Diffraction, Classification and Use by X-Ray Diffraction Patterns. Ind. Eng. Chem. Anal. Ed. **10**, 457-512.

Hess, J. B. (1951). A Modification of the Cohen Procedure for Computing Precision Lattice Constants from Powder Data. Acta Cryst. **4**, 209-215.

Hofer, L. J. E., Peebles, W. C. and Bean, E. H. (1963). X-Ray Powder Diffraction Patterns of Solid Hydrocarbons, Derivatives of Hydrocarbons, Phenols and Organic Bases, U. S. Bur. Mines Bull. **613**.

Huang, T. C. and Parrish, W. (1982). A new Computer Algorithm for Qualitative X-ray Powder Diffraction Analysis. Adv. X-ray Anal. **25**, 213-219, ibid., 221-229; IBM Res. Repts. RJ 3523-3526.

Hull, A. W. (1917). A New Method of X-Ray Analysis. Phys. Rev. **10**, 661-696.

Hull, A. W. (1919). A New Method of Chemical Analysis. J. Am. Chem. Soc. **41**, 1168-1775.

Hull, A. W. and Davey, W. P. (1921). Graphical Determination of Hexagonal and Tetragonal Crystal Structures from X-Ray Data. Phys. Rev. **17**, 549-570.

Jenkins, R. et al. (1980). A Qualitative Analysis Software Package for Use with the Computer Controlled Diffractometer. Norelco Reporter **27**, 11-19.

Jennings, L. D. (1969). Current Status of the I.U.Cr. Powder Intensity Project. Acta Cryst. **A25**, 217-222.

Jette, E. R. and Foote, F. (1935). Precision Determination of Lattice Constants. J. Chem. Phys. **3**, 605-616.

Johnson, G. G. (1977). Resolution of Powder Patterns, in Laboratory Systems and Spectroscopy, Chapter 3, edited by J. S. Mattson, H. B. Mack and H. C. MacDonald. New York: Dekker.

Ladell, J., Parrish, W. and Taylor, J. (1959). Interpretation of Diffractometer Line Profiles, Acta Cryst. **12**, 561-567.

Lang, A. R. (1956). Diffracted Beam Monochromatization Techniques in X-Ray Diffractometry. Rev. Sci. Instrum. **27**, 17-25.

Legalley, D. P. (1935). A Type of Geiger-Müller Counter Suitable for the Measurement of Diffracted MoK X-Rays. Rev. Sci. Instrum. **6**, 279-283.

Leroux, J., Lennox, X. D. H. and Kay, K. (1953). Direct Quantitative X-Ray Analysis by Diffraction-Absorption Technique. Anal. Chem. **25**, 740-743.

Lindemann, R. and Trost, A. (1940). Das Interferenz-Zählrohr als Hilfsmittel der Feinstrukturforschung mit Röntgenstrahlen. Z. Physik. **115**, 456-468.

Mack, M., Parrish, W. and Taylor, J. (1964). Methods of Determining Centroid X-Ray Wavelengths: CuKα and FeKα. J. Appl. Phys. **35**, 1118-1127.

Mantler, M. and Parrish, W. (1977). Energy Dispersive X-Ray

Diffractometry. Adv. X-Ray Anal. **20**, 171-186.

Marquart, R. G. et al. (1979). A Search-Match System for X-Ray Powder Diffraction Data. J. Appl. Cryst. **12**, 629-634.

Morris, M. et al. (1980). Standard X-Ray Diffraction Powder Patterns. Nat. Bur. Stand. Mon. 25-Section 17 (series of publications).

Nichols, M. C. (1966). A Fortran II Program for the Identification of X-Ray Powder Diffraction Patterns. UCRL-70078.

Parrish, W. (1949). X-Ray Powder Diffraction Analysis: Film and Geiger Counter Techniques. Science **110**, 368-371; and Hamacher, E. A. (1951). U.S. Patent 2,549,987.

Parrish, W. (1960). Results of the I.U.Cr. Precision Lattice Parameter Project. Acta Cryst. **13**, 838-850.

Parrish, W. (1962). Geiger, Proportional and Scintillation Counters. Int. Tables X-Ray Cryst. **III**, 144-156. Birmingham: Kynoch Press.

Parrish, W. (1965). X-Ray Diffraction Methods for Complex Powder Patterns. In "X-Ray and Electron Methods of Analysis," 1-35. New York. Edited by H. van Olphen and W. Parrish. New York: Plenum Press.

Parrish, W. and Cisney, E. A. (1948). An Improved X-Ray Diffraction Camera. Philips Tech. Rev. **10**, 157-167.

Parrish, W., Huang, T. C. and Ayers, G. L. (1976). Profile Fitting: A Powerful Method of Computer X-Ray Instrumentation and Analysis. Trans. Am. Crystallogr. Assoc. **12**, 55-73; (1980). Accuracy of the Profile Fitting Method for X-Ray Polycrystalline Diffractometry. Nat. Bur. Stand. Special Pub. **567**, 95-110.

Parrish, W. and Mack, M. (1967). Seemann-Bohlin X-Ray Diffractometry, I. Instrumentation. Acta Cryst. **23**, 687-692; II. Comparison of Aberrations and Intensity with Conventional Diffractometer. Ibid. **693-700**.

Parrish, W. and Wilson, A. J. C. (1959). Precision Measurement of Lattice Parameters of Polycrystalline Specimens. In Int. Tables X-Ray Cryst. **II**, 215-234. Birmingham: Kynoch Press.

Parsons, J., Holcomb, J. B. and Beher, W. T. (1968). X-Ray Diffraction Powder Data for Steroids. Henry Ford Hosp. Med. Bull., Suppl. VIII, **15**, 133-138 (and previous papers in same journal.

Reitveld, H. M. (1967). A Profile Refinement Method for Nuclear and Magnetic Structures. J. Appl. Cryst. **2**, 65-71.

Smith, D. K. (1968). Computer Simulation of X-Ray Diffractometer Traces. Norelco Reporter **15**, 57.

Smith, G. S. (1980). Advances in the Computer Indexing of Powder Patterns. Adv. X-Ray Anal. **23**, 295-303.

Smith, G. S. and Snyder, R. L. (1979). F_N: A Criterion for Rating Powder Diffraction Patterns and Evaluating the Reliability of Powder-Pattern Indexing. J. Appl. Cryst. **12**, 60-65.

Soller, W. (1924). A New Precision X-Ray Spectrometer. Phys. Rev. **24**, 158-167.

Sparks, C. J. (1981). Personal communication; (1967). Am. Crystallogr. Program Abstr. Ser. 2.

Sparks, C. J. and Gedcke, D. A. (1972). Rapid Recording of Powder Diffraction Patterns with Si(Li) X-Ray Energy Analysis System: W and Cu Targets and Error Analysis. Adv. X-Ray Anal. **15**, 240-253.

Straumanis, M. E. (1959). Absorption Correction in Precision Determination of Lattice Parameters. J. Appl. Phys. **30**, 1965-1969 (contains references to other papers by Straumanis).

Taylor, J., Mack, M. and Parrish, W. (1964). Evaluation of Truncation Methods for Accurate Centroid Lattice Parameter Determination. Acta Cryst. **17**, 1229-1245.

Warren, B. E. (1945). The Absorption Displacement in X-Ray Diffraction of Cylindrical Samples. J. Appl. Phys. **16**, 614-620.

Warren, B. E. (1959). X-Ray Studies of Deformed Metals. Prog. Metal Phys. **8**, 147-202.

Warren, B. E. and Averbach, B. L. (1950). The Effect of Cold-Work Distortion on X-Ray Patterns. J. Appl. Phys. **21**, 595-599.

Young, R. A. (1981). Application of the Rietveld Method for Structure Refinement with Powder Diffraction Data. Adv. X-Ray Anal. **24**, 1-23.

Zachariasen, W. H. (1948). Crystal Chemical Studies of the 5f-Series of Elements. I. New Structure Types. Acta Cryst. **1**, 265-268.

Zachariasen, W. H. (1963). Interpretation of Monoclinic Powder X-Ray Diffraction Patterns. Acta Cryst. **16**, 784-788.

CHAPTER 2

HISTORY OF THE POWDER DIFFRACTION FILE (PDF)

J. D. Hanawalt
Professor Emeritus, University of Michigan
Ann Arbor, MI 48109

The first publications which outlined the technique for X-ray powder diffraction and pointed out its important advantages as an analytical tool for phase identification of materials were by P. Debye and P. Scherrer (1916, 1917) in Europe and A. W. Hull (1917, 1919) in the United States. In the years following these publications there were no doubt many X-ray laboratories which were building private collections of reference diffraction patterns of the particular phases occurring in their work. Early literature references to such private collections were A. N. Winchell (1927) and J. D. Hanawalt and H. W. Rinn (1935, 1936) who reported on experience with a file of 1000 patterns and described a design of a search manual for identification of unknowns. A. W. Waldo (1935) published diagrams of the X-ray diffraction patterns of 51 copper ore minerals. At the Institute of Mines in Leningrad, A. K. Boldyev, V. I. Mikheev, V. N. Dubinina, and G. A. Kovalev (1938) published the X-ray powder diffraction data for 142 minerals along with a design of search manuals for the retrieval of these patterns. J. D. Hanawalt, H. W. Rinn and L. K. Frevel (1938) published the X-ray powder diffraction data for 1000 common inorganic chemicals and illustrated the system for retrieval of the components of a mixture of phases. The data for these 1000 phases were the starting point for the Powder Diffraction File (PDF) and in 1941 the data were reprinted on 3 x 5 cards by a working group formed by the National Research Council and by Committee E-4 of the American Society for Testing and Materials (ASTM).

The British Institute of Physics subsequently joined the group and furnished a significant portion of the data published in the second set of patterns in 1945. Later, the American Society of X-ray and Electron Diffraction (now the American Crystallographic Association) joined, to replace the National Research Council. During the next several years other groups joined, including the National Association of Corrosion Engineers, the American Ceramic Society, the Mineralogical Society of America, the Clay Minerals Society, the Mineralogical Society of Canada, the Mineralogical Society of Great Britain and Ireland, and Societe Francaise de Mineralogie et de Cristallographie. The group of volunteers and representatives from the sponsoring societies was known as the Joint Committee for Chemical Analysis by X-ray Powder Diffraction, later shortened to Joint Committee for Powder Diffraction Standards (JCPDS).

The work of the Joint Committee was first directed and housed in the office of Dr. Wheeler P. Davey at The Pennsylvania State University from 1941 until the mid-1950's. Before going to The Pennsylvania State University, Professor Davey worked with A. W. Hull at the General Electric Company. Davey's work there (Davey, 1921) made basic contributions to the developing science of X-ray powder diffraction. Professor Davey (1939) made an ideal leader for the practical chemical analysis objectives of the Joint Committee and served as the first Chairman of the Joint Committee (1941-1956). As chairman of Committee E-4 of ASTM, Dr. William L. Fink had a strong guiding hand in the operations of the Joint Committee and followed H. W. Rinn (1956-1961) as the third chairman of the JCPDS (1961-1974). As a charter member of the Joint Committee and as the astute treasurer for many years, LeRoy R. Wyman made a very major contribution to the success of the Joint Committee by helping the organization achieve operations in the "black" and insisting after the initial contributions enabled it to get started that the Joint Committee always pay its way. From the start the Joint Committee depended upon the voluntary and dedicated efforts of many people from many disciplines. About one-hundred voluntary scientists comprise the Joint Committee today.

The high standard of quality of the PDF, which is recognized world-wide, results from and continues to depend upon the contributions of physicists, chemists, mineralogists, crystallographers, computer experts, and others. While the major historical source of data for the PDF has been the scientific literature, a substantial proportion of the data comes from grants-in-aid for the generation of high quality data from well characterized materials. For example, since 1951 the JCPDS has joined the National Bureau of Standards (NBS) to maintain an associateship there in the crystallography section. Hundreds of experimentally produced patterns from high purity materials, as well as calculated patterns, are received from the group of six JCPDS employees at the NBS. This work was formerly under the leadership of Marlene Morris in collaboration with Dr. Camden R. Hubbard of NBS as coordinator.

In 1964 the Joint Committee moved to the headquarters of ASTM in Philadelphia. In 1969 the JCPDS became an independent non-profit organization incorporated under the laws of the Commonwealth of Pennsylvania. A new wholly-owned building designed for the purpose and located in Swarthmore, Pennsylvania, now serves as headquarters. Although originating in the USA, the organization has contacts and members from many nations of the world. The distribution of the PDF and related publications of the JCPDS to other parts of the world already exceeds the volume within the USA. In recognition of the opportunity and responsibility to serve the world-wide scientific community, the name of the organization was changed in 1977 to JCPDS-International Centre for Diffraction Data. Because the present active membership has expanded to include far broader representation than that of the original cooperating societies, it is likely that simply "International Centre for Diffraction Data" (ICDD) will become an accepted and less ambiguous name.

Since the publication of the first 1000 patterns in card form in 1941, the publications and scope of activities have increased and expanded year by year. The complete PDF comprises nearly 37,000 phases in 1980. Also available are sub-files of Inorganics, Organics, Minerals, Metals and Alloys, NBS generated patterns and a special file of about 2500 of the more commonly occurring phases. Other sub-files are currently under consideration, e.g., Ceramics, Forensics, etc. These data are available in card form, microfiche, in books and on computer disks and on magnetic tape. Computer programs are available which search the disks and magnetic tape for pattern matches. In addition, the PDF has been incorporated into the Chemical Information System (CIS). The CIS network has the capability to cross index with the other data bases on the system through the common compound code, the Chemical Abstracts Registry Number. Thus, one data base can augment the information contained in another data base. A more direct link is being established between the PDF and the Crystal Data File which will allow cross indexing through several data channels, including the crystal cell.

Like many scientific organizations, the International Center for Diffraction Data operates both as a professional society (having about 100 members drawn from industry, universities and government laboratories) and as a publishing house. The Board of Directors, elected by the membership, is responsible for the overall operations. A full-time general manager reports to the Board of Directors and, with a staff of about twenty persons at the Swarthmore headquarters, handles all aspects of the publishing business, which in fiscal 1979-80 exceeded one million dollars. Much of the volunteer work is broken down by the Technical Committee into subcommittees such as Editorial, Methods and Practices, Search Procedures, Computer, Education, Organic, Mineral, Metals and Alloys, Ceramics, Forensics, etc.

Of topmost priority in the publication of the PDF is that the data are correct, reliable, and of the highest quality in all respects. The quality control involves insuring that the sample is a chemically analyzed pure phase, that the highest precision X-ray diffraction equipment is correctly used and that the resulting diffraction data are crystallographically consistent if the crystal structure is known. This responsibility rests with the Editorial Staff. This work is not voluntary, and the editors are chosen from recognized authorities in the various fields, inorganics, organics, minerals, metals, etc., in which the phases lie. The history of the PDF is one of continual upgrading by replacing older patterns as newer, better data become available. For example, while the first 1000 patterns were prepared from C.P. grade samples, they were not chemically analyzed, and only for 300 of them were the lines completely checked against a unit cell by (*hkl*) indexing. These better characterized patterns were designated "starred" patterns. While a less than "starred quality" pattern may have practical value when used with discretion, the objective for the PDF is to include only the highest quality data possible for every phase. The editors have a standard and comprehensive set of specifications against which all new patterns are compared whether they are generated from grants-in-aid or are taken from the literature. A big advance followed with the developments in calculating powder diffraction data, as it provides a check on the experimental data [D. K. Smith (1963, 1967)]. Considerable attention has been directed to the questions of correct generation and publication of powder data [Int'l Union Cryst. Commission (1971); Am. Cryst. Assoc. Sub-Com., Final Report on Standards for Publication of Powder Patterns (1979)].

During the forty years of growth of the PDF there have naturally been many developments both in apparatus and experience which have contributed to a continuously rising standard of quality of powder diffraction data. In the early years the data were generated in the Debye-Hull type of non-focusing camera and intensities were measured by photometers or in many cases simply visually estimated. The first 1000 patterns of the PDF were produced on the original General Electric Company unit familiarly known as the "Ash-can." This unit used Mo radiation and exposure times were about ten hours. But it could handle 20 samples simultaneously in multiple cassettes of eight inch radius surrounding the X-ray tube. It was an ingenious and practical design which, with enough care, served its purpose very well. Improved quality of data followed with the advent and adoption of the Guinier type of crystal monochromator focusing camera [Guinier (1939)], and with the refinements in diffractometer design [William Parrish (1948, 1965)].

At the present time no published tables of powder diffraction data approach the volume and scope of the PDF of the International Center for Diffraction Data (ICDD). A compendium of data on clay minerals has recently been produced by Thorez (1975). Workers in Leningrad in a series of publications beginning in 1938 have tabulated powder diffraction data for more than a thousand minerals. In recent years workers at Leningrad and other places in the USSR have contributed many patterns to the PDF and are cooperating generously with the ICDD.

Concurrently with the build-up of the PDF, there has been extensive effort to develop retrieval systems designed to manually, or by mechanical or computer methods, match an unknown pattern to one in the PDF. In the main, such retrieval does not involve any information about crystal structure but simply makes use of the line spacings and intensity characteristics of the powder diffraction pattern. It is somewhat analogous to the matching of a "fingerprint" to one contained in the FBI files making use of the characteristic loops, whorls, arches, pockets, etc. The problem of pattern matching in X-ray powder diffraction is generally easy for patterns obtained from single phases, but it becomes more complicated for real unknowns where the pattern may represent a mixture of two or more phases.

Of course, even a simple alphabetical index based upon the name or the chemical formula of a material provides one practical means of very effectively using a large reference library of patterns. Its uses in the analytical chemical laboratory are myriad; – Is the material composed of the phases to be expected? Is the material a single phase or, on comparing its reference patterns, are there extra diffraction peaks indicating the presence of more than one phase? Has a reaction taken place between components of a mixture and what are the new phases present? Is a material anhydrous or are certain hydrated phases present, etc.? All such questions are answered by "matching" line for line with the appropriate PDF standard reference pattern. By going to the alphabetical index, which gives the three strongest lines, the analyst quickly checks whether these lines are present in the pattern of his unknown material. If a match of these three lines is found, the analyst uses the pattern number given in the alphabetical index to consult the complete pattern in the PDF.

It will be seen that there are two distinct steps involved in the solution of an unknown. The "search" consists of determining *which* standard pattern from the PDF to use for the comparison. The "match" consists of comparing all of the *d*'s and *I*'s of the standard pattern with *d*'s and *I*'s of the unknown. Often the analyst knows what to expect, so he checks the correctness of his information by the matching process. But sometimes the analyst may not know what to expect. The unknown material may contain a completely unexpected phase. It is then somewhat of a "challenge" and a "game" to be able to retrieve a pattern from the PDF having no information beyond the *d* and *I* diffraction data of the unknown. The development of retrieval methods historically parallels the development of the PDF itself. Successful retrieval can be accomplished by any one of the several forms of manual search systems. Naturally, all systems make use of the "*d*" values of the stronger lines of the pattern because these lines are the easier to measure and also because these lines may be the only ones observable for a minor component in a mixture. How many lines or parameters are needed to distinguish one pattern from every other one depends upon the number of patterns in the file. For a file of 1054 patterns it was observed that, making use of relative intensities, a measurement of three lines was sufficient to distinguish one pattern from all others for more than 98% of the patterns [J. D. Hanawalt and H. W. Rinn (1935)]. For a larger file, more parameters may be necessary to narrow the search down to a few patterns which may then require the checking of more lines.

The publication from Leningrad [Boldyev, Mikheev, Dubinina, Kovalev (1938)] listed the 5 strongest lines of the pattern in order of decreasing intensity as an entry in a table of continuously decreasing "*d*" values of the first line of the entry. The 1957 publication from Leningrad used the original format [Boldyev, Mikheev, Dubinina, Kovalev (1938)] but each pattern is given 5 entries as each of the 5 lines in turn is rotated into position as the first line of the entry.

In the USA, search systems published by Waldo (1935) used the three or four strongest lines of a pattern arranged in order of decreasing intensity for an entry. The line of strongest intensity determined the location of the entry in a table of continuously decreasing *d* values. Waldo also made charts of the diffraction patterns of his 51 minerals and showed them in two tables, in one according to the line of largest "*d*" value, in the other according to the "*d*" value of the line of strongest intensity.

The search manual which was put into use in 1934 in the X-ray laboratory at the Dow Chemical Company had a unique feature which proved useful and is now used in the design of the search manuals published by the JCPDS International Centre for Diffraction Data. The original search manual was characterized by dividing the range of "*d*" values from 0.8 to 20 Å into 77 Δd size "groups" so that all patterns with their strongest line falling within a particular Δd range would be found in this particular group. The location within the group, i.e., the sub-group, was determined by the "*d*" value of the second strongest line. With the growth in the number of patterns in the PDF the Δd range of the groups has been modified and two other features have been introduced. In the present search manuals of the ICDD the total Angstrom range of "*d*" values is divided into 48 groups, the groups are over-lapped to take care of experimental errors in determination of "*d*" values, and the patterns are ordered within sub-groups on the first line of the pair used for access to the search manual. [Hanawalt, J. D. (1970)].

The advantage of the concept of the group system making use of a pair of lines to position the pattern on a page rather than a sequential listing based on a single *d*-value is illustrated in the accompanying figure. This figure is a page from the PDF inorganic Hanawalt search manual which contains multiple entries for 26,000 phases. It shows how the group format brings together the patterns which should be examined when looking for a match to an unknown pattern. At intermediate *d*-values, the error in *d* may be $\pm$ 0.01 Å; hence, if an analyst is looking for a match for the line pair 3.04 with 2.96, he should look at all nine combinations of 3.04 $\pm$ 0.01 with 2.96 $\pm$ 0.01. If the pattern listings

3.04 – 3.00 (±.01)

										File No.	I/Ic
i	3.05_5	2.97x	3.20_5	2.84_5	2.78_5	1.93_5	2.89_4	2.14_4	$(Sm_2O_3)30N$	25- 749	
	3.05_4	2.97x	2.84_6	2.15_4	1.65_4	4.48_2	2.00_2	1.70_2	$K_3Gd_4(PO_4)_5$	21- 635	
c	3.05_3	2.97x	2.59_5	1.82_3	1.85_2	1.56_2	1.55_1	1.58_1	Lu_6MoO_{12}	20- 651	
i	3.05x	2.97_9	2.12_4	1.74_3	4.06_2	3.65_2	1.76_2	1.52_1	$Ba_3Y_4O_9$	27-1037	
	3.04x	2.97_9	2.95_7	2.09_5	1.97_4	1.93_3	3.34_3	1.96_3	$Sr_5Cl(VO_4)_3$	19-1269	
	3.04x	2.97_7	2.56_4	4.38_3	2.13_1	3.71_1	3.21_1	2.23_1	$((Zn_{0.80}Cd_{0.20})_2P_2O_7)$	17- 634	
*	3.04x	2.97_7	2.52_5	2.60_4	3.36_3	4.41_2	2.57_2	1.54_2	$NaCrGe_2O_6$	26-1483	1.50
*	3.03x	2.97_9	5.68_8	3.59_4	2.20_4	2.85_3	3.49_3	1.85_3	$Al_4Bi_2O_9$	25-1048	3.10
	3.03_8	2.97x	3.18_8	2.83_8	2.77_8	2.14_8	1.92_8	1.70_8	$(Eu_2O_3)30N$	12- 384	
	3.03x	2.97_8	2.94_8	2.79_6	2.02_5	2.91_5	2.82_3	2.75_3	$Sr_6Al_{18}Si_2O_{37}$	10- 25	
i	3.02_9	2.97_9	1.82x	5.98_7	5.20_6	2.61_6	3.64_5	10.1_4	$Ca_2Sb_2O_7$	26- 293	
*	3.01x	2.97x	6.01_7	5.18_6	1.82_6	3.64_4	3.09_3	1.57_2	Na_2NiCoF_7	24-1114	
*	3.01x	2.97_5	5.28_3	1.72_3	3.06_3	2.96_2	2.13_2	1.72_2	Sr_2CeO_4	22-1422	
	3.01x	2.97x	4.14_8	2.89_8	2.08_8	1.97_7	1.93_7	1.90_7	$3Pb_3(PO_4)_2.PbBr_2$	6- 365	
	3.01_8	2.97x	3.06_8	2.70_8	2.65_8	2.48_8	1.88_8	1.82_8	K_4SnO_4	16- 221	
	3.01x	2.97x	2.70_8	2.65_8	2.48_8	5.54_6	5.11_4	4.84_4	K_4SnO_4	20- 938	
*	3.01_8	2.97x	2.10x	4.12_6	2.29_6	2.17_6	1.72_6	1.51_6	$K_3Mn_2F_7$	19- 976	
	3.00_5	2.97x	5.21_7	3.63_4	5.94_4	1.82_3	5.13_3	3.09_3	$Na_5Ga_3F_{14}$	22-1376	
i	3.00_6	2.97x	3.31_9	1.70_6	3.13_6	1.64_6	2.89_5	1.70_5	$LiBa_2Nb_5O_{15}$	27-1215	
	2.99x	2.97x	3.14_8	2.64_5	3.83_4	3.36_4	2.01_4	1.58_3	$BaTi_4O_9$	8- 367	
	2.99x	2.97x	2.85_7	2.90_6	3.29_4	4.33_3	4.13_3	3.39_3	$Pb_5GeV_2O_{12}$	15- 435	
	3.05x	2.96_5	4.54_3	2.53_3	6.54_3	2.56_3	1.65_2	1.63_2	$NaScSi_2O_6$	21-1369	
i	3.05x	2.96x	3.94x	2.69_9	6.35_6	1.97_6	5.37_5	1.85_4	$Na_2ZrSi_3O_9.2H_2O$	14- 297	
i	3.05x	2.96x	2.89x	2.86x	3.48_7	2.46_7	3.80_6	2.51_6	$CuMoO_4$	26- 546	
	3.04x	2.96_9	2.52_9	3.34_8	1.43_8	2.60_7	1.55_7	3.08_6	$Fe_4Ge_{14}O_{20}$	15- 505	
*	3.04x	2.96x	2.09x	2.21_8	1.73_7	1.52_7	6.67_5	2.24_5	K_2MnF_4	19- 978	
*	3.03x	2.96_9	3.22_5	2.12_5	2.82_4	1.83_2	1.66_2	2.79_2	Pb_2O_3	23- 331	
i	3.03x	2.96_6	2.92_5	1.97_5	4.18_4	3.66_4	2.25_4	2.08_4	$(Ca,Pb)_5(AsO_4)_3Cl$	14- 213	
	3.03x	2.96_8	1.85_8	3.39_4	2.65_4	2.56_4	1.88_4	1.84_4	$Ga_2Y_4O_9$	22- 306	
*	3.03x	2.96_9	1.82_8	5.21_7	3.66_6	2.36_4	1.58_4	2.36_3	Na_2MgFeF_7	22- 895	
	3.02_6	2.96x	3.07_6	3.88_3	3.27_3	2.27_3	2.01_3	1.93_3	$KYMo_2O_8$	23-1352	
*	3.02x	2.96x	1.83_7	6.05_6	5.19_5	3.66_4	3.12_3	3.11_3	Na_2NiFeF_7	24-1115	
*	3.02x	2.96_9	1.83_7	5.19_6	3.66_6	2.36_4	1.58_3	6.04_3	Na_2MgVF_7	22- 896	
i	3.01x	2.96x	7.88x	2.23_7	4.91_6	3.26_6	3.11_6	1.78_6	K_3NbO_4	26-1325	
*	3.01_9	2.96_7	3.06x	3.36_4	2.11_3	1.99_2	1.91_2	4.44_2	$Pb_5(AsO_4)_3Cl$	19- 683	
	3.01x	2.96x	2.92x	2.17_5	6.53_4	2.03_3	3.99_3	2.19_2	$Gd_2O_2SO_4$	22- 293	
i	3.01x	2.96x	2.92x	6.50_8	1.74_8	1.71_8	1.69_8	4.00_6	$Cf_2O_2SO_4$	26-1174	
i	3.01x	2.96_5	2.44_5	2.09_5	1.72_4	1.50_4	4.89_3	2.16_3	$NaRb_2MnF_6$	27- 554	
*	3.00_9	2.96x	9.13_9	5.71_7	4.02_7	7.01_6	3.12_5	2.73_5	$(NH_4)_2Co_3(SO_4)_3(OH)_2.2H_2O$	23- 779	
*	3.00_6	2.96x	6.45_6	3.23_6	1.62_4	2.52_3	2.16_3	2.57_3	$CaNaFeMnZnMg(SiAl)_2O_6$	19- 1	
o	3.00_8	2.96x	3.27_8	2.83_8	2.75_5	2.80_3	2.69_3	2.47_3	$K_2Mg_6Al_2B_3P_3O_{20}F_4$	15- 352	
i	2.99x	2.96x	3.56_8	2.18_8	1.95_8	2.84_7	4.10_6	2.69_6	$Ce_2Ti_2Si_2O_{11}$	19- 302	
*	2.99_8	2.96_4	3.08x	2.48_3	2.29_3	2.60_2	2.57_2	2.13_2	K_2CrO_4	15- 365	
*	2.99x	2.96x	2.89_6	4.13_5	3.27_4	2.06_4	3.38_3	1.92_3	$Pb_5(PO_4)_3Cl$	19- 701	
	2.99x	2.96_9	1.89_7	3.12_4	5.92_3	3.30_3	3.17_2	1.97_2	$Sr_2B_2O_5$	19-1268	
*	3.05_9	2.95_7	2.24x	2.08_5	1.72_4	1.78_3	2.26_2	3.37_2	Ba_2HfO_4	22- 83	
	3.04_6	2.95_6	3.92x	1.96_6	1.85_6	1.74_6	6.31_5	5.36_5	Ba–Ti–Mn–Si–O–OH.Na–SO_4	15- 71	
*	3.04_8	2.95_5	2.97x	4.25_4	2.44_3	2.24_3	2.74_2	2.13_2	K_2HPO_4	25- 639	1.10
i	3.04_3	2.95_3	2.73x	2.49_3	1.63_2	3.37_2	2.27_2	2.01_2	(Na,Al,Ca,Fe)Mn(PO)(OH)	25- 774	
i	3.03_8	2.95x	3.16_9	3.59_7	2.97_6	2.62_4	3.70_4	3.98_3	Y_2GeO_5	23-1484	
	3.03x	2.95x	2.11x	1.75_8	1.73_8	2.04_7	1.69_7	1.51_7	$Ba_3Tm_4O_9$	23- 59	
	3.02x	2.95x	3.09_3	5.50_2	2.97_2	2.66_2	4.43_2	2.80_2	$(YCeNdTh)(NbTiTa)_2O_6$	18- 765	
	3.01_6	2.95x	4.13_7	2.47_6	3.58_6	4.19_5	2.62_5	3.75_5	$CaZn_2(PO_4)_2$	20- 248	
*	3.01x	2.95x	3.79_6	3.66_5	4.80_5	2.48_4	2.56_4	2.19_4	$ScTaO_4$	24-1017	5.10
i	3.01x	2.95x	1.73_9	2.11_8	2.63_8	4.70_7	1.74_7	1.54_7	$(K_3Sb)8H$	4- 643	

in the search manual were arranged sequentially on 3.03, 3.04 and 3.05, the entries would (at 80 entries per page) be scattered over eight pages of this manual. However, with the group format, as seen in the figure, all 25 cases of 3.04 ± 0.01 with 2.96 ± 0.01 lie on the same page. This is a considerable convenience for the analyst and also helps to insure that none of the possible matches to the unknown pattern will be overlooked. All search systems published by the ICDD have adopted the "group" format.

An ingenious punched card system, designed by F. W. Matthews (1954) was formerly supplied by the JCPDS. This mechanical system made possible the utilization of other properties of the standard in addition to the *d*-values of its X-ray diffraction pattern. However, the difficulties of retrieval increased with the number of components in a multiphase unknown and also it seemed impracticable to include the data for more than 10,000 phases on a card.

Two situations which may occur in practice and which may make difficulties in the process of matching and retrieval of standard patterns from the PDF are: *First;* anomalies in intensities caused, for example, by preferred orientation. Deviations from the relative intensity characteristics of the X-ray diffraction pattern are also often found in electron diffraction patterns. With the objective of making the PDF usable for electron diffraction and making retrieval less dependent upon exact intensities, W. C. Bigelow and J. V. Smith (1965) designed a new search manual based upon multiple entries using the strong lines of the pattern in order of decreasing d-values rather than in order of decreasing intensities. They named this system the "Fink" index in honor of W. L. Fink, then Chairman of the JCPDS. This system has since been modified by introducing the "group" format and by making more use of intensities and is presently also available from the ICDD. *Second;* Difficulty in the process of retrieval of standard patterns may occur because of shifting d-value due to solid solution changes of the lattice constant. Dr. William L. Fink has designed a graphical system in which the diagrams of the eight strongest lines of the diffraction pattern are plotted on a horizontal logarithmic scale of d-values and intensities are shown on the vertical scale. Further work is underway on this Fink Graphical Index (1975).

Another attempt to solve the difficulties created by solid solutions was proposed by B. D. Sturman (1976) who used the three strongest lines to prepare a series of charts. The strongest line determined into which "group," i.e., the Δd range of d-values covered by each chart, the pattern would be located. The point locating the pattern on this chart is determined by the d-values of the second strongest and third strongest lines plotted as the x and y coordinates. Intermediate solid solutions then are located on a line connecting the end points representing the known phases. This concept also needs further study.

Within the limitations discussed above, experience shows that manual searching, when carried out systematically using the search manuals published by the ICDD, is successful in identifying any phase present, even in a complicated mixture of unknowns if its reference pattern is contained in the PDF. No additional information beyond the X-ray diffraction data is required. While for some cases the process requires only minutes, in more difficult cases the procedure may require an hour or more and can become quite tedious for the eye and mind. This problem can obviously be alleviated somewhat by using smaller search manuals for those cases in which the sample is known to contain only phases limited to certain fields, for example, only minerals, or only metals, or only ceramics, etc. Such "minifiles" and accompanying search manuals are available from the ICDD. One helpful search manual contains only the patterns for about 2500 of the more commonly occurring phases. Further developments can be expected and manual searching will continue to be used to advantage in many circumstances.

It would be expected that use of the computer can do much to eliminate the human tedium and risk of mistakes involved in manual search. While much remains to be done in computer search/match considerable success has been attained in the past 15 years and astonishing results and also reduction in costs can be anticipated in the future [Nichols, M. C. (1966); Johnson, G. G. and V. Vand (1967); Frevel, L. K. and C. E. Adams (1968); Johnson, G. G. (1974); Marquart, R. G. ėt al. (1979) and Nichols, M. C. (1980)].

References

Am. Cryst. Assoc. Sub-Com. L. D. Calvert, J. L. Flippen-Anderson, C. R. Hubbard, Q. C. Johnson, P. G. Lenhert, M. C. Nichols, W. Parrish, D. K. Smith, G. S. Smith, R. L. Snyder, and R. A. Young (1980). Nat'l Bur. Standards, Special Publication No. 567.
Bigelow, W. C. and Smith, J. V. (1965). A.S.T.M.T.P. **372,** 54.
Boldyev, A. K., Mikheev, V. I., Dubinina, V. N., and Kovalev, G. A. (1938). Ann. Inst. Mines Leningrad **11,** 1.
Davey, W. P. (1921). Phys. Rev. **17,** 549 and (1939) J. Appl. Phys. **10,** 820.
Debye, P. and Scherrer, P. (1916). Physik. Z. 17, 277 and (1917) **18,** 291.
DeWolff, P. M. (1948). Acta Cryst. **1,** 207.
Frevel, L. K. and Adams, C. E. (1968). Anal. Chem. **40,** 1335.
Fink, W. L. (1937). C. R. Acad. Sci. **204,** 1115 and (1939) Ann. Phys. **12,** 161.
Guinier, A. (1939). Ann. Phys. **12,** 161-237.
Hanawalt, J. D. and Rinn, H. W. (1935). Am. Phys. Soc. Bull. and (1936) Ind. Eng. Chem. Anal. Ed. **8,** 244.
Hanawalt, J. D., Rinn, H. W., and Frevel, L. K. (1938). Ind. Eng. Chem. Anal. Ed. **10,** 457.
Hanawalt, J. D. (1970). JCPDS, Swarthmore.
Hull, A. W. (1917). Phys. Rev. **10,** 661 and (1919). J. Am. Chem. Soc. **41,** 1168.
I.U.Cr. Commission, Kennard, O., Hanawalt, J. D., Wilson, A. J. C., DeWolff, P. M. and Frank-Kamenetsky, V. A. (1971). J. Appl. Cryst. **4,** 81.
Johnson, G. G. and Vand, V. (1967). Ind. Eng. Chem. **59,** 19.
Johnson, G. G. (1974). JCPDS, Swarthmore.
Marquart, R. G., Katsnelson, I., Heller, S. R., Johnson, G. G. Jr. and Jenkins, R. (1979). J. Appl. Cryst. **12,** 629.
Matthews, F. W. (1954). A.S.T.M.T.P. **197,** 34 and (1962) A.S.T.M. Mat. Res. Stds. **2,** 643.
Nichols, M. C. (1966). YCRK-70078, 19 and (1980) Norelco Reporter **27,** No. 1.
Parrish, W. (1948). Am. Mineral. **33,** 770 and (1965) X-ray Anal. Centrex Publ. Co. Eindhoven.
Smith, D. K. (1963). Lawr. Rad. Lab. UCRL-7196 and (1967) UCRL-50264.
Sturman, B. D. (1976). JCPDS, Swarthmore.
Thorez, G. L. (1975). B4820 Dison, Belgium
Waldo, A. W. (1935). Am. Mineral. **20,** 575.
Winchell, A. N. (1927). Am. Mineral. **12,** 261.

CHAPTER 3

AUTOMATIC SINGLE CRYSTAL DIFFRACTOMETRY IN THE U.S.A.

Thomas C. Furnas, Jr.
Molecular Data Corporation
Cleveland, Ohio 44118

Walter Bond of the Bell Telephone Laboratories in Murray Hill probably built the first automatic single crystal diffractometer in the United States. In the early 1950's he built a horizontal axis diffractometer to use in a Weissenberg geometry and incorporated search routines with a backset provision to scan slowly across angular regions in which the search routine had sensed significant intensity. It worked. (Bond, 1954; Bond 1955; Benedict, 1955).

During this time (1950) Dr. David Harker started the Protein Structure Project at the Polytechnic Institute of Brooklyn and gathered a small group of people to tackle the several fearsome aspects of protein structure determination. Not least among those was accurate and high speed data collection. Dave had a strong antithesis to photographic methods but a deep love for the reciprocal lattice and appreciation for the corresponding beauty of precession films. Therefore, the first efforts toward new instrumentation at the Protein Structure Project centered around servo chart followers and drive mechanisms to simulate the precession geometry and the "simplicity of scanning rows of reciprocal lattice points" drawn on a chart for the particular crystal species being studied.

His associate, Tom Furnas, nearing completion of his Ph.D. in Physics at MIT and who had had considerable experience producing precision charts, insisted there *had* to be a better way. His slightly different view of the reciprocal lattice had been nurtured by Professor Bertram Warren and his thesis description of testing monochromator crystals lacked only the words from being among the earliest examples of X-ray topography.

It is difficult to pinpoint the moment or exact manner of invention, but it was decided: 1) to scrap the charts, 2) to depend solely and exclusively upon calculated coordinates of reciprocal lattice points, and 3) to measure reflections only in the equatorial plane (indeed all possible reflections *could* be measured in the equatorial plane and with only one orientation or mounting of the specimen). It was Dave Harker who sensed that the proposed coordinate system was the ancient set of Euler's angles. This and other mechanical aspects of the resulting device which Dr. Furnas designed and built with William G. (Bridgie) Weber account for its becoming known as the Eulerian Cradle. (Furnas, 1954; Furnas and Harker, 1955).

Initial "automation" of this device was people powered–i.e., hand setting the angles read from an IBM printout of *hkl* and coordinate angles with columns for handwriting the observed intensities and backgrounds. Some operators were able to collect more than 100 reflections an hour in this manner and with several operators taking turns it went on for weeks and weeks.

By 1954 a second instrument using the Eulerian geometry was well under way at the Protein Structure Project. It was designed with full automation in mind and incorporated a full circle instead of a "cradle" and had other features to solve the problems encountered with the initial instrument which had been built on a GE XRD-3 table and spectrogoniometer.

In the early spring of 1954, Dr. Harker participated in a "diffraction school" at General Electric Company in Milwaukee, Wisconsin. Just before he left Milwaukee, he was shown a "new instrument" which GE had built and was about to begin to market for single crystal data collection. It was a normal incidence equatorial and upper level Weissenberg geometry–essentially a copy of an instrument built in Professor Zachariasen's laboratory at the University of Chicago. Dave described this to his associates at the Protein Structure Project the following Monday morning. While on the Long Island Railroad en route home that evening and the next morning returning to work, Tom Furnas fumed at this terrible mistake about to be "foisted onto the unsuspecting crystallographer!!!" On Tuesday he sought and obtained permission from Dr. Harker to design the kind of instrument that ought to be made instead. A complete and detailed set of drawings were completed and mailed special delivery to Milwaukee before the end of the week. The following week was the ACA meeting at Harvard. Armed with several duplicate sets of drawings, a concerted effort was expended to convince Howard Pickett and others from GE that they should scrap the device they had already built and start a different approach.

Dr. Furnas, as chairman of the ACA Apparatus and Standards Committee, called an emergency meeting to discuss the impending creation and availability of automatic single crystal diffractometers and the corresponding need for standardization of the dimensions of goniometer heads. The meeting was well attended and the response almost overwhelming. Yes, GE dropped their Weissenberg geometry, built the Single Crystal Orienter, then requested Dr. Furnas to write an instruction manual (Furnas, 1957). The data and experiments described in that manual were actually performed on the full circle instrument at the Protein Structure Project. At that time they could not have been done on any other instrument in the world as it alone had all the degrees of freedom under micrometer control for ascertaining and evaluating delicate alignment parameters, etc.

With cooperation from W. Parrish, a member

of the I.U.Cr. Commission on Apparatus and other members of that body, the standardization of goniometer heads was approved and recommended in 1956 (I.U.Cr. 1956). That same year the ACA Apparatus & Standards Committee published its first Index of Products & Suppliers for Crystallographers, etc. through the combined efforts of T.C. Furnas, L.E. Alexander and Karl E. Beu. Their next project involving dovetail tracks for easy interchangeability of accessories was considered too expensive by most manufacturers so has appeared only sporadically on commercially produced X-ray equipment. (It is interesting to note that a manufacturer of laser optical accessories independently arrived at exactly the same dovetail dimensions for its track and base so a full line of optical devices and tracks are available now.)

From the late 50's there were feverish efforts to automate the task of single crystal data collection, especially for neutron diffraction applications; such as by Melvin Mueller and his associates at Argonne National Laboratory and several groups working at Brookhaven National Laboratory (Prince and Abrahams, 1959; Langdon and Frazer, 1959). The automation of X-ray equipment was tackled by numerous individuals (most of whose efforts were never published), principally at large research laboratories (Abrahams, 1962); by a few commercial concerns such as DATEX (using the GE single crystal orienter); and Philips with a totally different 'step backward' approach with PAILRED (described elsewhere with operations at their Irvington-on-Hudson and Briarcliffe Manor Research Laboratories) (Ladell and Lowitzsch, 1960).

Meanwhile, in 1958, Dr. Furnas had been plucked from the Protein Structure Project by Picker X-ray Corporation in Cleveland, Ohio. From his unique understanding of the needs and desires of the crystallographic research community, he invented a whole new array of X-ray diffraction equipment: The diffractometer which supported its own X-ray tube; the x-y-z translation device under the focal spot of the X-ray tube for controlled precise alignment and invariance of alignment while changing take-off angle; the shutter mechanism which with his new X-ray tube envelope provided a leak-proof joint through an entire 10° change of take-off angle; the main gear drive with differential for both angle bisection and independent insertion of "omega"; the multiple input shafts for easy manual as well as fully automatic operation, etc.

In 1963, PAILRED was "demonstrated" at the Rome, Italy, I.U.Cr. Congress along with other instruments from European manufacturers eager to meet the crying demand for automated data collection. Picker had only the best research diffractometer and an artist's conceptual drawing of its full circle goniostat. Although not offered commercially, IBM displayed a fully closed-loop automated 4-circle diffractometer using the Picker base and the GE single crystal orienter (Cole, Okaya, and Chambers, 1963).

Picker had decided to enter the single crystal diffractometry field primarily because of requests from neutron diffractionists. The Picker diffractometer was the only one available that could be used easily in any orientation and also easily support the heavy neutron detectors with their required shielding. The first such installation was at Argonne National Laboratory where they did their own construction of the goniostat and automation (Mueller, Heaton and Sidhu, 1963).

Up to this time most single crystal data had been collected photographically. That permanent record gave the investigator a sense of security and confidence for it was easy to go back and look at the films again or to "see" that something was amiss in the first place. But diffractometry was changing this. There was no "permanent" record, the measurements were made "blind" and a new sense of high precision prevailed with the counting of individual X-ray quanta.

A variety of sources of error were known but their impact upon measurements by diffractometric techniques needed to be evaluated both toward maximum utilization of the apparent accuracy and precision achievable and toward determination of acceptable methods for data collection by fully automatic equipment. Therefore the ACA organized a symposium on "Accuracy in X-ray Intensity Measurement" as part of its winter meeting held at Suffern, New York in February, 1965.

Dr. Furnas had spoken on "Pitfalls in the Measurement of Integrated Intensities" at the Montreal I.U.Cr. Congress in 1957 but had never published the work. Acutely aware that much of his work was unpublished, he insisted that at least his talk at Suffern be taped. The outgrowth of that request was not only the taping of the entire symposium but the establishment of a new publication: *"Transactions of the American Crystallographic Association"* which contains valuable contributions many of which would never have been published except for the taping of the original presentation.

Picker soon thereafter introduced an IBM punched card controlled single crystal goniostat and diffractometer. The coordinate calculations were based primarily upon computer programs written by Busing and Levy (1966; 1967) of the Oak Ridge National Laboratory. It was not until 1967, however, that a truly computer-controlled American-built single crystal diffractometry system capable of collecting a complete three dimensional set of X-ray diffraction data from a single mounting of a single crystal was commercially unveiled. It happened at a remarkably significant ACA meeting in Atlanta, Georgia.

Nobel Laureate W. L. Bragg gave a memorable lecture on his involvement with protein structure determination over the preceding thirty-odd years. Hal Wyckoff of Yale introduced Dr. David Harker who described that he and his continuing group from the earlier Protein Structure Project had finally, after sixteen (16) years, solved the structure of ribonuclease. After that same sixteen (16) years, Dr. Furnas finally demonstrated his computer controlled FACS-I system. It was a delightful blend of foresight – computer-control of an instrument still capable of manual operation and using the Eulerian geometry which is easily plotted by hand and a powerful pedagogical tool with reciprocal space.

Laboratories all over the world obtained the FACS-I system. Production could not keep up with the demand so delivery times stretched out. Potential customers besought manufacturers all over the world for new computer controlled instruments and the manufacturers responded. So fierce was the clamor that one manufacturer, in its efforts to combine with Picker, revealed its estimate of the potential market – it was nearly ten (10X) times the independent estimates made by Picker! The combine never took place but the stage was set for the most lavish and sophisticated instrument exhibit ever held in conjunction with an I.U.Cr. Congress. It happened in Stony Brook, New York in September, 1969.

At least nine different companies including four from the United States (GE, Philips, Picker and Syntex) exhibited computer-controlled single crystal diffractometry systems. They were, with one exception, a rehash of Weissenberg or Eulerian geometries. That one exception was by Nonius who demonstrated a Precession geometry which they have called "Kappa" after their designation of an inclined axis. They claimed a simpler, less expensive construction and a set of computer programs (initially written by Y. Okaya of IBM) to do all the necessary manipulations. Another obvious trend was toward instruments that could be used under computer control *only*.

But just then the 'Glorious Research Budgets of the 1960's" evaporated!!!

It has been said that during the four or five months immediately after the Stony Brook meeting only one of those nine companies sold any and that one sold only three units. Grant requests were being refused or only partially funded. Then, in an apparent effort to trim inventories in the USA, Syntex offered its system for apparently "whatever sum of money you have." Nonius matched the offer with their CAD III system even with an offer to later replace it with their 'Kappa" system and for a while it seemed a battle of who would give the most extras, such as precession cameras or other odds and ends, without altering the price.

No longer was it the mechanical features of the equipment but instead, the fancy computer programs, color displays, time sharing of data reduction while data collecting and the like became the center of the sales pitches. Yes, these new "black boxes" can turn out a lot of data, but it seems that too few students learn the X-ray optics, get the "feel of an instrument" to know that it is operating properly or what needs to be done to fix it mechanically, or to get the "feel of a reflection". The pedagogical loss with respect to the future of instrumentation and an improved understanding of the influence of instrumental parameters upon the ultimate accuracy of data and what it means, is traumatic.

Drs. Jack Ball and T. Furnas of Picker Corporation shared an I.R 100 Award in 1970 for their development of a Dynamic Diffraction System using fiber optic image intensifiers and a television readout. But the "bubble had burst" and within two years most of those companies that had exhibited in 1969 had gone out of the automatic single crystal diffractometry business, if not out of the diffraction business entirely. In the United States, only Syntex and Nonius offered any such equipment for many years and even now the only other activity seems to be new automation to upgrade the older mechanical systems or the sister activity of automated powder diffractometry systems.

Few of the developments since the very early 1970's have been commercially funded. Individuals in their respective laboratories have developed those accessories which they needed for their own research. Some are particularly exciting: for example, large area detectors especially useful for protein structure work (C. Cork, et al., 1973; R. Hamlin, et al., 1981), high temperature accessories (F. Lissalde, et al., 1978), low temperature accessories (S. Samson, et al., 1980), rotating anode sources (W. Massey, et al., 1976), etc.

Probably the most exciting new area of activity concerns the application of synchrotron X-ray sources (Philips, et al., 1980) regarding which a whole new flood of instruments, techniques, methods, papers and discoveries are sure to come along.

But let us not lose sight of some realities of the different times in which we live today. During the past three decades, hundreds of laboratories throughout the world have experienced and blossomed through their multiplied effectiveness at problem solving made possible by the contributions of relatively few dedicated and skilled artisans of instrument design nurtured (at least temporarily) by a free commercial market. Modern instruments are generally expensive in time and effort as well as in capital equipment and funds to develop, produce and maintain regardless of the value of money *per se*. It is only through the production of many identical items and their purchase by many different customers that high quality instruments can be produced at a reasonable cost in time, effort and money. We seem to have entered another period in which too many good research and results-oriented people are having to spend their time solving instrumental problems for which they are ill fitted (one side effect of the "black box" at educational institutions). Simultaneously, our promising instrument designers are trapped in one-of-a-kind situations with inadequate staff and funds. They are essentially unable to really help any other institution or laboratory to build or acquire the instruments which would appreciably improve their research capability.

It is a national disaster that most of the developments of this past decade, because they have been made at many independent laboratories primarily concerned with solving their own particular research problems, are suffering an involuntary entrapment and have been made at much greater *actual* cost than anyone is willing to admit. To achieve better utilization of our limited research funds, it is imperative that we all join together to devise ways in which good instrumentation can be made available to more laboratories and the talents there put to their most effective use.

References

Abrahams, S. C. (1962) Rev. Sci. Instrum. **33**, 973-977.

Ball, J. and Furnas, T. C. Jr. (1970) Am. Crystallogr. Assoc. Program Abstr. Ser. 2, Page 69.

Ball, J. and Furnas, T. C. Jr. (1970) Industrial Research 12 (December), 43.

Benedict, T. S. (1955) Acta Cryst. **8**, 747-752.

Bond, W. L. (1954) Acta Cryst. **7**, 620-621.

Bond, W. L. (1955). Acta Cryst. **8**, 741-746.

Busing, W. R. and Levy, H. A. (1966) Angle Calculations for 3-and 4-circle X-ray and Neutron Diffractometers, ORNL-4054, Oak Ridge National Laboratory, Tenn.

Busing, W. R. and Levy, H. A. (1967) Acta Cryst. **22**, 457-464.

Cole, H., Okaya, Y., and Chambers, F. W. (1963) Rev. Sci. Instrum. **34**, 872-876.

Cork, C., Fehr, D., Hamlin, R., Vernon, W., and Xuong, N. h., (1974). J. Appl. Cryst. **7**, 319-323.

Furnas, T. C. Jr. (1954). Acta. Cryst. **7**, 620.

Furnas, T. C. Jr. and Harker, D. (1955). Rev. Sci. Instrum. **26**, 449-453.

Furnas, T. C. Jr. (1957) Single Crystal Orienter Instruction Manual, Milwaukee: General Electric Company.

Furnas, T. C. Jr. (1957) Acta Cryst. **10**, 744-745.

Furnas, T. C. Jr. (1965) Trans. Am. Crystallogr. Assoc. **1**, 67-85.

Hamlin, R., Cork, C., Howard, A., Nielsen, C., Vernon, W., Matthews, D., and Xuong, N. h., (1981) J. Appl. Cryst. **14**, 85-93.

I.U.Cr. Commission on Crystallographic Apparatus (1956) Acta Cryst. **9**, 976.

Ladell, J. and Lowitzsch, K. (1960). Acta Cryst. **13**, 205-215.

Langdon, F., and Frazer, B. C. (1959) Rev. Sci. Instrum. **30**, 997-1003.

Lissalde, F., Abrahams, S. C., and Bernstein, J. L. (1978) J. Appl. Cryst. **11**, 31-34.

Massey, W. R. Jr., and Manor, P. C. (1976) J. Appl. Cryst. **9**, 119-125.

Mueller, M. H., Heaton, L., and Sidhu, S. S. (1963) Rev. Sci. Instrum. **34**, 74-76.

Philips, J. C., Cerino, J. A., Hodgson, K. O. (1979) J. Appl. Cryst. **12**, 592-600.

Prince, E. and Abrahams, S. C. (1959). Rev. Sci. Instrum. **30**, 581-585.

Samson, S., Goldish, E. and Dick, C. J. (1980). J. Appl. Cryst. **13**, 425-432.

CHAPTER 4

COMPUTING AIDS PRIOR TO THE USE OF DIGITAL COMPUTERS

Dan McLachlan, Jr.
Department of Geology & Mineralogy, Ohio State University
Columbus, OH 43210

The early work on structure determination was confined pretty much to the simple salts and minerals both in Europe and America but the applications of Fourier series to the computation of electron densities made possible the investigation of more complex inorganic structures as well as organic structures. Perhaps the first publications involving Fourier series was by Duane (1925) and Compton (1926). The first really difficult and detailed inorganic structure was that of diopside started by Warren and Bragg (1928) and published in detail by Bragg (1929). This paper by Bragg in which he depicted the electron density projected on planes through the crystal added understanding as well as importance to research on structure determination, but foretold some momentous problems to be faced by crystallographers. Among these are the measurement of accurate intensities, the phase problem and the time-consuming work of making the appropriate computations.

The first computing aids were the so called "strip methods" for handling the data. Beevers and Lipson (1934) produced a kit of strips that were used extensively in America and Europe. This was followed by the strips of Patterson and Tunell (1942). J. Monteath Robertson (1936) developed a set of strips which were named the "multiple slide rule" by Christ Torely, Cisney, Champaygne, and McLachlan who developed the same scheme (for the study of penicillin), independently.

The very important role that punched cards played in structure studies is supplied by the comments of E. W. Hughes : "At Caltech, in 1940, when an application was made for funds to build a mechanical Fourier machine, one of the referees, Dr. Eckert, suggested that punched cards could do the job better. As a result the International Business Machines Corporation granted Professor Pauling's group rent-free use, for about six years, of a key punch, a sorter, and a tabulator equipped with the maximum number of added controls."

Punched card sets were at once prepared for Fourier series summation of crystal diffraction data and for integrating Fourier integrals of electron diffraction data from gases (Shaffer, Schomaker, and Pauling, 1941). After a few years experience even faster card systems were developed. It was also found possible to adapt the machines to other types of calculation. Thus in the first least squares refinement of a crystal structure (Hughes, 1941) the reduction of the observational equations to normal equations was carried out by punched cards. Also structure factor calculations became possible. After the expiration of the rent-free grant, a more elaborate set of I.B.M. machines was rented by the Institute for a central computing facility; in addition to the machines listed above there were also a collator and an electronic multiplying punch. Dr. Verner Schomaker designed card sets for Fourier series summation which produced, for average sized jobs, 3-D series in about fifteen hours and 2-D series in about two hours. These machines served for several years until high speed digital computers became available. All told, punched card methods served this very active research group for about fifteen years.

Beginning about 1940, many crystallographers turned their attention to analogue computers, machines that did more than just handle the data; they actually carried out the mathematical procedures automatically. The time was ripe for such endeavors since, for many decades, classical physicists had not only been writing equations for every natural phenomenon they could understand, but they also were inventing machines that obeyed the same equations. A very good summary of such analogies is to be found in an 800 page book by William Hort (1922). A more recent book along the same lines is by Harry Falson in 1942 under the title *"Dynamical Analogies."* Like Hort's book it covers electrical, and mechanical analogies. Perhaps the greatest excitement in America derived from accomplishments in designing mechanical computers was a machine by Vannevar Bush (1931a). A book by an American author, Dayton C. Miller (1937) on *"The Science of Musical Sounds"* contains many beautiful analogue mechanisms for the portrayal of Fourier series including the famous machine of rolling sphera by Henrici (1894). The analogue machine that created the greatest sensation in America was "The Differential Analyzer, A New Machine for Solving Differential Equations" by Vannevar Bush (1931b).

With such a vast background of literature on analogies, the Americans were well prepared to apply them to crystallographic problems. McLachlan (1957) devoted fifty-five pages of his book to computing aids in structure determination and Fred Ordway (1950) described computers for crystallographers in Dr. Pepinsky's book.

One of the most interesting and instructive optical devices was Bragg's (1939) concept of the X-ray microscope or "Fly's Eye" based on the theoretical analysis of the ordinary light microscope by Abbé (1910). Over a period of ten years Buerger performed many beautiful experiments along the same lines [see for example Buerger (1950) and also Taylor and Lipson (1965)]. Among the first optical computers was Bragg's (1929) masks which consisted of a series of transparent films with dark and light streaks of sinusoidal density, each corresponding to a value of h and k. By successive exposures of these masks on print paper, a Fourier summation could be made. Huggins (1941) tried to put the Bragg masks on a commercial basis and then D. McLachlan and R. H. Wooley (1951) produced an optical set-up by which a hundred of the Bragg masks could be projected simultaneously.

Among optical computers, one that deserves mention is J. M. Robertson's (1943) method of getting theoretical Patterson projections. It was further developed by Philips and McLachlan (1954) to make shifted Patterson products.

Shimizu, Elsey and McLachlan (1950) devised an electrical analog computer which they called "Utah Computer." It made use of over two hundred selsyn motors operating together by shafts and gears in a manner suggested by A. D. Booth (1948) in his book. This machine was convenient to use and in trial runs produced beautiful projections of diopside and hexamethylbenzene for example.

It is impossible to describe all the American efforts to reduce the work associated with X-ray computations but we can mention the work of L. Alexander (1953), H. E. Montgomery (1938), J. Donohue and V. Schomaker (1949), E. G. Cox, G. L. Cross and G. A. Jeffrey (1949) and S. W. Mayer and K. N. Trueblood (1953).

While most analog computers can be classified as either mechanical, optical, or electrical, one was produced at the American Cyanamid Company by McLachlan and Champaygne (1946) that defies classification. It was called the "Sand Machine." It was made to conveniently incorporate the Miller indices, the F values, the phases and the planar angles on a two dimensional projection; but dwelling upon it here would be as superfluous as putting into a modern book of anatomy the details of the structure of an ancient extinct mammal.

The ultimate climax of all analogue computers for X-ray computations was devised and built by Ray Pepinsky (1947). Much has been written about this large machine. It would do almost anything desired for crystal structure determinations and attracted to Pennsylvania State University droves of guests, visiting professors and graduate students, for learning and research from the United States, Europe and the rest of the world. No attempt will be made to describe it here. A comprehensive discussion of it is to be found in the next chapter.

While the crystallographers were inventing and using analogue computers, great strides were being made in the development of electronic computers, particularly the digital computers based on the binary system of numbers. In a lecture by D. R. Hartree (1947) "Recent Developments in Calculating Machines" the ENIAC (Electronic Numerical Integrator and Calculator) is described in part. This machine by J. P. Eckert and J. W. Mauchly was America's greatest triumph in the field up to this time and started the age of digital computers. The crystallographers adapted this computer, and their use has made possible structure determinations of such complexity that without them the solutions would be impossible. The discussion of digital computers is to be discussed in a later chapter by Sparks and Trueblood.

References

Abbé, Ernst (1910), See book by O. Lummer and F. Reiche "Die Lehre von der Bildentschung in im Microscope von Ernst Abbe," Vieweg-Vertag, Brunswick, Germany (1910).
Alexander, L. (1953) Acta Cryst. **6**, 727-731.
Beevers, C. A. and Lipson, H. (1934). Phil. Mag. **17**, 855-859.
Booth, A. D. (1948) Fourier Techniques in X-ray Organic Structure Analyses. New York: Cambridge University Press, N.Y.
Bragg, W. L. (1929 a) Z. Kristallogr. **A70**, 475-492.
Bragg, W. L. (1929 b) Proc. Roy. Soc. (London) **A123**, 537-559.
Bragg, W. L. (1939) Nature (London) **143**, 678.
Buerger, M. J. (1950) Proc. Natl. Acad. Sci. U.S.A. **36**, 330-335.
Bush, V. (1931 a) J. Franklin Inst. **212**, 447.
Bush, V. (1931 b) Bull. Am. Math. Soc. **42**, 649-699.
Compton, A. H. (1926) X-rays and Electrons, New York: D. Van Nostrand Co., Inc. p. 151.
Cox, E. G., Cross, G. L. and Jeffrey, G. A. (1949). Acta Cryst. **2**, 251-255.
Donohue, J. and Schomaker, V. (1949) Acta Cryst. **2**, 344-347.
Duane, W. (1925) Proc. Natl. Acad. Sci. U.S.A. **11**, 489.
Hartree, D. R. (1947). J. Sci. Instrum. **24**, 172-176.
Henrici, O. (1894) Phil. Mag. **38**, 110.
Hort, W. (1922) Technische Schwingungslehre. Berlin: Verlag von Julius Springer.
Huggins, M. L., (1941) J. Am. Chem. Soc. **63**, 66-69.
Hughes, E. W. (1941) J. Am. Chem. Soc. **63**, 1737.
Mayer, S. W. and Trueblood, K. N. (1953). Acta Cryst. **6**, 427.
McLachlan, D. and Wooley, R. W. (1951) Rev. Sci. Instrum. **22**, 423-427.
McLachlan, D. and Champaygne, E. F. (1946) J. Appl. Phys. **17**, 1006-1014.
McLachlan, D. (1957) "X-ray Crystal Structure". New York: McGraw-Hill.
Miller, Dayton C. (1937), "The Science of Musical Sounds," New York: McMillan Company.
Montgomery, H. E. (1938) Bell Sys. Tech. Jour. **17**, 406-415.
Olsen, Harry F. (1943) Dynamical Analogies, New York: D. Van Nostrand.
Ordway, Fred (1950) In "Computing Methods and the Phase Problem," edited by R. Pepinsky. Penn State University Press.
Patterson, A. L. and Tunell, G. (1942), Am. Mineral. **27**, 655-679.
Pepinsky, R. (1947) J. Appl. Phys. **18**, 1001.
Philips, W. R. and McLachlan, D. (1954). Rev. Sci. Instrum. **25**, 123-128.
Robertson, J. M. (1936) Phil. Mag. **4**, 176-187.
Robertson, J. M. (1943) Nature (London) **152**, 411-412.
Shaffer, L. B., Schomaker, V. and Pauling, L. (1941). J. Chem. Phys. **14**, 648-659.
Shimizu, H. P., Elsey, P. J. and McLachlan, D. (1950) Rev. Sci. Instrum. **21**, 779-783.
Taylor, C. A. and Lipson, H. (1964), Optical Transforms. London: Bell and Sons Ltd.
Warren, B. and Bragg, W. L. (1928) Z. Kristallogr. **69**, 168-193.

CHAPTER 5

X-RAC AND S-FAC
(The Inventions of Ray Pepinsky)

Editorial note: Due to circumstances unforseen by the compilers of the chapters in this history book, Dr. Pepinsky was not able to get this chapter to the printers in time for inclusion in this volume. In view of the great contributions which have been made by Dr. Pepinsky to the progress of crystal structure determinations and also in view of the impact that his two major inventions, the X-RAC and the S-FAC have had upon the world-wide effort to speed up structure computations, we the editors could not face the thought of publishing a book without some recognition of his worthy contributions. We therefore are inserting the following paragraphs on this subject.

X-RAC was an analog computer used to calculate structure factors and Fourier (electron density or Patterson) maps. The results of the computation were displayed in contours on a cathode ray tube. It was possible, for example, for centrosymmetric structures, to flip the signs of certain reflections, and this was often done to get rid of very negative regions in electron density maps. Numerous crystallographers from all over North America came to use this computer which was originally at Auburn, Alabama, and then at Penn State University.

This famous computer, designed by Raymond Pepinsky and built by him, was first introduced to the public in a paper before the American Society for X-ray and Electron Diffraction, at Lake George, New York, June 11, 1946, and was soon followed by a publication "An Electronic Computer for X-ray Crystal Structure Analysis" in the *Journal of Applied Physics*, Vol. *18*, July 1947, pages 601-604. Recognition from abroad soon came from all sections of the world through published papers and references in books, and soon Ray's laboratory at Penn State University was swarming with visitors and people eager to come with their problems and to work with him.

In 1950 the first meeting of the merged societies, ASXRED and CSA to form ACA was held at Penn State and many of the American members saw fit to come early to attend a special meeting arranged by Ray Pepinsky April 6-8, 1950 on "Computing Methods and the Phase Problem." This was an especially impressive event partially because of the computers Ray had on display and fully in operation but also because of the rank and fame of the audience and speakers. Almost all noteworthy U.S. and Canadian crystallographers attended as well as many from overseas (several of whom were already there working with Ray.)

This meeting was reported in a book edited by Ray Pepinsky, *"Computing Methods and the Phase Problem,"* published by the X-ray Crystal Analysis Laboratory, Department of Physics, The Pennsylvania State College, State College, Pennsylvania in 1952. This 398 page book contains 166 pages of the texts presented by the twenty one authors at the meeting, a 116 page description of X-RAC and S-FAC and nine appendices (pages 283-361) largely co-authored by Pepinsky and related to X-RAC and S-FAC. The final appendix is a 36 page paper by David Sayre as "Fourier Transforms."

The importance of this meeting is partly borne out by the caliber of the people who attended, some of whose names are given below:

Introductory remarks by Mina Reese, Head of the Mathematical Sciences Division of the Office of Naval Research, Washington, were followed by papers by J. M. Bijvoet of Utrecht; P. P. Ewald of Brooklyn; C. A. Beevers of Edinburgh; A. L. Patterson of Philadelphia; M. J. Buerger of Cambridge, Massachusetts; C. H. MacGillavry of Amsterdam; J. A. Goedkoop of State College, Pennsylvania; B. Friedman of New York; F. R. Ahmed of London; M. Born of Edinburgh, W. H. Zachariasen of Chicago; J. M. Robertson of Glasgow; C. W. Bunn of Imperial Chemical Industries; C. A. Taylor, D. Rogers, I. W. Ramsay and H. Lipson of Manchester; E. G. Cox of Leeds; E. W. Hughes and V. Schomaker of Pasadena; F. Ordway of Washington; M. Perutz of Cambridge, England; and R. Pepinsky of State College, Pennsylvania. Several other leading crystallographers took part in the discussions. The conference was reported briefly by Prof. Ewald in *Acta Cryst.* **3**, 401-2 (1950).

It is impossible to cite all the literature in which these machines have been discussed, but we can mention that Dan McLachlan, Jr. thought enough of these achievements to devote pages 306-315 of his book *"X-ray Crystal Structure,"* McGraw-Hill Book Co., N. Y., Toronto London, (1957) to this subject.

A book that cannot fail to impress readers of history in North America is *"Computing Methods and the Phase Problem in X-ray Crystal Analysis"* (Report of a conference held in Glasgow in 1960) edited by Ray Pepinsky, J. M. Robertson and J. C. Speakman, Pergamon Press, Oxford, London, New York and Paris (1961). There were twenty-eight papers presented at this important conference of 114 participants and a fine group picture was taken of 98 of them. Speakers that would be of particular interest to crystallographers of North America and elsewhere are listed as follows:

A. Niggli, J. M. Robertson, G. A. Jeffrey, L. H. Jensen, D. W. J. Cruickshank, M. Wells, R. A. Jacobson, B. R. Penfold, W. N. Lipscomb, W. G. Sly, D. P. Shoemaker, W. R. Busing, H. Levy, D. H. Templeton, Q. C. Johnson, R. Pepinsky, J. van den Hende, V. Vand, R. A. Sparks, K. N. Trueblood, H. Hauptman, W. Cochran, M. M. Woolfson, J. C. Kendrew and M. G. Rossman. Pepinsky co-authored two of the papers himself.

This is an important book historically not only because it summarizes what was known among leading crystallographers of the time but it also pointed strongly to the prevalent trend for years to come in digital computers which were destined to replace all forms of analogue computers. IBM alone was involved in the computation referred to in seven chapters; and several other computers on the market were referred to.

This short chapter concludes what the editors have to say about Dr. Raymond Pepinsky's role in this important era of history of *Crystallography in North America.*

CHAPTER 6

DIGITAL COMPUTERS IN CRYSTALLOGRAPHY

Robert A. Sparks (*) and Kenneth N. Trueblood (**)
(*) California Scientific Systems, 574 Weddell Drive,
Sunnyvale, CA 94086
(**) Department of Chemistry, University of California,
Los Angeles, CA 90024

Early Computers

Digital computers have revolutionized the practice of crystallography in the last thirty years. Early computer development took place in England and the United States in the late 1940's and early 1950's. Almost as soon as computers first became available, they were used by crystallographers. The earliest repeated uses were by Ordway (1952) with the National Bureau of Standards Eastern Automatic Computer (SEAC) at NBS in Washington, by Bennett and Kendrew (1952) with the Electronic Delay Storage Automatic Calculator (EDSAC) at Cambridge, by Mayer and Trueblood (1953) with the National Bureau of Standards Western Automatic Computer (SWAC) at UCLA, and by Ahmed and Cruickshank (1953), with the Ferranti Mark I computer at Manchester University. These computers and all those that followed were of the stored-program type; both the data and the computer program were stored in a high-speed memory that could be modified under program control. Predecessors, such as the IBM or Hollerith card tabulator, did not have that feature. Thus, the tabulator could add many columns of numbers with control supplied by a wired plugboard. The IBM Card Programmed Calculator (CPC) used, in addition to a plugboard, IBM cards (one computer instruction per card) for its program. These cards had to be continuously fed through the card reader. The CPC also had the capability of multiplying numbers together.

Early users of stored-program computers were impressed with their vast improvement in speed over these predecessors. Thus, a UCLA student, John Bryden, spent six months calculating a three-dimensional Fourier synthesis by the IBM-card method (Shaffer, Schomaker and Pauling, 1946) with an IBM card tabulator, IBM card reproducer and IBM card sorter. On SWAC a large three-dimensional Fourier could be done in a few hours. The calculation of all of the (sin theta)/lambda values for the CuKalpha sphere for potassium chlorate took all night on the CPC. The same calculation on SWAC took 20 minutes.

The Ferranti Mark I computer had a fast electrostatic memory of 256 40-bit words and a magnetic drum memory that held an additional 16,384 words. It also had seven index registers (called B-lines in England). Addition or substraction took 1.2 msec, multiplications 2.2 msec. Input and output was on punched paper tape.

SWAC had 256 37-bit words stored electrostatically on the face of cathode-ray tubes (Williams tubes) and an additional storage of 8192 words on a magnetic drum. Additions and subtractions took 64 microseconds and multiplications 384 microseconds. There was no divide instruction. Input and output were on punched cards. SWAC was a four-address computer. Each instruction took its operands from addresses alpha and beta, placed its results in address gamma (and sometimes delta) and took its next instruction either from the next instruction in sequence or from that given in address delta. There were thirteen different instructions. Programs were written in machine language. Addresses were specified as 000-255 (or 00-zz in hexadecimal) and instruction types as 00-15 (or 0-z hexadecimal). Alphanumeric variable names and instruction mnemonics were never used.

SWAC, like all of the early computers, did all of its calculations in fixed-point arithmetic. The programmer had to keep track of the scale factor for all of the variables in his programs and was constantly worried about overflow or loss of significance.

This problem became so difficult for some types of programs that, at the expense of speed, a floating-point interpreter package of subroutines was written for SWAC. The programmer and user of SWAC was also the computer operator and sometimes the maintenance engineer. Crystallographers, who had some of the longest running programs, usually used the computer at night and had to know how to test and replace tubes, replace chasses, and lift and test diodes.

SWAC had an official mean error time of twenty minutes (many users thought this a gross overestimate). Some programs would occasionally sum those parts of memory that should not change to determine when errors were occurring. The technique sometimes used by crystallographers was to do every calculation at least twice, or until the results were identical on two separate runs. Programmers also became proficient at "programming around" hardware bugs that were hard to find.

One of the greatest triumphs of early computers was SWAC's contribution to the solution of the structure of the hexacarboxylic acid derivative of vitamin B_{12} (Hodgkin et al., 1955). Structure factor and three-dimensional Fourier calculations for the derivative were done on SWAC in 1954 and 1955; the excitement of the unraveling of the derivative structure is recounted by Trueblood in the Festschrift for Dorothy Hodgkin (1980). Later block-diagonal least-squares calculations for the derivative were done on SWAC, each cycle taking about 24 hours of computer time (Hodgkin et al., 1962).

Only one SEAC, one EDSAC and one SWAC were built. A second Ferranti Mark I was used by Ahmed and Barnes (1958) at the University of Toronto. With the placement of many IBM 650 computers in the United States and of many Ferranti Pegasus and Mercury computers in England in the late 1950's, crystallographic groups began to share programs (Jeffrey et al., 1961; Cruickshank, 1961; Mills and Rollett, 1961). Even though these computers were slower than SWAC, this ability to share programs soon overshadowed anything that could be done on one-of-a-kind computers simply because the writing and debugging of programs was such a very time-consuming task.

Large Computers

Computers in the late 1950's and early 1960's improved in many ways. Floating-point hardware, large (32,768 words) magnetic core memories, magnetic tape drives, magnetic disk drives, data channels for peripheral equipment, large versatile instruction sets, and higher speeds (e.g., a 12-microsecond machine cycle for the IBM 709) were now available to the crystallographer.

Software also improved dramatically. Assembly languages with alphanumeric variable names, mnemonics for instruction codes, macro-assembly and cross-assembly capability became available on all of the commercial computers. The most significant development was the use of higher level languages. FORTRAN, BASIC, ALGOL (mostly in Europe) and many other languages have been used by crystallographers. The first of the higher-level languages, FORTRAN, has become the universal language of crystallographic computing. It was developed by IBM Corporation in 1958, with a crystallographer, David Sayre, on the original team. FORTRAN programs were easier to write and debug than assembly language programs, but early programs written in FORTRAN were not as efficient of computer time as those written in assembly language by good programmers. By the late 1970's however, FORTRAN compilers had improved to the point where they generated very efficient code. Unfortunately there were enough differences in different manufacturers' FORTRAN compilers than FORTRAN programs that would run on one type of computer would often not run on another. For this reason, Stewart (1970) used "PIDGIN" FORTRAN (which could run on many different computers) when writing the crystallographic program package XRAY67. As a result, XRAY67 and its later versions became the most widely used general crystallographic software package.

In the 1960's hardware improvements continued to be made. The transistor replaced the vacuum tube. As a result the computer became faster and less expensive. It cost less to lease an IBM 7040 computer than to maintain the much less powerful IBM 650. The vacuum tube computers quickly became obsolete. Integrated circuits soon replaced individual transistors, again making computers less expensive. With the introduction of the IBM 360 series a complete range of computers from the least powerful and least expensive to the most powerful and most expensive had compatible instruction sets. By the end of the 1960's, almost every crystallographer in the United States could do calculations on a central computer at her (his) university, laboratory, or industrial establishment. Most of the computers used by crystallographers in the United States were manufactured by International Business Machines Corporation, Control Data Corporation, or UNIVAC Corporation.

Widely used programs other than XRAY67 were the least squares routines UCLASFLS (Gantzel, Sparks, and Trueblood, 1962) and ORFLS (Busing, Martin and Levy, 1962), the Fourier routine of Shoemaker and Sly (1961), the direct-methods programs REL (Long, 1965), MAGIC and MAGIA (Dewar, 1970) and MULTAN (Germain and Woolfson, 1968), a superposition program (Gorres and Jacobsen, 1964), a vector search program (Nordman, 1966), and a rigid-body thermal motion analysis program (Schomaker and Trueblood, 1968). For plotting of crystallographic structures ORTEP (Johnson, 1965) has been used almost universally, although recently a few other plotting programs have also been made available. The Cambridge Crystallographic Data File (Motherwell, 1976) of published structures could be searched on many central facilities and could be accessed by telephone link through the Chemical Information Service.

Minicomputers

When the minicomputer was introduced it was soon used to control single crystal diffractometers. The first computer-controlled diffractometers were built by Cole, Okaya and Chambers (1963) and Busing, Ellison and Levy (1965). The programs written by Busing and Levy (1968) were the models that were used by several commercial diffractometer manufacturers. Writing programs for the early minicomputers was much like writing programs for the earliest digital computers. Programs had to fit in a small amount of memory (4096 words) and had to be written in assembly language. FORTRAN compilers, more memory, floating-point processors, and all of the big computer peripherals became available for minicomputers in the early 1970s. New algorithms (Sparks, 1970) made it possible to mount a single crystal in an arbitrary orientation and within one or two hours obtain lattice parameters and an orientation matrix. Data collection then took typically one day to several weeks and was completely automatic. Minicomputers were also used to automate

X-ray film readers (Abrahamsson, 1966; Sparks, 1973a; Arndt, Gilmore and Wonacott, 1977). The most widely used minicomputers for crystallography were manufactured by Digital Equipment Corporation and Data General Corporation.

Syntex Analytical Instruments (Sparks, 1973b) introduced the XTL Structure Determination System, which was a set of software and hardware that made it possible to solve and refine crystal structures with up to 500 independent parameters on the same minicomputer that collected the data. In most places the dedicated minicomputer proved to provide better service and to be more economical than the large central computer facility. It was a great boon to crystallographers in those parts of the world where central facilities were not readily available. By 1977 (Sparks, 1977) it was estimated that about one fourth of all crystallographic calculations were being done on minicomputers; today the proportion is much higher. Okaya (1966), working at IBM Research, solved more than 24 structures in two years – a record at that time. In the year 1977, Struchkov (1978) in Moscow solved more than 100 structures using two computer-controlled diffractometers and one dedicated XTL Structure Determination System.

Because of the shortness of the word length (typically 16 bits) and the small size of address space (typically 32,768 words) of the minicomputer, a great deal of effort was required to take programs written for large computers and make them run on a minicomputer. For this reason many crystallographic groups (especially protein groups) with a large investment in crystallographic programs were initially reluctant to use a minicomputer. Recently the Digital Equipment Corporation introduced the VAX minicomputer with a 32-bit word length. FORTRAN programs written for large computers can easily be made to run on the VAX and as a result many groups are now beginning to use this computer.

Computing for Macromolecules

Protein and other macromolecular structures present problems not normally encountered with smaller structures. Special data-collection algorithms have been written (Sparks, 1980). Computer-controlled position-sensitive detectors have been built (Cork et al., 1975; Arndt, 1977; Abrahamsson, 1972) to collect many reflections simultaneously. Interactive three-dimensional computer displays are now used to help build molecular models to match electron density distributions (Morimoto and Meyer, 1976; Barry et al., 1975; Feldmann et al., 1974). Special Patterson-search algorithms (Rossman and Blow, 1962), isomorphous replacement and anomalous scattering programs (Kartha, 1970) and a three-dimensional Fast Fourier Transform program (Ten Eyck, 1977) have been developed.

Although approximations to the full-matrix least squares (FMLS) refinement technique were developed fairly early (Sparks, 1961; Rollett, 1965; Scheringer, 1963; Waser, 1963; Schomaker and Waser, 1958), most crystallographers preferred the FMLS method for small structures. Because the number of parameters in a macromolecule is so large, the FMLS method is impossible to use, requiring large amounts of memory and vast amounts of computer time on even the most powerful computer. Thus special refinement methods for macromolecules were developed (Diamond, 1971; Watenpaugh, 1973; Agarwal, 1978; Hendrickson and Konnert, 1980; Konnert, 1976). Macromolecular structure groups have not used shared programs as much as have small-molecule groups. Recently, however, Munn and Stewart (1980) have developed RATMAC, a structured FORTRAN preprocessor that can produce standard FORTRAN and that can run on any local computer with sufficient memory capacity. Under the auspices of the National Resource for Computation in Chemistry, Stewart and several protein crystallographers have cooperated to write in RATMAC a Multiple Isomorphous Replacement Phasing program (reported on by Watenpaugh, 1980) that promises to be used by many groups.

Computing for Powder Diffraction

Most powder diffractometers have not traditionally been computer-controlled, but because of the availability of low cost modern mini- and microcomputers, computer control has become popular in the last few years. Programs have been written to search the files of the Joint Committee for Powder Diffraction Standards (JCPDS) McCarthy and Johnson, 1979; Nichols, 1966; Frevel, 1977). Programs for the decomposition of powder patterns by profile analysis have been developed (Taupin, 1973; Huang and Parrish, 1978; Sparks, 1981). Powder patterns can be indexed (Visser, 1969; Werner, 1976; Shirley, 1978). Atomic parameters can be refined by the Rietveld method (Rietveld, 1969; Young, Mackie and von Dreele, 1977). Most of these techniques are now available on minicomputers (Parrish, Ayers and Huang, 1980; Sparks, 1981), as well as on large computers.

Recent Trends

Computer development has continued to increase the power of the computer while decreasing its cost. Large Scale Integrated (LSI) circuitry has produced the microprocessor and microcomputer ("computer on a chip"). These are now found in the California Scientific Systems and Nicolet diffractometers. The array processor (such as that available from Floating Point Systems) has not yet been widely used by crystallographers but promises to achieve computer speeds for mini- and microcomputers comparable to those for large computers (Furey, Wang and Sax, 1980).

Conclusion and Apology

Digital computers have had a revolutionary impact on crystallography. Hundreds of crystallographers have contributed to this revolution by developing algorithms, software and hardware for many different computers. The foregoing history of digital computers in crystallography, with emphasis on developments in the United States, is a personal and limited report, based in significant part on the experience of the authors. Doubtless many important contributions have been omitted, through a combination of our sheer ignorance, poor memories, and limited time to review the literature.

References

Abrahamson, S. (1966). J. Sci. Instrum. **43**, 931.
Abrahamsson, S. (1972) Acta Cryst. **A28**, S 248.
Agarwal, R. C. (1978) Acta Cryst. **A34**, 791.
Ahmed, F. R. and Barnes, W. H. (1958) Acta Cryst. **11**, 669.
Ahmed, F. R. and Cruickshank, D. W. J. (1953) Acta Cryst. **6**, 765.
Arndt, U. W. (1977) In "The Rotation Method in Crystallography", edited by U. W. Arndt and A. J. Wonacott. Chapter 17. Amsterdam: North-Holland.
Arndt, U. W., Gilmore, D. J. and Wonacott, A. J. (1977) In "The Rotation Method in Crystallography", edited by U. W. Arndt and A. J. Wonacott. Chapter 14. Amsterdam: North-Holland.
Barry, C. D., Bosshard, H. E., Ellis, R. A. and Marshall, G. R. (1975) In "Computers in Life Science Research," edited by W. Siler and D. A. B. Lindberg, p. 137. New York: Plenum Publishing Corporation.
Bennett, J. M. and Kendrew, J. C. (1952) Acta Cryst. **5**, 109.
Busing, W. R., Ellison, R. D., and Levy, H. A. (1965) Am. Crystallogr. Assoc. Program Abstr. Ser. **2**, p. 59.
Busing, W. R. and Levy, H. A. (1968) Oak Ridge National Laboratory ORNL-4143.
Busing, W. R., Martin, K. O. and Levy, H. A. (1962) ORFLS: A Fortran Crystallographic Least-Squares Program. Oak Ridge National Laboratory ORNL-TM 305.
Cork, C., Hamlin, R., Vernon, W., and Xuong, Ng. H. (1975) Acta Cryst. **A31**, 702.
Cole, H., Okaya, Y. and Chambers, F. W. (1963) Rev. Sci. Instrum. **34**, 872.
Cruickshank, D. W. J. and Pilling, D. E. (1961) In "Computing Methods and the Phase Problem in X-ray Crystal Analysis", edited by R. Pepinsky, J. M. Robertson, and J. C. Speakman, p. 32. New York: Pergamon Press.
Dewar, R. B. K. (1970) Crystallographic Computing, Proceedings of the 1969 International Summer School on Crystallographic Computing, edited by F. R. Ahmed, p. 63. Copenhagen: Munksgaard.
Diamond, R. (1971) Acta Cryst. **A27**, 436.
Feldmann, R. J., Heller, S. R. and Bacon, C. R. T. (1974) J. Chem. Soc. **12**, 234.
Frevel, L. K. (1977) Advances in X-ray Analysis, Vol. 20, edited by H. F. McMurdie, p. 15. New York: Plenum Press.
Furey, W., Jr., Wang, B. C. and Sax, M. (1980) Am. Crystallogr. Assoc. Program Abstr. Ser. **2**, p. 27.
Gantzel, P. K., Sparks, R. A., and Trueblood, K. N. (1962) ACA Computer Program No. 317.
Germain, G. and Woolfson, M. M. (1968) Acta Cryst. **B24**, 91.
Gorres, B. T. and Jacobsen, R. A. (1964) Acta Cryst. **17**, 1599.
Hendrickson, W. A. and Konnert, J. H. (1980) ACA Tutorial. Mathematical Tools in Crystallography. Am. Crystallogr. Assoc. Program Abstracts. Eufaula, Alabama, p. 60.
Hodgkin, D. C., Lindsay, J., Sparks, R. A., Trueblood, K. N. and White, J. G. (1962) Proc. Roy. Soc. (London) **A266**, 494.
Hodgkin, D. C., Pickworth, J., Robertson, J. H., Prosen, R. J. Sparks, R. A. and Trueblood, K. N. (1959). Proc. Roy. Soc. (London) **A251**, 306.
Huang, T. C. and Parrish, W. (1978) In "Advances in X-ray Analysis", Vol. 21, edited by C. S. Barrett, and D. E. Leyden, p. 275. New York: Plenum Press.
Jeffrey, G. A., Shiono, R., and Jensen, L. H. (1961) In "Computing Methods and the Phase Problem in X-ray Crystal Analysis", edited by R. Pepinsky, J. M. Robertson, and J. C. Speakman, p. 25. New York: Pergamon Press.
Johnson, C. K. (1965) ORTEP: A Fortran Thermal-Ellipsoid Plot Program for Crystal Structure Illustrations. Oak Ridge National Laboratory. ORNL-3794 revised.
Kartha, G. (1970) In "Crystallographic Computing. Proceedings of the 1969 International Summer School on Crystallographic Computing", edited by F. R. Ahmed, p. 132. Copenhagen: Munksgaard.
Konnert, J. H. (1976) Acta Cryst. **A32**, 614.
Long, R. E. (1965) Doctoral Thesis, University of California, Los Angeles
Mayer, S. W. and Trueblood, K. N. (1953) Acta Cryst. **6**, 427.
McCarthy, G. J. and Johnson, G. G. (1979) In "Advances in X-ray Analysis", Vol. 22, edited by G. J. McCarthy, p. 109. New York: Plenum Press.
Mills, O. S. and Rollett, J. S. (1961) In "Computing Methods and the Phase Problem in X-ray Crystal Analysis", edited by R. Pepinsky, J. M. Robertson, and J. C. Speakman, p. 107. New York: Pergamon Press.
Morimoto, C. N. and Meyer, E. F., Jr. (1976) In "Crystallographic Computing Techniques", edited by F. R. Ahmed, p. 488. Copenhagen: Munksgaard.
Motherwell, W. D. S. (1976). In "Crystallographic Computing Techniques", edited by F. R. Ahmed, p. 481. Copenhagen: Munksgaard.
Munn, R. J. and Stewart, J. M. (1980) Software-Practice and Experience **10**, 743.
Nichols, M. C. (1966). UCRL-70078, Lawrence Livermore Laboratory.
Nordman, C. E. (1966) Trans. Am. Crystallogr. Assoc. **2**, 29.
Okaya, Y. (1966) Acta Cryst. **21**, 726.
Ordway, F. D. (1952) In "Computing Methods and the Phase Problem in X-ray Crystal Analysis", edited by R. Pepinsky. Pennsylvania State College X-ray Crystal Analysis Laboratory.
Parrish, W., Ayers, G. L. and Huang, T. C. (1980) In "Advances in X-ray Analysis", Vol. 23, edited by J. R. Rhodes. p. 313. New York: Plenum Press.
Rietveld, H. M. (1969) J. Appl. Cryst. **2**, 65.
Rollett, J. S. (1965) In "Computing Methods in Crystallography", edited by J. S. Rollett, p. 73. New York: Pergamon Press.
Rossman, M. G. and Blow, D. M. (1962) Acta Cryst. **15**, 24.
Scheringer, C. (1963) Acta Cryst. **16**, 546.
Schomaker, V. and Trueblood, K. N. (1968) Acta Cryst. **B24**, 63.
Schomaker, V. and Waser, J. (1958) Private communication: see Sparks (1961).
Shaffer, P. A., Jr., Schomaker, V. and Pauling L. (1946) J. Chem. Phys. **14**, 648.
Shirley, R. (1978) In "Computing in Crystallography", edited by H. Schenk, R. Olthoff-Hazekamp, H. van Koningsveld, and G. C. Bassi, p. 221. Delft, Netherlands: Delft University Press.
Sly, W. G. and Shoemaker, D. P. (1961) In "Computing Methods and the Phase Problem in X-ray Crystal Analysis", edited by R. Pepinsky, J. M. Robertson, and J. C. Speakman, p. 129. New York: Pergamon Press.
Sparks, R. A. (1961) In "Computing Methods and the Phase Problem in X-ray Crystal Analysis", edited by R. Pepinsky, J. M. Robertson, and J. C. Speakman, p. 170. New York: Pergamon Press.
Sparks, R. A. (1970). Am. Crystallogr. Assoc. Program Abstr. Ser. 2, p. 20.

Sparks, R. A. (1973a) Syntex AD-1 Film Reader Manual. Syntex Analytical Instruments, Cupertino, California.

Sparks, R. A. (1973b) In "Computational Needs and Resources in Crystallography", p. 66. Washington, D.C.: National Academy of Sciences.

Sparks, R. A. (1977) In "Minicomputers and Large Scale Computations", edited by P. Lykos, ACS Symposium Series 57, p. 54. Washington, D. C.: American Chemical Society.

Sparks, R. A. (1980) In "Structural Studies on Molecules of Biological Interest: A Volume in Honour of Professor Dorothy Hodgkin", edited by G. G. Dodson, J. P. Glusker, and D. Sayre, Oxford, England: Oxford University Press.

Sparks, R. A. (1981) "Advances in X-ray Analysis", Vol. 24. Plenum Press, New York. To be published.

Stewart, J. M. (1970) In "Crystallographic Computing. Proceedings of the 1969 International Summer School on Crystallographic Computing", edited by F. R. Ahmed, p. 297. Copenhagen: Munksgaard.

Struchkov, Yu. (1978) Private communication.

Taupin, D. (1973) J. Appl. Cryst. **6**, 226.

Ten Eyck, L. F. (1977) Acta Cryst. **A33**, 486.

Trueblood, K. N. (1981) In "Structural Studies on Molecules of Biological Interest: A Volume in Honour of Professor Dorothy Hodgkin", edited by G. G. Dodson, J. P. Glusker, and D. Sayre. Oxford, England: Oxford University Press.

Visser, J. W. (1969) J. Appl. Cryst. **2**, 89.

Waser, J. (1963) Acta Cryst. **16**, 1091.

Watenpaugh, K. D. (1973) In "Computational Needs and Resources in Crystallography", p. 37. Washington, D. C.: National Academy of Sciences.

Watenpaugh, K. D. (1980) Am. Crystallogr. Assoc. Program Abstr. Ser. 2, p. 27.

Werner, P. E. (1976) J. Appl. Cryst. **9**, 216.

Young, R. A., Mackie, P. E. and von Dreele, R. B. (1977). J. Appl. Cryst. **10**, 262.

CHAPTER 7

THE FAST FOURIER TRANSFORM IN CRYSTALLOGRAPHY

Robert A. Jacobson
Ames Laboratory-DOE and Department of Chemistry
Iowa State University, Ames, Iowa 50011

Progress in crystallography has paralleled that of the digital computer. This is evident from even a casual perusal of the crystallographic literature. Prior to the 1960's, most crystallographic studies were done via two-dimensional projections and involved structural determination of crystals having relatively small unit cells. Computational devices such as the Beevers-Lipson strips were widely used in an attempt to ease the computational tasks involved even for such projections. (Indeed I've heard it said that a few crystallographers of that era could move their finger down a column of structure factor-cosine or sine products and immediately write down its sum.) The task of calculating Patterson and electron density functions was certainly greatly simplified by the advent of the digital computer. Before the ready availability of computers, it was a long and tedious task to calculate just a two-dimensional projection of a structure.

It should be noted though that just having a computer does not make the computation of the electron density function,

$$\rho(x,y,z) = \frac{1}{V} \sum_{h}\sum_{k}\sum_{l}{}_{-\infty}^{\infty} F(hkl)e^{-2\pi i(hx+ky+lz)} \quad (1)$$

a trivial proposition. Indeed even with today's high speed digital computers such a calculation is prohibited time-wise for structures of any complexity, if programmed in the straightforward way that the equation suggests, i.e. for any selected x, y, z, to compute the 2π(hx+ky+lz) product, determine its sine and cosine, multiply by $A(hkl)$ and $B(hkl)$, respectively, and then perform a sum with like products over all the available data. It has long been recognized that it is much more efficient to factor the expression so that partial summations can be stored and then reused as long as their associated coordinates are unchanged. In other words,

$$\rho(x,y,z) = \frac{1}{V} \sum_{h=-\infty}^{\infty}\left(\sum_{k=-\infty}^{\infty}\left(\sum_{l=-\infty}^{\infty} F(hkl)e^{-2\pi ilz}\right)e^{-2\pi iky}\right)e^{-2\pi ihx} \quad (1a)$$

can be rewritten by defining

$$R(hkz) = \sum_{l=-\infty}^{\infty} F(hkl)e^{-2\pi ilz},$$

$$S(hyz) = \sum_{k=-\infty}^{\infty} R(hkz)e^{-2\pi iky} \quad (1b)$$

and

$$V\cdot\rho(x,y,z) = \sum_{h=-\infty}^{\infty} S(hyz)e^{-2\pi ihx} \quad (1c)$$

Many of the first programs written tended to be space group specific; the next ten years saw the development of more general and faster programs, paralleling the advancements in computer capability, but saw no further major breakthroughs until the advent of the fast Fourier transform.

As complexities of molecules and sizes of unit cells increased in the late sixties, crystallographers began to explore other techniques to further speed up Fourier series calculation; at the same time mathematicians demonstrated the promise of a new algorithm – the fast Fourier transform (Cooley and Tukey, 1965; Borgland, 1969). The rationale of the fast Fourier transform has a number of similarities to the trigonometric approach, only this time the factoring is based on the properties of a number system, usually the binary number system. The approach used can be best illustrated by a one-dimensional example.

$$\rho(x) = \frac{1}{a} \sum_{h=0}^{N-1} F(h)e^{-2\pi ihx} \quad (2)$$

where $N=2^{n+1}$, a power of two.
Note that we have approximated an infinite one-dimensional Fourier transform by a finite Fourier transform of length N and therefore $F(-1)=F(N-1)$, etc. Letting $x = k/N$

$$\rho(k) = \frac{1}{a} \sum_{h=0}^{N-1} F(h)e^{-2\pi ih\cdot k/N} \quad (3)$$

and expressing h and k in terms of their binary representation:

$$h = h_0 + h_1 2^1 + h_2 2^2 + \ldots + h_n 2^n$$

and

$$k = k_0 + k_1 2^1 + k_2 2^2 + \ldots + k_n 2^n, \quad (4)$$

where
h_i and k_i can only take on values of 0 or 1.

Therefore

$$h\cdot k = h_0\left(\sum_{j=0}^{n} k_j 2^j\right) + h_1\left(\sum_{j=0}^{n} k_j 2^{j+1}\right) + \ldots + h_n\left(\sum_{j=0}^{n} k_j 2^{j+n}\right) \quad (5)$$

Although the righthand side of eq. 5 may seem much more formidable, some very helpful simplifications can now be made. Since $e^{-2\pi iN.p.q/N}=1$ if p and q are integers,

$$e^{-2\pi ihk/N} = e^{-2\pi i(h_0\sum_0^n k_j 2^j + h_1 \sum_0^{n-1} k_j 2^{j+1} + \ldots + h_n k_0 2^n)/N} \tag{6}$$

Therefore Eq. 3 can be written as

$$a \cdot \rho(k_n \cdots k_0) = \sum_{h_0=0}^{1} \sum_{h_1=0}^{1} \cdots (\sum_{h_n=0}^{1} F(h_n \cdots h_0) e^{-2\pi ih_n \cdot k_0/2}).$$

$$e^{-2\pi ih_{n-1}(\sum_0^1 k_j 2^{j+n-1})/N} \cdots e^{-2\pi ih_0(\sum_0^n k_j 2^j)/N} \tag{7}$$

The advantage of Eq. 7 is that we can define

$$F_0(k_0,h_{n-1},\cdots,h_0) = \sum_{h_n=0}^{1} F(h_n,\cdots,h_0) e^{-2\pi ih_n \cdot k_0/2}$$

$$F_1(k_0,k_1,\cdots,h_0) = \sum_{h_{n-1}=0}^{1} F_1(k_0,h_{n-1},\cdots,h_0) \cdot e^{-2\pi ih_{n-1}(\sum_0^1 k_j 2^{j+n-1})/N}$$

$$F_n(k_0,k_1,\cdots,k_n) = \sum_{h_0=0}^{1} F_{n-1}(k_0,k_1,\cdots,k_{n-1},h_0) \cdot e^{-2\pi ih_0(\sum_0^n k_j 2^j)/N} \tag{8}$$

This form now resembles that obtained by factoring as has been done with the trigonometric form (Eq. la, b, c) in that the first summation involves only h_n and yields F_o which is dependent on the value of k_o we have selected. Since there are two possible values of k_o (i.e. 0 or 1) and two values of h_n (again 0 or 1), the net effect has been to take two of the original coefficients and produce two new coefficients. Since the same considerations apply for F_1 through F_n, the fast Fourier transform can be said to have a net effect of changing a Fourier transform of length N into n transforms of length two.

By the time F_n has been obtained it is only necessary to reorder the binary terms and scale the results to obtain the desired function,

$$\rho(x) = (k_n,\cdots,k_0) = \frac{1}{a}(F_n(k_0,k_1,\cdots,k_n))_{\text{reordered}} \tag{9}$$

In order to carry out these calculations, approximately $N\log_2 N$ complex multiplications and additions are required; this can be compared to the $\sim N^2$ operations necessary for the direct evaluation of the transform. Therefore for a transform using 64 grid points, the fast Fourier transform algorithm means an order of magnitude increase in the speed of computation, (Hubbard, Quicksall & Jacobson, 1972). The complete three-dimensional Fourier series could be calculated via the fast Fourier transform approach; however a complete sphere of data would have to be stored in the high speed computer memory, significantly increasing the memory requirements to run the program. Also, in the typical electron density or Patterson calculation, the function is evaluated at many more grid points than one has *F(hkl)* coefficients. To do such calculations completely via a fast Fourier transform approach, a large number of zero coefficients would have to be added. Therefore a more typical approach is to use trigonometric factoring in the initial part of the three-dimensional transform followed by the fast Fourier approach in the later stages. In some cases, symmetry can be also used to reduce the amount of storage necessary (Bantz and Swick, 1974).

As noted above the advantages of the fast Fourier transform are especially evident as the size of the N increases. It is obvious then with the increased emphasis throughout the crystallographic community on structures of large biological molecules and on organic, inorganic and organometallic moieties of ever increasing size and complexity, that the fast Fourier transform will play an even more important role in the years to come. It should also be noted that it can be applied to the calculation of structure factors (Ten Eyck, 1977) and for the refinement (Agarwal, 1978) of large molecules.

References

Agarwal, R. C. (1978) Acta Cryst. **A34**, 791-809.
Bantz, D. A. and Zwick, M. (1974) Acta Cryst. **A30**, 257-260.
Borgland, G. D. (1969) IEEE Spectrum **6**, 41-51.
Cooley, J. W. and Tukey, J. W. (1965) Math. Comp. **19**, 297-301.
Hubbard, C. R., Quicksall, C. O. and Jacobson, R. A. (1972) J. Appl. Cryst. **5**, 234-235.
Ten Eyck, L. F. (1977) Acta Cryst. **A33**, 486-492.

CHAPTER 8

INTERACTIVE GRAPHICS IN THE STUDY OF MOLECULES OF BIOLOGICAL INTEREST

F. Scott Mathews
Department of Physiology and Biophysics and of Biological Chemistry
Washington University School of Medicine
4566 Scott Avenue, St. Louis, MO 63110

Introduction and Background

A. Crystal Structure representation

Prior to the introduction of computer-controlled graphic displays, graphic techniques have often been used by crystallographers to illustrate the results of their structural analyses. For relatively small molecules these have consisted of scale drawings of a molecule or unit cell contents in projection or of schematic representations of a structure with perspective to depict its three-dimensional nature. Such "hard copy" graphic reached its culmination in ORTEP written by Carroll Johnson (1965). This program can project a structure from any angle, produce stereoscopic pairs and depict anisotropic vibrations by shaded ellipsoids. Such plots have become the norm for publishing molecular structures in the crystallographic literature and can be used with structures as large as proteins to depict α-carbon atoms or limited structural features.

Ever since computers were able to control a CRT display, however, there has been a movement to depict structures in real time on the screen of a computer. The first goal was to provide a quick, easy way to view structures and understand their nature. This is particularly important for non-crystallographers such as organic chemists, biologists, students, etc. who are not accustomed to three-dimensional structures. It soon became apparent that complex structures such as proteins needed the speed and versatility provided by an on-line computer to dissect a structure and display selected portions of it instantaneously.

A second goal was to enable chemists, biochemists and crystallographers to manipulate parts of a molecule quickly and easily and be able to retain the modified atomic coordinates accurately. Such uses include fitting of a model to observed electron density, prediction of conformation from physical or chemical information and analyses of chemical reaction pathways. This latter goal represents the major area of growth for computer graphics and provides a research capability difficult or impossible to achieve by more conventional approaches.

B. Early graphic systems (1964-1970).

The first major use of computer graphics in chemistry and molecular biology was made by Cyrus Levinthal and Robert Langridge at MIT (Levinthal, 1966). In many respects their system was close to the state-of-the-art today. It consisted of a line drawing CRT with a refresh buffer which could hold about 400 short vectors for a static display. It was connected directly, as a time sharing terminal, to the IBM 7094 computer of Project MAC. This experimental system was very expensive but served as a prototype for future systems. A major difficulty was the delay caused by the time-sharing nature of the system, which caused jerkiness whenever global or bond rotations were performed. The system was used principally to display known molecular structures although some attempts at protein structure prediction were begun.

Another approach to computer graphics was taken somewhat later, by David Barry, Bob Ellis, and Garland Marshall at the Computer Systems Lab (C.S.L.) of Washington University, St. Louis. They used a LINC, which is a dedicated laboratory mini-computer, to display and manipulate small organic structures. The system could also display the α-carbon coordinate of a small protein. The software system they developed was called CHEMGEN.

C. Second generation systems (1970-1975)

(1) Stand alone systems. Computer graphics developed along two philosophically different pathways, the small (but powerful) stand-alone system and the satellite system connected to a large powerful time-sharing computer. This separation is gradually disappearing, with the stand-alone systems capable of communicating to larger systems and the satellites having a great deal of power on their own.

Representative of the stand-alone systems are the MMS-4 system (Barry, Bosshard, Ellis, & Marshall, 1974) of Washington University, the Vector General-PDP 11/40 system developed by Edgar Meyers (Meyers, 1971) at Texas A & M University and the Argus system with a coligraphic display developed in Oxford, England by Tony North and David Barry (Barry & North, 1971). The last two systems were not as powerful as the MMS-4, since they did not allow real time global rotations, and many fewer vectors could be displayed. They were used to display or manipulate molecular fragments or to fit protein ligand models to electron density difference maps. I will now describe the MMS-4 system in some detail (see Fig. 1) since I am most familiar with it and with its descendent, the MMS-X system.

The functions of a system can be divided into several tasks with different stringent requirements. One task is display. This involves computing the endpoints of lines and generating the connecting line segment on the

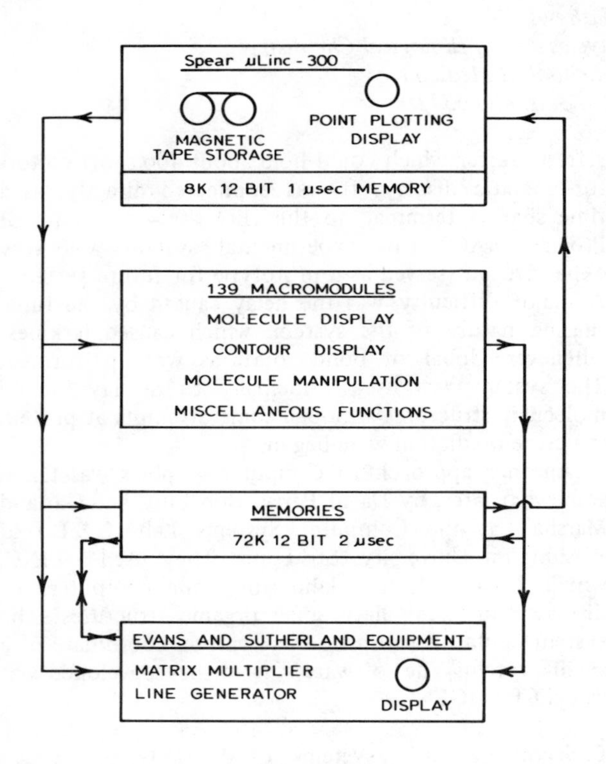

Fig. 1. Block diagram of the MMS-4.

display. To obtain 6,000 lines in 1/30 sec each line must be processed in less than 5 μsec. The second task involves computations associated with the static display, e.g., conversion of atomic coordinates to a display list and control of the display processor. This also must be rapid. The third task involves computations associated with the dynamics of a display, and includes scaling, translation and rotation of the display as a whole. Computation of the display matrix can be performed by a mini-computer sampling knobs or a keyboard, but the coordinate transformations require special hardware to match the speed of the display. The fourth task involves manipulation of atomic coordinates to generate different conformations of the molecule for display. This task is often carried out on a minicomputer.

(2) Large Computer Systems. Several graphics systems connected directly to large computers have been built. Examples of such systems are the Adage display computer connected to an IBM 360-91 computer at Columbia University (Katz & Levinthal, 1972), the Evans & Sutherland LDS-1 system linked to a PDP-10 at Princeton University (Langridge, 1974) and the GRIP-75 (Graphics Interaction with Proteins) system of the University of North Carolina. The last consists of a Vector General display system connected to an IBM 360-65 computer through a PDP 11/45 computer. Only the GRIP system has been used extensively by the crystallographic community. It represents the most recently developed of the time sharing systems and has been supported by a great deal of software development.

D. Third Generation Systems (1975 onwards).

The recent rapid development of cheap micro-processors promises the development of much cheaper, more powerful graphics systems. Two powerful graphics systems which have been developed recently and are located in a number of crystallographic laboratories are the Evans and Sutherland Picture System II and the MMS-X system. Since I am most familiar with the latter I will limit my discussion to it.

The expansion of the capability of the MMS-X is still under development but has potential for such things as dynamic contouring and hardware checking of hard sphere overlap of atoms. Physically, it is smaller and much more convenient than the MMS-4. The use of disk storage enables the computer memory to be loaded much more quickly and its FORTRAN capability greatly simplifies software development.

A National Collaborative Research Program (NCRP) has been established with funding from the NIH Biotechnology Research Program and is supporting the software development of the MMS-X in addition to program hardware maintenance and development. Naturally, each of the 9 laboratories has made its own program modifications and innovations to the supplied software to suit the local needs of the crystallographers. The NCRP has sponsored annual meetings of the users to coordinate programming activities for the MMS-X and serve as a forum for new techniques or requirements.

Illustrative graphics.

A. Static views.

If one wants to produce a good view for publication or illustration of a small molecule, protein fragment or α-carbon skeleton, then a plotter probably gives the best results, especially if augmented with hand coloring or other retouching. Tapered bonds, spheres of different sizes and shading, ellipses and lettering are very difficult to produce on a graphics screen because of size and resolution limitations. However, as a first step, the graphics system can provide the flexibility to view many orientations of the molecule quickly and easily, allow different portions to be selected, and provide the information needed to get the best plot most quickly.

A graphics system is very valuable also to enable one to "learn" a molecule, since particular views can be obtained easily which are not possible with a physical model. Structural features can be discovered and explored by selecting a number of subsets of the molecule and viewing them in different ways. For example, the first backbone structure we saw of cytochrome b_5 was produced on the CRT of a LINC computer. The lines were generated by connecting the dots representing approximate α-carbon coordinates measured directly from a minimap.

B. Motion Pictures.

One of the most striking uses of molecule graphics is the animation of motion pictures to present a protein structure. Various structural features can be emphasized and enzyme mechanisms illustrated. The most effective means of producing a movie is to have a computer-controlled shutter. Thus, the film can be shot frame by frame, each entire frame being composed by a computer program. If color filters are included, again under program control, then a color movie can be generated. The different colors can be used to represent various chemical features. The illusion of three-dimensional viewing is produced without the need for 3-D glasses by the motion programmed into the scenario.

From 1970-1975 my colleagues at the C.S.L. produced about a dozen motion pictures of various proteins, often in collaboration with the crystallographers who solved the structures. My own efforts yielded a 10 minute movies of cytochrome b_5. Other protein movies produced include insulin, chymotrypsin, lactate dehydrogenase, and ribonuclease. A "Protein Sampler" was also prepared illustrating a number of protein structures. Most of our motion pictures were in black and white because of the technical difficulties involved in producing color.

C. Shaded Surfaces.

Molecular models are most often represented by ball-and-stick models. The atoms are represented by small spheres or ellipsoids as in ORTEP or simply the intersection of two lines. This type of representation is useful for building models in an electron density map and for display and manipulation of a model on a graphics system. It also is useful for representing intermolecular interactions schematically such as a substrate with an enzyme. Often, however, a space-filling model is desirable both to indicate areas of solvent accessibility and to represent the shape of the protein surface available for interactions.

Organic chemists and biochemists are familiar with CPK models for representing the hard sphere interactions of atoms in a molecule. Such models are useful because they provide mechanical constraints to conformation flexibility similar to those of real molecules. However they are very heavy, cumbersome and expensive for large molecules. Two or three protein molecules have been built with CPK models, for example myoglobin, chymotrypsin and ribonuclease, but the task is very difficult.

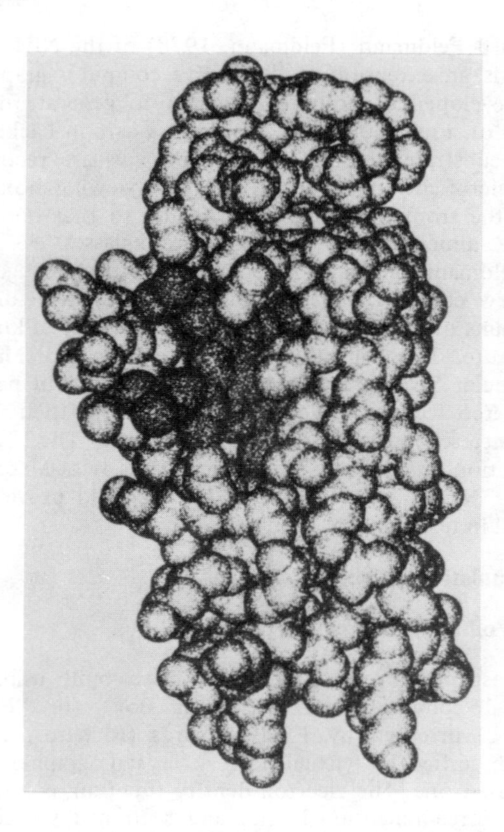

Fig. 2. Shaded surface diagram of cytochrome b_{562}. The heme group is shown with dark shading. The stick model was oriented on the MMS-X screen and the coordinate transmitted to the VAX computer for processing and plotting on a Printronics printer/plotter.

Recent developments of shaded surface and hidden line algorithms in computer graphics hardware and software make it possible to draw CPK models on a molecular graphics system. Richard Feldmann has a special Evans and Sutherland graphics colored surface display system with a frame buffer and raster monitor (Feldmann, Bing, Furie and Furie, 1978). His system is programmed to draw a sphere of given size around all atoms or subsets, and to assign different colors to different atoms. The surface of the spheres is sorted from back to front so that only the front surface at any point on the screen is in view. The color can be used to depict different atom or chemical group types or for more innovative illustrations such as thermal motion or dynamic flexibility. The algorithm can also be applied to a black and white matrix plotter at considerably less expense. Figure 2 shows a shaded surface view of cytochrome b_{562} produced on our VAX 11/780-Printronics printer/plotter.

D. Microfiche Atlas

An atlas of protein structures has been produced by

Richard Feldmann (Feldmann, 1978) of the NIH. This atlas is an extension of illustrative computer graphics. He developed a computer program to generate many views of a molecule such as protein α-carbon backbone or a substrate binding site. These views are recorded on microfiche in stereo. A polaroid viewing box fits over the front of a microfiche reader so that one gets a three-dimensional view using Polaroid glasses.

Feldmann and Bing (1979) have recently put together a very elegant collection of colored shaded surface diagrams of a large number of macromolecules of known structure. This collection of Teaching Aids for Macromolecular Structure (TAMS) includes a student packet of stereo views of selected structures and their component secondary structural elements. The TAMS collection is available commercially (Taylor-Merchant Corp., New York, N.Y. 10036) and should prove very useful in teaching.

Manipulative Graphics.

A. "Follies" and How They Work.

Most protein molecules have been built using an optical comparitor or "Richards Box" (or "Fred's Folly") invented by Fred Richards (Richards, 1968) and modified by virtually every crystallographer who has used one. The electron density function, computed on a three-dimensional grid, and sectioned serially in two dimensions, is represented as contour maps, with the contour levels proportional to density. If several sections are drawn on transparent material (such as plexiglass or window panes) they can be stacked to form a three-dimensional map.

A "Folly" is a vast improvement over earlier methods of model building. However, it is still very tedious and requires the drawing of many map sections. Furthermore, coordinates are difficult to measure and the model degrades with time through interaction with visitors.

The obvious thing to do is to display the electron density on a graphics computer and to adjust the molecule on the screen. There are several advantages of this: One is the ability to recontour the map to enhance or suppress details. Another is that coordinates of a molecule can be stored and retrieved easily without having to make inaccurate and tedious measurements. Also, the displays can be viewed in any orientation. Furthermore, storage space is greatly reduced. One graphics system can be used to fit and display several proteins and a scale model need never be built.

B. Problems of the Electronic Richards Box.

(1) Representation of Electron Density. Greater flexibility is offered for representing electron density by computer graphics than on transparencies. Although representations other than contours have been tested, such as a random collection of points whose density is proportional to the electron density (i.e., "clouds"), contouring still seems to be best. However, if contours are drawn and displayed in three intersecting sets of planes in a region of space, the contour lines join nicely to produce a three dimensional bird-cage like structure. One can still contour at several levels of density, but this is not necessary and tends to be confusing.

(2) Contour Editing. The easiest way to define a block to be contoured is to define the boundaries of a box parallel to the unit cell edges. This will usually give unwanted contours which obscure the areas of interest. For example, if an extended chain is passing diagonally through the parallelepiped, chains from nearby portions of the molecule will also be present and interfere with viewing the chain of interest.

Colin Broughton (Broughton & Bacon, 1980) of the University of Alberta has written a contouring program, BLOB, for the MMS-X. The coordinates of a fragment of a molecule are stored and used to define a volume in space for contouring. The resulting density block takes a few seconds to compute and requires little additional editing. A set of preliminary coordinates obtained from a minimap could be used for defining the volume.

C. Electron Density Fitting in Practice.

(1) MMS-4. The MMS-4 was very powerful in its display capability, but limited in its versatility since the LINC computer could only be programmed in machine language. Two serious attempts were made to use the MMS-4 to fit a protein model to electron density. Larry Webb attempted to build malate dehydrogenase in a 2.5A map and Sue and John Cutfield tried to fit insulin to a 1.9A map. Both attempts failed because the density fitting software was too primitive and no disk storage was available to the system.

Dave Barry and I did, however, experiment with several fragments of cytochrome b_5 to determine what was needed to achieve density fitting capability. From these experiments we learned that the single-level bird-cage contouring was adequate to represent density and that fitting a tetrapeptide using 6 active bonds was convenient.

The polypeptide model of cytochrome b_5 which we used had been refined by Bob Diamond's real space procedure (Diamond, 1971) to produce a good fit to the density. The heme group had not been refined in that manner, however. It was more convenient to construct a heme group model on the MMS-4 and to adjust it to fit the density. I used a planar heme group, placing its center on the iron atom, and adjusted the tilt of its plane and the orientation of its propionate side chains to maximize overlap.

The resulting fit was quite good, except that I could not get the propionate oxygen atoms into the center of the density. (Fig. 3a.) I decided to distort the heme group by folding it along the lines shown in Fig. 3b,

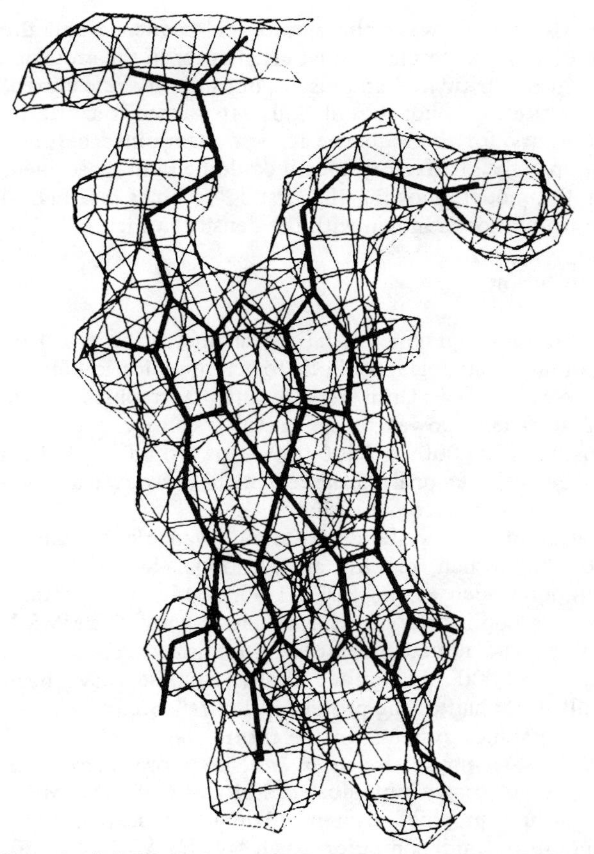

Fig. 3. (a) The planar heme of cytochrome b_5 superimposed on the 2.0Å electron density on the MMS-4 screen. The propionate groups at the top do not fit the density well even though the rest of the structure does.

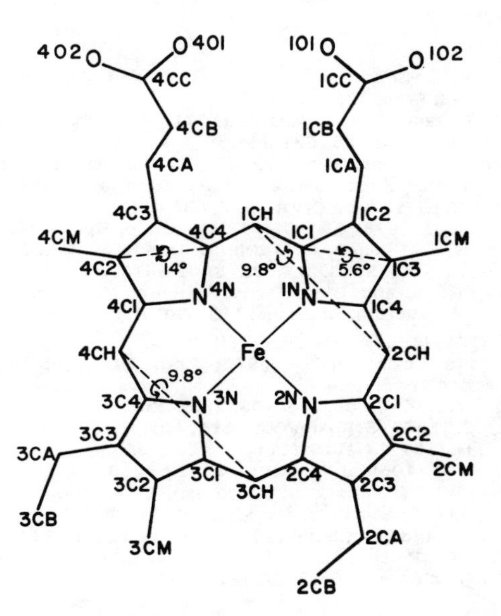

Fig. 3. (b) The heme skeleton showing how the model was "folded" to fit the density better.

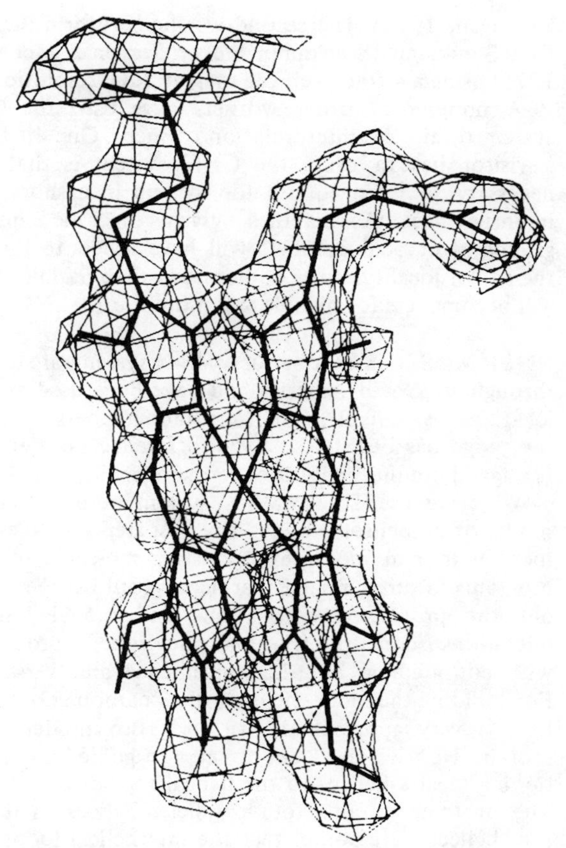

Fig. 3. (c) The non-planar heme adjusted to fit the 2.0Å electron density.

producing a non-planar heme group. The resulting fit is improved at the periphery of the molecule, as seen in Figure 3c. This folding or bending of the heme was rather arbitrary, and probably is distributed over several atoms.

(2) The GRIP System. The GRIP system at North Carolina has been used for the past several years by a number of crystallographers for density fitting. The system was designed and programmed after the MMS-4 had been built and the development of the GRIP software was aided by our experiences with MMS-4.

The GRIP system was used by Sung-Hou Kim (Sussman & Kim, 1976) to monitor the Fourier and difference Fourier refinement of tRNA. He also constructed a model of the system using the phosphate positions and targets for individual nucleotide fragments of the tRNA which he later fused together. In his refinement steps he omitted short polynucleotide fragments from the structure factor calculations and constructed improved models for them in the difference electron density.

At least two proteins have been constructed *ab inito* on the GRIP system. The first was a snake venom neurotoxin solved at 2.2Å resolution by Greg Petsko and Demetrius Tsernoglou (Tsernoglou, Petsko, Hermans &

McQueen, 1977). The second was hemerythrin built by Ron Stenkamp (Stenkamp, Sieker, Jensen & McQueen, 1978) using a 4-fold averaged map at 2.8Å resolution.

A number of other workers have used the GRIP system to aid the interpretation of maps. One difficulty a visitor has in using the GRIP system is that it is necessary to learn to use the system in a short, very intensive visit. As graphics systems become cheaper, groups of crystallographers will have access to them in their own locality, close to home, so that graphics usage will become less formidable and more routine.

(3) MMS-X. A number of MMS-X systems are located throughout North America and have been used at their locations by several protein crystallographers. A lot of the usage has been to analyze difference Fourier maps for ligand binding analysis or refinement.

We have used the MMS-X system here to build a model of cytochrome b_{562}. The first step was to experiment with and modify the existing molecule building programs to suit our particular needs. Paul Bethge carried out the program modifications and model building operations concurrently and developed procedures with considerable built-in convenience and versatility. For building the model he used the α-carbon coordinates from a very approximate Richards Box model of the protein. He used these coordinates to guide him in positioning ideal α-helices to the helical regions of the map. The protein contains four parallel α-helices. After the four helices were completed, the interhelical loops were added.

A major difficulty we encountered was a lack of consistency between the amino acid sequence and the density. In some places obvious changes in the sequence could be postulated and tested, but in others the correct structure was not so obvious. It turned out to be necessary for the sequence to be redetermined before a completely satisfactory model could be built. A segment of the model, showing the first 12 residues in a helical conformation is shown with the density in Figure 4.

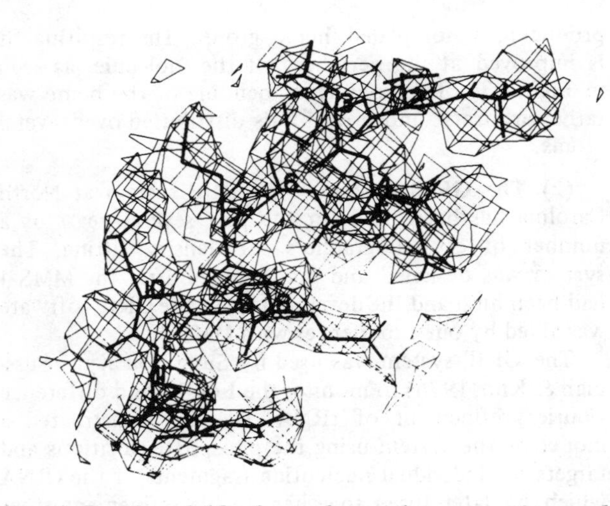

Fig. 4. Residues 1-12 of cytochrome b_{562} superimposed on the 2.5Å electron density. The conformation of cytochrome b_{562} was obtained by fitting it to the density on the MMS-X graphics system.

Conclusion

Computer graphics is having an increasing effect on protein crystallography both for communication and as a research tool. Until very recently a graphics system of sufficient power to be useful was very expensive, costing \$200,000 to \$300,000. This would limit them to being a regional resource which is awkward for a research worker to use easily.

Commercial systems are now available for about \$60,000 which require an additional \$40,000 minicomputer such as the PDP 11/45 or 11/70 to operate. Several laboratories now have such systems. The MMS-X offers stand-alone capability at a replication cost of less than \$50,000 but only a limited number have been built. Over half of the protein structure laboratories now have graphics systems. In the future, very sophisticated graphics terminals driven by one or more micro-computers may carry out most of the tasks of the MMS-X or other graphics system. When connected to a laboratory midi-computer, such as the VAX 11/780, the display and manipulation capabilities necessary for protein structure research will probably become widely available.

References

Barry, C. D., Bosshard, H. E., Ellis, R. A. and Marshall, G. R. (1974) Fed. Proc. **33**, 2368-2372.

Barry, C. D. and North, A.C.T. (1971) Cold Spring Harbor Symp. Quant. Biol. **36**, 577-584.

Broughton, C. G. and Bacon, D. J. (1980) Amer. Crystallogr. Assoc. Program Abstr. Ser. 2, Calgary, Alberta, No. PB34.

Diamond, R. (1971) Acta Cryst. **A27**, 436-452.

Feldmann, R. J. (1978) AMSOM, Tractor Jitco, Rockville, MD.

Feldmann, R. J., Bing, D. H., Furie, B. C. and Furie, B. (1978) Proc. Natl. Acad. Sci. U.S.A. **75**, 5409-5412.

Feldmann, R. J. and Bing, D. H. Teaching Aids for Macromolecular Structure, Bethesda, MD: D.C.R.T./N.I.H.

Johnson, C. K. (1965) ORTEP, ORNL-3794.

Katz, L. and Levinthal, C. (1972) Ann. Rev. Biophys. Bioengineering **1**, 465-504.

Langridge, R. (1974) Fed. Proc. **33**, 2332-2335.

Levinthal, C. (1966) Sci. American **214**, 42.

Meyers, E. F. (1971) Nature (London) **232**, 255.

Richards, F. M. (1968) J. Mol. Biol. **37**, 225-230.

Stenkamp, R. E., Sieker, L. C., Jensen, L. H. and McQueen, J. E. Jr. (1978) Biochemistry **17**, 2499-2504.

Sussman, J. L. and Kim, S. H. (1976) Science **192**, 853-858.

Tsernoglou, D., Petsko, G. A., McQueen, J. E. Jr. and Hermans, J. (1977) Science **197**, 1378-1381.

CHAPTER 9

HIGH ENERGY ELECTRON DIFFRACTION IN AMERICA

J. M. Cowley
Department of Physics, Arizona State University
Tempe, Arizona 85287

Introduction

"High energy," as applied to electron diffraction, refers to the use of accelerating voltages mostly in the range of 20 to 100kV, although extensions to 1MeV or more may be included since there is no fundamental difference in the phenomena observed or in the theory required over this extended range. The wavelengths of the electron radiation thus range from about 10^{-1} to 10^{-2}Å. For this high energy region the specimen thicknesses through which the electrons may usefully be transmitted rarely exceed 1000 Å.

The outstanding characteristic of electrons in relationship to diffraction is that they are much more strongly scattered by matter than X-rays or neutrons. Electron diffraction patterns can now be obtained routinely from specimen regions containing only a few thousand atoms, and the scattering from even a single heavy atom can be detected with sufficient strength to allow it to be unequivocally imaged. The technology which allows sufficient control of the incident beam to enable the selection of such minute specimen regions is a relatively recent development, coming with the application of the techniques of high resolution electron microscopy. In the early, pre-microscope days, when electron diffraction cameras contained only one long-focus lens, transmission diffraction was relatively rare. It was much more common to obtain diffraction patterns from flat surfaces of bulk specimens. Because the diffraction angles were typically 10^{-2} radians, the incident beam made a grazing angle with the surface. For a perfectly flat crystal surface the electron beam penetrated only 10-20Å. The method of reflection high energy electron diffraction (RHEED) thus became a major tool for studies of surface structure.

Following the development of low-energy electron diffraction, (LEED), the group at the Bell Laboratories applied the RHEED method very effectively to the study of the structures of monolayers of long-chain organic molecules deposited on flat metal surfaces (Germer and Storks, 1938). This work was followed up later by Jerry Karle and Lawrence Brockway (1947). Other applications of the method, mostly to rougher surfaces, included the extensive investigations of the oxidation of metal surfaces, carried out by Earl A. Gulbransen and coworkers at the Westinghouse Laboratories (for example, Hickman and Gulbransen, 1948).

Crystal structure analysis using electrons rather than X-rays appeared attractive in that it could be applied to very much smaller specimen volumes. In the late 1940's and the 1950's there was an energetic development of electron diffraction structure analysis methods, using mostly finely polycrystalline samples, in the Soviet Union by Pinsker, Vainshtein and colleagues (Pinsker 1953, Vainshtein 1964). In Australia attempts at structure analysis using single crystal electron diffraction patterns were made (Cowley and Rees, 1958) and some structure analyses were done in Japan. There were no parallel developments in America. In fact, faced by the prospect of being required to develop such methods, Lorenzo Sturkey of Dow Chemical Co. reformulated the theory of the dynamical scattering of electrons in crystals to demonstrate that electron diffraction intensities could not be interpreted in any reliable manner. The conviction developed in America that electron diffraction structure analysis was impossible. In spite of the successes of the Moscow group, the differing attitudes on the subject remained incompletely resolved for many years. It is only recently that there has been a sufficiently clear understanding of dynamical diffraction effects and an adequate development of instrumental techniques to allow a full appreciation of the possibilities in this direction.

Dynamical diffraction: phenomena and theory

The interaction of electrons with matter is so strong that the kinematical, single scattering approximation can fail for even a single heavy atom. The angular phase change of an electron wave passing through the potential field of an atom may be several radians and the atomic scattering amplitudes must be taken to be complex. Evidence for this was recognized in gas diffraction data by R. Glauber and Verner Schomaker (1953) at Cal Tech. Calculations of complex atomic scattering factors were made initially by Jim Ibers and Jean Hoerni. Complex atomic scattering amplitudes cannot usefully be applied for well ordered arrangements of atoms such as those in crystals (Hoerni, 1956). If an electron wave passes through two atoms in succession the effect is to give twice the phase change so the scattered amplitude can be very different from that obtained by taking twice the amplitude calculated for a single atom.

When an electron wave enters a crystal, diffracted waves of amplitude comparable to that of the incident wave can be produced within the first 100Å or less. The number of diffracted waves of appreciable amplitude produced in such a thin layer with short-wavelength radiation can be tens for small unit cell dimensions or hundreds for large unit cells. In the subsequent portion of the crystal all of these waves may be diffracted again. To understand the diffraction intensities one must consider the coherent interactions of large numbers of diffracted waves. For electron diffraction a many-beam dynamical theory is essential.

In his original treatment of electron diffraction, Bethe (1928) recognized the need for a many-beam theory and provided a formal statement of it. Lorenzo Sturkey developed the first formulation which could conceivably be applied to practical calculations, using a scattering matrix description. Unfortunately his work was published in 1949 and 1957 only in brief conference abstracts and was not presented in full until 1962 (Sturkey, 1962). By that time a variety of other formulations had been produced, including the multislice approach of Cowley and Moodie (1957), the Born series method of Fujiwara (1959), the matrix formulation by Fujimoto (1959) and others, and several more (see Cowley, 1975).

This theoretical work was stimulated by the great advances in electron optics, associated with the developments in electron microscopy, which allowed the sizes and shapes of crystals to be seen and permitted the physical parameters of electron diffraction experiments to be appreciated much more fully and even, in some cases, to be defined and controlled. Earlier electron diffractionists had used only the simple 2-beam approximation of Bethe's dynamical theory, similar to that used for X-rays or neutrons. For many situations involving simple crystal structures, this approximation does, in fact, give a reasonable qualitative representation of some electron diffraction phenomena and it has been used extensively in the interpretation of electron micrographs of crystalline materials.

Many of the well-known dynamical diffraction effects were first observed with electrons. These include the manifestations of the "pendellosung" of Ewald, the periodic variation of incident and diffracted beam intensities with crystal thickness. In direct space this produced the thickness fringes seen in electron microscope images of wedge-shaped crystals (Heidenreich 1942). Lorenzo Sturkey (1948) realized that the splitting of diffraction spots from small regularly shaped crystals was the reciprocal space equivalent.

In recent years the need for practical use of the many-beam dynamical diffraction approximations has become increasingly apparent. Computer programs based on either the multi-slice or matrix formulations have allowed quantitative comparison with experimental observations. The power of the electron diffraction methods has been correspondingly extended.

Theoretical analyses have confirmed to a considerable extent the assumptions of the Moscow group that, for the polycrystalline specimens they used for structure analysis, the averaging over thickness and orientation is effective for removing the more dramatic dynamical diffraction effects. The most important remaining deviations from simple kinematical theory are those arising from the one-dimensional "systematic interactions", the interaction of a reflection with its higher or lower orders. These effects can be calculated reasonably well. There is no reason why structure analysis based on electron diffraction patterns from suitable polycrystalline materials should not achieve accuracy comparable with that of most X-ray diffraction work.

The use of single crystal data is more difficult since the intensities vary strongly and often irregularly with crystal thickness and orientation. Progress can be made if crystal regions small enough to be uniform in thickness and orientation can be chosen or else if a well controlled integration over thickness and orientation is possible.

A particular area of interest for single crystal structure analysis is that of biologically significant materials. There are many cases where large molecules can form arrays well ordered in two dimensions or over small three dimensional volumes but crystals large enough for X-ray diffraction cannot be obtained. It has been considered (see Vainshtein, 1964) that for such materials dynamical scattering effects will not be appreciable because only light atoms are usually involved and also the large unit cell dimensions ensure that of the many reflections produced very few, if any, will be strong. The question remains as to whether a very large number of simultaneous reflections will collectively give strong many-beam dynamical diffraction effects. This question has been answered by Bob Glaeser and co-workers at Berkeley (e.g. Jap and Glaeser, 1980) by detailed dynamical calculations for particular crystals of organic compounds and biological molecules. Their conclusion is that for 100keV electrons it is possible to use the simple kinematical approximation for thicknesses of up to 100-300Å, depending on the accuracy desired. The effects of the ever-present bending of the very thin crystal sheets used for most such studies have been investigated in detail by Douglas Dorset (1978) at Buffalo.

Surface Studies

The use of RHEED tended to fall into disfavor as LEED was developed as the major diffraction tool for surface studies. In principle, for perfectly flat surfaces, RHEED has the same sensitivity to the structure of the topmost layers of atoms as LEED and also can be treated theoretically more simply because high energy scattering approximations can be used. A thorough appraisal of RHEED has been made on clean surfaces under ultra-high vacuum conditions and many-beam dynamical diffraction theory adapted for the reflection geometry (Colella and Menadue, 1972). In practice the need to use grazing angles of incidence for RHEED makes the intensities extremely sensitive to surface roughness. With the near-normal beam incidence used for LEED this sensitivity is minimized.

The sensitivity of RHEED to surface morphology may, on the other hand, become the foundation for new, more complete surface structure analysis methods. With high energy electrons it is possible to apply electron microscopy techniques to obtain images of the surface using the diffracted beams, with lateral resolution approaching that of normal transmission electron microscopy. The images are severely fore-shortened by the use of small grazing angles of incidence, but a vivid impression of the non-uniformity of most specimens giving surface diffraction patterns is provided. Exploratory work on this technique carried out at Arizona State University (Nielsen and Cowley, 1976) was limited by the poor vacuum conditions of standard electron microscopes. More recently an ultra-high vacuum transmission electron microscope has been used in Tokyo (Osakabe et al, 1980) and a high vacuum scanning transmission electron microscopy (STEM) instrument has been used in Arizona (Cowley, 1980). The images show clearly the configurations of atom-high surface steps and crystal defects, and provide important new insights on the formation of surface superlattices. The lateral resolution achieved in some cases has been about 10Å.

Electron Diffraction and Electron Microscopy

Observation of electron diffraction patterns forms an essential part of the electron microscopy of crystalline materials since the image intensities depend very strongly on the diffraction conditions. To a comparable extent, the observation of a specimen by electron microscopy is essential for the interpretation of electron diffraction patterns since the diffracton intensities are strongly influenced by the morphology of the diffracting crystals. It is therefore convenient to use the one instrument for both electron microscopy and diffraction and most transmission electron diffraction of solids in recent years has been carried out in electron microscopes.

The Selected Area Electron Diffraction (SAED) technique relies on the fact that, if an area of the specimen is selected by an aperture placed in the image plane of the objective lens, the diffraction pattern from that area may be obtained by imaging the diffraction patterns formed in the back-focal plane of the objective lens (see, for example, Heidenreich, 1964 or Cowley, 1975). The area selected cannot in general be less than about 1 μm in diameter for 100keV electrons (1000Å in diameter for 1MeV electrons) because of the effects of the spherical aberration of the objective lens. Even so, the possibility of obtaining diffraction patterns from areas of this size, chosen from specimens seen in the microscope with lateral resolution 10Å or less, represented an enormous advance.

A further advance in electron diffraction techniques has been associated with the development of scanning methods of electron microscopy. In scanning transmission electron microscopy (STEM), the electron beam from a small bright source is focussed on the object to give a fine electron probe. The diameter of the probe may be as small as 5Å for dedicated STEM instruments or 20-100Å for STEM attachments on conventional transmission electron microscopy (CTEM) instruments. For microscopy this probe is scanned across the specimen and the intensity of some part of the transmitted beam is detected to form an image displayed with a synchronized scan. If the probe is held stationary on the specimen, a diffraction pattern from the small illuminated area is formed on the detector plane. Diffraction patterns from regions of a diameter of about 5Å have been obtained. Patterns from regions of diameter 10-15Å have been used on a routine basis for the study of the structure of very small crystalline particles and of local ordering in nominally amorphous materials (Cowley, 1980). An alternative method for recording diffraction patterns from areas of diameter down to 20-30Å is that developed by Roy Geiss (1975) of IBM in which the incident beam direction in an electron microscope is scanned and the variation of intensity at a point in a high magnification image is recorded.

In principle the methods of electron microscopy and diffraction may be used in a complementary way to provide structural information in solids over the whole range of dimensions down to the sub-atomic level (Cowley and Jap, 1976). The electron micrographs provide information down to the limit of resolution, currently about 3Å. Electron diffraction patterns obtained from areas as small as 5Å diameter provide information on relative atom positions with a resolution of better than 1Å. In practice, of course, there are severe limitations due to the experimental difficulties of collecting sufficiently accurate data from the very small specimen regions. Also for very thin specimens for which the kinematical approximation is good the same phase problem applies as for the interpretation of X-ray diffraction data. For thicker specimens, and especially for crystals, there are the complications of dynamical scattering.

Convergent Beam Electron Diffraction and Microdiffraction

In order to get an electron probe of small diameter it is necessary to use an incident beam of correspondingly large convergence angle. The incident beam spot and the diffraction spots from a perfect crystal region are then spread into finite discs. The intensity distributions across the spots correspond to the variations of diffracted intensity with angle of incidence.

Convergent beam (CBEM) patterns of this type, from rather large specimen areas, were first observed by Kossel and Mollenstedt (1939). Goodman and Lehmpfuhl (1967) revived and perfected the technique which has since formed the basis for quantitative measurements of dynamical diffraction effects, leading to highly accurate determinations of structure amplitudes for simple crystal structures (Goodman, 1978). Also in the hands of Goodman (1978) and Buxton et al., (1976) the CBED technique has been shown to be a powerful tool for the determination of crystal symmetries since, from the dynamical diffraction effects, it is possible to make absolute determinations of space groups without the usual limitation of kinematical X-ray scattering which does not allow the detection of the presence or absence of a center of symmetry.

Use of the CBED technique in America has been limited almost entirely to practical applications as a method for obtaining microdiffraction from very small specimen regions in order to identify precipitate particles in alloys or ceramics or to determine local orientations of known phases. There have been indications, however, that a range of new and interesting diffraction effects may be observed in CBED patterns when a field emission electron gun is used to give such a small, bright source of electrons that the incident beam, converging on the specimen, may be regarded as a perfectly coherent convergent spherical wave (Cowley, 1979). A variety of interference phenomena, some analogous to those known in light optics (shearing interferometer fringes, Ronchigrams) appear in striking form.

Defects and Disorder

An outstanding characteristic of electron diffraction patterns which has been of great practical significance, is that streaks and diffuse spots due to crystal defects and disorder may be clearly and graphically displayed. Electron diffraction patterns are routinely used to obtain clear, qualitative indications of the diffuse scattering arising from thermal motions, from atomic displacements or replacements due to substitutional disorder or charge density waves and from the lattice distortions or disruptions dues to stacking faults, dislocations or other extended defects. One reason for their use is that they represent almost planar sections of reciprocal space so that the form of the scattering power distribution is readily comprehended. Another reason is that the sharp electron diffraction spots may easily be vastly overexposed even with exposure times of only a few seconds, so that diffuse scattering which is relatively very weak, appears strongly. Also, single crystal patterns may be obtained from very small areas of specimens which are highly disordered or have a high density of faults.

Outstanding examples of the use of electron diffraction patterns for the study of substitutional disorder and of the formation of long-period superlattices in alloys are provided by the work of Hiroshi Sato and Robert S. Toth (1965) at Ford Motor Co. More recently S. L. Sass and others (see Sass 1980) at Cornell have used the method in studies of periodic arrays of dislocations in grain boundaries and also in extensive investigations of the nature of the β-ω phase transformation in a variety of Ti and Zr-based alloys, in the course of which complicated wavy lines of diffuse scattering appear in the diffraction pattern (see, for example, Dawson and Sass, 1970).

Dynamical diffraction has profound but poorly understood effects on diffuse scattering in electron diffraction patterns and has prevented the interpretation of diffuse scattering intensities in any very quantitative manner. Calculations made using many-beam dynamical scattering approximations for idealized conditions suggest that the local configurations of diffuse scattering will not be greatly distorted but that relative intensities of widely separated diffuse features may be strongly affected by dynamical diffracton (Cowley and Fields, 1979). The calculations required for reliable interpretations of diffuse intensities are feasible but very laborious and usually require that the experimental parameters (ranges of thickness and orientation) be defined with much greater precision than is normally possible in a diffraction experiment.

Inelastic Electron Scattering

Fast electrons passing through matter can lose energy in inelastic scattering events in which the atoms or crystals are excited. The energy losses due to phonon excitation are detectable only under special circumstances. Magnetic spectrometers and other devices can be used to distinguish readily the electrons which have lost a few electrons volts or tens of electron volts by the excitation of plasmons, by exciting the outer-shell, valence electrons of atoms or by stimulating inter-band or intraband excitations of electrons in crystals. The same equipment allows the detection and measurement of larger energy losses occurring when the incident electrons excite electrons from the inner shells of atoms, giving rise to the emission of X-rays or Auger electrons.

The energy losses of a few tens of electron volts have been studied extensively, for example by L. Marton et al (1955) at the National Bureau of Standards and by John Silcox and associates (Silcox, 1977) at Cornell. Recently, however, the emphasis has been on the inner-shell energy losses which are characteristic of the elements present. Since electrons can be focussed to form small probes of high intensity, electron energy loss spectroscopy (ELS) offers a means for microanalysis of very small regions and so can provide information which is complementary to the crystallographic data from microdiffraction and the morphological data from electron microscope imaging. Semi-quantitative information on composition can be obtained from regions of the order of 100Å in diameter and groups of only a few thousand atoms of one type have given recognizable signals (Isaacson, 1980). Progress towards more quantitative ELS from somewhat larger specimen regions is being made, for example, by David Joy and Dennis Maher at Bell Laboratories, and Ray Egerton at Edmonton (see Joy et al., 1979).

The maxima in the energy loss spectra, corresponding to the excitation of K and L edges, show fine structure which is closely analogous to that observed in X-ray absorption. The analytical technique of EXAFS, using the fine structure of X-ray absorption edges, is proving valuable for the investigation of the local environments of particular atoms in solids. The corresponding technique using electrons, known as EXELFS, can give much the same information but can be applied to much smaller specimen volumes and is particularly useful for investigating the environments of the lighter elements, such as carbon, oxygen or nitrogen which have K-loss peaks in a readily accessible region of the spectrum (see Silcox, 1977).

The combination of energy-loss analysis with micro-diffraction offers some interesting possibilities which have scarcely been explored to date. Diffraction patterns obtained with electrons which have suffered particular characteristic energy losses could give information on the symmetries and the positions of particular atomic species in crystal structures. Experiments along these lines are now experimentally feasible.

References

Bethe, H. A. (1928). Ann. Phys. **87**, 55-129.
Buxton, B. F., Eades, J. A., Steeds, J. W. and Rackham, G. M. (1970) Phil. Trans. Roy. Soc. **281**, 171-194.
Colella, R. and Menadue, J. F. (1972) Acta Cryst. **A28**, 16-22.
Cowley, J. M. (1975) Diffraction Physics. Amsterdam: North Holland Publ. Co.
Cowley, J. M. (1979) Ultramicroscopy **4**, 425-450.
Cowley, J. M. (1980) in "Scanning Electron Microscopy/1980." Edited by Om Johari, O'Hare, Illinois: SEM Inc. In press.
Cowley, J. M. and Fields, P. M. (1979) Acta Cryst. **A35**, 28-37.
Cowley, J. M. and Jap, B. K. (1976 in "Scanning Electron Microscopy/1976." Edited by Om Johari. Chicago, Il.: IIT Res. Inst. pp. 377-384.
Cowley, J. M. and Rees, A.L.G. (1958) Rep. Progr. Physics **21**, 165-225.
Dawson, C. W. and Sass, S. L. (1970) Met. Trans. **1**, 2225-2233.
Dorset, D. L. (1978) Z. Naturforsch. **A33**, 964-982.
Geiss, R. H. (1975) Appl. Phys. Letters **27**, 174-176.
Germer, L. H. and Storks, K. H. (1938) J. Chem. Phys. **6**, 280-293.
Glauber, R. and Schomaker, V. (1953) Phys. Rev. **89**, 667-671.
Goodman, P. (1978) in "Electron Diffraction 1927-1977." Edited by P. J. Dobson, J. B. Pendry and C. J. Humphreys, Bristol: Institute of Physics, pp. 116-128.
Goodman, P. and Lehmpfuhl, G. (1967) Acta Cryst. **22**, 14-24.
Heidenreich, R. D. (1942) Phys. Rev. **62**, 291-292.
Heidenreich, R. D. (1964) Fundamentals of Transmission Electron Microscopy. New York: Interscience Publ.
Hickman, J. N. and Gulbransen, E. A. (1948) Anal. Chem. **20**, 158-165.
Hoerni, J. A. (1956) Phys. Rev. **102**, 1534-1542.
Ibers, J. A. and Hoerni, J. A. (1954) Acta Cryst, **7**, 405-408.
Isaacson, M. (1980) in "Scanning Electron Microscopy/1980." Edited by Om Johari, AMF O'Hare, Illinois: SEM Inc. In Press.
Jap, B. K. and Glaeser, R. M. (1980) Acta Cryst. **A36**, 57-67.
Joy, D. C., Egerton, R. F. and Maher, D. M. (1979) in "Scanning Electron Microscopy/1979 Vol. II." Edited by Om Johari, AMF O'Hare, Illinois: SEM Inc., pp. 817-826.
Karle, J. and Brockway, L. O. (1947) J. Colloid Sci. **2**, 277-287.
Kossel, W. and Mollenstedt, G. (1939) Ann. Phys. **36**, 113-140.
Marton, L., Leder, L. B. and Mendlowitz, H. (1955) in "Advances in Electronics and Electron Physics VII. New York: Academic Press.
Nielson, P. E. Højlund and Cowley, J. M. (1976) Surface Science **54**, 340-354.
Osakabe, N., Tanishiro, Y., Yagi, K. and Honjo, G. (1980) Surface Science **97**, In press.
Pinsker, Z. G. (1953) Structure Analysis by Electron Diffraction. London: Butterworths.
Sass, S. L. (1980) J. Appl. Cryst. **13**, 109-127.
Sato, H. and Toth, R. S. (1965) in "Alloying Behaviour and Effects in Concentrated Solid Solutions." Ed. by T. B. Massalski. New York: Gordon and Breach. pp. 295-419.
Silcox, J. (1977). in "Scanning Electron Microscopy/1977, Vol. I." Ed. by Om Johari. Chicago, Illinois: ITT Res. Inst. pp. 393-400.
Sturkey, L. (1948) Phys. Rev. **73**, 183.
Sturkey, L. (1962) Proc. Phys. Soc. **80**, 321-354.
Vainshtein, B. K. (1964). Structure Analysis by Electron Diffraction, Oxford: Pergamon Press.

CHAPTER 10

THE CRYSTALLOGRAPHIC ASPECTS OF ELECTRON MICROSCOPY

J. M. Cowley
Department of Physics
Arizona State University, Tempe, AZ 85287

Introduction

Of the various radiations used to investigate the structure of matter, electrons are unique in that they may be focussed to give images with a resolution approaching atomic dimensions and also are scattered so strongly by matter that even a single atom can give sufficient scattered intensity to allow it to be detected.

Although electron microscopy was first found to be feasible in the 1930s, it was only in the late 1940s that it became clear that the diffraction of electrons by crystals could have profound effects on the intensities of electron micrographs. For example, Bob Heidenreich (1942) at Bell Laboratories explained the appearance of thickness fringes in images of crystals in terms of the dynamical theory of electron diffraction. Before that and since then for perhaps 80 percent of electron microscopy, the wave nature of electrons and the possibility of diffraction effects could be conveniently ignored. For most biological samples, and many others without pronounced crystallinity, viewed with low resolution, it is sufficient to consider electron microscopy in terms of the geometric optics of particles which suffer an absorption effect in the specimen to give the image contrast. We will not concern ourselves with such work and so will ignore most of electron microscopy. Instead we will concentrate on the use of the technique for clearly crystallographic purposes: the determination of crystal structures, the study of crystal defects and the investigation of the configurations of large molecules and other small aggregates of atoms.

For all these crystallographic purposes it is essential to consider the interactions of electron waves with matter and to use the wave theory of image formation. In the simplest form, applicable for very thin films (10-50Å depending on the resolution being considered) of light-atom materials such as biological molecules, it is sufficient to consider the specimen as a weak phase object, changing the phase of the electron wave by a small amount. Then because the incident electron beam is well represented by a plane wave, it is possible to use coherent wave imaging theory. Because the scattering angles and the objective lens aperture sizes are all very small, it is possible to use the simplest of all phase contrast methods; that of defocussing the objective lens. Complications arise even for very thin films when heavier atoms are present. For crystal thicknesses of a few hundred Å or more, strong dynamical diffraction effects dominate the image intensities even for light-atom materials.

Two key developments were made in England in the 1950s. Hirsch and colleagues showed that dislocations and stacking faults in metal crystals could be clearly imaged and readily recognized: Menter showed that crystal periodicities in the 10Å range could be observed (see Hirsch et al. 1965). The former led to the widespread application of electron microscopy as the major tool in materials science for characterizing the extended defects of crystals and investigating their influence on physical and chemical properties of solids. The latter prompted the drive towards the imaging of crystal structures with ever-improving resolution until the goal was achieved, at least in part, of obtaining direct imaging of the arrangements of atoms in crystals, revealing the crystal structures and the form of the crystal defects in terms of atom positions.

Unfortunately for the biologists and organic chemists, the radiation damage produced by the incident electron beam is so severe that resolution of better than 10-20Å cannot, in general, be attained; but for many inorganic materials radiation damage is sufficiently low to allow imaging at the present-day resolution limits of 2-3Å.

Crystal defects with unresolved lattices

The information provided by electron microscopy on the nature and configuration of dislocations, stacking faults and other defects in crystals is quite unique. Most importantly, perhaps, it is possible to learn much about the interactions of the defects which are largely responsible for determining the mechanical properties of materials. The impact, initially on metallurgy and later on the whole of materials science, has been enormous. Following the demonstration in England of the power of electron microscopy in these fields, a number of research groups were set up to apply the methods in America. These include the groups led by Gareth Thomas at Berkeley, Bob Fisher at U.S. Steel, Ken Lawless and the Drs. Wilsdorf at the University of Virginia, Arthur Heuer at Case-Western Reserve, Ray Carpenter at Oak Ridge National Laboratory and others. An outline of the methods, the essential dynamical diffraction theory and the scope of the applications are given, for example by the book of Thomas and Goringe (1979). The applications in mineralogy and related areas are well described in the compilation edited by Wenk (1976).

In time all the groups named above acquired high voltage electron microscopes operating at 500keV to 1MeV. The main reason was that for the normal 100keV electrons the usable specimen thickness is limited to about 1000Å for medium-heavy elements. Especially for metals, the surfaces have such a strong influence on the dislocation configurations that the micrographs are not representative of bulk structure. The increase in penetration of higher energy electrons gives more reliable results.

A secondary use for high voltage microscopes developed from what was initially an inconvenience. As their energies increase, the incident electrons can cause radiation damage by knock-on collisions, displacing atoms from their lattice sites. It was soon realized that the radiation damage caused resembles closely that suffered by materials in nuclear reactors, but that in the electron microscope the damage rate could be thousands of times greater. The high voltage electron microscopes thus became primary tools for both creating and studying radiation damage effects. (Goringe, 1973).

Lattice fringe imaging

From the 10Å fringes observed by Menter, the progress in the detection of fringes due to crystal lattice periodicities has been steady until recently fringe spacings of less than 1Å have been recorded. Intersecting fringes, produced when several non-colinear Bragg reflections occur, have given two dimensionally periodic patterns showing detail on a comparable scale (Hashimoto et al. 1977). Such patterns are not images in the normal sense. They were obtained with microscopes of nominal resolution about 3Å. They may be regarded as interference patterns produced by a few sharply defined Bragg reflections given by a strictly periodic structure. Deviations from the periodicity can be imaged only with the normal (3Å) resolution.

Lattice fringe images of this sort normally show very little relationship to the crystal structure except to indicate periodicities. The positions of the intensity maxima and minima relative to the atomic planes depend strongly on dynamical scattering effects and vary with the crystal thickness and orientation and the defocus and aberrations of the microscope. However, if there is sufficient care taken in defining the experimental parameters and in interpreting the images, the fringe patterns can be used very effectively to reveal local changes of crystal structure and composition. In a series of studies of alloys and ceramics, Gareth Thomas and colleagues at Berkeley and Bob Sinclair at Stanford (see Sinclair 1978) have used lattice fringe imaging in the solution of problems of ordering and grain boundary structure.

By comparing observed image detail with that computed using many-beam dynamical theory it is possible to confirm conclusions drawn from images showing detail beyond the nominal resolution limit of the microscope. In this way John Spence and co-workers at Arizona State University (Spence et al., 1978) have been able to study the atom configurations within defects in silicon and other semiconductors.

Atomic resolution imaging of crystal structures

The possibility of determining the atomic arrangements in crystals directly from images is naturally of great significance for crystallographers. While X-ray diffraction structure analysis relies on the averaging over a vast number of unit cells, a direct imaging method can in principle show both the structure and the individual variations of the structure, revealing the atom configurations of particular defects, including those which occur only in low concentrations.

A major step towards this goal was made by Sumio Iijima (1971) of Arizona State University who showed that the heavy atom positions in some oxide crystals could be clearly represented with a resolution of about 4Å. Images from a sufficiently thin crystal, properly oriented, can be directly interpreted in terms of a projection of the structure, if obtained under well-defined instrumental conditions (Cowley and Iijima, 1977). The resolution limit has now been improved to better than 3Å for the usual 100keV or 200keV electron microscopes. By use of the special high resolution, high voltage microscopes operating at up to 1MeV and now in service in Japan and Europe, the resolution may approach 2Å or better. With the improvement of resolution a wide range of inorganic materials are accessible for the technique and a rapidly growing number of studies have been made on new crystal structures (especially superlattices) and on the form of defects in many types of material particularly for the non-stoichiometric oxides and sulfides (Anderson, 1979), a wide variety of minerals (Buseck, 1980), some alloys (Hiraga et al, 1979) and some semiconductors (Spence and Kolar, 1979).

Most of the materials contain medium weight atoms and for crystals oriented with the incident beam parallel to a principal axis the dynamical scattering effects become strong for thicknesses of 10-20Å so that the linear relationship between crystal potential and image intensity variation, expected for kinematical scattering, cannot be assumed. For thicknesses up to 50-100Å there is usually a clear, but non-linear relationship between image intensity and projected crystal structure but even this fails for very high resolution detail. The only way to ensure reliable interpretation of the image intensities is to compare the observations with intensities computed for the proposed structure, using many-beam dynamical theory.

For relatively large unit cells the number of simultaneous reflections which must be included in the dynamical calculations may be 500-1000. The calculations are then made most conveniently by use of computer programs based on the Cowley-Moodie multi-slice formulation of dynamic theory (see Cowley, 1975). For the calculation of image intensities for crystal defects, using an artifically imposed periodicity with a large unit cell, Michael O'Keefe and Sumio Iijima

(1978) have used calculations involving up to 12,000 beams.

For organic and biological materials, the radiation damage caused by the incident electron beam prevents the same sort of direct imaging of crystal structures. Here the main objective is to determine the periodic structure rather than the crystal defects so that it is possible to overcome the limitations of radiation damage to some extent by taking advantage of the periodicity. This is the basis for the method by which Unwin and Henderson (1975) obtained 7Å detail on the structure of the protein of the purple membrane. The diffraction amplitudes are obtained from electron diffraction patterns from large areas of thin crystal specimens while their relative phases are obtained from vastly underexposed images, from which the information on the content of one unit cell may be obtained by averaging the information from a very large number of unit cells. Bob Glaeser and associates at Berkeley (Glaeser, 1978) have worked for a number of years on application of such techniques. By the use of frozen sections and other methods to minimize radiation damage effects, they have obtained diffraction patterns from thin protein crystals showing information out to spacings of 4Å or less so that there is real hope of extending the effective resolution of images to this level. The particular value of this approach, as compared with the X-ray diffraction methods, is that it can be applied to the structure analysis of two dimensional crystals, such as occur in many membranes of biological significance, or of molecules which cannot form crystals more than 1 μm or so in diameter.

The imaging of molecules

The imaging of macromolecules which do not form periodic arrays is a potential advantage of electron microscopy which would be of great value to supplement the X-ray structure analysis results. The limitation of radiation damage is a severe one but various approaches have shown some signs of success in overcoming this.

Peter Ottensmeyer and associates (1979) of Toronto, have reported the recording of images of individual molecules showing recognizable structure on almost an atomic level of resolution. It appears possible that, although most biological molecules will be destroyed by the radiation level required for high resolution imaging, some may survive without major changes of structure. However these results are still not considered acceptable by many workers (Klug, 1979).

An alternative approach, aimed at locating the sites of particular reactive groups on a molecule, is that of tagging the group with heavy atoms in the hope that, even though the molecule may be destroyed by radiation damage, the heavy atoms will remain and be visible in the micrograph. The special ability of the scanning transmission electron microscope to show atom sites has been exploited for the purpose, especially by Michael Beer and associates (1979) at John Hopkins University. The results are promising but still limited by a number of factors, including the tendency of the heavy atoms to migrate under the electron irradiation.

At the level of resolution for which radiation damage is not a serious limitation or where staining techniques can give useful information on the outlines of molecular shapes, electron microscopy has provided a great deal of information on the conformation of biologically significant macromolecules. The methods of three dimensional reconstruction from multiple images obtained from different directions of the incident beam were developed by Klug and associates (e.g., Crowther et al., 1970) and have been applied very effectively to investigate structures, particularly of viruses and enzymes (see Frey, 1978) where repetition of a molecular unit in an ordered but non-periodic array provides many different views of the unit simultaneously in one image.

Scanning transmission electron microscopy

It was the development of the field emission gun as a practical electron source by Albert Crewe and associates (Crewe and Wall, 1970) which made the scanning transmission electron microscopy (STEM) instrument a useful high resolution device. STEM has provided a variety of new imaging modes and the serial form of its electronic image signal makes it ideally suited for quantitative image recording and the use of various techniques of image processing and image analysis.

The early applications of high resolution STEM were directed mainly towards biological problems but increasingly it is being applied to the crystallographic problem of solid state science. The correlation of imaging and diffraction information is very close and can be very powerful.

If the incident beam in a STEM instrument is stopped at any one point of a specimen, a convergent beam diffraction pattern coming from the illuminated region will be observed on the detector plane. Thus microdiffraction patterns from regions as small as 5Å in diameter can be recorded. In addition, by using detectors of suitable shape and size, any part or parts of the diffraction pattern can be selected to form the image when the beam is scanned over the specimen. With several detectors, several images can be recorded simultaneously. Special devices to capitalize on these possibilities have recently been developed (see Cowley and Spence, 1979).

The scanning scheme has also provided the basis for other developments. An electron beam striking a small specimen region can give a number of detectable signals. Some of these, such as the characteristic X-rays or Auger electrons given off, or the characteristic energy losses of the incident electrons, given information on the nature of the atoms present and so provide the basis for microanalysis of the irradiated region. Other signals, including the electron-induced conductivity or the cathodoluminescence provide data about the excited states of the specimen material. In each case, if the signal is collected as the incident beam is scanned over a specimen, an image is produced showing the spatial variation of the entity or interaction giving the signal. The image resolution, particularly for the microanalysis modes, is limited mainly by the relatively small cross-sections for the excitation processes involved. Currently the limits on determining distributions of individual atom types are of the order of tens of Angstroms.

While it is tempting to think that it may be possible to determine both the position and nature of each atom in a solid, the immediate prospects for achieving this goal are remote. In the meantime the combination of chemical information from microanalysis with the crystallographic information from diffraction and the morphological information from imaging is proving to be so valuable in practical applications in solid state science that its popularity is rapidly increasing. New electron microscopes are being designed specifically for the purpose and the subject of Analytical Electron Microscopy has been born and is flourishing (see Hren et al, 1979).

References

Anderson, J. S. (1978-9) Chemica Scripta **14**, 129-139.

Beer, M., Wiggins, J. W., Tunkel, D., and Stoeckert, C. J. (1978-9) Chemica Scripta **14**, 263-266.

Buseck, P. R. (1980) in "Electron Microscopy and Analysis 1979." Edited by T. Mulvey. Bristol: Institute of Physics. pp. 93-98.

Cowley, J. M. (1975) Diffraction Physics, Amsterdam: North Holland Publ. Co.

Cowley, J. M. and Iijima, S. (1977) Physics Today **30**, No. 3, 32-40.

Cowley, J. M. and Spence, J. C. H. (1979) Ultramicroscopy **3**, 433-438.

Crewe, A. V. and Wall, J. (1970) Optik **30**, 461-474.

Crowther, R. A., Derosier, D. J. and Klug. A. (1970) Proc. Roy. Soc. (London) **A317**, 319-340.

Frey, T. G. (1978) In "Electron Microscopy 1978 Vol. III." Edited by J. M. Sturgess, Toronto: Microscopical Soc. Canada. pp. 107-119.

Glaeser, R. M. and Hayward, S. B. (1978). in "Electron Microscopy 1978 Vol. III." Edited by J. M. Sturgess, Toronto: Microscopical Soc. Canada. pp. 70-77.

Goringe, M. J. (1973) J. Microscopie **16**, 169-182.

Hashimoto, H., Endoh, H., Tanji, T., Ono, A. and Watanabe, E. (1977) J. Phys. Soc. Jpn. **42**, 1073-1074.

Heidenreich, R. D. (1942) Phys. Rev. **62**, 291-292.

Hiraga, K., Hirabayashi, M. and Shindo, D. (1977) in "High Voltage Electron Microscopy 1977." Edited by T. Imura and H. Hashimoto, Tokyo: Japanese Soc. Electron Microscopy. pp. 309-312.

Hirsch, P. B., Howie, A., Nicholson, R. B., Pashley, D. W. and Whelan, M. J. (1965) Electron Microscopy of Thin Crystals, London: Butterworths.

Hren, J. J., Goldstein, J. I. and Joy, D. C., Eds., (1979) Introduction to Analytical Electron Microscoyp, New York: Plenum Press.

Iijima, S. (1971) J. Appl. Phys. **42**, 5891-5893.

Klug, A. (1978-9) Chemica Scripta **14**, 245-256 and 291-293.

O'Keefe, M. A. and Iijima, S. (1978) in "Electron Microscopy 1978, Vol. I." Edited by J. M. Sturgess. Toronto: Microcopical Soc. Canada. pp. 282-283.

Ottensmeyer, F. P., Bazett-Jones, D.P., Hendelman, R. M., Korn, A. P. and Whiting, R. F. (1978-9) Chemica Scripta **14**, 257-262.

Sinclair, R. (1978) in Electron Microscopy 1978 Vol. III., Edited by J. M. Sturgess. Toronto: Microscopical Soc. Canada. pp. 140-146.

Spence, J. C. H. and Kolar, H. (1979) Phil. Mag. **A39**, 59-62.

Spence, J. C. H., O'Keefe, M. A. and Iijima, S. (1978) Phil. Mag. **38**, 463-482.

Thomas, G. and Goringe, M. J. (1979) Transmission Electron Microscopy of Materials, New York: John Wiley & Sons.

Unwin, P.N.T. and Henderson, R. (1975) J. Mol. Biol. 94, 425-440.

Wenk, H. R. Ed. (1976) Electron Microscopy in Mineralogy. Berlin: Springer-Verlag.

CHAPTER 11

LOW-TEMPERATURE X-RAY DIFFRACTION

Reuben Rudman
Adelphi University
Garden City, NY 11530

Low-temperature X-ray diffraction (LTXRD) studies can be traced back to 1916, when St. John (1918) began his investigation of the crystal structure of ice in the Physics Laboratory of the Worcester Polytechnic Institute in Massachusetts. By 1922, the structure of mercury at 160K was reported (McKeehan and Cioffi, 1922), followed soon after by argon at 40K (Simon and Simson, 1924) and para-hydrogen at 1.65K (Keesom, de Smedt, and Mooy, 1930). More recently crystalline helium was studied at 60mk and 25 MPa (Heald and Simmons, 1977).

Instrumentation

LTXRD investigations require the use of a device for cooling the sample, while causing minimal loss of data, in addition to the three basic instrument components (source of X-rays, sample-positioning device, and detector). Since 1916, several hundred configurations have been devised in which the low-temperature system is either an integral part of the X-ray instrument or an accessory that can be installed or removed without affecting the normal operation of the instrument. Although the term *low-temperature*, as used in this chapter, includes the temperature range from just below ambient to just above absolute zero, any given low-temperature device is generally designed for optimum operation within a shorter temperature interval.

The low-temperature system itself consists of several components, which can be classified into three categories according to their function: provision for cooling the sample, control and measurement of the temperature, and frost prevention. A number of different approaches have been described in the literature for each of these components, and many successful modern low-temperature systems are based upon a combination of contributions from several sources.

The method used to cool the sample can be placed in one of three categories:

Gas Stream. A stream of cold gas (e.g., air, nitrogen, or helium), generated by boiling a liquefied gas or by passing warm, dry gas through a heat exchanger immersed in a cold bath, is directed over the sample. This method has the advantages of being relatively easy to set up and having minimum absorption problems. It suffers from the disadvantages of being most susceptible to frost formation and using relatively large amounts of cryogen for generating the cold gas stream. However, the latter disadvantage is largely off-set by recently developed mechanically refrigerated dewars and probes which can be used to replace the classical cryogens in many applications.

The first gas-stream system was described by Cioffi and Taylor (1922) and a similar system was used by Eastman (1924). However, this method did not become popular until a suitable method was devised for preventing frost-formation on the sample (Post, Schwartz, and Fankuchen, 1951). The development of this device, along with the basic principles of single-crystal growth, *in situ*, was presented in three papers appearing in 1949-1951 (Kaufman and Fankuchen, 1949; Abrahams, Collin, Lipscomb and Reed, 1950; Post, Schwartz and Fankuchen, 1951). A modern version of this device is shown in Figure 1. This device, used in conjunction with a suitable "cold-source" and heat exchanger, can be adapted easily to a variety of X-ray instruments.

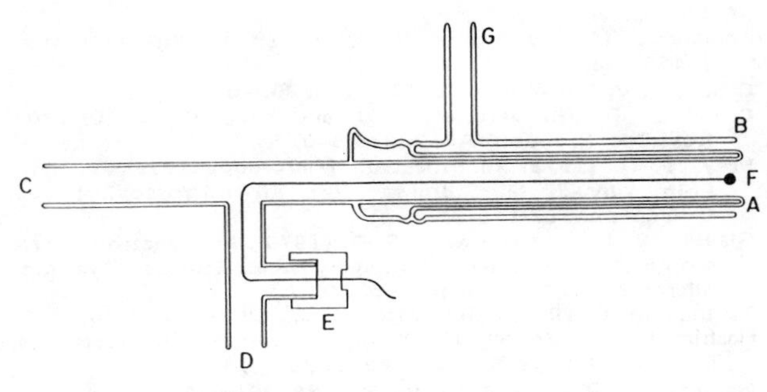

Fig. 1. The double-walled, evacuated and silvered tube (A) is surrounded by concentric tube (B). Cold dry gas enters through C and can be mixed with warm, dry gas entering through D. This permits fine control of the temperature at the outlet, thermocouple F. The thermocouple is inserted through a rubber serum cap (E) and lies along the inner tube. Dry gas at room temperature enters at G to form a protective sheath over the cold gas-stream thus preventing atmospheric vapors from condensing on the crystal. In some nozzles a heater over tube A is used in place of tube B to warm the outer portion of the cold-gas stream.

Conduction. The sample and sample chamber are cooled by means of a good thermal conductor in contact with a cooling unit which may be a cold bath, closed-cycle mechanical refrigerator, thermoelectric cooler, or Joule-Thomson expansion device. Advantages of this system include minimal use of cryogen, relatively good thermal equilibrium, and frost-free operation. Disadvantages are the need for a cooling chamber, difficulty in observing the sample, and errors introduced by absorption of the X-ray beam by the cooling chamber windows. The possibility of thermal gradients in the sample (due to poor conduction properties) must also be considered.

Immersion. Samples have also been cooled by immersing the entire camera in a cold bath or by dripping a cold liquid over the sample. This technique is not used very often, but has the advantage of maintaining a constant temperature (usually that of the cold liquid). Absorption problems and a lack of fine temperature control are some of the difficulties to be considered. A variation of this method is to place the entire apparatus in a cold room or cold chamber, with the temperature normally limited to a low of 220 K.

The temperature can be controlled by mixing a warm gas stream with the cold gas stream or by using a heater in the gas stream or near the conducting cold finger. In some applications precision control to within 0.01° is necessary while in others $\pm$5° accuracy is sufficient. A thermocouple, which may be attached to a recording potentiometer, is often used to monitor the temperature.

Atmospheric gases, most notably water and carbon dioxide, should be prevented from condensing on the sample and other points along the path of the X-ray beam. Gas-stream systems often employ an outer concentric stream of warm, dry gas which acts as a sheath around the inner cold stream, thus preventing atmospheric gases from reaching the cold sample. In the conduction method, a dry or evacuated chamber protects the sample. If vacuum insulation is not used, frost is prevented by directing a flow of warm, dry air over the cryostat windows or movable parts of the apparatus. For the immersion technique, the steady flow of cold liquid over the sample is generally sufficient to prevent condensation. In addition, it is often efficacious to construct a dry box or plastic tent over the diffraction apparatus so as to reduce the moisture content of the surrounding atmosphere.

Interestingly, in his study of ice, St. John (1918) appears to have used prototype versions of all three cooling methods. He initially cooled his sample by opening the laboratory windows and allowing the cold winter air to flow over his sample (gas-stream method). Since this was not satisfactory, he enclosed the camera in a chamber and cooled it with cans of ice and salt (cold-room immersion technique). Finally, he built a special spectrometer with a small ammonia refrigerating machine which was mounted in a well-insulated refrigerator box. No further details are given, but this may well have operated on the conduction principle*.

The early LTXRD measurements were often made in specially designed cameras and involved complex cooling techniques. Since the difficulties encountered were quite formidable and simple techniques for accurate investigations were not available, low-temperature studies were limited to only a few laboratories. However, during the late 1940's and early 1950's gas-stream cooling devices that could be assembled from common laboratory items and the techniques to use them were developed and popularized by several groups. These improvements led to the renaissance of low-temperature X-ray diffraction.

Many important chemical systems were studied (e.g. Lipscomb's work on the boron hydrides in the 1940s and 1950s, summarized in Lipscomb, 1977) and as more people became active in the field, a need arose for sophisticated and reliable apparatus. The current range of low-temperature instrumentation that is available includes an electronically controlled gas-stream device (useful for all types of powder and single-crystal studies to as low as 90K); closed-cycle liquid-helium systems for either powder or single-crystal diffractometers (temperatures between 10 and 20K can be maintained without the consumption of any liquid cryogens): and continuous-flow liquid-helium cryostats that have few restrictions on sample orientation. A detailed discussion of LTXRD apparatus and techniques, with a complete list of references is found in a recent book (Rudman, 1976).

No single system can be used for every low-temperature X-ray diffraction application, and, in fact, it would be a waste of resources to attempt to design such a system. The designer of any system must, therefore, know what the most likely applications will be and construct the system accordingly.

Some of the questions to be considered when

---

*In his pioneering study, St. John (1918) seemed to have encountered many of the problems that continued to plague future generations of LTXRD investigators. For the sake of historical interest, excerpts from his report are quoted here: "At first the apparatus already set up for another investigation was used, being kept cool by leaving the laboratory windows open. This method was very uncertain on account of the erratic weather and was otherwise unsatisfactory and was discarded after a single good photograph had been obtained. The spectrometer system was then enclosed and the chamber kept cool by cans of ice and salt. By this means the temperature could be kept reasonably constant but it was found virtually impossible to mount and maintain a specimen long enough to get a satisfactory photograph, probably on account of the presence of the salt vapor. Upon recommendation of Prof. A. G. Webster a grant was made from the Rumford Fund of the American Academy of Arts and Sciences, Boston, in aid of the investigation which made it possible to install a small ammonia refrigerating machine loaned by the Automatic Refrigerating Company and to build a specially adapted spectrometer mounted in a well-insulated refrigerator box. With this equipment the temperature could be maintained indefinitely and there was no further trouble from the melting of specimens. A marked tendency to sublimation, however, was troublesome until each specimen was mounted in a gelatine capsule when equilibrium was quickly established between the crystal and its vapor. Protected in this manner specimens were preserved for days . . . It was difficult to identify individual crystals and more difficult to isolate them for mounting but occasionally a reasonably good specimen was secured and mounted . . . Owing to the failure of the source of power the investigation was interrupted at this point and has not yet been renewed . . . The method is promising and is to be pursued further at a convenient season."

designing a system are:
- What is the minimum operating temperature?
- What are the sizes of the sample and sample holder?
- What is the maximum length of time that the apparatus will be operated continuously?
- Will the apparatus be used on one X-ray instrument only or must it be adaptable to several instruments?
- How often will it be used?
- What type of samples are to be studied?

Applications

The topics described in this section are representative of the areas in which LTXRD techniques have been applied. In some instances it is impossible to obtain the desired results without cooling the sample, while in others the use of low-temperature techniques serves to improve the quality of data which could also be obtained at room temperature.

Solidified gases and liquids. Methods have been developed for preparing single-crystal or polycrystalline samples of these materials directly on the diffraction apparatus or for growing them elsewhere and transferring them to the apparatus.

Although such studies are somewhat tedious, requiring skill in sample preparation as well as a carefully constructed and regulated apparatus, the data obtained from these investigations cannot be acquired by any other means. Important information concerning molecular structures, interatomic distances, and electron-density distributions in many simple molecules, fundamental to a proper understanding of molecular bonding theories, has been obtained in this way.

Examples include the crystal-structure determinations of mercury (McKeehan and Cioffi, 1922), ice (Barnes, 1929) chlorine (Collin, 1952; Stevens, 1979) diborane (Smith and Lipscomb, 1965), helium (Schuch and Mills, 1962), and carbon tetrachloride (Cohen, Powers, and Rudman, 1979).

Crystal-structure analysis. As the temperature of a sample is lowered, the thermal motion of the atoms within the sample is reduced. This factor is responsible for the preferred use of low-temperature data over room-temperature data in crystal-structure analyses, regardless of the physical state of the sample at room temperature. More data will be obtained, the peak-to-background ratio will be improved, the estimated standard deviations (esd) of the atomic co-ordinates will be lowered, and more accurate bond lengths and angles can be calculated.

A sophisticated treatment of thermal motion can improve the results obtained from room-temperature data. However, these correction factors are approximations which are most accurate for small librations. For a molecule which has a large amplitude of libration at room temperature, it is preferable to collect low-temperature data rather than attempt to correct room-temperature data, if accurate results are desired.

The advantages of using low-temperature data in crystal-structure analyses have been summarized by Coppens (1972). He shows that, in spite of the improved accuracy of the correction term for low-temperature data, it is also possible for a molecule to show a more pronounced thermal-motion anharmonicity at low-temperatures than at room temperature if the internal modes are less harmonic than the lattice modes.

Electron-density distribution. The electron distributions of atoms participating in chemical bonds differ from those of the free atoms. Theoretical evaluation of these differences shows that the calculated changes depend on the assumptions made for the wave functions of the molecules. Direct experimental information can, in principle, be obtained by X-ray diffraction as by this method the one-electron density functions in a molecule can be determined (Coppens, 1975).

Data should be collected at as low a temperature as possible so that more data can be collected (improved resolution in electron-density maps), thermal motion is reduced (less complex model in refinement), and thermal diffuse scattering is reduced (more accurate data).

Radiation damage. The interaction of X-radiation with a crystal often results in irreversible damage to the crystal. A steady decline in the intensity of the standard reflections measured during the course of data collection usually indicates that the crystal has decomposed (some reflections may even have their intensities increased). Radiation damage is strongly reduced or eliminated at low temperatures.

Protein crystallography. Petsko (1975) reviewed the many reasons for studying proteins at subzero temperatures and described techniques for cooling proteins to liquid nitrogen temperatures. The rate of radiation damage is markedly reduced. The rates of motion and exchange of loosely held groups are reduced, thus providing an improved image of "floppy" areas of the protein, and revealing details of conformational flexibility of backbone and side chains. In favorable cases, use of low temperatures may also provide a clear view of "bound" solvent molecules, giving valuable information about protein-water interactions and liquid structure. Another extremely important potential use for low-temperature protein crystallography is the direct observation of enzyme-substrate complexes and unstable intermediates. Such complexes can be stabilized in solution at subzero temperatures, and their lifetimes are long enough for high-resolution X-ray data collection.

Solid-solid phase transitions. A number of materials undergo transitions from one crystal structure to another as the temperature and/or pressure are changed. Many of these phases are unstable at room temperatures and can be studied only at low temperatures. Studies of this sort lead to an understanding of the mechanism of transition as well as a knowledge of the various structures.

Many investigations of plastic (orientationally disordered) crystals, metastable phases and cold-working of metals (Barrett, 1957) have been reported. High-pressure, low-temperature instruments have also been used and the transition to a superconductive state has been studied (Thorwarth and Dietrich, 1979).

Thermal expansion coefficients and precision lattice constants. The effect of temperature on the unit cell parameters and volume has been a fertile field of investigation for many years. Both powder and single-crystal samples have been examined. Many devices have been designed specifically for this purpose, and often a powder diffractometer or similar instrument is used with a large, platelike single crystal. Data on metastable, as well as high-pressure, modifications have been obtained.

For example, the lattice parameters and thermal expansion coefficients of aluminum, silver, and molybdenum to 30 K were determined using a symmetrical back-reflection vacuum camera with the sample holder attached to a closed-cycle, mechanically cooled refrigeration unit (Straumanis and Woodard, 1971).

Low-temperature powder diffraction. It is often simpler to obtain a poly-crystalline, rather than single-crystal, sample. Low-temperature techniques can be used to survey a group of isostructural compounds, to determine the occurence of major phase transitions, and to determine lattice parameter changes and coefficients of thermal expansion.

As an aid in indexing patterns of phases occuring at low temperatures, cameras have been developed for following the shifting of lines in the powder pattern as the temperature changes (e.g., Simon, 1971). For accurate work, the conduction method of cooling is generally used and over 100 different powder cryostats have been described in the literature (Table 4-2 in Rudman, 1976).

In conclusion, it is clear that low-temperature X-ray diffraction instrumentation has travelled a long road from the first studies of ice, in which the laboratory windows were opened during the winter, to the current use of a closed-cycle liquid-helium dilution refrigerator. Today, low-temperature techniques are routinely employed in many laboratories. The necessary apparatus can be assembled cheaply from common laboratory equipment or it can be custom-built for thousands of dollars. In the future, as miniaturized closed-cycle refrigeration devices are developed and electronic temperature regulators are improved, reliable LTXRD apparatus will become available at reasonable cost. It is instructive to remember that, for routine crystal-structure analysis, a set of data collected at 125 K can be satisfactorily refined using isotropic temperature factors. This tremendous saving of computer costs will pay for the initial investment in low-temperature equipment over a very short period of time. Eventually the use of LTXRD will be the normal means of collecting single-crystal data.

References

Abrahams, S. C., Collin, R. L., Lipscomb, W. N., and Reed, T. B. (1950) Rev. Sci. Instrum. **21**, 396-397.

Barnes, W. H. (1929) Proc. Roy. Soc. (London) **A125**, 670-693.

Barrett, C. S. (1957) Trans. Am. Soc. Metals **49**, 53-117.

Cioffi, P. P. and Taylor, L. S. (1922) J. Opt. Soc. Amer. **6**, 906-909.

Cohen, S., Powers, R., and Rudman, R. (1979) Acta Cryst. **B35**, 1670-1674.

Collin, R. L. (1952) Acta Cryst. **5**, 431-432.

Coppens, P. (1972) Proc. Adv. Study Inst. on Expt'l Aspects of X-Ray and Neutron Single Crystal Diffraction Methods, Aarhus.

Coppens, P. (1975) MTP Intn'l Rev. Sci., Ser. 2 - Phys. Chem., Chem. Cryst. **11**, 21-56 edited by J. M. Robertson. London: Butterworths.

Eastman, E. D. (1924) J. Am. Chem. Soc. **46**, 917-923.

Heald, S. M. and Simmons, R. O. (1977) Rev. Sci. Instrum. **48**, 316-319.

Kaufman, H. S. and Fankuchen, I. (1949). Rev. Sci. Instrum. **20**, 733-734.

Keesom, W. H., de Smedt, J., and Mooy, H. H. (1930) Proc. Acad. Sci. (Amsterdam) **33**, 814-819.

Lipscomb, W. N. (1977) Science **196**, 1047-1055.

McKeehan, L. W. and Cioffi, P. P. (1922) Phys. Rev. **19**, 444-446.

Petsko, G. A. (1975) J. Mol. Biol. **96**, 381-392.

Post, B., Schwartz, R. S., and Fankuchen, I. (1951) Rev. Sci. Instrum. **22**, 218-219.

Rudman, R. (1976) Low-Temperature X-Ray Diffraction: Apparatus and Techniques. New York: Plenum Press.

Schuch, A. F. and Mills, R. L. (1962) Phys. Rev. Lett. **8**, 469-470.

Simon, A. (1971) J. Appl. Cryst. **4**, 138-145.

Simon, F. and Simson, C. (1924) Z. phys. **25**, 160-164.

Smith, H. W. and Lipscomb, W. N. (1965) J. Chem. Phys. **43**, 1060-1064.

Stevens, E. D. (1979) Mol. Phys. **37**, 27-45.

St. John, A. (1918) Proc. Nat'l. Acad. Sci. U.S.A. **4**, 193-197.

Straumanis, M. E. and Woodward, C. L. (1971) Acta Cryst. **A27**, 549-551.

Thorwarth, E. and Dietrich, M. (1979) Rev. Sci. Instrum. **50**, 768-771.

CHAPTER 12

PHASE TRANSFORMATIONS AT LOW TEMPERATURES

Robert R. Reeber
Army Research Office,
Research Triangle Park, N.C. 27709

Phase transformations are complicated phenomena requiring spatial reorientation of atoms and/or chemical bonds relative to some original configuration. The driving force for change, a lowering of the Gibbs Free Energy in the transformed state, is a response to a change in intensive variables from initial conditions. The transformation process can be considered as the path resulting in either a change of symmetry and/or unit cell dimensions. Because transformation temperatures occur at greater than absolute zero, transformations and their associated kinetics must affect the lattice dynamics of atoms in the initial, intermediate, and end product crystal structure. One can readily appreciate that low temperatures have drastic effects on transformations that require a significant mobility or diffusion of atoms. For some systems true equilibrium can only be ascertained from studies of meteorites cooling through the eons of geological time (Scott and Clarke, 1979). At a given temperature, pressure, and other pertinent intensive variables a crystal is in dynamic equilibrium with its surroundings. At elevated temperatures, relative to the absolute melting point, energy dissipation within the crystal requires activation of relatively highly symmetrical optical vibrational modes. Upon cooling the crystal generally reduces its volume and energy dissipation may be accomplished less by optical modes and more by lower energy anisotropic longitudinal vibrational modes. One might expect for some materials as the anisotropy increases that alternate configurations may dissipate energy with lower vibrational frequencies. These may have varying degrees of metastability with respect to the one configuration with the lowest free energy. Unfortunately, or fortunately for the case of steel, no crystallographic demon exists to order each atom in its most appropriate place. If a mechanism does exist for a change in configuration by diffusion, shear, or to a metastable state by twinning and faulting, and such a process is energetically possible, such a transformation may occur at a rate dependent on the rate of the slowest component step of the process.

X-ray and neutron diffraction experiments have contributed perhaps the most significant part to our understanding of the structural stability and thermal behavior of solids. These have included low temperature experiments involving structural determinations, thermal expansion for chemical bonding information, and the determination of lattice vibrational spectra from neutron interaction experiments. The two earliest American X-ray studies that I have found reference to, that of ice by Ancell St. John (1918) and mercury by McKeehan and Cioffi (1922), were at the time exceptions rather than the rule. Perhaps this was to be expected, especially with so many interesting crystallographic problems to explore at room temperature. The added difficulties of reliably obtaining low temperatures could conceivably have discouraged all but the most resolute investigators. The fact that these difficulties were amenable to careful work is evident from the results obtained by Ellefson and Taylor (1934) at the University of Minnesota. In their work they calibrated two independent temperature sensors with a variety of low melting point organics and correlated small lattice parameter changes with reported changes in heat capacity for MnS, MnO, FeO, and Fe_3O_4. The 1930's saw only a limited amount of other phase change work at low temperatures. About the same time McFarlan (1936), working with Bridgman, determined structures for ice II and III. Some thermal expansion work, mercury by Hill (1935) and bismuth to -183°C with that one point below room temperature by Goetz and Hergenrother (1932), culminated the experimental investigations of what could perhaps be termed the first era of low temperature studies.

Concurrent with experimental work the review paper by Buerger and Bloom (1937) pointed out the need for quantitative relationships between crystal structure, free energy and "lattice energy". This work qualitatively addressed other fundamental concepts. These included dynamic thermodynamic stabilities and the tendency to higher symmetries at higher temperatures as a direct consequence of the fact that isotropic vibrational systems attain higher heat capacities than less symmetrical anisotropic ones. Volume contractions upon cooling in general lead to the latter. These ideas were perhaps building on those developing in the U.S.A. through Joseph Mayer and Levy (1933) from earlier contacts with Max Born in Europe. Understanding the lattice dynamics of different structures is important for phase changes since the vibrational nature of a solid determines feasible paths for transformation as well as the vibrational contribution to the total free energy at a given temperature. Theoretical developments, Fine's (1939) calculation of the frequency spectrum of a body-centered lattice by numerical integration and Montroll's (1941) approximate analytical methods, were American contributions that helped in the understanding of possible vibrational differences between different structures. Leighton (1948) extended this work to the face-centered lattice.

Significantly reduced activity during World War II lead to what can perhaps be called the second era of low temperature phase transformation studies. Interest in low temperature phase changes of the alkali metals and the experimental verification of theoretical predictions lead Barrett (1947, 1956) to design and use a low temperature X-ray apparatus for the discovery of strain-induced structural changes in lithium and sodium. In 1947 Buerger published another significant paper on derivative crystal structures. This paper and a later one in 1948 were important trends toward improving our classification of different types of transformations as well as indicating certain symmetry constraints to be expected at lower temperatures. The role of symmetry constraints has been more recently developed by de Fontaine (1975, 1979).

Concurrent with advances in theory, 1949 to 1953 showed a renewed interest in cryogenic X-ray methods initiated by Fankuchen, Abrahams, Lipscomb and their co-workers (Rudman, 1976, has a complete summary). Barrett's work is perhaps typical of the major efforts addressing low temperature phase transformations in the late 1940's through the decade of the 1950's. They were related more to the individual experimenter's special interests than to any concerted effort at understanding the interrelationships of solids at this temperature range vis-a-vis room temperature or above. Professor Lipscomb, for instance, at the University of Minnesota, unraveled many aspects of the molecular nature of boron chemistry with low temperature experiments while at the same time educating many of those destined to teach chemistry at other universities. Professor Post at the Polytechnic Institute of Brooklyn, Professor Templeton at Berkeley, both with a long succession of co-workers, applied crystallographic methods toward the investigation of phase transformations and bonding of a wide variety of chemicals. In the late 1950's and early 1960's Barrett continued his helium experiments that permitted the cold-working of metals below 10 K. Greater temperature stability and control electronics made the whole cryogenic range accessible on a routine basis. Black, Peiser, Bolz and others (1958) at the National Bureau of Standards were involved in this equipment development process that accompanied phase transformation studies and the structural studies of inert gas solids.

Just preceding this period perovskite transitions were being systematically evaluated by a number of workers including Matthias (1949) and Shirane working with Pepinsky and others (1954). Much if not all of this work is reviewed by Galasso (1969). This groundwork combined with higher temperature experiments led O'Leary and Wheeler (1970) to bring together ideas of Landau and the theory of lattice dynamics for partial explanations of observed transformation behavior. Some of these ideas can be related to Buerger's derivative crystal structures mentioned earlier. Increasingly in the 1960's neutron diffraction became an important tool for the study of displacive transformations. Axe (1971) has summarized this work. Shull and Wollan (1956) have reviewed neutron diffraction applied to ammonium halide transformations in a comprehensive paper that illustrates the role of this method for structural studies involving hydrogen.

Volume changes at a constant temperature are characteristic of first-order phase transformations. The measurement of thermal expansion and its anomalous behavior is therefore of interest in the study of phase transformations. The experimental and theoretical work on lattice dynamics lead to interesting predictions of negative thermal expansion with increasing temperature for some substances at low temperatures. LaPlaca and Post (1960) were able to show that ice was one of what turns out to be a number of solids that exhibit such behavior. During the 1960's many additional workers; including Bienenstock and Burley (1963) for AgI, Batchelder and Simmons (1964) for silicon and Reeber (1970) for ZnO (this summarizes CdS, ZnS and CdSe work also); have examined this phenomenon with X-ray methods. Straumanis and co-workers (1971) meanwhile were systematically working to obtain needed volume versus temperature data for a variety of other materials. This has been complemented by the development of a semiempirical method for accurately fitting thermal expansion data with a multi-frequency model of the material, Reeber (1975). The fit is excellent at all but the lowest extremes of the cryogenic range and can be extrapolated to 1.5 to 2 times the Debye temperature.

Also in the 1960's the continuing advances in computer, electronic, and cryogenic technology led to refinements and sometimes corrections of the structures done earlier by powder methods. These developments, discussed elsewhere in this volume, have continued through the decade of the 1970's until now we find real time Laue cameras, Bilderback (1979) and novel closed-cycle cooling goniometers, Samson et al., (1960). Such methods combined with modifications of the Laue developed earlier, Reeber (1975b), could give us continuous observations and records of low temperature phase transformations as they proceed.

Applications of simple lattice models to known crystal structures, Reeber and McLachlan (1971), have enabled the calculation of lattice characteristic temperatures. These have been useful for correlating low temperature phase transformations in crystallochemically and isostructurally similar compounds, Reeber (1976).

Some of the broad outlines of future directions are well within the realm of observation and imagination. More applications of the principles of lattice dynamics to practical examples, with better approximations to interatomic forces and defect configurations combined with continuing advances in computations, will lead to more accurate representations of existing structural relationships and volume changes as a function of temperature and pressure. In addition to these baseline studies, experiments where transformations are activated by higher intensities and/or more controlled energy spectral distributions will help differentiate the subtle differences between metastability and stability.

Such experiments will, when combined with new theoretical interpretations, lead directly to many practical benefits in the use of materials by mankind. The thirty to sixty year timing for such advances will depend equally on our ability to convince decision makers of the ultimate rewards from basic research and to convince our brightest students that the frontiers of crystallography and materials science will continue to be fascinating regions for exploration.

References

Abrahams, S. C., Collin, R. L., Lipscomb, W. N. & Reed, T. B. (1950). Rev. Sci. Instrum. **21**, 396.
Axe, J. D. and Shirane, G. (1973). Physics Today **26**, 32.
Barrett, C. S. (1947). Phys. Rev. **72**, 245.
Barrett, C. S. and Trautz, O. (1948). Trans. Metall. Soc. AIME. **175**, 579.
Barrett, C. S. (1956). Acta Cryst. **9**, 671.
Batchelder, D. N. and Simmons, R. O. (1964). J. Chem. Phys. **41**, 2324.
Bienenstock, A. & Burley, G. (1963). J. Phys. Chem. Solids, **24**, 1271.
Bilderback, D. H. (1979). J. Appl. Cryst. **12**, 95.
Black, I. A. Bolz, L. H. Brooks, F. P., Mauer, F. A. and Pelser, H. S. (1958) J. Res. Nat. Bur. Std. **61**, 367.
Buerger, M. J. and Bloom, M. C. (1937). Z. Kristallogr. **A96**, 182.
Buerger, M. J. (1947). J. Chem. Phys. **15**, 1.
Buerger, M. J. (1948) Am. Mineral. **33**, 101.
Ellefson, B. S. and Taylor, N. W. (1934). J. Chem. Phys. **2**, 58.
Fine, P. C. (1939). Phys. Rev. **56**, 355.
Fontaine, D. de, (1975). Acta Met. **23**, 553.
Fontaine, D. de (1979). Solid State Physics, Volume 34, pp. 73-272, edited by H. Ehrenreich, F. Seitz, and D. Turnbull. New York: Academic Press.
Galasso, F. S. (1969). Structure, Properties and Preparation of Perovskite-type Compounds. Volume 5. International Series of Monographs in Solid State Physics. Oxford: Pergamon Press.
Goetz, A. and Hergenrother, R. C. (1932). Phys. Rev. **40**, 643.
Hill, D. M. (1935). Phys. Rev. **48**, 620.
Kaufman, H. S. and Fankucken, I. (1949). Rev. Sci. Instrum. **20**, 733.
LaPlaca, S. and Post, B. (1960). Acta Cryst. **13**, 503.
Leighton, R. B. (1948). Rev. Mod. Phys. **20**, 165.
Matthias, B. T. (1949). Phys. Rev. **75**, 1771.
Mayer, J. E. & Levy, R. B. (1933). J. Chem. Phys. **1**, 647.
McFarlan, R. L. (1936). J. Chem. Phys. **4**, 60; ibid. **4**, 253.
McKeehan, L. W. and Cioffi, P. O. (1922). Phys. Rev. **19**, 444.
Montroll, E. W. (1942). J. Chem. Phys. **10**, 218.
Reeber, R. R. (1970). J. Appl. Phys. **41**, 5063.
Reeber, R. R. and McLachlan, D. Jr. (1971). Can J. Phys. **49**, 2287.
Reeber, R. R. (1974). Phys. Stat. Sol. **A26**, 253.
Reeber, R. R. (1975a). Phys. Stat. Sol. **A32**, 321.
Reeber, R. R. (1975b) Z. Kristallogr. **141**, 465.
Reeber, R. R. (1976). The Physics and Chemistry of Minerals and Rocks. pp. 469-477, edited by R. G. J. Strens. London, John Wiley, Proc. Conf. Newcastle upon Tyne. April 1974.
Rudman, R. (1976). Low-temperature X-ray Diffraction. New York: Plenum Press.
Samson, S., Goldish, E. and Dick, C. J. (1980). J. Appl. Cryst. **13**, 425.
Scott, E. R. D. and Clarke, R. S. Jr. (1979). Nature (London) **281**, 360.
Shull, C. G. and Wollan, E. O. (1956). Solid State Physics, Vol. 2, pp. 137-217, edited by F. Seitz and D. Turnbull. New York: Academic Press.
Straumanis, M. E. and Woodward, C. L. (1971). Acta Cryst. **A27**, 549.

CHAPTER 13

SMALL-ANGLE X-RAY SCATTERING

Paul W. Schmidt
Physics Department, University of Missouri, Columbia, MO 65211

Small-angle X-ray scattering is a technique which is useful for studying structures which are large compared to the wavelength of the scattered radiation. Since the X-ray wavelengths usually used in diffraction and scattering are of the same dimensions as atoms and small molecules, small-angle scattering can be employed to study structures on a larger-than-atomic scale.

The scattering pattern is the result of the interference effects produced by the superposition of the electric fields scattered by the inhomogeneities in the scattering sample. It often is convenient to express the scattering pattern as a function of $h = 4\pi\lambda^{-1}\sin\theta$ where λ is the X-ray wavelength, and 2θ is the scattering angle. For structures with a characteristic dimension D, the most important interval of scattering angles is specified by the condition $0 \leqslant hD \leqslant 2\pi$. Thus there is an inverse relationship between h and the size of the structure producing the scattering. In the small-angle region, which by convention has come to mean scattering angles 2θ no greater than a few degrees, θ can be considered to be proportional to h.

The reason that the scattering from structures with a characteristic dimension D much larger than the X-ray wavelength λ occurs primarily at *small* angles is that when $D/\lambda \gg 1$, $\sin\theta$ must be small in order to satisfy the condition $0 \leqslant hD \leqslant 2\theta$.

When the scattering centers scatter independently, the scattering pattern gives information about the form and dimensions of the scattering centers. Samples which have been studied by small-angle X-ray scattering include suspensions of biological macromolecules, inorganic colloids, and polymers. In my review I will mention some of the people and laboratories that have worked on these samples.

I learned about some of the earliest reports of small-angle X-ray scattering after Newell Gingrich recently told me that C. H. Slack (1926), working with a double-crystal spectrometer to determine the index of refraction of a number of materials at X-ray wavelengths, noticed that when the samples were in the form of finely divided particles, the rocking curve of the spectrometer was broadened by angles up to one minute of arc. Slack qualitatively interpreted his observations by suggesting that the X-rays were deviated by being refracted in a number of small particles. Von Nardroff (1926) soon published a calculation of the effects of refraction of X-rays by small spherical particles, and Davis (1927), at the end of a fairly detailed description of the double-crystal spectrometer which he and Slack used, listed Slack's 1926 scattering results and showed that within the experimental uncertainty, the broadening of the beam is proportional to the square root of the sample thickness, as von Nardroff predicted. The particles, which were large enough to be measured in an optical microscope, had dimensions in qualitative agreement with the sizes calculated from von Nardroff's theory.

As Beeman et al. (1957) point out in their review of small-angle X-ray scattering, the refraction theory developed by von Nardroff is valid for particles with dimensions greater than several microns, like the particles Davis and Slack worked with, but for the sub-microscopic particles and structures with which most small-angle X-ray scattering work is concerned, a different theory, often called the Rayleigh-Gans approximation, must be used. Thus, although for the relatively large particles considered by Slack and Davis, the refraction theory applies, the scattering, which is restricted to an interval of scattering angles much narrower than in most small-angle scattering investigations, really isn't quite the same effect as what is normally called small-angle X-ray scattering.

In a short letter, Raman and Krishnamurti (1929) mention the small-angle scattering of X-rays from natural graphite and attempt to explain the scattering by ascribing it to the effects of conduction electrons. Soon afterward, Krishnamurti (1930a; 1930b) described scattering studies of several amorphous carbons and some colloidal solutions and liquid mixtures. From the small-angle scattering pattern and the large-angle diffraction data, Krishnamurit concluded that the carbon particles were platelets. He calculated their thickness from the width of the first diffraction line and estimated the large dimension of the platelets by setting it equal to the Bragg distance corresponding to the scattering angle at which the small-angle scattering became negligible. For the eolloidal structures considered in his second paper, Krishnamurti (1930b), assumed the particles to be roughly spherical and found their linear dimensions by the same technique he used to obtain the large dimensions of the carbon platelets. Krishnamurti's interpretation of the scattering curves and his order-of-magnitude estimates of the particle sizes and molecular dimensions and weights were, I believe, essentially correct.

Gray and Zinn (1930) report that they and M. F. McDonald, working at Queen's University in Kingston, Ontario and using a set of slits to collimate the incident and scattered beams, measured the X-ray scattering from charcoals at scattering angles greater than about one-half degree. This interval of angles includes much of what is now called the "small-angle" region, and so the scattering discussed by Gray and Zinn can be considered the "usual" or "traditional" small-angle scattering. Gray and Zinn show scattering curves from two different charcoals and point out that the scattering patterns for the two charcoals are quite different.

In addition, Gray and Zinn present quite convincing evidence that for scattering angles between about 0.5 and 2.5 degrees, the scattered intensity from one of the charcoals was very nearly proportional to the inverse fourth power of the scattering angle. They thus observed what later experimental (Van Norstrand and Hack, 1953) and theoretical (Porod, 1951) investigations showed was a very important property of the outer part of the small-angle X-ray scattering curves from samples in which there is a sharp discontinuity in the electron density at the surface bounding the scattering centers. (The inverse-fourth-power scattering occurs at scattering angles satisfying the condition $hD_{min} \gg 1$, where D_{min} is the smallest dimension characterizing the scattering centers.) Though Gray and Zinn very cautiously point out that because of the possibility of errors in their techniques, they aren't especially confident of their results, I feel that it's clear that they noticed an effect which Van Nordstrand and Hack independently discovered over 20 years later in an experimental investigation of the small-angle scattering from some porous catalysts. Van Nordstrand and Hack, however, not only established the angular dependence in the outer part of the scattering curve but also found that at these scattering angles the intensity was proportional to the surface area bounding the scattering centers in the sample. These results, for which Porod (1951) had provided a theoretical basis, make it possible to employ small-angle X-ray scattering to determine the specific surface of materials with submicroscopic inhomogeneities, such as porous catalysts. The study of specific surfaces has become a very important application of small-angle X-ray scattering.

In a later paper, Gray (1935) gives a preliminary report of the Kingston group's later small-angle scattering studies, in which they examined more samples and found that the results could be classified in two groups. One of the two types of small-angle scattering was produced by coarse particles, with dimensions of some microns, and the second type was observed for samples with submicroscopic inhomogeneities. Gray et al. correctly, I believe, ascribe the first kind of scattering to refraction effects and use von Nardroff's theory (1926) to analyze their results. From the Debye scattering equation (Debye 1915), which gives the average scattering from randomly-oriented scatterers, they estimated the dimensions of the particles in samples giving the second type of small angle scattering, which corresponds to what nowadays is called small-angle scattering.

In a later paper, Penley and Gray (1937) provide more details about the work mentioned in Gray's brief preliminary report (1935) and show clearly that they correctly understood the sources and qualitative behavior of the small-angle scattering which they observed. At the end of their short paper, Penley and Gray say that they are continuing their scattering studies. Though they promise a more quantitative and detailed account, I have never seen any later work from the Kingston group. If they had published the more extended papers which they apparently intended to write, especially if their results had appeared in a journal with wider readership than the *Canadian Journal of Research*, their work probably would have received the attention I believe it deserves, and they would have been quite generally acknowledged to be among the first people who worked with small-angle X-ray scattering.

In a paper describing an X-ray study of the structure of carbon black, and also in an abstract of a paper presented at a meeting of the American Physical Society in 1936, B. E. Warren (Warren-1934; Warren 1936) mentions that he has observed X-ray scattering at small angles and states that this scattering is due to the small size of the particles. Though in his 1934 paper Warren does not attempt a quantitative description of the scattering, I think it's clear that by 1936 Warren was well on the way to understanding what small-angle X-ray scattering was and how it could be applied.

André Guinier, working in France during the late 1930's, devoted his Ph.D. thesis research to a detailed and thorough study of small-angle X-ray scattering. He obtained many of the fundamental equations employed for analysis of small-angle scattering data and applied these results in studies of the small-angle scattering from a number of samples. Though, as I have pointed out, Guinier was not the first person who worked with small-angle X-ray scattering, he was the first to have made an extended and detailed study of the subject. With his insights and thorough understanding of diffraction physics and his years of careful study, he was able to develop small-angle scattering into a technique which could be applied in many areas of research. I have always had the impression that Guinier is quite generally considered to be the person who discovered small-angle scattering, and I agree that he deserves this recognition.

Soon after Guinier began to work with small-angle scattering. O. Kratky became aware of the possibilities of this technique, and after he was appointed Professor of Physical Chemistry at the University of Graz, in Austria, he devoted most of his research efforts to developing the small-angle scattering laboratory which is now one of the world's foremost centers in this field.

Kratky's and Guinier's accounts of their early small-angle scattering work have recently been published (Kratky 1978, Guinier 1978).

While Guinier was working in France, Warren must still have been thinking about small-angle scattering, since in 1942 he and his student J. Biscoe (Biscoe and Warren, 1942) described an X-ray study of carbon black in which they made considerable use of small-angle scattering and employed some results from Guinier's work. Though small-angle X-ray scattering has never been a large part of Warren's research, between 1940 and 1960 he and his co-workers published a number of significant small-angle scattering papers.

A. S. Eisenstein, while studying the large-angle X-ray scattering from liquid and gaseous argon at the University of Missouri during the early 1940's, noticed that when the fluid was near its liquid-vapor critical point, there was a large increase in the X-ray scattering at small angles. (Under most conditions, liquids have no long-range structure and thus give very little small-angle scattering.) Eisenstein and his thesis advisor, N. S. Gingrich, spent considerable time trying to eliminate what they first believed must be an experimental artifact. After many efforts to adjust and improve their apparatus in order to eliminate the small-angle scattering, they concluded that the effect was real. In a paper reporting their work on argon, Eisenstein and Gingrich (1942) ascribe the small-angle scattering to a relatively long-range structure in the fluid and show in their discussions of this scattering that they were aware that they were measuring what now is known as small-angle X-ray scattering. They realized that this increased scattering was associated with the fact that the fluid was near its critical point, and they outline a technique for quantitatively estimating the dimensions of the scatterers, which they considered to be droplets of liquid in the vapor or bubbles in the liquid.

After his wartime assignments, Gingrich returned to the University of Missouri, and between 1945 and 1953, three of his Ph.D. students devoted their thesis studies to projects employing small-angle scattering. R. L. Wild (1950) investigated the small-angle scattering from nitrogen near the critical point and discussed many of the properties that now are recognized to be characteristic of the critical scattering from a fluid. About the same time, K. L. Yudowitch, another of Gingrich's students, began to investigate some applications and possibilities of the small-angle technique. Yudowitch showed considerable insight in his attempts to allow for collimation effects, which are present because the apparatus does not measure the radiation scattered only at a single scattering angle, as is assumed in most theoretical calculations, but instead averages the scattering over an interval of scattering angles around the nominal scattering angle. He also studied the scattering from some colloidal systems and observed the subsidiary maxima in the scattering pattern from spherical particles. In some of his measurements, Yudowitch obtained higher resolution by using wavelengths longer than the copper and molybdenum $K\alpha$ wavelengths usually employed in scattering and diffraction. After finishing his work at the University of Missouri, Yudowitch continued working with small-angle scattering and was active in the field through the mid-fifties. He prepared the small-angle-scattering bibliography which is at the end of the book by Guinier, Fournet, Walker, and Yudowitch (1955) on small-angle scattering. (This book was the first – and still is the only – textbook on small-angle scattering.)

J. W. Buttrey, Gingrich's last student to work with small-angle scattering, studied the critical scattering from four paraffin hydrocarbons – methane, ethane, propane, and butane (Buttrey, 1957).

G. H. Vineyard, now director of Brookhaven National Laboratory, joined the physics department at the University of Missouri in 1946. In one of his first publications there (Vineyard, 1948), he developed a quantitative method for analyzing the scattering from a fluid near its critical point in terms of the droplet model. With L. H. Lund and some of his other students, he also investigated other topics in small-angle scattering theory.

In the early 1940's, I. Fankuchen and others at the Brooklyn Polytechnic Institute began to work with small-angle X-ray scattering. Fankuchen had previously worked in England with J. D. Bernal on small-angle diffraction investigations of virus crystals. In a paper on the small-angle scattering from polymers, Fankuchen and Mark (1944) list the scattering angles for particles with different radii of gyration at which, according to the Guinier approximation, the scattering decreases to half its value for $2\theta = 0$. Between 1945 and 1949, Fankuchen, Jellinek and Solomon published a series of papers (Jellinek and Fankuchen 1945; Jellinek, Solomon and Fankuchen 1946; Jellinek and Fankuchen 1949) in which they describe the construction of a small-angle scattering system with Geiger-counter detection and use the apparatus to study the particle dimensions of alumina gels subjected to different heat treatments. They developed a quite sophisticated technique for finding not only the average particle dimensions but also the distribution of dimensions. In their investigation of alumina gels, rather than employing only small-angle scattering to obtain the dimensions of the particles, Jellinek and Fankuchen compared these dimensions with those calculated from the line widths of the large-angle powder diffraction patterns. The work of Jellinek, Solomon and Fankuchen was one of the first cases in which small-angle scattering, rather than being the subject of the investigation, was employed as a technique for obtaining new information.

By 1945 small-angle X-ray scattering had begun to move out of the universities and into industrial laboratories. At the Texas Company's laboratory in Beacon, N.Y., P. B. Elkin, C. G. Shull and L. C. Roess (1945) used small-angle scattering, nitrogen adsorption, and large-angle X-ray powder patterns to determine both the average dimensions and also the dimension distributions of particles. Shull and Roess (Shull and Roess, 1947; Roess and Shull, 1947) developed techniques for analysis of scattering curves and for collimation correction and also calculated the scattered intensity from polydisperse samples of ellipsoids of revolution with several diameter distribution functions. Roess, who I think could be called the theorist of the group, found (1946) an exact solution for the integral equation from which the diameter distribution for a system of spherical particles with uniform electron density can in principle be calculated from the small-angle scattering curve. During the 1950's the theoretical work of Shull and Roess was used frequently in the interpretation of small-angle scattering data.

About 1945, while investigating the fine structure of X-ray absorption edges, W. W. Beeman and some of his students at the University of Wisconsin in Madison noticed that when the crystals of a double-crystal spectrometer were placed in the parallel, or (1, -1) position, the X-ray beam was broadened when something with a macromolecular structure, such as a piece of paper, was placed between the crystals. Paul Kaesberg, who was ready to begin his thesis research and who was considering working with X-rays, at Beeman's suggestion looked through the literature to find out what was known about this effect. Kaesberg soon learned about the work that Guinier and Fankuchen's groups had done, and he and Beeman realized the possibilities of using small-angle scattering for investigating macromolecules in suspension. They became especially excited about using X-ray scattering to study the hydration of biological molecules. (These investigations are possible because the X-ray scattering is affected only by the internal hydration, or swelling, and does not "see" any water bound to the outside of the macromolecule.) Beeman (1978) has reviewed some of the early work in his laboratory.

Beeman and Kaesberg soon were joined by Harold N. Ritland, another doctoral student of Beeman's, and the group began to set up the necessary equipment for studying small-angle scattering.

Although they first planned to use the double crystal spectrometer, they soon found (Ritland, Kaesberg and Beeman, 1950a) that the background from the diffuse scattering was so great that the spectrometer could be used only for highly-scattering samples or at scattering angles no greater than a few seconds of arc.

To eliminate the high background that was unavoidable with the double crystal instrument, Beeman, Kaesberg and Ritland devised a system of slits which defined the incident and scattered X-ray beams. The unique feature of this collimation system, which is often called the Beeman four-slit system or simply the Beeman system, is a third slit placed between the sample and the slit just in front of the detector. This additional slit appreciably reduces the scattering from the two slit edges closest to the X-ray tube and thus makes possible the study of weakly scattering samples like suspensions of protein or viruses.

Ritland, Kaesberg and Beeman (1950b) used the new collimation system to obtain radii of gyration for several proteins in solution. By fitting theoretical scattering curves calculated by Porod, who then was the theorist in Kratky's group at Graz, Ritland, Kaesberg and Beeman used the scattering data from the outer parts of their scattering curves to estimate the elongation of the protein molecules. This study was the first of the Wisconsin group's many investigations of biological macromolecules.

The success of Beeman's group and the group's rapid growth in the 1950's is, I think, to a great degree the result of the fact that at this time conditions at the University of Wisconsin were especially favorable for the formation of a group interested in applying physical techniques in biological research. Without this ideal climate for biophysics, I don't think that even a person like Beeman, with his wide interests and abilities and his excellent physical insights, could have made such great progress. At Wisconsin, some of the best and most advanced instruments for biochemical and biophysical research, such as the ultracentrifuge and the electron microscope, were available. In addition, biochemistry had for a long time been a strong, active and productive department, and there were many biochemists, physical and protein chemists, and molecular biologists who recognized the possibilities of small-angle X-ray scattering and were willing to work with Beeman and his students to suggest important subjects for research and, especially during the first years of the small-angle scattering group's work, to help prepare the samples. Finally, during the 1950's, molecular biology was expanding quite rapidly, and research grants were relatively plentiful. The excellent reputation of the physics department and of biochemical and molecular biology research at the University of Wisconsin considerably helped, I'm sure, attract grants to support small-angle X-ray scattering.

To obtain a higher scattered intensity from the suspensions of proteins and viruses, which are quite weak scatterers, Bowen R. Leonard and Beeman designed and built a rotating-anode X-ray tube which provided 5 to 10 times the intensity available from a standard diffraction tube. Leonard et al. (1953) used their new tube to obtain scattering data from suspensions of spherical plant viruses.

John W. Anderegg, who joined the group not long after Leonard, made a detailed investigation of the small-angle X-ray scattering from solutions of bovine serum albumin (Anderegg et al., 1955). In addition to obtaining precise values of the radius of gyration, he measured the scattering at angles so large that the radius of gyration approximation does not apply. From this part of the small-angle scattering curve, he was able to learn something about the shape of the molecule and estimate the hydration of the molecule in solution. With David L. Dexter, Anderegg also made a quantitative study of collimation effects and applied the new correction techniques to his scattering data. One of the most important contributions from this work on collimation effects, I think, is the introduction of the concept of the weighting function, which is fundamental in the theory of collimation corrections.

In 1949 I began working in Beeman's laboratory, and in the fall of 1953, after finishing my thesis, I joined the faculty of the physics department at the University of Missouri. I arrived just a little before John Buttrey, the last student of Gingrich's to work with small-angle scattering, completed his studies of the scattering from hydrocarbons near their critical point. The small-angle scattering activity at the University of Missouri thus has continued without much interruption.

Harold D. Bale, my first Ph.D. Student, used small angle X-ray scattering to investigate the colloidal structure of aluminum hydroxides produced under different conditions (Bale and Schmidt, 1958; Bale and Schmidt, 1959). When he finished his work at Missouri, Bale joined the physics department of the University of North Dakota, where he set up a small-angle scattering laboratory and has used small-angle scattering in a number of investigations. Another of my students, James E. Thomas, who in his thesis research studied the critical small-angle scattering from argon and nitrogen (Thomas and Schmidt, 1963; Thomas and Schmidt, 1964), also has continued working with X-rays. In his thesis, H. Wu dealt with several topics in small-angle scattering theory, and he has collaborated with me on a number of later papers, including studies of chord distributions (Wu and Schmidt, 1971; Wu and Schmidt, 1974). To stay within the limits of the *short* history I was asked to write, I'm afraid I'll have to refrain from mentioning the work of any of my other students.

Anderegg, after completing his Ph.D. research in 1952, spent a postdoctoral year in England and then returned to the physics department at the University of Wisconsin. In his small-angle scattering research, he has concentrated on investigations of biological macromolecules. While Beeman has also been involved in this work, he has studied a number of other problems and has developed many new small-angle scattering techniques.

One of these investigations of Beeman's which I would like to discuss in more detail is the small-angle X-ray scattering from cold-worked metals. In France, Guinier and his associates observed in the early 1950's that when metal foils were plastically deformed, the small-angle scattering from the foils increased appreciably. This scattering was ascribed to the production of vacancies during cold-working. Further deformation was believed to make the vacancies cluster into large enough groups to produce noticeable X-ray scattering at small angles. In the early 1950's many people therefore became very interested in trying to use small-angle X-ray scattering in metallurgical research. The interpretation of the scattering as being caused by clusters of vacancies explained the results of Guinier's group very well and was published in the small-angle X-ray scattering text by Guinier et al. (1955).

Beeman was one of the people excited by the possibility of investigating the small-angle X-ray scattering from metals, and he and some of his students spent considerable time studying the properties of this scattering. After several years of work, they were able to show very clearly that the small-angle X-ray scattering from cold-worked metals was very rarely produced by clusters of vacancies but instead was almost always due to double Bragg scattering from the grains or subgrains in the metal. In this process, the X-rays were first diffracted through a large angle by ordinary Bragg diffraction and then diffracted again by a second grain which had a slightly different orientation from that of the first grain. Because the two grains had nearly but not exactly the same orientation, the net effect of the two diffractions was a scattering at small angles (Neynaber, Brammer and Beeman, 1959; Webb and Beeman, 1959). Since the information obtainable from measurements of the double Bragg scattering could also be found by other, simpler methods, there was little need to use small-angle X-ray scattering. The results of Beeman, Neynaber, Brammer and Webb were so convincing that they were quickly accepted by most people working with the small-angle scattering from metals, and almost all work on this subject was abandoned. However, since the cluster interpretation had become so well established as the result of the wide circulation of the book by Guinier et al., more than a year went by before the news of Beeman's group's work propagated to all small-angle scattering laboratories.

In 1967, after the Biophysics Laboratory had been built at the University of Wisconsin, Beeman and Anderegg moved the small-angle X-ray scattering equipment into the new building, although they still had joint appointments and a laboratory in the physics department. Anderegg's work continued to be directed toward biophysics and molecular biology, while Beeman's interests, as before, were oriented toward physics. Anderegg has continued his small-angle scattering studies of viruses and ribosomes and is now investigating conformational changes in viruses.

For the past few years, S. L. Cooper and his colleagues in the chemical engineering department at the University of Wisconsin have been using the small-angle scattering equipment in the biophysics laboratory for their polymer research. This work has been so promising that Cooper's group has joined with several other investigators on the Madison campus to buy their own small-angle scattering system, thus expanding the small-angle scattering facilities at the University of Wisconsin.

Peter Debye, who contributed so much to so many areas of physical chemistry and physics, also left his mark on small-angle X-ray scattering. The scattering equation which is often known as the Debye scattering equation (Debye, 1915) and which gives the average X-ray scattering at either large or small angles from a randomly-oriented assembly of scattering particles, has served as the starting point in many small-angle scattering investigations.

Debye was one of the people who, beginning about 1945, developed light scattering into an important technique in physical chemistry. The samples studied in this kind of light scattering usually have indexes of refraction that do not differ greatly from that of the solvent. Under these conditions, the scattering at optical wavelengths can be described by the equations used for small-angle X-ray scattering, provided the quantity h is evaluated for optical wavelengths. In other words, for a given value of h, light scattering at large angles and small-angle X-ray scattering have many similar properties and can be used in analogous ways.

A. M. Bueche and Debye expressed the intensity of light scattering in terms of the correlation function, often called the radial distribution function or the characteristic function (Debye and Bueche, 1949). Their scattering equation, though developed for use with light, is often convenient for analysis of small-angle X-ray scattering data.

Though Debye worked primarily with light scattering, some small-angle X-ray scattering studies also were carried out in his laboratory at Cornell. Debye, Anderson and Brumberger (1957) considered the scattering from materials in which there was a random distribution of pores. The scattering equation which Debye, Anderson and Brumberger developed for their work with these systems has been very useful in many investigations of catalysts and other porous materials.

Debye's interest in light scattering led him to investigate critical opalescence, which is the strong light scattering observed from fluids and binary liquid mixtures which are near their critical point. Though critical opalescence had been studied for many years, Debye's development of a convenient theory of analysis of the scattering data, together with his work on some of the equipment and techniques for making the scattering measurements, stimulated many studies of critical opalescence. These contributions of Debye and his associates were responsible for much of the renewed interest in both light and small-angle X-ray scattering studies of critical phenomena near critical points of pure fluids, binary liquid mixtures, and polymer solutions.

Benjamin Chu, who was a postdoctoral associate of Debye's, has often used scattering to investigate critical phenomena, both when he was at the University of Kansas and also after he moved to his present position at the State University of New York at Stony Brook. While most of Chu's work employed light scattering, he and his associates have also frequently employed small-angle X-ray scattering.

Small-angle X-ray scattering has also been important in many areas of materials science. I would especially like to mention the small-angle scattering work of R. W. Gould at the University of Florida and of H. Herman at the State University of New York at Stony Brook, as well as that which Robert W. Hendricks did at Oak Ridge National Laboratory.

In addition to having applied small-angle X-ray scattering in a number of studies of materials, Hendricks has made many contributions to the development of small-angle X-ray scattering instruments, both in the design of collimation systems and one-dimensional and two-dimensional position-sensitive detectors and also in what can be called the theory of instrumentation. For example, Hendricks and his co-workers have studied collimation corrections and some criteria of optimizing the arrangement of the slits in collimation systems.

Among the many people working in industrial laboratories who have employed small-angle X-ray scattering, there are two who have used this technique so much that I feel that they should be mentioned here. G. F. Neilson has investigated the small-angle scattering from glasses at Owens-Illinois laboratories in Toledo, Ohio. G. W. Brady frequently worked with small-angle X-ray scattering while he was at Bell Telephone laboratories. He has continued to use this technique after moving to the State University of New York at Albany, where he is concentrating on biological problems.

Studies of the structure of polymers are another important application of small-angle X-ray scattering. It sometimes seems to me that the work in this field has often tended to be somewhat independent of and removed from much of the rest of small-angle scattering. I expect that this tendency toward separation – assuming it really exists – is probably due to two things. First, X-ray studies of polymers, especially those with industrial applications, have dealt almost entirely with solid polymers, rather than with solutions. The methods and concepts needed for studies of solid polymers are rather different from those used with systems of independent or nearly-independent particles. Since most other small-angle scattering has tended to concentrate on systems of independent scatterers, the small-angle scattering from polymers often has more in common with large-angle diffraction than with other small-angle scattering. A second reason that the work with polymers has been somewhat isolated from other small-angle scattering research is that the polymer studies performed in industrial laboratories have often been discussed only in company reports or in patents, rather than being published in journals, and as a result may not be widely known.

W. O. Statton and his associates at DuPont Experimental Station in Wilmington, Delaware were among the first to use small-angle scattering in polymer research. These scattering investigations of polymers at DuPont have continued and are now under the direction of Frank C. Wilson.

By 1960, small-angle scattering had developed into an important tool for investigating polymer structure – both in fibers and in other polymers. Small-angle techniques have been and are being used so much in polymer research that in a short review like this, there isn't room to discuss the work of any individuals or groups. The brevity of my discussion of the small-angle X-ray scattering from polymers, however, certainly doesn't mean that I think this technique isn't important in polymer research. In fact, the existence of so much work that I can't discuss individual programs is an especially clear demonstration of how valuable this technique has become.

Small angle X-ray scattering became considerably more popular after ready-made small-angle collimation systems came on the market. Of the systems that have been or are now available, I would especially like to mention the Kratky system (Kratky and Skala, 1958), which is, I feel, an excellent general-purpose system with both sufficient resolution and enough intensity for most small-angle scattering work.

The arrival of high-intensity rotating-anode X-ray tubes on the market, including the Rigaku-Denki and Elliott tubes, has made it possible to examine many samples for which the scattered intensity is too weak to permit them to be studied by systems using an ordinary diffraction tube.

The automatic digital computer has changed small-angle scattering, just as it has affected so many other areas of physical science. The possibility of calculating the scattered intensity predicted by different theoretical models has greatly broadened the possibilities for interpretation of scattering patterns. Computers have also permitted the use of much more complicated and sophisticated techniques of data analysis, including collimation correction and estimation of errors, than would otherwise be possible. Small computers are now often replacing electronic programers for controlling small-angle scattering systems.

The first conference which I know about that was devoted especially to small-angle X-ray scattering was held at the University of Missouri in 1949. Yudowitch's bibliography (Guinier et al., 1955) lists the papers presented at this meeting.

A conference on the small-angle X-ray scattering from metals was sponsored by J. C. Grosskreutz and his associates at Midwest Research Institute in Kansas City, Missouri in September, 1958 (Robinson, 1959). At that time, Grosskreutz was interested in using small-angle scattering to predict fatigue in metals. Probably the most notable result of this conference was the rather negative one of spreading the information about the conclusions of Beeman and his group at the University of Wisconsin that small-angle scattering *wasn't* particularly useful for studying metals. Though of course the conference didn't stimulate further metallurgical applications of small-angle scattering, it served the very important purpose of making a number of people aware of the work of Beeman, Neynaber, Brammer and Webb at the University of Wisconsin. Propagation of information about this work was especially important because at the time of the conference, the group's results had not been published.

International small-angle scattering conferences have been held about every three to five years, beginning in 1965, when Harry Brumberger organized the first international conference at Syracuse, New York. The next two international conferences took place at Graz in 1970 and at Grenoble in 1973. Robert W. Hendricks was the chairman of the next conference, which was held at Gatlinburg, Tennessee in 1977. R. Hosemann and his colleagues organized the 1980 conference in West Berlin. The more recent international conferences have been quite deliberately called conferences on small-angle scattering, to help emphasize that both X-ray and neutron scattering are included. These international conferences have provided a forum where people with different backgrounds and with a common interest in the techniques and applications of small-angle scattering could meet to exchange information and discuss new developments.

It seems that most research specialties, after they reach a certain level of development and maturity, need some form of organization to help foster communication between different people and groups and to provide some coordinating activities. In the United States, small-angle scattering is fortunate to have found a home as an interest group of the American Crystallographic Association. The interest group, organized in 1975, has deliberately been called the Small-Angle Scattering Interest Group, with the word "X-ray" omitted, since from the beginning the group was intended to include both X-ray and neutron scattering. The business of the group is taken care of by a board of six directors. The group normally sponsors a small-angle-scattering session at an ACA meeting at least once every two years, though the sessions usually aren't held during years when there is an International Congress of Crystallography or an international small-angle scattering conference.

While I won't risk making very detailed predictions about the future of small-angle X-ray scattering, there are some fairly clear directions in which the field seems to be moving. I expect that soon more work will be done at large facilities available to users from different laboratories. Among these centers are the synchrotron laboratories at Stanford and Cornell Universities, where synchrotron radiation will provide a very intense X-ray source with a wavelength which can be quite easily varied. Facilities for small-angle X-ray and neutron scattering will be provided at the National Center for Small-Angle Scattering at Oak Ridge. But although these large centers will be used a great deal, I think that much important work will continue to be done at smaller laboratories. Since small dedicated computers are now available for control of the equipment and for data analysis, and as position-sensitive detectors make it possible to record scattering curves much more rapidly than was possible by step-scanning, small-angle scattering measurements and the interpretation of the scattering curves have become much easier. These advances will encourage people, especially those who may not be able to specialize in small-angle scattering, to acquire and use small-angle X-ray scattering equipment, in both university and industrial laboratories. While it's difficult to predict how far this growth will go, I certainly don't expect small-angle scattering to wither away. When interest in one application declines, other uses will be discovered, and with new techniques and equipment which are are now available, as well as with the further advances which can be expected, small-angle X-ray scattering should continue to be useful for many years.

Concluding Remarks and Acknowledgements

Because of space limitations, I've had to concentrate on the small-angle X-ray scattering from independent particles. I hope I haven't offended anyone by what I've said or left unsaid because of the fact that my choice of subjects has had to be somewhat restricted.

In organizing and collecting the material for this review, I was fortunate to have had some long, informative, and very pleasant discussions with Newell Gingrich, William Beeman, John Anderegg and Paul Kaesberg. I would like to take this opportunity to thank them for their willingness to review small-angle scattering history with me and for their helpful advice about the manuscript which I prepared after I talked with them. I also am very grateful to Edward S. Clark, Frank C. Wilson, Do Yoon, Richard S. Stein and Harry Brumberger for their assistance and suggestions.

References

Anderegg, J. W., Beeman,W. W., Shulman, S., and Kaesberg, P. (1955). J. Am. Chem. Soc. **77**, 2927-2937.

Bale, H.D., and Schmidt, P. W. (1958). J. Phys. Chem. **62**, 1179-1183.

Bale, H. D., and Schmidt, P. W. (1959). J. Chem. Phys. **31**, 1612-1618.

Beeman, W. W., Kaesberg, P., Anderegg, J. W., and Webb, M. B. (1957). Size and Shape of Particles from Small-Angle X-Ray Scattering; in "Handbuch der Physik" XXXII, pp. 321-389. Berlin, Göttingen, Heidelberg: Springer Verlag.

Beeman, W. W. (1978). In Remarks Delivered at the Awards Ceremony, Fourth International Conference on Small-Angle Scattering on X-Rays and Neutrons (Gatlinburg, Tenn., 1977), edited by H. Brumberger and R. W. Hendricks. Oak Ridge National Laboratory Technical Manual ORNL/TM 6317 (pub. June, 1978). Oak Ridge, Tenn., pp. 5-11.

Biscoe, J. and Warren, B. E. (1942). J. Appl. Phys. **13**, 364-371.

Buttrey, J. W. (1957). J. Chem. Phys. **26**, 1378-1381.

Davis, B. (1927). J. Franklin Inst., **204**, 29-39.

Debye, P. (1915). Ann Physik **46**, 809-823.

Debye, P., Anderson, H. R. and Brumberger, H. (1957). J. Appl. Phys. **28**, 679-683.

Debye, P. and Bueche, A. M. (1949). J. Appl. Phys. **20**, 518-525.

Eisenstein, A. S. and Gingrich, N. S. (1942). Phys. Rev. **62**, 261-270.

Elkin, P. B., Shull, C. G. and Roess, L. C. (1945). Ind. Eng. Chem. **37**, 327-331.

Fankuchen, I. and Mark, H. (1944). J. Appl. Phys. **15**, 364-370.

Gray, J. A. (1935). Canadian J. Res. **12**, 408-409.

Gray, J. A. and Zinn, W. H. (1930). Canadian J. Res. **2**, 291-293.

Guinier, A., Fournet, G., Walker, C. B. and Yudowitch, K. L. (1955). Small-Angle Scattering of X-Rays. New York: Wiley.

Guinier, A. (1978). In Remarks Delivered at the Awards Ceremony, Fourth International Conference on Small-Angle Scattering of X-Rays and Neutrons (Gatlinburg, Tenn., 1977), edited by H. Brumberger and R. W. Hendricks, Oak Ridge National Laboratory Technical Manual ORNL/TM 6317 (pub. June, 1978). Oak Ridge, Tenn., pp. 13-17.

Jellinek, M. H. and Fankuchen, I. (1945). Ind. Eng. Chem. **37**, 158-164.

Jellinek, M. H. and Fankuchen, I. (1949). Ind. Eng. Chem. **41**, 2259-2265.

Jellinek, M. H., Solomon, E. and Fankuchen, I. (1946). Ind. Eng. Chem. Anal. Ed. **18**, 172-175.

Kratky, O. (1978). In Remarks Delivered at the Awards Ceremony, Fourth International Conference on Small-Angle Scattering of X-Rays and Neutrons (Gatlinburg, Tenn., 1977), edited by H. Brumberger and R. W. Hendricks. Oak Ridge National Laboratory Technical Manual ORNL/TM 6317 (Pub. June, 1978). Oak Ridge, Tenn., pp. 19-27.

Kratky, O. and Skala, Z. (1958). Z. Elektrochem. **62**. 73-77.

Krishnamurti, P. (1930a). Indian J. Phys. **4**, 473-488.

Krishnamurti, P. (1930b). Indian J. Phys. **4**, 489-499.

Leonard, B. R., Anderegg, J. W., Shulman, S., Kaesberg, P. and Beeman, W. W. (1953). Biochim. Biophys. Acta. **12**, 499-507.

Neynaber, R. H., Brammer, W. G. and Beeman, W. W. (1959). J. Appl. Phys. **30**, 656-661.

Penley, H. H. and Gray, J. A. (1937). Canad. J. Res. **A15**, 45-47.

Porod, G. (1951). Kolloid Zeits. **124**, 83-114, Eqs. (42) and (43). (This journal is now called Colloid and Interface Science.)

Raman, C. V. and Krishnamurti, P. (1929). Nature (London) **124**, 53-54.

Ritland, H. N., Kaesberg, P. and Beeman, W. W. (1950a). J. Appl. Phys. **21**, 838-841.

Ritland, H. N., Kaesberg, P. and Beeman, W. W. (1950b). J. Chem. Phys. **18**, 1237-1242.

Robinson, W. H. (1959) Physics Today **12**, No. 5, 26-28, 30.

Roess, L. C. (1946). J. Chem. Phys. **14**, 695-697.

Roess, L. C. and Shull, C. G. (1947). J. Appl. Phys. **18**, 308-313.

Shull, C. G. and Roess, L. C. (1947). J. Appl. Phys. **18**, 295-307.

Slack, C. M. (1926). Phys. Rev. **27**, 691-695.

Thomas, J. E. and Schmidt, P. W. (1963). J. Chem. Phys. **39**, 2506-2516.

Thomas, J. E. and Schmidt, P. W. (1964). J. Amer. Chem. Soc. **86**, 3554-3556.

Van Norstrand, R. A. and Hack, K. M. (1953). Paper presented at Catalysis Club, Chicago, Il., May, 1953.

Von Nardroff, R. (1926). Phys. Rev. **29**, 240-246.

Vineyard, G. H. (1948). Phys. Rev. **74**, 1076-1083.

Warren, B. E. (1934). J. Chem. Phys. **2**, 551-555.

Warren, B. E. (1936). Phys. Rev. **49**, 885.

Webb, M. B. and Beeman, W. W. (1959). Acta Met. **7**, 203-209.

Wild, R. L. (1950). J. Chem. Phys. **18**, 1627-1632.

Wu, H. and Schmidt, P. W. (1971). J. Appl. Cryst. **4**, 224-231.

Wu, H. and Schmidt, P. W. (1974). J. Appl. Cryst. **7**, 131-146.

CHAPTER 14

HIGH PRESSURE CRYSTALLOGRAPHY

S. Block
and
G. Piermarini
Center for Materials Science, National Bureau of Standards
Washington, D.C. 20234

Introduction

High pressure crystallography is a very fitting topic for inclusion in this volume – not only for its scientific interest but because the genesis and development of this field parallels the growth of the ACA and is essentially an American story. Although he did not use X-ray diffraction techniques, the American Nobel Laureate P. W. Bridgman pioneered high pressure science (Bridgman, 1964). He discovered and characterized numerous phase transformations and determined the compressibility of a wide variety of substances. The first presentation of X-ray investigations at high pressure was at Berkeley, California at an American Physical Society meeting (Cohn, 1933). Other talks at that meeting included such pioneers as R. S. L. Mooney, W. H. Zachariasen, and I. Fankuchen. In 1938, R. B. Jacobs obtained Debye-Scherrer patterns on the 0.3 GPa* transition in AgI and showed for the first time that the transformation involved a structural change from the zinc blende to the halite form. Methods using sealed Pyrex capillary tubes or Be compact tubes as pressure vessels were also developed, and measurements of the shift of the interplanar spacings of CsI to 0.16 GPa (Frevel, 1935) and compressibility of many n-paraffins (Muller, 1941) were made. In 1949, A. W. Lawson and N. A. Riley used a large-grain-sized Be bomb to obtain transmission Debye-Scherrer patterns to 1.5 GPa. To improve on this design, Lawson and Tang, in 1950, substituted two single-crystal diamond plates in a split-bomb configuration. This is the first instance of the use of diamonds as load supporting components of a pressure vessel.

The two basic experimental methods of high pressure crystallography are presses and the diamond anvil cell (DAC). The presses are useable only for powder diffraction, but they have the advantages of larger sample sizes, and, currently, larger sustained temperature ranges. By using multi-anvils, the presses can also provide quasi-hydrostatic conditions. In comparison the DAC can be used for either single crystal or powder diffraction studies. Its advantages include small size, relatively low cost, optical access and applicability to a variety of experiments.

The DAC originated independently at the University of Chicago (Jamieson et al, 1959) and at the National Bureau of Standards (Weir et al, 1959). A modern version of the NBS cell is shown in Fig. 1.

An excellent indication of the state of the art of high pressure X-ray and neutron diffraction at the beginning of the past decade is provided by *Transactions of the American Crystallographic Association* (1969) along with a review article by M. D. Banus (1969). Introduction of the ruby fluorescence method for measuring pressure in the DAC (Forman et al, 1972) was a major development of the past decade and has generally replaced the use of the compressibility of internal calibrants such as NaCl. Pressure measurements with an accuracy of $\pm$ 0.05 GPa can be made in a matter of minutes over an extremely wide pressure range, while the degree of hydrostaticity of the pressure environment can be estimated from measurements of the R_1-R_2 line broadening. Such measurements have led to the discovery of pressure transmitting liquids such as a 4:1 mixture (by volume) methanol:ethanol with a hydrostatic range of 10.4 GPa (Piermarini, Block, and Barnett, 1973). Hydrostatic pressures are essential for the study of single crystals under pressure. Currently several laboratories are studying the rare gases as possible media for higher pressure ranges.

Single Crystal

The first pressure cell for single crystal studies, made at NBS in 1964 (Block, Weir, and Piermarini, 1965), was fabricated almost completely of Be metal and was used with a specially designed precession camera. However, because of the Be cell's large size and special camera requirements, and, at that time, absence of pressure measurement capability, the system was not used widely. With the advent of the ruby fluorescence method, further modification and miniaturization of the diamond cell has taken place and, today, several designs of varying degrees of complexity are available which can be used with standard precession cameras and automatic diffractometers (Fourme, 1968, Merrill and Bassett, 1974, Keller and Holzapfel, 1977, Schiferl, 1977).

Structural studies at high pressures are of three types: (1) those in which the high pressure polymorph is grown from the liquid or solid state by varying P and T and in which the crystal symmetry of the new phase is completely unknown; (2) those in which a crystal of

*The international system of units (SI) for pressure is the pascal (Pa) or newton per square meter,

1 Pa = 0.9869×10^{-5} atm
1 GPa = 0.9869×10^{4} atm

known structure undergoes either a second order phase transition or a first order transition in which ΔV, the change in volume, is so small that the crystal retains its integrity on transformation; and (3) those involving the measurement of cell constants (compressibility) of known structures as a function of pressure. For studies of the first type a device such as the original beryllium cell is needed. The spring-loaded lever arms and built-in heater are conducive to the precise control of pressure and temperature needed for growing single crystals of high pressure phases. This system was used to determine the structure of high pressure polymorphs such as benzene II, CCl_4-I, -II, -III, -IV, Cs-II and Ga-II, (Piermarini et al, 1969, Weir et al, 1971, Piermarini and Braun 1973). Studies of the second type are exemplified by the determination of the structure $CaCO_3$ -II (Merrill and Bassett 1975) and of Gillespite-II ($BaFeSi_4O_{10}$) (Hazen and Burnham, 1975). A single crystal of calcite remains intact on transition to $CaCO_3$ -II at a hydrostatic pressure of about 0.14 GPa; thus the high pressure phase which has undergone a solid state transition can be studied by single crystal methods. Although the $CaCO_3$-II lattice corresponds to a slightly distorted calcite lattice, the structure has not been uniquely determined by X-ray diffraction. The possible space groups are Pc or $P2_1/c$. The center of symmetry remains a problem. Gillespite I-II is an example of a first-order transition with a small volume change and a reduction in symmetry from $P4/ncc$ to $P2_12_12$. The I and II structures are distortions of the ideal gillespite structure of high symmetry.

Most high pressure X-ray measurements are of the third type. Using the beryllium cell with a precession camera, Weir et al, (1970) measured the anisotropic compressibility of several inorganic azides. Later, Mauer et al., (1974) used the same cell on a diffractometer with modified Bond geometry to measure very accurate cell parameters and anisotropic compressibilities on Si and α-$Pb(N_3)_2$. The accuracy was limited by line-width due to crystal imperfection and the uncertainty in the pressure measurement. The majority of studies on changes in bond lengths and bond angles with pressure have been on geologically interesting silicates and alumina (d'Amour et al, 1978, Finger and Hazen, 1978, Sato and Akimoto, 1979). Changes in the structure of alumina as a function of pressure and temperature are of particular interest because ruby (Cr^{+3} doped Al_2O_3) is used as the pressure sensor in most DAC experiments.

Reported R values for high pressure structures range from approximately 0.07 for film methods to 0.02 for diffractometers. In order to obtain such results, absorption effects due to the crystal and cell components (including the gasket), the centering of the crystal with respect to diffraction geometry, and the hydrostatic condition of the pressure transmitting medium all must be considered. A study of absorption effects in the Be-DAC (Santoro et al, 1968) showed that, in the case of Br_2, equivalent uncorrected intensities varied by as much as a factor of two. Because the DAC has limited X-ray access, centering of the crystal on the origin of the diffractometer circles is a problem. A procedure which utilizes Hamilton's centering method has essentially eliminated errors from this source (King and Finger, 1979). For meaningful results, the encapsulating fluid must retain its hydrostatic properties at all pressures as nonuniform stresses on the crystal can produce erroneous results.

Powder

Methods of generating high pressures for X-ray powder diffraction studies include the DAC, Bridgman anvils and various multi-anvil systems. The Bridgman opposed anvil method has been in use with few changes since the early 1960's. The multi-anvil system on the other hand has undergone considerable revision in design in recent years. Of particular note, is the work in Japan by T. Yagi and S. Akimoto (1976), who have, among other designs, developed a cubic anvil-type press in conjunction with a high power X-ray powder diffraction system. This press has reached pressures in the vicinity of 25 GPa. A. Ohtani et al, (1979) have developed a split octahedron type apparatus for X-ray powder diffraction studies to pressures in excess of 20 GPa. Both presses have the capability of simultaneous X-ray and electrical resistance measurements and use the lattice parameter of NaCl as the internal pressure calibrant.

Powder diffraction, ignoring nonhydrostatic effects, has been carried out to the highest pressures with the DAC. With an encapsulated sample, pressures of approximately 65 GPa can be obtained (Piermarini and Block, 1975). Over a megabar has been obtained (Mao et al, 1979) by pressing metal sheets and a ruby sensor between diamond anvils.

An important recent development in high pressure X-ray studies is the use of the energy dispersive powder diffraction method. Freud and La Mori (1969) were the first to adopt this method, but it was not until 1977 that it was combined with a DAC. Skelton et al, (1977) described a variable temperature cell for powder diffraction studies down to 2 K and used it in studying the phase diagram of Bi. As a means of overcoming the extremely low counting rates resulting from small sample volume, Buras et al., (1977) coupled energy dispersive diffraction with a synchrotron source to obtain patterns of TeO_2 at 8.0 GPa in only 1000 sec. The high intensity, extended energy range (3 to 70 KeV) and low divergence ($\Delta\theta = 10^{-4}$ rad) make the synchrotron a nearly ideal source for high pressure studies. Although the resolution of energy dispersive powder patterns is considerably lower than that of conventional patterns, the low divergence that can be attained with a synchrotron source results in a Gaussian peak shape that is amenable to curve fitting or the

Rietveld technique of total profile analysis. Synchrotron radiation is useful not only because of its high intensity, but also because selected wave lengths can be obtained by the use of monochromators. This latter property was used by O. Shimomura et al, (1978) in an EXAFS study of GaAs at 22 GPa. The tuneability will also be useful in the study of liquids, glasses and other amorphous materials at both ambient and pressure conditions. Such studies are currently underway. The combination of high intensity, tuneability and small beam size make the synchrotron source ideal for high pressure studies and the near future will likely see an explosion of such work.

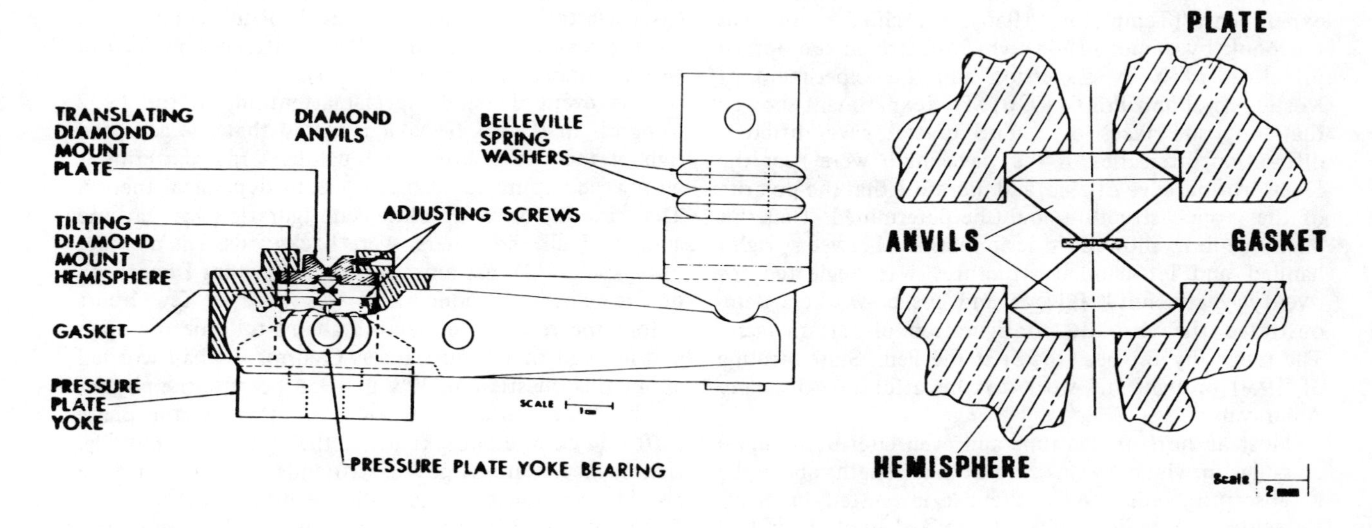

Figure 1. Cross-sections of the diamond anvil cell (left side) and diamond anvil assembly (right side).

References

Banus, M. D. (1969). High Temperatures-High Pressures. **1**, 483-515.

Block, S., Weir, C. E., and Piermarini, G. J., (1965). Science **148**, 947-948.

Bridgman, P. W. (1964) Collected Experimental Papers, Vols. 1-7, Cambridge, Massachusetts: Harvard University Press.

Buras, B. and Staun Olsen, J. (1977). J. Appl. Cryst. **10**, 431-438.

Cohn, W. M. (1933). Proc. Am. Phys. Soc. **63**, 326-327.

d'Amour, H., Schiferl, D., Denner, W., Schulz, H. and Holzapfel, W. B., (1978). J. Appl. Phys. **49**, 4411-4416.

Finger, L. W. and Hazen, R. M., (1978). J. Appl. Phys. **49**, 5823-5826.

Forman, R. A., Piermarini, G. J., Barnett, J. D., and Block, S., (1972). Science **176**, 284-285.

Fourme, R., (1968). J. Appl. Cryst. **1**, 23-30.

Frevel, L. K. (1935). Rev. Sci. Instrum. **6**, 214-215.

Hazen, R. M. and Burnham, C. W., (1975). Acta. Cryst. **B31**, 343-349.

Jacobs, R. B., (1938). Phys. Rev. **54**, 325-331.

Jacobs, R. B., (1939). Phys. Rev. **56**, 211-212.

Jamieson, J. C. Lawson, A. W., and Nachtrieb, N. D. (1959). Rev. Sci. Instrum. **30**, 1016-1019.

Keller, R. and Holzapfel, W. B., (1977). Rev. Sci. Instrum. **48**, 517-523.

King, H. E. and Finger, L. W., (1979). J. Appl. Cryst. **12**, 374-378.

Lawson, A. W. and Riley, N. A., (1949). Rev. Sci. Instrum. **20**, 763-765.

Lawson,A. W. and Tang, T. Y., (1950). Rev. Sci. Instrum. **21**, 815.

Mao, H. K., Bell, P. M., Dunn, K. J., Chrenko, R. M., and Devries, R. C., (1979). Rev. Sci. Instrum. **50**, 1002-1009.

Mauer, F. A., Hubbard, C. R., Piermarini, G. J., and Block, S., (1974). Adv. X-Ray Anal. **18**, 437-453.

Merrill, L. and Bassett, W. (1974). Rev. Sci. Instrum. **45**, 290-294.

Merrill, L. and Bassett, W. (1975). Acta. Cryst. **B31**, 343-349.

Muller, A. (1941). Proc. Roy. Soc. (London) **178**, 227-241.

Ohtani, A., Onodera, A., and Kawai, N., (1979). Rev. Sci. Instrum. **50**, 308-315.

Piermarini, G. J. and Block, S. (1975). Rev. Sci. Instrum. **46**, 973-979.

Piermarini, G. J., Block, S., and Barnett, J. D. (1973). J. Appl. Phys. **44**, 5377-5382.

Piermarini, G. J. and Braun, A. B. (1973). J. Chem. Phys. **58**, 1974-1982.

Piermarini, G. J., Mighell, A. D., Weir, C. E., and Block, S. (1969). Science **165**, 1250-1255.

Sato, Y. and Akimoto, S. (1979). J. Appl. Phys. **50**, 5285-5291.

Schiferl, D. (1977). Rev. Sci. Instrum. **48**, 24-30.

Shimomura, O., Fukamachi, T., Kawamura, T., Hosoya, S., Hunter, S., and Bienenstock, A. (1978). Jap. J. Appl. Phys. **17**, 221-223.

Skelton, E. F., Spain, I. L., Yu, S. C., Liu, C. Y., and Carpenter, Jr., E. R., (1977). Rev. Sci. Instrum. **48**, 879-883.

Weir, C. E., Block, S., and Piermarini, G. J. (1970). J. Chem. Phys. **53**, 4265-4269.

Weir, C. E., Lippincott, E. R., VanValkenburg, A., and Bunting, E. N., (1959). J. Res. Nat. Bur. St. **63A**, 55-62.

Weir, C. E., Piermarini, G. J., and Block, S. (1969). J. Chem. Phys. **50**, 2089-2093.

Weir, C. E., Piermarini, G. J., and Block, S. (1971). J. Chem. Phys. **54**, 2768-2770.

Yagi, T. and Akimoto, S. (1976). J. Appl. Phys. **47**, 3350-3354.

CHAPTER 15

ANOMALOUS SCATTERING

David H. Templeton
Department of Chemistry, University of California
Berkeley, CA 94720

It was in Häggs's laboratory in Uppsala during a visit in 1954 that I began to study anomalous X-ray scattering and its consequences in diffraction experiments (Templeton, 1955). My chief source was the book by James (1948) which described the optical principles of X-ray dispersion and the experiment of Koster, Knol and Prins (1930). This experiment showed that opposite faces of a ZnS crystal gave different diffraction intensities, if the wavelength were near the K absorption edge of zinc, and therefore that the polarity of the atomic structure could be determined. That this phenomenon allowed discrimination between right-handed and left-handed structures was neglected for twenty years, until Bijvoet and his co-workers demonstrated the absolute configuration of tartaric acid. The report by Bijvoet (1952) at the Penn State meeting of 1950 brought his work to the attention of many Americans.

Most authors at this time and even later used curves or tables in which f″ was zero at wavelengths above the K absorption edge. As a chemist interested in heavy elements and radioactivity, I realized that L and M electrons of many elements should affect X-ray dispersion at wavelengths used by crystallographers, and that significant effects should extend over considerable wavelength intervals. It struck me that the books and research papers gave too much emphasis to the importance of being "near" an absorption edge, and that this fact may have explained some of the neglect of this effect. A new table (Dauben and Templeton, 1955) was prepared to encourage and make more convenient the inclusion of dispersion in crystallographic calculations. This task was aided very much by the timely publication of L-edge calculations in Germany (Eisenlohr and Müller, 1954) and America (Parratt and Hempstead, 1954). Even so, a considerable amount of estimation and guess-work was necessary. It seemed better to use approximations than to neglect significant terms. These tables were later superseded by those of Cromer and Liberman (1970), which are based on more elaborate and consistent calculations.

Bijvoet pointed out that anomalous scattering could help solve the diffraction phase problem, and others both in America and abroad worked on ways to do this. Perhaps the first to formulate a method based purely on this effect were Okaya, Saito and Pepinsky (1955). Pepinsky (1956) emphasized the power of using the L-edge effects which he found in our tables, in a review which also gives a good picture of the best methods for solving crystal structures then in use. Over the next two decades work accelerated on the several fronts of more nearly correct diffraction calculations, absolute configuration, the phase problem, use of lighter atoms, and anomalous neutron scattering. This work is too extensive to itemize here, but much of it is summarized by the proceedings of the 1974 conference in Madrid (Ramaseshan and Abrahams, 1975).

One event demands special attention. At the 1972 Congress in Kyoto, Tanaka reported that the accepted sign of the phase shift in anomalous X-ray scattering is wrong, according to quantum electrodynamical theory. That is, Bijvoet's absolute configuration of tartaric acid, and all the others, were backwards. His abstract (Tanaka, 1972) was ambiguous enough that I had been unsure what his conclusion was until, a few hours before the session, Sidney Abrahams told me the news and insisted that I must act as Chairman. I had worried about this question already in 1954 because the physics textbooks I consulted, when they discussed the phase shift, did so in a context where the sign had no physical significance. Various physicist friends refused to give me absolute assurance, because they had not worked it out themselves, but the general opinion was that "of course it is right." By various arguments, based on macroscopic mechanics and the correspondence principle, I convinced myself that the sign was correct.

Had I gone wrong? Tanaka's talk did not help me much, because I failed to grasp the crucial step in his necessarily abbreviated exposition. A rather small attendance at the session left me wondering how much anyone else cared.

What followed was a scientific storm. Tanaka very graciously distributed copies of his unpublished manuscript. All over the world it was studied, and all sorts of alternate derivations were worked out. Some of these were described at Madrid. All agreed that the original sign was correct. Tanaka's mistake was that initial and final states were interchanged in an equation.

This incident is important because it revealed how precarious was our theoretical foundation. We were right, but nearly everyone trusted that someone else had determined that we were right. There is now a much more general understanding of why the phase shift is positive (in the conventional form of the equations). It is a paradox that an error did more to advance our understanding than did much work which was correct but less provocative.

The decade of the 70's brought a new tool, synchrotron radiation. This radiation, produced when very high energy electrons are deflected by strong magnetic fields, has a continuous spectrum so intense that diffraction experiments (and many others) can be carried out at whatever precise wavelength one chooses, rather than just at the characteristic lines of traditional X-ray tubes. It greatly enhances one's ability to study and apply anomalous scattering, as well as other techniques such as EXAFS spectroscopy which fall outside the present topic. The circumstance that a large electron accelerator had been constructed for high-energy physics research at Stanford University made that the first site in America for applications of synchrotron radiation to crystallography. Hodgson, whom we had known since his student days at Berkeley, was one of those who recognized early its potential for solving the phase problem (Phillips, Wlodawer, Goodfellow, Watenpaugh, Sieker, Jensen and Hodgson, 1977) and the value of combining a diffractometer with this radiation source. He influenced my wife Lieselotte and me to get involved in the exciting but often frustrating and inconvenient work in this laboratory where delays were routine and the accelerator operated best only late at night. Here we measured exceptionally large anomalous scattering effects which occur at L edges (Templeton, Templeton, Phillips and Hodgson, 1980; Templeton, Templeton and Phizackerley, 1980) and began to explore X-ray dichroism and the anisotropy of anomalous scattering which comes with it (Templeton and Templeton, 1980). Many others have also been studying various aspects of anomalous scattering at Stanford and in the several European synchrotron radiation laboratories. A major expansion of facilities is underway around the world, with a new source at Brookhaven as the largest American project. This work is so recent, and is developing so rapidly, that historical analysis seems premature. But I expect that it will merit major attention in a later volume like this one.

References

Bijvoet, J. M. (1952). In "Computing Methods and the Phase Problem in X-ray Crystal Analysis," edited by R. Pepinsky, pp. 84-89. State College: Pennsylvania State College.

Cromer, D. T. and Liberman, D. J. (1970) J. Chem. Phys. **53**, 1891.

Dauben, C. H. and Templeton, D. H. (1955) Acta Cryst. **8**, 841.

Eisenlohr, H. and Müller, G. L. J. (1954) Z. Phys. **136**, 491, 511.

James, R. W. (1948) The Optical Principles of the Diffraction of X-rays. London: Bell.

Koster, D., Knol, K. S. and Prins, J. A. (1930) Z. Phys. **63**, 345.

Okaya, Y., Saito, Y. and Pepinsky, R. (1955) Phys. Rev. **108**, 1231.

Parratt, L. G. and Hempstead, C. F. (1954) Phys. Rev. **94**, 1593.

Pepinsky, R. (1956). Record Chem. Progress (Knesge Hooker Science Library) **17**, 145.

Phillips, J. C. Wlodawer, A., Goodfellow, J. M., Watenpaugh, K. D., Sieker, L. C., Jensen, L. H. and Hodgson, K. O. (1977) Acta Cryst. **A33**, 445.

Ramaseshan, S. and Abrahams, S. C., editors (1975). Anomalous Scattering. Copenhagen: Munksgaard.

Tanaka, J. (1972) Acta Cryst. **A28**, S229.

Templeton, D. H. (1955) Acta Cryst. **8**, 842.

Templeton, D. H. and Templeton, L. K. (1980) Acta Cryst. **A36**, 237.

Templeton, D. H., Templeton, L. K., Phillips, J. C. and Hodgson, K. O. (1980) Acta Cryst. **A36**, 436.

Templeton, L. K., Templeton, D. H. and Phizackerley, R. P. (1980) J. Am. Chem. Soc. **102**, 1185.

CHAPTER 16

STRUCTURE DETERMINATIONS IN THE EARLY DAYS

Dan McLachlan, Jr.
Department of Mineralogy, Ohio State University,
Columbus, Ohio 43210

Any one who follows the history of the researcher into the structure of crystals is impressed by the accelerating increase year by year of the number of crystals of growing complexity that have been determined. This can be revealed by a perusal of the *Struktur Bericht.* The first volume edited by P. P. Ewald and C. Hermann, (1931) two men of great brilliance and foresight, covers the fourteen years 1913 to 1926 in 818 pages of text. The second volume by C. Hermann, O. Lahrman and H. Philipp (1937) covers only the five years from 1928-1932 in 963 pages, of which 128 pages were devoted to organic compounds. The word *"Struktur Bericht"* was changed eventually to *"Structure Reports"* and for years, A. J. C. Wilson was chief editor, publishing volumes of about one thousand pages per year. By 1964 the number of structures reported per year became so great that it was becoming almost impossible to keep up with them, in spite of the fact that the editors kept finding more ways to omit and abbreviate parts of the contributing authors' texts.

Of course modern crystallographers now have access to three highly developed achievements that were not available to the early researchers: automatic diffractometers, phase determining techniques and digital computers. The purpose of this chapter is to outline the methods used before these conveniences were available. Much can be learned by reading Bragg's (1933, '39 and 49) editions of his *"The Crystalline State,"* Davey's book (1934); and also pages 335 to 830 of Clark's (1955) *"Applied X-rays"* contain many of the older methods and references to apparatus. The very worthy theoretical foundations to X-ray studies were furnished in books by Compton and Allison (1935) and by R. W. James (1948). These extremely fundamental books were used almost as "bibles" by many young crystallographers to get themselves well grounded in the physical foundation of the diffraction process. A book that was put to daily use for many years by structure people was written by Buerger (1942) *"X-ray Crystallography".* This book not only contains the working principles of many of the cameras (some of which are still in use) but also furnishes many tables that people must constantly refer to.

A book that was written in an attempt to make the subject easily understood was *"X-ray Crystal Structure"* by McLachlan (1957) followed by two better books *"Vector Space"* by Buerger (1959) and *"Crystal Structure Analysis"* by Buerger (1960). In the meantime, Cullity (1978) and Klug and Alexander (1974) wrote books for direct daily use by crystallographers, and Azaroff and Buerger (1958) wrote about *"The Powder Method".*

The interest in the determination of atomic positions in crystals began almost immediately after the discoveries of von Laue in 1912. To assist in the thinking, handbooks of chemistry and physics already had furnished tables of the atomic numbers and atomic weights of all the known elements and the density of many pure crystalline compounds; and the "structural formulae" of many organic compounds had been established; and Ewald (1913) gave us the concept of the reciprocal lattice. The determination of the wavelength of X-rays from the various target materials was really performed as studies in the field of spectroscopy carried out initially by Moseley (1913) and Siegbahn (1921).

With the suggestion by W. H. Bragg (1915) that Fourier series could be used in the interpretation of X-ray diffraction patterns, it became apparent that a number of factors had to be considered before the intensity of the diffraction maxima could be correlated with the true electron density within the crystal. These are the Lorenz factor, polarization factor, temperature factors of Debye and Waller, absorption factor and the atomic scattering factors. These necessary factors are to be found in the International Tables and discussed in the books by Buerger (1942), Cullity (1956) and Klug and Alexander (1954). Using all these considerations, Warren and Bragg (1928) worked out the structure of diopside, a non-centrosymmetric crystal, and demonstrated their findings by means of two-dimensional contour maps.

In reviewing all these preliminary considerations that had to be given to the raw X-ray data before they could be applied to the determination of atomic positions, one cannot help but marvel at the hard work and dedication of the hundreds of researchers in chemistry, physics and mathematics on both sides of the Atlantic who sought, along separate paths, to converge on the single worthwhile objective of solving crystal structures. But even with all the tables made available in the literature and the equations readily at hand, the longhand computations required for Fourier syntheses were almost prohibitive. One labor-saving achievement that is frequently given less than its due credit is the theory of point groups and space groups.

Designers of wallpaper and carpets have long ago recognized that if a pattern has a mirror through it, then only half of the pattern need be original. The earliest crystallographers, even before the time of Haüy (1782), noticed that crystals exhibited their internal symmetry through their outside shapes. The crystallographers attacked symmetry on the basis of the arrangement of points in space and Schoenflies (1891) showed that there are 230 space groups while Fedorov (1895) had reached the same conclusion in a Russian Publication some years previously. The symbols of Schoenflies were modified by Hermann (1928) and Mauguin (1931). These symmetry concepts were very useful to people using the longhand methods of calculation. For example, if a crystal had the symmetry $P2_12_12_1$ it has four equivalent positions and the computations are reduced by a factor of four. In the cubic system there can be as many as 192 equivalent positions.

One more labor saving concept was furnished by Wyckoff (1930) when he published a book of 238 pages of which 44 pages were devoted to a picturization of the "special cases," that is cases where an atom might be located on an origin, a center, an axis, or a plane of symmetry. Only the older crystallographers with long memories can realize the impetus that this book had upon the hopes of structural researchers. Between the concepts of equivalent positions and special cases many of the structures of the simpler metals and minerals were deduced without Fourier computations because there were no structural parameters to consider.

The findings of Schoenflies, Hermann, Mauguin, Wyckoff and many others were incorporated in *"International Tables for Crystal Structure Determination"* in 1935 and carried forward to the *"International Tables for X-ray crystallography"* in 1960. These tables represent an attempt to furnish the researcher with everything he needs to know besides his own data in the location of atomic positions in a crystal.

In spite of all these fortifications benefitting the structure analyst there remained difficulties with one class of data: the Miller indices of the diffraction maxima. This problem existed for the two simpler cameras, the Laue camera and the powder camera and gave reasons to invent other cameras to help in the indexing. The Laue Camera presented some problems that were aided by the "ruler for gnomonic projections" developed by Wyckoff (1920) and a "saw" patented by Champaygne and McLachlan (1943). This latter instrument has never been used. The powder camera has caused the most trouble of all. Azároff and Buerger (1958) wrote an entire book on the subject referring to work as far back as the charts by Hull and Davey (1921). Azaroff has praise for the methods of Ito (1950). Perhaps the last paper ever to be published on the indexing of powder photographs without the use of computers was by McLachlan and Hsi-Che Lin (1971).

Some ease in indexing was achieved by *moving* the crystal while the diffraction pattern was being taken. Bernal (1926) was a promoter of the rotating crystal method and devised the Bernal charts. His interest was shared by many other scientists, including an American, John W. Gruner (1928) who, also worked with the oscillation method.

Great strides were made when diffraction patterns began to be made while moving *both* the crystal and the film. Three early cameras of this type are: the one by Weissenberg, the one by Schiebold-Sauter and the one by de Jong-Bouman. The Americans are greatly indebted to Buerger (1942) for his book in which he fully described these three European cameras giving a full account of how they operated and how to use the appropriate charts and tables. We are also indebted to Buerger (1944) for the precession camera which displays the diffraction spots in rows and columns that can be easily indexed. This camera is widely used today.

Among other cameras invented during those years were the gyrating Laue camera by Champaygne and McLachlan (1945), which enabled the estimation of some of the *d*-values from a Laue film, the Herzog and McLachlan (1954) camera for taking diffraction pictures on spheres and the optical machines by McLachlan and Wooley (1953) for transforming Weissenberg pictures to what they called "undistorted theta lattices". These last three machines won for their inventors almost exactly the amount of esteem and international recognition that they deserved.

The most hopeful attack on the structure problem during those days was made by Patterson (1934) in which he used the $|F|^2$ values to determine distance and direction of atoms from one another. Five years later he realized that it is possible for two or more structures to have the same interatomic distances but otherwise have different structures; he called these "homometric sets", Patterson (1939). Buerger (1978) has delved rather deeply into this problem under the subject of "Cyclotomic sets".

Immediately crystallographers set themselves the task of getting all the information possible out of Patterson projections. One of the first was Harker who took advantage of the fact that screw axes and glide mirrors made certain sections through the unit cell easier to handle (Harker, 1936). In a series of papers several authors including Wrinch (1939), Buerger (1950a and b), Clastre and Gay (1950) and Beevers and Robertson (1950) contributed to the solution of Patterson maps. Thomas and McLachlan (1952) made some critical tests of some of the method. Buerger's books (1959) and (1960) discuss some of the accepted terms such as vector sets, vector addition, vector multiplication, minimum function, vector convergency and implication theory, to all of which he made substantial

Along about this time, crystal structure analysis was on the verge of a revolution.* Probability methods were

*Regarding these busy days, Edward W. Hughes sent me the following note, "Volume 5, (1952) of *Acta Cryst.* is full of statistical papers, one after the other: Hauptman and Karle pp. 48-59, D. Sayre pp. 60-65, W. Cochran pp. 65-67, W. H. Zachariasen pp. 68-73. My own paper was in *Acta Cryst.* (1953) Vol. 6, p. 871."

beginning to creep into the literature as evidenced by an early paper by A. J. C. Wilson (1942) that showed that centrosymmetric and noncentrosymmetric structures have different probability distributions for the values of their structure factor magnitudes that could be used to distinguish them. Toward the end of the 1940's, phase relations in the form of inequalities among the structure factors were introduced by Harker and Kasper (1948) and the complete set was obtained by Karle and Hauptman (1950). The inequalities had a most important characteristic, first observed by Gillis (1948), namely, that the phase relationships they represented were likely to be correct even when the conditions on the structure factor magnitudes for them to be valid were not strictly obeyed. In other words, the phase relationships could be applied beyond the strict range of validity of the inequalities. This was later confirmed by probability theory and experience. With the introduction of the joint probability distribution among sets of structure factors and invariant and seminvariant theory (Hauptman and Karle, 1953), the stage was set for the development of a broadly effective procedure for phase determination for centrosymmetric and noncentrosymmetric crystals.

This new era is to be discussed in terms of their personal participation by Drs. Isabella and Jerome Karle in a later chapter in this book.

References

Azároff, L. V. and Buerger, M. J. (1958) The Powder Method. New York: McGraw-Hill.

Beevers, C. A. and Robertson, J. H. (1950) Acta Cryst. **3**, 164.

Bernal, J. D. (1926) Proc. Roy. Soc. (London) **A113**, 123-141.

Bragg, W. H. (1913) Proc. Roy Soc. (London) **17**, 43-47.

Bragg, W. L. (1949) The Crystalline State. London: G. Bell and Sons, Ltd.

Buerger, M. J. (1942) X-ray Crystallography. London: John Wiley and Sons, Inc.

Buerger, M. J. (1944) Monograph No. 1, ASXRED. Cambridge, MA: Murray Printing Company.

Buerger, M. J. (1950a) Acta Cryst. **3**, 87-97.

Buerger, M. J. (1950b) Proc. Natl. Acad. Sci. U.S.A. **36**, 376-382.

Buerger, M. J. (1959) Vector Space. New York: John Wiley and Sons, Inc.

Buerger, M. J. (1960) Crystal Structure Analysis. New York: John Wiley and Sons, Inc.

Buerger, M. J. (1978) Can. Min. **16**, 301-314.

Burke, J. G. (1966) The Origin of the Science of Crystals. Berkeley and Los Angeles: University of California Press.

Champaygne E. and McLachlan, D. (1943) U. S. Patent 2, 332, 391.

Champaygne E. and McLachlan, D. American Cyanamid Co., See McLachlan (1957). appendix.

Clark, G. L. (1955) Applied X-rays. New York McGraw-Hill.

Clastre, J. and Gay, R. (1950). C. R. Acad. Sci. **230**, 1876-1877.

Compton, A. H. and Allison, S. K. (1935) X-rays in Theory and Experiment. New York: D. Van Nostrand Co., Inc.

Cullity, B. D. (1978). Elementary X-ray Diffraction. Second edition. London: Addison-Wesley Publishing Co.

Davey, W. P. (1934) A Study of Crystal Structure and its Applications. New York: McGraw-Hill.

Ewald, P. P. (1913) Physik. Z. **14**, 405 and 1038.

Ewald, P. P. (1921) Physik. Z. **56**, 129-156.

Ewald, P. P. and Hermann, C. (1931) Structurbericht 1913-1928. Ann Arbor, Michigan: Edwards Brothers, Inc.

Fedorov, E. (1895) Z. Kristallogr. **24**, 209.

Garrido, J. (1950) C. R. Acad. Sci. **230**. 1878-1879.

Gillis, J. (1948) Acta Cryst. **1**, 174-179.

Gruner, J. W. (1928) Am. Mineral. **13**, 174-194.

Harker, D. (1936) J. Chem. Phys. **4**, 381-390.

Harker, D. (1947) J. Chem. Phys. **15**, 882-884.

Harker, D. and Kasper J. S. (1948) Acta Cryst. **1**, 70-75.

Hauptman, H. and Karle, J. (1953) American Crystallographic Association, Monograph No. 3 (Polycrystal Book Service, Pittsburgh).

Haüy, A. R. J. (1784), Essai d'une Theorie sur la Structure des Cristaux, Paris (See Burke (1966))

Hermann, C. (1928) Z. Kristallogr. **68**, 257.

Hermann, C., Lohrman, O., and Philippe, H. (1937) Structurbericht (1913-1928). Ann Arbor, Michigan: Edwards Brothers.

Herzog, A. and McLachlan, D. (1954) Am. J. Phys. **22**, 33-36.

Hull, A. W. and Davey, W. P. (1921) Phys. Rev. **17**, 549.

Ito, T. (1950) X-ray Studies in Polymorphism. Tokyo: Maruzen Co. Ltd. 187-228.

James, R. W. (1948) Optical Principles of the Diffraction of X-rays. London: G. Bell and Sons Ltd.

Karle, J. and Hauptman, H. (1950) Acta Cryst **3**, 181-187.

Klug, H. P. and Alexander, L. E. (1974). X-ray Diffraction Procedures. Second edition. New York: John Wiley and Sons, Inc.

McLachlan, D. and Harker, D. (1951) Proc. Natl. Acad. Sci. U.S.A. **37**, 846.

McLachlan, D. and Wooley, R. H. (1953) Rev. Sci. Instrum. **24**, 872-873.

McLachlan, D. (1957) X-ray Crystal Structure. New York: McGraw-Hill Co. Inc.

McLachlan, D. and Lin, H. C. (1971), Z. Kristallogr. **137**, 35-50.

Moseley, H. G. J. (1913). Phil. Mag. **26**, 1024.

Moseley, H. G. J. (1914) Phil. Mag. **27**, 203.

Patterson, A. L. (1935) Z. Kristallogr. **90**, 517-542.

Patterson, A. L. (1939) Nature (London) **143**, 939-940.

Sayre, D. (1952) Acta Cryst. **5**, 60-65.

Schoenflies, A. (1891) Kristallsysteme und Krisallstructure, Leipzig.

Siegbahn, M. (1921) C. R. Acad. Sci. **173**, 1350.

Siegbahn, M. (1922) C. R. Acad. Sci. **174**, 745.

Thomas, I. D. and McLachlan, D. (1952) Acta Cryst. **5**, 302-306.

Warren, B. and Bragg, W. L. (1928) Z. Kristallogr. **69**, 168-193.

Wilson, A. J. C. (1942) Nature (London) **150**, 152.

Wilson, A. J. C. (1949) Acta Cryst. **21**, 318.

Wilson, A. J. C. (1956) Structure Reports (1940). Utrecht: N. V. A. Oosthoek, Uitgevers M. I. J.

Wrinch, D. M. (1939) Phil. Mag. **27**, 490-507.

Wyckoff, R. W. G. (1920) Amer. J. Sci. **50**, 317.

Wyckoff, R. W. G. (1930) The Analytical Expression of the Results of the Theory of Space Groups. Washington, D. C.: Carnegie Institution of Washington.

Zachariasen, W. H. (1952) Acta Cryst. **5**, 68-73.

CHAPTER 17

STRUCTURE DETERMINING METHODS

D. Sayre
IBM Research Center, Yorktown Heights, NY 10598

One of the areas in which American crystallography has made a large contribution is that of structure determining methods, i.e. methods of passing from the diffraction intensities of a structure to the structure itself. An early example is the calculation of the first Fourier maps of electron density (Duane 1925; Havighurst 1925), carried out in the laboratory of physics at Harvard University. W. H. Bragg (1915) had suggested that it might be possible to pass from the intensities to the distribution of scattering power in the crystal by some Fourier process, but it remained for Duane to give the modern expression for the electron density as a Fourier sum with the structure factors as coefficients. Duane also discussed the problem of phasing the structure factors, which he did in the case of sodium chloride by symmetry arguments and by considering the chlorine atom to be at the origin, where it would make all coefficients positive. Using these techniques Havighurst calculated maps of NaCl, KI, NH_4I, NH_4Cl, and diamond. His maps are sometimes described as one-dimensional, but in fact they used full 3-dimensional data and are one-dimensional only in the sense that the density was evaluated along specific lines in the unit cell. It was later, in connection with the diopside structure, that W. L. Bragg (1929) showed that useful maps (projections of the structure along an axis) could be obtained from 2-dimensional data, which were then much more commonly available than 3-dimensional data. Incidentally, it was Duane who supplied one of the first high-precision methods of measuring Planck's constant by determining the high-frequency limit of the X-rays produced by electrons of known energy.

The Duane-Havighurst and Bragg maps may be considered as initiating the modern era of structure determination, for although they were of structures which had already been solved, it was evident that the preparation of similar maps could assist greatly in the solution of unknown structures. Indeed the present era of structure determination may be said to differ from the early period largely in the fact that it is based on the idea of first passing from structure-factor magnitudes to map, and then from map to structure, while in the pre-1930 technique the passage from magnitudes to structure was attempted essentially in one step (see Fig. 1). The vast increase in structure solving power today is basically due to the idea of dividing the problem into two relatively more tractable problems.

The first method of using maps in the solution of unknown structures was the Fourier refinement technique (Robertson 1933); once correctly initiated it usually leads rapidly to the successful completion of the structure, and it is still frequently used in the structure-completion process. The first examples of its use, due to Robertson, were with rigid-molecule structures, where it could be initiated by finding the approximate position and orientation of the molecule.

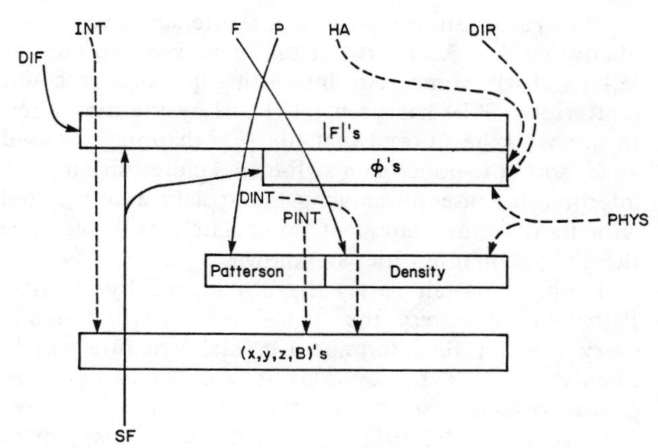

Fig. 1. Structure determination (schematic). Upper box is in reciprocal space and shows the diffraction pattern (magnitudes and phases). Lower boxes are in direct space and show images or maps (Patterson, electron density) and the structure itself (atom parameters (x, y, z, B)'s). Lines show structure determining techniques, in approximate historical order from left to right. Pre-1930: the magnitudes were found from the diffraction experiment (DIF), and a direct structural interpretation (INT) was attempted. (Dashed lines indicate techniques that may not succeed.) The interpretation could be checked by use of the structure factor formula (SF). Post-1930: introduction of maps into the technique. F is Fourier summation, DINT is density map interpretation, P is Patterson or $|F|^2$ summation, PINT is Patterson map interpretation, HA is heavy-atom methods, DIR is direct methods, PHYS refers to the possibilities noted in the last paragraph of the article. Not shown are the structure and phase refinement methods, or the processes of calculating a density map from a structure (inverse of DINT) or magnitudes and phases from a density map (inverse of F).

Subsequently, for structures of more general types, essentially three further structure determining methods have been found: in historical order they are: Patterson methods, heavy-atom methods, and direct methods. The first employs unphased maps; the second and third are means of obtaining phased maps, which are much easier to interpret than unphased maps; the first and second today are usually used in combination. The first, and to a large extent the third, were introduced by American crystallographers, and the second by British crystallographers. Here we shall concentrate principally on the first and third.

Patterson Methods

The Patterson or $|F|^2$-map (Patterson 1934) was discovered by A. L. Patterson, who was working at M.I.T. at the time. An interesting personal account (Patterson 1962) has been left to us by the discoverer. In this work he showed that the $|F|^2$-map may be used as a sort of poor man's F-map, more difficult to interpret because of showing the structure convoluted with its own inversion, but immediately available once the diffraction intensities are known.

I think it is fair to say that for some 30 years after Patterson's discovery the first thing done in virtually every attempt to determine a crystal structure was to calculate "the Patterson", as it was called by every crystallographer except Patterson; this was done to see if it would yield information on the atom positions or on the form or orientation of the molecules. If it did, an attempt to produce an interpretable F-map would be made, and if this succeeded the Fourier refinement process would begin.

Two tendencies can be discerned in the techniques of applying Patterson's idea in structure determination. In America, the theory of the $|F|^2$-map was stressed and an attempt was made to develop it into a general method of structure determination. This can be seen, for example, in the work of David Harker (1936), then at Cal Tech, in analyzing the effects of crystal symmetry on the map; of Patterson (1944), then at Bryn Mawr, on the existence of multiple map interpretations; of M. J. Buerger (1950, 1951), at M.I.T., on systematic methods for interpreting maps; and more recently in the work of C.E. Nordman (1966) at Michigan, M. G. Rossmann (1972) at the University of Cambridge and Purdue, and others, on methods for locating and orientating in the unit cell a known or approximately known structure via its $|F|^2$-map. In Britain, on the other hand, stress was laid on practical aspects of interpreting the maps. It was quickly noticed that the heavy-atom peaks in a Patterson map are generally rather easy to interpret. It was not long before this observation bore fruit with the discovery of the heavy-atom phasing method (Robertson 1936). Thus was born the combination of Patterson and heavy-atom techniques which over 40 years has registered an increasingly magnificent chain of successes on large structures, from penicillin in the 1940s, through the smaller proteins in the 1960s, to the viruses and other macromolecular biological systems under study today.

Direct Methods

It was after World War II that the American tendency to regard the process of structure determination from a mathematical point of view began to produce the third of the major structure determining techniques in use today, the direct methods. A direct method consists of a system of mathematical relations which involve the structure-factor amplitudes and phases (but not the atom coordinates), together with a method for solving the relations for the phases. There are three principal families of mathematical relation systems in use today: in decreasing order of current importance, they are the probabilistic linear relations providing estimates of the values of structure invariants and seminvariants, the convolutional relations, and the determinantal relations. Because of their convenience, direct methods are now usually the first technique tried for structures of up to 100-200 non-hydrogen atoms. The major relation systems will be discussed here in the order in which they appeared historically*.

Determinantal relations. The determinantal inequalities were discovered in 1950 by J. Karle and H. Hauptman at the Naval Research Laboratory in Washington, D.C. (Karle and Hauptman 1950). Earlier, D. Harker and J. S. Kasper, working at the research laboratory of the General Electric Company, had discovered the first system of mathematical relations among the structure factors to allow an unknown structure to be solved directly from the diffraction data (Harker and Kasper 1948; Kasper, Lucht and Harker 1950). Interest in the Harker-Kasper inequalities was considerable, and additional inequalities and improvements in the method of deriving inequalities were found by J. Gillis (1948) in Israel and Caroline MacGillavry (1950) in the Netherlands. It was at this point that the paper by Karle and Hauptman appeared. It showed that the Harker-Kasper inequalities are special cases of a more general inequality, the non-negativity of certain determinants constructed from the structure factors.

Were it possible to solve the full system of Karle-Hauptman inequalities, retaining only those sets of phases consistent with every inequality, the determinantal inequalities would probably provide more accurate phases on a wider range of real structures than the other direct-method relation systems currently available. Unfortunately, no good solution techniques for large systems of determinantal inequalities are known at present, and the possibilities for use of the inequalities in practice are accordingly much reduced. A partial step forward was taken when G. Tsoucaris (1970) in France showed, at least for the equal-atom case, that the phases which solve the full system can be

*In addition to the relation systems listed, one should mention the earliest such system, that of Ott (1927), later rediscovered by M. Avrami (1938) at the University of Chicago. The list here also omits the density modification methods, several of which (including non-crystallographic symmetry and solvent smoothing) have recently begun to be of great importance in macromolecular structure studies.

approximated by the phases which maximize the value of a single determinant, thus allowing the problem to be treated approximately as a maximization problem, for which reasonably good solution methods are available. In this form the determinantal relations are again a subject of research, both as a practical technique of phase refinement, and in recent work by Karle (1980) as a tool in the development of the probabilistic theory of the structure invariants and seminvariants.

Convolutional relations. The first convolutional relations were discovered by D. Sayre (1952), an American graduate student with Dorothy Hodgkin at Oxford. An account of the discovery has recently been written (Sayre 1980). His relations, the squaring-method equations, hold for structures composed of equal resolved atoms, and are exact (non-statistical). Equation systems of similar form, which are only of statistical validity but in which the requirements for atom equality and resolution are removed or reduced, have also been found: by W. H. Zachariasen (1952) at the University of Chicago (requirement on equality removed); by E. W. Hughes (1953) at Cal Tech (requirement on resolution reduced); by Hauptman and Karle (1953) at the Naval Research Laboratory (requirement on equality removed and on resolution reduced, centrosymmetric structures only); by Karle and Hauptman (1956) (as above, non-centrosymmetric structures); by Karle and Karle (1966) (as above but with requirement on resolution further reduced); and by H. Hauptman (1971), now at the Medical Foundation of Buffalo (similar to Karle and Karle, but with requirement on equality reinstated).

The convolutional relations derive their importance from the fact that it is considerably simpler to devise solution techniques for them than for the determinantal relations, even though the inherent phasing power of the two types of system is not greatly different. An important reason for this simplicity is that the convolutional equations express the value of each structure factor directly in terms of the full set of structure factors. This was observed by Karle and Hauptman (1956) and made the basis for tangent-formula refinement, a method of solving the convolutional equations by successive approximations once a starting set of phases is available. For many years direct-method systems have used this technique for the phase extension and refinement section of the system. A more reliable but costlier method of solution was introduced by Sayre (1972).

Probabilistic linear relations. The early history of these relations is somewhat complex. Gillis (1948) noted that the Harker-Kasper inequalities frequently held in cases where they were not strictly phase-determining, but did not place a probabilistic interpretation on this observation; also he did not at that time have the determinantal form of the inequalities, which under a probabilistic interpretation would have yielded the zero-estimate of the 3-phase structure invariant, to work with. Karle and Hauptman (1950) stated the determinantal form of the inequalities, but treated them purely algebraically. Sayre (1952) gave an informal probabilistic interpretation to both the determinantal inequalities and the convolutional relations, obtained the zero-estimate of the invariant as a result, and used it to solve the structure of hydroxyproline by a method somewhat similar to the modern multisolution technique. With knowledge of Sayre's result, Cochran (1952) and Zachariasen (1952) derived it by explicit probabilistic methods which were forerunners of the mathematical techniques used today, and also used it for the first time in the solution of unknown structures.

The value of the relations of this class is that they give rise to a system of equations which, being linear, can be solved for the phases *ab initio*. Thus a direct-method system can form an approximate solution using linear relations, and then extend and refine this solution, usually by means of a convolutional relation system.

Shortly after the work of Zachariasen and Cochran in 1952, Hauptman and Karle (1953) began their long series of contributions to the study of the probabilistic linear relations, introducing the important concept of structure invariant and seminvariant, and bringing more sophisticated mathematical techniques to the study of the relations than had been used heretofore. With this began the further story which ultimately led in 1975, due largely to the efforts of Schenk in the Netherlands, Giacovazzo in Italy, and Hauptman in the US, to the start of a major new series of discoveries of relations involving the higher-order structure invariants (quartets, quintets, etc.) and the structure seminvariants. In this volume J. Karle and H. Hauptman will tell the earlier and later parts of this story, respectively.

* * *

At the present time, structure determination is almost always carried out either by direct methods or by a combination of Patterson and heavy-atom methods, i.e. by methods which, although much refined since their inception, and made much much powerful by the advent of the computer, date back nearly 30 and 50 years, respectively. We are, in other words, in a period of considerable stability insofar as structure determining technique is concerned. But there are signs now that basic innovations may again be entering the field. It is too early to state it with positiveness, but it seems possible that the major developments may be physical, rather than mathematical and computational, in nature, and may involve imaging the structure directly in the physical experiment, or at least capturing the phases of the diffraction pattern in the experiment. Indications which suggest that the wind may be blowing in this direction, and indications also of the involvement of American scientists in the work, are the recent demonstration of the use of dynamical theory to provide an experimental means of measuring phases (Post 1977), the perceptible drawing-together of diffraction techniques and microscopy (McLachlan 1958, Kuo and Glaeser 1975), and the development of important new X-ray sources, optics, and detectors (Parsons 1980).

References

Avrami, M. (1938) Phys. Rev. **54**, 300.
Bragg, W. H. (1915) Phil. Trans. Roy. Soc. **A215**, 253.
Bragg, W. L. (1929) Proc. Roy. Soc. (London) **A123**, 537.
Buerger, M. J. (1950) Acta Cryst. **3**, 87 and 243.
Buerger, M. J. (1951) Acta Cryst. **4**, 531.
Cochran, W. (1952) Acta Cryst. **5**, 65.
Duane, W. (1925). Proc. Natl. Acad. Sci. U.S.A. **11**, 489.
Gillis, J. (1948) Acta Cryst. **1**, 76 and 174.
Harker, D. (1936) J. Chem. Phys. **4**, 381.
Harker, D. and Kasper, J. S. (1948) Acta Cryst. **1**, 70.
Hauptman, H. (1971) Am. Crystallogr. Assoc. Program Abstr. Ser. 2.
Hauptman, H. and Karle, J. (1953). ACA Monograph #3.
Havighurst, R. J. (1925) Proc. Natl. Acad. Sci. U.S.A. **11**, 502 and 507.
Hughes, E. W. (1953) Acta Cryst. **6**, 871.
Karle, J. (1980) Proc. Natl. Acad. Sci. U.S.A. **77**, 5.
Karle, J. and Hauptman, H. (1950) Acta Cryst. **3**, 181.
Karle, J. and Hauptman, H. (1956) Acta Cryst. **9**, 635.
Karle, J. and Karle, I. (1966) Acta Cryst. **21**, 849.
Kasper, J. S., Lucht, C. M. and Harker, D. (1950) Acta Cryst. **3**, 436.
Kuo, I. A. M. and Glaeser, R. M. (1975) Ultramicroscopy **1**, 53.
MacGillavry, C. H. (1950) Acta Cryst. **3**, 214.
McLachlan, D. (1958) Proc. Natl. Acad. Sci. U.S.A. **44**, 948.
Nordman, C. E. (1966) Trans. Am. Crystallogr. Assoc. **2**, 29.
Ott, H. (1927) Z. Kristallogr. **66**, 136.
Parsons, D. F. (1980) Ann. N.Y. Acad. Sci. 342.
Patterson, A. L. (1934) Phys. Rev. **46**, 372.
Patterson, A. L. (1944) Phys. Rev. **65**, 195.
Patterson, A. L. (1962) In "Fifty Years of X-Ray Diffraction" edited by P. P. Ewald, Utrecht: A. Oosthoek, Utrecht.
Post, B. (1977) Phys. Rev. Lett **39**, 760.
Robertson, J. M. (1933) Proc. Roy. Soc. (London) **A140**, 79.
Robertson, J. M. (1936) J. Chem. Soc. p. 1195.
Rossmann, M. (1972) The Molecular Replacement Method. New York: Gordon and Breach.
Sayre, D. (1952) Acta Cryst. **5**, 60.
Sayre, D. (1972) Acta Cryst. **A28**, 210.
Sayre, D. (1980) In "Structural Studies on Molecules of Biological Interest" edited by G. Dodson, J. P. Glusker, and D. Sayre. Oxford: Oxford University Press.
Tsoucaris, G. (1970) Acta Cryst. **A26**, 492.
Zachariasen, W. H. (1952) Acta Cryst. **5**, 68.

CHAPTER 18

RECOLLECTIONS AND REFLECTIONS*

I. L. Karle and J. Karle
Laboratory for the Structure of Matter, Naval Research Laboratory, Washington, D. C. 20375

In reviewing our participation in structure research, a number of aspects come to mind, personal anecdotes, interests and motivations, views of the subject and some highlights. The attempt to write a short article on these matters and to say something refreshing while avoiding repetitiveness presents some difficulties. Recently an article has been written concerning our work in electron diffraction (Karle and Karle, 1981), and reviews have been published on amorphous (Karle, J., 1977) and crystalline (Karle, J., 1978b) materials that contain much of the guiding philosophy that has pervaded our research activities.

The later parts of this article will concern a brief presentation of our views of some possible paths for future progress in structure research of the crystalline state. These views are based upon our personal experiences and perceptions to date and, therefore, one of the roles to be played by the recollections presented here is to provide the background for a discussion of the future. In this way, we have attempted to avoid the mere repetition of well-documented events.

It will be evident in this discussion that we present only our personal experiences in some areas of structure analysis by diffraction methods. We refer therefore mainly to our own publications, including only those of others that either played a role in our participation in the developments to be described or were closely related.

Electron Diffraction

During the 1940's, rather extensive changes took place in the practice of electron diffraction by gaseous molecules. These developments had their origins in technical and theoretical progress that was made before or early in World War II, but were not fully realized until the advent of renewed opportunities to pursue this type of research after the war.

In graduate school at the University of Michigan our specialty was electron diffraction by gases which we studied under Professor Lawrence O. Brockway. Our thesis work was completed during 1943. At that time in the United States, research on electron diffraction by gases was carried through with visual estimates of the total scattered intensity (Brockway, 1936). A theoretical paper by P. J. W. Debye (1941) pointed out the possibility of interpreting the Fourier transform of the molecular scattering in terms of the probability of finding various interatomic distances in a molecule. It was also indicated that such an interpretation could provide the opportunity to obtain information concerning the internal motion in molecules. Such possibilities were quite attractive to us and we had an excellent opportunity to attempt to fulfill this promise when we joined the Naval Research Laboratory in 1946. As matters turned out, the manner and underlying philosophy with which this program was pursued had a great influence on the future course of our research activities.

The very fine and extensive shop facilities at the Naval Research Laboratory made it relatively easy for us to design and have constructed an electron diffraction apparatus that incorporated the most advanced technical features of that time (Karle and Karle, 1955; Karle, 1973). Of particular interest to us in that regard was the potential for obtaining enhanced accuracies as compared to visual estimation by the introduction of a mechanical sector to level the background scattering and a microphotometer to measure the intensities. The mechanical sector had been introduced independently by Finbak (1937) and P. Debye (1939), and the Norwegian school had been already making microphotometer measurements of the scattered intensities (Finbak, 1941; Finbak and Hassel, 1941). We found, though, that in order to achieve the desired accuracy it was necessary to spin the photographic plates during the course of the measurements with the microphotometer (Karle, Hoober, Karle, 1947). In the development and assembly of the electron diffraction apparatus, we were greatly helped by our colleague John Ainsworth.

The achievement of enhanced accuracy for the values of the scattered intensities was only the first step in the development of a procedure that could provide interatomic distances of high reliability and information concerning internal motion. There were a number of crucial steps in the treatment of the data such as the elimination of termination errors from the Fourier transform of the molecular intensity and the derivation of a molecular intensity function from the total intensity in such a manner that it would represent the molecular scattering that would be obtained from atoms having essentially constant scattering factors.

A most interesting and, as it turned out, a most fruitful problem that had to be overcome was the development of an accurate representation of the background scattering. This is required in order to derive the relatively weak molecular scattering from the total intensity. Limitations of the photographic process and the existing theory made alteration of the tabulated values for the background scattering mandatory. A

*No pun intended.

method for achieving appropriate adjustments to the background scattering was derived on the basis that the Fourier sine transform of the molecular scattering had to be a non-negative function, since it represented probability distributions of values for interatomic distances (Karle, I. L. and Karle, J. 1949, 1950; Karle, J. and Karle, I. L. 1950). This worked out quite well and, in fact, we were able to achieve the promise contained in Debye's 1941 paper. Interatomic distances could be determined more accurately, and a considerable amount of information concerning internal molecular motion became available. Comparisons of the results from this newly developed technique with results from microwave and infra-red spectroscopic investigations of small molecules, e.g., CO_2 (Karle and Karle, 1949; Morino, 1950) and H_2CCF_2 (Karle, I. L. and Karle, J., 1950) showed that the agreement was quite satisfactory, thus implying that the new procedures were reliable.

We carried out a number of structural investigations by gas electron diffraction that included determinations of average amplitudes of vibration (Karle and Karle, 1955), gave indications of the shrinkage effect and provided evaluations of barriers that hinder internal rotation (Ainsworth and Karle, 1952). In addition, there had been earlier investigations made on monomolecular long-chain hydrocarbon films (Karle, 1946; Karle and Brockway, 1947; Karle, 1949) and, in recent years, the methods for gaseous molecules have been extended to the analysis of amorphous materials (D'Antonio et al. 1977).

It was apparent from the methods and procedures developed in our work in gas electron diffraction that the introduction of mathematical and physical constraints into a structure analysis could greatly enhance the quality of the results. This provided a guide to much of our future research in the field of structure determination (Karle, J., 1977, 1978).

Of particular interest was the effectiveness of the non-negativity criterion. This suggested that it might be useful to consider the applicability of this criterion to other fields of structure research. Soon crystal structure analysis was under consideration from the point of view that the electron density distribution in a crystal must be a non-negative function. Application of the non-negativity criterion to the electron density function resulted in a complete set of inequalities of increasing order among the structure factors (Karle and Hauptman, 1950), the inequalities becoming more restrictive with increasing order. The simple third order inequality, and its associated probability measures, contain several of the main formulas for present-day phase determination, e.g., sigma-2, the sum of angles formula and the tangent formula (Karle and Karle, 1966). This will be discussed in more detail later on.

Crystal Structure Analysis.

In ordinary X-ray diffraction experiments with crystals, phase information appears to be lost in the measured intensities. The phases, however, are in fact recoverable from the measured intensities, and the reason for this is the fact that electron distributions around atoms in crystals are known to quite a good approximation, differing little from those of free atoms. This, in effect, reduces the crystal structure problem to one concerning the location of point atoms, or atomic centers, a problem that is often highly overdetermined by the amount of measured intensity data. Since knowing the atomic positions is equivalent to knowing the phase values, these values are thus usually overdetermined by the intensity data (Karle, 1969).

It is important to emphasize, particularly in view of the forthcoming discussion, that overdeterminancy does not necessarily imply uniqueness. An overdetermined problem may still contain ambiguous answers, particularly if due regard is given to the inevitability of inaccuracies in the data and the fact that answers with quite unreasonable physical features may give a fairly good fit to the data or, at least, to the data that are used in the initial stages of a phase determination.

The importance of the application of the non-negativity criterion to the phase problem derives from the fact that it led to the general inequality theory, developed in collaboration with our colleague H. Hauptman (Karle and Hauptman, 1950), which with the associated probability theory (Hauptman and Karle, 1953; Woolfson, 1954; Cochran, 1955; Karle and Hauptman, 1956; Karle, J. and Karle I. L., 1966; Tsoucaris, 1970; Karle, 1971; Karle, 1978a; Heinerman, Krabbendam and Kroon, 1979) provides the foundation mathematics for the practical determination of phases. Related to this was the development of the theory of invariants and seminvariants associated with origin specification in crystals and the specification of the enantiomorph and/or axis direction, when appropriate. This theory was also developed in collaboration with H. Hauptman (Hauptman and Karle, 1953, 1956, 1959; Karle and Hauptman, 1961). The overdeterminancy of the phase problem permits the relatively simple, main phase determining formulas to have, under appropriate circumstances, a high probability of being correct.

The inequalities can be treated as phase relations that are probably correct even when the conditions for the application of the inequalities are not strictly obeyed. The importance of this point is that this greatly extends the range of applicability of inequality theory and provides phase formulas, for example, the phase relations contained in the third order determinantal inequality, that under proper circumstances have a high probability of being correct. We call this the probabilistic implications of inequality theory and it was first observed by Gillis (1948) with respect to the Harker-Kasper (1948) inequalities. The probabilistic implications were made quantitative by the development of associated probability formulas (Hauptman and Karle, 1953; Woolfson, 1954). It was later shown how the probability formulas associated with the inequalities could, with use of the central limit theorem, be read directly by inspection from the inequalities themselves (Karle, 1971). Various general forms of joint probability distributions associated with the inequalities have appeared in the literature (Tsoucaris, 1970; Karle, 1971; Karle, 1978a; Heinerman, Krabbendam and Kroon, 1979).

The role of the non-negativity criterion in leading to valuable relations for phase determination is illustrated by the effect in neutron diffraction of the presence of a proportionately significant number of atoms such as hydrogen that have negative scattering factors along with atoms that have positive scattering factors. Under these circumstances, the main phase determining formulas such as the sum of angles formula and the tangent formula may not be accurate enough to apply. A solution to this problem was developed by the formulation of a calculation that derives from the measured intensities the intensities that would have been obtained if all the atoms present scattered as if their scattering factors were squared (Karle, 1966). For the derived intensities, all effective scattering factors would be positive, permitting the formulas of the Symbolic Addition Procedure (Karle and Karle, 1963, 1966) to be applied.

The foundation mathematics for phase determination, namely, the general inequalities, their probabilistic implications and associated probability theory are valid whether the substance of interest has essentially equal atoms or atoms of rather disparate atomic number. Because of misunderstandings concerning the mathematical foundations for phase determination, students of this subject have often expressed surprise to us when they first learn that the procedure for phase determination is readily applicable to structures composed of atoms of disparate atomic number. In fact, probability theory and experimental results show that when the number of atoms is the same, the phase relations are more reliable for the case of disparate atoms than when only atoms of equal atomic number are present. Evidently, for a mathematical system to form a basic source for the phase determining formulas, it must be valid for both equal-atom and unequal-atom structures.

Despite the fact that our interest in pursuing the implications of the non-negativity of the electron density in crystals led eventually to the general inequality theory, it did not occur to us that this theory might include the Harker-Kasper (1948) inequalities, of which we were well aware, until the inequalities were at hand. It was subsequently straightforward to demonstrate in several examples that Harker-Kasper inequalities could be derived from the general set (Karle and Hauptman, 1950). Examination of the derivation of the Harker-Kasper inequalities showed that the non-negativity of the electron density was implicitly assumed in the derivation. Although the Harker-Kasper inequalities did not include the third order determinantal inequality of the complete set of inequalities, they served the valuable purpose of providing an opportunity to observe the characteristics of the inequalities such as their probabilistic features and usefulness in structure determination. By use of the Harker-Kasper inequalities, Kasper, Lucht and Harker (1950) determined the structure of decaborane.

Not long after the development of the general inequality theory, Sayre (1952) and Zachariasen (1952) gave a practical illustration of phase determination by use of the phase formula implied by the third order determinantal inequality. They each made an application to a centric structure and anticipated the future developments. Soon procedures were considerably enhanced by the introduction of probability measures and associated formulas (Hauptman and Karle, 1953) and by use of a number of strategies designed to improve reliability and maintain the ambiguousness of the problem at a tolerable minimum. Sayre (1952) also derived a formula that defines, in the equal-atom case, the value of a phase in terms of large numbers of other phases. This formula has found application in phase refinement and extension, particularly for macromolecules (Sayre, 1974).

While the mathematical developments have greatly facilitated crystal structure analysis, there are many structures that do not yield merely to the mathematics. In our experience, such structures require other analytical aids for their solution, such as chemical information and special intuitive skills that derive from experience in this field.

We started experimental work in crystal structure analysis during the latter part of the 1950's. In a fairly short time, it was possible to derive a procedure for phase determination, the Symbolic Addition Procedure (Karle, I. L. and Karle, J. 1963; Karle J. and Karle, I. L., 1966), that was efficient, had a wide range of applicability and, importantly, was able to handle the ambiguousness that normally characterizes a phase determination. The ambiguousness is expressed by the use of symbols corresponding to a variety of values for the phases. There is generally a high probability that the physical answer can be represented by an appropriate set of values for the symbols.

It is often a rather difficult matter to develop appropriate analytical procedures that can bridge the gap between theory and practical application. At the present time, when so many structures can be determined routinely or almost so with the use of computer programs, it is difficult to imagine that crystal structure analysis was ever a problem, particularly after phase relations from the theoretical investigations became available. However, this was the case and bridging the gap between theory and practice was not automatic. For example, it was quite valuable to recognize that the full set of phase values could be derived from only a very few symbolic specifications and that symbols representing acentric reflections need be assigned only discrete values separated by 90°. In addition to developing a procedure, it was also necessary to establish the conditions and standards for applying the procedure, for example, by formulating criteria for specifying the origin determining reflections, fixing the enantiomorph and estimating the acceptability of relationships that often develop among the symbols as the phase determination proceeds.

As applications of the procedure for phase determination, particularly for noncentrosymmetric crystals, increased during the 1960's, it became apparent that the procedure would often produce only part of the structure. Thus, it was evident that the range of application of the Symbolic Addition Procedure could be greatly extended if a procedure were introduced for developing partial structures. This led to the method for the development of partial structures by means of the tangent formula, a method that was found to be quite effective (Karle, 1968). If the partial structure is not properly located in the unit cell, a translation function can be used to relocate it (Karle, 1972). Other procedures such as the use of a lower space group, e.g., *P*1 is also effective in this regard (Karle and Karle, 1971).

With the development of a method for phase determination, the advent of commercially produced automatic diffractometers, and a considerable enhancement of the sophistication and availability of computing facilities toward the end of the 1960's, the opportunity to perform a wide range of structural investigations was at hand. In fact, soon after the Symbolic Addition Procedure was introduced, we applied it to solve the structure of the first noncentrosymmetric crystal determined by direct methods, L-arginine dihydrate (Karle and Karle, 1964) and to other early investigations of noncentrosymmetric structures, panamine (Karle, I. L. and Karle, J., 1966), 6-hydroxycrinamine (Karle, Estlin and Karle, 1967), reserpine (Karle and Karle, 1968) and digitoxigenin (Karle and Karle, 1969). As further examples, in our laboratory, we have investigated nucleic acid bases, photorearrangements, natural products, antibiotics, heart drugs, frog toxins, polypeptides, ionophores, steroids and plant growth regulators among others. Many examples of these studies are to be found in review articles (Karle, 1969; Karle and Karle, 1972; Karle, 1973b; Karle, 1976; Karle, I. L., 1977; Karle, 1980b; Karle, I. L. 1981).

In many instances, for example, in investigations of polypeptides and ionophores, new and stimulating insights have been obtained concerning conformational characteristics and special properties such as ion transport. In many natural products such as frog toxins and in complex photorearrangements, structures of unknown chemical composition and unexpected and unusual linkages were elucidated, thereby identifying the chemical nature of the substance of interest. X-ray analysis thus became a method of choice for identifying and characterizing reaction products and intermediates, especially when there were no models for interpreting other physical measurements such as n.m.r. spectra. By the latter part of the 1960's, X-ray analysis of essentially equal atom structures was largely transformed from a field in which there was little general confidence to one in which, in most instances, analyses could be made in a routine or fairly routine fashion, thus permitting the investigator to concentrate on the functional properties of substances of interest and correlate structure with function.

The considerable advances in structure determination facilitated by mathematical developments stimulated many such studies during the 1970's. They took many forms, for example, a generalization of the tangent formula (Karle, 1971) that has found application in phase refinement for proteins, an improved algebraic formula for triplet invariants (Karle, 1970), numerous investigations involving inequality and probability theory applied to triplet and higher phase invariants (e.g. Messager and Tsoucaris, 1972; Karle, 1978a, 1979, 1980a; Schenk 1973, 1974, Schenk and de Jong, 1973; Putten and Schenk, 1977; Hauptman, 1976, 1977, 1980; Hauptman and Fortier 1977a, 1977b; Giacovazzo 1976a, 1976b, 1977a, 1977b, 1980) and structure refinement for moderate size crystals (Flippen-Anderson, Konnert and Gilardi, 1982) and macromolecular crystals (Hendrickson and Konnert, 1980) that imposes restraints based on known stereochemistry.

It is too early to evaluate the impact of newer theoretical work in probability theory involving the higher order phase invariants noted above. An important measure of the value of new mathematical results is whether they can help solve structures more efficiently or, better, solve structures that could not be solved otherwise. An estimation of the extent to which this may ultimately be possible for formulas involving the higher order phase invariants awaits future developments and applications. In this connection, Jerome has recently developed a general set of formulas, recently published (Karle, 1982a,b), for computing high order phase invariants and embedded seminvariants that lend themselves to computation in any space group. To the extent that they represent the information content in present day joint probability calculations, they may afford a straight-forward means for assessing the value of this approach.

Computer programs written with the objective of automating phase determination began to appear toward the end of the 1960's and by the early 1970's became a very active enterprise in many laboratories. Some of the currently well-known ones have been developed in M. M. Woolfson's laboratory in York, G. M. Sheldrick's laboratory in Cambridge and Goettingen, H. Schenk's laboratory in Amsterdam and J. M. Stewart's laboratory at the University of Maryland. Extensive efforts are also currently under way by C. J. Gilmore in Glasgow and by D. Viterbo in Torino in collaboration with G. Giacovazzo in Bari. Various other laboratories, including our own, share special programs from time to time. These efforts are quite noteworthy and some of the older programs, such as MULTAN from Woolfson's laboratory or SHELX from Sheldrick's laboratory have facilitated the solution of large numbers of structures. Improvements are also continually finding their way into the programs.

To the extent that the programs facilitate rapid determination of structure so that more time is available to consider the functional implications of structure, they obviously play an important role. However, when they are used in such an exclusive way that their failure incapacitates the investigator, the science of crystallography is not being pursued at its optimal. It has been our view that the many facilities in the hands of the analyst that have not been well-programmed as yet can extend the range considerably in the complexity of the problems that can be addressed. We have tried to maintain a balance between the contribution of the analyst and that of the automated procedure – perhaps somewhat in favor of the analyst. It is more work and requires on the average more thought, but it can also be more rewarding.

It has seemed to us that this view is not the most popular one, although appearances may be deceptive. There has been a tendency, however, which we attribute to dependency upon computer programs, to overlook the fact that useful ways were developed years ago for handling many perceived problems. This is especially so for techniques or formulas that for one reason or another are not contained in some widely distributed programs, such as a particular method for enantiomorph specification in space group $P2_1$ (Karle, I. L. and Karle, J., 1966), or partial structure development by means of the tangent formula (Karle, 1968) or the use of the modified $B_{3,0}$ formula (Karle, 1970) for screening triplets (Karle, Gibson and Karle, 1970; Flippen, 1972) or the generalization of the Σ_3 formula for noncentrosymmetric crystals (Karle, 1969).

The development of the field of structure determination and its interrelationships with numerous associated disciplines in our scientific lifetimes has been most remarkable in terms of increased facility, productivity and meaningful insights into physical, chemical and biological processes. It is quite legitimate to ask, with anticipation, "what next?" An adequate response to this question could easily fill a large volume. In the following postscript, we touch briefly on a few points.

A Postscript – Past Impressions and Future Possibilities

Mathematical analyses of the phase problem have played a major role in facilitating structure determination by diffraction methods. And yet, today, there are numerous structure determinations in which the structure is greatly overdetermined by the measured data, but which nevertheless require chemical insight and the general fund of crystallographic knowledge to supplement the information derived from the application of mathematical relations. The question arises concerning whether the mathematics of this subject has simply not yet been brought to a high enough level of practicality or whether there are some inherent difficulties, such as ambiguousness with respect to the solutions obtainable, that would inhibit the realization of a unique, physically correct solution in many cases.

This question cannot be answered with certainty, but the existence of structures that are difficult to solve strongly suggests that ambiguousness may be inherent in the intensity data and no matter how precise our mathematical tools become, the ambiguousness will remain. In practice, the ambiguous solutions do not have to agree with the measured intensity data to precisely the same extent, just closely enough to cause confusion between the mathematical solutions and the physical one. This possibility has influenced our view of the character of future fruitful developments in crystal structure determination.

The argument is sometimes presented that evidence for the uniqueness of the mathematical system is afforded by the good physical answers that result from a structure determination, implying that we have reached a global minimum, in the agreement with the measured data, that no other solution could even reach approximately. We do not know if this is almost always true, or if or when it is true, but it is of interest to review how the ultimate fit to the data is obtained and how this relates to the present mathematical methods in phase determination. When performing a least-squares fit of computed structure factor magnitudes to the experimental data, much chemical information is introduced in terms of the number and type of atoms involved. In present joint probability distribution calculations, this type of information is included in a rather weak fashion. The chemistry appears only as collective sums over powers of the atomic scattering factors. Nowhere does there appear a constraint on the phase determinations that requires phase values such that the resulting Fourier maps have the correct number of peaks and correct heights to represent the known chemistry. This would be a very powerful constraint, indeed. It is missing from the phase determining relations but is effectively present in the final least-squares fit to the experimental data. This is a major distinction between the conditions imposed by the mathematics employed in phase determination and in structure refinement. This distinction may well explain the fact that calculations or procedures, especially for the more complex structures may be quite ambiguous, or partially valid, or sometimes simply wrong. It remains to be seen whether probability methods can be sufficiently strengthened to overcome these difficulties and, at the same time, it is worth considering other possibilities for enhancing phase determination.

When additional information can be introduced into a structure determination, it has been found to be very helpful. This is at the basis of the successful applications of techniques for partial structure development in which much chemical and structural information is introduced. Of particular value for future progress would be the development of the mathematics of structure determination in a way that encompasses more chemical and structural information than is possible to date.

It would evidently be quite valuable to have an experimental technique for obtaining phase information. Recent developments in the anomalous dispersion technique in our laboratory imply that such possiblities are quite promising for macromolecular research. It remains to be seen how far this may carry to the smaller structures. Recent solutions of the structures of the native proteins, crambin, by Hendrickson and Teeter (1981) and trimeric hemerythrin by Hendrickson and Smith (1980) directly by use of the anomalous dispersion technique have demonstrated in several ways the great potential of this technique. Crambin was solved by use of the anomalous scattering from the three disulfide bridges in the molecule. The experiment was a single wavelength experiment using CuKα radiation whose wavelength of 1.54Å is far from the absorption edge of sulfur at 5.02Å. The trimeric hemerythrin was solved by use of the anomalous scattering from the iron atoms present in the molecule, again in a single wavelength experiment using CuKα radiation. The absorption edge for iron is fairly close to that for copper at 1.74Å.

This work has shown that relatively light atoms can be employed as anomalous scatterers and that useful effects can be obtained quite far in wavelength from an absorption edge.

A theoretical analysis has recently been performed (Karle, J., 1980) in which equations have been derived for a multiple wavelength experiment involving anomalous dispersion. The simultaneous equations are valid without approximation for any number of anomalous scatterers of any type. They are also largely linear in character if the unknown quantities are chosen properly. Among the unknowns, for example, that can be evaluated by the simultaneous equations are individual terms representing intensities of scattering that would be obtained for each type of anomalous scatterer, if each type were isolated from the rest of the atoms present and each scattered nonanomalously. Given a set of intensities for one type of atom, it would be possible to perform phase determination in order to locate the atoms of this type. The advantage here is that locating the atoms for an anomalous scatterer may be a very much simpler problem than that posed by the structure as a whole. Phase determination incidentally may be necessary if the number of anomalously scattering atoms is too large for an analysis by use of the Patterson function. With isolated intensities, however, it is readily possible. Other unknowns definable by the simultaneous equations contain phase differences in such a way that once phases are obtained for an anomalous scatterer, sufficient information is at hand to compute the entire structure. This is briefly what is contained in the theory. It will be interesting to see what the range of applicability will be with current laboratory diffraction equipment and how much the range might be enhanced by use of tunable, high intensity synchrotron radiation.

References

Ainsworth, J. and Karle, J. (1952) J. Chem. Phys. **20**, 425-427.
Brockway, L. O. (1936) Rev. Mod. Phys. **8**, 231-266.
Cochran, W. (1955) Acta Cryst. **8**, 473-478.
D'Antonio, P., Moore, P., Konnert, J. H. and Karle, J. (1977) Trans. Am. Crystallogr. Assoc. **13**, 43-66.
Debye, P. (1941) J. Chem. Phys. **9**, 55-60.
Debye, P. (1939). Physik. Z. **40**, 66 and 404-406.
Finbak, C. (1937) Avhandl. Norske Videnskaps-Acad. Oslo, I. Mat.-Naturv. Kl., No. 13.
Finbak, C. (1941) Avhandl. Norske Videnskaps-Acad. Oslo, I. Mat.-Naturv. Kl., No. 7.
Finbak, C. and Hassel, O. (1941) Arch. Math. Naturvidensk. **45**, No. 3.
Flippen, J. L. (1972) Acta Cryst. **B28**, 3618-3624.
Flippen-Anderson, J. L., Konnert, J. and Gilardi, R. (1982). Acta Cryst. **B38**, in press.
Giacovazzo. C. (1976a) Acta Cryst. **A32**, 74-82.
Giacovazzo, C. (1976b) Acta Cryst. **A32**, 967-976.
Giacovazzo, C. (1977a) Acta Cryst. **A33**, 527-531.
Giacovazzo, C. (1977b) Acta Cryst. **A33**, 933-944.
Giacovazzo, C. (1980) Acta Cryst. **A36**, 711-715.
Gillis, J. (1948) Acta Cryst. **1**, 174-179.
Harker, D. and Kasper, J. S. (1948) Acta Cryst. **1**, 70-75.
Hauptman, H. (1976) Acta Cryst. **A32**, 877-882.
Hauptman, H. (1977) Acta Cryst. **A33**, 568-571.
Hauptman, H. (1980) Acta Cryst. **A36**, 624-632.
Hauptman, H. and Fortier, S. (1977a) Acta Cryst. **A33**, 575-580.
Hauptman, H. and Fortier, S. (1977b) Acta Cryst. **A33**, 697-701.
Hauptman, H. and Karle, J. (1953). American Crystallographic Association, Monograph No. 3. Pittsburgh: Polycrystal Book Service.
Hauptman, H. and Karle, J. (1956) Acta Cryst. **9**, 45-55.
Hauptman, H. and Karle, J. (1959) Acta Cryst. **12**, 93-97.
Heinerman, J. J. L., Krabbendam, H. and Kroon, J. (1979) Acta Cryst. **A35**, 101-105.
Heinerman, J. J. L., Kroon, J. and Krabbendam, H. (1979) Acta Cryst. **A35**, 105-107.
Hendrickson, W. A. and Konnert, J. (1980) In "Computing in Crystallograhy," edited by R. Diamond, S. Ramaseshan and K. Venkatesan. Bangalore: Indian Academy of Sciences.
Hendrickson, W. and Smith, J. (1980) Private communication

within our laboratory.
Hendrickson, W. and Teeter, M. (1981). Nature (London), **290**, 107-113.
Karle, I. L. (1976) In "Photochemistry and Photobiology of Nucleic Acids," Vol. 1, edited by S. Y. Wang. New York: Academic
Karle, I. L. (1977) Pure Appl. Chem. **49**, 1291-1306.
Karle, I. L. (1981). In "The Peptides," Vol. 4, edited by E. Gross and J. Meienhofer. New York: Academic.
Karle, I. L., Gibson, J. W. and Karle, J. (1970) J. Am. Chem. Soc. **92**, 3755-3760.
Karle, I. L., Hoober, D. and Karle, J. (1947) J. Chem. Phys. **15**, 756.
Karle, I. L. and Karle, J. (1949) J. Chem. Phys. **17**, 1052-1058.
Karle, I. L. and Karle, J. (1950) J. Chem. Phys. **18**, 963-971.
Karle, I. L. and Karle, J. (1963) Acta Cryst. **16**, 969-975.
Karle, I. L. and Karle, J. (1964) Acta Cryst. **17**, 835-841.
Karle, I. L. and Karle, J. (1966) Acta Cryst. **21**, 860-868.
Karle, I. L. and Karle, J. (1968) Acta Cryst. **B24**, 81-91.
Karle, I. L. and Karle, J. (1969) Acta Cryst. **B25**, 434-442.
Karle, I. L. and Karle, J. (1971) Acta Cryst. **B27**, 1891-1898.
Karle, J. (1946) J. Chem. Phys. **14**, 297-305.
Karle, J. (1949) J. Chem. Phys. **17**, 500.
Karle, J. (1966) Acta Cryst. **20**, 881-886.
Karle, J. (1968) Acta Cryst. **B24**, 182-186.
Karle, J. (1969) In "Advances in Chemical Physics," Vol. 16, edited by I. Prigogine and S. A. Rice. New York: Interscience.
Karle, J. (1970) Acta Cryst. **B26**, 1614-1617.
Karle, J. (1971) Acta Cryst. **B27**, 2063-2065.
Karle, J. (1972) Acta Cryst. **B28**, 820-824.
Karle, J. (1973a) In "Determination of Organic Structures by Physical Methods," Vol. 5, edited by F. C. Nachod and J. J. Zuckerman, New York: Academic.
Karle, J. (1973b) In "Modern Methods of Steroid Analysis," edited by E. Heftmann. New York: Academic.
Karle, J. (1977) Proc. Natl. Acad. Sci. U.S.A. **74**, 4707-4713.
Karle, J. (1978a) Proc. Natl. Acad. Sci. U.S.A. **75**, 2545-2548.
Karle, J. (1978b) Proc. Natl. Acad. Sci. U.S.A. **75**, 3540-3547.
Karle, J. (1979) Proc. Natl. Acad. Sci. U.S.A. **76**, 2089-2093.
Karle, J. (1980a) Proc. Natl. Acad. Sci. U.S.A. **77**, 5-9.
Karle, J. (1980b) Lipids **15**, 793-797.
Karle, J. (1980). Int. J. Quant. Chem. **7**, 357-367.
Karle, J. (1982a). Proc. Natl. Acad. Sci. U.S.A. **79**, 1337-1339.
Karle, J. (1982b). Proc. Natl. Acad. Sci. U.S.A. **79**, 2125-2127.
Karle, J. and Brockway, L. O. (1947) J. Chem. Phys. **15**, 213-225.
Karle, J., Estlin, J. A. and Karle, I. L. (1967) J. Am. Chem. Soc. **89**, 6510-6515.
Karle, J. and Hauptman, H. (1950) Acta Cryst. **3**, 181-187.
Karle, J. and Hauptman, H. (1956) Acta Cryst. **9**, 635-651.
Karle, J. and Hauptman, H. (1961) Acta Cryst. **14**, 217-223.
Karle, J. and Karle, I. L. (1950) J. Chem. Phys. **18**, 957-962.
Karle, J. and Karle, I. L. (1955) In "Determination of Organic Structures by Physical Methods," Vol. 1, edited by E. A. Braude and F. C. Nachod. New York: Academic.
Karle, J. and Karle, I. L. (1966) Acta Cryst. **21**, 849-859.
Karle, J. and Karle, I. L. (1972) In "Chemical Crystallography," MTP International Review of Science, Physical Chemistry, Vol. 11, edited by J. M. Robertson. London: Butterworth.
Karle, J. and Karle, I. L. (1981) In "50 Years of Electron Diffraction," edited by P. Goodman. Dordrecht: Reidel.
Kasper, J. S., Lucht, C. M. and Harker, D. (1950) Acta Cryst. **3**, 436-455.
Messager, J. C. and Tsoucaris, G. (1972) Acta Cryst. **A28**, 482-484.
Morino, Y. (1950) J. Chem. Phys. **18**, 395.
Putten, N. van der and Schenk, H. (1977) Acta Cryst. **A33**, 856-858.
Sayre, D. (1952) Acta Cryst. **5**, 60-65.
Sayre, D. (1974) Acta Cryst. **A30**, 180-184.
Schenk, H. (1973) Acta Cryst. **A29**, 480-481.
Schenk, H. (1974). Acta Cryst. **A30**, 477-481.
Schenk, H. and de Jong, J. G. H. (1973) Acta Cryst. **A29**, 31-34.
Tsoucaris, G. (1970) Acta Cryst. **A26**, 492-499.
Woolfson, M. M. (1954) Acta Cryst. **7**, 61-64.
Zachariasen, W. H. (1952) Acta Cryst. **5**, 68-73.

CHAPTER 19

THE PHASE PROBLEM DURING THE SEVENTIES

Herbert Hauptman
Medical Foundation of Buffalo, Inc.
73 High Street
Buffalo, NY 14203

The late forties and fifties saw the birth and vigorous development of direct methods; although the first applications were also made then, it was not until the sixties that these methods were widely and extensively applied. The decade of the seventies has seen a resurgence of activity in developing the theory and practice of direct methods, and some notable advances have been made. During this period three major themes were explicitly formulated and developed: (1) The fundamental principle of direct methods; (2) The neighborhood principle; and (3) The extension concept.

The Fundamental Principle of Direct Methods

There always exist certain linear combinations of the phases whose values are determined by the crystal structure alone and are independent of the choice of origin. These linear combinations of the phases are called structure invariants. For all space groups other than $P1$ the origin may not be chosen arbitrarily if one is to exploit fully the space-group symmetries. Those linear combinations of the phases whose values are uniquely determined by the crystal structure and are independent of the choice of permissible origin are known as the structure seminvariants. Thus the class of structure seminvariants includes, and is in fact much larger than, the class of structure invariants.

It is known that the values of a sufficiently extensive set of cosine seminvariants (the cosines of the structure seminvariants) lead unambiguously to the values of the individual phases (Hauptman, 1972). Magnitudes $|E|$ are capable of yielding estimates of the cosine seminvariants only or, equivalently, the magnitudes of the structure seminvariants; the signs of the structure seminvariants are ambiguous because the two enantiomorphous structures permitted by the observed magnitudes $|E|$ correspond to two values of each structure seminvariant differing only in sign. However, once the enantiomorph has been selected by specifying arbitrarily the sign of a particular enantiomorph sensitive structure seminvariant (i.e., one different from 0 or π), then the magnitudes $|E|$ determine both signs and magnitudes of the structure seminvariants consistent with the chosen enantiomorph. Thus, for fixed enantiomorph, the observed magnitudes $|E|$ determine unique values for the structure seminvariants; the latter, in turn, as certain well defined linear combinations of the phases, lead unambiguously to unique values for the individual phases ϕ. In short, the structure seminvariants serve to link the observed magnitudes $|E|$ with the desired phases ϕ of the normalized structure factors (the fundamental principle of direct methods). Thus the structure seminvariants play the central role in the solution of the phase problem.

The Neighborhood Principle

It has been stressed that, for fixed enantiomorph, the collection of all observed magnitudes $|E|$ determine, in general, the values of all the structure seminvariants. A major recent advance is the neighborhood principle: For fixed enantiomorph, the value of any structure seminvariant T is primarily determined, in favorable cases, by the values of one or more small sets of observed magnitudes $|E|$, the neighborhoods of T, and is relatively insensitive to the values of the great bulk of remaining magnitudes (Hauptman, 1975). The conditional probability distribution of T, given the magnitudes in any of its neighborhoods, yields an estimate for T that is particularly good in the favorable case that the variance of the distribution happens to be small.

Recent advances consist, first, in identifying the neighborhoods of the structure invariants and seminvariants and, second, in deriving their conditional probability distributions, assuming as known the magnitudes $|E|$ in their neighborhoods.

Systems of neighborhoods for all the general structure invariants are now known, and associated distributions, which have proved to be of great value in the applications, have been found in some cases, in particular for quartets (four-phase structure invariants), quintets, and sextets (Hauptman, 1975; Giacovazzo, 1975; Hauptman and Green, 1976; Hauptman and Fortier, 1977a, b; Fortier and Hauptman, 1977; Putten and Schenk, 1977, 1979). The analogous work for the special structure invariants associated with various space groups remains to be done.

The Extension Concept

By embedding a structure seminvariant T and its symmetry related variants in suitable structure invariants Q, one obtains the extensions Q of the seminvariant T (Giacovazzo, 1977; Hauptman, 1978). Owing to the space-group dependent relations among the phases, the value of T is simply related to the values of its extensions. In this way the probabilistic theory of the structure seminvariants is reduced to that of the structure invariants, which is well developed. In particular, the neighborhoods of the structure seminvariant T are defined in terms of the neighborhoods of its extensions. Details are given for the special case of the three-phase structure seminvariant in $P\bar{1}$ (Giacovazzo, 1978; Hauptman, 1979).

The Three-Phase Structure Seminvariant in $P\bar{1}$, The linear combination of three phases

$$T = \phi_h + \phi_k + \phi_l \quad (1)$$

is a structure seminvariant in $P\bar{1}$ if the three components of $h + k + l$ are even. Then the components of the eight reciprocal vectors $\frac{1}{2}(\pm h \pm k \pm l)$ are integers. One embeds the three-phase structure seminvariant T and its three symmetry-related variants

$$T_1 = \phi_{-h} + \phi_k + \phi_l \quad (2)$$

$$T_2 = \phi_h + \phi_{-k} + \phi_l \quad (3)$$

$$T_3 = \phi_h + \phi_k + \phi_{-l} \quad (4)$$

in the respective quintets (five-phase structure invariants)

$$Q = T + \phi_{-H} + \phi_{-H}, \quad (5)$$

$$Q_1 = T_1 + \phi_{-H_1} + \phi_{-H_1}, \quad (6)$$

$$Q_2 = T_2 + \phi_{-H_2} + \phi_{-H_2}, \quad (7)$$

$$Q_3 = T_3 = \phi_{-H_3} + \phi_{-H_3}, \quad (8)$$

where

$$\underset{\sim}{H} = \frac{1}{2}(h+k+l), \quad (9)$$

$$\underset{\sim}{H}_1 = \frac{1}{2}(-h+k+l), \quad (10)$$

$$\underset{\sim}{H}_2 = \frac{1}{2}(h-k+l), \quad (11)$$

$$\underset{\sim}{H}_3 = \frac{1}{2}(h+k-l). \quad (12)$$

In this way one obtains the extensions, Q, Q_1, Q_2, and Q_3 of T. In view of (1)-(4) and the space group dependent relationships among the phases, it is readily verified that (5)-(8) are in fact five-phase structure invariants and

$$T = Q = Q_1 = Q_2 = Q_3. \quad (13)$$

Thus the probabilistic theory of the three-phase structure seminvariant T is reduced to that of quintets. In particular, the neighborhoods of T are defined in terms of the neighborhoods of the quintet. In this way one finds that the first neighborhood of T consists of seven magnitudes $|E|$, the three "main terms"

$$|E_h|, |E_k|, |E_l|, \quad (14)$$

and the four "cross-terms"

$$|E_H|, |E_{H_1}|, |E_{H_2}|, |E_{H_3}|. \quad (15)$$

In view of the neighborhood principles, the value of T is, in favorable cases, primarily dependent on the seven magnitudes (14) and (15) in the first neighborhood. The conditional probability distribution of T, assuming as known the seven magnitudes in its first neighborhood, has been found (Hauptman, 1979). The distribution shows that the favorable cases here require first that the three magnitudes (14) all be large and second that either all four magnitudes (15) be large, in which case $T \approx 0$, or that precisely two of the four magnitudes (15) be large and the remaining two be small, in which case $T \approx \pi$. The methods devised for the estimation of the three-phase structure seminvariant in $P\bar{1}$ serve as the prototype for the structure seminvariants in general and carry over without essential change to all the space groups, noncentrosymmetric as well as centrosymmetric.

Recent Results

The results briefly described here have been generalized in several ways. First, the probabilistic theory of the n-phase structure seminvariants in $P\bar{1}$ and the restricted n-phase structure seminvariants in $P2_1$ and $P2_12_12_1$, n = 1,2,3,4,5, has been worked out. Finally, the probabilistic theory of both the enantiomorph insensitive (i.e. having the value 0 or π) and the enantiomorph sensitive (i.e. having the value $\pm \pi/2$) two-and three-phase structure seminvariants in $P2_1$ has been initiated (Hauptman and Potter, 1979; Putten, Schenk and Hauptman, 1980a,b). Owing to the enormous number of available structure seminvariants, it is anticipated, and some initial calculations confirm, that these seminvariants will prove to be of great importance in the applications.

The Applications

Owing to the discovery in recent years of methods for obtaining reliable estimates of large numbers of higher order structure invariants, in particular those whose values are equal to π, crystal structures in $P\bar{1}$ are rather routinely solvable nowadays. Of particular importance are those applications to difficult structures, with experimental data sets limited in quality and number, which had defied solution by traditional techniques (Sax *et al.*, 1976; Blank *et al.*, 1976; Gilmore *et al.*, 1977; Bonnett *et al.*, 1978; Glidewell *et al.*, 1979).

Successful applications of the higher-order structure invariants to structures in the noncentrosymmetric space groups, which had also defeated solution by other methods, are particularly noteworthy (Busetta, 1976; Gilmore, 1977; Freer and Gilmore, 1977; Silverton and Akiyama, 1978; Silverton and Kabuto, 1978). Finally a number of computer programs implementing the recent theoretical work on the estimation of the higher-order structure invariants have been written and are now in routine use [e.g. SHELX and QTAN (Langs and DeTitta, 1975)].

References

Blank, G., Rodrigues, M., Pletcher, J. and Sax, M. (1976) Acta Cryst. **B32**, 2970-2975.

Bonnett, R., Davies, J. E., Hursthouse, M. B. and Sheldrick, G. M. (1978) Proc. Roy. Soc. (London) **B202**, 249-268.

Busetta, B. (1976) Acta Cryst. **A32**, 139-143.

Fortier, S. and Hauptman, H. (1977) Acta Cryst. **A33**, 829-833.

Freer, A. A. and Gilmore, C. J. (1977) J. Chem. Soc. Chem. Commun. pp. 296-297.

Giacovazzo, C. (1975) Acta Cryst. **A31**, 252-259.

Giacovazzo, C. (1977) Acta Cryst. **A33**, 933-944.

Giacovazzo, C. (1978) Acta Cryst. **A34**, 27-30.

Gilmore, C. J. (1977) Acta Cryst. **A33**, 712-716.

Gilmore, C. J., Hardy, A. D. U., MacNicol, D. D. and Wilson, D. R. (1977) J. Chem. Soc. Perkin Trans. **2**, pp. 1427-1434.

Glidewell, C., Liles, D. C., Walton, D. J. and Sheldrick, G. M. (1979) Acta Cryst. **B35**, 500-502.

Hauptman, H. (1972) Crystal Structure Determination: The Role of the Cosine Seminvariants. New York: Plenum.

Hauptman, H. (1975) Acta Cryst. **A31**, 680-687.

Hauptman, H. (1978) Acta Cryst. **A34**, 525-528.

Hauptman, H. (1979) Proc. Natl. Acad. Sci. U.S.A., **76**, 4747-4749.

Hauptman, H. and Fortier, S. (1977a) Acta Cryst. **A33**, 575-580.

Hauptman, H. and Fortier, S. (1977b) Acta Cryst. **A33**, 697-701.

Hauptman, H. and Green, E. (1976) Acta Cryst. **A32**, 45-49.

Hauptman, H. and Potter, S. (1979) Acta Cryst. **A35**, 371-381.

Langs, D. A. andDeTitta, G. T. (1975) Acta Cryst. **A31**, S16, Abstr. 02.2-14.

Putten, N. van der and Schenk, H. (1977) Acta Cryst **A33**, 856-858.

Putten, N. van der and Schenk, H. (1979) Acta Cryst. **A35**, 381-387.

Putten, N. van der, Schenk, H. and Hauptman, H. (1980a). Acta Cryst. **A36**, 891-897.

Putten, N. van der, Schenk, H. and Hauptman, H. (1980b). Acta Cryst. **A36**, 897-903.

Sax, M., Rodrigues, M., Blank, G., Wood, M. K. and Pletcher, J. (1976) Acta Cryst. **B32**, 1953-1956.

Silverton, J. V. and Akiyama, T. (1978) Am. Crystallogr. Assoc. Program Abstr. Ser. 2, Abst. PB15.

Silverton, J. V. and Kabuto, C. (1978) Acta Cryst. **B34**, 588-593.

CHAPTER 20

NORMAL PROBABILITY ANALYSIS

Sidney C. Abrahams
Bell Laboratories, Murray Hill, NJ 07974

A characteristic feature of structural crystallography is the requirement that a measured set of observations be fitted by a corresponding set derived from a model that is varied in the fitting process. The quality of the resulting fit is a measure of the validity of the model. The traditional indicator used is the agreement factor $R = \Sigma||Fobs| - |Fcalc|| \div \Sigma|Fobs|$, where *Fobs* is the observed and *Fcalc* is the corresponding calculated structure factor, or is one of several other related single-valued indicators. Such indicators are useful in the course of the fitting process in which their minimum value is taken as corresponding to the best model. It is not possible to ascertain from the magnitudes of these indicators if a true minimum has been reached or if systematic bias is present either in the *Fobs* or *Fcalc* sets. The latter and, by inferring the presence of remaining systematic error in the model, the former may be determined by the use of normal probability analysis.

The possibility that different laboratories may independently determine the crystal structure of the same material has recently increased to the extent that at least one well-known journal has instituted special procedures for detecting unwitting duplications. When detected, the authors of the second determination are requested to undertake a quantitive comparison of the two sets of parameters, including the atomic coordinates and the thermal parameters, for the reader's benefit. Any valid comparison method is acceptable, but the method of normal probability analysis has many advantages.

The Probability Plot

Normal probability analysis was introduced into crystallography in 1971 by Abrahams & Keve. The method is based on the general properties of probability plotting for data analysis, see Wilk & Gnanadesikan (1968) for a discussion, in which the ordered weighted differences between the *i*-th values of two sets of data x_i and y_i are plotted against the order quantiles expected for the given probability distribution. The weighted differences are given by

$$\delta m_i = (x_i - ky_i)/\left[\sigma^2 x_i + k^2\sigma^2 y_i\right]^{1/2} \qquad (1)$$

where σx_i, σy_i are the standard deviations of the *i*-th values of x and y and k is the scale factor between the two sets of data.

The δm_i are ordered in increasing magnitude, forming a set of ranked deviates or order statistics. On plotting them against the expected ranked deviates for a given error distribution, the result is a linear array with zero intercept and unit slope if the two distributions match.

A normal probability distribution, corresponding to the familiar bell-shaped normal curve, is generally assumed to govern the distribution of error in crystallographic experiments.

The Normal Probability Plot

Values of the order quantiles to be expected for a normal probability distribution, i.e. one with zero mean and unit variance, are conveniently listed in the *International Tables for X-Ray Crystallography* (1974), under expected values $\xi(i|n)$ of the ranked normal deviates (Table 4.3.2B), for the range $2 \leqslant n \leqslant 101$ and $1 \leqslant i \leqslant 50$. It should be noted that $\xi(n-i+1|i) = -\xi(n|i)$. Values of $\xi(i|n)$ for larger n may be approximated by setting the cumulative normal distribution function

$$P(x_i) = \frac{1}{\sqrt{2\pi}} \int_{-\infty}^{x_i} \exp(-\alpha^2/2)\,d\alpha \qquad (2a)$$

$$\simeq [3(n-i)+2]/(3n+1) \qquad (2b)$$

and looking up the resulting value of $x_i = \xi(i|n)$ in a table of the normal probability function, such as that given by Abramowitz and Stegun (1964).

A plot is then made of the ordered values of δm_i against $\xi(i|n)$ which will result in a linear array with unit slope and zero intercept if the distribution of the δm_i order statistics is normal. A typical array is shown in Fig. 1, in which x_i and y_i are the structure factors measured on two different crystals of $3\,La(IO_3)_3.HIO_3.7H_2O$ and there are 2341 independent values of δm_i. The array is close to linear, except for 45 terms in the extrema, and is indicative of a normal distribution. The departure of these terms (less than 2% of the total) from normality shows that they may be regarded as outliers that contain excess error. In the case that such departures are large, their inclusion in subsequent calculations that are not both robust and resistant could introduce serious bias.

The array in Fig. 1 has zero intercept and a slope of 1.27. The zero intercept indicates that the scale factor k in Eq. (1) is well-determined and that the δm_i are rather free from systematic error. The difference from unity in the slope is most likely the result of a systematic underestimation in the standard deviation of the values of x_i, y_i by the factor 1.27.

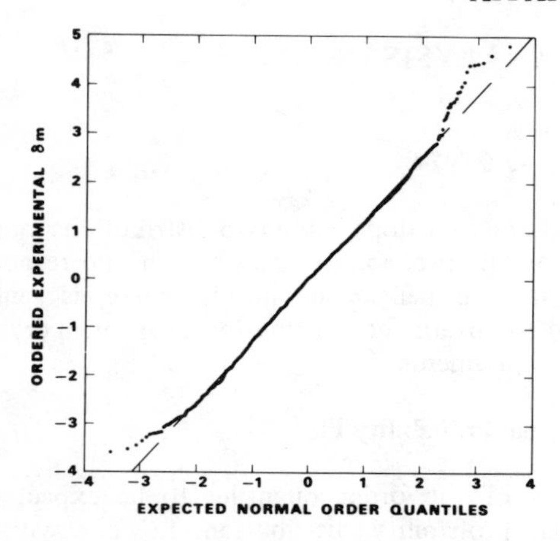

Fig. 1 Normal probability δm plot of 2341 ([Fobs (1) - k. Fobs (2)] ÷ [σ^2Fobs (1) + $\sigma^2 k^2$Fobs (2)]$^{1/2}$) terms common to the measurement sets made on two different crystals of $3La(IO_3)_3 \cdot HIO_3 \cdot 7H_2O$ in the same laboratory. The dash line is the least-squares linear fit to the δm array (after Abrahams & Bernstein, 1978).

If x_i = *Fobs* and y_i = *Fcalc* for each *i*-th value of *hkl*, the resulting δ*R* plot gives a direct comparison of the calculated with the experimental structure factors, as in Fig. 2. This δ*R* plot is based on 772 symmetry-independent structure factors measured on a crystal of KIO_2F_2: the corresponding *R* value is 0.046 and *S* (the goodness-of-fit) is 0.828. The array in Fig. 2 is very closely linear with very few outliers. The intercept is very small and the slope is 0.80. The distribution of

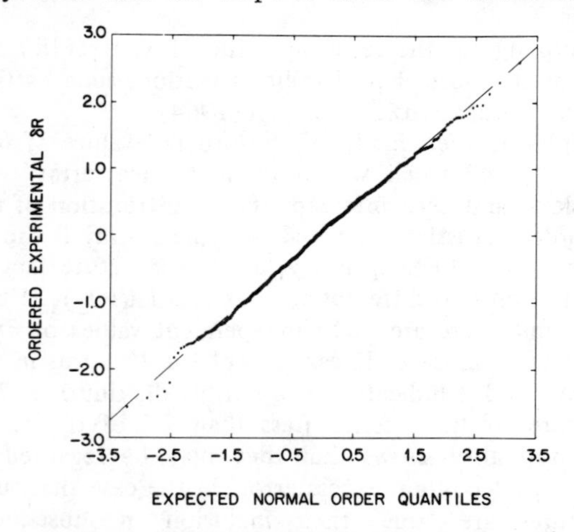

Fig. 2 Normal probability δR plot of 772 {[Fobs-Fcalc] ÷ [σFobs]} terms based on the structure determination of KIO_2F_2 by Abrahams & Bernstein (1976).

δ*R* terms is hence normal and the model and the experiment may be deduced to be free from systematic error, except for an overestimation of the σ*Fobs* by about 22%. The slope of the δ*R* array should be very similar to the value of *S*, as is generally found.

The presence of systematic error, other than a uniform over- or under-estimation of σ*Fobs*, either in the atomic parameters or in the experiment will result in a nonlinear δ*R*-array. This test is so simple and revealing that all crystal structure determinations should be subjected to it: major departures from linearity should be accounted for or, preferably, the cause of the error determined and then eliminated.

The Half-Normal Probability Plot

The sign of δm_i (Eq. 1) may be redundant in some cases, as in the comparison of atomic parameters for which the position *xyz* is transformed by a symmetry operation that introduces a sign change for one or more coordinates, e.g. from *xyz* to $\bar{x},\bar{y},z$. It is thus generally appropriate to compare atomic parameters by means of the half-normal probability plot, in which the $|\delta m_i|$ are ordered from maximum value to zero and plotted against the order quantiles given by the two-tailed normal probability function, Eqs. 3*a* and 3*b*,

$$P(x_i) = \frac{1}{\sqrt{2\pi}} \int_{-x_i}^{+x_i} \exp(-\alpha^2/2)\, d\alpha \qquad (3a)$$

$$\simeq [3(n-i)+2]/(3n+1) \qquad (3b)$$

and listed in Tables 4.3.2D and 4.3.2E of the *International Tables for X-Ray Crystallography* (1974) for $2 \leqslant n \leqslant 500$.

A powerful example of the δ*p* plot is given in Fig. 3, in which two sets of atomic parameters were determined

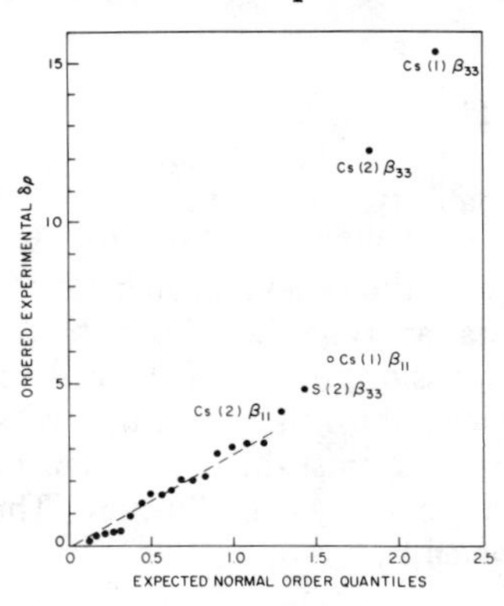

Fig. 3 Half-normal probability δp plot of 23 independent atomic parameters in $Cs_2S_2O_6$, measured on a crystal with minor radiation exposure p(1) and on another crystal p(2) that had been heavily exposed, with $\delta p_\iota = \| p(1)_\iota - |p(2)_\iota \| \div \{\sigma^2 p(1)_\iota + \sigma^2 p(2)_\iota\}^{1/2}$. Plotted from the data in Tables II and III of Liminga, Abrahams & Bernstein (1980).

in two different laboratories using different crystals.

The first crystal had received only 2h exposure before intensity measurements were commenced. The second crystal had been irradiated over 60h before intensity measurements started. The crystals of $Cs_2S_2O_6$ suffered from radiation damage as indicated by the intensity variation in the standard reflections. The primary effect of the radiation exposure is thought to be the partial redistribution of the Cs and S atoms along the trigonal axes on which they are located.

Examination of Fig. 3 shows that the principal outliers are the anisotropic components of the apparent thermal vibrations of the Cs atoms along the trigonal axis direction, as expected for the model. The remaining δp are nearly linear with zero intercept, indicating an essentially normal distribution for these ranked moduli of the deviates. The slope of 2.80 indicates that the pooled standard deviations of the atomic parameters have been underestimated by the factor 2.8.

It is well-known that the method of least squares may result in values of the estimated standard deviations for atomic parameters that are too small. Underestimation of positional parameters by factors of $\sqrt{2}$ on average, and of thermal parameters by $2\sqrt{2}$ on average, were found by Hamilton & Abrahams (1970) in the IUCr Single Crystal Intensity Measurement Project. The use of half-normal probability δp plots, such as that in Fig. 3, allows quantitative corrections for underestimation to be made in addition to providing a simple determination of the possible presence of systematic error in one or both parameter sets.

Further Uses of Probability Plot Analysis.

Normal probability analysis can provide valuable insight into the results of any two quantitative sets of independent observations or determinations. The method is readily applied to the comparison of the molecular geometry (bond lengths, bond angles and torsion angles) in chemically-equivalent but crystallographically-independent molecules (or parts of molecules). De Camp (1973) has shown that the technique gives a highly sensitive probe for the analysis of conformational differences. The distribution of interatomic distances has been shown to be normal (Albertsson and Schultheiss, 1974), as required for the application of normal probability analysis to molecular geometry measurements.

References

Abrahams, S. C. & Keve, E. T. (1971). Acta Cryst. **A27**, 157-165.

Abrahams, S. C. & Bernstein, J. L. (1976). J. Chem. Phys. **64**, 3254-3260.

Abrahams, S. C. & Bernstein, J. L. (1978). J. Chem. Phys. **69**, 2505-2513.

Abramowitz, M. & Stegun, I. A. (1964). Handbook of Mathematical Functions, Nat. Bur. Stds. Washington, D. C.

Albertsson, J. & Schultheiss, P. M. (1974). Acta Cryst. **A30**, 854-855.

DeCamp, W. H. (1973). Acta Cryst. **A29**, 148-150.

Hamilton, W. C. & Abrahams, S. C. (1970). Acta Cryst. **A26**, 18-24.

International Tables for X-Ray Crystallography (1974). Vol. IV. Birmingham: Kynoch Press.

Liminga, R., Abrahams, S. C. & Bernstein, J. L. (1980). J. Chem. Phys. **73**, in press.

Wilk, M. D. and Gnanadesikan, R. (1968). Biometrika **55**, 1-17.

SECTION E

INTERNAL PROPERTIES OF MATTER

CHAPTER 1

PHYSICAL PROPERTIES AND ATOMIC ARRANGEMENT

S. C. Abrahams
Bell Laboratories, Murray Hill, NJ 07974

Many physical properties are a direct function of atomic arrangement in the given material; others are more directly related to the electronic structure. The latter include such important optical properties as color and birefringence, such transport properties as electronic band gap and mobility, such thermal properties as conductivity and thermoelectricity, and such mechanical properties as elasticity and cohesive strength. Among the former are thermal expansivity, specific heat, superconductivity, and ferromagnetism in addition to optical activity, superionic conductivity, pyroelectricity, piezoelectricity, ferroelectricity and ferroelasticity. Limitations of space allow a consideration here only of the final group of six properties. A discussion of other properties may be found in *"Structure - Property Relations"* by Newnham.

Optical activity and crystal chirality

Relationships between the sense and magnitude of the optical rotatory power and the structure of disymmetric molecules have been extensively investigated: a review has been given by Caldwell and Eyring (1971). Recently, the relationship between the sense of rotation of plane polarized light and the chirality or handedness of the atomic arrangement in enantiomorphous crystals has started to receive attention again. The unexpected discovery of Beurskens-Kerssen et al. (1963) that crystals of isomorphous $NaClO_3$ and $NaBrO_3$ with identical chirality rotate plane polarized light in opposite senses was not followed by comparable studies until their results were confirmed by Abrahams, Glass and Nassau (1977). The optical rotatory power of neither crystal is large, ranging from 3.6 to 1.4 deg min^{-1} for wavelengths between 5461 and 7188Å. Each of the corresponding atoms in the two crystals differ in position by at least 0.1Å, sufficient to reverse the sense of the optical rotation.

The only other pair of isomorphous and optically active crystals to have been investigated, that are not composed of disymmetric molecules, is $Bi_{12}SiO_{20}$ and $Bi_{12}GeO_{20}$. The optical rotatory power of these materials ranges from 60 to 20 deg mm^{-1} for wavelengths between 4500 and 6500Å, an order of magnitude larger than in the previous pair. The effect of replacing Si by Ge is a displacement of about 0.01Å or less for all atoms except that oxygen which forms a tetrahedron about Si or Ge and which is displaced by 0.11Å. A result of the small differences in position between most atoms in $Bi_{12}SiO_{20}$ and $Bi_{12}GeO_{20}$, which are an order of magnitude less than in $NaClO_3$ and $NaBrO_3$, is that crystals of the former pair with identical chirality rotate plane polarized light in the same sense (Abrahams, Svensson and Tanguay, 1979).

Further investigation of other pairs of optically active isomorphous crystals with intermediate rotatory power should enhance our understanding of the interaction between chiral atomic arrangements and plane polarized light.

Superionic conductivity

A superionic conductor is generally taken to be a solid with an ionic conductance at temperatures well below that of melting or decomposition greater than 10^{-2} Ω^{-1} cm^{-1}. Such conductivity is always associated with the rapid diffusion of at least one of the ionic species from its normal position in the unit cell. Among the best conductors at room temperature are compounds that contain Ag^+ and I^- in addition to other ions or molecules, such as $RbAg_4I_5$ and $[C_5H_5NH]_5Ag_{18}I_{23}$, in which the Ag^+ ions diffuse very easily. At higher temperatures, β-alumina is of importance as an outstanding conductor of Na^+ ions and as a potential battery material. A number of other ions have also been investigated for their conductive properties. Useful reviews of the field may be found both in *"Superionic Conductors"* and in *"Physics of Superionic Conductors."*

The structural contribution of superionic conductors, first noted in α-AgI, was emphasized by Geller's (1967) study of the atomic arrangement in $RbAg_4I_5$, which has the highest room temperature conductance at 0.27 Ω^{-1} cm^{-1}. The Rb and I atoms are fully ordered in this cubic crystal, with the Ag^+ ions distributed over three sets of inequivalent sites each of which is only partially filled. These sites occupy face-sharing tetrahedra of iodide ions. Nearest-neighbor available sites are unoccupied if a given site is filled by an Ag^+ ion, since they are less than 1.9Å distant, unless occupied by a Rb^+ ion. The ionic conductivity is hence postulated to involve Ag^+ ions diffusing rapidly along paths that pass through shared tetrahedral faces. Disorder among the current-carrier ions, with easy passage through the crystal by way of face-sharing anion polyhedra, hence appears to be essential for superionic conductors of this type.

The sodium aluminates beta and beta″ alumina are superionic conductors at room temperature, with increased conductivity at higher temperatures. Beta alumina has nominal composition $Na_2O{\cdot}11Al_2O_3$, with $Al_{11}O_{16}$ spinel-like blocks separated by NaO layers. The high transport-properties are associated with excess sodium charge-compensated in the conduction plane by interstitial oxygen ions. Aluminum vacancies are also present, compensated by aluminum interstitials in the form of "Roth" defects. Peters et al. (1971) showed that the sodium ions are smeared into a complex disordered arrangement in the basal plane. Roth et al. (1976) have proposed a model for the disorder in the nonconducting state at liquid nitrogen temperature in which the sodium ions are distributed between two of three possible sites: the third site becomes gradually filled, with rising temperatures, as the ionic density at the mid-oxygen site becomes less localized. The added magnesium in beta " alumina increases the conductivity by increasing the number of sodium ions in the conduction layers, which are highly non-localized even at room temperature, without charge compensation by interstitial oxygen. The degree of ionic order in the conduction layers has been shown by diffuse scattering studies to be correlated with the conductivity.

Pyroelectricity

The characteristic property of a pyroelectric crystal is the development of an electric polarization with change of temperature. Pyroelectric behavior is possible only in the ten noncentrosymmetric point groups that possess a unique direction: any polarization generated is confined to these directions. All pyroelectric crystals contain permanent dipoles, usually formed by a separation between ions or originating in polar molecules. Primary pyroelectricity is caused by changes in the magnitude or orientation of the permanent dipole as the temperature is changed, for the case in which the shape and volume of the unit cell remains invariant. Secondary pyroelectricity, which will not be considered further here, originates in a piezoelectric contribution caused by thermal expansion or contraction.

$Ba(NO_2)_2.H_2O$, which has recently been found to have a moderately large pyroelectric coefficient (Liminga et al., 1980), may be used to illustrate the close relationship between atomic arrangement and pyroelectricity. Layers of NO_2^- and Ba^{2+} ions form normal to the polar direction and are connected by hydrogen bonds to the water molecule. The total permanent dipole or spontaneous polarization is the sum of the polarization associated with the Ba^{2+} ion which is about 1/4Å distant from the nearest mean NO_2^- ion plane, the NO_2^- ion dipoles each of which is inclined to the polar axis, and the H_2O dipole which is also inclined to this direction. An increase in temperature of 1K results in a negative polarity of 14×10^{-5} C m^{-2} being generated on (00.1). This polarization could be caused by a displacement of 5.7×10^{-4} Å by the Ba^{2+} ions toward $(00.\bar{1})$, for the NO_2^- ions remaining stationary. A rotation by a NO_2^- ion or an H_2O molecule through about 7.7 arc min would produce the same polarization. Alternatively, an electronic redistribution could give a similar effect. Definitive diffraction experiments to separate these typical possible mechanisms have still to be performed.

Piezoelectricity

A piezoelectric crystal characteristically develops an electric polarization when mechanically stressed along an appropriate direction and, conversely, undergoes a mechanical distortion under the application of an appropriately directed electric field. All noncentrosymmetric crystal classes except 432 are piezoelectric: all pyroelectric crystals are also piezoelectric. A spontaneous polarization is not a requirement for a piezoelectric crystal although dipoles that at least cancel by symmetry must be present. It is the application of stress in a direction such that the contribution of the individual dipoles either changes or no longer cancels that gives rise to the generation of an electric polarization.

An example of a nonpyroelectric but piezoelectric crystal is given by a ternary chalcopyrite such as $ZnSiP_2$. Each atom in this structure with point group $\bar{4}2m$ is tetrahedrally bonded (Shay and Wernick, 1975): in case, the tetrahedra are regular, interatomic dipoles cancel exactly; otherwise each tetrahedral group has a small residual dipole which is exactly cancelled by another corresponding dipole within the unit cell. Shear stress in the basal plane deforms all tetrahedra so as to produce a net dipole along the direction f the tetragonal axis. The atomic displacements in a number of piezoelectric crystals have recently been shown to lie in the range 0.1 to 3×10^{-4} Å per 10^6 N m^{-2}. In case dipole rotations are allowed by symmetry, these are on the order of arc mins per 10^6 N m^{-2} (Abrahams, 1978).

Ferroelectricity

A crystal is ferroelectric if it has a spontaneous polarization, or permanent dipole, that can be reversed in sense or reoriented by the application of an electric field larger than the coercive field. All ferroelectric crystals are necessarily pyroelectric, but not the converse In practice, other solid state properties such as the defect distribution within the crystal and the conductivity may affect ferroelectric reversal: in addition, the temperature and pressure and electrode conditions may be critical. In most ferroelectrics, the spontaneous polarization tends to zero as the Curie temperature, at which a higher symmetry phase transition occurs, is approached. In all ferroelectrics, the reversal or reorientation of the spontaneous polarization is the result of atomic displacements.

The magnitude of the atomic displacements resulting from ferroelectric reversal has now been studied in many materials. Typical is the one-dimensional ferroelectric lithium tantalate in which the oxygen atoms form a sequence of distorted face-sharing octahedra along the polar direction. Within these oxygen octahedra are cations in the sequence Li, Ta, vacancy, Li, Ta, vacancy, . . . At room temperature, the Ta atom is about 0.2Å from the center of its octahedron. The Ta atom moves closer to the center as the temperature rises, following a temperature dependence of location very similar to that of the spontaneous polarization, finally occupying the central position at the Curie temperature. The ionic dipole, hence the spontaneous polarization, given by the displacement of Ta from its central position and by that of Li from the nearest octahedral face is very close to the measured value at all temperatures. A general discussion of this structure and that of other representative ferroelectrics may be found in *"Principles and Applications of Ferroelectrics and Related Materials"* by Lines and Glass.

Ferroelasticity

A crystal is ferroelastic if two or more equally stable orientational states exist in the absence of mechanical stress and if the application of stress reproducibly transforms one state into another. As found with ferroelectric crystals, other solid state properties may affect ferroelastic reorientation. Ferroelasticity is invariably associated with small distortions from a higher symmetry phase: no other symmetry restrictions are applicable, hence the property may be unaccompanied by other structure-dependent properties or may be accompanied by or even coupled to such other properties as ferroelectricity or magnetic ordering. The distortion from higher symmetry is given by the spontaneous strain, which is generally on the order of parts per thousand.

Among the simplest examples of a hypothetical ferroelastic crystal is one with a monoclinic unit cell, space group $P2_1$, with $\beta = (90 + \delta)^\circ$ and $\delta < 5$ are mins. In the case that each atom at x_1, y_1, z_1 is related to another of the same element in the unit cell at x_2, y_2, z_2 such that $x_1, y_1, z_1 = (x_2, y_2, 1\text{-}z_2) + D$, and D is a displacement vector less than about 1Å, the crystal will be ferroelastic unless the crystal cohesive strength is less than the coercive stress. Compressive stress applied along [101] in such a crystal will hence reorient the c-axis by $2\delta^\circ$, converting the stress direction into the new $[10\bar{1}]$ direction. For low cohesive strength crystals the application of stresses as small as 10^4 N m^{-2} may cause crystal damage. The normal coercive stress range in ferroelastic crystals is 10^8 to 10^4 N m^{-2}.

The temperature dependence of the atomic displacements necessary for reorienting the spontaneous strain in $K_2Cd_2(SO_4)_3$, measured between room temperature and the first order transition to the paraelastic phase at 432K (Lissalde et al., 1979), may be taken as a typical example of a ferroelastic crystal. The largest ferroelastic displacement at 298K of about 1.2Å in $K_2Cd_2(SO_4)_3$ is reduced to 0.9Å at 417.5K, and the magnitude of all other displacements becomes significantly smaller on approaching the phase transition temperature. The angular orientation of the SO_4^{2-} ions changes less than 5° between 298 and 417.5k, with the remaining rotations of as much as 18° taking place only when very close to the transition temperature. It is possible to reorient the spontaneous strain in $K_2Cd_2(SO_4)_3$ only by applying a small stress to a crystal that is heated to the transition temperature and cooled again before the stress is removed. The large ionic rotations necessary for orientation in $K_2Cd_2(SO_4)_3$ thus result in a frozen ferroelastic at temperatures lower than about 5° below the phase transition. Ferroelastic reorientation in other materials is readily induced well below their transition temperature.

References

Abrahams, S. C. (1978). Mater. Res. Bull. **13**, 1253-1258.

Abrahams, S. C., Glass, A. M. and Nassau, K. (1977). Solid State Commun. **24**, 515-516.

Abrahams, S. C., Svensson, C. and Tanguay, A. R. (1979). Solid State Commun. **30**, 293-295.

Beurskens-Kerssen, G., Kroon, J., Endeman, H. J., van Laar, J. and Bijvoet, J. M. (1963). In "Crystallography and Crystal Perfection," Edited by G. N. Ramachandran. London: Academic Press.

Caldwell, D. J. and Eyring, H. (1971). "The Theory of Optical Activity." New York: Wiley.

Geller, S. (1967). Science **157**, 310-312.

Liminga, R., Abrahams, S. C. and Bernstein, J. L. (1980). J. Appl. Cryst. **13**, 516-520.

Lines, M.E. and Glass, A.M. (1977) "Principles and Applications of Ferroelectrics and Related Materials." Oxford: Clarendon Press.

Lissalde, F., Abrahams, S. C., Bernstein, J. L. and Nassau, K. (1979). J. Appl. Phys. **50**, 845-851.

Newnham, R. E. (1975. "Structure-Property Relations." New York: Springer-Verlag.

Peters, C. R., Bettman, M., Moore, J. W. and Glick, M. D. (1971). Acta Cryst. **B27**, 1826-1834.

"Physics of Superionic Conductors" (1979). Edited by M. B. Salamon, New York: Springer-Verlag.

Roth, W. L., Reidinger, F. and LaPlaca, S. (1976). In "Superionic Conductors," Edited by G. D. Mahan and W. L. Roth, pp. 223-241. New York: Plenum Press.

Shay, J. C. and Wernick, J. H. (1975). "Ternary Chalcopyrite Semiconductors: Growth, Electronic Properties and Applications." New York: Pergamon Press.

CHAPTER 2

DISLOCATIONS IN CRYSTALS

J. P. Hirth
Metallurgical Engineering Department
Ohio State University, Columbus, OH 43210

The Early Years

The elastic theory of dislocations in continua had been developed by the Italian School of mathematicians early in the twentieth century, but this work was not appreciated by those concerned with crystal properties. The concept of dislocations in crystals originated in Europe with the postulation of the edge dislocation by Taylor (1934) in England and of Orowan (1934) and Polanyi (1934) in Germany. Orowan later spent a large portion of his career in Mechanical Engineering at M.I.T. The screw dislocation originated in the work of Burgers (1939) who also developed the vector field theory for the elastic properties of dislocations. There was then a spate of activity in evolving the geometrical and elastic theory of dislocations with significant contributions in America, particularly from Seitz' group at Carnegie Institute of Technology and later at the University of Illinois and from Bell Laboratories.

The concept of a partial dislocation and its associated stacking fault was introduced by Shockley (1948) and Heidenreich (1948). General applications of dislocations in the area of crystal physics were discussed by Seitz and Read (1941) including the concept of dislocation climb by emission and absorption of vacancies and interstitials. Seitz (1950) also noted that jogs on dislocations carried a net charge in ionic crystals so that dislocations could act as an extrinsic charge source. The important concept of a virtual thermodynamic force on a dislocation produced by stresses acting at its core was forwarded by Peach and Koehler (1950).

Again in Europe, Frank (1949) demonstrated the importance of screw dislocations as sources for self-perpetuating monatomic ledges on singular low-index surfaces. This idea was the basis for the modern theory of crystal growth (Burton et al., 1951). One of the key ingredients in the description of crystal plasticity, the Frank-Read source (1950), evolved. Frank and Read jointly published this work since they determined that they had independently conceived the idea at the same time, Read at Murray Hill and Frank while strolling on the green while visiting Princeton. Read (1953) later contributed substantially to the geometric theory for dislocations in grain boundaries. The concept of special coincidence-lattice boundaries was introduced by Kronberg and Wilson (1949).

Further Developments and Observations

Throughout the decade of the 1950's, numerous experimental methods were developed which confirmed the presence of dislocations in crystals and led to considerations of detailed and complex dislocation configurations. Strong contributions were made in America from the General Electric Research Laboratories and from several Universities as well as the groups mentioned earlier. Heidenreich (1949) considered dynamical diffraction theory for thin films and observed grain boundaries and dislocation cell walls with transmission electron microscopy of deformed aluminum. Single dislocations were observed in TEM by Whelan et al., (1957) at Cambridge, spurring great growth in dislocation study by this technique. The Hirsch groups at Cambridge and later Oxford were active in the evolution of the kinematic and dynamical theory of diffraction by dislocations, with a significant contribution by Thomas (1962) at Berkeley.

Dislocations decorated with copper were observed in infra-red transmission by Dash (1956), while decorated dislocations in alkali halide crystals were studied optically by Hedges and Mitchell (1953) and Amelinckx (1958) in Europe. Mitchell later moved to the Univerity of Virginia. The Berg-Barrett (1945) technique for dislocation study by X-ray topography was evolved, with applications to dislocation structures and dislocation motion by Newkirk (1958) and by Lang (1959). Etch pits at dislocation-surface intersections in lithium fluoride were used for extensive studies of dislocation motion by Gilman and Johnston (1957). Direct observations of dislocations by field-ion-microscopy were performed by Mueller (1958). Examples of complex arrays revealed by these techniques include faulted prismatic loops in hcp metals (Price, 1961) and stacking fault tetrahedra (Silcox and Hirsch, 1959) produced by quenching in and precipitating vacancies. These observations led to theoretical models such as the concept of a zonal dislocation whose motion produces both shear and local shuffles in crystals with a basis (Kronberg, 1959; Westlake, 1961) and the elaboration of extended dislocation barrier reactions in fcc metals (Hirth, 1961). Models were developed for mechanical twinning and for martensitic phase transformations produced by partial (or zonal) dislocation motion, with some experimental support (reviewed in Reed-Hill et al., 1964). Also, the anistropic elastic theory for dislocations was developed by Eshelby et al., (1953).

Recent Work

The development at Cambridge, U.K. of the weak beam technique for transmission electron microscopy (Cockayne et al., 1969) has led to near atomic scale resolution of configurations such as jogs on extended dislocations. This method has made possible measurements of dislocation extension leading to more accurate estimates of stacking fault energy, as evidenced, for example, by the extensive work at Oxford of Carter (1975), now at Cornell. Improved TEM methods have also made possible the direct study of grain boundary dislocations, with notable contributions by Balluffi (see Schober and Balluffi, 1971) while at Cornell. These observations confirmed the refined O-lattice theory of Bollmann (1967) based on the earlier ideas on coincidence lattices. Electron diffraction effects associated with local periodicities in grain boundaries have also been revealed (Sass et al., 1975). Direct lattice imaging of dislocations has been achieved in high resolution TEM, for example in Cowley's group, now at the Arizona State University (Spence et al., 1978).

Head and co-workers in Australia (1967) developed numerical methods for computing TEM dislocation images from diffraction theory with the anistropic displacement field of dislocations as input, greatly aiding the interpretation of dislocation configurations. A new formation for the anistropic elastic theory of dislocations which facilitates numerical calculations has been presented by Barnett and Lothe (1973) and co-workers (Asaro et al., 1973). In particular the method facilitates atomic calculations of dislocation core configurations (Sinclair et al., 1978).

Atomic calculations have been performed for a number of metals, extending dislocation study to the nonlinear elastic core region. Perhaps most important in such work has been the finding by several groups, in Ottawa (Basinski et al., 1971) and at Battelle Laboratories (Gehlen, 1970), that screw dislocations in bcc metals have a local core dissociation which hinders screw dislocation motion, rationalizing the observed properties of screw dislocations which differ markedly from those in other crystal structures. Also, such studies have revealed the crystal volume expansion (of nonlinear elastic origin) associated with a dislocation (Sinclair et al., 1978).

The description, alternative to discrete dislocation models, of grain boundaries by disclinations has been achieved (Li, 1972). Single disclinations, having too large an energy to exist in metals or inorganic crystals, have been observed in polymers, in liquid crystals and in noncrystalline lattices such as superconducting flux-line lattices and block wall lattices, as reviewed by Harris (1974).

The elastic and geometric study of dislocations and of their influence on mechanical and physical properties of crystals remains a very active field today. Perhaps the greatest obstacle to further advances is the lack of a first principles atomic interaction potential which would permit more accurate atomic calculations to be achieved.

References

Amelinckx, S. (1958). Acta Met. **6**, 34.

Asaro, R. J., Hirth, J. P., Barnett, D. M., and Lothe, J. (1973). Phys. Stat. Sol. **B60**, 261.

Barnett, D. M., and Lothe, J. (1973). Phys. Norvegica **7**, 13.

Barrett, C. S. (1945). Trans. A.I.M.E. **161**, 15.

Basinski, Z. S., Duesbery, M. S., and Taylor, R. (1971). Can. J. Phys. **49**, 2160.

Bollmann, W. (1967). Phil. Mag. **16**, 363.

Burgers, J M. (1939). Proc. Kon. Ned. Akad. Wet. **42**, 293.

Burton, W. K., Cabrera, N. and Frank, F. C. (1951). Phil. Trans. Roy. Soc. **243**, 299.

Carter, C B. and Holmes, S. M. (1975). Phil. Mag. **32**, 599.

Cockayne, D J. H., Ray, I. L. F. and Whelan, M. J. (1969). Phil. Mag. **20**, 1265.

Dash, W. C. (1956) J. Appl., Phys. **27**, 1193.

Eshelby, J. D., Read, W. T. and Shockley, W. (1953). Acta Met. **1**, 251.

Frank, F. C. (1949). Discuss. Faraday Soc. **5**, 48.

Frank, F. C. and Read, W. T. (1950). in "Symposium on Plastic Deformation of Crystalline Solids," p. 44. Pittsburgh, Pennsylvania: Carnegie Institute of Technology.

Gehlen, P. C. (1970). J. Appl. Phys. **41**, 5165.

Gilman, J. J. and Johnston, W. G. (1957). in "Dislocations and Mechanical Properties of Crystals," edited by J. C. Fisher et al. p. 116. New York: Wiley.

Harris, W. F. (1974). Surface and Defect Properties of Solids, vol. 3, p. 57, London: The Chemical Society.

Head, A. K., Loretto, M. H. and Humble, P. (1967). Phys. Stat. Sol. **20**, 505.

Hedges, J. M. and Mitchell, J. W. (1953). Phil. Mag. **44**, 223.

Heidenreich, R. D. (1949). J. Appl. Phys. **20**, 993.

Heidenreich, R. D. and Shockley, W. (1948). Report of a Conference on Strength of Solids, p. 57, London: The Physical Society.

Hirth, J. P. (1961). J. Appl. Phys. **32**, 700.

Kronberg, M. L. (1959). J. Nucl. Mater. **1**, 85.

Kronberg, M. L. and Wilson, F. H. (1949). Trans. A.I.M.E. **185**, 501.

Lang, A. R. (1959). Acta Cryst. **12**, 249.

Li, J. C. M. (1972). Surface Science **31**, 12.

Mueller, E. W. (1958). Acta Met. **6**, 620.

Newkirk, J. B. (1958). Phys. Rev. **110**, 1465.

Orowan, E. (1934). Z. Phys. **89**, 634.

Peach, M. O. and Koehler, J. S. (1950). Phys. Rev. **80**, 436.

Polanyi, M. (1934). Z. Phys. **89**, 660.

Price, P. B. (1961). Phys. Rev. Lett. **6**, 615.

Read, W. T. (1953). Dislocations in Crystals. New York: McGraw-Hill.

Reed-Hill, R. E., Hirth, J. P. and Rogers, H. C. (1964). Deformation Twinning, New York: The Met. Soc. A.I.M.E.

Sass, S. L., Tan, T. Y. and Balluffi, R. W. (1975). Phil. Mag. **31**, 559.

Schober, T. and Balluffi, R. W. (1971). Phil. Mag. **24**, 165.

Seitz, F. (1950). Phys. Rev. **80**, 239.

Seitz, F. and Read, W. T. (1941). J. Appl. Phys. **12**, 100.

Shockley, W. (1948). Phys. Rev. **73**, 1232.

Silcox, J. and Hirsch, P. B. (1959). Phil. Mag. **4**, 72.

Sinclair, J. E., Gehlen, P. C., Hoagland, R. G. and Hirth, J. P. (1978). J. Appl. Phys. **49**, 3890.

Spence, J. C. H., O'Keeffe, M. A. and Iijma, S. (1978). Phil. Mag. **38A**, 463.

Taylor, G. I. (1934). Proc. Roy Soc. (London) **145A**, 362.

Thomas, G. (1962). Transmission Electron Microscopy of Metals. New York: Wiley.

Westlake, D. G. (1961). Acta Met. **9**, 327.

Whelan, M. J., Hirsch, P. B., Horne, R. W. and Bollman, W. (1957). Proc. Roy. Soc. (London) **240A**, 524.

CHAPTER 3

SOME STATISTICAL ASPECTS OF CRYSTAL SYMMETRY

Sterling B. Hendricks
Silver Spring, MD 20910

Introduction

Many phenomena displayed by crystals were first recognized in the 1920-40 period. The resulting principles are basic not only to crystallography, but to development of such diverse areas of knowledge as metallurgy, molecular biology, petrology, and the physics of the solid state. Initial aspects of a connecting thread in this knowledge are presented in the following article.

The space group formalization of unit cells repeated by translation in infinite lattices accords with crystal symmetry. As X-ray diffraction methods for analysis of crystal structure developed, Darwin (1922) and Ewald (1925) pointed out that a structure with a perfect lattice would give much lower intensities of diffraction than do real crystals in which units or groups of units are slightly misaligned in a mosaic. The real crystal approaches the ideal array only in a statistical or average sense. The early recognition of some irregularities which markedly influence the properties of real crystals are discussed here. Primitive cases are emphasized rather than later elaborations which led to many books on special aspects of the subject.

Isomorphism

Breakdown of perfect order in crystals relaxes the constraints of constant composition, the issue of conflict between Berthollet and Proust in the 19th century. Conversely, deviation from ideal composition indicates a non-periodic deviation from order. Variation of atoms by substitution in units of structure is an expression of Mitscherlich's concept of isomorphous replacement based on his observations in 1819 on crystals of alkali dihydrogen phosphates and arsenates. Thus substitution of some K^+ by $(NH_4)^+$ or $(PO_4)^{-3}$ by $(AsO_4)^{-3}$ in the KH_2PO_4 structure (Hendricks, 1927) leads to averaging of equivalent position occupancy. The substitution is constrained by atomic size (ionic radii), charge, and temperature. The electrical dipoles arising from the hydrogen bonds in the KH_2PO_4 unit are randomly oriented at room temperature but align at a low temperature to give a ferroelectric state (Busch and Scherrer, 1935).

Replacement between atoms of low and high scattering power for X-rays aids in determining the phase of the structure factor. Early examples of such use were with the isomorphous pairs Al_2O_3 (corundum) Fe_2O_3 (hematite) (Pauling and Hendricks, 1925) and $K_2CuCl_4 \cdot 2H_2O$ and the corresponding Rb salt (Hendricks and Dickinson, 1927). Locating the heavy atoms facilitates accurate analysis for the parameters of atoms of low atomic number in the alternate compound. This procedure is of particular value in structure analysis of proteins and other complex compounds.

Isomorphism, based on the criterion of forming solid solutions, is shown by crystals of fluoro- and chloroapatite. The large Cl^- ions with six fold coordination are displaced by $1/4\ c_o$ along the hexagonal axis from the position of the smaller F^- ions which have three fold coordinations with respect to Ca^{+2} (Hendricks et al., 1932a). A Ca^{+2} ion in the apatite lattice can also be replaced by $2H^+$ ions forming hydrogen bonds about Ca^{+2} positions (Posner et al., 1960).

Partial atomic replacement is random in apatites and many other crystals. This is the case for small amounts of Mg^{+2} replacing Ca^{+2} in the calcite lattice. The replacement, however, can be ordered leading to dolomite ($CaMg(CO_3)_2$). Replacement in dolomite in turn is random. The ordering is a result of the considerable difference in ionic radii of Mg^{+2} and Ca^{+2}.

Group Orientation

Some pure compounds having non-symmetrical groups of atoms form crystals in which the groups are randomly oriented in two or more directions. This is the case for the linear cyanate anion $(CNO)^-$ in the tetragonal KCNO crystals (Hendricks and Pauling, 1925). The group can have the orientation $(CNO)^-$ or $(ONC)^-$ along the four fold axis. The KCNO crystals are isomorphous with those of KN_3 in which the three atom azide group is symmetrical. The situation arises because of the low difference in lattice energy for the ordered and random structures and the requirement of a high energy of activation for inverting $(CNO)^-$. The randomness is fixed or "frozen" into the lattice leading to an entropy of mixing of R 1n2 at O°K for the extreme of complete disorder.

The thermodynamic consequences of the random orientation in crystals of many simple molecules was examined by Giauque and his students (1950) in work supporting the third law of thermodynamics. The method used involved measuring heat capacities to low temperatures and comparison with the free energies calculated from spectroscopic observations. The compounds studied in this way included nitric oxide $(NO)_2$ (Johnston and Giaque, 1929) for which the residual entropy at O°K was 2.94 J/(mol.°K). Measurements of residual entropy were also made on nitrous oxide (NNO), carbon monoxide (CO) and propylene (CH_3-CH=CH_2) all of which showed residual entropies, within experimental error, of R 1n2 at O°K (Giauque 1950). The carbonyl sulfide molecules (COS) (Kemp and Giauque, 1937), in contrast, are fully ordered in the lattice because of the large sulfur atom. Random mixing of isotopes in molecular crystals is expected in all cases including deuterium and tritium in ice.

An example in which the randomized molecules have atoms sufficiently different in scattering power to affect intensities of X-ray reflections is afforded by parabromochlorobenzene ($1,4BrC_6H_4Cl$) (Hendricks, 1932). The crystals are isomorphous with the corresponding dichloro- and dibromo- compounds. In the BrC_6H_4Cl crystals the molecules have an apparent center of symmetry, that is, a lack of distinction between the Cl and Br atoms.

The structure of ice I is of particular interest in showing random orientation of H_2O molecules in crystals. Each H_2O molecule has four neighboring H_2O molecules at the corners of a tetrahedron. The H_2O molecule shares four hydrogen bonds with its neighbors. The hydrogen bond (shown as ---) between two of the neighbors can be HO-H--X---OH_2 or H_2O---X--H-OH about the central position (X). Pauling (1935) calculated the residual entropy at 0°K to be R ln(3/2) or 3.37 J/ (mol.°K) in comparison with the measured value 3.43 (Giauque and Stout, 1936). Neutron diffraction shows an apparent hydrogen atom in a position half way (at (X)) between two oxygen atoms in agreement with the complete random orientation of the hydrogen bonds (Wollan et al., 1949). In some hydrates, e.g., $ZnSO_4.7H_2O$, residual entropy is absent, indicative of the H_2O molecules being in fixed orientations.

Equivalent Occupancy of Lattice Positions

Many oxides of fixed composition when quenched from their melts form crystals with the metal atoms distributed randomly in equivalent positions between the oxygen atoms in approximate close-packing. First recognition occurred in studies in $LiFeO_2$ (Posnjak and Barth, 1931) in which Li^+ and Fe^{+3} ions occupy an equivalent set of positions in an NaCl type structure. The difference in charge between the ions must lead to some local order but long range order was not observed. A tetragonal modification of $LiFeO_2$ in which the cations are ordered can be made by suitable annealing. A similar situation was found with some spinels of the general formula (XY_2O_4) (Barth and Posnjak, 1932). Thus in $MgFe_2O_4$ and $TiFe_2O_4$ the cation distribution is (Fe) $(FeMg)O_4$ and (Fe) $(FeTi)O_4$ where 16 (FeMg) or (FeTi) cations are randomly mixed in an equivalent set. In $MgAl_2O_4$ the site occupancy is ordered apparently due to the constraints of cation charge.

Spinels also show disorder in which vacancies are present. Examples are the series of solid solutions of spinel ($MgAl_2O_4$) as the composition varies towards Al_2O_3 and in the structure of maghemite (Fe_2O_3) which corresponds to $Fe_{2.67}O_4$ as a spinel. Thus 11% of the cation positions are vacant (Welo and Baudisch, 1925).

Randomness of atoms in equivalent positions is further illustrated by the structures of silver tetraiodomercurate Ag_2HgI_4 (Ketelaar, 1934a). The two Ag^+ and Hg^{+2} cations occupy distinct positions in a tetragonal unit with Ag^+ and Hg^{+2} in definite positions in an approximate cubic close-packed arrangement of large I^- anions. At temperatures above 50°C the structure is cubic with the cations randomly distributed in a face centered unit. Three cations are averaged over the four positions of an equivalent set (Fig. 1). The color changes

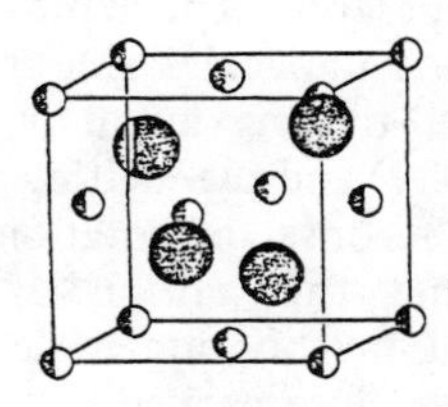

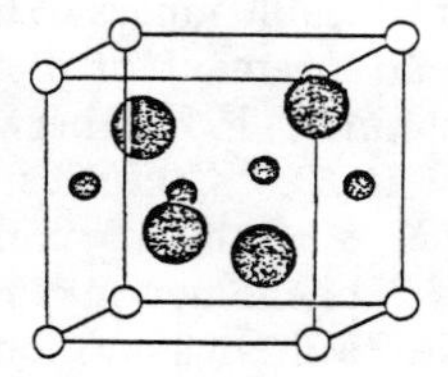

Fig. 1. Ag_2HgI_4 structures. Right hand view shows high-temperature form with cations randomly arranged.

from yellow to red as the temperature is advanced through the transition region indicative of a change in distribution of electrons. The electrical conductivity increases, indicative of greater freedom of movement of the cations. Such dependency of properties on vacancies proved to be an important factor in development of the physics of the solid state.

Random Vacancies and Non-Stoichiometric Compounds

Substitutions in some crystals include vacancies with deviations from constant composition. The first instances were in the structure of spinels as discussed earlier and in wurstite ($Fe_{1-x}O$), which has a NaCl type structure (Jette and Foote, 1933). The limits of x values are 0.02 and 0.30, which vary with temperature and pressure. A number of sulfides and other oxides are non-stoichiometric. The compounds TiO_x and VO_x are examples (Banus and Reed, 1970). About 15% of the cation and oxygen positions in these substances can be vacant at the TiO and VO compositions. The homogeneity range is between x values of 0.75 and 1.30. The NaCl type structure of TiO_x can be retained by quenching from above 1000°C. A monoclinic stoichiometric form, TiO, is stable below 950°C. Compositions are $TiO_{0.7}$ when the Ti positions are filled in the quenched material and $TiO_{1.25}$ when the oxygen positions are filled. The vacancies can show ordering in restricted ranges of composition (Watanabe et al., 1970).

The iron sulfides troilite and pyrrhotite with compositions of $Fe_{1-x}S$ are similar to wurstite. The pyrite and marcasite structures of FeS_2 have exact compositions. Ordered superlattice structures appear in the $Fe_{1-x}S$ sulfides (Parthé and Hohnke, 1970).

Group and Molecular Rotation

Rotation and extreme oscillation of groups of atoms in crystals are the limiting aspects of statistical averaging by thermal motion. Pauling (1930) pointed out that para H_2 molecules are rotating at O°K as required by the lowest spin state. He indicated that maxima in specific heat changes with temperature in other molecular crystals such as those of CH_4, N_2, O_2, and the hydrogen halides also are due to rotation. The transition extends over considerable ranges of temperature. The cubic close-packed arrangement of methane molecules (McLennan and Plummer, 1928) between the maximum in heat capacity at 20.4°K (Clusius, 1929) and the melting point at 90.6°K is of this nature. The onset of rotation is a collective phenomenon in that the motion of one molecule facilitates motion of neighboring molecules. Nitrogen (N_2) and O_2 also have close packed structures above transitions at 35.4°K and 43.7°K respectively.

Gradual onset of rotation was found for the nitrate anions $(NO_3)^-$ in $NaNO_3$ crystals as the temperature was raised (Kracek, 1931; Kracek et al., 1931). Gradual enhancement of heat capacity and increase in volume without a change in phase are also observed. The enhancements are completed at 275°C. X-ray reflections to which oxygen atoms alone contribute decrease to zero intensity at 275°C. These changes are expected for rotation of the planar $(NO_3)^-$ anion about the trigonal axis which is normal to the plane of the anion.

Sharp onset of the $(NO_3)^-$ anion rotation with a change in phase is shown by crystals of NH_4NO_3 (Hendricks et al., 1932b). Ammonium nitrate has a cubic unit of structure (CsC1 type) above the phase transition at 125.2°C. The $(NO_3)^-$ groups are rotating with essentially cubic holohedral symmetry (Fig. 2). Limited solid

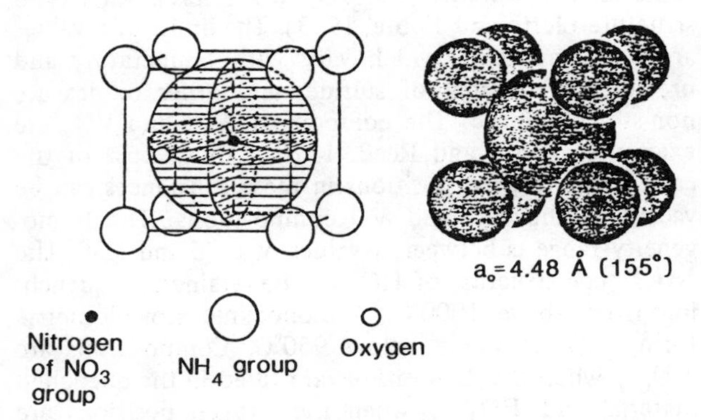

Fig. 2. NH_4NO_3 with rotation of the nitrate group at elevated temperatures.

solutions are formed between NH_4NO_3 and NH_4C1 compatible with the structural equivalence of $C1^-$ and rotating $(NO_3)^-$ ions. The nitrate groups have fixed orientations in the four polymorphic forms of NH_4NO_3 below 125.2°C. The $(NO_3)^-$ ion is also rotating in some manner in crystals of $RbNO_3$ and $CsNO_3$ at temperatures above 210°C and 170°, respectively.

The $(CN)^-$ ion has the two fixed orientations CN^- and NC^- in the orthorhombic lattice of LiCN at room temperature. Its motion is strongly perturbed in cubic or pseudo cubic lattices of the K^+, Rb^+, and Cs^+ cyanides. Ammonium cyanide has a deformed CsC1 type structure with strong aniostropic vibrations (Lely and Bijvoet, 1944). In the numerous coordination compounds containing $(CN)^-$ and other linear groups such as cyanates and azides the orientations probably are restricted to the linear inversion.

Rotation of $(NH_4)^+$ is present in crystals of many ammonium salts at room temperature. The rotation probably starts at gradual transitions in the region of 220°K to 240°K in the ammonium halides (Simon et al., 1927). The diffraction pattern is unchanged except for about a 1.0% change in volume of the salts in the transition region. The chloride is piezoelectric at temperatures below 243°K (Hettich and Hendricks, 1933). The NH_3 group in many coordination compounds probably has spherical symmetry at room temperature.

In substituted ammonium salts such as the tetragonal primary amyl ammonium chloride ($NH_3C_5H_{11}C1$) the $NH_3C_5H_{11}$ group rotates about its long axis and thus appears to have a linear sequence of carbon atoms (Hendricks, 1934). Measurement of the heat capacity, C_p, shows broad maxima at 225°K and 248°K (Southard et al., 1932). The transitions probably occur with rotation of the NH_3 and C_5H_{11} moieties. The tetragonal unit of structure, stable at room temperature, contains $2(NH_3C_5H_{11})^-$ ions with the axis of rotation along the tetragonal axis.

Random Sequences of Layers

Parallel layers of atoms or groups of atoms are present in many crystals. A number of these show irregularities in the sequence of layers. The irregularities affect intensities of X-ray diffraction and often lead to patterns of diffuse scattering. Examples of the extremes in succession of layers are shown by structures of cobalt metal and montmorillonite.

Cobalt metal has a cubic close-packed sequence of layers when prepared by reduction of the oxide at temperatures below 380°C. This can be represented by the layer successions ABCABCABC---. The metal obtained by reduction above 830° approached hexagonal close packing with the sequence ABABAB---. Reflections from *hkl*) $l \neq n.3$, however, are broad and have lower intensities than calculated for the regular sequence (Hendricks et al., 1930; Edwards and Lipson, 1942). Wilson (1942) showed that intensities calculated for an irregular sequence, ABABABCAB---, agreed with observed values if one layer in 7 to 10 of the hexagonal sequence was in the cubic close-packed sequence. Crystals of $CdBr_2$ and $NiBr_2$ in which the large Br^- ions form a cubic close-packed lattice can have random occupancy by the cations of octahedral coordinated positions between the layers (Bijvoet and Nieuwenkamp, 1933; Ketelaar, 1934).

Some carbon blacks have layers of carbon atoms in the graphite arrangement oriented randomly about the normal to the layers (Warren, 1941; Wilson, 1949). The randomness leads to the presence of only *hk*0 reflections with diffuse scattering to larger angles from the intense maxima.

The micas and other silicates with layer lattices afford many examples of randomness in succession of layers (Bijvoet and Nieuwenkamp, 1933; Ketelaar, 1934). order to the lattice energy. Diffuse scattering of X-rays in some zones from crystals of biotite was suggested as arising from irregularities in layer sequences. The randomness generally comes from displacement of layers by 1/3 b_o, if the crystals are monoclinic, in the plane of the layers. Such a shift does not change the relationship of atoms in the top of one layer relative to the atoms in the bottom of the next layer. The randomness leads to diffuse scattering of X-rays in zones with $k \neq n.3$ giving the asterism noted in extreme cases.

The structure of the kaolin mineral dickite ($Al_2Si_2(OH)_4O_5$), in which Al occupies 2/3 of the octahedrally coordinated positions with respect to O^{-2} and $(OH)^-$ is an example in which the number of displaced layers is very low. The asymmetry of the layer imparts long range order. Nevertheless random successions of layers is indicated by failure of agreement between observed and calculated intensities for (*kkl*) $k \neq 3.n$ (Hendricks, 1938). In the structurally related mineral cronstedite (idealized to $(Fe^{+2})_3$ (Fe^{+3}, Si^{+4}) $(OH)_5O_4$) the approach to complete filling of the octahedrally coordinated positions leads to a more symmetrical layer with a lower contribution to long-range order. The diffuse scattering is prominent (Hendricks, 1939).

The mica muscovites $(Al_2(Al,Si_3)O_{10}(OH,F)_2K)$ provide an example of a lattice without randomness in layer sequence because of the distortion of octahedral coordination. In biotites (with FeMgAl in octahedral coordination) and phlogopite $(Mg)^{+2}$ random superposition of layers and diffuse scattering are prominent. Nevertheless super-lattices in layer sequences are found indicative of order over long ranges (Hendricks and Jefferson, 1939).

An extreme in random succession of layers is present in montmorillonites, chlorites, and vermiculites. These minerals have biotite mica-like layers, with somewhat lower negative charge than the micas, and layers of water molecules and/or brucite type layers $(Mg,Al(OH)_2)$. The several types of layers in montmorillonite are random in number, sequence, and mica type displacement. X-ray diffraction leads to non-integral orders of maxima, diffuse scattering and broadening of all maxima. The pattern of scattering was calculated by Hendricks and Teller (1942). The general case of such sequences were also treated by Landau (1937) and Lifshits (1937).

Order-Disorder

Initiation of order in a disordered lattice as temperature is decreased is expected when the energy of activation for the ordering process is not too great with respect to kT. Rates of ordering decrease rapidly with temperature through an annealing region. The differences in charge and size of the exchanging atoms are the prominent factors both in forming a new structure, which might be a super lattice, and in determining physical properties. The ordering process can be abrupt at a change in phase as was the case for Ag_2HgI_4 and is generally true in congruent melting of crystals. The ordering process is well expressed in many alloys.

Tammann (1919) suggested that the change in resistivity of several alloys on annealing resulted from segregation of component atoms from random into ordered positions. This concept was later supported by observations on Au-Cu alloys (Johansson and Linde, 1925). When alloys near the AuCu composition are quenched from above 400°C the atoms are randomly distributed in a cubic close packed arrangement. Annealing near 300° leads to a tetragonal lattice with $a_o : c_o = 1.078$ with alternate close packed layers of Au and Cu atoms. A similar situation exists for compositions near $AuCu_3$.

The theory of the ordering process in a binary alloy with equal numbers of the two kinds of atoms was worked out by Bragg and Williams (1934) on the basis of the Weiss, i.e., mean field theory of ferromagnetism; namely, the ordering at a particular point is determined by the average order in the lattice. The resistivity of the AuCu alloy drops sharply as the temperature is decreased below the critical temperature (θ) for order and the specific heat increases after a "jump" at the critical value. Bethe (1935) pointed out that the ordering process depends strongly on the interaction of nearest neighbors. The long range order determines the short range order only on the average. Also, fluctuations of configuration need to be considered. The long-range order vanishes with a vertical tangent at the critical temperature (θ). Physical constants (resistivity, etc.) have "kinks" but no discontinuity at θ. The present state of the subject is presented in books by Christian (1965) and by Krivoglaz and Smirnov (1965).

The order-disorder phenomena in ionic crystals depend strongly on the electrostatic interaction of the ions. It was touched on earlier for the phase change in Ag_2HgI_4 and for the random succession of layers in the micas and related minerals in which the disorder is long-range. A much more involved situation exists for the feldspars where the near equivalence of Al^{+3} and Si^{+4} in scattering power limits the use of X-ray diffraction methods. A guiding principle to the structure of the alumino-silicates was advanced by Pauling (1960) as the electrostatic valence principle; namely, shared edges or faces in a coordinated structure decrease its stability. The effect is large for Si^{+4} and Al^{+3}.

In anorthite ($CaAl_2Si_2O_8$) Taylor et al. (1934) noted the presence of weak X-ray reflections indicative of a larger unit of structure than required by the stronger reflections (as superlattices). Goldsmith and Laves (1954, 1955) found that the Al^{+3} and Si^{+4} ions are orderly distributed in alternate coordination tetrahedra of oxygen ions. Variation of the composition from the Al:Si = 1 ratio results in the appearance of long range order. The complexities in the feldspar structures have been treated in detail by Megaw (1959) who distinguished between unit cell and lattice disorder. The unit cells can have atoms deviating in positions of an equivalent set and positions occupied by different atoms (Al and Si, Na and Ca). The lattice can consist of domains of many units separated from other domains by faults, the mosaic crystal. Elaboration of the basic principles are presented in the book by Smith on the nature of the feldspars (Smith, 1974).

Acknowledgments

I enjoyed discussions with Tom F. W. Barth (deceased) Linus Pauling, and Edward Teller during the formative period of the subject of lack of order in crystals. Current conversations with Richard A. Robie of the U.S. Geological Survey greatly facilitated preparation of this article.

References

Banus, M.D., and T. B. Reed (1970). Structural, Magnetic, and Electrical Properties of Vacancy Stabilized Cubic 'TiO' and 'VO'. In Eyring and O'Keeffe (Ed.) pp. 488-522.

Barth, T. F. W. and E. Posnjak (1932). Spinel structures with and without variate atom equipoints. Z. Kristallogr. **82**, 325-340.

Bethe, H. (1935). Statistical theory of superlattices. Proc. Roy. Soc. (London) **A150**, 552-575.

Bijvoet, J. M. and Nieuwenkamp, W. (1933). The variable structure of cadmium bromide. Z. Kristallogr. **86**, 466-470.

Bragg, W. L. and Williams, E. J. (1934). The effect of thermal agitation on atomic arrangements in alloys. Proc. Roy. Soc. (London) **A 145**, 699-730.

Busch, G. and Scherrer, P. (1935). A new seignettoelectric substance. Naturwiss. **23**, 737.

Christian, J. W. (1965). The Theory of Transformation in Metals and Alloys. Oxford, Pergamon, 973 pp.

Clusius, K. (1929). The specific heat of some condensed gases between 10° absolute and their triple points. Z. Phys. Chem. **B3**, 41-79.

Darwin, C. G. (1922). The reflection of X-rays from imperfect crystals. Phil. Mag. **43**, 800-820.

Edwards, O. S. and Lipson, H. (1942). Imperfections in the structure of cobalt. I. Experimental work and proposed structure. Proc. Roy. Soc. (London) **A 180**, 268-277.

Ewald, P. P. (1925). Reflection of Rontgen rays. Physik. Z. **26**, 29-32.

Eyring, L. and O'Keeffe, M. (1970). (editors) The Chemistry of Extended Defects in Non-Metallic Solids. New York: Elsevier. 669 pp.

Giauque, W. F. and Stout, J. W. (1936). The entropy of water and the third law of thermodynamics. The heat capacity of ice from 15° to 273°K. J. Am. Chem. Soc. **58**, 1144-1150.

Giauque, W. F. et al. (1950). The Scientific Papers of W. F. Giauque I. (1923-1949). New York: Dover Publications. pp. 641.

Goldsmith, J. R. and Laves, F. (1954). Potassium feldspars structurally intermediate between microcline and sanidine. Geochim. et Cosmochim. Acta **6**, 100-118.

Hendricks, S. B. (1927). The crystal structure of potassium dihydrogen phosphate. Amer. J. Sci. **23**, 269-287.

Hendricks, S. B. (1932). Para-bromochlorobenzene and its congeners: various equivalent points in molecular lattices. Z. Kristallogr. **84**, 85-96.

Hendricks, S. B. (1934). The crystal structure of primary amyl ammonium chloride. Z. Kristallogr. **74**, 29-40.

Hendricks, S. B. (1938). On the structure of the clay minerals dickite, halloysite, and hydrated halloysite. Am. Mineral. **23**, 295-301.

Hendricks, S. B. (1939). Random structure of layer minerals as illustrated by cronstedite. Am. Mineral. **24**, 529-539.

Hendricks, S. B. and Dickinson, R. G. (1927). The crystal structures of ammonium, potassium, and rubidium cupric chloride dihydrates. J. Am. Chem. Soc. **48**, 2149-2162.

Hendricks, S. B. and Jefferson, M. E. (1939). Polymorphism of the micas. Am. Mineral. **24**, 729-771.

Hendricks, S. B. and Pauling, L. (1925). The crystal structures of sodium and potassium azides and potassium cyanate and the nature of the azide group. J. Am. Chem. Soc. **47**, 2904-2920.

Hendricks, S. B. and Teller, E. (1942). X-ray interference in partially ordered layer lattices. J. Chem. Phys. **10**, 147-167.

Hendricks, S. B., Jefferson, M. E. and Mosley, V. M. (1932a). The crystal structures of some natural and synthetic apatite-like substances. Z. Kristallogr. **31**, 352-369.

Hendricks, S. B., Jefferson, M. E. and Shultz, J. F. (1930). The transition temperatures of cobalt and nickel. Some observations on the oxides of nickel. Z. Kristallogr. **73**, 376-380.

Hendricks, S. B., Posnjak, E. and Kracek, F. C. (1932b). Molecular rotation in the solid state. The variation of the crystal structure of ammonium nitrate with temperature. J. Am. Chem. Soc. **54**, 2766-2786.

Hettick, A. and Hendricks, S. B. (1933). Molecular rotation in solid ammonium chloride. Naturwiss. **21**, 467.

Jette, E. R. and Foote, F. (1933). An X-ray study of wustite solid solutions. J. Chem. Phys. **1**, 29-36.

Johansson, C. H. and Linde, J. D. (1925). A rontgengraphic determination of the atomic arrangement in mixed crystals of gold-copper and palladium-copper. Ann. Phys. **78**, 439-463.

Johnston, H. L. and Glauque, W. F. (1929). The heat capacity of nitric oxide from 14°K to the boiling point and the heat of vaporization. Vapor pressures of solid and liquid phases. The entropy from spectroscopic data. J. Am. Chem. Soc. **51**, 3195-3214.

Kemp, J. D. and Giauque, W. L. (1937). Carbonyl sulfide. The heat capacity, vapor pressure and heats of fusion and vaporization. The third law of thermodynamics and orientation equilibrium in the solid. J. Am. Chem. Soc. **59**, 78-84.

Ketelaar, J. A. A. (1934a). The crystal structure of the high temperature modifications of Ag_2HgI_4 and Cu_2HgI_4. Z. Kristallogr. **87**, 436-445.

Ketelaar, J. A. A. (1934b). The crystal structure of nickel bromide and iodide. Z. Kristallogr. **88**, 26-34.

Kracek, F. C. (1931). Gradual transition in sodium nitrate. I. Physiochemical criteria of the transition. J. Am. Chem. Soc. **53**, 2609-2624.

Kracek, F. C., Posnjak, E. and Hendricks, S. B. (1931). Gradual transition in sodium nitrate. II. The structure at various temperatures and its bearing on molecular rotation. J. Am. Chem. Soc. **53**, 3339-3348.

Krivoglaz, M. A. and Smirnov, A. A. (1965). The Theory of Order and Disorder in Alloys. New York: Elsevier Publishing Company.

Landau, L. D. (1937). Dispersion of X-rays by crystals with alternating structure. J. Exptl. Theoret. Phys. (USSR) **12**, 26-31.

Laves, F. and Goldsmith, J. R. (1955). The effect of temperature and composition on the Al-Si distribution in anorthite. Z. Kristallogr. **106**, 227-235.

Lely, J. A. and Bijvoet, J. M. (1944). Crystal structure and temperature vibration of NH_4CN. Rec. Trav. Chim. Pays-Bas. **63**, 39-43.

Lifshits, E. M. (1937). Scattering of X-rays by crystals of variable structure. Phys. Z. Sowjet Union. **12**, 623-643.

Megaw, H. D. (1959). Order and disorder. I. Theory of stacking faults and diffraction maxima. Proc. Roy. Soc. (London) **A 259**, 59-78. II. Theory of diffraction effects in the intermediate plagioclase feldspars. ibid 159-183. III. The structure of the intermediate plagioclase feldspars. ibid 184-202.

McLennan, J. C. and Plummer, W. G. (1928). The crystal structure of solid methane. Nature (London) **122**, 571-572.

Parthé, E. and Hohnke, D. (1970). The Ordered Vacancy Arrangement in Pyrite Defect Structures. In Eyring and O'Keeffe (editors) pp. 220-222.

Pauling, L. (1930). The rotational motion of molecules in crystals. Phys. Rev. **36**, 430-443.

Pauling, L. (1935). The structure and entropy of ice and of other crystals with some randomness of atomic arrangement. J. Am. Chem. Soc. **57**, 2680-2684.

Pauling, L. (1960). The Nature of the Chemical Bond. Ithaca: Cornell University Press, 644 pp.

Pauling, L. and Hendricks, S. B. (1925). The crystal structure of hematite and corundum. J. Am. Chem. Soc. **47**, 781-790.

Posjnak, E. and Barth, T. F. W. (1931). A new type of crystal fine-structure: lithium ferrite ($Li_2O.Fe_2O_3$). Phys. Rev. **38**, 2234-2239.

Posner, A. S., Stutman, J. M. and Lippincott, E. R. (1960). Hydrogen bonding in Ca-deficient hydroxyapatites. Nature (London) **188**, 486-487.

Simon, F., Simson, C. V. and Ruheman, M. (1927). An investigation of specific heat at low temperature. The specific heat of the halides of ammonia between -70° and room temperature. Z. Phys. Chem. **129**, 339-348.

Smith, J. V. (1974). In "Feldspar Minerals I Crystal Structure and Physical Properties." New York: Springer-Verlag. Chap. 3. Order-Disorder, pp. 50-84.

Southard, J. D., Milner, R. T. and Hendricks, S. B. (1932). Low temperature specific heats III. Molecular rotation in primary normal amyl ammonium chloride. J. Chem. Phys. **1**, 95-102.

Tammann, G. (1919). Chemical and galvanic properties of mixed crystals and their atomic properties. Z. Allg. Anorg. Chem. **107**, 1-239.

Taylor, W. H., Darbyshire, J. A., and Strunz, H. (1934). An X-ray investigation of the feldspars. Z. Kristallogr. **87**, 464-498.

Warren, B. E. (1941). X-ray diffraction in random-layer lattices. Phys. Rev. **59**, 693-698.

Watanabe, D., Terasaki, O., Justin, A. and Castle, J. R. (1970). Electron microscopic studies of the structure of low temperature modifications of titanium monoxide phase. In Eyring and O'Keeffe (editors) pp. 238-258.

Welo, L. A. and Baudisch, O. (1925). Two stage transformation of magnetite into hematite. Phil. Mag. **50**, 399-408.

Wilson, A. J. C. (1942). Imperfections in the structure of cobalt. II. Mathematical treatment of proposed structure. Proc. Roy. Soc. (London) **A 180**, 277-285.

Wilson, A. J. C. (1949). X-ray optics. The Diffraction of X-rays by Finite and Imperfect Crystals. London: Methuen and Co., 127 pp.

Wollan, E. O., Davidson, W. L. and Shull, C. G. (1949). Neutron-diffraction study of the structure of ice. Phys. Rev. **75**, 1348-1352.

CHAPTER 4

X-RAY ABSORPTION BY METALS AND ALLOYS

Leonid V. Azároff
Institute of Materials Science and Physics Department
University of Connecticut, Storrs 06268

Before Max Laue and Paul Ewald even began their respective speculations about the suitability of natural crystals as diffraction gratings for X-rays, other physicists had been busy trying to understand their properties and origins. Most notable among them was C. G. Barkla (1906) who demonstrated that X-rays become polarized upon scattering and behaved, therefore, like transverse light waves. Even more dramatic was Barkla's discovery that, in the course of scattering incident X-rays, metal scatterers produced a form of fluorescent radiation whose hardness (penetrability) increased with the metal's atomic weight. He did this by measuring the attenuation (absorption) of X-rays scattered by a progression of metals after passage through a, say, copper foil and an aluminum foil. A plot of the ratio of the two attenuation coefficients showed a more rapid decline in the attenuation in copper than in aluminum as the scatterer's atomic weight (number) increased with a sharp discontinuity occurring whenever the scatterer and first absorber were the same metal (Barkla, 1908, 1911). For discovering this energy dependence of what we now call absorption edges and later relating them to the Bohr atomic model, Barkla received the Nobel prize in 1917.

Because X-ray emission and absorption spectra are relatively uncomplicated, at least within early twentieth-century detection limits, they provided some of the best foils for testing the emerging quantum theory of matter. By approximately 1930, however, it became apparent that a kind of standoff had been reached between the ability to detect changes in metal and alloy spectra, attributable to such variables as structure (polymorphism) and composition (phase changes), and the theoretical limitations on calculating the expected spectral shapes. Thus it is not surprising that the amount of activity in this once very fertile area of research in physics began to decline leaving very few active centers in the U.S., notably Cal. Tech. (J.W.M. Dumond), Cornell (L.G. Parratt), Johns Hopkins (J.A. Bearden), and Ohio State (C.H. Shaw).

By 1930, moreover, a great deal was known about X-ray spectra including their wavelength dependence, general shape, relation to the electronic structure of the emitting or absorbing atoms, and so forth. Thus it was believed that the absorption discontinuity occurring at one of the characteristic frequencies (wavelengths) involved the ejection of an inner bound electron to an outer unoccupied energy state so that the fine structure evident at this edge was relatable to the possible availability of such states.

In 1931, a simple one-dimensional periodic potential array was shown to predict that the energy states for an electron moving along such an array were segregated into bands of allowed and forbidden energy values. One of the progenitors of this model, R. de L. Kronig, applied it that same year to demonstrate that an X-ray-ejected inner electron can undergo transitions only to energy values contained in these allowed bands so that an absorption edge should be followed by a series of absorption maxima alternating with minima (Kronig, 1931). Order-of-magnitude calculations quickly showed a good correspondence between the predictions of this simple band model and actually observed variations in the extended X-ray absorption fine structure. What is now known in our acronym-prone society as EXAFS, thus was first known as Kronig structure.

Making use of somewhat more sophisticated energy-band calculations emanating from J. C. Slater's group at M.I.T., J. A. Bearden, W. W. Beeman, and H. Friedman showed in a series of papers around 1940 that the fine structure at the main edge, visible in high-resolution absorption spectra of pure metals and their alloys, was related to the unfilled density-of-states distribution deduced from the corresponding band model (Azároff and Pease, 1974). Probably because some of the recorded spectra showed little if any change upon alloying and the band calculations were not sufficiently precise, these physicists subsequently turned their attention to other applications of X-rays.

Following the technological advances that resulted from the outpouring of funds for research during World War II, the availability of more sensitive and precise detectors and, probably most significantly, high-speed inexpensive computers, a virtual renaissance of activity in X-ray spectroscopy began to develop. Thus the availability of electron-probe microanalyzers after 1951 stimulated interest in emission spectra and their utilization in characterizing the emitter's environment. Similarly, it became apparent that proliferating X-ray powder diffractometers could be readily modified to serve as spectrometers for recording the Kronig structure of absorbing foils. While in Japan and the U.S.S.R. new theoretical models were being developed to explain the extended fine structure, in the U.S.A. two chemists were chiefly responsible for stimulating interest in this field: R. A. Van Nordstrand in Harvey, Ill. and F. W. Lytle in Seattle, Wash. In collaboration with two physicists, E. A. Stern and D. E. Sayers, Lytle has succeeded in making significant progress in developing EXAFS for characterizing an absorbing atom's environment and related applications (Sayers et al., 1970).

Because X-ray absorption spectra can selectively

sample the electronic structure in the vicinity of one atomic species at a time, they offer unique information about binary and ternary alloys especially. The availability of increasingly sophisticated and reliable energy-band calculations for metals prompted Azároff to reexamine the previous measurements of Beeman et al. around 1960 and led to a series of studies of the electronic structures of binary and ternary solid solutions (Azároff et al., 1964, 1965, 1966, 1967, 1973). By 1973, direct theoretical calculations of X-ray K and L absorption spectra began to appear for metals so that it is now possible to correlate changes caused by alloying to observed changes (or lack of change) in their X-ray absorption edges (Szmulowicz and Pease, 1978). In the case of transition metals, where changes typically take place in the small number of unfilled states in the valence or d band, absorption spectroscopy is a particularly sensitive tool since it depends entirely on the allowed state density in this portion of the band.

Improvements in energy-band calculations for alloys coincided with advances in X-ray detectors and sources, notably the development of synchrotron sources that will be utilizable in a broader X-ray energy region. The study of directional properties that the plane-polarized synchrotron radiation enables is also beginning to receive increased attention. Thus the future of X-ray absorption spectroscopy in the United States looks to be very bright indeed.

References

Azároff, L. V. and Das, B. N. (1964). Phys. Rev. **143**, A747.
Azároff, L. V. and Das, B. N. (1965). Acta Met. **13**, 827.
Azároff, L. V. and Yeh, H. C. (1965). Appl. Phys. Letters **6**, 207.
Azároff, L. V. (1966). Science **151**, 785.
Azároff, L. V. (1967). J. Appl. Phys. **38**, 2809.
Azároff, L. V. and Donahue, R. J. (1967). J. Appl. Phys. **38**, 2813.
Azároff, L. V. and Yeh, H. C. (1967). J. Appl. Phys. **38**, 4034.
Azároff, L. V. and Brown, R. S. (1973). In "Band Structure. Spectroscopy of Metals and Alloys," edited by D. J. Fabian and L. M. Watson, p. 491. London: Academic Press.
Azároff, L. V. and Pease, D. M. (1973). J. Appl. Phys. **44**, 3419.
Azároff, L. V. and Pease, D. M. (1974). X-Ray Spectroscopy, edited by L. V. Azaroff, p. 284. New York: McGraw-Hill Book Co.
Barkla, C. G. (1906). Proc. Roy. Soc. (London) **77**, 247.
Barkla, C. G. (1908). Phil. Mag. **16**, 550.
Barkla, C. G. (1911). Phil. Mag. **22**, 396; ibid **23**, 987.
Kronig, R. deL. (1931). Z. Phys. **70**, 317.
Sayers, D. E., Lytle, F. W. and Stern, E. A. (1970). Adv. X-Ray Anal. **13**, 248.
Szmulowicz, F. and Pease, D. M. (1978). Phys. Rev. **17**, B3341.

CHAPTER 5

TWENTY YEARS OF ELECTRON DENSITY STUDIES IN NORTH AMERICA

Philip Coppens
State University of New York at Buffalo
Buffalo, NY 14221

The idea of studying the electron density by means of X-ray diffraction was conceived very soon after von Laue's 1912 discovery. The knowledge of electron distributions at that time was confined to the existence of the Bohr electron rings proposed in 1913. Debye deduced that the electron rings might give an interference effect and together with Scherrer constructed the well known camera to verify this theory. Somewhat later scattering measurements on mono-atomic gases such as mercury definitively showed the absence of maxima and minima in agreement with wave mechanics (Scherrer, 1962).

Though chemical theory based on electron distributions was developed extensively in North America by G. N. Lewis, Linus Pauling and others, early attempts to use diffraction methods for locating electrons seem to be confined mainly to Europe.

The somewhat startling success of the free-atom model in accounting for observed X-ray intensities must have discouraged many investigators. It certainly created an ingrained skepticism towards claims that bonding density had been observed, which became increasingly frequent as the accuracy of the experimental measurements improved.

In the U.S. some of the early work was centered at Watertown Arsenal by a group headed by Dick (R.J.) Weiss, who also organized at very short notice the first Sagamore conference, which has since become institutionalized by the I.U.Cr. commission on Charge, Spin and Momentum densities of which Dick was the first chairman. Sagamore conferences have since been held at Aussois, France; Minsk, Russia; Kiljava, Finland; Mont Tremblant, Canada with a 1982 conference being planned for Japan. The origin of the name Sagamore has become almost forgotten by the success of the following meetings. As related by Dick Weiss, the first initiative was conceived at short notice to relieve the U.S. Army of the Sagamore Conference Center on Racquette Lake in the Adirondacks. The following quote is from the April 1965 issue of *Physics Today*.

> SPECIFICATIONS-US ARMY TANK-JULY 2064
> . . . medium tank, 4 passenger, armour plated steel, maximum speed 120 mph . . .
>
> Wave functions must be specified to within 1% at all points in the atomic polyhedra. One electron observables to be evaluated. . .

The above excerpt might be taken from specifications for an Army tank in the year 2064 since high-speed computers may then be capable of calculating all pertinent observables (strength, hardness, thermal conductivity, etc.) from the wave functions. However, projected military requirements a hundred years hence had little to do with the Army's sponsorship of the Sagamore conference on charge and spin density (Aug. 17-21, 1964), which resulted rather from an active interest of the research group of the US Army Materials Research Agency at Watertown, Mass., in the measurement and calculation of charge and spin density.

In the fifties much work in the U.S. was done on relatively simple structures like diamond, silicon and several metals, some of it producing very puzzling results, such as the observed sharp contraction of the core electrons in Al and Fe (Weiss, 1966). Even today it is difficult to study the charge distribution in such small unit cells, which give very few Bragg reflections in the low angle region where valence scattering is concentrated. This difficulty led many of the physicists involved in the field to concentrate on the measurement of momentum densities by Compton scattering rather than on the charge density, and it culminated in the statement that "both the physics and the nature of the experiments dictate the use of the Compton-scattering technique to measure anisotropies in electron distributions" (Berggren, Martino, Eisenberger and Reed, 1976).

A more balanced view must take into account that nature is often more complicated than is the case for LiF, in the course of which study the above statement was made. More complex solids with larger unit cells and more low order Bragg reflections are especially suited for the X-ray diffraction technique, but yield no information pertaining to individual atoms or bonds when their Compton scattering is analyzed.

The study of electron densities in molecules was taken up in the early sixties by theoretical chemists such as Ransil, Cade, Bader and Lipscomb and their collaborators. Cade and Bader and co-workers (Cade and Huo, 1966, Cade, Bader, Henneker and Keaveny, 1969, Cade, 1972) published charge density difference maps of unusual quality for a series of diatomic molecules, while Stevens, Switkes, Laws and Lipscomb (1971) showed the peculiar charge accumulation in BHB triangles associated with two electron-three center bonding in electron deficient compounds. The work by Smith and Richardson (1965, 1967) on the basis set dependence of the deformation density in CO and N_2 should be mentioned as it was well ahead of extensive discussions published later on the subject.

For crystallographers the most persistent reminder that not everything was accounted for by the simple model was the consistent shortening of bonds to hydrogen atoms. The subject was discussed by Jensen and Sundaralingam (1964) in an article in *Science* and by Jones and Lipscomb in *Acta Crystallographica* (1969). But it was Stewart who together with Davidson and Simpson (1965) was able to bridge the gap between the theoretical distribution around an H atom and the crystallographic practice. Stewart calculated the scattering of the best spherical hydrogen atom that could be fitted to the theoretical density for the H_2 molecule and also derived an aspherical form factor curve for hydrogen. The bonded hydrogen scattering factor has been so widely used that the corresponding paper published in *J. Chem. Phys.* in 1965 has become one of the most quoted papers in all scientific literature. It foreshadowed later development of atom-centered generalized scattering formalisms.

A much better understanding of the apparent success of the spherical atom model was obtained by comparison of X-ray and neutron results. It became very clear that X-ray least squares parameters can be systematically biased by an incorrect (i.e., spherical-atom) assumption in the model when X-N difference ellipsoids and X-N density maps were plotted. In the latter, first calculated for s-triazine and published in *Science* in 1967 (Coppens, 1967), spherical atoms with neutron diffraction parameters were subtracted from the total density. The resulting map was not at all flat, as is common after a successful least squares refinement, but showed density accumulations in the bonds and in the regions associated with the lone pair electrons.

The progress in the field was recognized by a symposium held at the ACA meeting in Albuquerque in March 1972 which resulted in the publication of a volume of the *ACA Transactions* (Vol. 8). In addition to some of the work mentioned above it includes articles on electron diffraction studies of charge densities by L. S. Bartell, on spin densities by R. H. Moon and a landmark study by Hanson, Sieker and Jensen on sucrose, which is still among the largest molecules that have been studied.

The period after the Albuquerque symposium saw a rapid acceptance of modified scattering formalisms. They were not the bond form factors originally pioneered by McWeeny, which proved to be too strongly correlated with density functions centered on the atoms, but one-center formalisms in which all functions are centered on the atoms. Such functions were used by Weiss (1966) to describe the electron distribution in diamond, and subsequently generalized with major contributions by Kurki Suonio, Dawson, Stewart and Hirshfeld. The post-Albuquerque period also witnessed an increasing interaction between crystallographers and theoreticians with many theoretical maps being published precisely because the electron density, or at least the thermally smeared electron density, became recognized as an experimental observable. Combined studies by crystallographers and theoreticians, stimulated by 1978 and 1980 Gordon Conferences, became more and more common, with the experiment being used as a calibration of the theory, or in other cases being scrutinized for evidence of molecular associations such as hydrogen bonding. In addition to the charge density, physical properties including molecular moments, electric field gradients and the electrostatic potential have become a focus of interest.

Whereas the early work was confined to laboratories in a few centers like Boston, Buffalo and Pittsburgh, charge density analysis is now a more generally applied technique which may eventually become as routine as structure determination. It is increasingly being applied to series of related compounds such as in the work on silicates by Gibbs, Ross and co-workers at Virginia Polytechnic Institute, supported by theoretical contributions from M. D. Newton and J. Tossell. Furthermore, it is becoming evident that much heavier atoms can be studied than was assumed earlier. Especially transition metal complexes provide a fertile ground as many have complex electronic structures which are only partially understood. They are of importance as models for enzymes and oxygen transport proteins and play a prominent role in many catalytic processes. With the study of such complicated molecules the electron density technique has come of age, and become a tool of increasing importance for our understanding of molecular processes.

The history of electron density work in North America cannot be described in its isolation. Throughout the past decades there has been a steady exchange of ideas across the Atlantic and with scientists in countries like Japan, Australia and Israel, resulting in many joint publications. The rapid developments of the 1960-1980 period would not have been possible without this extensive international collaboration.

For further detail, the reader is referred to a number of reviews which are listed at the end of the following bibliography.

References

Berggren, K.-F., Martino, F., Eisenberger, P. and Reed, W. A. (1976) Phys. Rev. B. **13**, 2292-2304.

Cade, P. E. (1972) Trans. Am. Crystallogr. Assoc. **8**, 1-30.

Cade, P. E. and Huo, W. M. (1966). J. Chem. Phys. **45**, 1063-1065.

Cade, P. E., Bader, R. F. W., Henneker, W. H. and Keaveny, I. (1969). J. Chem. Phys. **50**, 5315-5333.

Coppens, P. (1967). Science **158**, 1577.

Jensen, L. H. and Sundaralingam, M. (1964). Science **145**, 1185.

Jones, D. S. and Lipscomb, W. N. (1969). Acta Cryst. **A28**, 635-645.

Scherrer, P. (1962). In "Fifty Years of X-ray Diffraction,"

Edited by P. P. Ewald. Utrecht, Netherlands: Oosthoek.
Smith, P. R. and Richardson, J. W. (1965). J. Phys. Chem. **69**, 3346.
Smith, P. R. and Richardson, J. W. (1967). J. Phys. Chem. 71, 924.
Stevens, R. M., Switkes, E., Laws, E. A. and Lipscomb, W. N. (1971). J. Am. Chem. Soc. **93**, 2603-2609.
Stewart, R. F., Davidson, E. R. and Simpson, W. T. (1965). J. Chem. Phys. **42**, 3175.
Weiss, R. J. (1966). X-ray Determination of Electron Distributions. New York: John Wiley and Sons.

Reviews

Transactions of the American Crystallographic Association Vol. 8 (1972). Edited by P. Coppens.
Israel J. of Chemistry Vol. 6 (1977). Edited by F. L. Hirshfeld.
Electronic and Magnetisation Distributions in Molecules and Solids (1979). Edited by P. Becker, Plenum.
Electron Distributions and the Chemical Bond. (1982). Edited by P. Coppens and M. Hall. Plenum.
Physica Scripta Vol. 15, (1977). Edited by K. Kurki Suonio.

CHAPTER 6

THE INTRODUCTION OF SOME IMPORTANT CONCEPTS AND PRINCIPLES

Maurice L. Huggins
Woodside, CA 94062

Introduction

In this article I briefly discuss the American history of some of the main concepts and principles of crystallography, emphasizing those to which I contributed, with no desire to belittle the many important contributions of others.

The Periodic Table, Valence, and the Numbers and Orientations of Chemical Bonds

Gilbert N. Lewis (1916, 1923) made a great contribution to the progress of science when he showed how the valences of the elements and the periodic table can be related to the numbers of valence electrons in the atoms and the tendencies of these electrons to form stable valence shells, often containing 8 electrons, by the transfer of electrons from one atom to another, and by the sharing of electron pairs between pairs of atoms.

At first he thought of the eight electrons in the stable valence shells of many atoms as being located at corners of a cube. He considered the formation of single bonds and double bonds as consisting of the sharing of one or two cube edges between two atoms. Adding the assumption that the two electrons of a shared pair could somehow be drawn close together, and (in this case) abandoning the cubical valence shell assumption, he could explain the existence of triple bonds – in acetylene, for example.

Returning to the University of California in Berkeley early in 1919, after my service in World War I, I learned about this theory and also about various chemical observations and experiments that had not been satisfactorily explained in terms of earlier theories of atomic and molecular structure. For my own satisfaction I tried to use the Lewis theory, with some modifications and extensions of my own, to obtain reasonable explanations for the anomalous phenomena (Huggins, 1919).

I tried the idea that the electrons in a stable 8-electron valence shell are *all* paired – even those that are not shared between atoms. The cubical valence shell is then replaced by a tetrahedral valence shell. (Like Lewis, I then considered the electrons or electronpairs as having definite *locations.* Later, "locations of electronpairs" became replaced by "orientations of electron orbits" or "orbitals.")

Hydrogen Bonds

The distributions of valence electron pairs in ammonia and water molecules and in ammonium and hydroxyl ions could be simply represented by Lewis dot formulas as follows:

$$\begin{matrix} \mathrm{H} \\ \mathrm{H}{:}\underset{\cdot\cdot}{\ddot{\mathrm{N}}}{:} \\ \mathrm{H} \end{matrix} \qquad \mathrm{H}{:}\underset{\cdot\cdot}{\ddot{\mathrm{O}}}{:}\mathrm{H}$$

$$\left(\begin{matrix} \mathrm{H} \\ \mathrm{H}{:}\underset{\cdot\cdot}{\ddot{\mathrm{N}}}{:}\mathrm{H} \\ \mathrm{H} \end{matrix}\right)^{+} \qquad ({:}\underset{\cdot\cdot}{\ddot{\mathrm{O}}}{:}\mathrm{H})^{-}$$

It seemed to me that the previously puzzling fact that a solution of ammonia in water sometimes behaves as if it is composed of ammonia and water molecules and sometimes as if it contains ammonium and hydroxyl ions could be explained by the simple assumption that a proton (a hydrogen kernel or nucleus H) attached to one electronegative atom (N or O) should have an attraction for a "lone electron pair" (one not serving to bond two atoms together) in the valence shell of another electronegative atom. If the attraction is strong enough, a "hydrogen bridge," perhaps persisting for a long time in spite of the intermolecular collisions, should be formed:

$$\begin{matrix} \mathrm{H} \\ \mathrm{H}{:}\underset{\cdot\cdot}{\ddot{\mathrm{N}}}{:}\mathrm{H}{:}\underset{\cdot\cdot}{\ddot{\mathrm{O}}}{:}\mathrm{H} \\ \mathrm{H} \end{matrix}$$

The bridge could be considered a weak bond (Huggins, 1922a,b,c).

Then and later I applied this idea to various other situations. For example, I predicted that the structure of ice would be one in which each oxygen atom is hydrogen-bonded tetrahedrally to four others. This prediction was confirmed by W. H. Bragg (1922).

Not long after my speculations about the hydrogen bond, Latimer and Rodebush (1920), apparently independently, had essentially the same idea. (See Lewis, 1923, p. 109). They used it to explain a variety of chemical observations. Many later researches have confirmed the existence and importance of hydrogen bonds in crystals.

Lewis, in his valence theory, had dealt with only a few of the elements–those in which the usual valences could be related to the presence of stable valence shells containing eight electrons (or two, in the case of hydrogen). I extended his theory to the rest of the elements in the periodic table, postulating concentric tetrahedral, octahedral and cubic shells, with pairs and triplets of electrons at the polyhedron corners (Huggins, 1922a,b 1926a). Langmuir (1919) and Bury (1921) made other proposals, but my proposals, I believe, were better. They accounted for the observed valences and coordination numbers of the elements and they could reasonably be related to the (later) requirements of the Pauli exclusion principle (Pauli, 1925).

Secondary Valence and Complex Ion Structures

According to the Lewis theory, an electron pair bond can be formed either by the combination of one electron from each of two atoms or by the interaction of an atom having a lone pair in its valence shell with an atom which, by adding two electrons, can obtain a more stable valence shell:

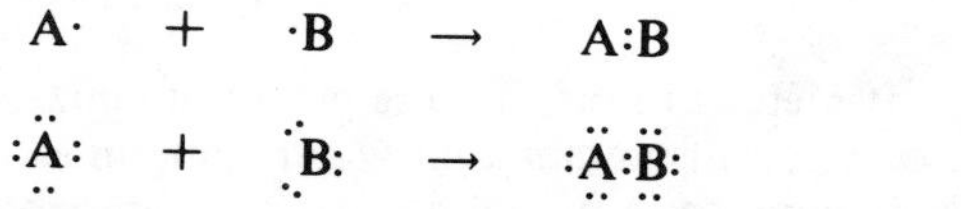

A· + ·B → A:B

:Ä: + B: → :Ä:B̈:

For example, a fluoride ion can add to boron trifluoride to form a fluoborate ion:

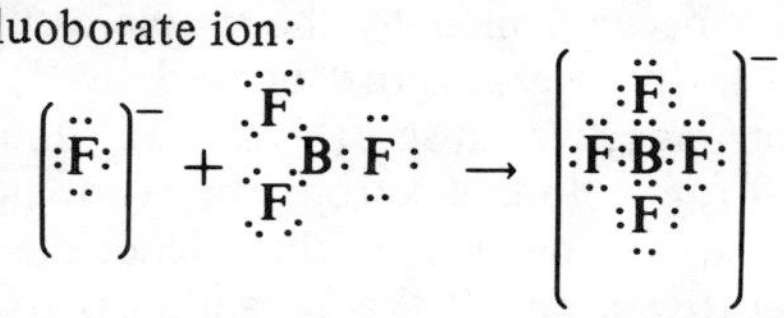

$[:\ddot{F}:]^-$ + F:B:F (BF$_3$) → $[\text{:F:B:F:}$ with F above and below$]^-$

I showed (Huggins, 1922a,b; 1926b) how this concept leads simply to the structures of many complexes (often ions), in which the number of bonds connecting an atom to others does not conform to the usual valence of that atom. My theory also predicted the geometrical arrangements of the atoms in these complexes, agreeing with the results of many X-ray structure determinations, especially by R. W. G. Wyckoff and R. G. Dickinson and their co-workers.

The Benzene Ring

In 1919, after learning about various attempts that had been made to modify the old ideas of valence to account for the chemical behavior of benzene and other "conjugated systems," I tried to apply my tetrahedral atom modification of Lewis' valence theory to the problem. I soon arrived at the concept that the benzene molecule oscillated between two Kekulé structures, passing through an intermediate puckered ring structure, having 6 tetrahedron corners at or near the center of the molecule. Studying the literature, I found that the intermediate structure had previously been proposed by Körner (1874). At Lewis' suggestions, I concentrated (in my lectures and publications) on the intermediate structure, which I designated "the centroid structure." Of course I considered, not only benzene, but naphthalene and various other aromatic compounds (Huggins, 1922d,e; 1923a).

I also postulated that each sheet of carbon atoms in graphite has a puckered structure, like a composite of many centroid rings.

These models accounted nicely for many experimental facts and I looked for a definite experimental way of testing their correctness. One way of doing so required the synthesis of certain naphthalene derivatives. A friend and fellow student undertook the job and showed conclusively that my predictions were wrong (Fuson, 1924).

Learning, in 1919, about the new science of crystal structure determination, I decided that the benzene structure problem (and others in which I was interested) could be solved by X-ray diffraction studies. I could not undertake research in this field immediately, but Lonsdale (1929) and Brockway and Robertson (1939) later showed that the centers of the twelve carbon atoms in hexamethylbenzene, $C_6(CH_3)_6$, are in (or very close to) a plane – at least as a time and space average. For a photographic Fourier series projection of the molecule structure, made from Brockway and Robertson's X-ray data, see Huggins (1944). This result also showed that my benzene model was wrong.

At the California Institute of Technology in 1923-1925, Linus Pauling and I discussed many things, including the structures of benzene and other conjugated systems. Not long thereafter he found a way to improve on the centroid model by substituting, for my idea of oscillation of tetrahedral carbon atoms between two Kekulé structures, the concept of "resonance" between different patterns of electron paths (Pauling, 1939):

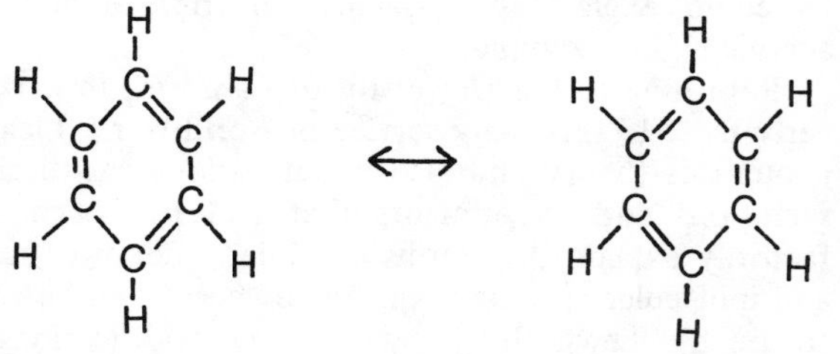

Application of the Lewis-Huggins Theory to Crystal Structures

It seemed to me in 1919 that the Lewis theory of valence and molecular structure, as modified and extended by me, should be as applicable to crystals as to small molecules and ions. I was therefore delighted when the Braggs' book (Bragg and Bragg, 1916) appeared in the chemistry library, because I could then compare experimentally determined structures with my theoretical predictions of the numbers and relative orientations of the bonds connecting each atom to its neighbors (Huggins, 1922f; 1932). I also found good examples illustrating the fact that atoms can be held together by electron pair sharing. Coulombic attractions between ions, or both (if the electron pair bonds are polar).

Looking for examples of hydrogen bonding, I soon found some, first in ammonium chloride and ice.

I found that in some substances (diamond, SiC, ZnS, CuC1, $CuFeS_2$, FeS_2, etc.) each crystal is a single molecule, held together by shared electron pairs. Other crystals (e.g., As, Sb, Bi) are assemblages of puckered sheets of macromolecules. Still others (Se, Te) are composed of (helical) macromolecular chains of atoms.

In my graduate years (1919-1922) I was eager to determine crystal structures experimentally, but there was no appropriate apparatus at Berkeley and I could not take the time to set up an X-ray laboratory. I decided, however, to see if I could use my ideas, plus crystallographic and other pertinent information from the literature, to deduce new crystal structures.

The Braggs had obtained and published (1916) some results of X-ray measurements on quartz, SiO_2, but they had been unable to find a structure accounting for their data. (Their chemical colleagues told them that the crystal should be composed of rodlike molecules corresponding to the formula O=Si=O.) I believed that each silicon atom should be bonded tetrahedrally by single bonds to four oxygens, and each oxygen atom should be connected by single bonds to two silicon atoms, with the angle between these bonds a little larger than 110°. Using crude models, I deduced the structure and showed that it agreed with the Braggs' measurements and the published symmetry and density data. The fact that the structure showed spirals was satisfying, since it gave a reasonable explanation for the optical rotation properties (Huggins, 1922g).

Using the predictions of our molecular structure theory with densities and non-X-ray data in the crystallographic literature, I predicted the structures of many crystals for which X-ray diffraction analyses had not yet been published.

For example, I deduced the structure of marcasite, the orthorhombic modification of FeS_2. The atomic arrangement in the cubic modification (pyrite) had previously been determined by the Braggs (1916). In that structure, according to our theory, each sulfur atom is tetrahedrally bonded to one sulfur and three iron atoms, and each iron atom is octahedrally bonded (by semipolar bonds) to six sulfur atoms. Using models, I found a structure (Huggins, 1922h) for marcasite that agreed with the density and axial ratio data (Groth, 1906). I also assumed that other crystallographically similar minerals, such as arsenopyrite (FeAsS) and loellingite ($FeAs_2$), have similar structures. (In order to have the same pattern of electron pair bonds, the iron atoms should have valences of two in FeS_2, three in FeAsS, and four in $FeAs_2$.)

In calcite, the rhombohedral modification of $CaCO_3$ (Bragg and Bragg, 1916), each carbon atom is at the center of an equilateral triangle of three oxygens. I proposed (Huggins, 1922c,f) that each oxygen has a tetrahedral 4-electronpair valence shell, so oriented as to form a double bond to the central carbon atom and a (quite polar) single bond to each of the two neighboring calcium atoms.

Pauling (1939) prefers to describe the carbonate group as a resonance hybrid of three structures, each with tetrahedral bonding around the carbon, but I disagree (Huggins, 1936b).

Proceeding as for marcasite, I deduced a structure for aragonite, the orthorhombic modification of $CaCO_3$ (and for other carbonates isomorphous with it), that had a bond pattern similar to that in calcite and agreed with the (non X-ray) crystallographic data (Huggins, 1922i). Later X-ray analyses (Bragg, 1924; Wyckoff, 1925a,b) showed that there is a calcite-like arrangement of atoms within each CO_3 group, but that there are 9 instead of 6 of these groups around each calcium. (I had not allowed for the fact that aragonite, which is more stable at high pressures, should be expected to have a more close-packed structure than calcite.)

Studying the (non X-ray) crystallographic literature, I found many other examples of apparent isomorphism with crystals whose structures had already been determined by X-ray diffraction (Huggins, 1923b,c,d; 1926c). Interatomic distances calculated for such substances were helpful in deducing atomic radii.

Atomic Radii

Long before the use of X-rays for crystal structure analysis, scientists (Barlow, 1883, 1898; Barlow and Pope, 1906, 1907; Lord Kelvin, 1889) had used the concept that crystals are composed of spherical atoms or ions in mutual contact, but they had no good way of determining the relative sizes of the spheres or the patterns of their arrangement. Later, after these patterns became known, it was natural for scientists working in this field to try to relate the then known distances between the centers of neighboring atoms to the sums of radii of hypothetical atomic spheres. W. L. Bragg (1920) did this, calculating and comparing the sizes of atomic spheres in many elements and compounds of known structure.

Reading his paper on this subject, it seemed to me that his assumption of atomic radii, each characteristic of an atom of a particular element, was reasonable for atoms connected by (single) electronpair bonds, if the environment of each kind of atom is sufficiently similar in the structures being compared. For two atoms held together by an electronpair bond, I thought of the atomic radii of these atoms as the distances from the atomic nuclei to the center of the electronpair.

As appropriate interatomic distance data became available in the literature, I calculated atomic radii, concentrating on crystals in which (according to my theory) there are valence electronpairs on the centerlines between adjacent atoms (Huggins, 1921, 1922c). I first calculated a set of "standard radii" for atoms tetrahedrally surrounded by others, with the sum of the kernel charges of the bonded pairs of atoms equal to eight. Then, assuming the radii of the more electronegative atoms to be constant, I computed radii for relatively electropositive atoms in other situations (e.g., surrounded by six electronegative atoms).

If the difference in kernel charges is large, the interactions between the atoms are largely coulombic. For some purposes it is then better to use "ionic radii," related to the magnitudes of the ionic charges and to the coordination numbers. See the next section.

In 1926 I published the results of my atomic radii calculations up to that time (Huggins, 1926d). Several years later Pauling and I reviewed the data then available and published a revision and extension of my earlier tables. In addition to the crystal structure data, we considered interatomic distances in small molecules, deduced from electron diffraction and infrared data (Pauling and Huggins, 1934).

Much later I correlated departures from strict additivity of atomic radii with bond energies and polarities (Huggins, 1953).

Ionic Radii and Ionic Crystal Energies

Davey (1923, 1925) published a set of "ionic radii" which, assuming additivity, gave approximate agreement with most of the then known distances between closest ion centers in alkali halide crystals. His radii were computed on the assumed approximation that

$$r_{Cs^+} = r_{I^-}$$

Radii calculated for Li^+, Na^+ and F^-, assuming the radii adopted for the other ions, departed considerably from constancy.

Cuy (1927) in papers submitted for publication in 1924 and early 1925, used interatomic distance data for many more crystals and introduced the idea that a structure type is unstable if the ratio of the radii of the component ions approximates or exceeds the value which would give contact between spherical ions of like charge, as well as between those of opposite charge.

Pauling (1927, 1928, 1929, 1939) deduced sets of "crystal radii" and "univalent crystal radii" of ions, relating them with the aid of inverse power law terms for the coulombic and overlap energies. This treatment gave a semiquantitative explanation for Cuy's radius ratio effect.

Pauling's calculations of ionic radii were paralleled by those of scientists in other countries, including Landé (1920), Wasastjerna (1923), Goldschmidt (1926), and by those of Zachariasen (1931), soon after coming to this country from Norway. For a comparison of Pauling's and Zachariasen's ionic radii, see Huggins (1932).

The bonding between neighbor atoms, as I have mentioned, is often partly of the covalent (shared electronpair) type and partly of the ionic type. Comparison of the interatomic distances for the same pairs of atoms, computed by my atomic radii with those computed from Pauling's or Zachariasen's ionic radii, using appropriate corrections for structure type, etc., showed (Huggins, 1932) that in all (or nearly all) cases where the bonding is chiefly of one type or the other, the interatomic distance calculated on the basis of that assumption is the smaller one.

The introduction of an exponential form for the overlap repulsion energy (Born and Mayer, 1932), plus other refinements (Huggins and Mayer, 1933; Huggins, 1937), led to the calculation of interatomic distances and lattice energies with great accuracy for all 20 alkali halides, using just nine empirical "basic radii."

The alkaline earth chalcogenides (Huggins and Sakamoto, 1957) and the silver halides (Huggins, 1943a, 1951) were later treated in a similar manner. The silver halide calculations were used as a basis for a theory of the photographic latent image (Huggins, 1943b, 1954).

Metallic Radii

The structures and distances between closest atom centers were known for many metals in the late 1920s and early 1930s (Goldschmidt 1928, 1929; Neuberger 1931; Huggins 1932), but little was then known about the numbers and distributions of the valence electrons, especially for the transition element metals.

Empirical "metallic radii," defined as half the shortest interatomic distances, were not very different from the "atomic radii" discussed above. It is not surprising that W. L. Bragg (1920) and Goldschmidt (1928, 1929) used the same radii for both classes.

Pauling (1938, 1939) later related some of the variations of the interatomic distances (and hence the empirical radii) in metals to atomic orbital assignments and the ferromagnetic properties of the elements.

Van der Waals Radii

Reviewing the crystal structures of organic compounds, Hendricks (1930) compared the distances between two halogen centers in different molecules, in crystals of several organic compounds. Soon thereafter, I (Huggins, 1932) published a table of "standard radii for nonbonded atoms with 12 like neighbors," noting that the radii for atoms with only six neighbors appeared to be about 5 or 10% smaller. Pauling (1939), using the results of many later structure determinations, published a revised table of nonbonded atom radii – which he preferred to call "van der Waals radii."

Since his radii were deduced from (and intended for use in) crystals in which the number of nonbonded contacts is often less than 12, his tabulated values were generally smaller than those in my standard set.

Pauling also included a van der Waals radius for hydrogen which, in my opinion, is too small – at least for contacts between hydrogen atoms in *different* molecules. For contacts between hydrogen atoms not directly bonded together *in the same molecule,* these atoms are often forced (by bond angle or other structural factors within the molecule) to be closer together than they would otherwise be. I have discussed this subject in some detail (Huggins, 1966, 1967, 1968a).

Although the van der Waals radii are much less constant than the covalent radii and ionic radii, they have proved very useful, especially for limiting parameters to be considered in investigations of complex structures.

Summaries of Structures and Structural Principles

In the early days of crystal structure research in this country, I had many occasions to describe typical structures and to explain the structural principles involved.

In one lecture (Huggins, 1931b) I presented and discussed the structural principles that seemed most important in determining the atomic arrangement in a crystal of given composition.

The *first principle* mentioned was that the influence of an atom on other atoms decreases very rapidly with distance, so that the effect on all but those atoms immediately adjacent to a given atom is almost negligible.

The *second principle* was that atoms of the same kind, crystallizing in the same environment, tend to be surrounded similarly. (Similar surroundings may be impossible for like atoms in molecules whose structures were determined before the crystallization.)

The *third principle* was that the whole crystal (except, in some cases, the crystal surfaces) must be electrically neutral.

The *fourth principle* was that an ion of a given sign tends to be surrounded by ions of opposite sign.

The *fifth principle* was that, in a crystal containing anions of more than one kind, those having the largest negative charge tend to be in the positions of greatest positive potential. Those having a smaller charge tend to be in positions of lower positive potential. This is the basis for Pauling's (1929) "electrostatic valence principle."

The *sixth principle* was that the atoms (or ions or molecules) tend to form as close-packed a structure as possible, consistent with other requirements.

The *seventh principle* was that the electron pairs in the valence shells of electronegative atoms or ions are oriented around the atomic kernels as if they repelled each other, often with shared electron pair bonds connecting pairs of atoms, as prescribed by the Lewis (1923) theory of valence, with modifications (Huggins, 1922b,f). This theory often determines (or reduces to a few possibilities) the close neighbor arrangements in crystals.

The *eighth principle* was that the interatomic distances between close neighbor atoms are usually approximately the sums of the appropriate atomic (or ionic or metallic or van der Waals) radii.

These principles are, I believe, still valid. Another principle, not pertinent to the structures being considered in 1931, proved to be very important later. It may be expressed as follows: hydrogen bonds will be formed to the maximum extent possible, whenever the groups appropriate for their formation are present.

Silicates, Borates, and the Structon Theory

After deducing the structure of quartz, I had planned to study silicate structures, expecting each silicon atom to be surrounded by (and connected by semipolar electron pair bonds to) four oxygens, with some or all of these oxygens connected by more polar electronpair bonds (or, perhaps, by purely ionic forces) to electropositive atoms.

I abandoned this plan, however, since I was busy with other problems and because good progress was being made by Sir Lawrence Bragg and his associates and by Linus Pauling and his co-workers. Both groups treated silicate crystals as assemblages of ions (Si^{+4}, Na^{+}, O^{--}, etc.).

Much more recently, I have been concerned with developing and applying my "structon" theory to glasses. In this theory I concentrate attention on the types and relative numbers of close neighbor arrangements ("structons"), as functions of the overall composition. To check my assumptions and to help in predicting *glass* structures, I have studied the structon types present in *crystalline* silicates and borates and some related compounds of known structure (Huggins, 1968b, 1970, 1971). The structon theory relationships should, I believe, be useful in understanding and predicting new (crystalline and glass) structures, including those for many compositions other than those I have dealt with.

Chain Molecules

One of the unsolved problems that I pondered about in my student days was an explanation for the fact that plots of the melting points and certain other properties of normal alkanes and their simple derivatives against the number of chain atoms showed an alternation of properties. Points for the compounds having an even number of chain atoms followed one smooth curve; points for the compounds having an odd number of chain atoms followed another, roughly parallel, curve. The proposed explanations in the literature did not seem reasonable to me.

I speculated that this behavior resulted from a slight polarity of each C-H bond, probably giving each hydrogen atom a small positive charge and each carbon atom a small negative charge. If so, there would not be strictly free rotation around each C-C single bond, as assumed by classical organic chemists, but there would be a slight preference for a zigzag chain structure. Chains with odd numbers of carbon atoms would pack together in crystals in a different manner from those with even numbers of carbon atoms. This reasoning explained the observed behavior.

I presented a paper dealing with this and related problems at a meeting of the American Association for the Advancement of Science (Huggins, 1922j), but I never submitted the paper for publication, because I discovered a paper by Pauly and Stark (1921), giving essentially the same idea, with new evidence for its correctness.

When I went to Stanford University in 1925, I met O. L. Sponsler, a botanist working temporarily in the physics department, and learned from him about his X-ray research (Sponsler and Dore, 1926, 1928) on the structure of cellulose. He showed it to be an assemblage of what I then called "long string molecules." This result interested me very much because the concept fitted well with my knowledge of Emil Fischer's work on polypeptides, my interpretation of the structures of crystalline selenium and tellurium, and my conclusions in 1923 about synthetic condensation polymers (see Huggins, 1972, p. 479).

Some years later, after becoming much involved in research on linear polymers, I published some papers dealing with the structures of those that are crystalline (Huggins, 1945). Some, such as polyethylene, have (like the shorter *n*-alkanes) a planar zigzag molecular structure. If such a structure is not possible, I predicted that (in order to place like groups in like surroundings) the chains should form spirals (helices). The evidence then available (and much more since then) showed agreement with this prediction.

Studying the structures of fibrous proteins (Huggins 1943c), I deduced various hypothetical helical structures, making use of published experimental data, plus the principles that I have just mentioned. The experimental data then available were insufficient to make decisions as to the detailed structure of any fibrous protein, but my main conclusions about the structures have since been amply verified (Huggins, 1952, 1958).

I have made further progress more recently, using some new structural principles pertinent to the packing of helical chain molecules (Huggins 1966, 1967, 1975, 1977, 1980a,b). I have presented evidence that the long-accepted concept that alpha-keratin and collagen are assemblages of ropelike groups of three helical chains coiled around a common axis is wrong. There are groups of three chains, but they have straight parallel axes. Like amino acid residues are helically arranged around the "3-stack" axis.

It seems likely that the detailed structures of the chief fibrous proteins will soon be known.

Conclusion

We marvel at the great progress that has been made in the development of methods for determining crystal structures. Good progress has also been made in finding and understanding the principles on which these structures are based.

References

Barlow, W. (1883). Nature (London) **29**, 186, 205, 404.
Barlow, W. (1898). Z. Kristallogr. **29**, 433-588.
Barlow, W. and Pope, W. J. (1906). J. Chem. Soc. **89**, 1675-1744.
Barlow, W. and Pope, W. J. (1907). J. Chem. Soc. **91**, 1150-1214.
Born, M. and Mayer, J. E. (1932). Z. Phys. **75**, 1-18.
Bragg, W. H. (1922). Proc. Phys. Soc., London, **34**, 98-103.
Bragg, W. H. and Bragg, W. L. (1916). X-rays and Crystal Structure. London: G. Bell and Sons.
Bragg, W. L. (1920). Phil. Mag. **40**, 169-189.
Bragg, W. L. (1924). Proc. Roy. Soc. (London) **105**, 16-39.
Bragg, W. L. (1929). Z. Kristallogr. **70**, 475-492.
Brockway, L. O. and Robertson, J. M. (1939). J. Chem. Soc. pp. 1324-1332.
Bury, C. R. (1921). J. Am. Chem. Soc. **43**, 1602-1609.
Cuy, E. J. (1927). J. Am. Chem. Soc. **49**, 201-215.
Davey, W. P. (1923). Phys. Rev. **22**, 211-220.
Davey, W. P. (1925). Chem. Rev. **2**, 349-367.
Fuson, R. C. (1924). J. Am. Chem. Soc. **46**, 2779-2788.
Goldschmidt, V. M. (1926). Geochemische Verteilungsgesetze der Elemente, Skrifter det Norske Videnskaps-Akad. Oslo I. Matem.-Naturvid Klasse.
Goldschmidt, V. M. (1928). Z. Phys. Chem. **133**, 397-419.
Goldschmidt, V. M. (1929). Trans. Faraday Soc. **25**, 253-283.
Groth, P. (1906). Chemische Krystallographie, Vol. 1, Leipzig: Engelmann.
Hendricks, S. B. (1930). Chem. Rev. **7**, 431-477.
Huggins, M. L. (1919). Some Speculations Regarding Molecular Structures, Term paper for advanced inorganic chemistry course, University of California (Berkeley). Unpublished.
Huggins, M. L. (1921). Phys. Rev. **18**, 333.
Huggins, M. L. (1922a). Science **55**, 459-461.
Huggins, M. L. (1922b). J. Phys. Chem. **26**, 601-625.
Huggins, M. L. (1922c). Phys. Rev. **19**, 346-353.
Huggins, M. L. (1922d). Science **55**, 679-680.
Huggins, M. L. (1922e). J. Am. Chem. Soc. **44**, 1607-1617.
Huggins, M. L. (1922f). J. Am. Chem. Soc. **44**, 1841-1950.
Huggins, M. L. (1922g). Phys. Rev. **19**, 363-368.
Huggins, M. L. (1922h). Phys. Rev. **19**, 369-373.
Huggins, M. L. (1922i). Phys. Rev. **19**, 354-362.
Huggins, M. L. (1922j). Disposition of the Atoms in Molecules of Simple Aliphatic Compounds, Paper presented at Am. Assoc. Adv. Sci. Meeting, Boston, Mass. Unpublished.
Huggins, M. L. (1923a). J. Am. Chem. Soc. **45**, 264-278.
Huggins, M. L. (1923b). Phys. Rev. **21**, 509-516.
Huggins, M. L. (1923c). Phys. Rev. **21**, 719-720.
Huggins, M. L. (1923d). Phys. Rev. **21**, 211-212.
Huggins, M. L. (1926a). Phys. Rev. **27**, 286-297.
Huggins, M. L. (1926b). J. Chem. Educ. **3**, 1110-1114.
Huggins, M. L. (1926c). Phys. Rev. **28**, 1086-1107.
Huggins, M. L. (1931). J. Phys. Chem. **35**, 1270-1289.
Huggins, M. L. (1932). Chem. Rev. **10**, 427-463.
Huggins, M. L. (1936a). J. Am. Chem. Soc. **58**, 694-695.
Huggins, M. L. (1936b). J. Chem. Educ. **13**, 160-165.
Huggins, M. L. (1937). J. Chem. Phys. **5**, 143-148.
Huggins, M. L. (1943a). J. Chem. Phys. **11**, 412-419.
Huggins, M. L. (1943b). J. Chem. Phys. **11**, 419-426.

Huggins, M. L. (1943c). Chem. Rev. **32**, 195-218.
Huggins, M. L. (1944). J. Chem. Phys. **12**, 520.
Huggins, M. L. (1945). J. Chem. Phys. **13**, 37-42.
Huggins, M. L. (1951). Phase Transformation in Solids. Ed. by R. Smoluchowski, J. E. Mayer and W. A. Weil. New York: Wiley, pp. 238-256.
Huggins, M. L. (1952). J. Am. Chem. Soc. **74**, 3963.
Huggins, M. L. (1953). J. Am. Chem. Soc. **75**, 4126-4133.
Huggins, M. L. (1954). J. Chem. Phys. **22**, 1389-1396.
Huggins, M. L. (1958). J. Polym. Sci. **30**, 5-16.
Huggins, M. L. (1966). Makromol. Chem. **92**, 260-276.
Huggins, M. L. (1967). Pure Appl. Chem. **15**, 369-389.
Huggins, M. L. (1968a). Structural Chemistry & Molecular Biology, edited by A. Rich and N. Davidson. San Francisco and London: Freeman, pp. 761-768.
Huggins, M. L. (1968b). Inorg. Chem. **7**, 2108-2115.
Huggins, M. L. (1970). Acta Cryst. **B26**, 219-222.
Huggins, M. L. (1971). Inorg. Chem. **10**, 791-798.
Huggins, M. L. (1972). Br. Polymer J. **4**, 465-489.
Huggins, M. L. (1975). Acta Cryst. **A31**, S37.
Huggins, M. L. (1977). Macromolecules **10**, 893-898.
Huggins, M. L. (1980a). Macromolecules **13**, 465-470.
Huggins, M. L. (1980b). Contemporary Topics in Polymer Science, 4. Edited by W. J. Bailey. New York: Plenum, in press.
Huggins, M. L. and Mayer, J. E. (1933). J. Chem. Phys. **1**, 643-646.
Huggins, M. L. and Sakamoto, Y. (1957). J. Phys. Soc. Jpn. **12**, 241-251.
Kelvin, Lord (1889). Proc. Roy. Soc. Edinburgh **16**, 693.
Körner, W. (1874). Gaz. Chim. Ital. **4**, 437-445.
Landé, A. (1920). Z. Phys. **1**, 191-197.
Langmuir, I. (1919). J. Am. Chem. Soc. **41**, 868-934.
Latimer, W. M. and Rodebush, W. H. (1920). J. Am. Chem. Soc. **42**, 1419-1433.
Lewis, G. N. (1916). J. Am. Chem. Soc. **38**, 762-795.
Lewis, G. N. (1923). Valence and the Structure of Atoms and Molecules. New York: Chemical Catalog Co.
Lonsdale, K. (1929). Proc. Roy. Soc. (London) **A123**, 494-515.
Neuberger, M. C. (1931). Z. Kristallogr. **80**, 103-131.
Pauli, W., Jr. (1925). Z. Phys. **31**, 765-783.
Pauling, L. (1927). J. Am. Chem. Soc. **49**, 765-792.
Pauling, L. (1928). Z. Kristallogr. **67**, 377-404.
Pauling, L. (1929). J. Am. Chem. Soc. **51**, 1010-1026.
Pauling, L. (1938). Phys. Rev. **54**, 899-904.
Pauling, L. (1939). The Nature of the Chemical Bond. Ithaca: Cornell University Press.
Pauling, L. and Huggins, M. L. (1934). Z. Kristallogr. **A87**, 205-238.
Pauly, H. and Stark, J. (1921). Z. Anorg. Allg. Chem. **119**, 271-298.
Sponsler, O. L. and Dore, W. H. (1926). Fourth Colloid Symposium Monograph. New York: Chemical Catalog Co.
Sponsler, O. L. and Dore, W. H. (1928). J. Am. Chem. Soc. **50**, 1940.
Wasastjerna, J. A. (1923). Soc. Sci. Fennica, Comm. Phys.-Math. 1, No. 38, 1-25.
Wyckoff, R. W. G. (1925a). Amer. J. Sci. **9**, 145-175.
Wyckoff, R. W. G. (1925b). Z. Kristallogr. **61**, 425-451.
Zachariasen, W. H. (1931). Z. Kristallogr. **80**, 137-153.

CHAPTER 7

HYDROGEN BONDING: SOME GLIMPSES INTO THE DISTANT PAST

Jerry Donohue
Department of Chemistry and Laboratory for Research on the Structure of Matter
University of Pennsylvania, Philadelphia, PA 19104

It should come as no surprise that there is confusion as to who first invented the hydrogen bond as we know it today. Credit for this is often given to Latimer and Rodebush (1920), although Pauling (1940, p. 285) cites a mostly ignored paper by Pfeiffer (1913) in which he "introduced the bond into organic chemistry," and an earlier paper by Moore and Winmill (1912). As this chapter is part of an historical volume, let us examine in detail what was really said in these above papers of 1912, 1913, and 1920.

The paper by Moore and Winmill is entitled "The State of Amines in Aqueous Solution," is 35 pages long, and is divided into two parts, part I, largely experimental, by Winmill, and part II, largely theoretical, by Moore. There are only two structural formulas in part II:

```
CH3  CH3        / CH3  CH3 \
  \  /         |    \  /    |
   N           |     N      |—OH
  /  \         |    /  \    |
CH3   H—OH      \ CH3  CH3 /
```

"thick strokes mean strong unions and thin strokes mean weak unions"

In trimethylammonium hydroxide a hydrogen bond (as now called – Moore did not use the term) between H_2O and $N(CH_3)_3$ was used to account for the fact that it was a weak base, compared to tetramethylammonium hydroxide.

The paper by Pfeiffer is entitled "Zür Theorie der Farblacke, II", and is 59 pages long. He gives the following structural formula.

```
          CO
     /\  /  \  /\
    |  ||    ||  |
     \/  \  /  \/
          C     |
          ‖     |
          O     O
           ` . /
             H
```

to explain the weaker bonding strength of compounds with C=O and –OH groups placed as shown with amines and metallic hydroxides. This, according to Pauling, "introduced the bond into organic chemistry."

In the paper by Latimer and Rodebush which is entitled "Polarity and Ionization" and is 14 pages long, we find the formula

```
  H
  ..  ..
H:N:H:O:H
  ..  ..
  H
```

in which "the hydrogen nucleus held between two octets constitutes a weak 'bond'." If the formula is strictly interpreted, the hydrogen atom appears to be forming two shared pair bonds of equal strength, in violation of the Pauli exclusion principle (which had not yet been postulated).

Although it might be argued that in all three of these papers hydrogen bonds appear, their impact on structural chemistry at the time was, in modern parlance, zilch.

Thus, shortly thereafter, there appeared a spate of papers on urea, which, as we know now is well-endowed with N-H-----O interactions, by Mark and Weissenberg (1923), Hendricks (1928), Wyckoff (1930, 1932), and Wyckoff and Corey (1934), just for starters on this much investigated substance. In none of these was hydrogen bonds between molecules mentioned. In Wyckoff's second (1932) paper, an electron density projection down the c-axis was presented, in the form of a photograph of an elegant and meticulously constructed plaster of Paris three dimensional model of $\rho(xy)$ - ah, those were the days! It was also stated that there was "no indication of the separate existence of hydrogen atoms outside of NH_2 groups." The work by Wyckoff and Corey (1934) was based on single crystal data obtained with a spectrometer, and said to be more accurate than previous work. Bond distances were given, but *no* intermolecular distances.

Usage of the term "hydrogen bond," or sometimes "hydrogen bridge" began to spread slowly in the early 1930's. One important early paper which was largely responsible for popularizing the term is that of Pauling (1935) who freely used it in accounting for the residual entropy of ice. Pauling cited the 1920 paper of Latimer and Rodebush "as the discovery of the hydrogen bond," but cites neither the 1912 paper of Moore and Winmill nor the 1913 paper of Pfeiffer. One is led to suspect that between 1935 and 1938, the date of preface to Pauling (1940), somehow he discovered or was led to the two earlier papers. Full flower came a bit later, especially in publications from CalTech on such complicated molecules as diketopiperazine (Corey 1938) and glycine (Albrecht and Corey, 1939) both with "hydrogen bonds," acetamide (Senti and Harker, 1940) with "N-H-O bridges," and DL-alanine (Levy and Corey, 1941) with "N-H-----O hydrogen bonds."

Since then hardly a year passes without a symposium or a book entitled "Hydrogen Bonding" or the like, dealing mostly, with N-H-----O and O-H-----O interactions, and in each issue of *Acta Cryst.* may be found more papers on them. The energies of these are of the order of 5 kcal/mole as compared with an order of magnitude larger for covalent bonds, and are thus not to be ignored, especially in complex natural products. Spectacular success was soon achieved by Pauling and Corey (1950) with their formulation of the α-helix for polypeptides. For a while thereafter *everything* was thought to consist of hydrogen-bonded helices, the culmination being the discovery by Watson and Crick (1953) of a helical structure for DNA, which, in a sense, is the illegitimate descendant of the α-helix.

Among the early symposia was a memorable one, part of the American Chemical Society meeting in 1951 in Cleveland, in which I was a participant. Just before the scheduled time of the start of the symposium I arrived at the appointed room to be confronted by a real mob scene: dozens of reporters and other media people, hundreds of the curious off the streets, plus a few chemists. It turned out that this session had been listed in the official program as "Symposium: Hydrogen Bombs." The chemists remained following the exodus when the chairman corrected the error, and we settled down to a cozy symposium which has been much quoted ever since, even now.

In closing, let me remark that many people have done very well working on hydrogen bonds and the complexities thereof. (Not as well as those into cancer, but pretty well nevertheless.) Even now there are many features which deserve further study, among them the stringency of linearity, the less frequently encountered bonds such as C-H-----O, symmetrical O-H-----O systems, bifurcation, disorder, and many more. Because of the importance of hydrogen bonding in biologically interesting compounds I do not foresee any diminution of research in this area for some time.

This work was supported in part by the National Institutes of Health, Grant GM 20611-04 and in part by The National Science Foundation MRL Program Grant DMR-7923647.

References

Albrecht, G. and Corey, R. B. (1939). J. Am. Chem. Soc. **61**, 1087.

Corey, R. B. (1938). J. Am. Chem. Soc. **60**, 1598.

Hendricks, S. B. (1928). J. Am. Chem. Soc. **50**, 2455.

Latimer, W. M. and Rodebush, W. H. (1920). J. Am. Chem. Soc. **42**, 1419.

Levy, H. A. and Corey, R. B. (1941). J. Am. Chem. Soc. **63**, 2095.

Mark, H. and Weissenberg, K. (1923). Z. Phys. **16**, 1.

Moore, T. S. and Winmill, T. F. (1912). J. Chem. Soc. **101**, 1635.

Pauling, L. (1935). J. Am. Chem. Soc. **57**, 2680.

Pauling, L. (1940). The Nature of the Chemical Bond, 2nd edition. Ithaca, N.Y.: Cornell University Press.

Pauling, L. and Corey, R. B. (1950). J. Am. Chem. Soc. **72**, 5349.

Pfeiffer, P. (1913). Ann. **398**, 137.

Sentl, F. and Harker, D. (1940). J. Am. Chem. Soc. **62**, 2008.

Watson, J. D. and Crick, F. H. C. (1953). Nature (London) **171**, 737.

Wyckoff, R. W. G. (1930). Z. Kristallogr. **75**, 529.

Wyckoff, R. W. G. (1932). Z. Kristallogr. **81**, 102.

Wyckoff, R. W. G. and Corey, R. B. (1934). Z. Kristallogr. **89**, 462.

CHAPTER 8

NEUTRON DIFFRACTION STUDIES OF ASYMMETRY IN SHORT HYDROGEN BONDS*

Jack M. Williams
Chemistry Division, Argonne National Laboratory
Argonne, Illinois 60439

Single crystal neutron diffraction studies were first reported nearly 30 years ago in the United States and since that time this method has served as the cornerstone for obtaining precise metrical information about the hydrogen bond (H-bond) in crystalline materials. Neutron diffraction played a seminal role in the investigations of the shortest and strongest H-bond known, i.e., in the bifluoride ion, $(F\text{-}H\text{-}F)^-$, as first reported by Peterson and Levy (1952). In this study the F---F distance was observed to be 2.26 Å and, more importantly, it was concluded that within the errors of the measurement the H-bond was symmetrical (centered). The symmetrical nature of the H-bond in $(F\text{-}H\text{-}F)^-$ has become accepted as intrinsic to this system and has significantly influenced later thought regarding the position of H-atoms in very short H-bonds. Thus, until the early 1970's, it was generally thought that most very short $(X\text{-}H\text{-}X)^{\pm}$ H-bonds were likely to be centered. However, one of the frequent problems that existed in diffraction studies of very short H-bonds was that the $(X\text{-}H\text{-}X)^{\pm}$ bond of interest is frequently positioned across a crystal symmetry element. The symmetry element limited the possible proton distributions and made it difficult to distinguish between an unsymmetric bond with superimposed disorder and the symmetric or centered proton bond. It should be pointed out that early investigators were very much aware that while the symmetry of an H-bond might be influenced by its environment, there were either insufficient materials known at that time on which to test this hypothesis, or available thermal neutron fluxes were insufficient in the 1950's to allow studies of more complex molecules. In addition, data collection was extremely tedious and time consuming. The latter problems were obviated with the development of high flux nuclear reactors and automated data collection techniques.

In the 1950's and 1960's neutron diffraction studies of H-bonds focussed on those of the exceptionally short O-H-O type (O---O less than 2.5Å) because of their rather unbiquitous nature. It was, of course, well understood that *long* H-bonds should not be symmetric, but difficulties arose when attempting to understand very short O-H-O bonds. From the results of many X-ray and neutron diffraction studies it had been observed that linear O-H-O bonds possessed O---O distances ranging from approximately 2.42 - 3.0Å. It also appeared that there was an almost predictable lengthening of the O-H distance as the O---O distance shortened until, it seemed, the H-atom appeared to be centered between the oxygen atoms. Many of these results were derived from X-ray studies from which it was not possible to precisely determine H-atom locations. A further complication was that O---O bonds were often nonlinear and, in addition, the true bond distances were often obscured by the thermal motions of the atoms. Thus, the rather smooth inverse variation of O---O and O-H distance led scientists in the field to attempt to establish *correlations* which relied largely on *measured* O---O distances. Probably what was the most recent of these correlations, but possibly the last, was published by Pimentel and McClellan (1971). They presented a plot of O-H distance vs O---O separation for those cases in which the H-atom had been located (see Fig. 1). From Fig. 1 it is clear that

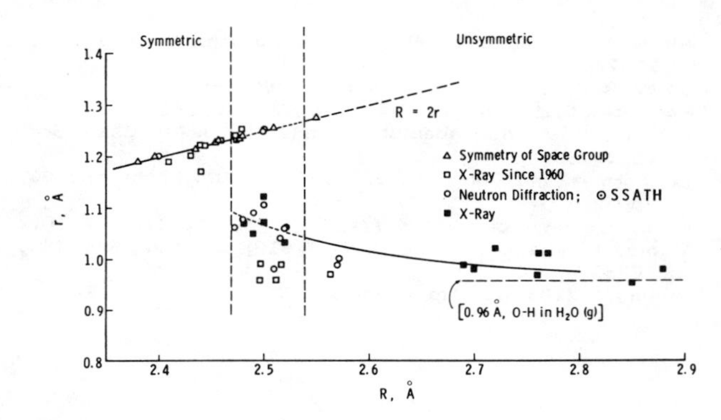

Fig. 1. A plot of O-H (ordinate) vs O---O distance for O-H-O bonds which indicated that "very short" (O---O ⩽ 2.5 Å) hydrogen bonds usually contained centered hydrogen atoms (From Pimentel and McClellan, 1971).

the points define two types of O-H---O bonds. The great majority of the hundreds of studies which had been reported, but not necessarily shown in Fig. 1, had O---O separations between 2.5 and 3.0 Å, with the average being 2.76 Å. Most of those had unsymmetrical H-bonds, i.e., the H-atom was closer to one oxygen atom than the other. In the group with O---O distances between 2.40-2.54 Å, it was suggested that the results consistently indicated that the "hydrogen atom is apparently symmetric between the two oxygen atoms" (Pimentel and McClellan, 1971).

When Fig. 1 was published in 1971 there was little reason to suspect that very short H-bonds might not be symmetric even when, for example, an O-H-O bond was situated in a very asymmetric environment in a crystal. In this regard, an important milestone neutron diffrac-

*Work performed under the auspices of the Division of Material Sciences, Office of Basic Energy Sciences, U. S. Department of Energy.

tion study had been that of potassium hydrogen chloromaleate, KHCM, by Ellison and Levy (1965). Although KHCM was chosen specifically because it possessed *no* intrinsic molecular symmetry, the neutron diffraction study revealed that the O-H distances were essentially identical at 1.206 and 1.199 $\pm$ 0.005 Å (see Fig. 2).

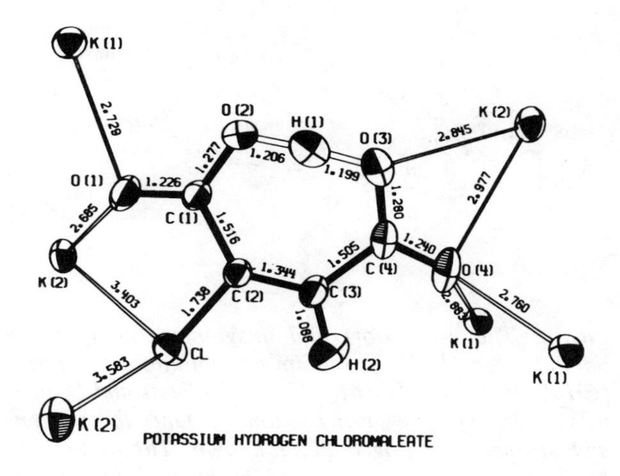

Fig. 2. The molecular structure of potassium hydrogen chloromaleate, $KH_2C_4O_4Cl$, (From Ellison and Levy, 1965). The atomic nuclei are represented by ellipsoids of thermal motion. Even though the molecule contains no inherent symmetry, and the O-H-O bond "sees" different environments, the hydrogen atom is symmetrically located in the hydrogen bond.

Thus, even in the absence of the mirror plane of symmetry, which was present in analogous potassium hydrogen maleate (Peterson and Levy, 1958), the bridging proton appeared to be located at the bond midpoint. At this point a detailed discussion of the neutron diffraction study of KHCM is most instructive. It should be stated at the outset that it is very difficult to assess in any *quantitative* manner the structural consequences of the differences in chemical environment about O (2) and O(3) (see Fig. 2), and the resultant effect on the potential energy function of the bridging H-atom. It is true that the Cl atom and all O atoms except O(2) have K^+ ion neighbors at contact distances nearly equal to or larger than the sum of the appropriate van der Waals radii, but all C–O bond distances that should be *chemically* equivalent are such when bond distances are compared. In order to determine if the potential energy function for the bridging H(1) is of the symmetric single minimum type one must turn to a detailed consideration of the thermal motion of H(1). In such an analysis one attempts to answer the question of whether or not the component of the mean square displacement of H(1) along the O---O bond line is due to *thermal* oscillation about a single point or alternately to the existence of equilibrium positions slightly displaced (disordered) about the bond center and randomly occupied in the lattice. From a detailed comparison of the mean square stretching amplitudes of the O-H bond derived from both diffraction and spectroscopic data, it was concluded that zero-point energy alone was sufficient to cause H(1) to move from one of the (assumed) equilibrium points to the other. The implication was that such a low barrier exists between the two sites in the double minimum potential that the former assumption of a static disorder was meaningless. Hence, it was finally concluded that H(1) was effectively centered and resides in an apparently symmetric potential well. Thus, the neutron study of KHCM and other very short H-bonds suggested that they may be symmetric even when the H-bonding environment about the bridging proton is asymmetric. Clearly, a meaningful test of this hypothesis required the investigation of H-bonds in which the H-bonding environment about an $(X\text{-}H\text{-}X)^+$ H-bond was very different.

The first clear-cut test of the effect of an exaggerated asymmetry about a very short H-bond came with the neutron diffraction study of sulfosalicylic acid trihydrate, SSATH, (Williams, Peterson and Levy, 1972). In this case the hydrated proton specie appeared to be an $(H_5O_2^+ \cdot H_2O)$ ion with an unsymmetrical H-bond in the $H_5O_2^+$ ion (see Fig. 3). In this compound there is a

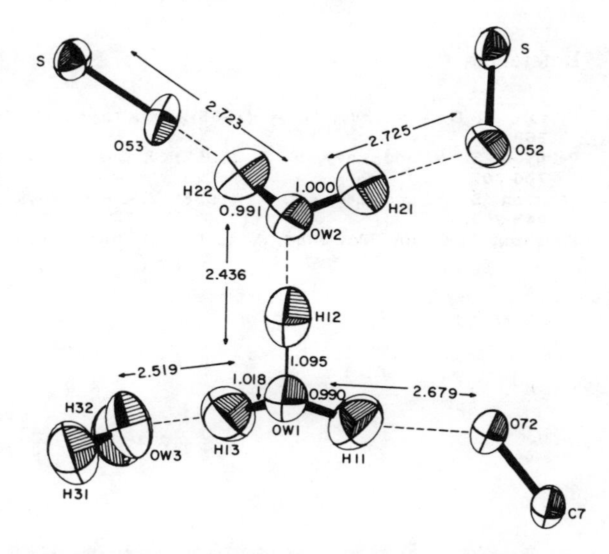

Fig. 3. The hydrogen bonds in the $H_5O_2^+ \cdot H_2O$ ion in sulfosalicylic acid trihydrate, C_6H_3 (COOH) (OH)SO_3 $H \cdot 3H_2O$ (From Williams, Peterson and Levy, 1972). The very different environments of water molecules W1 and W2 causes the hydrogen bond between them [OW1→H12→OW2] to be very unsymmetrical even though the O---O bond length is very short (2.436 Å).

very short and unsymmetrical O-H-O bond of length 2.436 Å while the H-----O bond length is 1.341 Å (O-H

1.095 Å). The hydrogen atom is located between two water molecules, the constituent assembly of which is an $H_5O_2^+$ ion, but the *external environments* of the two water molecules are markedly different. This was the first study in which a very short O-H-O bond was observed to be unsymmetrical largely because of environmental factors.

The most decisive test of the effect of environmental asymmetry on an $(X\text{-}H\text{-}X)^+$ bond came with the study of the $(F\text{-}H\text{-}F)^-$ bond in *p*-toluidinium bifluoride (Williams and Schneemeyer, 1973). In this compound the F---F bond length is 2.260(4) Å, the F-H distance is 1.025(6) Å, and the H-----F distance is 1.235(6) Å. More importantly, the H-bonding environments about F(1) and F(2) are extremely different with F(1) more rigidly bound by two N-H---F H-bonds of 1.607 and 1.675 Å length whereas F(2) is involved in only one N-H---F H-bond of 1.777 Å length (see Fig. 4).

Each of the hydrogen bonding studies discussed herein contributed greatly to our understanding of the effects of environmental asymmetry on the position of a H-atom in a H-bond. The quantitative assessment of these effects is a remaining challenge to this day.

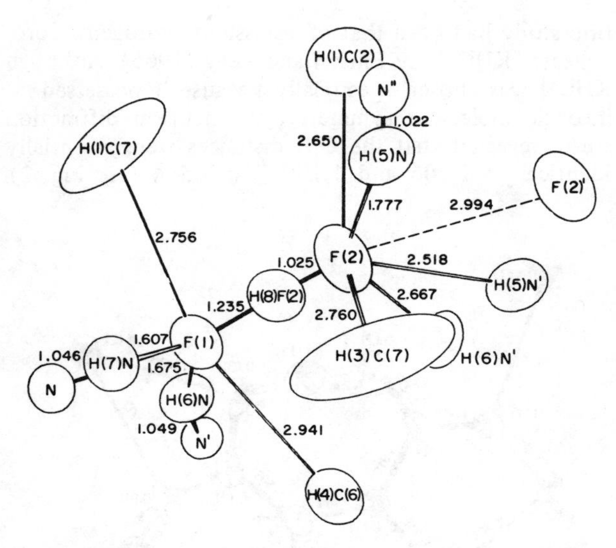

Fig. 4. The very short and unsymmetrical hydrogen bond in the (F-H--F)⁻ ion in p-toluidinium bifluoride $(CH_3C_6H_4NH_3)^+$ $(F\text{-}H\text{-}F)^-$ *(From Williams and Schneemeyer, 1973). The (F-H-F)⁻ ion contains the shortest and strongest hydrogen bond known. The asymmetry in the hydrogen bond is caused, in this case, by the very different N-H---F hydrogen bonding environments about the two fluorine atoms F(1) and F(2).*

References

Ellison, R. D. and Levy, H. A. (1965). Acta Cryst. **19**, 260-268.

Peterson, S. W. and Levy, H. A. (1952). J. Chem. Phys. **20**, 704-707.

Peterson, S. W. and Levy, H. A. (1958). J. Chem. Phys. **29**, 948-949.

Pimentel, G. C. and McClellan, A. L. (1971). Ann. Rev. Phys. Chem. **22**, 347-385.

Williams, J. M., Peterson, S. W. and Levy, H. A. (1972). Am. Crystallogr. Assoc. Program Abstr. Ser. 2, **17**, p. 51, Albuquerque, N.M.

Williams, J. M. and Schneemeyer, L. F. (1973). J. Am. Chem. Soc. **95**, 5780-5781.

SECTION F

APPLICATIONS TO VARIOUS SCIENCES

CHAPTER 1

CRYSTALLOGRAPHY AND GLASS STRUCTURE

Doris L. Evans
Research & Development, Corning Glass Works, Corning, New York 14830

Introduction

Glass, one of our oldest and most useful families of materials, occupies a very special place in the history of science. Optical properties provided the magnification, resolving power and dispersion which launched modern physics and astronomy while chemical inertness gave the unreactive containers and substrates required for chemistry and biology. Dimensional stability yields reliable thermometers as well as space-shuttle windows.

"Sweet" glasses are conforming and forgiving; with heat, mistakes are erased and shapes altered. Such glasses, with low permeability or high dielectric strength, begin a parade of technological progress: -vacuum systems, the incandescent and TV bulb as well as X-ray tubes being followed by glass capacitors and the like. The parade is a long one. Not-so-sweet glasses have a parade of their own. Being unforgiving their shapes are unaltered by heat. They generate the glass-ceramic family of fine-grained polycrystalline materials in which nucleation and sequential crystallization are controlled by composition and heat treatment.

It is therefore one of history's more pleasant ironies that glass continues today to elude precise structural definition for the same reasons that generated consternation and controversy in the earliest days of X-ray investigation.

We attempt first not to review the glass structure literature but to put it, and the major American contributions, in historical perspective. A few of the phenomena which have been controlled in the glassy state to yield new and useful materials are then briefly outlined. The hope is that the inherent limitations and bias of a structural approach that is restricted to diffraction will be clarified and the exciting possibilities that glass systems offer for reintegrating fragmented disciplines, controlling materials properties and extending the crystallographer's range, both conceptually and operationally, will be recognized.

Structure History

Dame Kathleen Lonsdale has recalled the surprise of X-ray pioneers when they learned that silk, rubber, paper and hair were "more crystalline" than glass. Their chagrin can readily be imagined for these pioneers were leaping from one scientific success to another, rapidly cracking the codes which translate observed diffraction patterns into the spatial distribution of atoms and molecules.

The liquid-like glass halo, first observed by Debye and Scherrer in 1916, was difficult to reconcile with the obvious 3-dimensional solidity of glasses. It must also have been frustrating to the point of disbelief to those who were beginning a new chapter of X-ray achievements with their study of fibrous substances. The discovery that even very thin fibers were bundles of crystals, all with the same axis parallel to the fiber axis but otherwise randomly oriented, marks the beginning of understanding fibrous minerals, long-chain polymers and biological substances. Glass however persisted in escaping structural definition; attempts to orient the glass halo, by drawing fibers, were not successful.

Reasoning exclusively from diffraction evidence, the structure of the silica polymorphs known at the time and the Scherrer formula which relates diffraction line breadth to crystallite size led Randall, Rooksby and Cooper (Randall et al., 1930) to conclude, in 1930, that the pattern of vitreous silica is accounted for by cristobalite crystallites about 15 Å in size. The problem of vitreous silica structure has since been a perennial favorite for diffraction analysis as experimental and mathematical techniques improved and structural knowledge expanded. As the high and low temperature structures of silica were determined, and the number of crystalline polymorphs increased, diffraction from different crystals was broadened to model the vitreous silica pattern. Depending on technique, models tried and the criteria of fit selected, alternative "best fits" abound in the literature. Fringe benefits from vitreous silica's intractability include structure determinations of more polymorphs, improved recognition of sources of experimental error and uncertainty, and the development of mathematical techniques to minimize spurious detail in the radial distribution function. The microcrystalline "best fits" remain however far from perfect, a not unexpected result if one considers the properties and behavior of vitreous silica. The dangers inherent in structural studies which ignore all properties except fitting the diffraction pattern to a microcrystalline model can be appreciated by doing a similar exercise with other properties. For example the average thermal expansion is fit best by β-quartz and the density best by β-cristobalite while the improved diffraction pattern is best fit by tridymite.

A rather fundamental but as yet unrecognized problem concerning accuracy in determining interatomic separations is raised if one examines the discrepancies between recent X-ray and neutron studies. These results show that, for reasons yet to be explained, we do not generate data that are free of distortion and sufficiently accurate to define the region of order intermediate between the short range order of the SiO_4 tetrahedron and the long range liquid-like continuum of statistically uniform density. The difference between the values of the Si-O separation is too large (Konnert and Karle, 1974; and Leadbetter and Wright, 1972). It corresponds to a physical density difference in excess of 5%, a difference that precludes accurate definition of intermediate range order. The scattering factor bias of X-rays against O-O pairs argues in favor of neutron studies. However adding high Q linac data to lower Q reactor data shifts the Si-O separation to larger values and also distorts the Si-O peak somewhat (Wright, 1979). The structure of vitreous silica is clearly mirroring for us unrecognized limitations of our methods. We have been somewhat slow to look into that mirror objectively, recognize the sources of error, uncertainty and distortion, and initiate the cooperation required to define and improve undistorted resolving power. Further diffraction studies, directed toward defining our resolving power, are suggested if crystallography is again to play the vital role American crystallographers assumed in the 1930's.

A structural interpretation of the scattering, alternative to microcrystals, was the random network hypothesis and glass building rules of Zachariasen (1932). The random network reflects inductive structural reasoning based on all the physical, chemical and diffraction data. A strong inference of the polymorphism of silica is that spatially continuous non-periodic networks, which include but need not be restricted to the ring topologies of the polymorphs, could account for all the properties of vitreous silica: the density, expansion, isotropy, liquid-like diffraction pattern and that region of the volume-temperature characteristic that distinguishes the glassy from the crystalline and liquid states. This region is the transformation range where glass structure "sets-up" during cooling and can be "fine-tuned" by annealing. It is the temperature region where a glass melt becomes rigid as the viscosity rapidly increases.

Further inferences originate in the low coordination, high charge and small size of the cation in the classical glass formers (SiO_2 and B_2O_3) and the 2-connectedness of the oxygen. It is inductive reasoning, seeking to weave together all the known facts to find the building rules governing glass formation, that characterizes the enormous contribution made by American crystallographers in the thirties to the science of glass. The rules of Zachariasen are similar to those of Goldschmidt and Pauling. While these rules are not the whole story they provide a structural basis for predicting glass forming systems. They also provide a conceptual framework for understanding the seeming enigmas of the glassy state.

An idealized network of vitreous silica would consist of tetrahedral building units, identical with those of the crystals, each making a single link at every vertex with another unit to form a completely connected space-filling 3-dimensional structure without lattice periodicity. The randomness originates in the geometries of mutual arrangement. In such networks the short range order can be everywhere identical, as it would be in the chemically simple glasses SiO_2 and B_2O_3, built of tetrahedral and triangular units respectively. The randomness of mutual arrangement is due to the small number of bonding constraints at the oxygen vertices. In terms of radial distributions both the number and kind of atom neighbors, and the directions in which they are located, become increasingly indeterminate with increasing separation of atom pairs.

Such a network was swiftly given structural support by experiments of B. E. Warren (Warren, 1933; and Warren, 1934). Practical support was already available. For example the property changes in vitreous silica caused by small amounts of Na_2O, in particular the more fluid viscosity-temperature characteristic, are accounted for by "breaking" the required number of tetrahedral vertex connections (non-bridging oxygen).

More support was forthcoming as the conception was put to practical use in predicting composition regions favoring glass formation. For example, adding Al_2O_3 to Na_2O-SiO_2 yields glasses with a less fluid viscosity-temperature characteristic. This also is as expected if the networks formed had increasing numbers of vertex-connected tetrahedral building units, AlO_4 and SiO_4, replacing those in which a SiO_4 link had been broken by the addition of Na_2O.

One aspect of the historic irony which characterizes the glass - X-ray determinations of structure relationship is that the crystallographic "wealth", accessible in the property measurements for a very large range of glass compositions, has been essentially ignored by contemporary crystallographers. Had the structural (i.e., coordination and connectivity) implications of these properties been recognized, it seems likely that silicate crystal chemistry as well as our understanding of the dissolution processes which result in a melt, glass forming compositions and non-periodic structures in general would have benefited.

Crystallography of the thirties and forties was a rapidly developing and interdisciplinary science equally concerned with the physics of interaction between X-rays and matter and discovering the rules by which structures were built and the properties of real materials determined.

The American contributions were fundamental, and far reaching. They reflect the urgency and excitement of those who could see that mysteries would be solved once the proper concepts and formalism were worked out. It is no accident that Patterson's (Patterson, 1934) recognition of what can be learned directly from $|F(hkl)|^2$ values and the application by Warren (Warren et al., 1936) of an approximate Fourier analysis method to vitreous SiO_2 and B_2O_3 diffraction are parallel developments.

The physics of X-ray scattering and the limitations on

our ability to clearly distinguish, or confidently correct for, signals that are not related to the cooperative scattering of structural significance was a most vital part of crystallography's development. Intrinsic limitations like the burying of structurally significant signals in Compton "noise" were recognized. Loss and distortion of information due to limitations in data collection and analysis, like the inevitability of spurious ripples due to series termination, were also recognized.

Structurally "messy" materials, like glasses and work hardened metals, present a challenge and some similar problems. The problems demand rigorous objective definition, constant recognition of wherein the tools (experimental and theoretical) are too blunt for unambiguous solution and development of improved tools and renewed attack. The contributions of Warren and his school are continuous and span a very wide range of disordered materials. Unambiguously extracting structurally significant information was always difficult. Critical awareness of sources of error, uncertainty and ambiguity was always demanded. The possibility of defining structures more clearly, with different or sharper methods, had always to be recognized and welcomed. Warren and co-workers have re-explored the structures of chemically simple glasses as they improved methods of detection, resolution and interpretation (Mozzi, 1967; and Warren and Mavel, 1965).

Their evaluation has been comprehensive and critical. The unsatisfactory situation for boron scattering factors, as well as the possibility that X-ray measured peaks of geometrically averaged electron density may be slightly displaced from the internuclear distances, has been noted (Mozzi, 1967).

Specific models proposed over the years have been systematically ruled out while geometrically random networks of chemically ideally ordered building units, BO_3 triangles and SiO_4 tetrahedra, have always survived. In vitreous B_2O_3, it seems possible that the 3-membered rings of BO_3 triangles, the $(B_2O_3)^{-3}$ boroxyl ring of Krogh-Moe, are a fundamental unit of network construction, these units being linked to each other by BO_3 triangles.

The difficulty with random networks is that they must be constructed physically. The models must contain a large number of building units to confirm that space-filling random construction could be continued indefinitely and also yield reasonable statistics for the various geometries of mutual arrangement. The random rule must be specified as, for example, the range of Si-O-Si angle in vitreous silica. The scale and regularity of the building units should be such that model co-ordinate measurement has the accuracy required for radial distribution and density calculations. Until the 1960's, when computer capabilities justified model building (Ordway, 1964; Bell and Dean, 1966; and Evans and King, 1966), the random network was a hypothesis. The model distribution of Evans and King (1966) was used by Mozzi (1967) to interpret his X-ray R.D.F. He found "a remarkably good match." This model has since been manipulated, by computer, with constraints imposed to generate density differences while maintaining tetrahedrality of the units and the 4:2 connectivity scheme (Evans and Teter, 1977).

Scientific Significance

The broad scientific importance of glasses lies in the experimental bridge they offer to span the gulf between melts or liquids and the crystalline solid state. The large number of materials families that have been derived from the glassy state illustrate a few of the fundamental phenomena that can occur.

The glass formers SiO_2 and B_2O_3 are miscible in all proportions while Na_2O-B_2O_3-SiO_2 glasses are the starting point for Vycor® brand silica (96%). The process depends upon control of phase separation with spatial continuity of the Na_2O-B_2O_3 phase. Being more soluble than SiO_2 this phase can be leached out by acid leaving a material with interconnected channels. The channels can then be filled with something else, used for chromatographic separation, or heated such that the channels collapse and the material shrinks, without change of shape, to yield the dense 96% vitreous silica body called Vycor™.

Opal glasses, either translucent or opaque, demonstrate increasing supersaturation upon cooling with the dispersed phase separated as fine fluorine rich droplets. These may crystallize to NaF or CaF_2 upon further cooling. Another class of opal glasses requires reheating to develop the phase separation while yet another class is generated by fine, uniformly dispersed gas bubbles.

There is a very large family of light or radiation sensitive glasses whose uses extend from chemical machining to eye glasses whose transmission of visible light decreases with light intensity. Light induced nucleation of dendritic Li_2SiO_3 crystals, which are more soluble in HC1 than the residual glass, makes very fine chemical machining possible using photographically reduced masks. Another family with a fine dispersion of silver halide crystallites makes possible, in glass, what is essentially the chemistry of a reversible latent image formation in photography. These are the photochromic glasses. The local trapping of reaction products makes the reaction reversible when irradiation stops. The dynamics of the glass-silver halide interface is such that light generates a silver rich surface and heat dissipates it.

Glass ceramics are generally chemically complex aluminosilicate glasses whose crystallization sequence is controlled through composition and heating schedule. They yield a uniformly fine-grained polycrystalline body that is more than 50% (and often more than 90%) crystalline. The crystal phases are often unnatural minerals, being solid solutions of silica in a mineral silicate structure. The crystallization sequence may also be unnatural, reflecting both the energy available and the structural paths accessible at any temperature.

The rapid development of these and other materials is due in part to the interweaving of short-range order or coordination rules determined by crystal structure analysis. It is also due to the recognition that the same rules may not be operative at high temperature or when a particular chemical species is either very dilute, physically constrained by the surrounding matrix or chemically constrained by species competitive with it. The challenge presented is clearly demonstrated by the phenomena and property changes. It is a lively one which should force us to evaluate and sharpen our resolving power and also invent the missing discipline called, by J. D. Bernal, "statistical geometry."

The future offers possibilities as exciting as those of X-ray crystallography's early days.

References

Bell, R. and Dean, P. (1966). Nature (London) **212**, 1351.

Evans, D. and King, S. (1966). Nature (London) **212**, 1353.

Evans, D. and Teter, M. (1977). In "The Structure of Non-Crystalline Materials," pp. 53-57, Edited by P. H. Gaskell, London: Taylor and Francis, Ltd.

Konnert, J. H. and Karle, J. (1974). Trans. Am. Crystallogr. Assoc. **10**, 29-43.

Leadbetter, A. J. and Wright, A. C. (1972). J. Non-Cryst. Solids **7**, 141-155.

Mozzi, R. L. (1967). Doctorate Thesis MIT, Physics.

Ordway, F. (1964). Science **141**, 800.

Patterson, A. L. (1934). Phys. Rev. **46**, 372.

Randall, J. T., Rooksby, H. P. and Cooper, B. S. (1930). J. Soc. Glass Tech. **14**, 219.

Warren, B. E. (1933). Z. Kristallogr. **86**, 349.

Warren, B. E. (1934). J. Amer. Cer. Soc. **17**, 249.

Warren, B. E., Krutter, H. and Morningstar, O. (1936). J. Amer. Cer. Soc. **19**, 202.

Warren, B. E. and Mavel, G. (1965). Rev. Sci. Instrum. **36**, 196.

Wright, A. C. (1979). Private communication.

Zachariasen, W. H. (1932). J. Am. Chem. Soc. **54**, 3841.

CHAPTER 2

LIQUID CRYSTALS

Adriaan de Vries
Liquid Crystal Institute, Kent State University, Kent, Ohio 44242

Introduction

Before going into the actual "history" of liquid crystal research, it seems appropriate to first review briefly what liquid crystals are. Liquid crystals, or mesophases, are phases exhibited by certain compounds or mixtures of compounds, in between the solid crystalline phase and the isotropic liquid phase. The "in between" of the previous sentence can refer to temperature, concentration, or to any other parameter used to effect the transition from solid to liquid. For instance, sodium stearate is a solid at room temperature, but at a higher temperature it becomes a liquid crystal, and at a still higher temperature it becomes an isotropic fluid. Or, one can obtain a liquid crystal from sodium stearate by adding water: at a certain water concentration the mixture becomes a liquid crystal, and at a higher water concentration it becomes an isotropic fluid. In the first example the liquid crystal phase obtained is called thermotropic, in the second example it is called lyotropic.

Lyotropic liquid crystals are generally formed by mixing an amphiphilic substance with water or oil (additional components may also be introduced). With a large amount of water, e.g., one obtains a true solution. If the water concentration is decreased, micelles are formed, and one has a micellar solution. At a still lower water concentration, the micelles become larger and pack in a regular manner, forming a liquid crystalline phase. The main water-containing liquid crystalline phases are: lamellar phases (bilayers of the amphiphilic molecules alternating with water layers), hexagonal phases (hexagonal arrangements of rod-like micelles with water in between), and cubic phases (e.g., a cubic arrangement of spherical micelles with water in between).

Thermotropic liquid crystals are obtained by changing the temperature of certain compounds (e.g., para-azoxyanisole) or mixtures of such compounds. Again, additional components may be introduced to modify the properties of the material. The most important feature of compounds giving thermotropic liquid crystal phases is that the molecules have to be pronouncedly anisotropic in shape. Originally, one only used rod- or lath-like molecules, but more recently it has been found that disc-shaped molecules also can give liquid crystal phases. In both cases one uses virtually exclusively organic molecules. According to Gray (1962a, p. 11), approximately one in every two hundred organic compounds has one or more liquid crystal phases. Long-range order in the alignment of these anisotropically shaped molecules (i.e., parallelism of their long axes or their planes) is the basic feature of all thermotropic mesophases.

The thermotropic phases of rod-like molecules are classified as nematic, cholesteric, or smectic, depending on the arrangement of the molecules. In nematic phases, the only restriction on the molecular arrangement is the above mentioned parallelism of the long axes. In cholesteric phases, the long axes are twisted into a helical pattern. In smectic phases, the molecules are parallel and arranged in layers. Different ways of arranging molecules in layers (e.g., the average direction of the molecular long axes can be perpendicular or tilted with respect to the plane of the layer, and the molecules can be packed in a regular or irregular manner) lead to different types of smectic phases; these are most commonly identified as smectic *A*, smectic *B*, smectic *C*, etc.

Broad overview of historical development

Liquid crystal research started with the discovery of "liquid crystalline behavior" in 1888 by the Austrian botanist F. Reinitzer, and a discussion of this behavior in 1890 by the German chemist O. Lehmann. Interest in this field of research remained first limited to Europe, with O. Lehmann (Germany, 1890-1923)*, D. Vorländer (Germany, 1906-1938), G. Friedel (France, 1907-1931), F. Grandjean (France, 1911-1921), T. Svedberg (Sweden, 1914-1918), and L. S. Ornstein (The Netherlands, 1917-1940) as some of the main people involved in the initial period. The first publication originating from the United States appears to be one by E. C. Bingham and G. F. White (Richmond College, 1911),** but this paper deals with liquid crystals only in a peripheral manner and the authors appear not to have published any further papers on liquid crystals. The first real research publications on liquid crystals from the U.S. that I have been able to locate are two papers by J. S. van der Lingen (Johns Hopkins University, 1921), who had earlier published a paper on liquid crystals in Switzerland (1913).

Of particular interest to the readers of this book will be that several noted crystallographers from the "early years" have published papers dealing with liquid crystals:

---

*The years indicate the period during which the author published papers on liquid crystals.

**To identify who has done the work, and where and when, and to allow the reader to find the corresponding reference fairly easily in the literature, we have always given authors' names, place, and year of publication. To distinguish these entries from those pointing to references at the end of the paper, we have marked the latter by a lower case *a, b,* or *c* directly after the year of publication.

G. Friedel (1907-1931), who formulated Friedel's Law; C. Mauguin (1911-1913) and C. Hermann (1931), originators of the Hermann-Mauguin symbols which have been adopted as the standard space-group symbols in the *"International Tables for X-ray Crystallography"*; J. D. Bernal (1933-1941); D. Crowfoot (1933), now known as D. Hodgkin; W. Bragg (1934); and I. Fankuchen (1937-1941).

Figure 1a illustrates the growth of liquid crystal research over the years, as measured by the number of papers, reports, patents, etc., per year. This growth shows a quite interesting behavior. After an initial brief period with 0-6 entries per year (1888-1904), there is first a significant growth (to 20 entries in 1908), but then there comes an extremely long period (1908-1959) of no sustained growth and fairly low output (the average over the period 1905-1959 is less than 13 entries per year). Next, there is a brief period (1960-1963) of somewhat greater output (averaging at about 27), and then comes a decade of extremely sharp growth: from 43 entries in 1964 to 1111 in 1973! After that (1973-1978), the output appears to have remained fairly constant.

Some items that probably have contributed to the phenomenal growth of liquid crystal research from 1964 to 1973 are the following. (*1*) The publication of a review paper by Brown and Shaw (1957*a*, University of Cincinnati) with 475 literature references, which made the liquid crystal literature through 1955 easily accessible. (*2*) The increasing awareness of the unique electro-optical properties of nematic liquid crystals, evidenced, e.g., in a government report by J. M. Ruhge and D. Green in 1959. (*3*) The publication of extensive tables of transition temperatures by W. Kast (Germany, 1960), which made available a comprehensive list of compounds and their phases. (*4*) The book by Gray (1962*a*, England), which provided an additional extensive review of the literature, in particular information about the relationship between molecular structure and mesomorphic behavior. (*5*) The discovery of the usefulness of the color changes of cholesteric liquid crystals for temperature measurement, e.g., a patent by J. L. Fergason, T. P. Vogel, and M. Garburg (Westinghouse Electric Corporation, 1963). (*6*) The increasing interaction between academic and industrial researchers on an international level, e.g., in the International Liquid Crystal Conferences, started at Kent State University in 1965.

The apparent leveling-off of the output of liquid-crystal-related work after 1973 (see Fig. 1a) probably represents a saturation effect. The second set of data in Fig. 1a, however, the full circles, suggests that after 1974 we have entered another period of steady growth. A reason for this may be the growing appreciation of the relationship of liquid crystals to human diseases such as atherosclerosis, gallstones, and sickle cell anemia, and to general life processes (Brown and Wolken, 1979*a*; Small, 1977*a*; Stewart, 1974*a*; see also section on Membranes).

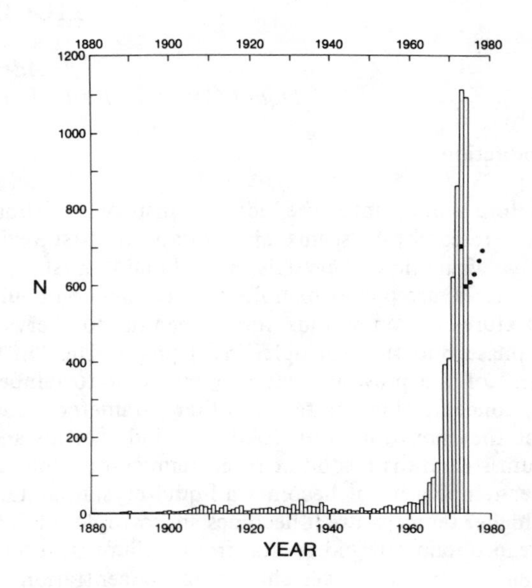

Fig. 1a.

Further significant increases might be expected if, e.g., important industrial applications for smectic liquid crystals are found.

Figure 1b illustrates the growth of *X-ray* work on liquid crystals (please note the large difference in the

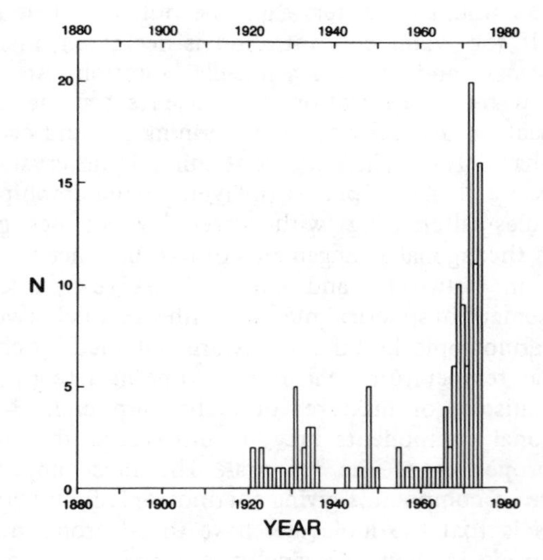

Fig. 1b.

vertical scales of the two graphs, and that Fig. 1b only refers to papers which have the word "X-ray(s)" in the title). The first entries in Fig. 1b appear in 1921, some 30 years after the first entry in Fig. 1a, but, just as in Fig. 1a, we find in Fig. 1b an extended period without sustained growth (1921-1965) followed by a sharp increase (1966-1972). The increase is much less dramatic, however, than in Fig. 1a, and the number of X-ray contributions as a percentage of the total declines drastically, from about 7% over the period 1921-1959, to only 1.5% for the years 1972-1974. The less dramatic rise in Fig 1b may be attributed to the fact that X-ray work on liquid crystals has generally been in the area of "basic research," and as such has been less affected by the increased interest in practical industrial applications of liquid crystals.

Early work on X-ray diffraction

As noted above, the United States trailed Europe in liquid crystal research by about 30 years. In the more narrow field of X-ray studies of liquid crystals, however, the U.S. got off to a much better start: Of the first two papers (1921; Fig. 1b), one is from the United States (J. S. van der Lingen, Johns Hopkins University), the other from Germany (E. Hückel, Göttingen). After that, limiting ourselves now to authors with three or more papers, come J. W. McBain (University of Bristol, England, 1924; Stanford University, California, 1943-1948), W. Kast (Germany, 1927-1934), K. Herrmann and A. H. Krummacher (Berlin-Charlottenburg, Germany, 1930-1935), and G. W. Stewart (University of Iowa, 1931-1936). No others appear until 1955!

Although the *number* of X-ray publications has always been only a small fraction of the total (see above, and Fig. 1), their *influence* on the understanding of the structure of liquid crystals has always been considerable: E. Hückel (1921) found that the X-ray diffraction patterns of the nematic and isotropic liquid phases are very similar, leading to the conclusion that the short-range "structure" of the two phases must be nearly identical, and that the nematic phase has no regular spatial arrangement of the molecules. M. de Broglie (1923) and G. Friedel (1923, 1925) presented strong evidence confirming the idea that the molecules in smectic phase and D. Vorländer (organic chemist) who showed that these layers have a thickness approximately equal to the length of the molecules. K. Herrmann (1935) presented conclusive evidence that there are *at least two* types of smectic phases, one in which the molecules in each layer are arranged in a random manner, and one in which they are arranged in a regular two-dimensional lattice presumably corresponding to a hexagonal close-packing of cylinders. Herrmann's work was an important factor in resolving the dispute going on at that time between G. Friedel (physicist) who on theoretical grounds argued there could be only *one* smectic phase, and D. Vorländer (organic chemist) who on the basis of experimental observations concluded there were *more* smectic phases.

As far as diffraction studies on liquid crystals in the United States are concerned, the first "series" of papers (i.e., three or more) came from G. W. Stewart and co-workers (Laboratory of Physics, University of Iowa, 1931-1936). It is of interest to note here that Stewart appears to have been the only American invited to participate in the famous Discussion of The Faraday Society in London in 1933, on "Liquid Crystals and Anisotropic Melts." This meeting was a high point of the "early years," and its occasion probably caused the maximum in Fig. 1a for 1933, which was not equalled or surpassed until 1964! The main subject of Stewart's work was the comparison of the diffraction patterns of the nematic and isotropic phases of para-azoxyanisole. The results confirmed the earlier conclusion of Hückel (see above) and others, that these diffraction patterns differ only slightly.

The second series of papers is from J. W. McBain and co-workers (Chemistry Department, Stanford University, 1943-1948). Their interest was in lyotropic systems, in particular those of soaps and detergents, and part of their work involved the use of small-angle scattering (in lyotropic systems one frequently encounters spacings of more than 50 Å). Various hexagonal and lamellar phases were found by them.

In more recent years, liquid crystal X-ray diffraction papers from the United States and Canada have become numerous and varied (in accord with the overall trend; see Fig. 1b), and it appears best to deal with them in a number of separate categories: thermotropic phases, disc-like molecules, lyotropic phases, biological membranes, polymers, and crystal structures.

Thermotropic liquid crystals

The first work to appear after Stewart's (1931-1936), referred to above, is that of G. H. Brown and co-workers (University of Cincinnati, 1959; Kent State University, 1968). These authors again studied the diffracted intensity as function of the diffraction angle for the nematic and isotropic phases of a number of compounds. The main result of their studies appears to be the "herringbone packing" model proposed by L. W. Gulrich and G. H. Brown (1968): neighboring molecules, viewed in the direction of their long axes (which are parallel), pack in such a way that the molecular planes are arranged in a herringbone pattern. Although this model—proposed for a nematic phase—does not appear to have found general use for *nematic* phases, it is now widely accepted for most *smectic* phases which have an ordered arrangement of the molecules within the layers, viz., for smectic *E, H.* and *B.* Also, it has recently been used by DeVries (1979*a*, 1981*a*) as the basis for a complete structural classification of all smectric liquid crystals.

Next is A. J. Mabis (Procter and Gamble Company, 1962), with a beautifully illustrated paper on the classification scheme for mesophases published by C. Hermann in 1931. C. C. Gravatt and G. W. Brady (Bell Laboratories, 1969) did the first small-angle studies of thermotropic liquid crystals, comparing the isotropic and liquid crystal (nematic or cholesteric) phases of two compounds. They found no difference, except for one compound when the samples were not very pure.

A. de Vries and co-workers (Kent State University, 1969-1981) are the oldest still active group in the U.S., and, consequently, have published more papers than any other group in the country. De Vries's first two papers (1969, 1970) contributed to a significant change in the understanding of the structure of smectic phases. Up to that time, it was quite generally assumed that in *all* smectic phases (with the possible exception of those of thallium palmitate and stearate; K. Herrmann, 1933) the molecules are, on the average, *perpendicular* to the smectic layers. An X-ray paper from the U.S.S.R. (I. G. Chistyakov, L. S. Schabischev, R. I. Jarenov, and L. A. Gusakova, 1969), the X-ray work of De Vries, and three papers on microscopic studies from Kent State University (S. L. Arora, J. L. Fergason, A. Saupe, 1970; T. R. Taylor, J. L. Fergason, S. L. Arora, 1970; T. R. Taylor, S. L. Arora, and J. L. Fergason, 1970), however, showed conclusively that in phases classified as smectic *C* the molecules are *tilted.* The same papers by De Vries also showed that certain nematic phases (skewed cybotactic nematic phases) have a very strong smectic-*C*-like short-range order that can persist over a very long temperature range (further detailed results have been published in 1980 by De Vries and Qadri, and by Sethna, De Vries, and Spielberg). In another paper in 1970, De Vries presented the first accurate data on the thickness of smectic layers and on the average intermolecular distance in smectic *A* phases. He showed that in the smectic *A* phase the layer thickness was not equal to the molecular length, as previously believed, but somewhat less. The reason for this was not found until much later (1979; see below). The paper also stressed that the low value for the intermolecular distance means that in none of the phases can there be free rotation of the molecules around their long axes. A similar paper on a series of other compounds (1973) confirmed the data from the earlier paper, and also pointed out that in the isotropic phase, just above the nematic-isotropic phase transition, the molecules appear to be predominantly in the fully extended conformation.

In 1972, De Vries and Fishel gave the first detailed and extensive evidence of the existence of a three-dimensional lattice in a smectic phase: *forty* different *hkl* reflections were observed and indexed. This phase is now classified as smectic *G.* Also in 1972, De Vries published two papers showing that the calculation of molecular cylindrical distribution functions from X-ray diffraction data on liquid crystals, a procedure used in many Russian and French papers since 1963, is not allowed. This conclusion appears to have been tacitly accepted, for very few papers have used the method since. In 1977 and 1979, De Vries and co-workers pointed out that comparisons of layer thickness with molecular length suggest that there are smectic *A* phases in which the molecules are tilted rather than perpendicular. S. Diele, P. Brand, and H. Sackmann (East Germany, 1972) had earlier noted such discrepancies between layer thickness and molecular length, but had not drawn a definite conclusion. Further studies (including work by A. J. Leadbetter, England) led De Vries, Ekachai, and Spielberg (1979) to conclude that in *all* smectic *A* phases the molecules are tilted (with an average angle of about 20°), and that these phases owe their optical uniaxiality to the randomness of the direction of this tilt rather than to an absence of tilt. This "orientational disorder" model was subsequently expanded to the smectic *C* phase (De Vries, 1979*b*).

In 1972 and 1973, W. L. McMillan (Bell Laboratories) published four papers on the results of very detailed and precise measurements of X-ray diffraction intensities as a function of scattering angle for a number of liquid crystal phases. The first paper presents data at several temperatures in the smectic *A*, cholesteric, and isotropic phases of two cholesterol derivatives. The intent was to test McMillan's theoretical model for the smectic *A* phase, but no definite conclusions could be drawn. Short-range-order effects were observed in the cholesteric and isotropic phases. The second paper describes the determination of smectic-*A*-phase order-parameter fluctuations from X-ray diffraction data in a nematic phase near a second-order nematic-smectic *A* phase transition. The third paper does the same for a *first*-order nematic-smectic *A* phase transition. The measurements were fitted by the Landau-theory expression and confirmed both the temperature and the momentum dependence of the Landau free energy. In the fourth paper, the smectic-*C*-phase order-parameter fluctuations in a skewed cybotactic nematic phase are described and analyzed in terms of a Landau theory. McMillan, now at the University of Illinois at Urbana, appears to have published no further papers on X-ray diffraction of liquid crystals.

In 1973 several new names appear in the literature. W. R. Krigbaum, J. C. Poirier, and M. J. Costello (Duke University) presented a very intriguing analysis of the diffraction intensity data of three smectic *A* phases. They conclude that the standard smectic *A* model—i.e., molecules essentially perpendicular to the planes—does not fit any of their phases, even though for two of them the smectic layer thickness is approximately equal to the molecular length. For one compound, a lack of sufficient data made further conclusions impossible, but for the other two a good fit was obtained with dimer units making a considerable angle with the normal to the planes. For one compound, thallium stearate, this model is in basic agreement with the earlier model of Herrmann (1933), and the apparent conflict with the standard smectic *A* model can be removed by noting that the thallium stearate phase might not be a smectic *A* phase. For the other compound, however, the conflict remains. So far, Krigbaum's work has neither been confirmed nor disproved.

M. V. King and M. Young (Massachusetts General Hospital and Harvard Medical School, 1973) describe a specimen cell for studying oriented liquid crystals by X-ray diffraction and polarized light. J. H. Wendorff and F. P. Price (University of Massachusetts, 1973) studied the mesophases and isotropic phases of several cholesteryl esters. They conclude that in the isotropic phase, just above the transition point, the molecules are still essentially extended, that the cholesteric phases have significant smectic-*A*-like short-range order, that in the smectic phase the thermal vibrations of the molecules perpendicular to the layers are very large, and that in all phases the molecules pack in an antiparallel fashion. A. de Vries (Kent State University, 1973) proposes a preliminary classification of smectic phases (*A* through *H*) on the basis of their X-ray diffraction patterns, and discusses (1975) the diffraction patterns of all liquid crystal phases with reference to classification.

R. Pynn (Brookhaven National Laboratory, 1975) discusses X-ray and neutron diffraction data from the nematic and isotropic phases of para-azoxyanisole; this appears to be the only neutron study done in the United States. B. M. Craven and G. T. DeTitta (University of Pittsburgh, 1976) studied the smectic and cholesteric phases of cholesteryl myristate. Based on their determination of the structure of this compound in the solid crystalline state they propose, contrary to conventional understanding and the conclusions of Wendorff and Price (see previous paragraph) and those of Loomis, Shipley and Small (see section on Lyotropic Liquid Crystals), that the small-angle diffraction maximum found in the smectic phase (and, to a much lesser degree, also in the cholesteric phase) is not caused by a periodic layer structure but rather by pairs of antiparallel molecules with overlapping myristate chains. This quite novel idea has neither been confirmed nor disputed, but further studies by Craven and co-workers on other cholesteryl esters (see section on Crystal Structures) do not appear to lend any support to this hypothesis.

Beginning in 1977, R. J. Birgeneau, J. D. Litster, and co-workers (Massachusetts Institute of Technology) have published a series of papers on "high-resolution X-ray studies." The first three papers (1977-1979) discuss the critical behavior associated with the (nearly) second-order nematic-smectic *A* phase transitions in three cyano compounds. The authors conclude that the pretransitional behavior in the nematic phase is essentially consistent with the superfluid He^4 analogue proposed by de Gennes, and that the anomalous behavior shown by the elastic constants in the smectic phase is probably associated with logarithmically divergent phase fluctuations. In the fourth paper (Safinya et al., 1980*a*), tilt angle and layer thickness are measured near a second-order smectic *A*-smectic *C* phase transition, and it is concluded that the results support a simple molecular-tilt model for the transition. The authors appear not to have taken into account, however, the evidence presented by De Vries and co-workers in their studies of several smectic *A* phases and one smectic *A*-smectic *C* phase transition (De Vries, 1979*b*, and another paper in 1979), which indicates that in smectic *A* and *C* phases the deviations from such a "simple model" are quite significant.

A short note by A. C. Griffin and J. F. Johnson (University of Southern Mississippi and University of Connecticut, 1977) draws our attention to another area of current interest. The authors point out that mesogenic compounds with a hydrocarbon chain at one end of the molecule and a CN or NO_2 group at the other end show the following unusual properties: *(1)* The layer thickness in the smectic *A* phase is *longer* than the molecular length, rather than shorter, as is usually the case; G. W. Gray and J. E. Lydon (England, 1974) have proposed that these phases have a bilayer structure with partially overlapping molecules. (*2*) Mixtures of these compounds exhibit "re-entrant nematic" phases (P. E. Cladis, Bell Laboratories, 1975), i.e., phase diagrams with the phase sequence nematic-smectic *A*-nematic (Note: more recently also pure compounds have been found to show this behavior). (*3*) Mixtures with other mesogenic compounds may give enhanced smectic phases, i.e., smectic phases with a far longer temperature range than for the pure components (J. W. Park, C. S. Bak, and M. M. Labes, Temple University, 1975). We may point out here that the first three papers from Birgeneau et al. (see above) also dealt with compounds of the type discussed by Griffin and Johnson. D. Guillon, P. E. Cladis and J. Stamatoff (Bell Laboratories, 1978) report X-ray data on the re-entrant nematic phase of a mixture of two such compounds, and suggest that this nematic phase is similar to the classical nematic phase but may coexist with crystalline fluctuations. Another paper on such compounds, by P. E. Cladis, D. Guillon, W. B. Daniels, and A. C. Griffin (Bell Laboratories, University of Delaware, and University of Southern Mississippi, 1979), confirms that the layer thickness in the smectic *A* phase exceeds the molecular length, and proposes a new model for the structure of the bilayers. D. Guillon, P. E. Cladis, D. Aadsen and W. B. Daniels (Bell Laboratories and University of Delaware, 1980) report the first measurements of the smectic *A* layer spacing at high pressure for a compound of this type. They find that at relatively high temperatures the smectic layers are virtually incompressible. Another 1980 paper from the Bell Laboratories (Stamatoff, Cladis, Guillon, Cross, Bilash and Finn, 1980*a*) on a compound in this category measures the intensity of the second-order diffraction from the smectic layers relative to the first order. The authors conclude to an unusually large short-range *disorder* combined with a well-defined long-range *order*.

A. Blumstein and L. Patel (University of Lowell, 1978) studied a series of p-*n*-alkoxybenzoic acids. They conclude that in the crystalline state the aromatic and aliphatic parts of the molecules have the same tilt angle, but that in the smectic *C* phase the tilt angle of the aliphatic part is much smaller. A. C. Griffin, D. L. Wertz, and A. C. Griffin, Jr. (University of Southern Mississippi, 1978) studied the effect of the CF_3 group in the nematic phase, and conclude to strongly repulsive electronic interactions. A. C. Griffin, M. L. Steele, J. F. Johnson and G. J. Bertolini (University of Southern Mississippi and University of Connecticut, 1979) note that their study of Siamese-twin liquid crystals in the smectic *C* phase suggests intramolecular entanglement of the alkoxy chains. D. Guillon, P. E. Cladis, J. Stamatoff, D. Aadsen, and W. B. Daniels (Bell Laboratories and University of Delaware, 1979) report that in a chiral smectic *C* phase, near the *C*-to-*A* phase transition, the tilt angle varies with pressure according to a power law with a critical exponent of about 0.5. A. de Vries (1979*c* Kent State University) wrote a review on structure and classification of liquid crystals, with 126 references, and L. V. Azároff (1980*a*, University of Connecticut) wrote a review on X-ray diffraction by liquid crystals, with 128 references (other reviews of relevant literature are: Brown and Shaw, 1957*a*; Falgueirettes and Delord, 1974*a* Chistyakov, 1975*a*). Azároff (1980) also published a paper proposing an explanation for the diffraction pattern of the skewed cybotactic nematic phase that is quite different from the generally accepted one given in 1970 by De Vries.

D. E. Moncton and R. Pindak (1979*a*, 1980*a*; Bell Laboratories) report on some very interesting studies on freely suspended films of smectic *B* phases. In the first paper, they conclude that *thick* films of the smectic *B* phase of 4-*n*-butyloxybenzylidene-4′-*n*-octylaniline have three-dimensional long-range order, even though the energy of interlayer order must be weak; *thin* films are found to be two-dimensional crystals. In the second paper, preliminary data are presented on the smectic *B* phase of *n*-hexyl-4′-*n*-pentyloxybiphenyl-4-carboxylate. The authors conclude that this *B* phase is different: it has long-range "bond-orientational correlation" (i.e., correlation with respect to the orientation of the lattice in the smectic layer), but exponential decay of in-plane positional correlations. It is suggested that this phase may be related to models which involve the stacking of interacting hexatic layers (Note: the date presented on this second "*B* phase" suggest that it may be better to call it a "non-tilted smectic *F* phase").

L. Gunther, Y. Imry, and J. Lajzerowicz (Tufts University, 1980) analyzed the X-ray scattering power-law singularities in smectic *A* phases. Very close to the Bragg points, the finite size of the sample appears to become important.

No publications in this area of research appear to have originated from Canada.

Mesophases of disc-like molecules

This subsection of the thermotropic mesophases has been the subject of considerable study in recent years, in a large part due to the significance of such mesophases in the carbonization of coal tar and in the fabrication of high-strength carbon fibers. Most of the basic X-ray work has been done in France and India, but a few applied papers have come from the United States.

L. D. Wakeley, A. David, R. G. Jenkins, G. D. Mitchell, and P. L. Walker, Jr. (Pennsylvania State University, 1979) investigated the conditions for the formation of mesophase semicoke from solvent-refined coal. L. S. Singer, F. Delhaes, J. C. Rouillon, and G. Fug (Union Carbide Corporation, 1979) studied the structure of an oriented mesophase pitch obtained by heat-treating a mesophase petroleum pitch. They propose an average structure consisting of small aromatic regions connected by aryl and aliphatic bridges. J. J. Friel, S. Mehta, G. D. Mitchell, and J. M. Karpinski (Bethlehem Steel Corporation, 1980) observed diffuse rings in the electron diffraction pattern of hot coal samples containing mesophase regions, and discuss the relation between mesophase formation and the coking properties of coal.

Lyotropic Liquid Crystals

Even though X-ray work on lyotropic liquid crystals got off to such a good start in the U.S. (McBain and co-workers, 1943-1948, see section on Early Work on X-ray Diffraction), further research in. this area has mainly been done in Europe, in particular by the groups of V. Luzzati and co-workers (France) and of P. Ekwall and co-workers (Sweden) (see Fontell, 1974*a*). So, our review of the American activities can be brief.

The Procter and Gamble Company (Cincinnati) has, of course, always shown considerable interest in lyotropic liquid crystals (for soaps, emulsions, etc.), and their X-ray work in this area has resulted in three papers. H. Nordsieck, F. B. Rosevear, and R. H. Ferguson (1948) conducted an X-ray investigation of the stepwise melting of anhydrous sodium palmitate (strictly speaking, this is not a lyotropic liquid crystal system, but the phases are similar to lyotropic phases, so we include this study in this section). They found a pronounced break, in the curve of long spacing versus temperature, between the lower-temperature mesophase ("waxy phase") and the higher-temperature mesophase ("neat phase"). Also, there were differences in the diffuse diffraction rings corresponding to the short spacings. The authors conclude to structural restraints on molecular position and motion in the waxy phase. E. S. Lutton (1966) reported measurements of the long spacing of the lamellar phase and the hexagonal phase of a two-component system. K. D. Lawson, A. J. Mabis, and T. J. Flautt (1968) made a detailed X-ray study of the same system and reported discontinuities in the spacings at the phase boundaries. The data agree with a cylindrical model for the "middle phase," a face-centered cubic structure for the viscous isotropic phase, and a lamellar structure for the "neat phase."

S. E. Friberg, formerly director of the Swedish Institute for Surface Chemistry, has, starting in 1976, established an active research group at the University of Missouri. Friberg, Thundathil, and Stoffer (1979) and Thundathil, Stoffer, and Friberg (1980) used low-angle X-ray diffraction to study polymerization in a lyotropic liquid crystal (sodium undecenoate and water), during which the structure changed from a hexagonal closepacking of cylinders to a lamellar phase. Moucharafieh and Friberg (1979) found shorter interlayer spacings in a lamellar structure with ethylene glycol as solvent, than in the same system with water. The same authors with Larsen (1979) reported extremely small changes in the interlayer spacing of a lamellar phase when aromatic hydrocarbons were added, but larger changes when aliphatic hydrocarbons were added (see also McIntosh et al. in section on Membranes).

Another currently active group is that of G. G. Shipley and D. M. Small at Boston University. In connection with their interest in the biological aspects of liquid crystals (see Small, 1977*a*, and section on Membranes), Janiak, Small and Shipley (1979) studied the ternary phase diagram of cholesteryl myristate (I)-dimyristoyl lecithin (II)-water. Above 23°C, II formed a lamellar phase with disordered hydrocarbon chains, into which limited amounts of I were incorporated. Below 23°C, II formed ordered-chain structures which did not incorporate I. Loomis, Shipley, and Small (1979) found that cholesterol with excess water has a smectic bilayer phase from 123-157°C. This smectic phase helps them to explain the ability of cholesterol to exist in high concentrations in biological membranes. Calhoun and Shipley (1979) determined the repeat distance of synthetic sphingomyelin-lecithin bilayers.

Biological membranes, model membranes, and related systems

Because of the biological importance of membranes, multilayer membranes are an especially interesting group of lyotropic lamellar liquid crystals (see also Chapman, 1979*a*). For this reason, we discuss here membranes and related systems separately.

We will limit ourselves to the most recent work. Several reviews of the earlier research have appeared in 1973 and 1974, and references to these may be found in a paper by Schwartz, Cain, Dratz, and Blasie (1975*a*, University of California and University of Pennsylvania). These authors also describe a generalized Patterson function analysis of lamellar X-ray diffraction data from disordered membrane multilayers. With this method, both lattice disorder and substitution disorder can be eliminated from the electron density profile. The method is applied to data from retinal rod outer segments. L. Herbette, J. Marquardt, A. Scarpa, and J. K. Blasie (University of Pennsylvania, 1977) use the same method in an analysis of lamellar X-ray diffraction data from oriented multilayers of fully functional sarcoplasmic reticulum. They obtained the profile structure, showing marked asymmetry, to a resolution of 10 Å. G. Zaccai, J. K. Blasie, and B. P. Schoenborn (Brookhaven National Laboratory and University of Pennsylvania, 1975) phased lamellar neutron diffraction data from oriented multilayers of dipalmitoyl lecithin by isomorphous H_2O-D_2O exchange and swelling techniques. Bound-water sites were located and a 6 Å resolution structure is proposed for the bilayer.

At Carnegie-Mellon University, C. R. Worthington and co-workers also did studies on the structure of multilamellar systems. Khare and Worthington (1977) recorded series of X-ray reflections from oriented samples of sphingomyelin, of cholesterol, and of their mixtures, and Worthington and Khare (1978) describe a method for determining the phases of X-ray reflections from oriented model membrane systems, at low resolution; using this method, they derive phases for the first five or six orders from phosphatidylethanolamine and lecithin. Khare and Worthington (1978) present an analysis of the X-ray data (up to 14 orders of diffraction) from oriented stacks of sphingomyelin bilayers, giving electron density profiles at resolutions of about 6 Å and 2.5 Å. Worthington and Wang (1979) describe a theory for small-angle X-ray data processing for planar multilayered structures. Most recently, Worthington (1979) and Worthington, McIntosh and Lalitha (1980) report on X-ray studies of frog sciatic nerve and nerve myelin.

At Boston University, D. M. Small, G. G. Shipley, and co-workers have done much work on phase transitions in membranes and related systems. A physiologically very important phase transition in membranes is considered to be a change from a state with ordered hydrocarbon chains (often called the gel phase, sometimes called crystalline) to a state with disordered chains (often called the fluid phase or the liquid crystal phase). Shipley, Avecilla, and Small (1974) found in sphingomyelin a disordered lamellar phase above 47°C and an ordered lamellar phase below 47°C. The layer thickness was determined at different temperatures as a function of water concentration. Janiak, Small, and Shipley (1976) investigated the nature of the so-called pretransition (which occurs at a temperature below the chain-melting transition) in synthetic lecithins. They conclude that below this transition the hydrocarbon chains are fully extended and tilted with respect to the plane of the bilayer; the tilt angle decreases with increasing temperature. Above the transition temperature, the chains remain tilted but the lipid lamellae are now distorted by a periodic ripple (for another hypothesis, see Rand et al., 1975, below). At the chain-melting transition, the hydrocarbon chains assume a liquid-like conformation and the ripple structure disappears. In a study of the cholesterol esters in plasma low-density lipoproteins from monkeys fed an atherogenic diet,

Tall, Small, Atkinson, and Rudel (1978) found that these esters formed an *ordered* smectic-like structure at body temperature, whereas those from monkeys on a control diet formed a more *disordered* structure. In other work, Tall, Small, and Lees (1978) conclude that in tendon xanthomas there is a smectic liquid crystalline phase of cholesterol esters, layered between the collagen fibrils.

Elsewhere, R. S. Khare, R. K. Mishra, W. H. Falor, and P. H. Geil (Akron City Hospital, 1974) determined by X-ray diffraction the changes caused by a number of psychoactive drugs in the lamellar structure of sphingomyelin. S. W. Hui and D. F. Parsons (Roswell Park Memorial Institute, 1974) made the first electron diffraction patterns from single wet phospholipid bilayers, and noted the existence of semicrystalline domains and a structural phase transition with change in temperature. R. P. Rand, D. Chapman, and K. Larsson (Brock University, England, and Sweden, 1975) studied the phase transitions in dipalmitoyl lecithin with X-rays. They conclude that, below the pretransition, the chains are fully extended, packed in a hexagonal lattice, and tilted with respect to the plane of the bilayer. Between the pretransition and the main transition, the chains are similarly packed but oriented perpendicular to the plane (for another hypothesis, see Janiak, Small, and Shipley, above). Above the main transition, the chains are disordered. A. Hybl (University of Maryland, 1976) concluded from electron microscopic and X-ray data that the myelin lamellar structure is paracrystalline. C. D. Linden, J. K. Blasie, and C. F. Fox (University of California and University of Pennsylvania, 1977) confirmed by X-ray diffraction the existence of a broad order-to-disorder phase transition in cytoplasmic membrane lipids. They also calculated Patterson functions based on data from partially oriented cytoplasmic membranes, indicating a decrease in membrane thickness upon chain melting.

At the University of Waterloo, J. E. Thompson and co-workers investigated phase changes in biological systems with wide-angle X-ray diffraction. Several of their papers deal with senescence. Studies in chloroplast and microsomal membranes of leaves, by McKersie and Thompson (1978), revealed that portions of the lipid become crystalline (i.e., gel phase) as the tissue senesces, and that the gel-to-liquid crystal transition temperature increases with age. The same authors (1978) found similar changes in microsomes from bean cotyledons and also (1979) in liposomes prepared from total lipid extracts of the membranes; liposomes prepared from purified phospholipid fractions showed little change with age. Lees and Thompson (1980) confirm the results on bean cotyledons and also report crystallinity in senescent membranes attributable to sterol-sterol interactions. Thompson, Mayfield, Inniss, Butler, and Kruuv (1978) studied the microsomal membranes from algae. For young cultures, the membrane lipid is entirely liquid crystalline at physiological temperatures, but as the cultures age, portions of the lipid become crystalline. In no-senescence-related work, Thompson, Fernando, and Pasternak (1979) examined the effects of coccidial infection on chick intestinal cells. They found that, during the later stages of parasite maturation, the host cell plasma membrane acquired increasing proportions of gel-phase lipid. By contrast, purified membrane from isolated parasites was in a liquid-crystalline state. Buhr, Carlson, and Thompson (1979) investigated the phase behavior of microsomal membranes from corpora lutea of rats. During periods of optimal progesterone secretion, all of the membrane lipid was in the liquid-crystalline phase at physiological temperatures, but in animals undergoing regression, mixtures of liquid-crystalline and gel-phase lipid were observed. This was accompanied by a parallel rise in the lipid phase-transition temperature.

At Duke University, T. J. McIntosh and co-workers studied the influence of additional components in phosphatidylcholine bilayers. McIntosh (1978) used electron density profiles at 5 Å resolution, and chain-tilt and chain-packing parameters. He concluded that, for 12-16 carbon-chain bilayers, the addition of cholesterol *increases* the width of the bilayer, by removing the chain tilt from the gel-phase lipids, and by increasing the transconformations for liquid-crystalline lipids. For 18 carbon-chain bilayers, however, adding cholestrol *reduces* the width of the bilayer in the gel phase, by causing chain deformations. McIntosh, Simon, and McDonald (1980) found that 14 and 16 carbon alkanes align parallel to the lipid acyl chains, *in*creasing the phase transition temperatures. Alkanes with only 6 or 8 carbons, on the other hand, formed an alkane region in the center of the bilayer, and *de*creased the transition temperatures (see also Friberg et al. in section on Lyotropic Liquid Crystals).

Elsewhere, J. Stamatoff, D. Guillon, L. Powers, P. E. Cladis, and D. Aadsen (Bell Laboratories, 1979) made the first X-ray measurements of the lamellar periodicity of phosphatidylcholine as a function of pressure. Below the gel transition, the spacing *de*creased with increasing pressure, above the transition it *in*creased significantly. W. F. Graddick, J. B. Stamatoff, P. Eisenberger, D. W. Berreman, and N. Spielberg (Bell Laboratories and Kent State University, 1979) studied the interaction of calcium ions with phosphatidylcholine multilayers, as a function of temperature and calcium concentration, and found order-disorder-order transitions. R. D. Frankel and J. M. Forsyth (University of Rochester, 1979) obtained nano-second-exposure X-ray diffraction patterns from spinal nerves and cholesterol powder, with a laser-produced plasma source. T. N. Estep, W. I. Calhoun, Y. Barenholz, R. L. Biltonen, G. G. Shipley and T. E. Thompson (University of Virginia and Boston University, 1980) report that the stable gel-state of stearoylsphingomyelin is more highly ordered than the corresponding form of *other* phospholipids.

Liquid-crystalline polymers

Like research on other types of liquid crystals, research on liquid-crystalline polymers has been stimulated considerably by the discovery of industrial applications; in this case, the fabrication of high-modulus and tensile-strength fibers (e.g., Kevlar). An overview of the general state of research on polymer liquid crystals may be obtained from the proceedings of a symposium on this subject in 1977 (Blumstein, 1978*a*).

In the X-ray literature before 1975, we found only two U.S. contributions. J. B. Stamatoff (Edgewood Arsenal, 1972) made X-ray diffraction photographs of oriented solutions of poly-γ-benzyl-L-glutamate, the longest-known liquid-crystalline polymer. The patterns indicated that the polypeptide maintains its α-helical conformation in the electric field. E. M. Friedman and R.-J. Roe (Princeton University and Bell Laboratories, 1974) calculated the effect of a cholesteric twist on the diffraction pattern of a pack of rod-like polymer molecules.

The oldest still-active group in the U.S. is that of A. Blumstein, S. B. Clough, and co-workers at the University of Lowell, Blumstein, Blumstein, Clough, and Hsu (1975) studied three cases of oriented polymer growth in mesomorphic and potentially mesomorphic media, and found that neither the presence of a mesomorphic matrix nor large amounts of cross-linking agents are necessary for the development and locking-in of mesomorphic superstructures (see also Newman et al., 1978, below). Clough, Blumstein, and Hsu (1976) report that the mesomorphic organization of a polymer is not necessarily that of the monomer: a nematic monomer can yield a polymer with smectic order. They also give X-ray data on the anisotropic thermal expansion of the polymer. Blumstein, Clough, Patel, Blumstein, and Hsu (1976) show on two examples that under favorable circumstances the organization of the side groups of a polymer can proceed beyond smectic order to crystallization; hydrogen bonding appears to be essential for this. Clough, Blumstein, and De Vries (University of Lowell and Kent State University, 1978) report X-ray diffraction results for a number of polymers in which the side chains have nematic- or smectic-like order. Sivaramakrishnan, Blumstein, Clough, and Blumstein (1978) present preliminary X-ray data on some polymers with mesogenic elements and flexible spacers, indicating that these polymers are crystalline at room temperature.

Elsewhere, C.-Y. Chen and I. Piirma (University of Akron, 1978) report X-ray and electron diffraction data suggesting that rigid-chain polyacenaphthylene is paracrystalline. B. A. Newman, V. Frosini, and P. L. Magagnini (Rutgers, The State University, and Italy; 1978) conclude from X-ray data that it is not necessary to have mesogenic monomers or polymerization in mesomorphic phases in order to obtain polymers with mesomorphic structure (see also Blumstein et al., 1975, above). M. Jaffe (Celanese Research Company, 1978) used X-ray diffraction to investigate structural differences between wet-spun and dry-jet wet spun aramid fibers; wet-spun fibers were less crystalline and gave patterns reminiscent of lamellar structures.

S. B. Warner, D. R. Uhlmann, and L. H. Peebles, Jr. (Massachusetts Institute of Technology, 1979) propose that, in the partially ordered regions of polyacrylonitrile fibrils, the polymer molecules assume a contorted helical shape, forming rods which are ordered into a liquid-crystal-type array. W. W. Adams, L. V. Azároff, and A. K. Kulshreshtha (University of Massachusetts, 1979) report that the diffraction pattern from a nematic polybenzothiazole fiber is a composite of what has been predicted for isolated periodic cylinders and for parallel arrays of infinite cylinders. A. Garton, D. J. Carlsson, R. F. Stepaniak, and D. M. Wiles (National Research Council of Canada, 1979) obtained a smectic X-ray diffraction pattern from a rapidly quenched polypropylene fiber. S. M. Aharoni (Allied Chemical Corporation, 1979, two papers) discusses the X-ray diffraction of certain isocyanate rigid-backbone polymers which are mesomorphic in concentrated solutions at ambient temperatures and in bulk at elevated temperatures.

Y. Onogi, J. L. White, and J. F. Fellers (University of Tennessee, 1980) investigated the structures of the liquid-crystalline phases of solutions of aromatic polyamides in sulphuric acid, of hydroxypropyl cellulose in water, and of poly-γ-benzyl-L-glutamate in dioxane. The poly-amide solutions were nematic; the other two compounds gave cholesteric solutions. In related work, R. S. Werbowyj and D. G. Gray (Pulp and Paper Research Institute of Canada, 1980) showed that hydroxypropyl cellulose formed an ordered liquid-crystalline phase in concentrated (30-40%) solutions in water, acetic acid, and acetic acid anhydride. In the latter two solvents, high-pitch cholesteric structures were observed. S. B. Warner and M. Jaffe (Celanese Research Company, 1980), in their study of quiescent crystallization in thermotropic flexible-chain polyesters, found that the structure of the crystal closely resembles that of the nematic melt. Friberg and co-workers investigated the polymerization of a lyotropic liquid crystal (see section on Lyotropic Liquid Crystals).

Crystal structures of mesogenic compounds

Even though work in this area ("mesogenic compounds" means "compounds which on heating give mesophases") started fairly early (Bernal and Crowfoot, 1933), further studies had for many years been very few: as late as 1970, only *four* complete crystal structures had been determined, all of them in Europe. Now, far more structures have been done, and many of them in the United States.

The first structures determined in the U.S. are from

W. R. Krigbaum and co-workers at Duke University. Krigbaum, Chatani, and Barber (1970) reported the structure of p-azoxyanisole. The non-planar molecules are parallel and form an imbricated packing, as predicted for nematogenic compounds. In the first determination of the structure of a smectogenic compound, Krigbaum and Barber (1971) found the nearly-planar molecules packed in a parallel array, again as predicted.

At Kent State University, Lesser, De Vries, Reed, and Brown (1975) also found an imbricated packing for a nematogenic compound, with the non-planar molecules arranged in herringbone fashion in planes perpendicular to the long axes. Chung, Carpenter, De Vries, Reed, and Brown (1978) found again a layered structure for a smectogenic compound. The non-planar molecules form a herringbone packing, as in the previous structure, and are tilted relative to the layers.

J. H. Wendorff and F. P. Price (University of Massachusetts, 1973 and 1974) determined the unit-cell dimensions of three cholesteryl esters. They concluded that, for the two esters that have mesophases (the myristate and stearate), the molecules are arranged in some sort of antiparallel fashion which has no similarity to the structure in the mesophase. In a pretransition region below the melting point, the order in directions *normal* to the long axis decreases markedly with increasing temperature. For the ester without mesophase (the acetate), a decrease in the order *along* the long axis is the main effect in the pretransition range.

Complete crystal structures of cholesterol and several of its esters were determined by B. M. Craven and co-workers at the University of Pittsburgh, in an effort to gain more insight in the structure of their thermotropic and lyotropic phases. In the myristate (Craven and DeTitta, 1976), the molecules pack in an antiparallel fashion (as concluded by Wendorff and Price, see above), in bilayers, with overlapping myristate chains. The molecules are almost fully extended. Cholesterol (Craven, 1976 and 1979) also has a stacking of bilayers, but the molecules are not fully extended. In the laurate (Sawzik and Craven, 1979), the chains are again almost fully extended, but the molecules are arranged in *mono*layers in which the ester chains are packed with the cholesteryl ring systems. In the acetate (Sawzik and Craven, 1979), the C(17) chains are again almost fully extended. The molecules are packed in separate stacks, antiparallel in each stack. All molecular long axes are parallel. In the oleate (Craven and Guerina, 1979), the oleate chains are almost straight. The molecules are parallel, and are packed in such a way as to form a structure with alternating layers of completely interdigitating oleate chains and layers of overlapping cholesteryl parts. The angle between the molecular long axes and the plane of the layers is only about 27°. The authors suggest that it is unlikely that the structure of the liquid crystalline phase is closely related to the crystal structure. The decanoate (Pattabhi and Craven, 1979) and the nonanoate (Guerina and Craven, 1979) are isostructural with the laurate (see above). The molecules in both structures have almost fully extended conformations and are arranged in antiparallel order in monolayers. The molecular long axes make an angle of about 67° with the layer plane in the decanoate, about 61° in the nonanoate.

At the University of Virginia, R. F. Bryan (who published the first complete crystal structure determination of a liquid-crystal-related compound in England in 1967) and his co-workers studied the relationship between crystal structure and liquid crystallinity. Bryan, Forcier, and Miller (1978) conclude from the study of a potentially mesogenic substance, that for this compound mesophase formation is inhibited by a crystal structure with infinite interlocking hydrogen-bonded zig-zag chains of molecules. In another study of a potential mesogen, Bryan and Forcier (1980) assume that mesophase formation is prevented by a crystal structure of hydrogen-bonded sheets of molecules combined with an interlocking of adjacent sheets by alternating ring orientations. Comparing the structures of two nematogens, Bryan and Forcier (1980) find that whereas in one structure the molecular long axes are all more or less *parallel*, the other consists of bimolecular sheets with the molecular axes in successive sheets *perpendicular* to each other, which is extremely unusual for a nematogen.

Acknowledgment

Searching the literature for relevant papers has been aided significantly by the data available from Brown and Shaw (1957*a*), Eastman Kodak Company (1975*a*), Falgueirettes and Delord (1974*a*), Fehrenbach, Dreher, and Meier (1975*a*, 1977*a*, 1978*a*, 1979*a*, 1980*a*) and Fontell (1974*a*), and from the CA Selects on Liquid Crystals.

References

Azaroff, L. V. (1980). Mol. Cryst. Liq. Cryst. **60**, 73-98.

Blumstein, A. (1978). Mesomorphic Order in Polymers and Polymerization in Liquid Crystalline Media. ACS Symposium Series Vol. 74. Washington: American Chemical Society.

Brown, G. H. and Shaw, W. G. (1957). Chem. Rev. **57**, 1049-1157.

Brown, G. H. and Wolken, J. J. (1979). Liquid Crystals and Biological Structures. New York: Academic Press.

Chapman, D. (1979). Liquid Crystals, The Fourth State of Matter. pp. 305-334. New York: Marcel Dekker.

Chistyakov, I. (1975). Advances in Liquid Crystals. Vol. 1, pp. 143-168. New York: Academic Press.

De Vries, A. (1979*a*). J. Chem. Phys. **70**, 2705-2709.

De Vries, A. (1979*b*). J. Chem. Phys. **71**, 25-31.

De Vries, A. (1979*c*). Liquid Crystals, The Fourth State of Matter. pp. 1-72. New York: Marcel Dekker.

De Vries, A. (1981). Mol. Cryst. Liq. Cryst., **63**, 215-230.

Eastman Kodak Company (1975). Liquid Crystal Bibliography. Kodak Publication No. JJ-193. Rochester: Eastman Kodak Company.

Falgueirettes, J. and Delord, P. (1974). Liquid Crystals and Plastic Crystals. Vol. 2, pp. 62-79. Chichester: Ellis Horwood.

Fehrenbach, W., Dreher, R. and Meier, G. (1975). Bibliography on Liquid Crystals. 1973 and 1974. Freiburg: Institut für Angewandte Festkörperphysik.

Fehrenbach, W., Dreher, R. and Meier, G. (1977). Mol. Cryst. Liq. Cryst. **43**, 103-174.

Fehrenbach, W., Dreher, R. and Meier, G. (1978). Mol. Cryst. Liq. Cryst. **48**, 53-126.

Fehrenbach, W., Dreher, R. and Meier, G. (1979). Mol. Cryst. Liq. Cryst. **55**, 251-329.

Fehrenbach, W., Dreher, R. and Meier, G. (1980). Mol. Cryst. Liq. Cryst. **61**, 79-161.

Fontell, K. (1974). Liquid Crystals and Plastic Crystals. Vol. 2, pp. 80-109. Chichester: Ellis Horwood.

Gray, G. W. (1962). Molecular Structure and the Properties of Liquid Crystals. London: Academic Press.

Moncton, D. E. and Pindak, R. (1979). Phys. Rev. Lett. **43**, 701-704.

Moncton, D. E. and Pindak, R. (1980). Ordering in Two Dimensions. pp. 83-90. New York: Elsevier North Holland.

Safinya, C. R., Kaplan, M., Als-Nielsen, J., Birgeneau, R. J., Davidov, D., Litster, J. D., Johnson, D. L. and Neubert, M. E. (1980). Phys. Rev. B **21**, 4149-4153.

Schwartz, S., Cain, J. E., Dratz, E. A. and Blasie, J. K. (1975). Biophys. J. **15**, 1201-1233.

Small, D. M. (1977). J. Colloid Interface Sci. **58**, 581-602.

Stamatoff, J., Cladis, P. E., Guillon, D., Cross, M. C., Bilash, T. and Finn, P. (1980). Phys. Rev. Lett. **44**, 1509-1512.

Stewart, G. T. (1974). Liquid Crystals and Plastic Crystals. Vol. 1, pp. 308-326. Chichester: Ellis Horwood.

CHAPTER 3

CRYSTALLOGRAPHY OF WATER STRUCTURES

Barclay Kamb
Division of Geological and Planetary Sciences
California Institute of Technology, Pasadena, California 91125

The crystallography of the various forms of ice and crystalline hydrates is interesting both in its own right and because of its implications for the bonding properties of water molecules and for the structure and properties of liquid water and aqueous solutions. In writing about the crystallography of water from an American historical perspective, my main focus is on the ices and ice-like hydrates, but I stray from strict crystallography enough to recall how crystallographic approaches and conclusions have influenced thinking about water more generally.

The artistically most beautiful crystallography ever done for water, and perhaps indeed for any substance, was Wilson A. Bentley's photography of snowflakes that fell in Vermont winters in the early 1900's (Bentley and Humphreys, 1931). Much has been done subsequently to understand how the forms of ice crystals are controlled (Hobbs, 1974; La Chapelle, 1969), but Bentley's photographs remain unsurpassed as visually delightful crystallography.

Ice was of course studied from almost the earliest days of X-ray crystallography, and a couple of American contributions figure significantly, in particular the first powder pattern, the first unit cell determination, and the suggestions that the structure consisted of a "wurtzite-like" arrangement of oxygen atoms or a hexagonal-close-packed arrangement of "H_4O_2 molecules" (St. John, 1918; Dennison, 1921). Only later was it realized that ice was built up of H_2O molecules little altered from their form in the vapor phase and held together by asymmetric hydrogen bonds (Bernal and Fowler, 1933).

The most important American contribution to the modern crystallography of water was Linus Pauling's far-reaching paper on the configurational entropy of ice (Pauling, 1935). It introduced the concept of orientational disorder in molecular crystals, a concept which, as Pauling showed, went far beyond the particular example that prompted its discovery – ice. The phenomenon of water-molecule orientation disorder, often called proton disorder, is now known to be intimately involved in most of the physical properties of ice. Although proton disorder is essentially a crystallographic concept, involving partial occupancy of atomic sites, Pauling established it not by crystallographic means but by thermodynamics and statistical mechanics. He showed that the rules governing proton arrangements in a tetrahedral framework of asymmetric hydrogen bonds (the so-called "ice rules") could be satisfied by a statistical arrangement involving short range order without long-range order; he devised a simple approximate way of calculating the configurational entropy of the statistical arrangement, and he showed that this entropy agreed with the residual entropy that had been found experimentally for ice close to absolute zero. Years later, sophisticated methods for improving on the exactness of the statistical calculation were developed (Nagle, 1966); they changed the result only slightly from Pauling's value.

Crystallographic proof of the proton arrangement in ice had to await the development of neutron diffraction, by which the proton positions could be detected strongly enough to distinguish between ordered and disordered structures. Powder methods were first used (Wollan et al., 1949), but the decisive test was with single-crystal neutron diffraction, by Peterson and Levy (1957) at Oak Ridge. They showed that the proton-disordered structure, with proton site occupancies of 1/2, is correct.

Experimentally, the next major step beyond single-crystal diffraction was investigation of the proton-position correlation function for the short range order by means of single-crystal neutron diffuse scattering. This was undertaken by John Axe and Walter Hamilton at Brookhaven National Laboratory in the early 1970's (see Kamb, 1973, p. 39). They showed that the diffuse scattering in the prism zone could be accounted for reasonably well by a simple model of short range order consistent with the Pauling concept of proton disorder.

A further contribution to the crystallography of ice was the accurate determination of its lattice constants as a function of temperature, by La Placa and Post (1960).

The discovery by Pauling and Marsh (1952) of the structure of chlorine hydrate opened a new chapter in the crystallography of water: the structures of the clathrate hydrates, which are built of ice-like, tetrahedrally-linked frameworks of hydrogen-bonded water molecules forming polyhedral cavities that enclose hydrophobic molecules such as Cl_2, N_2, O_2, CH_4, etc. Prominent among these polyhedra is the pentagonal dodecahedron. Larger guest molecules, such as CH_3Cl, are accommodated in expanded modifications of the pentagonal dodecahedron, such as the tetrakaidecahedron and hexakaidecahedron. In subsequent years, crystallographic studies of a panoply of clathrate hydrate structures were carried out by George Jeffrey and his colleagues at the University of Pittsburgh (Jeffrey and McMullan, 1967). When hydrophilic guest molecules, capable of hydrogen bonding, are incorporated into these structures, they are found often to bond into the tetrahedral frameworks of water molecules, a modification that makes the frameworks no longer fully ice-like. "Semi-clathrate" hydrates of this kind are removed from the ice-like hydrates by one step in the direction toward the broad classes of salt hydrates and organic hydrates

(Baur, 1965, 1972; Hamilton and Ibers, 1968; Jeffrey, 1969, 1982; Clark, 1963), which, while important and fascinating, have to be left outside the scope of this historical sketch.

Air hydrate, with the same clathrate structure as chlorine hydrate, is probably present below a depth of about 1200 m in the Antarctic and Greenland ice sheets (Miller, 1973). Most of the crystalline water in glaciers is, however, ordinary ice I. Its crystallographic study has focussed on the size and peculiar geometry of the crystals (Bader, 1951), their modifications with depth and time (Gow, 1971), and their patterns of spatial orientation in relation to glacier flow (Rigsby, 1960; Kamb, 1959, 1972a; Gow and Williamson, 1976; Hooke and Hudleston, 1980).

The dense polymorphs of ice, formed under high pressure, were first studied crystallographically by McFarlan (1936a, b), working with samples made by Bridgman. Results obtained for ice II and ice III were interpreted as ionic structures containing independent H^+ and O^{--} ions, but they were based only on powder diffraction data, which proved inadequate to the task.

In 1958 Pauling interested me in taking up crystallographic study of the dense ice phases. This would have been a technically formidable task if it had had to be done working at the pressures of 2 to 22 kbar required to stabilize them. But an easier approach was possible, thanks to the fact that these forms of ice could be released to atmospheric pressure and retained metastably if first cooled to 77 K. They could then be studied by techniques of low temperature crystallography. The only requirement was that the crystals be maintained at all times below their temperature of inversion, which lay in the range 125-170 K. For the purpose, I used a cold nitrogen gas stream in combination with a precession camera. During the period 1958-67, my colleagues and I did X-ray studies of a succession of ice phases made at progressively higher pressures. We of course obtained primarily the arrangement of the oxygen atoms in each structure, but could reason about the hydrogen atoms by stereochemical arguments and from certain detailed features of space group symmetry that turned up. The results indicated that though the structures show great variations of detail, in overall features they all represent tetrahedrally linked frameworks of asymmetrically hydrogen-bonded water molecules, and in this respect are fundamentally similar to ice I and the clathrate hydrates. The densification consequent upon their origin at high pressure is achieved primarily by distortion of the hydrogen bonding from the nearly perfect tetrahedral geometry represented by ice I, which makes it possible for the molecules to accommodate nonbonded neighbors at distances much shorter than in ice I. The bending of the hydrogen bonds and the insertion of nonbonded neighbors at distances where intermolecular repulsion is significant causes the energies of the dense ices to be raised relative to ice I, which explains why they are stable only at high pressure. These features have been reviewed in several papers (Kamb, 1968, 1972b, 1973).

The first high-pressure-ice structure determined was that of ice III (Kamb and Datta, 1960; Kamb and Prakash, 1968), but the study that opened up the most interesting new ideas and set a pattern that reappeared several times in later work was that of ice II (Kamb, 1964); A pseudostructure was found in space group $R3c$, while the actual structure was a symmetry-degraded distortion from it, in space group $R\bar{3}$. To obtain this structure it was necessary to make the assumption that the observed $\bar{3}m$ Laue symmetry of the crystals was caused by twinning of individuals with $\bar{3}$ Laue symmetry. There were two possible explanations of the distortion: (1) it could be analogous to the symmetry-degrading distortion that occurs in the $\beta \rightarrow \alpha$-type transitions in silica structures, which were known ice-analog frameworks; (2) it could be due to proton ordering. From the X-ray crystallographic evidence I was able to choose alternative (2) and to derive the actual ordered arrangement of the protons in the H bonds.

As my conviction about the validity of the crystallographic arguments grew, so did my excitement over the idea that here, for the first time, was proton-ordering in ice – something long sought but never found in ice I. The proton order in ice II was in fact reflected in its thermodynamics: its entropy, calculated by Bridgman from the H_2O phase diagram, was 0.8 entropy units (cal deg^{-1} mole^{-1}) lower than that of ice I. This was just the amount expected from Pauling's calculation of the configurational entropy of proton disorder in ice I. Here was a direct manifestation of Pauling's entropy, entirely independent of the zero-point entropy of ice I, and free of the mystery that always seemed to pervade that quantity. The transition entropies also indicated that all the other forms of ice known at that time should be proton disordered, which to some extent hid from us an expectation of what was to come.

About six years later, in 1968, I had the satisfaction of being able to prove by neutron diffraction that these ideas were correct. It was Walter Hamilton, at Brookhaven, who made our neutron diffraction work possible. In 1967 I went to see him about the possibility, and he received my proposal for collaboration enthusiastically. In the ensuing years, as I got to know him better through this collaboration, I realized that everything he did – and he did a lot – he did with graceful enthusiasm and unassuming competence. This collaboration was the high point of my crystallographic life, and Walter's untimely death in 1973 was its greatest tragedy. Walter was one of the finest products and exponents of American crystallography.

Our neutron diffraction work on ice II decisively confirmed the proton-ordered arrangement deduced earlier, and it provided a rather direct experimental proof of the twinning hypothesis (Kamb, Hamilton, La Placa, and Prakash, 1971).

Ice III presented a contrasting situation. Its space group symmetry did not constrain proton order-disorder, so that a continuous and progressive transition from proton disorder to proton order would be possible without any change in unit cell or space group along the way. In fact, such a transition, occurring progressively over the temperature range from 200 to 165 K, was discovered by Whalley et al., (1968) in Canada, by means of dielectric measurements under high pressure. They named the proton-ordered phase ice IX. This is the only instance where a named distinction has been created between two ice phases that are not distinguished by any qualitative difference in crystallography. The crystallographic distinction between them must be made in terms of the proton site occupancy parameters, which should be reflected in differences in neutron diffraction intensities, coupled, perhaps, with slight differences in lattice constants (over and above what could be expected from thermal expansion). A group at Los Alamos looked for such differences by neutron powder-diffraction experiments under high pressure and reported that they could see them, although the basic unit cell of ice III, as manifested by its powder pattern, remained almost unchanged in the transformation to ice IX (Arnold et al., 1971). Our single-crystal neutron work on this phase gave a proton-ordered structure (La Placa et al., 1973), as did an earlier powder-diffraction study (Rabideau et al., 1968). Evidently our experimental procedure had not quenched in the disordered arrangement of ice III, and what we had made was ice IX. The orientations chosen by the water molecules in the ordered arrangement were the best ones available, in the sense that the corresponding O· · ·O· · ·O angles were nearest to 104.5° from among the angles available for occupancy by H-O-H. Because the neutron diffraction data were particularly good, the ice IX structure provided the most accurate determination then available of the geometry of asymmetrically hydrogen bonded water molecules in the solid state. The good quality of the neutron data let us see an unexpected detail: although the structure was basically proton ordered, protons were actually present in the "unoccupied" sites to the extent of about 4% (± 1%). A little of the disorder in ice III actually had been retained after all. The power of crystallography was shown here, because infrared spectroscopy had revealed the proton order but failed to detect the slight disorder.

In the summer of 1965, while I was doing X-ray work on quenched crystals of ice VI, Block et al. (1965) at the National Bureau of Standards published a study of ice VI with the new diamond-anvil high-pressure cell for the precession camera, which allowed single crystals to be observed visually and studied by X-ray diffraction under high pressure, *in situ*. The published cell was orthorhombic, but I found that my crystals of ice VI were tetragonal, space group $P4_2/nmc$. The discrepancy doubtless resulted from the limited flexibility for manipulating crystal orientation in the diamond-anvil cell, which kept the N.B.S. group from finding the 4-fold axis. Otherwise, the *in situ* diffraction data were mostly compatible with mine, showing that ice VI under pressure had basically the same crystallography as the quenched material. The tetragonal structure was easily solved (Kamb, 1965a), and presented a curious new structural feature, which seems to be characteristic of the densest ice phases, since it also appeared in ice VII and ice VIII. Instead of a single tetrahedrally linked framework of hydrogen bonds, the structure contained two independent frameworks, not connected to one another by bonding. The molecules of one framework occupied void space in the other, and vice versa, so that the frameworks interpenetrated one another, but did not interconnect. The enclosure of one set of molecules by a tetrahedrally bonded framework of another was a clathrate-type feature, and since the enclosing and enclosed frameworks were identical (crystallographically equivalent), I called this type of structure a "self clathrate." Pauling coined the term *bireticulate* for a two-framework structure of this kind, and discovered a triretriculate hydrogen-bonded structure, formed by $H_3Co(CN)_6$ (Pauling and Pauling, 1968). A related structural feature is seen in the interlocking sheets of H-bonded trimesic acid molecules discovered by Duchamp and Marsh (1969).

The quenched crystals of ice VI actually showed ever so slight a departure from space group $P4_2/nmc$, in which the c glide was violated, as indicated by two or three very faint X-ray reflections. The situation was reminiscent of the symmetry degradation from $R\bar{3}c$ to $R\bar{3}$ in ice II, but the effects were very much weaker. Neutron diffraction showed that the protons were partially ordered, in space group *Pmmn* (Kamb, 1973). The crystallogrpahic change from a necessarily proton-disordered structure in $P4_2/nmc$ at high temperature to a proton-ordered structure in *Pmmn* at low temperature gave an unambiguous basis for recognizing the ordered form as a new phase of ice, which, according to previous practice, should be named ice X, although temporarily I called it ice VI′ (Kamb, 1973).

Ice V is crystallographically the most complex of the ice phases, having a 28-molecule structure in space group $A2/a$. X-ray study (Kamb et al., 1967) suggested that all bonds were proton disordered, so when we did neutron diffraction (Hamilton et al., 1969) we got a surprise: most of the proton positions turned out to have occupancy probabilities differing significantly from 1/2, 0, or 1. The range was from 0.07 to 0.95. We could be sure that these numbers were meaningful, because they satisfied the "ice rules" – that is, the sum of the proton probabilities in each bond was 1, and around each oxygen atom 2, to a good approximation. As in ices IX and VI′, this state of partial proton order had not been detected by spectroscopic or other methods, so the power of crystallography was once again well shown. In 1973 I discovered a few X-ray reflections that violated the *A* centering in ice V; these had apparently developed subsequent to our earlier studies. Neutron diffraction showed that reflections of this type were generally observable and indicated that they were due to proton ordering in space group $P2_1/a$, which drops the *A* centering but retains the *a*-glide of $A2/a$. Dropping the *A* centering eliminates a symmetry center and a 2-fold axis that had constrained two of the H-bonds in the $A2/a$ structure to be proton disordered, so that partial or full order was then possible in all bonds. Discovery of the ordered structure in $P2_1/a$ provided a crystallographic basis for recognizing and defining yet another new phase of ice – ice V′ or ice XI. In the course of neutron diffraction work on ice V′ we discovered that a reversible order-disorder transition between the $P2_1/a$ and $A2/a$ structures occurs over the temperature range 115 to 125 K at atmospheric pressure (Kamb and La Placa, 1974).

The densest forms of ice proved to have the most novel features. Their crystallography was basically quite simple, but a little tricky to obtain and interpret. To make ice VII required pressures of above 21 kbar, which our original apparatus for making and extracting quenched samples was unable to reach. I approached the structure first from a theoretical standpoint, and in 1964 developed a line of reasoning that suggested that the oxygen atoms would be in a body-centered cubic arrangement. Since this proposition could be easily tested by powder diffraction, I got in touch with Briant Davis at U.C.L.A., who had developed a high-pressure cell for powder diffractometry, capable of working up to 30 kbar. Our only problem in studying high-pressure ice by this technique was that we had to keep the temperature of the high-pressure cell below –21° C, the minimum melting point in the H_2O system, otherwise the ice sample was violently expelled from the cell, like a rifle shot. Working at -50°C and 22 kbar, we got a powder pattern of a body-centered cubic cell with a = 3.30 Å (Kamb and Davis, 1964). At about the same time, Bertie et al. (1964) in Canada got a powder pattern of quenched ice VII, which was more complicated but whose stronger lines fitted a body-centered cubic cell with a = 3.45 Å. Whatley and Van Valkenburg (1966), using the diamond anvil cell, observed a rhombic dodecahedral crystal of ice VII growing from liquid water at the freezing temperature of about 100°C, and Weir et al. (1965) obtained single-crystal diffraction patterns at 25°C and 25 kbar showing a body-centered cubic cell with a = 3.40 Å. Shortly thereafter, Whalley et al. (1966) discovered by high-pressure dielectric measurements that upon cooling to 4°C, ice VII transformed to a proton-ordered phase, which they named ice VIII. The transformation ran rapidly, so that the disordered form (ice VII) was not quenchable. This meant that the phase from which we had obtained the cubic powder pattern was actually ice VIII. When finally we were able to make quenched single crystals of ice VIII, we found that while the strong reflections conformed to a body-centered cubic cell with a = 3.42 Å (at atmospheric pressure), there were weak reflections indicating a doubling of the cell edges; in addition, a few strong reflections were resolved into doublets. Because of this splitting, the diffraction pattern, in spite of its cubic symmetry, could not be explained by a cubic cell, but a tetragonal cell with c/a = 1.03 could account for it if the sample consisted of a mosaic of tetragonal crystals with their *c* and *a* oriented in three mutually perpendicular directions at random. Our high-pressure powder pattern of ice VIII had failed to detect the weak superlattice reflections and had missed the peak splitting because of accidental interferences from powder peaks from the beryllium of the pressure vessel and from diamond used as an internal standard.

The hydrogen-bonded structure made visible by the crystallography of ice VII and VIII is perhaps the neatest, and certainly the simplest of the dense ice phases. Illustrated in Figure 1, it is a bireticulate structure, in which each of the two frameworks is that of ice

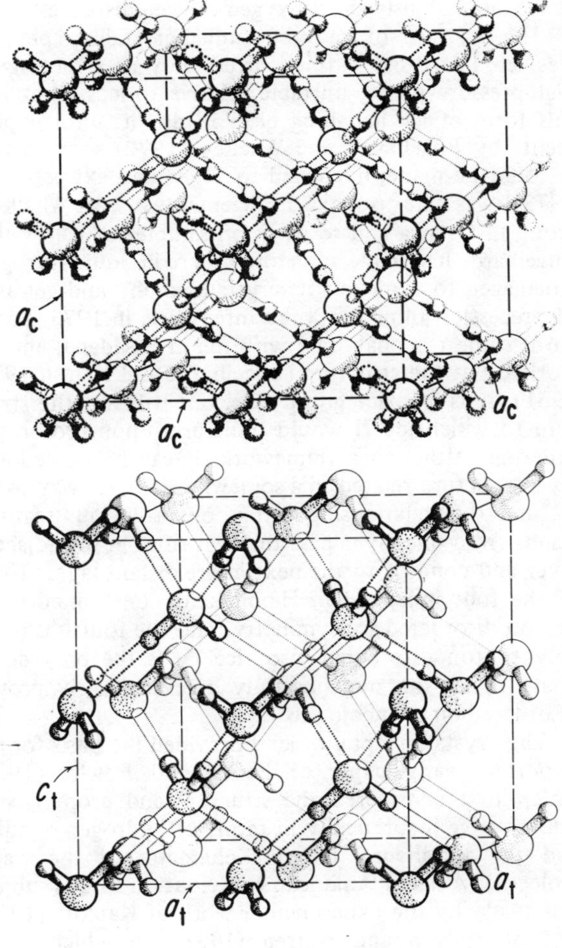

Figure 1

Ic, the low-pressure cubic form of ice that has the diamond arrangement for the oxygen atoms. Each water molecule in ice VII has eight neighbors at equal distances; it is tetrahedrally H bonded to four of these, and is in non-bonded (repulsive) contact with the other four, which causes a large stretching of the hydrogen bonds (Kamb, 1965b). In the transition from ice VII to VIII, all the water molecules in one framework line up with their dipole-moment vectors along an *a* axis pointing in one direction, and those in the other framework line up pointing in the opposite direction, so that the net polarization is zero and the transition is antiferroelectric. The pointing direction becomes the *c* axis of the tetragonal cell. The two frameworks shift slightly (0.30 Å) relative to one another parallel to the *c* axis. (A similar antiparallel polarization and shift occurs in ice VI′.) The framework shift causes two of the non-bonded contacts of each water molecule to lengthen and two to shorten; as a result, two non-bonded contacts become 0.17 Å shorter than the hydrogen bonds. This remarkable feature can be explained by formation of a cyclic hydrogen bond between pairs of water molecules at the anomalously short oxygen-oxygen distance.

The last form of ice to be studied crystallographically was ice IV. Because it occurred only as a metastable high-pressure phase, unstable relative to ices III and V, this form of ice had long been in doubt, until experiments by Engelhardt and Whalley (1972) showed that it could be made, quenched to 77 K, and extracted. In 1972 I was able to make arrangements to visit Whalley's group in Ottawa and to work on samples of ice IV that Engelhardt had made and stored there in liquid nitrogen. I managed to work out its crystallography and get a set of precession films for X-ray intensities. In 1973, Engelhardt came to collaborate with me in Pasadena, and we worked out the structure (Engelhardt and Kamb, 1978, 1981). It is in space group *R*3*c*, but it is not the structure to which ice II would transform upon proton disordering. Although a framework, it can be viewed as a layer structure related in a somewhat indirect way to ice Ic, and this relationship brings out a curious feature: bonds from one layer pass through rings in the adjacent layer and connect to the next, more distant layer. Three of the four independent H-bonds are constrained to be proton disordered by symmetry, and the fourth is probably proton disordered also. Ice IV is the only dense phase of ice that remains fully or even largely proton disordered on quenching to 77 K.

The crystallography of ice I provided the basis for the important early paper of Bernal and Fowler (1933) which first considered the structure and properties of liquid water theoretically in terms of hydrogen bonding and the tetrahedral bonding character of the water molecule. A major American contribution to this subject was made by the experimental work of Katzoff (1934) and of Morgan and Warren (1938), in which X-ray diffraction was first extensively used to determine the radial distribution function (r.d.f.) for liquid water. The results showed that the water molecules in the liquid are basically 4-coordinated, with nearest neighbors lying at normal H-bond distances, as in ice. In comparing the observed r.d.f. with calculated r.d.f.'s based on the ice I structure, Morgan and Warren noted that (1) the nearest-neighbor peak is shifted from 2.76 Å in ice I to about 2.9 Å in water near 0°C, (2) the next peak, around 4.4 - 5.0 Å, is much broader in water than in ice I, and (3) in the interval between 3.0 and 4.0 Å in the water r.d.f. there is a considerable number of oxygen-oxygen distances that are absent in the ice I structure and that correspond to a broad peak containing about 3.4 distances centered at about 3.6 Å. More detailed and extensive X-ray diffraction study of liquid water in the 60's by Narten, Danford, and Levy (1967) at Oak Ridge confirmed the general features of Morgan and Warren's r.d.f. and indicated an average H-bond length of about 2.84 Å at 0°C (Narten and Levy, 1971; Narten, 1972).

What did the results of the subsequent work on ice crystallography have to add to the understanding of liquid water structure? First and perhaps foremost, they showed that the features (1) - (3) identified by Morgan and Warren (1938) in the water r.d.f. are in fact features that would be expected from the oxygen-oxygen distances found in the dense ice phases if such molecular arrangements occurred to a significant extent in the liquid. Thus in the dense ice structures the hydrogen bond lengths are universally increased over the bond length in ice I, by amounts varying from little up to 0.2 Å; the topologically next-nearest neighbors, as counted outward through the bond network, lie at distances ranging from 3.5 to 5.3 Å; and there is a significant number of non-bonded neighbors that are topologically more remote but lie at distances in the interval 3.0 to 4.0 Å. The water r.d.f. could therefore be explained by models in which contributions from regions with various different ice-like structures are combined (Kamb, 1968, 1970). Contributions from molecular arrangements like those in the dense ice phases could account in a natural way for the increase in density on melting of ice I. More generally, the relationships among spectroscopic properties, thermodynamic properties, and structure (particularly bond bending, bond stretching, and non-bonded interaction), developed from the various crystalline ice phases, provided a comparative basis for considering those aspects of water structure in which the molecules are fully hydrogen bonded (Kamb, 1968; Eisenberg and Kauzmann, 1968; Sceats and Rice, 1980a).

Of the many and diverse models for water that were proposed over the years, those that had a close basis in ice crystallography include the hydrate model based on the clathrate hydrate structures (Pauling, 1959), the ice I-based model of Narten and Levy (1969), and the multiple-ice-structure model (Kamb, 1968). The first two of these models had the common feature of including unbonded water molecules in cavities in the structure. This feature proved to be noncrystallographic in the sense that such interstitial water molecules have not been found in crystallogrpahically detectable amount in any ice structure.

In the mid 70's, Stuart Rice and his colleagues at the University of Chicago and Oak Ridge introduced an important new source of information for thinking about water structure by their experimental work on the amorphous form of ice, a substance that may be regarded as an intermediate between crystalline ice and liquid water in the sense that it has the non-crystalline disorder but not the labile structure of the liquid. Amorphous ice samples formed by vapor deposition at 10 K and at 77 K gave X-ray r.d.f.'s that were similar in overall features to that of liquid water near 0°C, but showed differences of detail that indicated differing densities of 1.1 and 0.94 g cm^{-3} for the 10 K and 77 K deposits (Narten et al., 1976). For the less dense form, the interval of the r.d.f. from 3 to 4 Å was almost empty, as would be the case for an ice I-like structure, while for the denser form the r.d.f. showed a well developed small peak centered at 3.3 Å. Alternative structures suggested for the denser form of amorphous ice were (1) an ice I model with interstitials, like the Levy-Narten model mentioned above, and (2) a model containing contributions from ice II- and ice III-like molecular arrangements in addition to ice I.

The work on amorphous ice provided a stimulus for development by Rice and his co-workers of a new random network model for water. Based on potential functions for bond bending, bond stretching, and non-bonded interaction, and on concepts as to the different time scales of vibrational and diffusional motions, the model led to calculation of thermodynamic quantities for liquid water as a function of temperature (Sceats et al., 1979; Rice and Sceats, 1980; Sceats and Rice, 1980 a,b,c; Rice and Sceats, 1981). The theory succeeded in accounting for the thermodynamic functions with a structure in which all or most of the molecules are fully hydrogen bonded, with little or no bond breakage, and in which most of the energy storage is in bond bending. Such a structure has many of the qualitative attributes of the dense ice phases. The most important parameter that defines the structure in this model is the mean square bending angle of the hydrogen bonds; for water at 0°C the theory uses an experimentally-based value of about 20° for the r.m.s. diviation of the O· · ·O· · ·O angles from 109.5°, which is intermediate between the corresponding deviation values 18.5° in ice V and 23° in ice VI, and somewhat smaller than the value 26° obtained earlier by Pople (1951) in a geometrically similar model of water structure.

An excursion of large amplitude but short duration in the structural chemistry of water was generated in 1969-70 by the famous hubbub over "polywater." For this supposed form of H_2O, with density 1.4 g cm^{-3} and a variety of anomalous properties, no less than twelve structural models were put forward, most of which bore little or no relationship to the structures of the dense forms of water as known from crystallography. These theoretical models provided considerable food for thought as to what water molecules could and could not be expected to do in condensed matter within the framework of existing understanding and evidence (Kamb, 1971). However, the models soon evaporated along with the elusive liquid they were concocted for (Chemical & Engineering News, 1973).

References

Arnold, G. R., Wenzel, R. G., Rabideau, S. W., Nereson, N. G. and Bowman, A. L. (1971). J. Chem. Phys. **55**, 589.
Bader, H. (1951). J. Geol. **59**, 519.
Baur, W. H. (1965). Acta Cryst. **19**, 909.
Baur, W. H. (1972). Acta Cryst. **B28**, 1456.
Bentley, W. A. and Humphreys, W. J. (1931). Snow Crystals. New York: McGraw Hill. (Republished in 1962 by Dover Publications, New York).
Bernal, J. D. and Fowler, R. H. (1933). J. Chem. Phys. **1**, 515.
Bertie, J. E., Calvert, L. D. and Whalley, E. (1964). Can. J. Chem. **42**, 1373.
Block, S., Weir, C. W. and Piermarini, G. J. (1965). Science **148**, 947.
Chemical & Engineering News (1973) July 16, p. 13.
Clark, J. R. (1963). Aust. J. Pure Appl. Phys. **13**, 50.
Dennison, D. M. (1921). Phys. Rev. **17**, 20.
Duchamp, D. J. and Marsh, R. E. (1969). Acta Cryst. **B25**, 5.
Eisenberg, D. and Kauzmann, W. (1969). The Structure and Properties of Water. Oxford: Oxford University Press.
Engelhardt, H. and Kamb, B. (1978). J. Glaciol. **21**, 51.
Engelhardt, H. and Kamb, B. (1981). J. Chem. Phys., in press.
Engelhardt, H. and Whalley, E. (1972). J. Chem. Phys. **56**, 2678.
Gow, A. J. (1971). Cold Regions Res. Eng. Lab., Res. Report 300.
Gow, A. J. and Williamson, T. (1976). Geol. Soc. Am. Bull. **87**, 1665.
Hamilton, W. C. and Ibers, J. A. (1968). Hydrogen Bonding in Solids. New York: Benjamin, p. 204-220.
Hamilton, W. C., Kamb, B., La Placa, S. J. and Prakash, A. (1969). In "Physics of Ice," edited by N. Riehl, B. Bullemer, and H. Engelhardt. New York: Plenum Press, p. 44.
Hobbs, P. V. (1974). Ice Physics. Oxford: Oxford University Press, pp. 526-568.
Hooke, R. L. and Hudleston, P. J. (1980). J. Glaciol. **25**, 195.
Jeffrey, G. A. (1969). Accounts Chem. Res. **2**, 344.
Jeffrey, G. A. (1982). In "Inclusion Compounds." Edited by J. L. Atwood, J. E. Davies, and D. D. MacNicol. New York: Academic Press.
Jeffrey, G. A. and McMullan, R. K. (1967). Prog. Inorg. Chem. **8**, 43.
Kamb, B. (1959). J. Geophys. Res. **64**, 1891.
Kamb, B. (1964). Acta Cryst. **17**, 1437.
Kamb, B. (1965a). Science **150**, 205.
Kamb, B. (1965b). J. Chem. Phys. **43**, 3917.
Kamb, B. (1968). In "Structural Chemistry and Molecular Biology." Edited by A. Rich and N. Davidson, San Francisco: p. 507.
Kamb, B. (1969). Trans. Am. Crystallogr. Assoc. **5**, 61.
Kamb, B. (1970). Science **167**, 1520.
Kamb, B. (1971). Science **172**, 231.
Kamb, B. (1972a). Am. Geophys. Union, Geophys. Monograph **16**, p. 211.
Kamb, B. (1972b). In "Water and Aqueous Solutions." Edited by R. A. Horne, New York: Wiley, p. 9.
Kamb, B. (1973). In "Physics and Chemistry of Ice." Edited by E. Whalley, S. J. Jones and L. W. Gold, Ottawa: Royal

Society of Canada, p. 28.
Kamb, B. and Datta, S. (1960). Nature (London) **187**, 140.
Kamb, B. and Davis, B. L. (1964). Proc. Natl. Acad. Sci. U.S.A. **52**, 1433.
Kamb, B., Hamilton, W. C., La Placa, S. J. and Prakash, A. (1971). J. Chem. Phys. **55**, 1934.
Kamb, B. and La Placa, S. J. (1974). Am. Geophys. Union (EOS) **56**, 1202.
Kamb, B. and Prakash, A. (1968). Acta Cryst. **B24**, 1317.
Kamb, B., Prakash, A. and Knobler, C. (1967). Acta Cryst. **22**, 706.
Katzoff, S. (1934). J. Chem. Phys. **2**, 841.
La Chapelle, E. R. (1969). Field Guide to Snow Crystals. Seattle: University of Washington Press.
La Placa, S. J., Hamilton, W. C., Kamb, B. and Prakash. A. (1973). J. Chem. Phys. **58**, 567.
La Placa, S. J. and Post. B. (1960). Acta Cryst. **13**, 503.
McFarlan, R. L. (1936a). J. Chem. Phys. **4**, 60.
McFarlan, R. L. (1936b). J. Chem. Phys. **4**, 253.
Miller, S. L. (1973). In "Physics and Chemistry of Ice," Edited by E. Whalley, S. J. Jones and L. W. Gold. Ottawa: Royal Society of Canada, p. 46.
Morgan, J. and Warren, B. E. (1938). J. Chem. Phys. **6**, 666.
Nagle, J. (1966). J. Math. Phys. **7**, 1484.
Narten, A. H. (1972). J. Chem. Phys. **56**, 5681.
Narten, A. H., Danford, M. D. and Levy, H. A. (1967). Discuss. Faraday Soc. **43**, 97.
Narten, A. H. and Levy, H. A. (1969). Science **165**, 447.
Narten, A. H. and Levy, H. A. (1971). J. Chem. Phys. **55**, 2263.
Narten, A. H., Venkatesh, C. G. and Rice, S. A. (1976). J. Chem. Phys. **64**, 1106.
Pauling, L. (1935). J. Am. Chem. Soc. **57**, 2680.
Pauling, L. (1959). In "Hydrogen Bonding." Edited by D. Madzi, New York: Pergamon Press, p. 1.
Pauling, L. and Marsh, R. E. (1952). Proc. Natl. Acad. Sci. U.S.A. **38**, 112.
Pauling, L. and Pauling, P. (1968). Proc. Natl. Acad. Sci. U.S.A. **60**, 362.
Peterson, S. W. and Levy, Henry, A. (1957). Acta Cryst. **10**, 70.
Pople, J. A. (1951). Proc. Roy. Soc. (London) **A205**, 155.
Rabideau, S. W., Finch, E. D., Arnold, G. P. and Bowman, A. L. (1968). J. Chem. Phys. **49**, 2514.
Rice, S. A. and Sceats, M. G. (1980). J. Chem. Phys. **72**, 3236.
Rice, S. A. and Sceats, M. G. (1981). J. Phys. Chem. **85**, 1108.
Rigsby, G. P. (1960). J. Glaciol. **3**, 589.
Sceats, M. G. and Rice, S. A. (1980a). J. Chem. Phys. **72**, 3248.
Sceats, M. G. and Rice, S. A. (1980b). J. Chem. Phys. **72**, 3260.
Sceats, M. G. and Rice, S. A. (1980c). J. Chem. Phys. **72**, 6183.
Sceats, N. G., Stavola, M. and Rice, S. A. (1979). J. Chem. Phys. **70**, 3927.
St. John, A. (1918). Proc. Natl. Acad. Sci. U.S.A. **4**, 193.
Weir, C. E., Block, S. and Piermarini, G. (1965). J. Res. Nat. Bur. Std. **69C**, 275.
Whalley, E., Davidson, D. W. and Heath, J. B. R. (1966). J. Chem. Phys. **45**, 3976.
Whalley, E., Heath, J. B. R. and Davidson, D. W. (1968). J. Chem. Phys. **48**, 2362.
Whatley, L. S. and Van Valkenburg, A. (1966). In "Advances in High Pressure Research." Edited by R. S. Bradley. Vol. 1, New York: Academic Press, p. 327.
Wollan, E. D., Davidson, W. L. and Shull, C. G. (1949). Phys. Rev. **75**, 1348.

CHAPTER 4

CLAY MINERALOGY – STRUCTURAL DEVELOPMENT

Ralph E. Grim
Geology Department, University of Illinois
Urbana, IL 61801

A review of the older literature (Grim, 1968) shows that through the years many concepts were suggested to portray the fundamental and essential components of clay materials and to explain their various properties. The diversity of the concepts was a consequence of the lack of adequate analytical tools to determine, with any degree of certainty, the exact nature of the fundamental building blocks of most clay materials. The components of clays and shales were in particles too small and variable for available methods of study.

Le Chatelier in 1887 and Lowenstein in 1909 suggested that clays were composed of extremely small particles of a limited number of crystalline minerals. This is the clay mineral concept which has now attained universal agreement. But it was not until 1923 when Hadding in Sweden and a year later Rinne (1924) in Germany performed X-ray diffraction analyses on clays that the concept was established. The work by Ross and his colleagues of the U.S. Geological Survey, beginning with superb optical studies with the petrographic microscope, and later expanded to include X-ray diffraction, thermal, and chemical data, led to a monumental series of reports (e.g., Ross and Kerr, 1931) which firmly established the clay mineral concept.

The properties of soils, clays, shales, in fact all argillaceous materials (i.e., clay materials), are largely a consequence of their clay mineral composition and ultimately of the atomic structure of the clay minerals. As a consequence, a vast amount of research has been done, and a voluminous literature has developed on the structure of the clay minerals. Within a reasonable amount of space it will be possible only to refer to the studies describing the general atomic structure of the various clay minerals, and to recent investigations revealing details and variations in their structures.

The generalizations of Pauling in 1930 for the structure of mica and related layer minerals suggested that two structural units were involved in the atomic structure of such minerals. One unit consisted of two planes of closely packed oxygens or hydroxyls in which aluminum, iron and/or magnesium atoms were embedded in octahedral coordination so that they were equidistant from six oxygens or hydroxyls. The second unit is built of silica tetrahedrons. In each tetrahedron a silicon atom is equidistant from four oxygens or hydroxyls, if needed to balance the structure, with the silicon atom at the center. The tetrahedral groups are arranged to form a hexagonal network which is repeated indefinitely to form a sheet with the composition $Si_4O_6(OH)_4$.

Kaolinite

The structure of this clay mineral, as suggested by Pauling (1930a) and confirmed by Gruner (1932), is composed of a single tetrahedral sheet and a single alumina octahedral sheet combined in a unit so that the tips of the silica tetrahedron and one of the planes of the octahedral sheet form a common layer. Numerous investigators, notably Brindley (1961), have studied the kaolinite structure in detail and have pointed out distortions in the structure. Noteworthy are the electron diffraction studies of Zvyagin (1960) in which he pointed out rotations within both the octahedral and tetrahedral part of the structure and slight displacement of both the silicon and aluminum atoms. Bailey (1963) presented a concept of the variation of the stacking sequence in kaolinites, and the variation of the position of the aluminum atoms in the three possible octahedral sites, only two of which are occupied in kaolinite. Plançon and Tchoubar (1977) recently have concluded that the major defect in kaolinite is not the 1/3 translation but the displacement from one layer to another (or from one domain to another) of aluminum vacancies.

Careful chemical analyses of purified samples of kaolinite have suggested that small amounts of titanium, iron, magnesium, and possibly other elements, can be present in the kaolinite structure. Angel and Hall (1972) established the presence of iron and magnesium in some kaolinite structures on the basis of electron spin resonance studies. Using this same technique, Mead and Molden (1975) confirmed the presence of iron and other transition metal ions in kaolinite and pointed out the influence of these elements on the kaolinite structure. Very recently Mestdagh et al. (1980), using electron paramagnetic resonance, have indicated the presence and position of iron in the kaolinite structure and its relation to the crystallinity.

Murray and Patterson (1976) have considered the variation of the crystallinity of kaolinite in kaolins, fireclays, and Ball Clays and its influence on properties.

Halloysite

There are two forms of halloysite, one with the composition of kaolinite $(OH)_8Si_4Al_4O_{10}$ and the other, its tetrahydrate, with the composition $(OH)_8Si_4Al_4O_{10}.4H_2O$. The latter form dehydrates to the former at relatively low temperatures. Under ambient conditions there is a slow transition to the lower hydration form with a drastic change in plasticity and other properties. It is generally agreed that the dehydrated form has a structure similar to that of poorly crystalline kaolinite, and that the hydrated form has a single layer of water molecules between the silicate layer (Hendricks and Jefferson, 1938). It is also generally agreed that the halloysite structure is considerably more random than the kaolinite structure. Chukhrov and Zvyagin (1966) have discussed in detail the difference in degree of randomness between kaolinite and halloysite. The extensive discussion of this paper by many investigators emphasized the fact that the peculiar morphology of some halloysites, for example, the tubular forms and those showing sawtoothed edges (Hofmann et al, 1962) have not been adequately explained on the basis of atomic structure.

Smectite

The structure of smectite first proposed by Hofmann, Endell, and Wilm (1933) is composed of units of two silica tetrahedral sheets with a central alumina octahedral sheet. The population of the octahedral positions, which may consist of aluminum, magnesium, iron, and some other atoms, and to a lesser extent the tetrahedral positions is somewhat variable, but always results in a structure with a net negative charge which is balanced by exchangeable cations between and around the silicate unit. An outstanding feature of the smectite structure is that water, other polar molecules, and certain organic molecules can enter the structure causing it to have a variable *c*-axis dimension.

Numerous investigators, especially Zvyagin and Pinsker (1949), Conley and Goswami (1961), Radoslovich (1962), Besson et al., (1974), and Glaeser and Mering (1976), have presented detailed information on translations and rotations within the smectite structure. Very recently Rozenson and Heller-Kallai (1977) have used Mossbauer spectra data to investigate the position of ferric iron and ferrous iron in dioctahedral smectites and accompanying structure distortions. Guven in the volume by Grim and Guven (1978), evaluated the major factors affecting the electron diffraction pattern of mica-type layer silicates. He considered in detail the interference function for undeformed lattices and then the effect of bending of the lattice on this function. A detailed study by selected area electron diffraction of muscovite preceded an investigation of the crystal structure of beidellite and montmorillonite. This volume presents data on the morphology and structure of about 75 samples of smectite in various bentonites from around the world.

Many investigators (see Grim, 1968) have presented evidence to show that the water held directly on the surfaces of the clay particles is in a physical state different from that of liquid water. Not the least of this evidence is that the nature of this water is important in understanding many of the physical properties of clay-water systems (Grim and Cuthbert, 1945). Thus the interlayer water in smectites is believed to have some sort of organization. At the present time there is no general agreement regarding the exact nature of the organization of this water (Ravina and Low, 1972).

Illite

The micaceous clay minerals have a structure like that of smectite, except there is less variation in the composition of the octahedral positions, and there is substantial substitution of aluminum for silicon in the tetrahedral positions. Further, the net negative charge is balanced by potassium ions between the silicate sheets binding them together so that the *c*-axis dimension is fixed. These clay minerals occur in extremely small particles, and often precise diffraction data to reveal structural detail cannot be obtained, e.g., to determine whether they are dioctahedral or trioctahedral. Grim, Bradley, and Bray (1937) suggested that illite be used as a general term, not as a specific mineral name, to describe these minerals.

Radoslovich (1962) has considered the translations and other structural variations in micaceous minerals and proposed (1963) a new ideal structure of layer silicates in which the tetragonal surfaces are ditrigonal, and the octahedral units are stretched and flattened, and the interlayer cations are effectively six-coordinated. Giese (1971) has shown that it is feasible to calculate the lattice energy of layer silicate structures and to derive thereby information on the bonding mechanism. Giese later (1973) considered the orientation of the hydroxyl ion in such phyllosilicates. Farmer (1974) has studied the relation of infrared spectra to these and other clay minerals. Velde and Weir (1979) claimed to have synthesized illite, and have studied the structural implications of such synthesis.

Many investigators have considered the possible development of illite from smectite by the fixation of potassium during diagenesis following the burial of sediments (Grim, 1968). Mamy and Gaultier (1976) have investigated the fixation of potassium by montmorillonite by successively wetting and drying montmorillonite saturated with potassium at 80°C. They found that on successive treatments the exchange capacity decreases, the interlayer distance collapses to about 10Å, and swelling is lost. Further electron diffraction data show that the initial turbo-stratic structure changes to regular stacking. The structural changes involved in the transition from smectite to mixed layer structures, and to illite, have also been considered in detail recently by Jonas (1976), Plançon and Tchoubar (1979), and Eberl (1979).

Chlorite

The structure of chlorite suggested by Pauling (1930b) and confirmed by McMurchy (1934) consists of alternate mica-layers and brucite-like layers $(Mg.Al)_6(OH)_{12}$. Various polymorphic forms of chlorite are known to exist (Brindley and Robinson, 1961) but the extremely small particle size of the mineral in clay materials makes the identification of the form about impossible. In clay mineral analyses the name chlorite is applied to a component with a *c*-axis dimension of about 14Å which does not expand and which does not collapse on moderate heating.

Vermiculite

According to Gruner (1939) the structure of vermiculite consists of layers of trioctahedral mica or talc separated by layers of water molecules occupying a space of 4.98Å which is about the thickness of two water molecules. The structure is unbalanced chiefly by substitutions of aluminum for silicon which results in a net charge deficiency that is satisfied by cations which occur chiefly between the mica layers and are largely replaceable. The balancing cation is commonly Mg^{2+}. On moderate heating the water leaves the structure with a reduction in the *c*-axis dimension.

There have been numerous studies of the morphology of vermiculite and variations in its structure (see Walker, 1961). In clay mineral analyses vermiculite is identified as a component with a 14Å spacing which collapses on moderate heating and does not expand again beyond about 14Å. As in the case of chlorite, its extremely small particle size and intimate mixing with other minerals makes any further identification impossible.

Attapulgite, Palgorskite, and Sepiolite

Fersman in 1913 and later Caillere in 1936 and Nagelschmidt in 1938 pointed out that there are clay materials which are composed of elongated fibrous particles rather than the usual flake-shaped units. It is now known that the fundamental unit of these fibrous particles are double silica chains linked together through oxygens at their longitudinal edges. The apices of the tetrahedra in successive chains point in opposite directions, thus forming a kind of double-ribbed sheet. The ribbed sheets are arranged so the apices are held together by aluminum and/or magnesium in octahedral coordination between the apex oxygens of successive sheets, (deLapparent, 1938, Bradley 1940, Brauner and Preisinger, 1956, Zvyagin et al., 1963). The structure yields lath shaped units with a gutter and channel surface. The names attapulgite or palygorskite (the names are used interchangeably with palygorskite being favored by European investigators) and sepiolite are applied to these minerals with the sepiolite units having wider lath dimensions.

Recently Rautureau and Tchoubar (1976) have studied sepiolite with selected-area electron diffraction and have confirmed the Brauner and Preisinger model with the added detail that the magnesium at the edges of the laths occupy two sites and the H_2O bound to the magnesium also occupies two positions.

Imogolite and Allophane

Following the general acceptances of the clay mineral concept, it was considered that clay materials were composed totally of crystalline components, i.e., they contained no amorphous material. It is now known that material too poorly organized to yield distinct diffraction effects by either X-rays or electron beams is present in some instances. Allophane is the name applied to such material. Wada (1967) first proposed a chain-like structure for allophane with a Si/A1 ratio of one in which a silica tetrahedral chain shared corners with an aluminum octahedral chain. Wada believed infra-red data supported his structure. Later Udagawa, Nakada, and Nakahira (1969) suggested a sheet structure rather than a chain structure in which substantial percentages of aluminum (40%) are in tetrahedral rather than octahedral coordination. Various investigators have suggested that allophane has a defect kaolinite structure and Wada (1979) has recently summarized concepts of allophane structures and concluded that the major features are the defects in both the tetrahedral and octahedral sheets and the position of the water molecules bonded to aluminum in tetrahedral sheets. Earlier Wada and Wada (1977) described spherical units with walls having the structural attributed noted above.

Only slightly better organized than allophane is a material with a tubular morphology observed especially in Japan in the form of gel-like films coating weathered pumice particles which has been called imogolite (Wada and Howard 1974). Cradwick et al. (1972) on the basis of electron diffraction data, suggested that the basic structure of an imogolite tube was a single gibbsite sheet bent round in the form of a cylinder with orthosilicate groups attached to the inside of the cylinder, each group replacing three (OH) groups around an empty octahedral site. Farmer and Fraser (1979) have succeeded in synthesizing tubular material which seems to confirm the suggested structure.

Mixed-Layer Structures

So-called mixed-layer minerals are interstratifications of layer silicate minerals in which the individual layers are of the order of a single or a few alumina-silicate sheets. These structures are a consequence of the fact that the layers of the minerals have very similar structures and dimensions. The interstratification may be regular with the stacking along the *c*-axis a regular repetition of the different layers with the consequence that the *c*-dimension is equivalent to the sum of that of the unit layers and a regular order of reflections is obtained.

It is well known now that a common component of many clay materials are random interstratifications of two and sometimes three different layer silicates. Interstratification of smectite and illite components are the most common mixed-layer minerals, and much material reported in the literature as illite is such a mixed-layer structure. Mixed-layer structures containing kaolinite have been reported (Cradwick and Wilson, 1972) but are less frequent than those composed of the 2:1 layer silicates.

In many cases diffraction data for clay mineral components permit the identification of the individual components, but occasionally such identification is very difficult. As a consequence, the nomenclature and classification of the clay minerals is not as straightforward and unequivocal as that of minerals which occur in large enough units so that single crystal investigations are possible.

Random mixed-layer structures do not give an integral series of basal orders. Hendricks and Teller (1942) and Bradley (1945) considered the diffraction characteristics of random mixed-layers. MacEwan in 1956 presented a Fourier transform method for studying the scattering of random interstratification. In 1965, MacEwan extended the Fourier transform method to crystallites with irregularities in two and three dimensions. In recent years many investigators have studied the scattering effects of random mixed-layers and suggested procedure for their evaluation. Thus in 1970, Reynolds and Hower suggested a computer procedure and Tettenhorst and Grim (1975) presented a computational method based on a somewhat different transform method which allows the examination of individual layer transforms or of the mean for any given set of layers. Cradwick and Wilson (1978) presented a calculation of the diffraction effects for a three component system.

Farmer and Russell (1971) have studied the infrared absorption characteristics of interlayer complexes and in 1976 Drits and Sakharova published a comprehensive survey of the basic theoretical and practical problems of X-ray structural analyses of mixed-layer minerals of any type of interstratification and composition. These authors used statistical probability coeffcients to present a complete scheme of classification of mixed-layer structures. Hopefully, this volume will be translated from the Russian.

Clay Minerals - Organic Reactions

Students of soils in the field of agriculture appear to have been the first to suggest that the inorganic components of clay materials may form definite complexes with organic compounds. In recent years it has been shown that a large variety of organic compounds, but especially polar and ionic organics can form definite complexes with the clay minerals. Investigations were first directed to the expanding clay minerals, but recently have included kaolinite and halloysite. The research has been concerned with the nature of the bonding of the organic molecules to the clay mineral surface and the structural orientation of the molecules on the clay mineral surface. For information on this matter up to about 1965 see Grim (1968). Theng (1974) has admirably summarized recently the chemistry of clay-organic reactions.

It has long been known that various clay minerals may serve to catalyze certain organic reactions (Robertson, 1948). Much information on this subject is to be found in the patent literature, e.g., Hauser and Kollman (1960). In recent years, Johns and his colleagues (see Johns and Shimoyama, 1972) have been studying the catalytic action of the clay minerals as a factor in the origin of petroleum. Of special interest from a structural point of view is the production of commercial catalysts by processing either smectite or kaolinite by alkali or acid treatment in such a way that the atomic structure is modified or a new structure is produced. An example of the latter is the production of zeolite-type catalysts by treatment of kaolinite with an alkali. Again, the patent literature needs to be consulted for information on this subject. (Haden and Dzlerzanowski, 1968).

Structural Changes in the Clay Minerals at Elevated Temperature

The fact that clay minerals used in the ceramic industry are subjected to elevated temperatures has led to extensive studies of the successive changes that take place when the clay minerals are heated up to their fusion point. In general, the clays first lose pore and adsorbed water and then at temperatures of 500-700°C lose their hydroxyl water. The loss of hydroxyls may or may not be accompanied by a loss of structure. If the clay material contains substantial iron, alkalies, and/or alkaline earths, fusion may begin at temperatures of the order of 900-1000°C. In the absence of such fluxes, as in kaolinite for example, a series of crystalline forms develop in the interval of from about 900°C up to the fusion temperature of about 1700°C. This matter has been much investigated by thermal analyses and X-ray diffraction. Particularly revealing are diffraction data obtained while the mineral is at elevated temperatures.

Kaolinite experiences an intense and very sharp exothermic reaction at about 950°C. Considerable differences of opinion exist as to the cause of this thermal reaction even though it has been extensively studied. It has been attributed to the nucleation of mullite, the formation of α-alumina, and/or a spinel type structure. Above about 1100°C, kaolinite yields mullite and cristobalite. DeKeyser (1965) who has made the most thorough study of the changes in the structure of kaolinite on heating, has shown, for example, that a very small amount of water (0.4%) remains at temperatures above 800°C and that the loss of this water accompanies the slight endothermic peak that immediately precedes the large exothermic peak at about 950°C. He believed that a spinel-type structure preceded the formation of mullite and that its formation accompanied the 950°C exothermic reaction. A difficulty in accepting this conclusion results from finding Comfero et al. (1948) that the hexagonal morphology of kaolinite is preserved and needles of mullite are present on heating to about 1000°C. Brindley (1976) has recently summarized information on the thermal transformation of the clay minerals.

Of interest and importance is the fact that small amounts of various elements present as additives to the clay mineral or present in the atmosphere during calcining may very substantially change the thermal transition. Reductions in temperature of the order of several hundred degrees, for the formation of mullite and cristobalite may take place in the case of kaolinite (Wahl, 1962).

Investigation of the 2:1 minerals has shown a considerable number of high temperature phases formed between dehydration and fusion. Grim and Kulbicki (1961) have emphasized the point that in smectites, the structure immediately following the loss of hydroxyls shows inheritance characteristics of the original structure. Phases formed at higher temperatures are largely the result of the chemical composition of the sample. Brindley (1963) considered in some detail the relation of the structure of parent material to reactions at elevated temperatures in clay and related materials.

With the continued importance of clay minerals for use in ceramics, catalysis, masonry, fillers for dynamite, extenders for glues, and dozens of other uses, we are assured that research on clay minerals will continue in North America for many years.

References

Angel, B. R. and Hall, P. L. (1971). Proc. Int. Clay Conf., 1972, Madrid, **47**.

Bailey, S. W. (1963). Am. Mineral. **48**, 1196.

Besson, G., Mifsud, A., Tchourbar, C. and Mering, J. (1974). Clays and Clay Minerals **22**, 379.

Bradley, W. F. (1940). Am. Mineral. **25**, 405.

Bradley, W. F. (1945). Am. Mineral. **30**, 704.

Brauner, K. and Preisinger, A. (1956). Tschermaks Mineral. Petrog. Mitt. **6**, 120.

Brindley, G. W. (1961). X-ray Identification and Structures of Clay Minerals. Chapter II, p. 32. London: Monograph Mineral. Soc. Gr. Britain.

Brindley, G. W. and Robinson, K. (1961). X-ray Identification and Structures of Clay Minerals. Chapter VII, p. 199. London: Monograph. Mineral. Soc. Gr. Britain.

Brindley, G. W. (1963). Proc. Int. Clay Conf., Stockholm, **1**, 37.

Brindley, G. W. (1976). Proc. Int. Clay Conf. 1975, Mexico City, **1**, 119.

Caillere, S. (1936). Bull. Soc. Franc. Mineral, **59**, 353.

Chukhrov, F. V. and Zvyagin, B. B. (1966). Proc. Int. Clay Conf., Jerusalem, **I**-11.

Cibula, D. J., Thomas, R. K., Middleton, S. and Ottewill, R. H. (1979). Clays and Clay Minerals, **27**, 39.

Comfero, J. E., Fischer, R. B. and Bradley, W. F. (1948). J. Amer. Cer. Soc. **31**, 254.

Conley, J. M. and Goswami, A. (1961). Acta Cryst. **14**, 1071.

Cradwick, P. D. and Wilson, M. J. (1972). Clay Minerals, **9**, 391.

Cradwick, P. D., Farmer, V. C., Russell, J. D., Masson, C. R., Wada, K. and Yoshinaga, N. (1972). Nature Phys. Sci.. (London) **240**, 187.

Cradwick, P. D. and Wilson, M. J. (1978). Clay Minerals. **13**, 53.

DeKeyser, W. L. (1963), Proc. Int. Clay Conf., 1963, Stockholm, **2**, 91.

deLapparent, J. (1938). Bull. Soc. Franc. Mineral. **61**, 253.

Drits, V. A. and Sakharova, B. A. (1976). X-ray Structural Analysis of Mixed-Layer Minerals. Trans. Acad. Sci. U.S.S.R. 295.

Eberl, D. (1970). Proc. Int. Clay Conf 1978, Oxford, U.K., 375.

Farmer, V. C. (1974). The Infrared Spectra of Minerals. London: Mineral. Soc. Gr. Britain.

Farmer, V. C. and Russell, J. D. (1971). Trans. Farady Soc. **67**, 2737.

Farmer, V. C. and Fraser, A. R. (1979). Proc. Int. Clay Conf., 1978, Oxford, U.K., 547.

Fersman, A. (1913). Mem. Russian Acad. Sci. **32**, 377.

Giese, R. (1971). Science **172**, 263.

Giese, R. (1973). Clays and Clay Minerals **21**, 145.

Glaeser, R. and Mering, J. (1975). Proc. Int. Clay Conf. 1975, Mexico City, 175.

Grim, R. E. Bradley, W. F. and Bray, R. M. (1937). Am. Mineral **22**, 813.

Grim, R. E. and Cuthbert, F. L. (1945). J. Amer. Cer. Soc. **28**, 90.

Grim, R. E. and Kulbicki, G. (1961). Am. Mineral. **46**, 1329.

Grim, R. E. (1968). Clay Mineralogy. New York: McGraw-Hill Book Company.

Grim, R. E. and Guven, N. (1978). Bentonite; Mineralogy, Properties, Uses. Amsterdam; Elsevier Scientific Pub. Co.

Gruner, J. W. (1932). Z. Kristallogr. **83**, 75.

Gruner, J. W. (1939). Am. Mineral. **24**, 428.

Hadding, A. (1923). Z. Kristallogr. **58**, 108.

Haden, W. L. and Dzierzanowski, F. J. (1968). U.S. Patent, 3, 367, 887.

Hauser, E. A. and Kollman, R. C. (1960). U.S. Patent 2, 951, 087.

Hendricks, S. B. and Jefferson, M. E. (1938). Am. Mineral. **23**, 863.

Hendricks, S. B. and Teller, E. (1942). J. Chem. Phys. **10**, 147.

Hofmann, U., Morcos, S. and Schembra, F. W. (1962). Ber. Deut. Keram. Ges. **39**, 475.

Hofmann, U., Endell, K. and Wilm, D. (1933). Z. Kristallogr. **86**, 340.

Johns, W. D. and Shimoyama, R. (1972). Bull. Am. Assoc. Petrol. Geol. **56**, 2160.

Jonas, E. C. (1976). Proc. Int. Clay Conf. 1975, Mexico City, **1**, 3.

LeChatelier, H. (1887). Bull. Soc. Franc. Mineral. **10**, 204.

Lowenstein, E. (1909). Z. Anorg. Allg. Chem. **63**, 69.

Mamy, J. and Gaultier J. P. (1976). Proc. Int. Clay Conf. 1975, Mexico City, **1**, 149.

MacEwan, D. M. C. (1956). Kolloid Zeitz. **149**, 96.

MacEwan, D. M. C. and Sutherland, H. H. (1965). Proc. Int. Clay Conf.. 1963, Stockholm, **2**.

McMurchy, R. C. (1934). Z. Kristallogr. **88**, 420.

Mead, R. E. and Molden, P. J. (1975). Clay Minerals **10**, 317.

Mestdagh, M. M., Vielvoye, L. and Herbillon, A. J. (1980). Clay Minerals **15**, 1.

Murray, H. and Patterson, S. (1976). Proc. Int. Clay Conf. 1975, Mexico City, **1**, 511.

Nagelschmidt, G. (1938). Nature (London) **143**, 114.

Pauling, L. (1930a). Proc. Natl. Acad. Sci. U.S.A. **16**, 123.

Pauling, L. (1930b). Proc. Natl. Acad. Sci. U.S.A. **16**, 578.

Plançon, A. and Tchoubar, C. (1977). Clays and Clay Minerals **25**, 436.
Prost, R. (1976). Proc. Int. Clay Conf. 1975, Mexico City, **1**, 351.
Radoslovich, E. W. (1962). Am. Mineral. **47**, 617.
Radoslovich, E. W. (1963). Proc. Int. Clay Conf., Stockholm, **3**.
Rautureau, M. and Tchoubar, C. (1976). Clays and Clay Minerals. **24**, 43.
Ravina, I. and Low, P. F. (1972). Clays and Clay Minerals, **20**, 109.
Reynolds, R. C. Jr. and Hower, J. (1970). Clays and Minerals. **18**, 25.
Rinne, F. (1924). Z. Kristallogr. **60**, 55.
Robertson, R. H. S. (1948). Clay Minerals Bull. **2**, 38.
Ross, C. S. and Shannon, E. V. (1926). J. Amer. Cer. Soc. **9**, 77.
Ross, C. S. and Kerr, P. F. (1931). U.S. Geol. Surv. Prof. Paper. 165E.
Rozenson, I. and Heller-Kallal, L. (1977). Clays and Clay Minerals **25**, 94.
Tettenhorst, R. and Grim, R. E. (1975). Am. Mineral., **60**, 49 and 60.
Theng, B. K. G. (1974). Chemistry of Clay-Organic Reactions London: Adam Hilger Ltd.
Udagawa, S., Nakada, T. and Nakahira, M. (1969). Proc. Int. Clay Conf. Vol. 1, 1978, Tokyo, 151.
Velde, B. and Weir, A. H. (1979). Proc. Int. Clay Conf. 1978, Oxford, Gr. Britain, 395.
Wada, K. (1967). Am. Mineral. **52**, 690.
Wada, K. and Howard, M. E. (1974). Adv. Agron. **26**, 211.
Wada, S. I. and Wada, K. (1977). Clay Minerals, **12**, 289.
Wada, K. (1979). Proc. Int. Clay Conf. 1978, Oxford, Gr. Britain, 537.
Wahl, F. M. (1962). Advances in X-ray Analysis. 10th Conf. p. 264. New York: Plenum Press.
Walker, G. F. (1961). X-ray Identification and Structures of Clay Minerals. Chapter VII, p. 199. London: Monograph. Mineral. Soc. Gr. Britain.
Zvyagin, B. B. and Pinsker, Z. G. (1949). C. R. Acad. Sci. USSR **68**, 65.
Zvyagin, B. B. (1960). Kristallografiya **5**, 40.
Zvyagin, B. B., Mishchenko, K. S. and Shitov, V. A. (1963). Kristallografiya **8**, 201.

CHAPTER 5

METALS

Charles S. Barrett
Denver Research Institute, University of Denver, Denver, CO 80208

American developments in the field of crystallography of metals and alloys are so numerous and so varied in type that it is impossible to condense a comprehensive history of them into a small space, even if details regarding them are omitted. It is necessary to choose a few and let the reader refer to the major reviews and books, and also to subjects covered elsewhere in this volume for the many subjects not included here. My choices include some topics that I have always found especially interesting, including of course some in which I have been involved.

Metal crystallography in this country made a quantum jump ahead in 1917 when Hull developed the powder method of X-ray diffraction independently and almost simultaneously with Debye and Scherrer (Hull, 1917) and solved the crystal structure of most of the common metals with this method. Another product of General Electric scientists (in addition to W. D. Coolidge's improvement in X-ray tubes) was the development of Hull-Davey charts for indexing powder photographs (Hull and Davey, 1921). These were widely used for years, then were modified to an improved form, the Bunn charts.

It is scarcely possible to overstate the importance of the powder method to metallurgical research, especially in the early years when there were so many important phases of alloys that were relatively simple in structure and readily solved by the powder method. Lattice constants could be measured with precision, phase changes found, composition limits of the phases determined, quantitative phase analyses made, the types of solid solutions determined, preferred orientations could be disclosed, and the effects of imperfections, cold working and annealing could be seen, all without taking the trouble to grow single crystals. Successively improved powder cameras and diffractometers appeared on the market, so that increased precision in lattice constant and intensities became possible; also correction formulas were improved, for example Cohen's least square method for lattice constants (Cohen, 1935) and others which became widely used in computer programs which also were used to compute intensities and angles when given known parameters of any crystal structure.

The subject of phase diagrams of alloys, so basic to nearly all metallurgical research, should be opened by citing J. Willard Gibbs' fundamental work *(Collected Works,* Gibbs, 1948). The amount of research on phase diagrams has been tremendous with X-rays being just one of many methods used in determining them. We must mention Bain's pioneer X-ray diffraction work on alloys (Bain, 1923) in which he deduced order in the arrangement of atoms in some phases, and then merely cite some of the most widely used summaries of the structures of alloy phases that were written in this country: *Crystal Structures* (Wyckoff, 1963); *Structure Reports* (many volumes, 1951 to date, following the earlier volumes of *Structurbericht*); the JCPDS files (Joint Committee on Powder Diffraction Standards) which grew out of the files of J. D. Hanawalt and coworkers at the Dow Chemical Company (Hanawalt, Rinn and Frevel, 1938) and the subsequent committee work of the ASTM and the JCPDS; *Constitution of Binary Alloys* (Hansen and Anderko, 1963 and *Supplement* (Elliot, 1965); *The Metals Handbook* in numerous editions (American Society for Metals); *Structure of Metals* (first edition in 1943, third edition by Barrett and Massalski, 1966); *Electronic Structure and Alloy Chemistry of the Transition Elements* (Beck, ed., 1963); *Alloying Behavior and Effects in Concentrated Solid Solutions* (Massalski, ed., 1965); *Rare Earth Alloys* (Gschneider, 1961); *Crystal Chemistry of Tetrahedral Phases* (Parthé, 1964); *Phase Stability in Metals and Alloys* (Rudman and Stringer, ed., 1966); *Intermetallic Compounds* (Westbrook, ed., 1967).

A series of important results has been obtained with P. Duwez's method of extremely rapid cooling of liquids, by splashing the liquid drops against a metal surface of high thermal conductivity (Duwez, Willens and Klement, 1960), (Duwez and Willens, 1963). New metastable phases were formed and solid solution composition ranges were extended; also some phases that were expected to form under equilibrium conditions were in fact obtained only after the very fast cooling, for example the CsCl type structures of Au with Y, Pr, Nd, and Gd (see Duwez in Westbrook, 1967). Although vacuum deposition on substrates at cryogenic temperatures had been known to produce glassy thin films, these were very metastable, whereas rapid cooling from the melt produced many different foils of metallic glasses that in several cases did not recrystallize below 125°C or even higher. Many amorphous alloys involved Te with Ge, Ga and In (Luo and Duwez, 1963); research on the subject has continued actively at Caltech, with expansion into multicomponent alloys. The subject is of obvious interest to industry since glassy metallic ribbons that have unusual properties can now be produced and are being marketed.

Another active field of metals research in this country has been crystallography at high pressures, which started with the pioneer work of P. Bridgman (Bridgman, 1949). Gray tin, of the diamond structure (A4 type), was transformed to the higher density white tin structure under pressure. Similar transformations were found in silicon and germanium, and transformations in several III-V and II-VI compounds (see lists in Westbrook, 1967, and Barrett and Massalski, 1966).

Plastic flow in the form of extreme amounts of shear under pressure were explored by Bridgman with interesting results, and even moderate plastic flow was found to act as a substitute for thermal agitation in overcoming barriers to phase transformations, changing metastable phases to more stable ones, as in lithium at low temperatures (Barrett and Trautz, 1948) and also sodium and some alloys (Barrett and Massalski, 1966). When the transformation is merely a sliding of close-packed layers over each other in face centered cubic or hexagonal closepacked phases, the curve on a phase diagram representing the curve along which the phases have equal free energy can be determined by this strain-induced transformation method. Also with martensitic transformations the curve representing the highest temperature at which strain initiates the transformation (the so-called M_d curve) can be determined; it may lie at much higher temperatures than the M_s curve for the beginning of martensite in unstrained material, as for example in β-brass (Massalski and Barrett, 1957).

Phase transformations in alloys are treated, with many references, in *Phase Transformations* (1970) and more briefly in Barrett and Massalski (1966). The summaries discuss the many branches of the subject, including for example the notable theory of Wechsler, Lieberman and Read (1953) for the change of shape. orientation, and habit plane in martensitic transformations. Earlier approaches to this subject had been made as early as 1924 by E. C. Bain and later by A. B. Greninger and A. R. Troiano, by E. S. Machlin and Morris Cohen and by J. S. Bowles and numerous further contributions have come from D. S. Lieberman and from C. M. Wayman at the University of Illinois. Other branches of the subject include bainite, order-disorder of both long range and short range, clustering in solid solutions, spinodal transformations, and massive transformations. We became used to expecting continuing contributions to the field from MIT, especially from Morris Cohen and co-workers; from Northwestern University where J. B. Cohen continued research on local order; from H. Sato and R. S. Toth on superlattices; from R. F. Mehl and co-workers on transformations involving long range diffusion, also from H. I. Aaronson in this area; from J. W. Cahn on spinodal decomposition; from T. A. Read and others on thermoelastic martensite; and from T. B. Massalski on massive transformations.

The transformation from cubic to tetragonal in In-Tl alloys with a very slight change in axial ratio was of special interest to some of us (Bowles, Barrett and Guttman, 1950); a similar one, with even smaller axial ratio change was found in V_3Si (Batterman and Barrett, 1966). The importance of V_3Si and other examples of the A15 structure is that many of them are superconducting at low temperatures, so they have been extensively studied by B. T. Matthias and others (see Westbrook, 1967 for a summary). There is a connection between these In-Tl and A15 transformations that can be credited to an early proposal of C. Zener's (1947). He suggested that the body centered cubic structure is soft with respect to shear on (110) planes in the $[1\bar{1}0]$ direction, and that unusually large thermal vibrations of this shear type should tend to make that structure unstable at low temperatures. This suggestion led to the finding of such changes in Li and Na. The cubic form of In-Tl and V_3Si likewise has lowered stiffness for (110) $[1\bar{1}0]$ shear as temperature is lowered, as shown by elastic constants measurements. This shear is the right one for causing the structure change, so the implication is that amplitudes of thermal vibration become comparable with distortions of the structural transformation and provide nuclei for it as the temperature is lowered toward the starting point for martensite formation. Pre-martensitic effects have been seen in other phases by X-ray, electron and neutron diffraction experiments which appear to indicate precursers of the low temperature phase, and which have formed an active field of research in the last decade. A brief review appears in Barrett (1976).

References

Bain, E. C. (1923). Chem. Met. Eng. **28**, 21, 65; Trans. AIME **68**, 625.

Barrett, C. S. (1976). Trans. Jap. J. Metals, **17**, 465.

Barrett, C. S. and Massalski, T. B. (1966). Structure of Metals. New York: McGraw Hill, New York; reprinted in 1980 by Pergamon Press, Fairview Park, Elmsford, New York, 10523.

Barrett, C. S. and Trautz, O. R. (1948). Trans. AIME **175**, 579.

Batterman, B. W. and Barrett, C. S. (1966). Phys. Rev. **145**, 296.

Beck, P. A. (editor) (1963). Electronic Structure and Alloy Chemistry of the Transition Elements. New York: Wiley-Interscience.

Bowles, J. S., Barrett, C. S. and Guttman, L. (1950). Trans. AIME, **188**, 1478.

Bridgman, P. W. (1937). J. Appl. Phys., **8**, 328.

Bridgman, P. W. (1949). The Physics of High Pressure. London: Bell and Sons

Cohen, M. U. (1935). Rev. Sci. Instrum. **6**, 68.

Donnay, J. D. H. (1963). Crystal Data. Am. Crystallogr. Assoc. 3rd ed. pub. by JCPDS-ICDD, Swarthmore, PA 19081.

Duwez, P. and Willens, R. H. (1963). Trans. AIME **277**, 362.

Duwez, P., Willens, R. H. and Klement, W. (1960). J. Appl. Phys. **31**, 1136.

Elliot, R. P. (1965). First Supplement. New York: McGraw Hill.

Gibbs, J. W. (1948). Collected Works. New Haven, Connecticut: Yale University.

Gschneidner, K. A. (1961). Rare Earth Alloys. New York: Van Nostrand.

Hanawalt, J. D., Rinn, H. and Frevel, L. K. (1938). Ind. Eng. Chem. Anal. Ed. **10**, 457.

Hansen, M. and Anderko, K. (1958). Constitution of Binary Alloys. New York: McGraw Hill.

Hull, A. W. (1917). Phys. Rev. **10**, 661.

Hull, A. W. and Davey, W. P. (1921). Phys. Rev. **17**, 549.

Jamieson, J. C., Lawson, A. W. and Wentorf, R. W. (editors) (1962). Modern Very High Pressure Techniques. Washington, D.C.: Butterworth.

Jamieson, J. C. (1963). Science **139**, 762.
Luo, H. L. and Duwez, P. (1963). Appl. Phys. Letters, **2**, 21.
Massalski, T. B. Ed. (1965). Alloying Behavior and Effects in Concentrated Solid Solutions. New York: Gordon and Breach.
Massalski, T. B. and Barrett, C. S. (1957). Trans. AIME **209**, 455.
Matthias, B. T., Gebelle, T. H. and Compton, V. B. (1963). Rev. Mod. Phys. **35**, 1.
Metals Handbook. Am. Soc. Metals, Metals Park, Ohio (numerous eds.); 8th ed., vol. 8, (1973): Metallography, Structures and Phase Diagrams.
Mueller, M. H., Heaton, L. and Miller, K. T. (1960). Acta Cryst. **13**, 828.
Parthé, E. (1964). Crystal Chemistry of Tetrahedral Structures. New York: Gordon and Breach.
Phase Transformations (1970). Am. Soc. for Metals, Metals Park, Ohio.
Rudman, P. S. and Stringer, J. (editors) (1966). Phase Stability in Metals and Alloys. New York: McGraw Hill.
Wechsler, M. S., Lieberman, D. S. and Read, T. A. (1953). Trans. AIME **197**, 1503.
Westbrook, J. H. (editor) (1967). Intermetallic Compounds. New York: John Wiley.
Wyckoff, R. W. G. (1963). Crystal Structures 2nd Ed., Vols. 1-6. New York: Wiley Interscience.
Zener, C. (1947). Phys. Rev. **71**, 846.

CHAPTER 6

STUDIES OF THE STRUCTURES OF ORGANOMETALLIC AND COORDINATION COMPOUNDS IN THE U.S.

James A. Ibers
Northwestern University, Evanston, Illinois 60201

The growth of determinations of organometallic and coordination structures has been phenomenal and has been more rapid than the determination of structures of other classes of small molecules. We will take as a definition of organometallic and coordination compounds those included in the Cambridge Crystallographic Database. Table 1, derived from that database, indicates the early rise in determinations of the structures of these compounds.

Two events led to this growth. The first, which is not peculiar to this class of compounds, was the increased availability of diffractometers and powerful computers, beginning around 1962-1965. The second was the discovery of ferrocene in 1951 (Kealy and Pauson, 1951) and the subsequent synthesis of entirely new classes of chemical compounds involving organic moieties and transition metals. Coupled with the rising importance of metal complexes in homogeneous catalysis, as embodied in the hydroformylation reaction, catalyzed by $Co_2(CO)_8$, organometallic chemistry experienced an explosive growth. Since most of the compounds synthesized were of unprecedented composition and architecture, spectroscopic methods were of limited use in their characterization, and increasing reliance was placed on diffraction methods.

About half of the 32,000 structures in the current Cambridge Crystallographic Database contain one or more transition metals. If we include those compounds containing main-group metals, such as Sn or Ge, then the number of known organometallic structures exceeds the number of known organic structures. Today, perusal of the number of American journals, including *Journal of the American Chemical Society* and *Inorganic Chemistry*. will reveal the significant contributions that American structural chemists are making to organometallic chemistry. Of course, there are significant contributions by non-Americans as well, just as there were to those papers cited in Table 1. But it is interesting to detail who was involved in early work on organometallic and coordination compounds and what types of structures were determined.

It is clear from reading some of the early papers that the Second World War interrupted or delayed a number of structure determinations. The earliest published work by an American appears to be the determination of the structure of gadolinium formate by Pabst at Berkeley (1943). In the same year Griffith of Eastman Kodak Company published the structure of silver oxalate (Griffith, 1943). Another early work by an American was the study of the structure of nickel(II) glycinate dihydrate by Stosick at Caltech in 1945 (Stosick, 1945). This was part of the amino acid project initiated by Pauling at Caltech. In fact studies at Caltech dominate these early years. This is not surprising in view of the tradition for structure determinations established at Caltech by Dickinson and Pauling and by the presence at Caltech in the 1945-1955 period of Pauling, Sturdivant, Hughes, and Schomaker and their many visitors, post-doctorals, and students, including Davies, Donohue, Dunitz, Hamilton, Ibers, Lipscomb, Marsh, Merritt, Samson, Sly, Trueblood, Rich, Rollett, and Waser. Other notable results from Caltech during this period include the structure determination of trimethylplatinum chloride by Rundle and Sturdivant in 1947 (Rundle and Sturdivant, 1947) and that of hexamethylenetetramine manganous chloride dihydrate (Tang and Sturdivant, 1952).

During the 1950's a number of former Caltech students were carrying out structure determinations on organometallic compounds. Notable among these were Rundle at Iowa State, Hoard at Cornell, and Lipscomb at Minnesota. Notable among Rundle's work were the study of silver perchlorate-benzene with Smith (Smith and Rundle, 1958) and the studies of the Ni and Pd dimethylglyoximates with Williams and Wohlauer (Williams et al., 1959). A number of Rundle's students, including Dahl and Baenziger, went on to establish their own schools of organometallic structure determination. Among Hoard's contributions was the determination of the structure of silver perfluorobutyrate dimer with Blakeslee (1956). Lipscomb contributed some important structure determinations, including that of cyclooctatetraene silver nitrate with Mathews (Mathews and Lipscomb, 1959). Other seminal determinations from these early days include those on ferrocene itself, first in a preliminary way (Eiland and Pepinsky, 1952) and later by Dunitz, Orgel, and Rich at Caltech (Dunitz et al., 1956) and the structures of ruthenocene by Hardgrove and Templeton at Berkeley (1959) and of diphenylacetylene dicobalt hexacarbonyl by Sly at Caltech (1959). This last structure appears to be first of what is now a very large class in which an organic group, in Sly's case an acetylene, bridges more than one metal.

Those early days were indeed exciting, because few were able to predict structures from composition. I remember Norm Davidson returning to Caltech sometime around 1953 or 1954 from an ACS meeting where he had just learned about a vast collection of new organometallic compounds. He quizzed Verner Schomaker on what he would expect for their structures. Verner, as usual, made some extraordinarily good guesses, but there were some structures that could not be guessed.

Today, some 20 years and 16,000 organometallic structures later, there is still great excitement in the field as new architectural marvels are discovered and characterized by diffraction methods. There is no doubt that in the general area of organometallic chemistry the influence of the structural chemist has been of extreme importance. Without characterization of these new compounds, as provided by those early pioneers, a number of Americans among them, and the generation of active workers that followed, organometallic chemistry would still be in the witchcraft stage.

References

Blakeslee, A. E. and Hoard, J. L. (1956). J. Am. Chem. Soc. **78**, 3029.

Dunitz, J. D., Orgel, L. E. and Rich, A. (1956). Acta Cryst. **9**, 373.

Eiland, P. F. and Pepinsky, R. (1952). J. Am. Chem. Soc. **74**, 4971.

Griffith, R. L. (1943). J. Chem. Phys. **11**, 499.

Hardgrove, G. L. and Templeton, D. H. (1959). Acta Cryst. **12**, 28.

Kealy, T. J. and Pauson, P. L. (1951). Nature (London) **168**, 1039.

Mathews, F. S. and Lipscomb, W. N. (1959). J. Phys. Chem. **63**, 845.

Pabst, A. (1943). J. Chem. Phys. **11**, 145.

Rundle, R. E. and Sturdivant, J. H. (1947). J. Am. Chem. Soc. **69**, 1561.

Sly, W. G. (1959). J. Am. Chem. Soc. **81**, 18.

Smith, H. G. and Rundle, R. E. (1958). J. Am. Chem. Soc. **80**, 5075.

Stosick, A. J. (1945). J. Am. Chem. Soc. **67**, 365.

Tang, Y.-C. and Sturdivant, J. H. (1952). Acta Cryst. **5**, 74.

Williams, D. E., Wohlauer, G. and Rundle, R. E. (1959). J. Am. Chem. Soc. **81**, 755.

Table 1. Organometallic Structure Determinations[a]

Year	Number
1935-1939	2
1940-1944	2
1945-1949	4
1950-1954	19
1955-1959	53
1960-1964	207
1965	105
1966	146
1967	215

a) Defined by classes 71-86 of the Cambridge Crystallographic Database.

CHAPTER 7

NEUTRON CRYSTALLOGRAPHY AND THE STUDY OF 1-DIMENSIONAL PLATINUM-CHAIN METALS*

Jack M. Williams
Chemistry Division, Argonne National Laboratory
Argonne, Illinois 60439

An interesting and illuminating chapter in X-ray and neutron crystallography in the U.S.A. began in the early 1970's with structural studies of "1-dimensional" electrical conductors of which the Pt-chain metal $K_2[Pt(CN)_4]Br_{0.3}.3H_2O$, "KCP", was the inorganic prototype. Salts of this type were of especially high interest at that time because of the suggestion by Little (1964) that certain organic derivatives containing highly polarizable side groups bound to a metal-chain spine might be high temperature superconductors. Although this prospect has not yet materialized, primarily because of the difficulty in the chemical synthesis of such a system, calculations still suggest that room temperature superconductivity is possible in a real chemical system.

Inorganic complexes of the KCP type contain closely stacked square planar $Pt(CN)_4$ groups (see Fig. 1) having

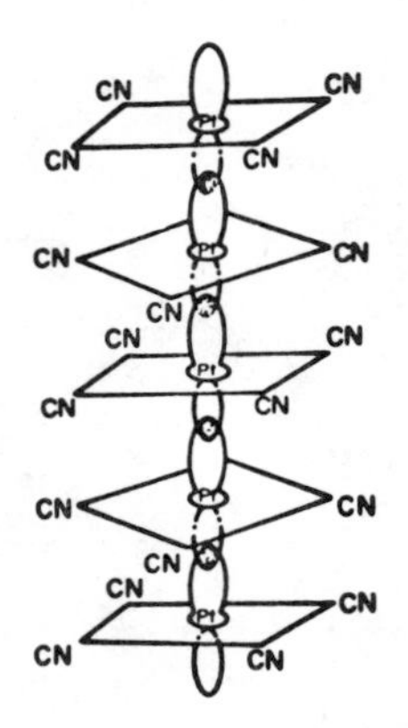

Fig. 1. Illustration of overlapped d_{z^2} orbitals of platinum which result in metal "chain" formation.

short Pt-Pt separations and extensive d_{z^2} orbital overlap, which gives rise to electron delocalization along the metal-atom chain and concomitant metallic conductivity at room temperature. At this point it should be noted that the origin of metallic properties in these salts is a partially filled band which is achieved by partial oxidation of the metal atom to a non-integral valence state. Recent studies (Williams and Schultz, 1979) have established that the degree of partial oxidation (DPO) in these salts varies in the range 0.2-0.4 thereby producing complexes containing $Pt^{2.2\text{-}2.4}$.

Actually, the KCP saga began nearly 140 years ago with the discovery of this unusual salt by Knop (1842). However, the lustrous metallic bronze colored salt received very little attention until the 1960's, at which time an X-ray structure determination was published by Krogmann and Hausen (1968). Although it was not known at that time, a parallel X-ray study of KCP was also reported by Piccinin and Toussaint (1967), but on a material thought to be $K_2[Pt(CN)_5]\cdot 3H_2O$. Our entry into this field occurred while performing a neutron structure study of KCP (Williams, Petersen, Gerdes and Peterson, 1974) from which we learned (vide infra) that while Piccinin and Toussaint (1967) had the wrong compound in the correct space group, Krogmann and Hausen (1968) described the correct analytical composition of KCP in a more symmetrical but incorrect space group, *viz. P4/mmm* rather than noncentrosymmetric *P4mm* (see Fig. 2). At the same time the neutron

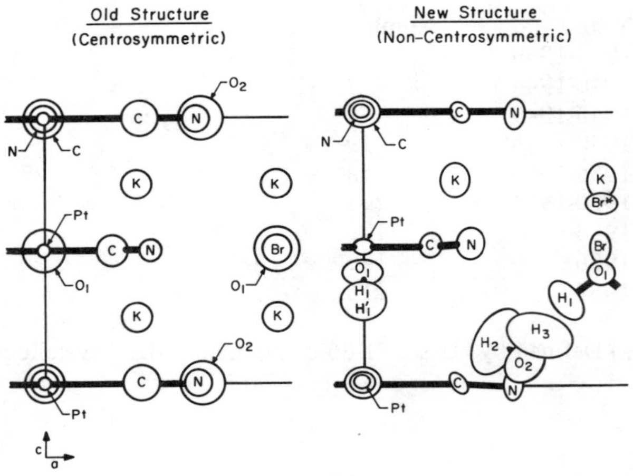

Fig. 2. The original X-ray structure (centrosymmetric) of KCP (Krogmann and Hausen, 1968) and revised neutron diffraction determined structure (Williams, Petersen, Gerdes and Peterson, 1974). In the revised structure the atom labelled as Br is likely a highly disordered H_2O molecule (see text).*

structural study was reported, an independent X-ray study was also reported (Deiseroth and Schulz, 1974). The results of the parallel neutron and X-ray studies were essentially identical except that in the neutron investigation an additional site thought to be occupied by a halide ion or a disordered water molecule was discovered. Recent combined low temperature X-ray and neutron studies (Peters and Eagen, 1976; Deiseroth and Schulz, 1978) have confirmed the existence of the

*Work performed under the auspices of the Division of Materials Sciences, Office of Basic Energy Sciences, U.S. Department of Energy.

additional site and that it most likely contains a highly disordered water molecule. Unoccupied halide sites also appear to be occupied by water molecules. However, it should be noted that the essential structural features of KCP, *viz,* the stacked $Pt(CN)_4$ groups and the closely spaced Pt-atom chains, were originally recognized by Krogmann (1969) and discussed in a classic review article on the subject. The choice of an incorrect centrosymmetric space group by Krogmann and Hausen (1968) occurred because although they observed "weak" 001 (1 odd) reflections, they incorrectly ascribed them to a "superstructure." This problem arose mainly because of the identical repeat periodicity of the square planar $Pt(CN)_4$ groups ($d_{Pt\text{-}Pt}$ = *2.88* Å) which stack along the chain direction which in turn corresponds to the tetragonal *c* axis of the crystal. In X-ray studies this precise stacking causes the 001 (1 even) reflections to be very intense while the 1 odd reflections are very weak (but measurable!). This is, of course, *not* the case in the neutron diffraction experiment where all of the atoms have nearly similar scattering amplitudes and all orders of the 001 reflections are easily observed. The minimal absorption corrections in the neutron case (μ_c = 1.24 cm^{-1}) also simplified data interpretation. Thus the initial neutron diffraction study (Williams, Petersen, Gerdes and Peterson, 1974) led to a reinterpretation of the crystal structure of KCP with the following significant findings:

(a) The K^+ ions are *not* disordered between the planes of the $Pt(CN)_4$ groups as was previously supposed (Krogmann and Hausen, 1968), but are asymmetrically located in one-half of the unit cell;

(b) The intrachain Pt-Pt distances are essentially identical although they are *not* required to be so;

(c) Water molecules serve to connect the Pt-atom chains via hydrogen bonds to nitrogen or halogen atoms;

(d) An extra or "defect" site exists in KCP which is only partially occupied; and

(e) The $Pt(CN)_4$ groups are not planar but rather cant toward the K^+ ions with which they interact (see Fig. 3).

In summary, the original neutron diffraction single crystal study established incisively the crystal and molecular structure of KCP which later led to a vastly improved understanding of the structure-electrical conductivity relationships in KCP type salts (Williams and Schultz, 1979). Thus the new field of "synthetic metals" (a new journal bearing this name was established in 1979) grew immensely in the 1970's and diffraction studies played a key role in providing an understanding of structure property relationships in these unusual materials. With the discovery of the first man-made 'organic" superconductors such as $(TMTSF)_2X$ (TMTSF = tetramethyltetraselenafulvalene and X = PF_6^- or ClO_4^-) (Bechgaard *et al.*, 1980) the absolute need for 3-dimensional diffraction data at very low temperatures (∿1°K) and high pressures (∿12kbar) was established. Thus, the field of synthetic metals has provided a continuing and growing challenge, especially to crystallographers, because it requires new and improved crystallographic research tools and data analysis techniques. There is no doubt in my mind that crystallographers engaged in this research area will meet that challenge.

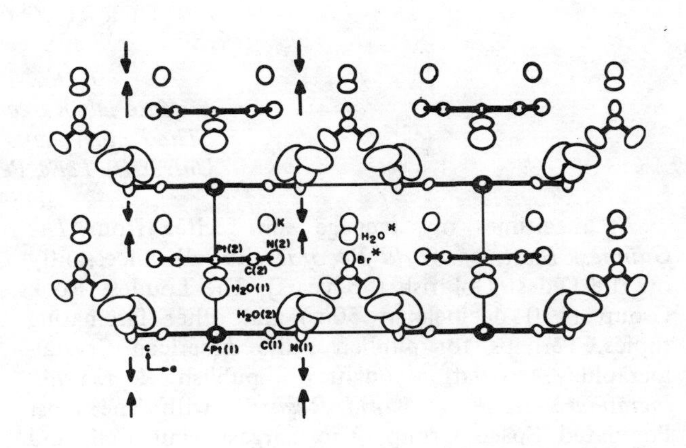

Fig. 3. A b-axis half-cell projection of the structure of KCP, $K_2[Pt(CN)_4]Br_{0.3}{\cdot}3H_2O$. Only independent atoms in the asymmetric unit are identified and the Pt atom at the origin is overlapped by C and N atoms. The disordered Br^- and H_2O sites are labelled as Br and H_2O*. The allowed longitudinal distortions of the Pt-atom chain caused by the K^+. . . N≡C interactions are indicated with arrows.*

References

Bechgaard, K., Jacobsen, C. S., Mortensen, K., Pedersen, H. J. and Thorup, N. (1980). Solid State Commun. **33**, 1119.

Deiseroth, H. J. and Schulz, H. (1974). Phys. Rev. Lett. **33**, 963-965.

Deiseroth, H. J. and Schulz, H. (1978). Acta Cryst. **B34**, 725-731.

Knop, A. (1842). Justus Liebig's Ann. Chem. **43**, 111-115.

Krogmann, K. and Hausen, H. D. (1968). Z. Anorg. Allg. Chem. **358**, 67-81.

Krogmann, K. (1969). Angew. Chem. Int. Ed. Eng. **8**, 35-42.

Little, W. A. (1964). Phys. Rev. **A134**, 1416-1424.

Peters, C. and Eagen, C. F. (1976). Inorg. Chem. **15**, 782-788.

Piccinin, A. and Toussaint, J. (1967). Bull. Soc. Roy. Sci. Liege **36**, 122-126.

Williams J. M. Petersen, J. L., Gerdes, H. M. and Peterson, S. W. (1974). Phys. Rev. Lett. **33**, 1079-1081.

Williams, J. M. and Schultz, A. J. (1979). Molecular Metals, NATO Conference Series VI. Edited by W. E. Hatfield, New York: Plenum Press. pp. 337-368.

CHAPTER 8

FERROIC CRYSTALS

R. E. Newnham
Materials Research Laboratory
The Pennsylvania State University
University Park, Pennsylvania 16802

At mealtimes my teenage son recites from *The Guinness Book of World Records* and talks incessantly of The Oldest Goldfish (36 years), The Loudest Rock Group (120 decibels at 50m) and other fascinating topics. Perhaps for publicity, the American Crystallographic Association ought to publish *A Crystallographer's Book of World Records,* with The Most Populated Space Group, The Largest Unit Cell, and The Toughest Phase Problem. And if physical properties were included, there would be categories such as The Biggest Dielectric Constant, The Highest Elastic Compliance, and The Largest Saturation Magnetization.

Most of the record-setting materials with outstanding physical properties are ferroic crystals with movable twins or domains: ferroelectrics, ferromagnetics, ferroelastics, and a host of secondary ferroics. Such crystals are very responsive to external fields, forces, and temperatures changes, and unlike ordinary crystals which just lie there, ferroics quiver, shake, and polarize under the least provocation. These are the kind of crystals that turn me on.

American crystallographers have made a number of important contributions to the study of ferroic crystals. The phenomenon of ferroelectricity was first reported by Joseph Valasek (1921), and one of his students—Ray Pepinsky—went on to discover many other new and interesting ferroelectrics (Shirane, Jona and Pepinsky, 1955). Arthur von Hippel and Howard Evans at M.I.T. were two of the pioneer investigators of barium titanate, and at Bell Laboratories important discoveries were made by Bernd Matthias, Elizabeth Wood, Walter Bond, W. P. Mason, Joseph Remeika, and Robinson Burbank. Sidney Abrahams and Joel Bernstein continue this tradition with their work on piezoelectric and ferroelastic crystals (Abrahams, 1971).

In the field of magnetism, the determination of magnetic spin structures by neutron diffraction was pioneered at U.S. national laboratories, revolutionizing our understanding of ferrimagnetic and antiferromagnetic crystals (Koehler, 1967). Other highlights include the work of Axe (1971) and Shirane on soft modes in ferroic crystals; a structural classification of phase transitions by Buerger (1971); and the investigations of Donnay (1967) on symmetry and twinning, and the list goes on.

Proper and Improper Ferroics

Recent attention has focused on the causes of displacive phase transitions, particularly on ferroics with incommensurate structures, those with one- and two-dimensional behavior, and on improper ferroics. The Pennsylvania Hex diagram in Fig. 1 illustrates many of

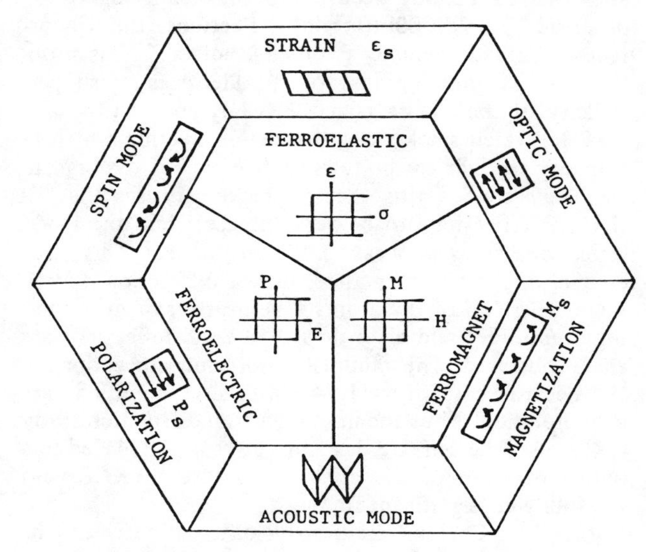

Fig. 1. Order-parameter diagrams for the three types of primary ferroics: ferroelasticity, ferroelectricity and ferromagnetism. Ferroic behavior can be caused by any of the six order parameters grouped around the outside of the diagram, giving rise to a wide range of physical properties.

the origins of ferroic phenomena. The three types of primary ferroics lie in the center of the drawing, and grouped around them are six different classes of order parameters.

Because of their instability, ferroics exhibit both proper and improper behavior; in phase transition terminology, propriety refers to the order parameter of a phase transition, rather than to the social graces. Polarization is the order parameter for a proper ferroelectric, strain for ferroelasticity, and magnetization for ferromagnetism. If the order parameter of a ferroelectric is anything other than polarization, it is called improper.

Gadolinium molybdate is an interesting example of impropriety in which the order parameter is a condensed optic mode, an atomistic order parameter which doubles the size of the unit cell. The optic mode causes a spontaneous strain which in turn causes a small spontaneous polarization through piezoelectric coupling. Thus gadolinium molybdate is both an improper ferroelastic and a very improper ferroelectric.

In recent years there has been increasing awareness of

cross-coupled domain phenomena caused by electric, elastic, and magnetic interactions. Lithium ammonium tartrate, for instance, is a type of elastoferroelectric in which mechanical strain is the order parameter at the 98 K phase transition. Strain gives rise to ferroelectricity through piezoelectric coupling with the polarization (Sawada, et al., 1977).

Other cross-coupled effects arise when magnetic phenomena are included. There are a number of examples of elastoferromagnetism and magnetoferroelasticity, and at least one good example of electroferromagnetism: nickel iodine boracite (Ascher, et al., 1966). At room temperature, NiI boracite is cubic, point group $\bar{4}3m$. Below room temperature at 120 K, it undergoes a transition to an antiferromagnetic state as the Ni^{2+} moments align; at this stage, the material is an antiferromagnetic piezoelectric, but it is neither ferromagnetic nor ferroelectric. On further cooling, a second phase transition to an orthorhombic ferroelectric state takes place at 64 K. As the crystal structure develops a spontaneous polarization, the magnetic structure is also altered, destroying the balance of spins in the antiferromagnetic state and producing a weak ferromagnetism. The ferromagnetic effect is therefore of electric origin and can be referred to as an electroferromagnet.

In the opposite effect–magnetoferroelectricity–a reversible spontaneous electric polarization appears on passing through a magnetic phase transition. The one-dimensional model in Fig. 2 illustrates the principle of a magnetically induced ferroelectric. At high temperatures, the system is paramagnetic and non-polar with the paramagnetic cations located on centers of symmetry. On cooling through the Néel point, the system becomes antiferromagnetic with spins aligned along the chain direction. The spins order in such a way that each magnetic ion has one neighbor with parallel spin and one with antiparallel spin. The inversion centers are destroyed by magnetic order, making the chain a polar axis. Small atomic movements take place because of the differences in interatomic forces between atom pairs with parallel spin and atom pairs with antiparallel spin.

If the cations move in the same direction, as in Fig. 2, an electric polarization is produced. The atomic displacements destroy the centers of symmetry and make the material pyroelectric, piezoelectric, and potentially ferroelectric. The expected polarization is small because most magnetic transitions show second-order behavior with very small magnetostrictive effects.

Magnetoferroelectricity was recently discovered in chromium chrysoberyl (Cr_2BeO_4) on the basis of symmetry arguments (Newnham, et al., 1978), establishing an interesting connection between crystallographic symmetry and magnetic symmetry. In teaching crystal physics, one finds that there are very few cross-connections between crystallographic symmetry and magnetic symmetry. Matrices describing elasticity,

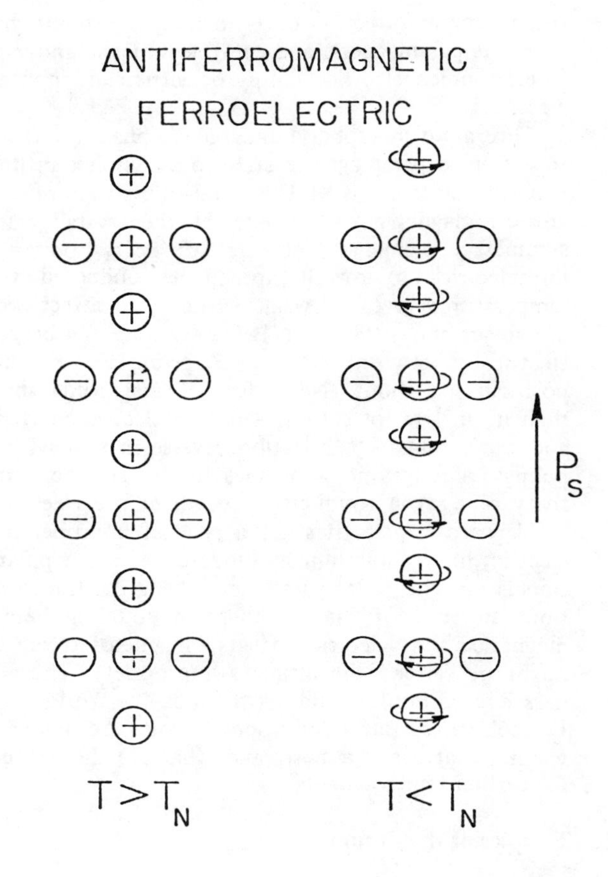

Fig. 2. Paramagnetic and antiferromagnetic states in a hypothetical magnetoferroelectric in which arrows indicate spin directions. Centers of symmetry are destroyed at the magnetic transition, converting the chain to a polar axis and causing a spontaneous polarization P_s to appear.

piezoelectricity, and other traditional topics (Nye, 1957) are developed in terms of the 32 crystallographic point groups, whereas the 90 magnetic point groups are introduced separately to explain magnetic properties such as magnetoelectricity and piezomagnetism (Birss, 1964). But after introducing magnetic symmetry, the electric, thermal and mechanical properties are seldom re-examined with regard to the symmetry change accompanying magnetic transitions. The discovery of magnetoferroelectricity establishes a very satisfying link between magnetic symmetry and pyroelectricity, one of the standard topics in crystal physics.

Magnetoferroelectric Cr_2BeO_4 is orthorhombic and isostructural with the minerals chrysoberyl (Al_2BeO_4) and forsterite (Mg_2SiO_4). The crystal structure consists of a close-packed lattice of oxygen ions with Cr^{3+} in octahedral sites and Be^{2+} in tetrahedral positions. At room temperature, Cr_2BeO_4 belongs to point group *mmm*, is paramagnetic and centrosymmetric and, therefore, nonpiezoelectric, nonpyroelectric, and nonferroelectric.

Chromium chrysoberyl undergoes a phase transformation from a paramagnetic state to a complex antiferromagnetic state at 28 K. The spiral spin structure of the antiferromagnetic state violates all the crystallographic symmetry elements, making Cr_2BeO_4 potentially ferroelectric. Chynoweth experiments conducted at low temperatures reveal a weak pyroelectric effect which disappears above 28 K. Cr_2BeO_4 ceramics can be poled electrically between 24 and 28 K, giving rise to remnant polarizations about five orders of magnitude smaller than normal ferroelectrics. The pyroelectric coefficient and the remnant polarization reverse in sign with the poling field, but no anomalies in the electric permittivity or electric conductivity occur at the Néel point.

Magnetoferroelectrics are a type of improper ferroelectric, like gadolinium molybdate, in which polarization is not the order parameter driving the transformation. Because of the weakness in coupling between magnetic and electric effects, magnetoferroelectrics might be termed the ultimate impropriety. The Guinness Record book would list them as The World's Worst Ferroelectrics. But if improper ferroics are among the worst, what are the best, and what are the prospects for further improvements?

Ferroics of the Future

Progress in ferroic materials–like progress in most fields–follows an S-shaped curve of history (Fig. 3).

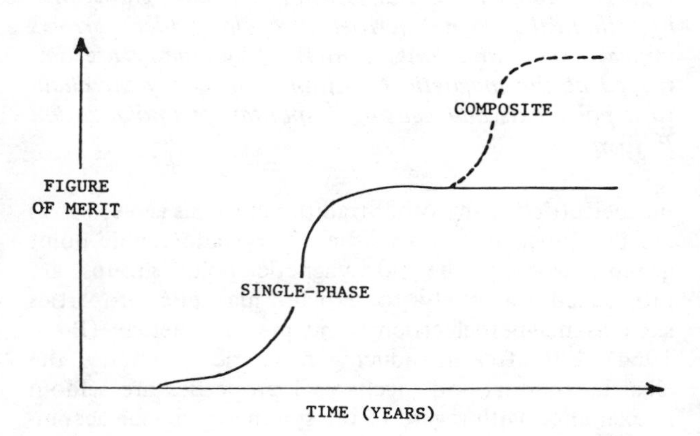

Fig. 3. Curve of history for practical materials. The discovery of a potentially useful property is followed by a period of rapid development during which many new materials are examined. Eventually the best materials are selected and progress temporarily slows. For many applications, further advances can be made with composite materials designed to optimize a given figure of merit.

When a new effect such as ferroelectricity is discovered, scientific development is rather slow at first, until its importance is recognized. Then follows a period of rapid growth when practical applications and many new materials are discovered. During this period, rapid changes take place in selecting the "best" material for each application, but eventually the field matures as the choices are made, and the curve of history saturates.

We see this saturation effect in many branches of applied crystallography. Among electroceramics, lead zirconate-titanate (PZT) has been the best transducer material, and barium titanate the best capacitor material for the past twenty years. Similar trends can be noted in magnetic materials, semiconductors, and superconductors. Despite intensive search for new compounds, relatively few important new materials have been made in the past decade.

Led by the semiconductor industry, materials science now appears to have entered a new era, the age of carefully patterned inhomogeneous solids designed to perform specific functions. Examples of heterogeneous systems optimized for particular applications include semiconductor integrated circuits, fiberreinforced metals, and barrier-layer ferroelectric capacitors. No longer as much concerned with the properties of the best single-phase materials, many scientists now search for the best combination of materials and ways to process them. In a very real sense, the field has matured from materials science to materials engineering just as electrical science changed to electrical engineering many years ago.

In many applications there are several phases involved and a number of material parameters to be optimized. An electromechanical transducer, for example, may require a combination of properties such as large piezoelectric coefficient (d or g), low density, and mechanical flexibility. A pyroelectric detector might require large pyroelectric coefficient, low thermal capacity, and low dielectric constant. In general, the task of materials design may be considerably simplified if it is possible to devise a figure of merit which combines the most sensitive parameters in a form allowing simple intercomparison of the possible "trade offs" in property coefficients. In certain pyroelectric systems, for example, a useful figure of merit is p/ϵ where p is the pyroelectric coefficient and ϵ the electric permittivity.

Unfortunately, the figure of merit often involves property coefficients which are conflicting in nature. To make a flexible electromechanical transducer it would be desirable to have the large piezoelectric effects found in poled piezoelectric ceramics, but ceramics are brittle and stiff lacking the required flexibility, while polymers having the desired mechanical properties are at best very weak piezoelectrics. Thus, for such an application a composite material combining the desirable properties of two different phases might be vastly superior. The main problem is to effect the combination in such a manner as to exploit the desirable features of both components and thereby maximize the figure of merit.

A knowledge of crystallography and of structure-property relationships is very useful in designing composites. As an example, it is instructive to examine the structure-property relationships for the hydrostatic piezoelectric effect, and to exploit the relationships in transducer design.

Hydrostatic pressure coefficients for a number of piezoelectric materials is given in Table I. Since the symmetry requirements for pyroelectricity and hydrostatic piezoelectricity are identical, all the materials are also pyroelectric. For puposes of discussion, they can be divided into ferroelectric pyroelectrics and ordinary (non-ferroelectric) pyroelectrics. As shown in Table I, ferroelectrics have substantial d_h (polarization) coefficients but the g_h (voltage) coefficients are not very big because of their large dielectric constants.

Among ordinary pyroelectrics, those with the wurtzite structure have very small hydrostatic piezoelectric effects. The wurtzite crystal structure is based on a hexagonal close-packed anion lattice with cations in tetrahedral interstices. Compared to the other pyroelectrics, the atomic bonding in the wurtzites is very isotropic. It is not surprising, therefore, that under hydrostatic pressure they deform isotropically, leading to very small piezoelectric effects.

Silicate pyroelectrics have somewhat larger hydrostatic coefficients than the wurtzite group. Tourmaline is a complex borosilicate mineral containing tetrahedral SiO_4 groups. The silica tetrahedra are arranged in Si_6O_{18} rings oriented perpendicular to the pyroelectric axis. This imparts an anisotropy to the structure not found in the wurtzite group, but the silicate and borate groups are linked together by Al^{3+} and Mg^{2+} ions which also form fairly strong chemical bonds and hence tourmaline is not as anisotropic as some other crystals.

More anisotropic structures are found among the water-soluble pyroelectrics. Lithium sulfate monohydrate ($Li_2SO_4.H_2O$) is an important example with an extremely large g_h coefficient, so large that the crystals have been used as hydrostatic pressure sensors. The crystal structure of lithium sulfate contains Li^+ cations, tetrahedral SO_4^{2-} anions, and water molecules. Ionic bonds between cations and anions extend in all directions in the crystals but the hydrogen bonding between water molecules extends only along *b*, the unique polar axis. Tensile stress in this direction produces a large electric polarization. Short hydrogen bonds like those in lithium sulfate make an important contribution to the piezoelectric effect because the proton position changes as the oxygen-oxygen is stretched. Mechanical stress affects the dipole moments of the water molecules, producing electric polarization along the *b* axis and producing an unusually large d_{22} coefficient in lithium sulfate. The large piezoelectric effect together with a small dielectric constant gives it the largest hydrostatic voltage coefficient (g_h) of any material, including ferroelectrics.

Because of their large polarizabilities, ferroelectrics also have large piezoelectric constants but the hydrostatic coefficients are not large for those with symmetric crystal structures. $BaTiO_3$, and other perovskites

TABLE I

Hydrostatic piezoelectric coefficients for a number of materials. For a given pressure, d_h measures the electric polarization, and g_h the open-circuit electric field. d_h is expressed in units of 10^{-12} C/N and g_h in $10^{-3}m^2/C$.

	d_h	g_h
Water-Soluble Pyroelectrics		
Ethylene diamine tartrate (EDT) $C_2H_4(NH_3)_2$ $C_4H_4O_6$	1.0	15
Lithium sulphate monohydrate (LH) $Li_2SO_4.H_2O$	16.4	180
Others	<4	<100
Pyroelectric Silicate Minerals		
Tourmaline (Na,Ca) $(Mg,Fe)_3B_3Al_6Si_6(O,OH,F)_{31}$	2.5	38
Others	<3	<30
Wurtzite-Family Pyroelectrics		
BeO,ZnO,CdS,CdSe	<0.2	<3
Ferroelectric Single Crystals		
Barium titanate $BaTiO_3$	16.6	11
Triglycine sulfate (TGS) $(NH_2CH_2COOH)_3.H_2SO_4$	8.0	30
Antimony sulfur iodide (10°C) SbSI	1100	14
Lithium niobate $LiNbO_3$	14.5	57
Poled Ferroelectric Ceramics		
Barium titanate $BaTiO_3$	34	2
Lead niobate $PbNb_2O_6$	67	34
Lead zirconate titanate (PZT) $Pb(Ti,Zn)O_3$	20-50	2-9
Sodium potassium niobate $(Na,K)NbO_3$	40	10

contain closed-packed arrays of oxygens and large cations. The $LiNbO_3$ structure is also based on close-packing. Hydrostatic piezoelectric coefficients for these materials are small compared to antimony sulfur iodide (SbSI) which has the largest d_h coefficient in Table I.

The structure of SbSI (Fig. 4) is very anisotropic

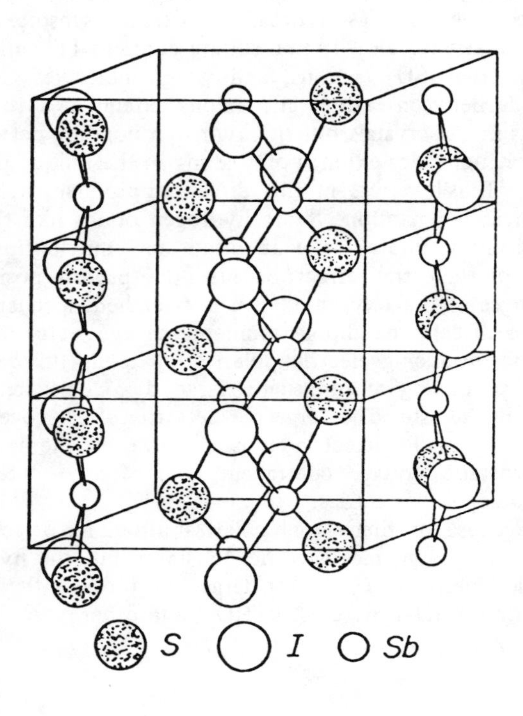

Fig. 4. Crystal structure of antimony sulfur iodide (SbSI). The polar chain structure leads to very large hydrostatic piezoelectric coefficients.

with covalently bonded chains parallel to the polar axis. Neighboring chains are only weakly bonded by ionic or van der Waals forces. Crystals of SbSI cleave readily parallel to the polar *c* axis. Under tensile force parallel to *c*, the crystals develop a large piezoelectric polarization similar in origin to that in lithium sulfate. The antimony cations displace relative to the anions causing the polarization. Piezoelectric effects in the perpendicular directions are much smaller because of the loose packing of chains.

In summary, we see that the best piezoelectric crystals for hydrostatic sensors are those with anisotropic structures and chain-like molecular mechanisms for piezoelectricity. This concept has been used to design ceramic-plastic composites with very large piezoelectric coefficients (Newnham, Skinner, and Cross, 1978). An interesting design for hydrostatic applications is one with a parallel arrangement of stiff piezoelectric PZT fibers in a compliant polymer matrix (Fig. 5), which is a

Fig. 5. Composite piezoelectric transducers consisting of poled ceramic rods in a soft polymer matrix.

macroscopic imitation of the SbSI structure. Depending on composition and various geometric factors, these composites give piezoelectric coefficients 10-100 times larger than the best single-phase materials (Klicker, et al., 1981).

Conclusion

We are entering a new age in crystallography—the age of crystallographic engineering. No longer content with the structures Nature has given us, we begin to design and build structures of our own; at first rather crude macroscopic composites—like the PZT-plastic transducers—but gradually progressing to smaller and smaller scale until we achieve control at the Angstrom level, and build crystal structures of our own design. Recent work on GaAs-AlAs epitaxial superlattices are an indication of things to come (Esaki and Chang, 1975). Guiding us in this great enterprise will be our knowledge of Nature's crystal structures, our understanding of structure-property relationships, and our dreams of the future.

References

Abrahams, S. C. (1971). Mater. Res. Bull. **6**, 881-890.
Ascher, H., Rieder, H., Schmid, H. and Stossel, H. (1966). J. Appl. Phys. **37**, 1404-1409.
Axe, J. D. (1971). Trans. Am. Crystallogr. Assoc. **7**, 89-103.
Birss, R. R. (1964). Symmetry and Magnetism. Amsterdam: North Holland Publishing Co.

Buerger, M. (1971). Trans. Am. Crystallogr. Assoc. **7**, 1-20.
Donnay, J. D. H. (1967). Trans. Am. Crystallogr. Assoc. **3**, 74-95.
Esaki, L. and Chang, L. L. (1975). Crit. Rev. Solid State Sci. **6**, 195-205.
Klicker, K. A., Biggers, J. V. and Newnham, R. E. (1961). J. Amer. Cer. Soc. **64**, 5-9.
Koehler, W. C. (1967). Trans. Am. Crystallogr. Assoc. **3**, 53-73.
Newnham, R. E., Kramer, J. J., Schulze, W. A. and Cross, L. E. (1978). J. Appl. Phys. **49**, 6088-6091.
Newnham, R. E., Skinner, D. P. and Cross, L. E. (1978). Mater. Res. Bull. **13**, 525-536.
Nye, J. F. (1957). Physical Properties of Crystals. Oxford: Oxford University Press.
Sawada, A., Udagawa, M. and Nakamura, T. (1977). Phys. Rev. Lett. **39**, 829-831.
Shirane, G., Jona, F. and Pepinsky, R. (1955). Proc. I.R.E. **43**, 1738-1793.
Valasek, J. (1921). Phys. Rev. **17**, 475-481.

CHAPTER 9

LANTHANIDE AND ACTINIDE OXIDES, HYDROXIDES AND OXIDE HYDROXIDES

W. O. Milligan, G. W. Beall and D. F. Mullica
The Robert A. Welch Foundation, Houston, TX 77002

Introduction

There are many points of view in physical solid-state chemistry from which a historical development of the lanthanide and actinide oxides, hydroxides, and oxide hydroxides can be discussed. Since a significant amount of formidable knowledge related to physical properties (electrical, magnetic, and optical), crystal growth, thermodynamics and X-ray diffraction studies (other than structural) are readily accessible in the literature, this presentation considers only available X-ray diffraction structural investigations of oxides, oxyhydroxides, and hydroxides of both the lanthanide and actinide series. These elements constitute approximately 30% of the periodic table.

The lanthanide (L^n) oxides and hydrous oxides are dominated by the tendency of the metal to be trivalent across the entire series. This results in a reasonably small number of solid phases. Fig. 1 illustrates in general the

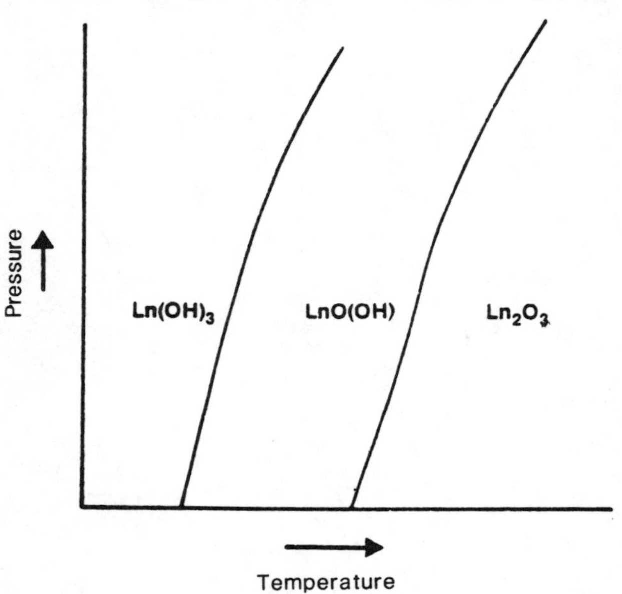

Fig. 1. Major chemical compounds that predominate in the trivalent rare earth oxide-water system.

major phases that predominate in the trivalent rare earth aqueous system. The trihydroxides of rare earth metals occur in the hexagonal form, prototype UCl_3, with the exception of the end member, lutetium, which is cubic. The lanthanide oxide hydroxides are predominately in the monoclinic form across the entire series at lower temperature. But, a high temperature tetragonal form has been reported for a few of the rare earth oxyhydroxides. The lanthanide oxides have three crystalline phases, cubic, hexagonal, and monoclinic. The only other crystalline phases of importance are the rare earth dioxides and and monoxides. There are, however, several isolated mixed valent oxides that exist under special conditions.

The case is much the same for the trivalent actinides. The dominant phases observed in the trivalent lanthanides are also found for the trivalent actinides. However, this simple one-to-one relationship is broken for the actinide oxides and hydroxides by the fact that higher valence states in the actinide system are found to be more stable. This leads to a much greater number of compounds in the actinide (Ac) series.

The Rare Earth and Actinide Oxides

The predominant stoichiometry for lanthanide oxides and a great portion of the actinide oxides is the sesquioxide (M_2O_3). There are three major crystalline phases – hexagonal, monoclinic and cubic – in the lanthanide and actinide oxides (Goldschmidt, Ulrich and Barth, 1925), illustrated in Figs. 2 and 3. As shown in Fig. 2,

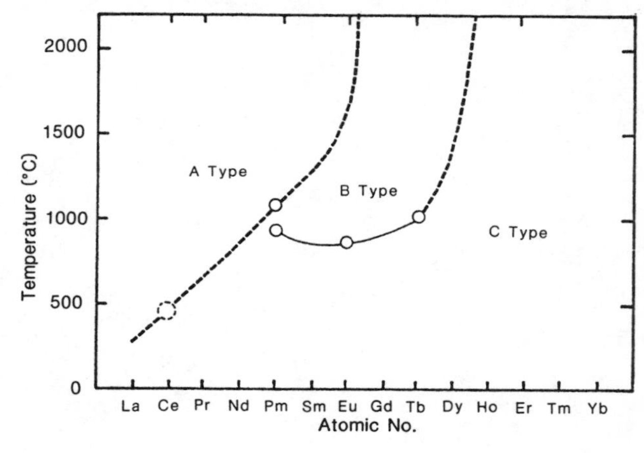

Fig. 2. Phase diagram for the lanthanide sesquioxides showing hexagonal (A type), monoclinic (B type) and cubic (C type) forms.

the dominant low temperature form is the cubic phase. At higher temperatures, the monoclinic phase exists for the mid-atomic elements in the lanthanide series and finally, a high temperature hexagonal form is observed for only the lower atomic number lanthanide oxides. The same characteristic trends can be seen in Fig. 3 for the actinide oxides, although a limited set of data exists.

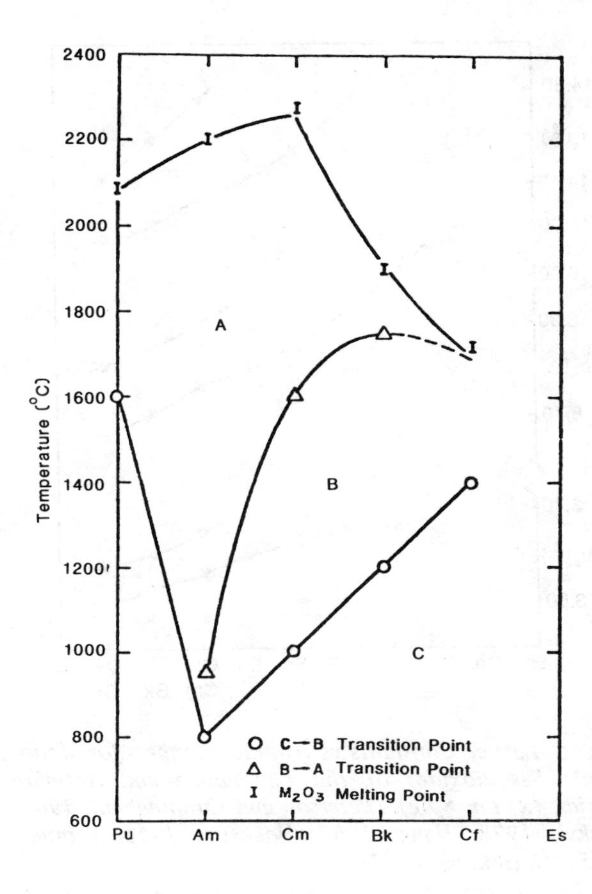

Fig. 3. Phase diagram for the actinide sesquioxides showing the A type (hexagonal), B type (monoclinic) and C type (cubic) forms.

Cubic Lanthanide Sesquioxides (M_2O_3)

The cubic lanthanide and actinide sesquioxide phase is in the space group *Ia*3 (T_h^7), No. 206, with 16 molecules per unit cell (O'Connor and Valentine, 1969). The structure of Er_2O_3 is analogous to bixbyite (Fe, Mn)$_2$ O_3 (Pauling and Shappell, 1930). The lanthanide metal atom is surrounded octahedrally by six neighboring oxygen atoms, each a distance of approximately $0.214a_o$. A neutron diffraction study of the crystal structure of the C-form of yttrium sesquioxide (Y_2O_3) has yielded very accurate metal and oxygen atom positions (O'Connor and Valentine, 1969). The investigation has also shown that there are six equivalent Y-O distances in the 8(b)-octahedron (2.284Å) and three significantly different independent distances for the 24(d)-octahedron (2.243, 2.274, and 2.331Å). The lattice constant used in this work was obtained from an X-ray diffraction analysis (Paton and Maslen, 1965). The cubic form is the dominant phase at low temperature and exists for all the rare earths and all the trivalent actinides that have been studied. Information on the lattice constants of lanthanide and actinide C-type oxides is graphically displayed in Fig. 4. The figure shows a smooth uniform change in lattice constants that is only interrupted by a small cusp at Gd and Cm. If one assumes a radius of 1.380Å for the oxygen ion in these structures, a set of crystal radii can be derived for the rare earths and actinides (Templeton and Dauben, 1954). The set of crystal radii thus derived is given in Table 1. Small differences (0.01-0.020Å) between successive elements in the lanthanide series (the lanthanide contraction) explain the historical difficulty in the chemical separation of these elements and the strikingly uniform chemistry of the series.

Table 1. Lanthanide and actinide crystal radii

Atomic No.	Element	r_o(Å)	Atomic No.	Element	r_o(Å)
57	La[h]	1.061	68	Er[a]	0.881
58	Ce	1.034	69	Tm[a]	0.869
59	Pr[i]	1.013	70	Yb[a]	0.858
60	Nd[i]	0.995	71	Lu[a]	0.848
61	Pm[c]	0.979	94	Pu[d]	0.989
62	Sm[a]	0.964	95	Am[b]	0.985
63	Eu[a]	0.950	96	Cm[g]	0.978
64	Gd[a]	0.938	97	Bk[e]	0.954
65	Tb[i]	0.923	98	Cf[f]	0.942
66	Dy[a]	0.908	99	Es[c]	0.928
67	Ho[a]	0.894			

Radii were derived from lattice constants used in Fig. 4 a(Templeton and Dauben, 1954); b(Templeton and Dauben, 1953); c(Baybarz and Haire, 1976); d(Gardner, Markin and Street, 1965); e(Peterson and Cunningham, 1967); f(Baybarz, Haire and Fahey, 1972); g(Asprey, Ellinger, Fried and Zachariasen, 1955); h(Koehler and Wollan, 1953); i(Zachariasen, 1926).

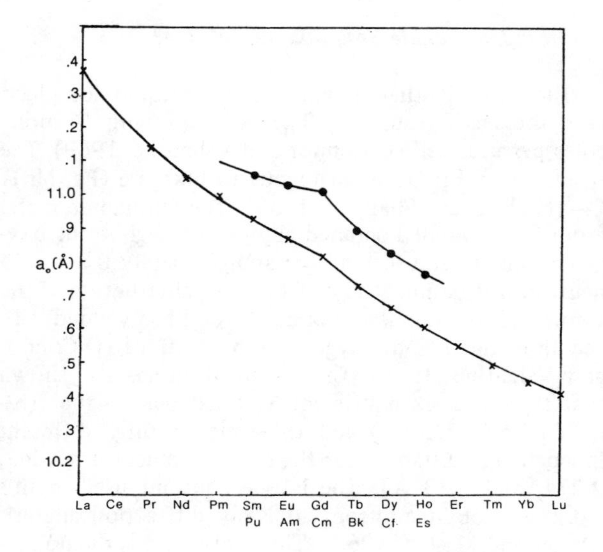

Fig. 4. Lattice constants for the Cubic Lanthanide and Actinide Sesquioxides vs. Atomic Number (x, Rare Earths. Actinides).

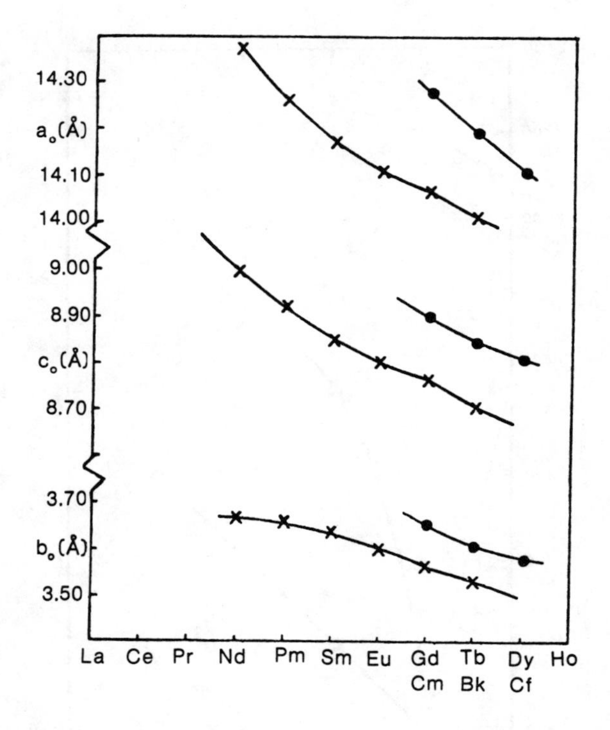

Fig. 5. Lattice Constants vs Atomic Numbers for Monoclinic Sesquioxides of the Lanthanide and Actinide Series, (x, Ln, • Ac). (Peterson and Cunningham, 1967; Yakel, 1978; Hang, 1967; Hoekstra, 1966; Cromer, 1957; Glushkova, 1967).

Monoclinic Lanthanide and Actinide Sesquioxides

The monoclinic lanthanide and actinide sesquioxides are in space group *C*2/*m* (No. 12) with six formulas per unit cell. A recent single crystal structural refinement (Yakel, 1979) confirmed this. Each of three independent europium ions in the unit cell are seven-coordinated with two types of average bond distances. Six of the seven Eu-O bonds in such hepta-polyhedron lie in the range 2.23-2.54Å; the seventh distance is greater than 2.6Å. In all cases, the bond that is greater than 2.6Å involves the oxygen atom that caps the polyhedral coordination sphere. In the case of two europium ions, the coordination polyhedra are monocapped trigonal prisms. For the third europium ion the coordination polyhedron is a monocapped octahedron. The overall packing of the structure has been described as layers of oxygen atoms, tetrahedrally coordinated with europium atoms. Lattice constants for the monoclinic lanthanide and actinide sesquioxides are plotted against atomic number in Fig. 5. The trend of decreasing lattice constants, as seen in the cubic form is apparent. The small cusp at Gd is only evident in a_o *and* c_o.

Hexagonal Lanthanide and Actinide Sesquioxides

The hexagonal sesquioxides of the lanthanide and actinide series are in space group $P\bar{3}m1$, No. 164 (Zachariasen, 1949). The metal atom is seven-coordinated having four M-O bond lengths less than 2.4Å and three ranging from 2.4 to 2.7Å. The unit cell contains one formula unit with each metal atom having a polyhedron described as a monocapped trigonal antiprism. This structural phase exists mainly for the lower atomic numbered lanthanides and intermediate actinides. Lattice constants are plotted against atomic number in Fig. 6. It can be seen that a_o decreases continuously with

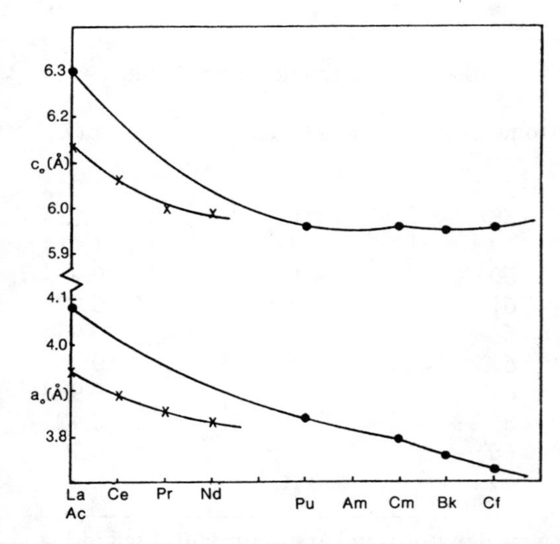

Fig. 6. Lattice Constants for the Hexagonal Sesquioxides of the Lanthanide and Actinide Series. (•, Ac, x, Ln). (Baybarz and Haire, 1976; Gardner, Markin and Street, 1965; Zachariasen, 1926; Zachariasen, 1949).

increasing atomic number, however, the c_o values tend to be constant for most of the actinide series. A single crystal analysis of hexagonal-La_2O_3 and Nd_2O_3 has yielded another hexagonal space group after considering submicro twining [Müller-Buschbaum and Schnering (1965) and Müller-Buschbaum (1966)].

The Rare Earth and Actinide Dioxides

The second most common oxides of the rare earth and actinide series are the dioxides. The dioxide form is much more prevalent in the actinide series than the lanthanide series, due primarily to the greater stability of the tetravalent state found in the actinide series. The initial precipitation of a dioxide from solution yields an amorphous gel that will age to the crystalline dioxide (Haire, Lloyd, Beasley and Milligan, 1971). The structure of the lanthanide and actinide dioxides is that of fluorite, space group *Fm3m* (No. 225). The metal atom is located in the 4(a) position with point symmetry *m3m* at the cell origin (0,0,0) and the oxygen atoms are located in the 8(c) position with point symmetry $\bar{4}3m$ at 1/4,1/4,1/4. Lattice constants are graphed against atomic numbers in Fig. 7. The graph clearly shows a smooth change in lattice constant for the actinide compounds.

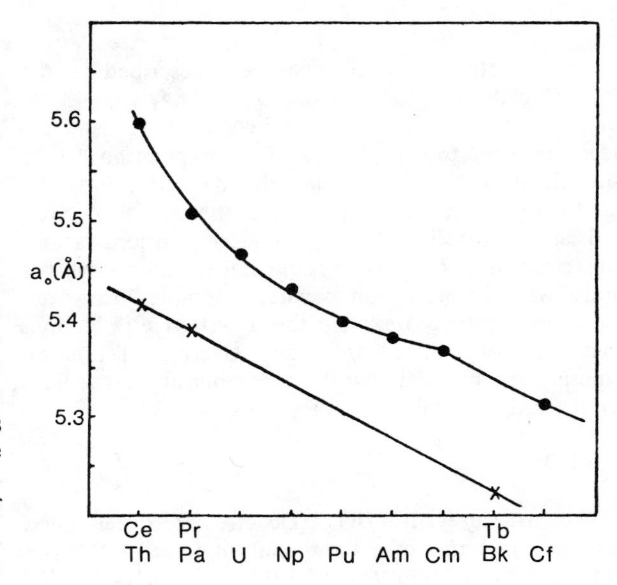

Fig. 7. Lattice Constants vs. Atomic Number for the Cubic Dioxides of the Lanthanide and Actinide Series (x, Ln;•, Ac). (Templeton and Dauben, 1953; Baybarz, and Haire 1976; Lloyd, Beasley and Milligan, 1971; Brauer and Willaredt, 1971; Mulford and Ellinger, 1958; Baenziger, Eick, Schuldt and Eyring, 1961; Goldschmidt and Thomassen, 1923; Zachariasen, 1949).

Lanthanide and Actinide Lower Oxides

A few of the lanthanide and actinide elements have a monoxide phase. This structure is cubic face-centered (*Fm3m*, No. 225; halite, NcCl type) with the metal atom located in the 4(a) position at the origin (0,0,0) and the oxygen atom in the 4(b) position at 1/2,1/2,1/2. The stereochemistry about both the metal and oxygen ions is octahedral (O_h). EuO (Eick, Baenziger and Eyring, 1956) and YbO (Fishel, Haschke and Eick, 1970) are examples. Prepatory methods of these lower-oxides can be found in the literature (McCarthy and White, 1970). Table 2 lists the lattice constants for this rock-salt structure type. Eu_3O_4 (Bärnighausen and Brauer, 1962) is another example but, this lower-oxide best fits the orthorhombic space group *Pnam* (No. 62).

UO_3 Polymorphs

Uranium trioxide has at least five different polymorphs (α-ϵ). Of these five, only three (α, β, γ) have detailed structural data. Each will be described.

Table 2. Lattice constants for the Monoxides (MO) of the lanthanides and actinides.

Atomic Number	Element	a_o(Å)
62	Sm[d]	4.9883
63	Eu[b]	5.1439
70	Yb[f]	4.86
92	U [e]	4.92
94	Pu[c]	4.959
95	Am[a]	5.05

a(Zachariasen, 1949); b(Domange, Flahaut and Guittard, 1959); c(Ball, Greenfield, Mardon and Robertson, 1958); d(Ellinger and Zachariasen, 1953); e(Rundle, Wilson, Baenziger and Tevehaugh, 1958); f(Archard and Tsoucaris, 1958).

α-UO_3

The structure of α-UO_3 has been described on the basis of orthorhombic unit cell, a_o = 6.84, b_o = 43.45, and c_o = 4.157Å (Greaves and Fendes, 1972) and has been compared to L-Ta_2O_5 (Roth and Stephenson, 1971). The alpha form of uranium trioxide was refined by employing the U_3O_8 structure, to be described later, with approximately 12% uranium atom positions vacant. This structural form requires further research since the study was conducted on powdered samples. However, the refinement does explain the superlattice reflections observed both in neutron and electron diffraction experiments and the low observed density as well as peculiar infra-red vibrational frequencies.

β-UO_3

The structure of β-UO_3 (Debets, 1966) has been described in terms of a monoclinic unit cell ($P2_1$, No. 4) with a_o = 10.34(1), b_o = 14.33(1), c_o = 3.910(4)Å and β = 99.3(2). This β-form has been described as a layered structure connected by uranyl groups. These layers are 7.16Å apart. The idealized symmetry of the polyhedra around two of the five types of uranium atoms is a pentagonal bipyramid (D_{5h}), whereas the hexacoordination system associated with the other three types of uranium atoms is a highly distorted octahedron.

γ-UO_3

The structure of γ-UO_3 (Siegel and Holkstra, 1971; Ergmann and DeWolff, 1963) appears to be pseudo-tetragonal ($I4_1/amd$ No. 141) with a_o = 6.89 and c_o = 19.94Å (Ergmann and DeWolff, 1963). All the uranium atoms in this structure are six-coordinated with oxygen atoms in a distorted octahedral arrangement. The octahedra around one of the uranium atoms share edges, thus forming strings that run in the a and b directions at different z levels. These strings are not immediately connected, but each octahedron surrounding the other crystallographically independent uranium atom shares corners with the strings in both directions.

Mixed Valence State Oxides of the Lanthanides and Actinides

There are many of the oxides in the lanthanide and actinide series that have mixed valence states. It is beyond the scope of this work to describe each in detail.

α-U_3O_8

α-U_3O_8 (Loopstra, 1964) crystallizes in the orthorhombic system a_o = 4.14(8), b_o = 11.96(6), and c_o = 6.71(7)Å, two molecules per unit cell (z = 2). The hepta-coordination around the uranium atoms is in the form of a pentagonal bipyramid (D_{5h}). The interatomic bond distances between the uranium atom and six of the seven coordinated oxygen atoms range from 2.07 to 2.23Å. The seventh bonded oxygen atom is at a distance of 2.44Å when associated with one U atom and 2.71Å when related to the second U atom.

β-U_3O_8

The structure of β-U_3O_8 (Loopstra, 1970) appears to be very closely related to the alpha form, orthorhombic (*Cmcm*, No. 63). The lattice constants are a_o = 7.069(1), b_o = 11.445(1), and c_o = 8.303(1)Å with four molecules per unit cell. The U-O bond distances range from 1.888 to 2.398Å. The polyhedra about the uranium atoms are divided into two systems, heptacoordination (pentagonal bipyramid) and hexacoordination (a distorted octahedron).

U_4O_9

The crystal system of U_4O_9 is cubic (space group $I4_132$, No. 214) with a lattice constant of 21.76Å (Masaki and Doi, 1972). The crystallographic make-up appears to be a modification of the fluorite type structure (four times the UO_2 arrangement). For this mixed-valence compound (U_4O_9), the fluorite structure is transformed by the addition of 64 oxygen atoms at the centers of interstices in the basic structure. The remaining 24 atoms are displaced from the centers of the interstices by 0.52Å along the face diagonal. Of the 256 uranium atoms constituting the fluorite type structure, 16 are found to be displaced by 0.45Å along the body diagonal directions from their normal positions. This arrangement results in U-O bond distances of 2.37Å and 1.92Å for the two types of uranium atoms respectively.

Lanthanide Oxide Hydroxides

The compounds of most interest after considering oxides of the lanthanide and actinide series are the oxide hydroxides of these elements. These oxyhydroxide compounds occur in two crystalline phases, monoclinic and tetragonal. The monoclinic is by far the most common at room temperature and therefore, will be treated in more detail.

The monoclinic form of the lanthanide oxide hydroxides crystallizes in space group $P2_1/m$ with 2 molecules per unit cell (Gondrand and Christensen, 1971; Christensen, 1965; Christensen, 1966; Milligan, Mullica and Hall, 1980; Klevtsov and Sheina, 1965; Bärnighausen, 1965). The lattice constants for the known LnOOH series are plotted against atomic numbers (z) in Fig. 8. It can be seen that the lattice constants decrease linearly with increasing atomic numbers. The

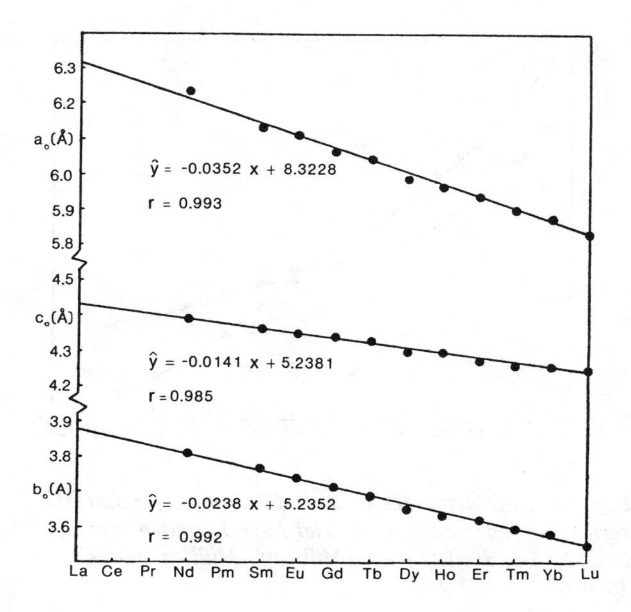

Fig. 8. Lattice Constants vs. Atomic Number for the Monoclinic form of LnOOH. (Christensen, 1966; Milligan, Mullica and Hall, 1980; Klevtsov and Sheina, 1965; Bärnighausen, 1965).

cusp, seen at Gd in the case of the lanthanide and actinide oxides, is not evident in the lattice constants of the LnOOH. Linear regressional analyses of the experimental unit cell parameters and consequential correlation analyses (r's $>$ 0.98), see Fig. 8, verify linearity. The values for DyOOH seem to be slightly low; perhaps re-investigation of this compound is needed. The structure consists of Ln atoms surrounded by seven oxygen atoms forming a monocapped trigonal prism. There are four O^{2-} ions and three hydroxyl groups in the coordinating sphere. There is no hydrogen bonding evident in the structure. The determined Lu-O bond lengths from the structural refinement of lutetium hydroxide oxide range from 2.13 to 2.42Å.

The tetragonal form of the LnO(OH) (Gondrand and Christensen, 1971) has been assigned space group $P\bar{4}2_1m$ (No. 113). This phase is prepared by high pressure-high temperature methods. The lattice constants (Gondrand and Christensen, 1971) for the tetragonal oxyhydroxide form of the lanthanide series are graphed against the corresponding atomic number in Fig. 9. As seen previously, the lattice constants uniformly decrease in the familiar manner with increasing atomic number. The cusp at Gd is quite evident for c_o values but not a_o values. Further investigations of this modified form of LnO(OH) may be fruitful. Little detailed structural data are available for this modification. By analogy with $LiLnO_2$, the Ln atoms appear to be seven coordinate as in the case with the monoclinic form.

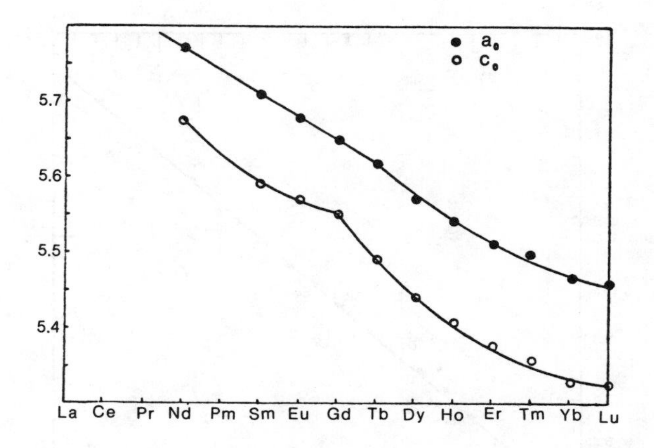

Fig. 9. Lattice Constants vs. Atomic Number for the Tetragonal Form of LnO(OH).

Lanthanide and Actinide Trihydroxides

The most common compound formed by precipitation of trivalent rare earths or actinides from aqueous solutions are trihydroxides (Haire and Willmarth, 1960; Beall, Milligan and Wolcott, 1977; Mullica, Milligan and Beall, 1979). These trihydroxides are primarily hexagonal with only lutetium exhibiting a cubic modification and not crystallizing in the hexagonal form as yet. It has been possible, from a knowledge of the structural relationships between the two series, to develop the chemistry of the actinides. This can be demonstrated by electron micrograph and diffraction patterns of $Am(OH)_3$ (Milligan, Beasley, Lloyd and Haire, 1968). Such experiments have also been carried out for $Cm(OH)_3$ (Haire, Lloyd, Milligan and Beasley, 1977) and $Pm(OH)_3$ (Haire, Dillin, Milligan and Beasley, 1977). These investigations are only possible after having extensive supporting data from the lanthanide series. In these studies, the effects of self irradiation were carried out.

The hexagonal form of the heavy-metal lanthanide and actinide trihydroxides has been assigned space group $P6_3/m$ (No. 176). This series has been studied extensively (Beall, Milligan and Wolcott, 1977; Mullica, Milligan and Beall, 1979). The lattice constants for the lanthanide and actinide trihydroxides are plotted against one another in Fig. 10. A regressional analysis has been applied to the data and it is obvious that the data points are best represented by a straight-line curve. The correlation coefficient (R) which has been calculated from the independent variables (a_o and c_o) is 0.9996. The computerized analysis has been accompanied by a computation of homoscedasticity (standard deviations 0.108 and 0.096). An empirical relationship has been derived for calculating the lattice constants from the crystal radii of the lanthanide or actinide of interest (see Milligan, Mullica and Oliver, 1979 for complete

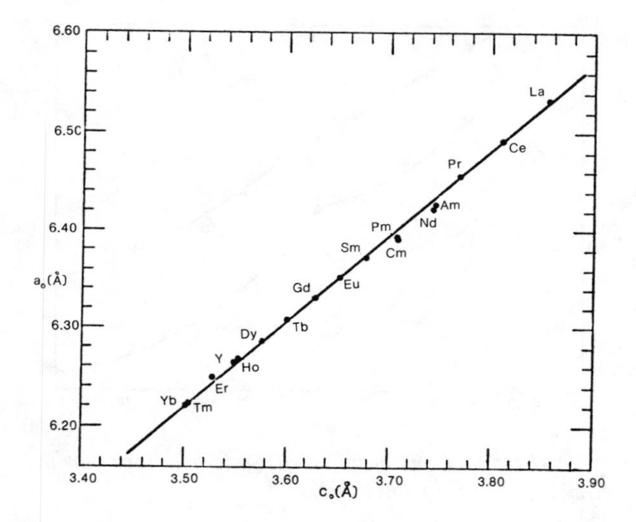

Fig. 10. Lattice Constants a_o vs. c_o for Hexagonal heavy-metal trihydroxides. (Milligan, Mullica and Oliver, 1979; Milligan, Beasley, Lloyd and Haire, 1968; Roy and McKinstry, 1953; Haire, Lloyd, Milligan and Beasley, 1977; Dillin, Milligan and Williams, 1973; Beall, Milligan and Wolcott, 1977; Fricke and Durrwachter, 1949).

details). The atom positions for the $M(OH)_3$ compounds (where M = La, Ce. . .Yb) are M at 1/3, 2/3, 1/4) and the oxygen atom at approximately (0, 0.39, 0.31). The hydrogen position has been identified as (0.2734, 0.1436, 1/4) in a recent neutron diffraction study of $Tb(OH)_3$ (Beall, Milligan, Korp and Bernal, 1977). The hexagonal structure consists of infinite chains of nine-coordinate metal ions running along the *c* axis. The coordination polyhedron is a tricapped trigonal prism. The structure has two unique metal to oxygen bond distances that have been designated as apical, and equatorial respectively. The positional parameters in the lanthanide trihydroxides are essentially the same, and hence the metal-oxygen distances are a function of the lattice constants. Therefore, if the lattice constants of compounds of interest are known (for example, from a powder X-ray diffraction photograph), the metal-oxygen and other distances may be calculated without solving the complete structure by single crystal methods. This calculation can be applied, for example, to $Yb(OH)_3$ and many of the actinide trihydroxides. The apical and equatorial bond distances are graphed against atomic numbers in Fig. 11. Values for M-O bond lengths in Angstrom units may be calculated from the derived empirical equations, M-O(1) = 0.3899 a_o and M-O(2) = $[(0.2627a_o)^2 + (0.5c_o)^2]^{1/2}$. Respective standard deviations for each equation are 0.004 and 0.005N and the percent average deviation between calculated and observed metal to oxygen bond distances is 0.2%. As can be seen, the plotted apical and equatorial M-O bond distances exhibit some rather peculiar trends. The most striking feature is the reversal in relative bond distance at Gd.

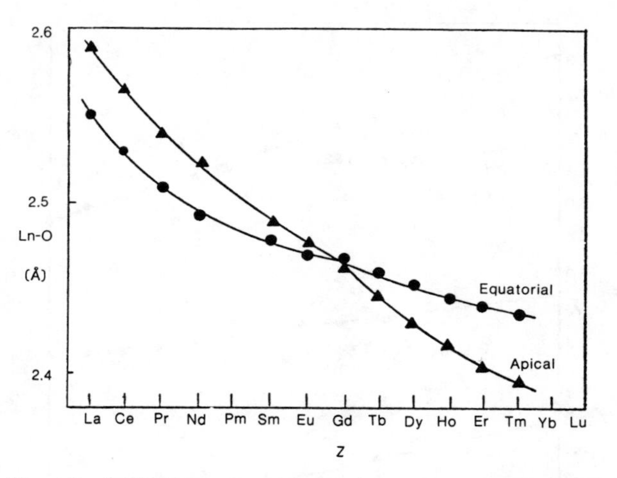

Fig. 11. M-O distances vs. atomic number for $Ln(OH)_3$ series • represents equatorial M-O(1) and ▲ represents M-O(2) bond distances. (Milligan, Mullica and Oliver, 1979).

The divergence of the bond distances at the end of the series may be responsible for the change to the cubic phase at lutetium. The apical M-O bond distances have been employed to derive crystal radii for the lanthanide and actinide trihydroxides by assuming a radius of 1.487Å for the hydroxyl ion (details can be found in Milligan, Mullica and Oliver, 1979). The agreement between crystal radius values obtained from the experimental M-O(2) apical bond distances and other reliable literature values (Templeton and Dauben, 1954; Ergmann and DeWolff, 1963; Milligan, Mullica and Oliver, 1979) is quite sound; a 0.08% average deviation is observed.

The other lanthanide trihydroxide modification, $Lu(OH)_3$ (Mullica, 1977) is cubic (space group *Im3*, No. 204), a_o = 8.222Å. The Lu atom is at 1/4,1/4,1/4 and the oxygen atom is at 0.0, 0.319(1), 0.166(1). This results in a Lu-O bond distance of 2.243(3)Å (Mullica and Milligan, 1980). Each Lu atom is octahedrally (O_h) coordinated with six oxygen atoms and each oxygen atom is coordinated by two lutetium atoms. The structure of $Lu(OH)_3$ is a polyhedral framework consisting of two slightly distorted pentagonal dodecahedra and twelve 7-hedra per cell in the body-centered cubic (bcc) arrangement. The eight Lu atoms and the twelve O atoms occupy the 20 vertices of the dodecahedra. This type of arrangement has been observed in a neutron diffraction study of $In(OH)_3$ (Mullica, Beall, Milligan, Korp and Bernal, 1979) and is thoroughly discussed by Jeffrey and McMullan, 1967.

Acknowledgements

Much of the most recent work related to the lanthanide oxide hydroxides and trihydroxides is attributed to equipment and support, in part, by the U.S. Atomic

Energy Commission (AEC-AT-3955), M.D. Anderson Foundation, National Science Foundation (GH-34513) and the Robert A. Welch Foundation (AA-668).

References

Archard, J. C. and Tsoucaris, G. (1958). C. R. Acad. Sci. **246**, 285-288.

Asprey, L. B., Ellinger, F. H., Fried, S. and Zachariasen, W. H. (1955). J. Am. Chem. Soc. **77**, 1707-1708.

Baenziger, N. C., Eick, H. A., Schuldt, H. S. and Eyring, L. (1961). J. Am. Chem. Soc. **83**, 2219-2223.

Ball, J. G., Greenfield, P., Mardon, P. G. and Robertson, J. A. L. (1958). At. Res. Establ. (Great Britain) M/R 2416.

Bärnighausen, H. (1965). Acta Cryst. **19**, 1047.

Bärnighausen, H. and Brauer, G. (1962). Acta Cryst. **15**, 1059.

Baybarz, R. D. and Haire, R. G. (1976). Proc. Moscow Symp. Chem. Transuranium Elem. 7-12.

Baybarz, R. D., Haire, R. G., and Fahey, J. A. (1972). J. Inorg. Nucl. Chem. **34**, 557-565.

Beall, G. W., Milligan, W. O., Korp, J. and Bernal, I. (1977). Acta Cryst. **B33**, 3134-3136.

Beall, G. W., Milligan, W. O. and Wolcott, H. A. (1977). J. Inorg. Nucl. Chem. **39**, 65-70.

Brauer, G. and Willaredt, B. (1971). J. Less Common Met. **24**, 311-316.

Christensen, A. N. (1965). Acta Chem. Scand. **19**, 1391-1396.

Christensen, A. N. (1966). Acta Chem. Scand. **20**, 896-898.

Cromer, D. T. (1957). J. Phys. Chem. **61**, 753-756.

Debets, P. C. (1966). Acta Cryst. **21**, 589-593.

Dillin, D. R., Milligan, W. O. and Williams, R. J. (1973). J. Appl. Cryst. **6**, 492-494.

Domange, L., Flahaut, J. and Guittard, M. (1959). C. R. Acad. Sci. **249**, 697-699.

Eick, H. A., Baenziger, N. C. and Eyring, L. (1956). J. Am. Chem. Soc. **78**, 5147-5149.

Ellinger, F. H. and Zachariasen, W. H. (1953). J. Am. Chem. Soc. **75**, 5650-5652.

Ergmann, R. and deWolff, P. M. (1963). Acta Cryst. **16**, 993 996.

Fishel, N. A., Haschke, J. M. and Eick, H. A. (1970). Inorg. Chem. **9**, 413-414.

Fricke, R. and Durrwachter, W. (1949). Z. Anorg. Allg. Chem. **259**, 305-308.

Gardner, E. R., Markin, T. L. and Street, R. L. (1965). J. Inorg. Nucl. Chem. **27**, 541-551.

Glushkova, V. B. (1967). (Nauka: Leningrad). 133 pp. 61.

Goldschmidt, V. M. and Thomassen, L. (1923). Skrifter Norske Videnskaps-Akad. Oslo, I. Mat. Naturv. Kl. No. 2.

Goldschmidt, V. M., Ulrich, F. and Barth, T. (1925). Skrifter Norske Videnskaps-Akad. Oslo, Mat. Naturv. Kl. No. 5.

Gondrand, M. and Christensen, A. N. (1971). Mater. Res. Bull. **6**, 239-243.

Greaves, C. and Fendes, B. E. F. (1972). Acta Cryst. **B28**, 3609.

Haug, H. O. (1967). J. Inorg. Nucl. Chem. **29**, 2753-2758.

Haire, R. G., Dillin, D. R., Milligan, W. O. and Beasley, M. L. (1977). J. Inorg. Nucl. Chem. **39**, 53-57.

Haire, R. G., Lloyd, M. H., Beasley, M. L. and Milligan, W. O. (1971). J. Elect. Mic. **20**, 8-16.

Haire, R. G., Lloyd, M. H., Milligan, W. O. and Beasley, M. L. (1977). J. Inorg. Nucl. Chem. **39**, 843-847.

Haire, R. G. and Willmarth, R. (1960). ORNL report TM-2387.

Hoekstra, H. R. (1966). Inorg. Chem. **5**, 754-757.

Jeffrey, G. A. and McMullan, R. K. (1967). Prog. Inorg. Chem. **8**, 43-108.

Klevtsov, P. V. and Sheina, L. P. (1965). Izv. Akad. Nauk. SSSR Neorg. Mater. **1**, 912-917.

Koehler, W. C. and Wollan, E. O. (1953). Acta Cryst. **6**, 741-742.

Loopstra, B. O. (1964). Acta Cryst. **17**, 651-654.

Loopstra, B. O. (1970). Acta Cryst. **B26**, 656-657.

Masaki, N. and Doi, K. (1972). Acta Cryst. **B28**, 785-791.

McCarthy, G. J. and White, W. B. (1970). J. Less Common Met. **22**, 409-417.

Milligan, W. O., Beasley, M. L., Lloyd, M. H. and Haire, R. G. (1968). Acta Cryst. **B24**, 979-980.

Milligan, W. O., Mullica, D. F. and Hall, M. A. (1980). Acta Cryst. **B36**, 3086-3088.

Milligan, W. O., Mullica, D. F. and Oliver, J. D. (1979). J. Appl. Cryst. **12**, 411-412.

Mulford, R. N. R. and Ellinger, F. H. (1958). J. Phys. Chem. **62**, 1466-1467.

Müller-Buschbaum, H. (1966). Z. Anorg. Allg. Chem. **343**, 6-10.

Müller-Buschbaum, H. and Schnering, H. G. (1965). Z. Angor. Allg. Chem. **340**, 232-245.

Mullica, D. F. (1977). Thesis, Baylor University, Waco, Texas.

Mullica, D. F., Beall, G. W., Milligan, W. O., Korp, J. D. and Bernal, I. (1979). J. Inorg. Nucl. Chem. **41**, 277-282.

Mullica, D. F. and Milligan, W. O. (1980). J. Inorg. Nucl. Chem. **42**, 223-227.

Mullica, D. F., Milligan, W. O. and Beall, G. W. (1979). J. Inorg. Nucl. Chem. **41**, 525-532.

O'Connor, B. H. and Valentine, T. M. (1969). Acta Cryst. **B25**, 2140-2144.

Paton, M. G. and Maslen, E. N. (1965). Acta Cryst. **19**, 307-310.

Pauling, L. and Shappell, M. L. (1930). Z. Kristallogr. **75**, 128-142.

Peterson, J. R. and Cunningham, B. B. (1967). Inorg. Nucl. Chem. Lett. **3**, 327-336.

Roth, R. S. and Stephenson, N. G. (1971). The Chemistry of External Defects in Non-Metallic Solids. Amsterdam: North Holland.

Roy, R. and McKinstry, H. A. (1953). Acta Cryst. **6**, 365-366.

Rundle, R. E., Wilson, A. S., Baenziger, N. C. and Tevehaugh, A. D. (1958). U.S.A.E.C. Rept. TID-5290 Book 1, 67.

Siegel, S. and Holkstra, H. R. (1971). Inorg. Nucl. Chem. Lett. **7**, 455-457.

Templeton, D. H. and Dauben, C. H. (1953). J. Am. Chem. Soc. **75**, 4560-4562.

Templeton, D. H. and Dauben, C. H. (1954). J. Am. Chem. Soc. **76**, 5237-5239.

Yakel, H. L. (1978). Acta Cryst. **B35**, 564-569.

Zachariasen, W. H. (1926). Narsk. Geol. Tidssk. **9**, 310-316.

Zachariasen, W. H. (1949). Acta Cryst. **2**, 60-62.

Zachariasen, W. H. (1949). Acta Cryst. **2**, 388-390.

CHAPTER 10

THE ROLE OF CRYSTALLOGRAPHY IN URANIUM MINERALOGY

Deane K. Smith, Jr.
Department of Geosciences
The Pennsylvania State University 16802

Introduction

If one examines the *"Glossary of Mineral Species 1980"* (Fleischer, 1980, 1981) over 160 mineral species can be identified which contain uranium as an essential element. In addition there are several dozen other species in which uranium is known to play a significant substitutional role. Table 1 summarizes the distribution of uranium minerals by major chemical subgroups. It is at first surprising that uranium minerals comprise over 5% of the known mineral species. However, the large amount of work on uranium chemistry and the large number of uranium mines which have been developed help account for this abundance. The complex chemistry of uranium is also a contributing factor.

TABLE 1

Uranium Mineralogy

Reduced forms	**Known species**
U^{4+} minerals	11
U^{4+} - U^{6+} minerals	7
Niobates, tantalates and titanates	18
U^{4+} substituted (› 20)	
Oxidized forms [as $(UO_2)^{2+}$ ion, mostly hydrated]	
Uranyl oxide hydrates	5
Alkali, alkali-earth uranyl oxides	19
Uranyl silicates	11
Uranyl phosphates and arsenates	54
Uranyl vanadates	17
Uranyl molybdates	7
Uranyl sulfates	11
Uranyl carbonates	15
Uranyl selenites and tellurites	7
Total U minerals	162
Total minerals	∿ 2800

Uranium minerals are usually divided into two main groups, the so-called "primary" and "secondary" minerals. This classification was based on the initial belief that all uranium was initially deposited in the U^{4+} state then subsequent oxidation and ground water transport allowed the formation of U^{6+} minerals. Better terminology now would be to use the terms "reduced" and "oxidized" minerals as many examples of primary U^{6+} minerals are known.

The understanding and classification of uranium minerals are a combination of chemical and crystal structure knowledge which probably began with the recognition by Goldschmidt and Thomassen (1923) that uraninite had the CaF_2, fluorite, structure. The chemistry of uraninite, however, is not so simple as to imply that it is a simple stoichiometric UO_2. Not only does it form extensive solid solution with other compounds which take the fluorite structure such as CeO_2, cerianite, ThO_2, thorianite and many rare-earth and actinide oxides; it also usually shows marked deviations in stoichiometry. Fig. 1 (see next page) is a composite T-X phase diagram of the U-O system compiled from sources too numerous to list (Smith, 1982). The geologically important region of this phase diagram is between $UO_{2.0}$ and $UO_{2.33}$ and below 1000°C. Although oxides higher than $UO_{2.33}$ could exist in nature, they evidently do not form, and hydrated uranyl oxides form in their place. Grønvold (1955) showed that the excess oxygen in UO_2 is interstitial and that the formula should be considered UO_{2+X}. The limiting value for X is 0.25 and both disordered $UO_{2.25}$ and ordered U_4O_9 (Masaki, 1972: Masaki and Doi, 1972 and Naito, 1974) exist. Many studies have been made on the ordered U_4O_9, and Willis (1978) has reviewed the structural data. Between $UO_{2.25}$ and $UO_{2.33}$ the cubic fluorite structure distorts to a tetragonal derivative. Only one of the several forms known has been found in nature, αU_3O_7 (Voultsidis and Clasen, 1978).

U^{4+} Mineralogy

Table 2 (see page 372) shows the ionic radii and crystal chemical behavior of uranium in its major oxidation states. The U^{5+} ion has been reported in some compounds using spectroscopic methods, but its effect in mineral structures is apparently minor. The U^{4+} ion is similar in size to Ca^{2+} and the rare-earth ions; consequently uranium often plays a substititional role in

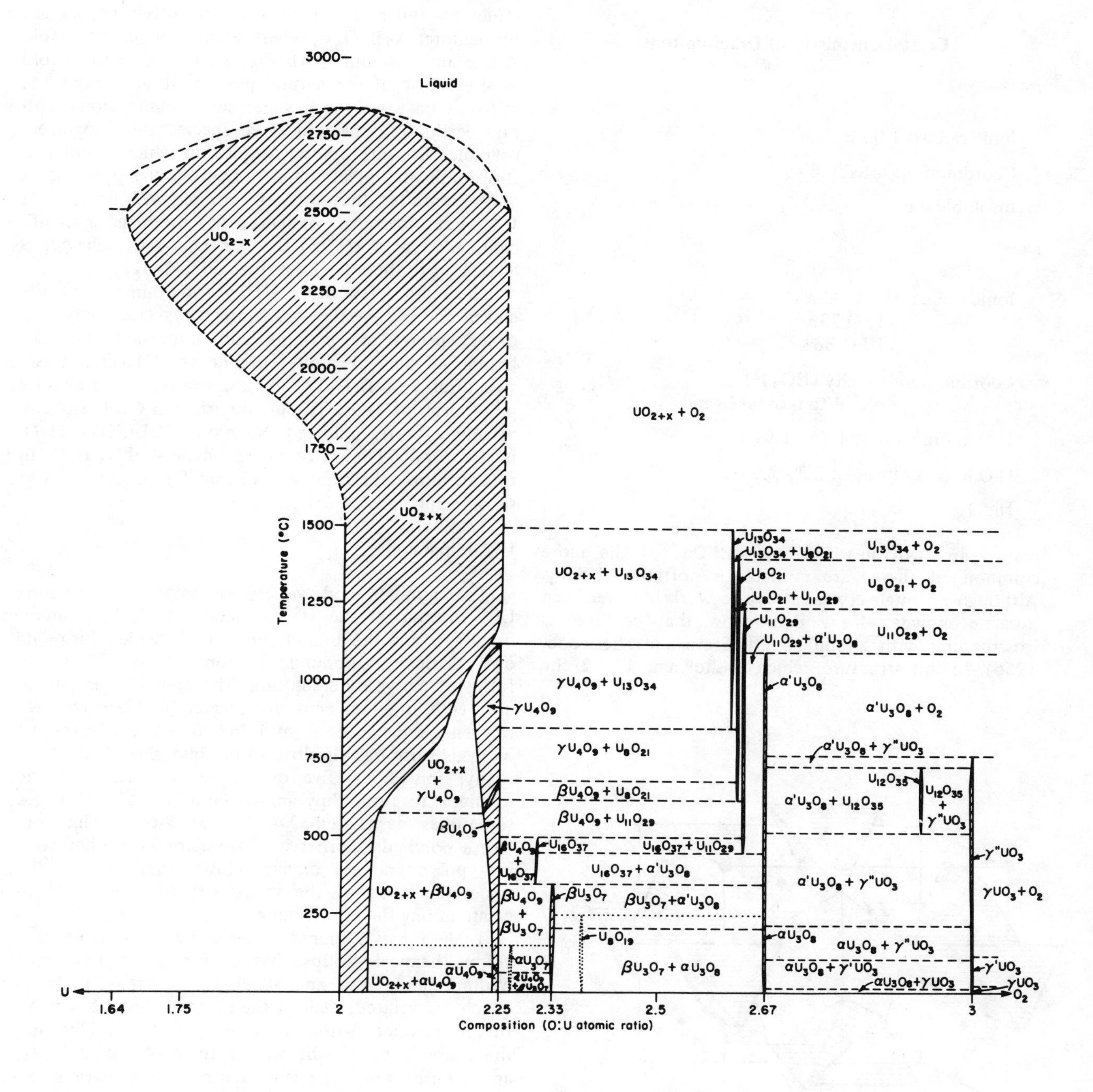

Figure 1. Phase relations in the uranium-oxygen system.

TABLE 2

Crystal Chemistry of Uranium Ions

U^{4+}

Ionic radius: 1.00Å

Coordination: usually 6 to 8

Insoluble ion

U^{6+}

Ionic radius: II 0.45Å
VI 0.73Å
VIII 0.88Å

Coordination: usually $(UO_2)^{2+}$
plus 4 to 6 other ligands

U-O in uranyl ion: 1.76 - 1.91Å

U-O in other ligands: 2.3 - 2.6Å

Highly soluble ion

many rare-earth bearing minerals. One of the more common of these rare minerals is coffinite, $USiO_4$. Although no single-crystal structure work has ever been possible, powder X-ray data show that coffinite is isostructural with zircon (Stieff, Stern and Sherwood, 1956). In this structure, which is shown in Fig. 2, the

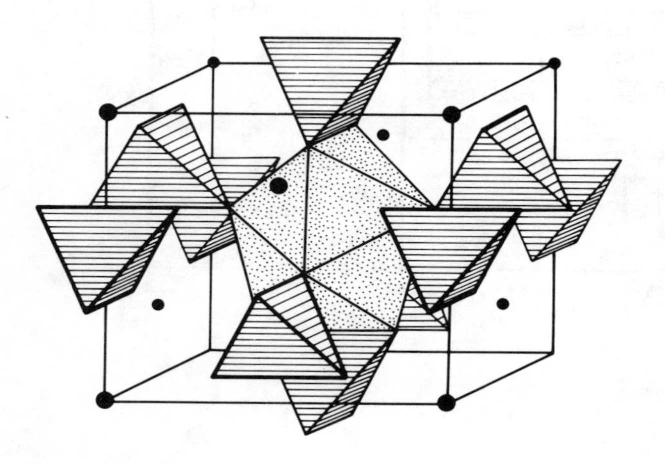

Figure 2. The structure of coffinite, $USiO_4$. The uranium is in 8-fold coordination shown stippled in the figure. The SiO_4 tetrahedra are ruled.

uranium is in 8-fold coordination. Zr, Th and Hf may substitute for the uranium and PO_4 and evidently $(OH)_4$ may substitute for the SiO_4. Two other common structure types in which uranium often occurs are pyrochlore, $A_2B_2O_6F$, where uranium is in the 8-fold A site and columbite, AB_2O_6, where it is in the 6-fold A site. Most of the natural pyrochlore and columbite minerals have only a few percent uranium along with rare earths and thorium. Two species are recognized, uranmicrolite and uranpyrochlore, in which uranium is the dominant ion in the A-site. In all the pyrochlores and columbites, they probably formed with uranium as U^{4+}, but subsequent oxidation has produced a significant fraction of U^{6+} in the minerals. This behavior is analogous to the oxidation observed in UO_{2+x}.

There are a few other U^{4+} bearing minerals. Of the eleven specific species which are recognized, only two other species are economically important minerals, brannerite and ningyoite. Brannerite, UTi_2O_6, is isostructural with $ThTi_2O_6$ and a derivative of the anatase polymorph of TiO_2. The uranium is in a 6-fold site and usually is partly oxidized. Ningyoite, $CaU(PO_4)_2.H_2O$, is structurally related to rhabdophane, $CePO_4.H_2O$. In this structure the uranium is in an 8-fold site and also shows significant oxidation.

U^{6+} Mineralogy

The number and variety of minerals containing U^{6+} outnumber the U^{4+} minerals. As U^{6+}, uranium usually forms the uranyl ion, UO_2^{2+}, which forms its own distinct compounds in contrast to U^{4+} which forms extensive solid solution. The uranyl ion is a linear unit with uranium centrally located between the two oxygens at distances from 1.76Å to 1.91Å. Additional oxygens form ligands close to the bisecting plane of the uranyl ion. Four, five or six oxygens surround the uranium to form dipyramidal coordination polyhedra sometimes designated 2-4, 2-5 or 2-6 coordination. These polyhedra corner- or edge-share with other uranium polyhedra and oxyanion polyhedra to form the structural motif of the many structure types which occur among the uranyl minerals.

Table 1 lists a uranyl mineral classification by oxyanion chemical groups. Except for the uranyl oxide hydrates and alkali and alkaline-earth uranyl hydrates which are related, each of the chemical groups is structurally distinct. Within most of the chemical groups, there are distinct subgroups with similar uranyl:oxyanion ratios which are isostructural or otherwise structurally related. Even when uranyl: oxyanion ratios are different within a chemical group, structural similarities exist.

The first structure of a uranyl compound was proposed by Beintema (1938) for the phosphates, autunite and meta-autunite. These two minerals are the structure types for the two most populated subgroups as shown by the lists in Tables 3 and 4. In these structures 2-4 uranyl dipyramids corner-share with PO_4 tetrahedra to form a tetragonal sheet which is the basic structural unit for both mineral sub-groups. Intersheet cations and water molecules complete the structure and control the manner in which the sheets stack and the polyhedra articulate to minimize packing and bond energies. The number of water molecules is variable and sensitive to humidity conditions and controls the stacking that is the main distinction between the autunite and meta-autunite subgroups. The cations easily exchange with other monovalent or divalent cations, and the phosphate groups may be replaced by arsenate groups which accounts for the many mineral species.

TABLE 3

The Autunite Family

Mineral	**Formula**
Arsenuranospathite	$HAl(UO_2)_4(AsO_4)_4 \cdot 40H_2O$
Autunite	$Ca(UO_2)_2(PO_4)_2 \cdot 8\text{–}12H_2O$
Fritzcheite*	$Mn(UO_2)_2(VO_4)_2 \cdot 10H_2O$
Heinrichite	$Ba(UO_2)_2(AsO_4)_2 \cdot 10\text{–}12H_2O$
Kahlerite	$Fe(UO_2)_2(AsO_4)_2 \cdot 10\text{–}12H_2O$
Novacekite	$Mg(UO_2)_2(AsO_4)_2 \cdot 12H_2O$
Sabugalite	$HAl(UO_2)_4(PO_4)_4 \cdot 16H_2O$
Saleeite	$Mg(UO_2)_2(PO_4)_2 \cdot 10H_2O$
Threadgoldite	$Al(UO_2)_2(PO_4)_2(OH) \cdot 8H_2O$
Torbernite	$Cu(UO_2)_2(PO_4)_2 \cdot 8\text{–}12H_2O$
Uranocircite	$Ba(UO_2)_2(PO_4)_2 \cdot 12H_2O$
Uranospathite	$HAl(UO_2)_4(PO_4)_4 \cdot 40H_2O$
Uranospinite	$Ca(UO_2)_2(AsO_4)_2 \cdot 10H_2O$
Xiangjiangite	$(Fe,Al)(UO_2)_4(PO_4)_2(SO_4)_2(OH) \cdot 22H_2O$
Zeunerite	$Cu(UO_2)_2(PO_4)_2 \cdot 40H_2O$

*More likely a member of the carnotite group

TABLE 4

The Meta-Autunite Family

Mineral	**Formula**
Abernathyite	$K_2(UO_2)_2(AsO_4)_2 \cdot 8H_2O$
Bassettite	$Fe(UO_2)_2(PO_4)_2 \cdot 8H_2O$
Meta-ankoleite	$K_2(UO_2)_2(PO_4)_2 \cdot 6H_2O$
Meta-autunite	$Ca(UO_2)_2(PO_4)_2 \cdot 6H_2O$
Meta-autunite II	$Ca(UO_2)_2(PO_4)_2 \cdot 4\text{–}6H_2O$
Metaheinrichite	$Ba(UO_2)_2(AsO_4)_2 \cdot 8H_2O$
Metakahlerite	$Fe(UO_2)_2(AsO_4)_2 \cdot 8H_2O$
Metakirchheimerite	$Co(UO_2)_2(AsO_4)_2 \cdot 8H_2O$
Metalodevite	$Zn(UO_2)(AsO_4)_2 \cdot 10H_2O$
Metanovacekite	$Mg(UO_2)_2(AsO_4)_2 \cdot 4\text{–}8H_2O$
Metatorbernite	$Cu(UO_2)_2(PO_4)_2 \cdot 8H_2O$
Meta-uranocircite	$Ba(UO_2)_2(PO_4)_2 \cdot 8H_2O$
Meta-uranocircite II	$Ba(UO_2)_2(PO_4)_2 \cdot 6H_2O$
Meta-uranospinite	$Ca(UO_2)_2(AsO_4)_2 \cdot 8H_2O$
Metazeunerite	$Cu(UO_2)_2(AsO_4)_2 \cdot 8H_2O$
Przhevalskite	$Pb(UO_2)_2(PO_4)_2 \cdot 2H_2O$
Ranunculite	$(H_2O)Al(UO_2)(PO_4)(OH)_3 \cdot 3H_2O$
Sodium meta-autunite	$(Na_2,Ca)(UO_2)_2(PO_4)_2 \cdot 8H_2O$
Sodium uranospinite	$(Na_2,Ca)(UO_2)_2(AsO_4)_2 \cdot 5H_2O$
Troegerite	$(H_3O)_2(UO_2)_2(AsO_4)_2 \cdot 6H_2O$
Troegerite - (P) (=Hydrogen meta-autunite)	$(H_3O)_2(UO_2)_2(PO_4)_2 \cdot 6H_2O$
Uramphite	$(NH_4)_2(UO_2)_2(PO_4)_2 \cdot 4\text{–}6H_2O$

Some of the members of the meta-autunite subgroup have been studied in detail. Fig. 3 (see next page) shows the structure of abernathyite, $K_2(UO_2)_2(AsO_4)_2 \cdot 8H_2O$, which was reported by Ross and Evans (1964). Subsequent studies have shown that although the general topology of the structure is identical for all meta-autunites, ordering effects of the interlayer cations and related bonding to the sheets result in structural distortions which reduce the symmetry below the ideal

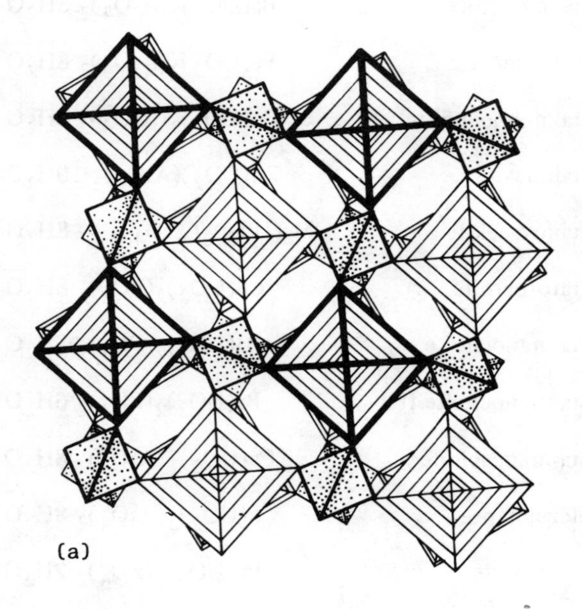

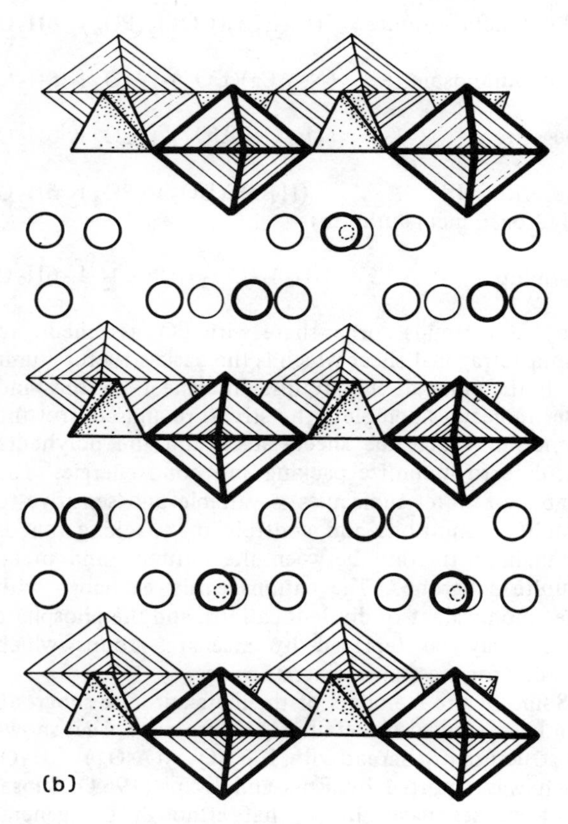

Figure 3. The structure of abernathyite and meta-torbernite showing the deviations from the ideal structure (a) The structure of two superimposed [$(UO_2)(AsO_4)$] sheets showing the alternating articulation of the polyhedra in adjacent sheets. (b) The stacking of the sheets in metatorbernite showing interlayer water and cations. The Cu atom is the small circle. In abernathyite, the Cu site is not occupied and the K is disordered on the H_2O sites, (after Ross and Evans, 1964).

tetrahedral symmetry shown by some of the compounds. Evidently, thermally induced transformations which change the symmetry occur in most of the species.

In the autunites and meta-autunites, the UO_2:TO_4 ratio is 1:1. Other arsenates and phosphates having ratios of 4:2, 3:2, and 2:4 are known. The 3:2 compounds are essentially isostructural with a structural motif based on a chain of edge-shared 2-5 and 2-6 uranyl polyhedra and edge-shared PO_4 or AsO_4 tetrahedra. This chain unit corner-links to adjacent chains to form a sheet unit. These sheets then stack with interlayer cations and water molecules like in the autunites and meta-autunites. Most of the 4:2 compounds appear to be isostructural with the 3:2 compounds and may actually be mistakenly characterized. Phosphuranylite is an example that was first given the formula $Ca(UO_2)_4(PO_4)_2(OH)_4 \cdot 7H_2O$ but shown by Shashkin and Sidorenko (1974) to be $Ca(UO_2)_3(PO_4)_2 \cdot 6H_2O$ by structure analysis.

Structural studies among the uranyl silicates show the individual subgroups are closely related. Three subgroups are known with UO_2:SiO_4 of 2:1, 1:1 and 1:3. The known minerals are listed in Table 5. The first structure

TABLE 5

Uranyl Silicates

(UO_2):(TO_4)	Mineral	Formula
2:1	Soddyite	$(UO_2)_2SiO_4 \cdot 2H_2O$
1:1	Betauranophane	$(H_3O)_2Ca(UO_2)_2(SiO_4)_2 \cdot 3H_2O$
	Boltwoodite	$K_2(UO_2)_2(SiO_3OH)_2 \cdot 5H_2O$
	Cuprosklodowskite	$(H_3O)_2Cu(UO_2)_2(SiO_4)_2 \cdot 4H_2O$
	Kasolite	$Pb_2(UO_2)_2(SiO_4)_2 \cdot 2H_2O$
	Sklodowskite	$(H_3O)_2Mg(UO_2)_2(SiO_4)_2 \cdot 4H_2O$
	Sodium boltwoodite	$(H_3O)_2(Na,K)_2(UO_2)_2(SiO_4)_2 \cdot 2H_2O$
	Uranophane	$(H_3O)_2Ca(UO_2)_2(SiO_4)_2 \cdot 3H_2O$
1:3	Haiweeite	$Ca(UO_2)_2Si_6O_{15} \cdot 5H_2O$
	Weeksite	$K_2(UO_2)_2Si_6O_{15} \cdot 4H_2O$
	Haiweeite - (Mg)	$Mg(UO_2)_2Si_6O_{15} \cdot 9H_2O$

of this group to be solved was for uranophane, $Ca(H_3O)_2(UO_2)_2(SiO_4)_2 \cdot 3H_2O$) (Smith, Gruner, and Lipscomb, 1957). This structure first revealed the 2-5 coordination for uranium and showed that the basic structural unit is a chain of edge-shared pentagonal dipyramids which also share with SiO_4 tetrahedra as shown in Fig. 4.

Figure 4. The uranyl silicate chain and sheet structure as found in the 1:1 uranyl silicates. The uranium 2-5 polyhedra are ruled, and the SiO_4 tetrahedra are stippled.

Since this study, almost all the 1:1 compounds have been refined by single crystal methods, and all the minerals have this chain unit. The chain corner-shares with adjacent chains to form a sheet and cations, oxonium ions and water molecules occupy the intersheet volume. A variety of monovalent and divalent cations can substitute in the structure accounting for the variety of minerals observed.

Although the structure of the 1:3 compounds have not been completely solved, the presence of the same chain unit as in the 1:1 compounds have been verified by Stohl and Smith (1981) and Anderson (1980). The unit is shown in Fig. 5. It does not corner share with adjacent chains; rather the extra SiO_4 tetrahedra evidently act as a bridge between two or more units. The 2:1 compounds have never been studied directly, but their structure is proposed by analogy with $(UO_2)_2GeO_4 \cdot 2H_2O$ (LeGros and Jeannin, 1975). Again the

Figure 5. The uranyl silicate chains in the 1:3 uranyl silicates. The uranium 2-5 polyhedra are ruled, and the SiO_4 tetrahedra are stippled.

basic structure unit is the chain as found in the 1:1 compounds; however, the structure is composed of adjacent chains lying at 90° angles and sharing the same SiO_4 tetrahedra. This structure is shown in Fig. 6. (See page 376). Thus, the uranyl silicate minerals are an interesting example of a chemical group with interrelated structural motifs and crystal chemistry.

Other uranyl compounds with tetrahedral oxyanions include the sulfates and some of the molybdates. The sulfates have yet to receive adequate study but several structure types exist as implied by studies on synthetic compounds. Although, most of the minerals are fibrous implying a chain-like structural unit, the unit must be distinctly different from those found in silicates and phosphates. The chain in $UO_2SO_4 \cdot 3.5H_2O$ (Brandenburg and Loopstra, 1973) contains uranium in 2-5 coordination but shows all corner-shared polyhedra. There are a large number of sulfate minerals with $UO_2:SO_4 = 2:1$ that have closely related structures (Frondel et al., 1976). No structural information is available on any of these species, but they evidently share a basic structural motif with extensive substitution of the characteristic cation. The molybdates are even less well studied and considerably more varied in their structural chemistry. They are complicated by the potentially

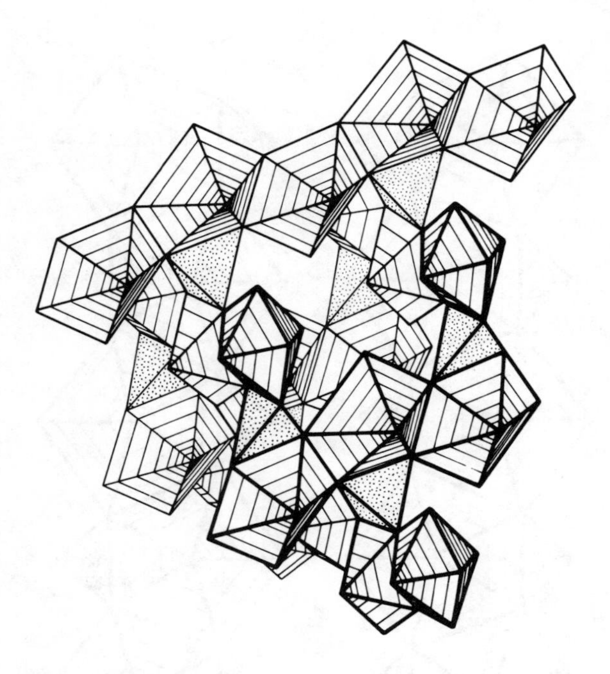

Figure 6. The structure of soddyite, $(UO_2)_2SiO_4 \cdot 2H_2O$. The uranium 2-5 polyhedra are ruled, and the SiO_4 tetrahedra are stippled. Chains, as found in the other uranyl silicates, lie 90° to each other and share SiO_4 tetrahedra.

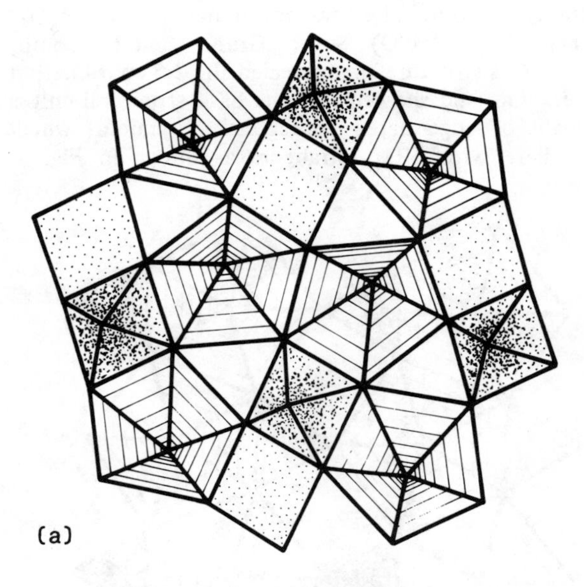

Figure 7. The structure of francevillite, $Ba(UO_2)_2V_2O_8 \cdot 5H_2O$. (a) The structure of the $[(UO_2)_2V_2O_8]$ unit. (b) The stacking of the sheets showing interstitial cations and water molecules. The uranium 2-5 polyhedra are ruled, and the V_2O_8 groups are stippled, (after Shashkin, 1974).

variable valence of both U and Mo. To date no systematic structural characteristics have been revealed.

The vanadate minerals are not VO_4 compounds but are composed of dimeric or higher polyanions. The most commonly encountered anion is V_2O_8 which takes the form of an edge-shared pair of square VO_5 pyramids which edge-share with U_2O_{12} dimers to form an infinite sheet. This sheet is shown in Fig. 7. The structure was first found in anhydrous carnotite by Appleman and Evans (1965), and since found in all other members of the carnotite group of minerals listed in Table 6. (See page 377). Examination of this table shows the similar ity of the a and b values, which correspond to the sheet dimensions, and the different c values and symmetries, which imply the many ways in which this sheet can stack. The stacking reflects the interlayer bonding and limits the types of cation substitutions which may occur in each structure type.

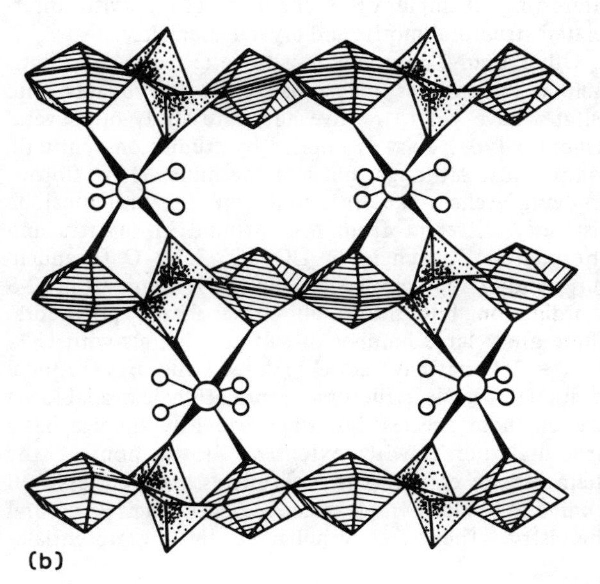

The carbonate minerals show very little similarity among the species which are known. In those structures which have been solved, the uranium is in 2-6 coordination, and one 2-6 polyhedron edge-shares with 1 to 3 carbonate groups. In the $UO_2:CO_3 = 1:1$ compounds, a layer structure results (Christ *et al*, 1955). In the 1:3 compounds, a distinct ionic unit $[(UO_2(CO_3)_3]^{4-}$ occurs which forms complex arrangements linked through cations and water molecules. The selenite and tellurite minerals, like the carbonates, show no structural systematics. The uranium is either in 2-5 or 2-6 coordination and edge- or corner-shares with SeO_3 or TeO_4 groups to form the structural motif. As more new minerals are recognized and described this picture may well change.

The uranyl oxide hydrates and alkali and alkaline-earth uranyl hydrates are based on sheet structures with uranium in 2-4, 2-5, or 2-6 coordination. The uranium polyhedra edge- or corner-share to form sheets, and the

TABLE 6

Uranyl Vanadates

Mineral	**Formula**	**Symmetry**	**Lattice Constants (Å)**
Carnotite	$K_2(UO_2)_2V_2O_8 \cdot 3H_2O$	$P2_1/a$	a=10.471, b=8.41, c=6.59, β=103°50'
Curienite	$Pb(UO_2)_2V_2O_8 \cdot 5H_2O$	*Pcan*	a=10.40, b=8.45, e=16.34
Francevillite	$Ba(UO_2)_2V_2O_8 \cdot 5H_2O$	*Pcan*	a=10;41, b=8.51, c=16.76
Fritzcheite	$Mn(UO_2)_2V_2O_8 \cdot 10H_2O$	*Pnam*	a=10.59, b=8.25, c=15.54
Margaritasite	$Cs(UO_2)_2V_2O_8 \cdot 1.5H_2O$	$P2_1/a$	a=10.51, b=8.45, c=7.32, β=106°5'
Metatyuyamunite	$Ca(UO_2)_2V_2O_8 \cdot 3\text{-}5H_2O$	*Pnam*	a=10.54, b=8.49, c=17.34
Metavanuralite	$Al(UO_2)_2V_2O_8(OH) \cdot 8H_2O$	$P\bar{1}$	a=10.46, b=8.44, c=10.43 α=75°53', β=102°50', γ=90°
Sengierite	$Cu_2(UO_2)_2V_2O_8(OH)_2 \cdot 6H_2O$	$P2_1/a$	a=10.62, b=8.10, c=10.11, β=103°36'
Strelkinite	$Na_2(UO_2)_2V_2O_8 \cdot 6H_2O$	*Pnmm*	a=10.64, b=8.36, c=32.72
Tyuyamunite	$Ca(UO_2)_2V_2O_8 \cdot 8H_2O$	*Pnan*	a=10.36, b=8.36, c=20.40
Vanuralite	$Al(UO_2)_2V_2O_8(OH) \cdot 11H_2O$	$A2/a$	a=10.55, b=8.44, c=24.52, β=103°
Vanuranylite	$(H_3O)_2(UO_2)_2V_2O_8 \cdot 4H_2O$		a=10.49, b=8.37, c=20.30, β=90°?

cations, water molecules and oxonium ions lie in the interlayer region. There are many structural similarities among these two groups as discussed by Evans (1963) and Sobry (1971). The number of detailed structural studies in these groups is unfortunately small, and most of the proposed structures are based on poor data which permit only the uranium atoms to be located with certainty. Fig. 8 shows the sheet structures as found in the various forms of $UO_2(OH)_2$ and in αU_3O_8 (Taylor, 1971; Roof *et al*., 1964; Siegel *et al*., 1972; and Loopstra, 1964). None of these compounds are known as minerals,

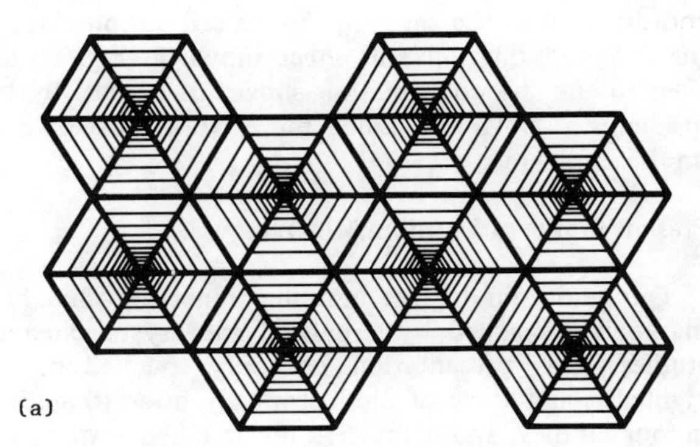

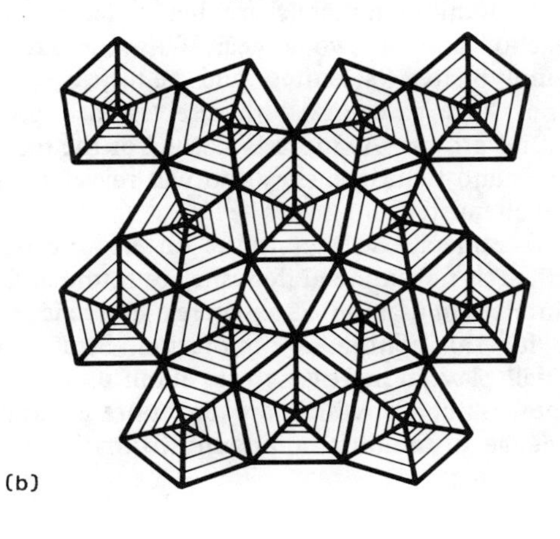

Figure 8. Possible structures of U(O,OH) layers in the hydrated uranyl oxides. (a) Arrangement with 2-6 coordination as found in $\alpha UO_2(OH)_2$. (b) Arrangement with 2-5 coordination as reported for U_3O_8. (c) Arrangement with 2-4 coordination as reported for β and $\gamma UO_2(OH)_2$.

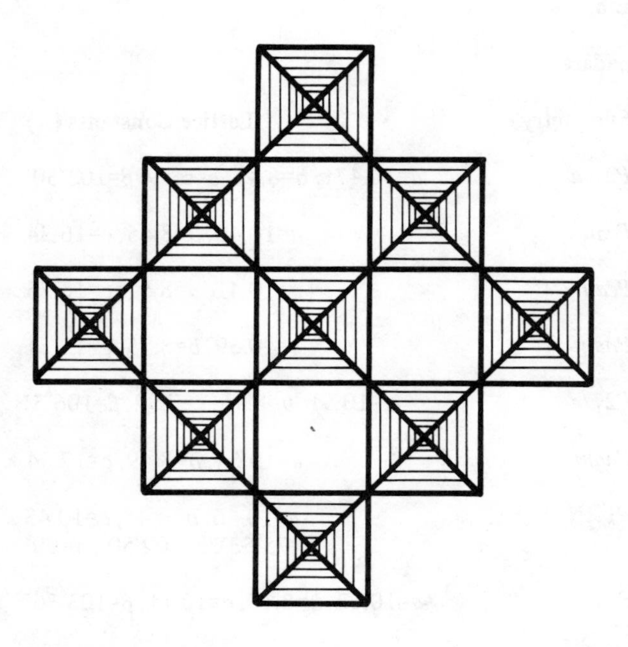

Figure 8 (Cont'd.)

but the sheet structures or derivatives thereof probably form the basic structural units in these two groups. Sobry (1971) argues that the layers are all as shown in Fig. 8a and that all the differences between many mineral series are due to the interlayer cations, oxonium and water molecules. The structure reported by Mereiter (1979) for curite, $Pb_2U_5O_{17}\cdot 4H_2O$ disputes this premise and reveals a sheet of edge-shared 2-5 and 2-4 polyhedra. This structure is more compatible with the ideas of Evans (1963) who proposes that the major differences occur in the bridging oxygen atoms in the sheet and that two (OH) ions shared by the two uraniums can be replaced by one O^{2-} ion shared by two uraniums. The 2-6 coordination of the sheet in Fig. 8a can be converted to the 2-5 coordination and sheet shown in Fig. 8b and even to the 2-4 coordination shown in Fig. 8c by this mechanism. Obviously much more work must be done on these minerals.

Present Status of Uranium Mineralogy

Our understanding of uranium mineralogy has been markedly enhanced by structural and crystal chemical studies. This presentation has only touched on the highlights and some of the historically interesting stages in our studies, and a much more detailed review is in preparation (Smith, 1982). Systematics of uranium in nature have proved very useful in understanding the many known minerals and in predicting new minerals which may be found and described. The mineralogy is important in understanding the formation of economically important ore deposits and in the location of these deposits.

The chemistry of U^{4+} is that of an ion which can form a small number of discrete compounds in nature but can easily substitute in a variety of other minerals. Uranium as U^{4+} is often associated with rare-earths and thorium. As such it is usually found in pegmatites and granites. In these high-temperature environments, the uranium concentrations are low, and solid solution phases form such as uraninite with ThO_2 and CeO_2 solid solution or the pyrochlore and columbite-type minerals.

As U^{6+} uranium is soluble and easily transported. Oxidation of U^{4+} from igneous sources makes it available for transport and it is geochemically separated from the associated rare-earths and actinides. As the U^{6+} is dispersed in the ground water it may deposit as one of the many uranyl minerals or it may encounter a reducing environment and precipitate as a U^{4+} mineral. These U^{4+} minerals are usually the oxide or silicate but may rarely form phosphates or titanates. These "second cycle" "primary" minerals are chemically simpler than are their high temperature counterparts. As more U deposits are found and studied, the number of such minerals is slowly increasing, but the number of U^{4+} minerals is still quite small.

The number of U^{6+} minerals, however, is quite large. These minerals usually occur in small to trace concentrations, and their chemistry depends on the other ions encountered during migration. Structural studies have shown that the anions are very important in determining the type of mineral which forms. Oxyanions like V_2O_8, AsO_4, PO_4 and SiO_4 are very effective in fixing U^{6+}, as the resulting minerals are very insoluble. The variety of minerals also depends on available cations and the state of hydration. Structural studies have shown that these cations are in interlayer sites and often are easily exchanged with other cations. Even the water which may occur as oxonium plays a significant role in this exchange.

New uranium minerals are being described on the average of one or two a year. With the exception of the molybdates, selenites and tellurites, these new minerals have categorized themselves into pre-existing structural groups. As more is learned of the molybdates, selenites and tellurites, they too will reveal a systematic crystal chemistry.

New minerals will continue to be discovered and described as new mineral deposits are found and studied. The interest in uranium as an energy source is the driving force for this continued exploration. Structural chemistry will play an increasing role in our understanding of the new minerals which are encountered and in our knowledge of uranium compound formation in general.

References

Anderson, C. A. F. (1980). The Crystal Structure of Weeksite. MS Thesis, Pennsylvania State University.

Appleman, D. E. and Evans, H. T., Jr. (1965). The crystal structures of synthetic anhydrous carnotite, $K_2(UO_2)_2V_2O_8$ and its cesium analogue, $Cs_2(UO_2)_2V_2O_8$. Am. Mineral., **50**, 825-842.

Beintema, J. (1938). On the crystallography of autunite and the meta-autunites. Rec. Trav. Chim. Pays-Bas., **57**, 155-175.

Brandenburg, N. P. and Loopstra, B. O. (1973). Uranyl sulfate hydrate, $UO_2SO_4 \cdot 3\frac{1}{2}H_2O$. Cryst. Struct. Commun., **2**, 243-246.

Christ, C. L., Clark, J. R. and Evans, H. T., Jr. (1955). Crystal structure of rutherfordine, UO_2CO_3. Science, **121**, 472-473.

Evans, H. T., Jr. (1963). Uranyl ion coordination. Science, **141**, 154-157.

Fleischer, M. (1980). Glossary of Mineral Species, 1980. Mineralogical Record Inc., P.O. Box 35565, Tucson, Arizona 85740, USA.

Fleischer, M. (1981). Additions and corrections to the Glossary of Mineral Species, 1980. Min. Record., **12**, 61-63.

Frondel, C., Ito, J., Honea, R. M. and Weeks, A. M. (1976). Mineralogy of the zippeite group. Can. Min., **14**, 429-436.

Goldschmidt, V. M. and Thomassen, L. (1923). Die Kristallstrüktur natürlicher, und synthetischer Oxyde von Uran, Thorium und Cerium. Norske vidensk. Kristiana Skr. Mat.-naturvidensk. Kl. **(2)**, 1-48.

Grønvold, F. (1955). High temperature X-ray of uranium oxides in the UO_2-U_3O_8 region. J. Inorg. Nucl. Chem., **1**, 357-370.

Legros, J. P. and Jeannin, Y. (1975). Coordination de l'uranium par l'ion germanate. II Structure du germanate d'uranyle dihydrate $(UO_2)_2GeO_4(H_2O)_2$. Acta Cryst., **B31**, 1140-1143.

Loopstra, B. O. (1964). Neutron diffraction investigation of U_3O_8. Acta Cryst., **17**, 651-654.

Masaki, N. (1972). Structure of U_4O_9 below and above the transition temperature. Acta Cryst., **A28**, S54.

Masaki, N. and Doi, K. (1972). Analyses of the superstructure of U_4O_9 by neutron diffraction. Acta Cryst., **B28**, 785-791.

Mereiter, K. (1979). The crystal structure of curite $[Pb_{6.56}(H_2O \cdot OH)_4]\frac{1}{2}[(UO_2)_8O_8(OH)_6]_2$. Tschermaks Min. Petr. Mitt., **26**, 279-292.

Naito, K. (1974). Phase transitions of U_4O_9. J. Nucl. Mater., **51**, 126-135.

Roof, R. B., Jr., Cromer, D. T. and Larson, A. C. (1964). The crystal structure of uranyl dihydroxide, $UO_2(OH)_2$. Acta Cryst., **17**, 701-705.

Ross, M. and Evans, H. T., Jr. (1964). Studies of the torbernite minerals (1): The crystal structure of abernathyite and the structurally related compounds $NH_4(UO_2ASO_4) \cdot 3H_2O$ and $K(H_3O)(UO_2AsO_4)_2 \cdot 6H_2O)$. Am. Mineral., **49**, 1578-1602.

Shashkin. D. P. and Sidorenko, G. A. (1974). Crystal structure of phosphuranylite $Ca[UO_2)_3(PO_4)_2] \cdot 6H_2O$. Dokl. Akad. Nauk SSSR, **220**, 1161-1164.

Siegel, S., Hoekstra, H. R. and Gebert, E. (1972). The structure of γ-uranyl dihydroxide, $UO_2(OH)_2$. Acta Cryst., **B28**, 3469-3473.

Smith, D. K. (1982). Uranium Mineralogy in "Handbook on Uranium" edited by F. Ippolito and B. De Vivo. Amsterdam: Elsevier Scientific Publishing Co. (in preparation).

Smith, D. K., Gruner, J. W. and Lipscomb, W. N. (1957). The crystal structure of uranophane $[Ca(H_3O)_2](UO_2)_2(SiO_4)_2 \cdot 3H_2O$. Am. Mineral, **42**, 594-618.

Sobry, R. (1971). Water and interlayer oxonium in hydrated uranates. Am. Mineral., **56**, 1065-1076.

Stieff, L. R., Stern, T. W. and Sherwood, A. M. (1956). Coffinite, uranous silicate with hydroxyl substitution: a new mineral. Am. Mineral., **41**, 675-688.

Stohl, F. V. and Smith, D. K. (1981). The crystal chemistry of the uranyl silicates. Am. Mineral, **66**, 610-625.

Taylor, J. C. (1971). The structure of the α form of uranyl hydroxide. Acta Cryst., **B27**, 1088-1091.

Voultsidis, V. and Clasen, D. (1978). Problems and boundary ranges of uranium mineralogy. Erzmetall., **31**, 8-13.

Willis, B. T. M. (1978). The defect structure of hyperstoichiometric uranium dioxide. Acta Cryst. **A34**, 88-90.

CHAPTER 11

MORPHOLOGY AND CRYSTAL GROWTH

Cecil J. Schneer
University of New Hampshire

Introduction

Morphological crystallography, considered as the relationship between form and structure, goes back to the Pythagoreans of the 6th Century B.C.

> The first principle of all things is the One. From the One came an indefinite Two . . . . From the One and the Indefinite Two came numbers; and from numbers points; from points, lines; from lines, plane figures; from plane figures, solid figures; from solid figures, sensible bodies. The elements of these four are: fire, water, earth, air,."
>
> Diogenes Laertius (Cornford, 1957)

The elements were identified with the five regular solids of Plato, the tetrahedron, octahedron, cube, icosahedron, and the fifth element (space) with the pentagonal dodecahedron.

Max von Laue placed the birth of modern crystallography in the 17th Century with Johann Kepler's association of mineral forms with the Platonic solids of Euclid's Book 13. In the *"New-Year's Gift of a Hexagonal Snowflake"* (1611), Kepler compared close-packing arrays of spherical particles with Pythagorean patterns for numbers, distinguishing hexagonal and cubic arrays. This association of an internal micro structure with the plane faces of crystals and the derivation of a periodic internal geometry from the external forms, remained central to crystallography until the 20th Century. The Greek atomism revived in the 17th century as the corpuscular philosophy remained hypothetical until von Laue's discovery of the diffraction of X-rays in 1912. Nevertheless because of the inverse relationship between structure and diffraction the existence of a well developed body of structure theory at the turn of the century (Barlow and Pope, 1907) proved essential for the elucidation of the first X-ray patterns. Although morphological studies, aided by the rapid spread of X-ray structure determinations were to reach new heights of achievement in the decade before World War II, the problem had been changed. From the derivation of structure from morphology, it had become the derivation of form from structure - a problem which was dismissed by Max Born (1923) with a theorem proving that the properties of crystals are independent of size and shape. With this dismissal, morphology reverted to mineralogists, mathematicians, technicians of crystal growth and a growing array of neo-Platonists drawn from every field of thought. Yet the most recent work of morphologists relating form, surface energies and growth converges with the conclusions of the physicists studying the kinetics of crystal growth, namely the dependence of growth and habit on the crystallography of the surface. The complexity and richness of detail of Fedorov's empire of crystals continue to engage the attention of amateurs and professionals alike.

Morphology as Evidence of Structure

By 1666 in the *"Micrographia"* Robert Hooke noted the invariance of 90° or 120° angles throughout extremes of form development in halite and quartz, in snowflakes and gem crystals. Hooke took this as confirmation of corpuscularian mechanism and explained it by diagrams of close-packed spheres. The impetus to this work may be attributed to the introduction of the microscope in mid-seventeenth century. The identification of the three Platonic solids with the five elements led 17th century mechanists to the idea of associating physical and chemical properties with geometric form, for example René Descartes represented magnetic force by corkscrew-shaped effluvia, Nicolas Lemery and Pierre Gassendi attributed the action of acids to spear and razor-shaped particles; or more crystallographically obvious, Robert Boyle explained cleavage by a structure of thin-bladed lamellae, and Domenico Guglielmini used morphology in 1705 as the basis of a system of stereochemistry (Burke, 1966).

The inference of atomism was not a necessary conclusion, however. The constancy of the relative orientation of crystal faces was also interpreted by René Descartes and later by Nicolas Steno as a growth phenomenon and a response to the vectorial distribution of a central adhesive force. The mechanistic chemistry of the 17th Century provided essentially modern interpretations of the growth of crystals from aqueous solutions and by extension, for the growth of minerals. It was in fact the relatively advanced understanding of the processes of crystal growth and recrystallization combined with a paucity of geological theory which made for the fossil controversy and held back the development of a science of paleontology for so long. Hooke distinguished between morphologies reflecting the kind of material (mineral) and the shapes of fossils reflecting the organism, a distinction developed by Steno in his *"Prodromus to a dissertation on solids contained within solids"* (1669). The distinction was that the growth of minerals is by the addition of layers to the outer surface, and in the case of organic growth, from within.

Christian Huyghens (1690) replaced spherical corpuscles by oblate ellipsoids to account quantitatively for both the birefringence and the rhombohedral form of calcite. Steno, who had discarded the corpuscles entirely, printed templates for the construction of models of complex hematite crystals from the Isle of Elba, thus carrying the principle of the constancy of interfacial angles beyond the squares and triangles dictated by simple packing models or the angles of the Platonic solids (Schneer, 1971). In order to construct the templates, Steno had to determine the interfacial angles, but how? The contact goniometer was not invented until 1782. This invention by Arnould Carangeot in the process of preparing models for Romé de l'Isle, ushered in an era of quantitative mineralogy, and led rapidly by way of Romé de l'Isle's and Torbern Bergmann's cleavage molecules to René Just Haüy's law of simple rational intercepts (ca 1783).

Mineral morphology had been from earliest times, handmaiden to practical mineralogy, mining and gemology, and a means to identification (Anselm Boece de Boodt in 1609; Domenico Guglielmini in 1705; Maurice Cappeller in 1723; Carolus Linnaeus in 1768; Abraham Gottlob Werner in 1774; cf. Burke, 1966). With the work of Haüy morphology became the empirical basis of atomistic physics. Cuvier wrote of Haüy that he had made of mineralogy a science as exact as astronomy. Haüy's nuclei, lamellae, and molecules were an empirically derived internal *structure.* Morphological investigation and theory rendered ever more exact by such developments as William H. Wollaston's invention of the reflecting gioniometer (1809) applied extensively by William Phillips, now turned on the problem of structure. But these first structural concepts were essentially geometric and had little bearing on the processes of growth of crystals or connections to the materials techniques of the day. Together with parallel developments in the mathematics of symmetry associated with the names of Weiss, Miller, Mohs, Naumann, Seeber, Delafosse, Cauchy and Hessel, (Mauskopf, 1976), this period culminated with the appearance of the concept of the space lattice (Frankenheim and Bravais). Haüy's shaped molecules had been replaced by shaped spaces defined by periodic distributions of geometric points (lattices).

Auguste Bravais's concern went beyond abstract mathematics. He spoke of his points as the centers of gravity of molecules of repetition, and prepared the way for the theory of space groups by the association of morphological polarity and reduced symmetry with molecular polarity and symmetry. The physical properties of crystals and especially the cleavage, were his means of deriving what he called the *structure* and we call the space lattice i.e., hexahedral, dodecahedral, octahedral (*P, I, F* in modern notation). Reasoning that the forces holding the molecules together would be weaker with increasing interplanar spacing, Bravais proposed as a rule of cleavage (and by extension, of morphology) that the directions of cleavage should parallel the planes of greatest net reticular density which would be the lattice planes of greatest separation. Therefore octahedral cleavage would mean that the mineral species should be referred to the *octahedral mode;* dodecahedral cleavage, the dodecahedral mode, hexahedral cleavage, the hexahedral mode, etc. While recognizing that the morphology of individual specimens reflected the circumstances of their growth, Bravais nevertheless considered that dominant morphologies would lead to a similar determination of the structure (lattice). An example was the predominantly dodecahedral morphology of garnet, a mineral with no cleavage. The garnet structure or elementary parallelopipehedron [unit cell] was therefore cubic dodecahedral (I) after the morphology.

Greene and Burke (1979) place the date of American achievement of standards in mineralogy parallel to the European, as 1822 with the publication of the second edition of Parker Cleaveland's *"Mineralogy."* Some impression of the high level of morphological investigations carried out in the United States in the first half of the 19th Century may be gleaned from Lewis Beck's *"Mineralogy of New-York"* of 1842. Cleaveland's theory was Haüy's, essentially unmodified. Beck was less concerned with theory, more with morphological and chemical methodology. He considered himself as following Haüy but using the methodology of Henry Brooke's *"Crystallography"* of 1823 (England) and William Phillips' *"Mineralogy"* published in N.Y. by Samuel Mitchell in 1818. James Dwight Dana's *"System of Mineralogy"* of 1837 derived its theoretical base from Haüy's concept of both the geometrical and the chemical characterization of the mineral species. With his several manuals, in successive editions it came to dominate American mineralogy. As a practical system of classification of minerals and instruction of mineralogists, it set directions and standards which were universally adopted.

The Equilibrium Theory of Crystal Morphology

Modern theory associating crystal morphology and growth began with J. Willard Gibbs' *"On the Equilibrium of Heterogeneous Substances"* published in the Transactions of the Connecticut Academy of Sciences 1875-1878. Gibbs had spent his life at Yale (his father before him had also been a professor) where science had been dominated by geological and mineralogical concerns since Benjamin Silliman began the Sheffield School. In fact Silliman had acquired the country's premier mineral collection for Yale from Colonel George Gibbs of Newport, and the senior scientist at Yale during most of Gibbs' career would have been the geologist-mineralogist Dana. Gibbs' *"Equilibrium"* was a definitive treatment of the principle of entropy. Even now, a century later, there are few articles on theoretical morphology or crystal growth which do not refer to Gibbs' *"Equilibrium."*

The reasons for Gibbs' choice of the Connecticut Transactions for publication are obscure since Silliman's *American Journal of Science* was uniquely associated with Yale. Yet although his work was supposedly neglected for its difficulty and his students few, Wilhelm Ostwald translated the *"Equilibrium"* into German, Henry Louis Le Châtelier into French, and James Clerk Maxwell wrote to him in admiration.

In the section of the *"Equilibrium"* on the theory of capillarity, there is a chapter on "Surfaces of discontinuity between solids and fluids" introducing the ideas of specific surface energies and vectorial growth, and establishing the principal that the equilibrium form will be that for which the total surface energy is a minimum, cf. below. Seven years later (1885) Pierre Curie independently arrived at the corollary to Gibbs' rule, derived from Gauss' principle of virtual velocities. The areas of faces would be inversely proportional to their specific surface energies. G. W. Wulff (1901) in Petersburg proposed an extension of Curie's idea citing Leonhard Sohncke as demonstrating that the Bravais and the Curie theories lead to a mutually concurrent result. Let the specific surface energy of the ith face be designated as σ_i., and the area of the face by F_i. Then for the equilibrium crystal there is a central point, the Wulff point from which the perpendiculars on each face h_i are directly proportional to σ_i. Wulff's paper included experimental measurements of capillary constants which he considered to confirm his theorem, but the proofs of this theorem were not obvious and remained an interesting if minor problem for theoretical physics culminating in a lengthy discussion by Max von Laue during the war years. As an aside, it is worth recording that von Laue, an outspoken opponent of the Nazis, was dismissed from his positions in German physics during the war, and returned to pure crystallography, considered irrelevant by official science.

Although Wulff considered that his theorem was essentially the application of the physics of Curie's principle to the phenomenological theory of Bravais, the Wulff-Curie work was not accepted by Georges Friedel, whose whole approach to science was in Baconian contrast to the Cartesianism of theoretical physics. He conceived of his task as constructing principles of mineralogy drawn directly and firmly from nature, and restricting a priori constraints solely to the definition of the crystalline state. Nature alone is sufficient to establish Haüy's law and the lattices of Bravais, he wrote, but these lattices are a "geometric intermediary suitable to express the law and nothing more" (Friedel, 1907). Friedel used morphological data derived from statistical studies to determine the lattices and the proportionate cell constants for a number of mineral species. From the densities and the chemical proportions, he was able to fix the absolute values of the cell constants (off in these examples by a factor of two since he erred in his assumptions of the numbers of formula units per cell).

Of the many exhaustive studies of morphology, the work of Victor Goldschmidt of Heidelberg deserves especial notice. Goldschmidt took his Pythagoreanism literally, publishing *"Ueber Harmonie und Complication"* privately in 1901. In this work inspired by Kepler's "Harmonies of the World" of 1619, Goldschmidt showed that from the assumption of forces of attraction between the periodically distributed particles composing the crystal, he could derive what he called the 'form system' of the crystal. This derivation based on a harmonic series, he termed the 'law of complication.' Extending this to a consideration of musical harmony, he found a direct parallel. As Kepler had before him, he extended this to all of the natural world and the human psyche. He considered the physiological and psychological foundations for harmony in music, in light and color, the brain, the solar system, art, aesthetics in general, astronomy and biology. It was all the more remarkable then, that the private research institute he founded at Heidelberg with his wife, the Portheim Stiftung, became the world center of morphological studies. Among others, it attracted A. E. Fersman from Russia. With the Americans, Charles Palache and M. A. Peacock, Goldschmidt demonstrated that the complex system of 92 forms of calaverite, $AuTe_2$ could be brought into excellent agreement with the development sequence of his law of complication. The calaverite morphology had appeared to contradict the law of rational intercepts and had defied the efforts of morphologists for thirty years. W. T. Schaller also applied the law in the morphological analysis of mercury minerals from Texas, but it was dismissed by Friedel as an involved method of restating the laws of Haüy and Bravais.

The efforts of Goldschmidt and his students and foreign visitors led to the elaboration of the Goldschmidt method of morphological analysis. It was a system of measurements of angles and a representation of morphology on the gnomonic projection which brings out the internal periodicity. The method was applied to the compilation of a massive *"Atlas der Krystallformen"*-18 volumes in which nothing less than the republication and reconciliation of every morphological mineral study ever made was projected and essentially carried out. No description of a mineral since the time of Carangeot and Romé de l'Isle, had been considered complete without such a morphological study which was crowned by exact isometric projections. The purpose of the *Atlas* was to provide the data base for the determination of crystal structures.

The other European school of morphological study was at Zurich under Paul Niggli. Like Fedorov and the crystallographers of the 19th century, Niggli considered structure as the core of crystallography, and morphology as fundamental reality with which methods of structure analysis such as X-ray diffraction must concur. Niggli was a powerful personality whose contributions to petrology and mineralogy are available in English translation and still influence American geologists. He had in fact worked with Norman Bowen at the Geophysical Laboratory in Washington. (One of Bowen's early papers dealt with the petrologic implications of morphology). As geologist and petrologist, Niggli used a statistical approach for the characterization of the morphology of mineral species in nature; comparing the relative frequency of occurence of the forms of a species in localities (Fundorts-persistenz), in persistence in habits and combinations of forms (P-Werte), and in size (G-Werte). Yet oddly enough for Niggli who developed sophisticated analytic and graphic techniques for the reconciliation of the far more intractable data of petrology, neither he nor his school (R. L. Parker, F. Braun, T. Gebhardt), in their analysis of these persistence values, went beyond Friedel's method of listing forms in the order of their relative persistences to establish the "morphological aspect" of a mineral species.

This use of mineralogy, in particular the collection of the data defining the statistical distribution of the forms of a mineral species in geological time and space, marks Niggli with Friedel and Goldschmidt as naturalists in their approach and distinguishes them from those who like Wulff (and Fedorov) looked primarily to the laboratory growth of crystals for their data.

The Donnay-Harker Extension of the Bravais Law

Although in the original formulation of his lattice theory, Bravais had understood that the degeneracy of the point-group symmetries, that is to say, the phenomenon of merohedry, implied a corresponding degeneracy to the symmetry of the molecule, rigorous modification of the Bravais-Friedel Law was not accomplished for nearly a century. Niggli, who was responsible for transforming the theory of space groups "from a skeletal structure to a helpful friend" (P. P. Ewald, quoted by C. G. Amstutz, 1978) and independently J. D. H. Donnay and David Harker (1937), saw that the space-group symmetry operations were modifications of the lattice equivalent to the introduction of 'basis' points to the primitive lattice. The morphological effects should therefore be similar. This concept was developed independently and with rigor by J. D. H. Donnay and David Harker (1937) in a paper which ranks with the original work of Bravais for its lasting and pervasive influence. According to the Donnay-Harker Law, the effective interplanar spacings and therefore the morphological order of the corresponding forms are decreased according to the systematic space-group extinctions of corresponding X-ray reflections. The weakness of the law as noted by Donnay and Harker in the original paper was that it was purely crystallographic and involved no physical or chemical considerations. Yet as Friedel had claimed for the rule of Bravais, it was a geometrical convenience transformed into a law of nature by observation. Friedel had demonstrated the preponderance of the influence of the lattice over that of the motif. Donnay and Harker showed that the symmetry of the space group in the large majority of cases, so outweighed all other influences, as to predominate in the determination of the morphology of a species in nature. Nevertheless, the reason for the attractiveness of the new formulation was precisely its independence of all physical and chemical considerations. For Friedel in 1907, "all scientists consulted on this matter [the reality of the lattice in the crystal and real existence of the chemical atom] would answer, without hesitation, that such an assumption is actually useless..."

The power of the Donnay-Harker approach was precisely in its total freedom from *ad hoc* assumptions and in its demonstration of the rule of considerations of symmetry alone.

Morphology and Growth After World War II

Referring to the period before the transformation of physics during World War II, I. I. Rabi called it the era of string and sealing wax. The entire equipment budget for the Columbia University Physics Department had been $1000 per year. The dramatic applications of sciences to the technology of warfare, in particular the development of nuclear weapons, radar and rocketry led directly to a transformation in the nature of the scientific enterprise. The post-war years saw the emergence of solid state physics as a major discipline with an equivalent reorientation within physical and inorganic chemistry and related shifts in metallurgy, ceramics, and electronics. Crystallography and crystal growth were no longer largely relegated to mineralogists in departments of geology. *The Journal of Crystal Growth* began in 1967. Within the earth sciences the study of crystal growth transformed geophysics and petrology. With the invention of the transistor, the synthesis of crystals and the growth of single crystals became a major industry. The crystallographic approach to morphology and growth with its traditional mineralogical base came under critical scrutiny.

The Donnay-Harker papers were dissected by A. F. Wells in a series of papers "On Crystal Habit and Internal Structure" (1946) and by M. J. Buerger in the following year (1947). From the point of view of inorganic chemistry, and with laboratory experimental evidence, Wells demonstrated that a definite equilibrium shape could be defined for specified conditions, this equilibrium shape being a function of both the internal structure and the interactions of the various crystal surfaces with the solute. This topic, the modification of habit by selective adsorption had been studied by C. Frondel (1935, 1940) and others cf. Buckley (1951) in general concluded that theories of morphology not taking account of the chemical environment of the growing crystal were of limited validity. Wells criticized the Donnay-Harker Law for its limitation to space-group equipoints, and in the translation to actual structures, its restriction to coplanar atoms. Wells suggested that not only would the relative stability of crystal faces be a function of space group extinctions but also should be related to the intensities of corresponding X-ray reflections. Buerger (1947) also noted that the Donnay-Harker ordering is ". . . exactly that of the order of appearance of the corresponding lines on the [X-ray] powder photograph. . ." The difficulties of X-ray studies and calculations in the pre-computer, pre-diffractometer era, delayed any attempt to study these relationships for more than a dozen years.

The limitation of the Donnay-Harker Law to symmetry and its independence of considerations of the physical environment of growth was nowhere more apparent than in the studies of the habits of ice (snowflakes). The early experiments at Harvard of Percy Bridgman that revealed the polymorphism of ice at high pressures, were models for the foundation of solid state physics. No account of morphology would be complete without mention of the 202 plates reproducing 2453 stunning photographs of snow crystals by W. A. Bentley of Jericho, Vermont (Bentley and Humphreys, 1931). Meteorologists had long connected the 22- and 46 degree solar and lunar haloes to the simple needle and/or plate habits of snowflakes and the cirrus and/or cirrostratus clouds characteristic of high altitude conditions. During and after the war, the habits of snow were related to cloud types and temperatures by direct observations in flights principally in Germany and in Canada by L. W. Gold and B. A. Power in 1952 and in New England by J. Kuettner and R. J. Boucher in 1958; (B. J. Mason, 1963). Synthesis of snowflakes was of major concern in Japan and the relationships of habit to temperature and saturation defined by U. Nakaya (1954). The development of cloud seeding by I. Langmuir and V. J. Schaeffer of the General Electric Laboratories in 1946 led to intensive laboratory, field, and theoretical studies of the growth of ice crystals, in particular variations of habit with physical conditions and nucleation. (B. J. Mason, 1963). In 1957 D. McLachlan, Jr. attributed the dihexagonal growth coordination of the arms of the snowflake dendrite to standing thermal and acoustical waves.

The absence of kinetic and energy considerations from the Bravais-Donnay-Harker Laws was noted by Buerger. His qualitative discussion emphasized the effects of the geometry and bonding of the particles arriving at the surface of the growing crystal. Buerger and Wells agreed that only in the simplest structures would the symmetry alone play a predominant role in the determination of the habit.

There already was an extensive body of physical treatment of the kinetics of crystal growth in 1946, but the applicability of this theory to mineralogy and its connections with crystallography were tenuous. W. Kossel (1927) had proposed a simple cubic model for a growing crystal and first demonstrated the theoretical dependence of growth on molecular steps and kinks, that is, on disorder, imperfections and dislocations.

The kinetics of crystal growth was of considerable importance to metallurgists, ceramicists, materials scientists and technologists even in Gibbs' time and the topic was pursued with considerable intensity in central and eastern Europe in the period between the wars in papers by J. Frenkel, I. N. Stranski, R. Kaischew and many others. It is essentially beyond the scope of this article and the reader should refer to one of the numerous review volumes on the subject of crystal growth. Buckley (1951) is particularly recommended although out of date. More recent volumes include Hartman, 1973; Kaldis and Scheel, 1977; Pamplin, 1979; see also the article by McLachlan (1978).

In 1951 in work stemming from Bristol University, W. K. Burton (of I.C.I.) N. Cabrera (who later moved to the Physics Department of the University of Virginia), and F. C. Frank used the purely classical physics of the Ising model to show that when dislocations with a component parallel to the line of dislocation assumed a screw form (Burgers, 1939), the need for two-dimensional nucleation was eliminated and perpetual growth of the crystal face "up a spiral staircase" (Frank, 1949), was provided. Evidence for spiral growth was visible on so many crystals that the rigorous mathematical theory seemed to most scientists not only valid but uniquely explanatory of all crystal growth. (Cabrera, 1952; Verma, 1953. These kinetic papers distinguished between flat faces defined as close-packed, (with a physicist's disregard for crystallographic usage)* stepped, and kinked; demonstrating the requirements for growth on each (kinked or K-faces, for example were self-nucleating).

Dan McLachlan, Jr. at the University of Utah in 1952 extended the Donnay-Harker Law to take into account the spatial configuration of the actual motif. The Donnay-Harker Law had been criticized by Wells (1946) as implying that only atoms which were exactly coplanar contributed to the reticular density. McLachlan's equations calculated 'laminarity' as the measure of morphological significance of a plane in place of net reticular density. Laminarity was the product of the interplanar spacing times the 'congestion,' a function introduced by McLachlan which allowed for the chemical difference of each atom as well as the positions of the atoms in the detailed structure. The paper, published in a bulletin of the Engineering Experiment Station of the University of Utah, attracted little notice. Yet it was the first paper to derive morphological equations for complete structures (to take into consideration the motifs of the equipoints).

---

*"When I use a word it means just what I choose it to mean, neither more nor less." Humpty Dumpty.

P. Hartman and W. G. Perdok at Groningen undertook to calculate the energies of the strongest vectors in the crystal which they called periodic bond chain vectors (PBC). In an influential set of three papers in 1955, Hartman and Perdok defined F- or flat faces as those parallel to two or more PBC vectors; S- or stepped faces as parallel to at least one; and K- or kinked faces as all others. The occurrence of faces was determined by their F-, S-, or K- character in nominal agreement with the kinetics of Burton and Cabrera (1949). Niggli had pointed out the significance of a *Hauptbindungsrichtung* for prominent zone axes. Hartman and Perdok attempted to calculate the bonds for real crystals (in place of the simplified Kossel crystals of the Bristol physicists). In later papers they calculated *attachment energies,* of slices d_{hkl} in thickness, inversely proportional to the lengths of the bonds, so that the theory becomes a reformulation of the Bravais-Donnay-Harker relationship (Hartman, 1978). These calculations of the energies have become increasingly detailed in recent years. Eric Dowty (1976) at Princeton has undertaken detailed calculations of the energies associated with particular faces of crystals by means essentially parallel to the more recent methods employed by Hartman and his students, that is calculating the strengths of selected sets of bonds within layers or slices, (Hartman, 1977; Dowty, 1977). His calculations take into consideration bond types as well as detailed structural information for each crystal species. Y. Aoki (1979) at Kyushu calculates linkages of coordination polyhedra called LCP vectors.

Meanwhile, a physical treatment of equilibrium form was published by Conyers Herring of the Bell Telephone Laboratories in 1951. Pointing out that the flat plane-sided polyhedron was but one of five possible cases for the shape of the crystal, Herring analyzed the Wulff diagram or polar plot of the surface free energy vectors (γ- or Wulff vectors) as a function of direction from the center of the crystal (Wulff point).

The consequence of a periodic internal distribution of centers of force would be that the surface of the γ-plot would be made of intersecting spheres of various sizes with major cusps at orientations of simple indices. The minimum polyhedron bounded by the planes perpendicular to the Wulff vectors at the cusps, would be the equilibrium polyhedron of Wulff. Herring derived relationships between the relative free energies of different surface configurations of the γ-plot, demonstrating in particular the conditions for smoothly rounded as well as sharply cornered faces. This analysis was directly applicable to the morphology of the sharp metal points used in field emission spectroscopy. The points are single crystals only microns in size which are rounded at temperatures half that of the melting point, with flats at major crystallographic directions. Herring's demonstrations of the effects of bonding and atomic assumptions on the shapes of the γ-plot and therefore on the equilibrium form were the point of departure for Wolff and Gualtieri (1962) (then with the U.S. Army Signal Corps research facility at Fort Monmouth, New Jersey) and N. Cabrera. Wolff and Gualtieri calculated the modifications of the cusps of the Wulff diagrams for simple structure types and bonding assumptions, demonstrating striking agreement with observations of micro-cleavage. For the crystal growth physicists, figuratively in the shadow of Massachusetts' Route 128, or California's Silicon Valley, the problem had become one of finding a phenomenological theory applicable to kinetic processes (Cabrera and Coleman, 1963). Morphology was largely ignored and structure was defined as the physical configuration of the surface of growth.

Surface Energies From Morphology

In 1902, J. W. Gibbs published a systematic exposition of the mechanical principles which make the rational foundation of thermodynamics. In particular, he defined as *canonical,* "ensembles of systems in which the index or logarithm of probability of phase is a linear function of the energy. [Boltzmann's Law]. At the 1966 meetings of the International Mineralogical Association at Cambridge, England, structure density maps derived from morphology were displayed. C. J. Schneer of the University of New Hampshire proposed that "the equilibrium crystal form is a function of the electron density and the electron density is the Fourier transform of the morphological probability distribution." Assuming Niggli's *Kombinations-persistenz* values approached a Gibbsian canonical ensemble (statistical equilibrium), Schneer calculated empirical specific surface energies for Fourier coefficients. Comparing these with the structure factors (F_i) used as coefficients for Fourier synthesis of electron density calculated by standard methods, residuals (anomaly factors) as low as 10% were obtained. At these meetings which were attended by J. D. H. and G. Donnay, P. Hartman, R. Kern, and I. Sunagawa among others, Donnay pointed out that the morphology of polar crystals used for such a synthesis would allow for the determination of the absolute polarity of acentric crystals. The significance of the correspondence of McLachlan's laminarity equations to the structure factor equations of diffraction was then explored by McLachlan (now at Ohio State University by way of Stanford Research Institute) in an addendum to the reprinting of his 1952 paper (1974).

Although the Fourier syntheses using morphologically derived energies as coefficients, required *a priori* structural assumptions for the determination of phases, Patterson-Harker syntheses (requiring no assumptions other than symmetry derived from morphology) were shown at the meetings of the International Union of Crystallography at Stony Brook and again in 1970 at the Kyoto session of the meetings of the International Mineralogical Association. The conclusion that ". . .the electron distribution is probably a basic criterion describing the crystal face-formation as a function of its structure and composition," was drawn by Leonyuk *et al.* (1980).

Assuming then that the statistical equilibrium form is a function of the structure, the problem of the definition of the Wulff equilibrium form still remained. Consider only the confusion between equilibrium and growth forms. The Eastern Europeans (I. Vesselinov, 1979) have formulated the problem in geological terms as a distinction between the normative and the modal form. I. Kostov of Bulgaria has undertaken a massive compilation of the normative and modal forms of the mineral species. In his definitive study *"On Growth and Form"* D'Arcy Thompson has written ". . . in the representation of form. . . we see a diagram of forces in equilibrium. . ." and ". . . in the comparison of kindred forms, we . . . discern the magnitude and the direction of the forces which have sufficed to convert the one form into the other." Morphology requires a 3rd Law, a 0- or base form against which the forms of real crystals may be measured. What are the ideal polyhedra corresponding to the periodicity of the solid state?

According to Leonyuk *et al.* (1980) Wulff had observed that the Bravais Law required that a crystal take "the form of the Dirichlet region of a reciprocal lattice point, i.e., a polyhedron in which each inner point is closer to the given lattice point than to any other point of the lattice." B. N. Delone, a Soviet mathematician, referred to the Dirichlet regions as "Wulff ideal habits," an approach elaborated in detail by W. Nowacki but one which attracted little attention in the West.

Faced with the profligacy of habits and forms of natural crystals, Friedel had defined the problem as that of obtaining structural information from morphology. It would not be possible to ". . . give me the parameters of a crystal and I will tell you what are its forms and their relative development. I will draw the crystal for you." (Friedel, 1907). But the reciprocal relationship of X-ray diffraction patterns and morphology was known from the first experiment of Friedrich and Knipping in 1912. With the development of the band theory of solids resting on a theorem of F. Bloch's extending the Schrodinger wave mechanics to solids, Leon Brillouin (1946) used the morphology of the Dirichlet regions or domains in the treatment of wave propagation in solids. The Brillouin zones were three-dimensional regions in reciprocal space bounded by planes of discontinuity. That the first Brillouin zones were the Wulff polyhedra derived from the Bravais lattices was pointed out by Leonyuk *et al.* (1980).

Wulff had found 17 such ideal polyhedra, Delone 24. By contrast Donnay and Harker tabulated 97 morphological aspects. Schneer (1978) classified the polyhedral complements of periodic structures according to their complexity as first; Brillouin zones for those complementary to Bravais lattices, next; Donnay-Harker zones for those complementary to the arrays of space-group equipoints; and finally Wulff zones for those obtained from the Wulff or γ-plots of Herring, (1951), Bennema and Gilmer (1973) and Wolff and Gualtieri (1962).

Convergence of Kinetic and Phenomenological Analyses

The success of Burton, Cabrera and Frank's spiral growth theory confirmed a majority of the physicists and engineers of crystal growth in their rejection of mineralogical studies relating morphology and growth. It had at first seemed as if continuous growth could only occur through a spiral or related mechanism, originating in a dislocation. Yet with more detailed models of interface kinetics, it became clear that faceting (crystallographic morphology), impurities, temperature, structure, surface configuration and growth rates were all related in a complex way. Growth rates could not be determined from a series of independent functions.

G. H. Gilmer who moved from Washington and Lee to the Bell Laboratories, with K. A. Jackson, (1977) and J. D. Weeks (1976) used computer simulations to try to take into consideration the multitude of factors influencing and effecting the growth of crystals. Other mechanisms than spiral growth were shown to account for significant growth rates. The same kinetic Ising model used by Burton, Cabrera, and Frank could allow for 2-dimensional nucleation even on dislocation-free surfaces. The impingement and evaporation of atoms at each site of the crystal was determined by a Monte Carlo process of randomization. Chance could provide a finite probability of nucleation, even at low temperatures. At temperatures above a critical temperature determined by the crystal structure, nucleation is no longer required and growth is continuous. This is referred to as the roughening transition. Gilmer represented the various possibilities of growth graphically by computer-generated color movies, dramatically illustrating the effects of the variation of the factors of growth. These films must be seen by anyone seriously considering the subject of crystal growth.* Although cubic closest packing is the most complex structure analyzed, Gilmer's work illustrates the significance of the crystallographic surface structure by comparing growth rates for ideal (100), (111), and (110) surfaces, and the effects of temperature, screw dislocations, and impurities. Since the basis for the simulations is statistical, the equations are of the same form as the phenomenological equations deriving surface energies from morphological persistences.

Mineralogists have also used computer graphics. J-M Sempels and J. Raymond (1976) of the University of Ottawa, and E. Dowty at Princeton (1980) cf. Keester and Giddings (1971), and Prewitt (1973) have programmed machine drawings of crystals but mineralogists have

*Film Library, Bell Laboratories, Murray Hill, N. J. 07974: Gilmer, 1980.

yet to take advantage of their unique familiarity with the details of structure and morphology. There has been little interest in any computer simulations of morphology that would be comparable to the kinetics simulations by Gilmer.

The National Science Foundation does not even have a division for mineralogy, and few students now understand the difference between a structure and a lattice, but the consideration of geometric order continues to fascinate biologists, artists, crystallographers, architects and others. The metallurgist and historian of science Cyril S. Smith (theory of shapes of grains in polycrystalline aggregates) in his introduction to Arthur L. Loeb's *"Space Structures, Their Harmony and Counterpoint"* (1976) refers to the informal Boston group of *Philomorphs* ". . . brought together by a common interest in the underlying patterns of interaction between things." Form follows structure as form follows function. Alan Holden in *"Shapes, Space and Symmetry"* (1971) wrote that "the best way to learn about these objects [polyhedra] is to make them." Perhaps the best way to learn about crystals is to grow them and the best way to learn to grow crystals could be to study their forms which are the record of their origin and their interactions with their environment throughout their history. Of his solids, Plato concluded in the Timaeus, "So their combinations with themselves and with each other give rise to endless complexities, which anyone who is to give a likely account of reality must survey."

I apologize to the very large number of scientists both here and abroad, living and dead, whose significant contributions to the subject were not mentioned in this article. Those known to me were omitted because of the pressures of time and space; those unknown to me, because the topic is both larger and grander than my experience and abilities can encompass.

References

Amstutz, C. G. (1978). Paul Niggli in "Dictionary of Scientific Biography, X." pp. 124-127.

Aoki, Y. (1979). Morphology of crystals grown from highly supersaturated solution. Mem. Fac. Sci., Kyushu Univ., Serv. D. Geol. **24**, 75-108.

Barlow, W. and Pope, W. J. (1907). The relation between the crystalline form and the chemical constitution of simple inorganic substances. Trans. Chem. Soc. **91**, 1150.

Bennema, P. and Gilmer, G. H. (1973). Kinetics of crystal growth. In P. Hartman, (editor) 1973.

Bentley, W. A. and Humphreys, W. J. (1931). Snow Crystals. New York: McGraw-Hill. Reprinted (1962) New York: Dover.

Born, M. (1923). Atomtheorie des festen Zustandes. Leipzig: B. G. Tuebner.

Braun, F. (1932). Morphologische, genetische und paragenetische Tracht studien an Baryt. Neues Jahrb. Mineral., Bell Band, 65A, pp. 173-222.

Bravais, A. (1849). The Crystal Considered as an Assemblage of Polyatomic Molecules Translated from Etudes cristallographiques, Part II, Paris, pp. 194-195, 197-202, in Schneer, 1977.

Bravais, A. (1849). The Crystal Considered as a Simple Assemblage of Points. Translated from Etudes cristallographiques, Part I, Paris, pp. 165-170, 193, in Schneer, 1977.

Brillouin, L. (1946). Wave Propagation in Periodic Structures. New York: Dover Publications, 1953.

Buckley, H. E. (1951). Crystal Growth. New York: John Wiley.

Buerger, M. J. (1947). The relative importance of the several faces of a crystal. Am. Mineral, **32**, 593-606.

Burgers, J. M. (1939). Some considerations on the fields of stress connected with dislocations in a regular crystal lattice. Proc. Kon. Ned. Akad. Wet. **42**, 293-325.

Burke, J. G. (1966). Origins of the Science of Crystals. Berkeley: University California Press.

Burton, W. K. and Cabrera, N. (1949). Crystal Growth and Surface Structure. Discuss. Faraday Soc. **5**, 33-39.

Burton, W. K., Cabrera, N. and Frank, F. C. (1951). The Growth of Crystals and the Equilibrium Structure of their Surfaces. Phil. Trans. Roy. Soc. **A243**, 299.

Cabrera, N. (1952). Discussion. In Gomer and Smith, 1953.

Cabrera, N. and Burton, W. K. (1949). Crystal Growth and Surface Structure. Part II. Discuss. Faraday Soc., **5**, 40-48.

Cabrera, N. and Coleman, R. V. (1963). Theory of Crystal Growth from the Vapor. In Gilman, 1963.

Cornford, F. M. (1957). Plato and Parmenides. New York: Liberal Arts Press.

Curie, P. (1970). On the Formation of Crystals and on the Capillary Constants of Their Different Faces. J. Chem. Educ. **47**, 636-637, translated from Bull. Soc. Franc. Mineral., **8**, 145-150 (1885).

Donnay, J. D. H. and Harker, D. (1937). A new law of crystal morphology extending the law of Bravais. Am. Mineral. **22**, 446-467.

Donnay, J. D. H. and Harker, D. (1962). Center of charges inferred from barite morphology. Sov. Phys. Crystallogr. **6**, 679-684.

Dowty, E. (1976a). Crystal structure and crystal growth: I. The influence of internal structure on morphology. Am. Mineral. **61**, 448-459.

Dowty, E. (1976b). Crystal structure and crystal growth: II. Sector zoning in minerals. Am. Mineral. **61**, 460-469.

Dowty, E. (1977). Crystal structure and crystal growth: I. The influence of internal structure on morphology: a reply. Am. Mineral., **62**, 1036-1037.

Dowty, E. (1980). Computing and drawing crystal shapes. Amer. Min. **65**, 465-471.

Fedorov, E. S. von (1920). The Crystal Empire (Das Krystallreich). Mem. Acad. Sci. Russ., VIIIe Ser., **36**, rev. in Schneer, (editor) 1977.

Frank, F. C. (1949). The influence of Dislocations on Crystal Growth. Dis cuss. Faraday Soc. **5**, 48-54.

Friedel, G. (1907). Studies on Bravais' Law. Bull. Soc. Franc. Mineral. pp. 347-349, 355-357, 454-455.

Frondel, C. (1935). Oriented intergrowth and overgrowth in relation to the modification of crystal habit by adsorption. Amer. J. Sci. **30**, 51-56. (1940). Amer. Min. **25**, 69-87.

Gibbs, J. W. (1906). On the Equilibrium of Heterogeneous Substances. The Scientific Papers of J. Willard Gibbs, Vol. I, Thermodynamics. New York: Longmans, Green & Co.

Gibbs, J. W. (1902). Elementary Principles in Statistical Mechanics, New Haven: Yale University Press.

Gilman, J. J. (1963). (editor). The Art and Science of Growing Crystals. New York: John Wiley.

Gilmer, G. H. (1980). Computer Models of Crystal Growth, Science, **208**, 355-363.

Gilmer, G. H. and Jackson, K. A. (1977). Computer Simulation of Crystal Growth, in Kaldis and Scheel, 1977. pp. 80-144.

Goldschmidt, V. (1913-1923). Atlas der Kristallformen. 9 Vols. & 9 vols. ill. Heidelberg: Winter Universitatsverlag Gmbh.

Gomer, R. and Smith, C. S. (1953). (editors) Structure and Properties of Solid Surfaces. Chicago: University of Chicago Press. Chicago.

Greene, J. C. and Burke, J. G. (1979). The Science of Minerals in the Age of Jefferson. Trans. Amer. Phil. Soc. **68**, 113.

Hartman, P. (editor) (1973). Crystal Growth: An Introduction. North Holland, Amsterdam.

Hartman, P. (1977). Crystal structure and crystal growth: I. The influence of internal structure on morphology: a discussion. Am. Mineral. **62**, 1034-1935.
Hartman, P. (1978). On the validity of the Donnay-Harker law. Can. Mineral. **16**, 387-391.
Hartman, P. and Perdok, W. G. (1955a,b,c.). On the relations between structure and morphology of crystals. Acta Cryst. **8**, 49-52; 521-524; 525-529.
Herring, C. (1951). Some Theorems on the Free Energies of Crystal Surfaces. Phys. Rev. **82**, 87-93.
Holden, A. (1971). Shapes, Space and Symmetry. New York: Columbia University Press, N.Y.
Huyghens, C. (1690). Traité de la lumière. Leiden: P. Van der Aa.
Kaldis, E. and Scheel, H. J. (editors) (1977). Crystal growth and materials. Amsterdam: North-Holland.
Keester, K. L. and Giddings, G. M. (1971) Morphological crystallography via IBM 2250 interactive graphic display terminal. Am. Crystallogr. Program Abstr. Ser. **2**, p. 45.
Kepler, J. (1611, 1966). The Six-cornered snowflake. Edited and translated by C. Hardie. London: Oxford University Press.
Kossel, W. (1927). The theory of crystal growth. Nachr. Ges. Wiss. Gottingen Jahresber. Math-Physik. Klasse. pp. 135-143.
Laue, M. von (1943). The Wulff Theorem for the Equilibrium Form of Crystals. Translated from Z. Kristallogr. **105**, 124-133 in Schneer, 1977.
Leonyuk, N. I., Galiulin, R. V., Al'shinskaya, L. I. and Delone, B. N. (1980). Practical Determination of Perfect Habits of Crystals. Z. Kristallogr. **151**, 263-269.
McLachlan, D., Jr. (1952). Some Factors in the Growth of Crystals: Part I. Extensions of the Donnay-Harker Law. Bull. Univ. Utah Eng. Exptl. Sta. **57**, Addendum (1974) in Schneer, 1977.
McLachlan, D., Jr. (1957). The symmetry of dendritic snow crystals. Proc. Natl. Acad. Sci. U.S.A. **43**, 143-151.
McLachlan, D., Jr. (1978). Progress in crystal-growth theory. Can. Mineral. **16**, 415-425.
Mason, B. J. (1963). Ice. In Gilman, 1963, pp. 119-150.
Mauskopf, S. H. (1976). Crystals & Compounds. Trans. Amer. Phil. Soc. **66**, 86.
Nakaya, U. (1954). Snow Crystals. Cambridge: Harvard University Press.
Niggli, P. (1941). Lehrbuch der Mineralogie und Kristalchemie. Vol. I. Berlin: Borntraeger.
Niggli, P. (1941). Lehrbuch der Mineralogie und Kristallchemie. tinuums. Leipzig: Borntraeger.
Niggli, P. (1920). Beziehung zwischen Wachstumformen und Structur der Kristalle. Z. Anorg. Allg. Chem. **110**, 55-81. Reviewed in Schneer, 1977.
Pamplin, B. R., ed. (1979). Progress in Crystal Growth and Characterization. Oxford: Pergamon Press.
Peacock, M. A. (1932). Calaverite and the law of complication. Am. Mineral. **17**, 317-337.
Prewitt, C. T. (1973). XTLCm A Program to Plot Crystal Morphologies. Unpublished.
Schneer, C. J. (1968). Crystal form and crystal structure. Helv. Phys. Acta, **41**, 1151-1155.
Schneer, C. J. (1971). Steno: on crystals and the corpuscular hypothesis. In Dissertations on Steno as geologist, G. Scherz, (editor) Acta Historica Scientiarum Naturalium et Medicinalium, **23**, 293-307. Copenhagen: Odense University Press.
Schneer, C. J. (1977). (editor) Crystal Form and Structure. Stroudsburg, PA: Dowden, Hutchinson and Ross.
Schneer, C. J. (1978). Morphological complements of crystal structures. Can. Mineral. **16**, 465-470.
Sempels, J-M and Raymond, J. (1976). A computer program for the cartesian translation of crystallographic data. Computers and Geosciences **2**, 417-435.
Stranski, I. N. (1928). Zur Theorie des Kristallwachstums. Z. physik. Chem. **136**, 259-278.
Steno, N. (1669). Prodromus to a Dissertation on Solids contained within Solids. In Steno; Geological Papers edited by G. Scherz and translated by A. J. Pollock. Copenhagen: Odense University Press.
Thompson, D'Arcy W. (1948). Growth and Form. New York: Cambridge University Press.
Verma, A. R. (1953). Crystal Growth and Dislocations. London: Butterworth.
Verma, A. R. and Krishna, P. (1966). Polymorphism and Polytypism in Crystals. New York: John Wiley.
Weeks, J. D., Gilmer, G. H. and Jackson, K. A. (1976). Analytical theory of crystal growth. J. Chem. Phys. **65**, 712-720.
Wells, A. F. (1946). Crystal habit and internal structure. Phil. Mag. **37**, 184-199.
Wolff, G. A. and Gualtieri, J. G. (1962). PBC Vector, Critical bond energy ratio and crystal equilibrium form. Am. Mineral. **47**, 562-584.
Wulff, G. (1901). Zur Frage der Geschwindigheit des Wachstums und der Auflosung der Kristallflachen. Z. Kristallogr. **34**, 449-530, (On the Question of the Rate of Growth and of Dissolution of Crystal Faces) Transl. in C. Schneer (editor) 1977.

CHAPTER 12

CRYSTALS IN INDUSTRY

William R. Cook, Jr.
Cleveland Crystals, Inc., Cleveland, Ohio 44117

Unlike most of the reviews in this volume, this one is concerned with the uses of crystals. In some of the earlier uses, mostly in Europe, crystals were part of optical components such as lenses and prisms (quartz, calcite, fluorite, and alum), watch bearings (corundum, as "sapphire" and "ruby"), and electrical or heat insulators (mica). Many of us remember the old Franklin stoves with "isinglass" windows, made from large mica cleavages. In all but one of the above examples (alum), the crystal used was solely a natural product. About the turn of the century materials became insufficient and syntheses were frequently employed in Europe. In this chapter, we shall cover those syntheses only as they came to the United States in the third and fourth decades of this century. Four developments will be covered, two of which originated in Europe, and two here. Each of these founded an industry and was in turn brought about by the needs of another industry. Each involved a different growth technique, and inspired growth of other crystals by the same technique. Almost all of the well over a hundred crystals grown commercially today use variants of one of these methods. Exceptions are mica, silicon carbide, diamond, and garnet-structure oxides. The first three of these are not completely dependent on *single* crystals, and the last is twenty years newer than any of the others.

One of the early non-jewelry applications of crystals was as bearings in watches and other precision instruments. An early method of crystal growth was that of Verneuil (1902) before the turn of the century for growing sapphire. This involved dropping aluminum oxide powder through an oxyhydrogen flame onto a seed at just below its melting point so that the molten material crystallizes on the seed. Interestingly, this early method may well be the most difficult of the commercial growth techniques.

The first attempt to acquire the process in this country occurred during World War I, when shipments were cut off from Germany. The details of the process were closely guarded by the European manufacturers. Nevertheless, Harris Calorific was successful in growing sapphire, but the project died after the war as material again became available from Europe.

World War II came along, and again the government asked for bids for growing sapphire. The Linde division of Union Carbide obtained the contract, and set up a plant near Chicago in 1940, which produced 20 tons a year during the war. Perhaps it was the better technology available this time, or a longer war that permitted more development of the process, but this time the business survived the end of the war and renewed competition from overseas. The synthetic star sapphire and ruby business grew out of the accidental discovery of a star in crystals that had been grown from contaminated material, and was commercialized in 1947.

While the "first" development did not really get started in this country until World War II, and came from overseas, the next crystal venture started in America.

Piezoelectricity (a linear relationship between mechanical pressure and electrical charge) was discovered by the Curie brothers in 1880. Among the materials studied were quartz, tourmaline and Rochelle salt. Although the piezoelectric constants were measured as early as 1894 by Pockels, real commercial interest dates from World War I and the beginnings of military sonar and radio. Pockels, under the influence of earlier incorrect work on the dielectric properties of Rochelle salt, had failed to fully solve the properties along the ferroelectric *a*-axis. (Ferroelectricity consists of a polar axis in a crystal which is switchable between at least two positions by an electric field. Ferroelectricity is frequently called seignettelectricity in Europe after an alternate name for Rochelle salt.) It was left to Valasek in 1921-1922 to fully describe the unusual properties along the *a*-axis and to explain the new phenomenon, in analogy to ferromagnetism.

After graduate school, two young friends, Charles Francis Brush, Jr. (son of the arc-light pioneer, who was also first president of Linde Air Products Co. in 1905) and Charles Baldwin Sawyer, started a small commercial laboratory in the senior Brush's carriage house. Business being slow for the new enterprise, each did some research on a subject of interest. For Sawyer, it was beryllium metal (which developed into Brush Beryllium Co., now Brush Wellman Co.); for Brush it was Rochelle salt, which became Brush Development Co., later part of Clevite Corp. It is likely that young Brush's interest was aroused by Valasek's 1921 paper.

The younger Brush died tragically in 1927, and his work was carried on by Dr. Sawyer and Bengt Kjellgren. One pioneer paper, dating from 1930, the year that Brush Development was formed, described a means of displaying a ferroelectric hysteresis loop on an oscilloscope, and is still regularly quoted in the literature. (Sawyer and Tower, 1930).

The crystals were grown by placing a rod of Rochelle salt in a slot in the bottom of a tray; a slightly supersaturated solution was added, and the solution was slowly cooled. Crystals 18 to 24 inches long were grown. In developing the growth process, however, all three crystal axes were tried as the direction for seed elongation, and large crystals from these early experiments still exist, now badly dehydrated on their surfaces. Variants of this method are still used for growing many water-soluble crystals, and are well-described by Singer and Holden (1960).

The availability of large piezoelectric crystals of uniform quality opened up a host of new uses; the first and largest volume was the phonograph pickup element found in most phonographs for over three decades. Rochelle salt was the strongest piezoelectric known, and at exactly 24°C it still is (as defined by the efficiency of conversion of mechanical to electrical energy or vice versa). Unfortunately, the high piezoelectric effect is due to a transition at 24°C between two crystal forms, both piezoelectric and one ferroelectric. As little as 2° change in temperature cuts the piezoelectric effect in half. Also, the crystal becomes progressively less stable in both low and high humidity with increasing temperature, eventually dissolving in its own water of crystallization at 55°C. Brush Development once received a letter from a customer in Arizona saying that his record player, which he kept on a closet shelf when not in use, would no longer play. The problem seemed to be with the packet of white powder (enclosed); would they please send another packet of white powder that would work? His Rochelle salt had actually dehydrated completely in the hot Arizona climate, the normal preventative sealing having failed on the hot closet shelf.

Because of these problems, in 1940 Brush Development hired a refugee German scientist, Dr. Hans Jaffe, to find a substitute for Rochelle salt. Three developments resulted: ADP, $NH_4H_2PO_4$, in 1942 (with the first major use in Navy sonar); followed by barium titanate, $BaTiO_3$, in 1949; and lead titanate-zirconate, $Pb(Ti,Zr)O_3$, in 1957. The latter two are ferroelectric ceramics, and act like single crystals after treatment with a strong electric field. In its major uses, ADP was replaced by $BaTiO_3$ after the Korean war.

ADP crystals were grown on flat seeds cut perpendicular to the *c*-axis, in a series of trays on a rack which rocked all the trays together. The crystals grew by slowly cooling the solution, with virtually all the growth occuring in the *c*-direction. Since the whole room was the solution temperature, the growth began in exceedingly hot rooms, and the workers, mostly female, dressed accordingly. The scenery, I am told, often resembled a summer beach.

Dr. Jaffe also developed the use of ADP and its isomorphs as linear electro-optic (Pockels) modulators. (Jaffe, 1949)*. As frequently happens, this development was ahead of its time, and found little use until after the invention of the laser. For laser modulation, KD_2PO_4, the deuterated potassium analog of ADP, has proven most useful. Crystal elements as large as ten inches across are now available, considerably larger than the two inch elements sold thirty years ago.

*A parallel effort occurred in Switzerland (Zwicker and Scherrer, 1944).

Another part of the prolific crystal growth effort at Brush Development Co. was the growth of synthetic quartz under Dr. Danforth R. Hale. In 1905, Spezia obtained crystals of a few grams in a high pressure sodium silicate solution. Under U.S. Army Signal Corps sponsorship a program was started in 1946 that by 1955 had produced a commercial process. Basically it involves growth from Na_2CO_3 solution in a temperature gradient on a seed elongated perpendicular to the prism faces. To achieve reasonable solubility pressures of about 5000 psi and temperatures near 350°C were used. Many engineering problems were solved before it became a reliable production process. For instance, in order to monitor crystal growth without waiting months for the end of a run, some way was needed to "watch" the crystals as they grew. Windows were impractical in a heavy steel autoclave with high-pressure corrosive liquids. Radiography was considered, but according to the experts, there would not be sufficient contrast with ^{60}Co radiation through the thick steel walls to detect the low-absorbing quartz. Dan Hale tried it anyway, carefully aligning the thickest direction of the seed parallel to the radiation, and it worked well. Again, long seeds were almost unavailable, and growth on the prism face is very slow. Wilfred Charbonnet, manager of the pilot plant, tried to abut two seeds end to end and grow one extra-long crystal. The extremely critical orientation needed made the effort unlikely to succeed, but after several tries, a long joined crystal was obtained.

The time came to start production. The Board of Directors of Clevite Corp. (successor to Brush Development Co.) helped by the decision of the Brazilians to halve the price of natural quartz (the timing was unlikely to have been a coincidence) chose not to set up production. Dr. Sawyer felt they were wrong, took a license, and started a new company, Sawyer Research Products, to manufacture quartz. This, the largest independent manufacturer of quartz, recently became a part of Brush-Wellman, joining the successor company of the beryllium branch of the original Brush Laboratories.

We should point out that Bell Laboratories, inspired by the needs of the telephone system, had embarked on a closely parallel effort in crystal growth. Although most of these did not get beyond the research stage, there is evidence (Jaffe, 1949) that ethylenediamine tartrate (EDT) was used commercially for a short while, probably as a substitute for quartz. The many publications of Warren P. Mason testify to the Bell activity. The only piezoelectric crystal from the 1940's that is produced in volume by Western Electric Co. is synthetic quartz. The major differences in the Bell process are use of NaOH rather than Na_2CO_3, a higher pressure, and *c*-plates as seeds. Dr. A. C. Walker, the head of the early effort at Bell Laboratories, acknowledged how helpful the Brush effort was to the Bell development.

The next crystal type to appear on the commercial market was a halide. Harshaw Chemical Co. started in the 1890's as a producer of ceramic colors and plating chemicals. By the 1930's the company was interested in diversification, and William J. Harshaw, a son of the founder, decided to try the halides for ultraviolet optics. While there had been limited growth of crystals such as alum, NaCl, NaF, LiF, and in 1936 KRS-5 (thallium iodide-bromide) in Europe for such applications as microscope lenses, crystals were not generally available. In 1937 a group under Dr. H. C. Kremers began producing lithium fluoride using the method developed by Prof. Donald Stockbarger (1936, 1937) at MIT. A conical-tipped platinum crucible filled with the material is lowered through a two-zoned furnace with a tight baffle between the zones. The two zones are separately controlled, permitting a steep temperature gradient at the baffle, a necessary condition in obtaining high quality crystals. Proper control of conditions will cause one nucleus to take over the crystallization surface and propagate up the crucible while the crucible is lowered into the cooler zone. The Stockbarger method is a modification of the classical Bridgman (1925) technique. The author recalls seeing one of their then-standard Pt crucibles in 1955, which was about a foot across and very impressive. Current sizes can be much larger.

The original application for LiF was as a replacement for natural fluorite in ultraviolet optics. While the market was quickly satisfied, new needs appeared as prisms for infrared spectroscopy, for which NaCl and KBr were better. By 1939, NaI, CsBr, and CsI had also been added to the list. The availability of very large prisms (4" to 8") immensely speeded the development of infrared spectroscopy. While prisms have now been superseded by gratings, the uses of halides continues to expand. About 1950, thallium-activated NaI was introduced as a nuclear detector, quickly followed by other activated halides. During World War II, LiF was introduced as an analyzing crystal for the early X-ray fluorescence units, and in the early 1960's it was bent for focusing analyzers. Recently interest has returned to the ultraviolet, with LiF and MgF_2 supplied for UV space applications.

An interesting sidelight to how technology in one area affects other technology is illustrated by the following experience. The original LiF was grown from chemicals purified using standard techniques (Kremers, 1940). The government plant at Savannah River began processing lithium ores for nuclear needs, and so very pure inexpensive raw materials for LiF growth became available. They were used and much to everyone's surprise the crystals, which had been brittle, were now ductile, making them difficult to handle. In order to process them properly, the defects which had been removed with the purification, needed to be restored!

The fourth crystal product resulted from the invention of the transistor at Bell Laboratories, probably the world's leading research establishment. The transistor was invented in late 1947, and resulted in a Nobel prize for three Bell scientists, John Bardeen, Walter Brattain, and William Shockley. The transistor used single crystal germanium at first, changing in the 1950's to silicon. In order to commercialize this invention, and to obtain the sophisticated devices and perfect crystals made today, many other inventions were also necessary. Probably the most important, the invention of zone refining by William G. Pfann in 1950, was not even made with transistor technology in mind, but was an attempt to solve variability of impurity content in copper from differing sources.

The original transistor was a point contact device, and needed only a small area to work, but with the advent of purer materials, whole wide layers of npn junctions could be introduced during growth, first by alternate doping by Ga and Sb, then later by taking advantage of the variation of segregation coefficient with growth rate. The elemental material was zone refined by passing multiple hot zones down a long bar of the material. At melting, some impurities would tend to segregate in the liquid and others in the solid, and would thus migrate to one end or the other of the bar. After purification a seed of the material at just below the melting point would be dipped in a molten crucible at just above the melting point. Heat conducted along the seed would cool material near it and start crystal growth. The crystal would be slowly withdrawn, and the crystal increased in length. This is the Czochralski method. Perfection of silicon crystals has reached a level that would have been thought impossible a few years ago. While this accomplishment has been of tremendous scientific and commercial importance by itself, spawning a host of multimillion dollar companies, it has produced additional benefits: a large number of other crystals grown by the same method. These range from other semiconductors such as GaAs and piezoelectric crystals such as $LiNbO_3$ to laser hosts such as YAG (yttrium aluminum garnet $Y_3Al_5O_{12}$). The existence of experienced crystal growers undoubtedly spurred the development of the laser industry as well as others dependent upon crystals grown by this method.

Acknowledgments

The author was a member of Dr. Jaffe's group at Brush starting in 1951. Special thanks are due Dr. Carl Swinehart of Harshaw, Dr. Dennis Holt of Union Carbide, and Dr. C. B. Sawyer, Jr. formerly of Sawyer Research Products and Bell Laboratories, for their information. None of these people should be blamed, however, for any errors that may have been inadvertently introduced.

References

Bridgman, P. W. (1925). Proc. Am. Acad. Arts Sci. **60**, 305.
Busch, G. (1938). Helv. Phys. Acta **11**, 269-98.
Busch, G. and Scherrer, P. (1935). Naturwiss. **23**, 737.
Curie, J. and Curie, P. (1880). Bull. soc. min. France **3**, 90-3.
Jaffe, H. (1949, Sept.). Physics Today, 14-19.
Kremers, H. C. (1940). Ind. Eng. Chem. **32**, 1478-83.
Pockels, F. (1894). Abh. Gött. **39**, 1-204.
Sawyer, C. B. and Tower, C. H. (1930). Phys. Rev. **35**, 269-73.
Singer, P. and Holden, A. (1960). Crystals and Crystal Growing. Columbus, Ohio: Anchor Books.
Spezia, G. (1905). Acad. Sci. Torino Atti **40**, 254.
Stockbarger, D. C. (1936). Rev. Sci. Instrum. **7**, 133.
Stockbarger, D. C. (1937). J. Opt. Soc. Amer. **27**, 416.
Valasek, J. (1921). Phys. Rev. **17**, 422-3, 475-81.
Valasek, J. (1922). Phys. Rev. **19**, 478-91; 20, 639-64.
Verneuil, A. (1902). C. R. Acad. Sci., **135**, 791.
Zwicker, B. and Scherrer, P. (1944). Helv. Phys. Acta **17**, 346-73.

CHAPTER 13

ROLE OF X-RAYS AND CRYSTALLOGRAPHY IN THE MANUFACTURE OF QUARTZ OSCILLATORS IN WORLD WAR II

William Parrish
IBM Research Laboratory
San Jose, California 95193

Background

X-rays and crystallography played an important role in the development of the industry for the mass production of quartz oscillator plates in World War II. The thin postage stamp-size wafers were used for frequency control in virtually all military radio communications in aircraft, tanks, artillery units, walkie-talkies, naval vessels and many other uses.

Early in 1942 I was called to Washington by the Office of the Chief Signal Officer to set up a program to support an enormous expansion of the industry. The small number of manufacturers then in operation were producing crystals by hand-crafted methods. Suddenly millions of crystals were needed to meet much stiffer military specifications. This required setting up new plants, and designing improved methods and equipment to get into mass production in a minimum of time. A number of my mineralogist friends soon joined: Clifford Frondel, Samuel Gordon and Richard Stoiber; there were many other civilian and military personnel involved. After our first visit to a crystal plant we realized we knew virtually nothing about the methods and even the terms used were strange; e.g., Brazil twinning was called "optical" twinning in the trade and Dauphiné "electrical." The small existing literature was of little help in evolving mass production processes (see for example Booth, 1941).

We began an intensive crash program to learn the fundamentals and improve the processes. There was no laboratory to work in and only later were facilities set up at Fort Monmouth, New Jersey. The testing had to be done in operating plants while avoiding interruptions in their production. We received help from managers and engineers who were experienced in certain phases of the processes. Methods were developed for virtually all aspects of production and we wrote detailed manuals which we combined with lecture demonstrations to speed the introduction of X-rays and crystallographic methods. Production rose from about 100,000 plates in 1941 to a rate of over 60 million a year at the end of the war, and prices dropped to a small fraction of prewar prices (Frondel, 1945). A special enlarged issue of *American Mineralogist* contains a series of papers written by authors who were actively involved in various aspects of this huge scientific/engineering/production task (see Parrish and Hamacher, 1945). The X-ray method of orientation is still widely used in cutting quartz, silicon, garnet and similar processes requiring accurate orientations control.

Virtually all the quartz used was mined in the states of Minas Gerais, Goiaz and Baia in Brazil. A crisis occurred when a large tonnage of quartz was lost in a submarine attack, and thereafter it was shipped by air. The crystals were 100 to 2,000 grams and later in the war only the smaller poorer quality was available. At least one-half of the crystal had to be "eye-clear" and later this restriction was reduced.

Crystallographic Orientation

The manufacturing process required the orientation of "raw" quartz (usually known as "mothers" in the industry) and also the reorientation of bars, sections, wafers and diced blanks at various stages. This was a formidable task because the methods had to be used by people with little or no technical or crystallographic training, and the methods had to work fast to handle the large number of crystals to be processed. Errors in the orientation and inspection caused a large amount of useless processing, greatly reduced output and a loss of quartz crystals.

Prior to processing, the raw quartz was immersed in an oil bath between crossed polaroids to determine the extent of optical twinning, and in a bright spotlight to see cracks, inclusions and other defects. The Brazil twins occur as multiple thin lamellae of one hand within a host crystal of the opposite hand. They are often triangular, parallel to r and have remarkably straight edges. Dauphiné twins could not be detected optically, and usually occurred as large areas with irregular boundaries. The crystals were classified into usability groups for processing.

Twinning caused two major problems: 1) the final oscillator plate had to be completely free of twinning, 2) Dauphiné twinning could cause the cut to be made on the wrong side of the optic axis. Virtually all the oscillators were the low temperature coefficient type, e.g., AT plates were inclined +35°15' from the c-axis and roughly parallel to z (01.1), BT plates inclined -49° toward r (10.1).

Ideally four parameters determined the correct orientation for cutting the crystal into oscillator plates: the hand of the crystal which could be determined opitcally as right or left, and the polarity of the piezoelectric a-axes which could be determined with a piezometer. Because of the ubiquitous nature of Dauphiné twinning these tests were usually wrong or misleading. It was therefore necessary to develop orientation and cutting methods which showed the twinning within the crystal, and to take it into account in deciding how to cut it.

The best and most widely used method was the X-block method in which the cutting was done in two stages (Gordon and Parrish, 1945). In the first stage the crystal was mounted on a prism or rhombohedral face and a cut was made on each side of the crystal parallel to (11.0). This required locating the c-axis which could be done from the morphology, or if the faces were absent with a conoscope (Bond, 1943).

The X-block was then deeply etched and the twinned portions marked. When the block was placed on a pin-hole light source, a figure appeared from reflections from the etch pits and showed the orientation of the etched plane, Fig. 1 (Parrish and Gordon, 1945a). The parallelogram or arrow light figures directly gave the r and z

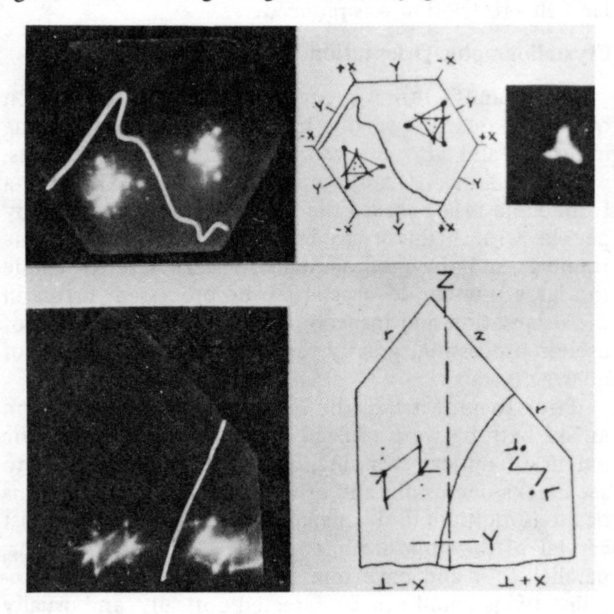

Figure 1. Light figures on etched quartz blocks. Top: Z-section (00.1). Connecting lines drawn between the bright outer spots form an equilateral triangle with sides parallel to Y(11.0). The triangles on both sides of the irregular Dauphiné twin boundary are related by a mirror plane. Bottom: X-section (11.0). The long fainter sides of the parallelogram show the slope of r(10.1) and the shorter brighter sides are parallel to the c-axis. The arrow figure on the opposite side of the Dauphiné twin is parallel to (11.0) and the short ends parallel to z(01.1).

directions regardless of the Dauphiné twinning. The block could be cut along a twin boundary and each part wafered separately. If the entire block was wafered, one part would have been useless because of its incorrect orientation. The pinhole light figures were found to be the only practical infallible method for obtaining the correct orientation.

They were used in remounting the X-blocks on the selected (11.0) surface for the second stage of wafering. Methods for orienting crystals without faces were also developed.

Stauroscopic and conoscopic optical methods were developed to locate the optic axis direction in unfaced raw quartz and for checking the sawed sections or wafers prior to X-ray measurements or dicing. There were many other important steps in the manufacture such as automatic machine lapping to frequency, final frequency adjustment, aging of the oscillator plates, etc., which are described in the Symposium see Parrish and Hamacher, (1945).

X-ray Orientation

Once the general orientation was known, the X-block had to be cut within 10' to 15' of the correct angle into thin wafers. Optical methods were found to be unsatisfactory. Laue patterns had been used but the long exposure and film development time (this was prior to the development of Polaroid film) and low precision were not satisfactory for mass production. A Bragg reflection method appeared to be the best solution, but goniometers were not available.

The urgency of the situation required a new type of goniometer that could be quickly designed, mass-produced and used by operators who had no knowledge of X-rays and crystals. The readings had to be fast and accurate with no calculations or interpretation. The danger of operator exposure to X-rays required the use of safety shutters and shielding that could not be easily defeated in the plant environment.

The horizontal goniometer developed for the quartz industry was somewhat similar in principle to the Braggs' ionization spectrometer (Parrish and Gordon, 1945b; Parrish, 1950; Parrish and Hamacher, 1952) and became the prototype of the powder diffractometer, Fig. 2. The detector was fixed at the selected 2θ-angle

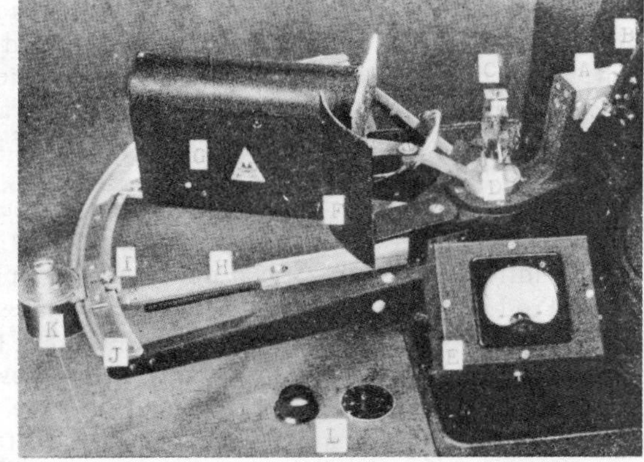

Figure 2. X-ray goniometer developed for precision orientation of quartz crystals. Various other crystal holders were used for large pieces.

and only the crystal was rotated. The flat surface of the wafer might be parallel to the reflecting plane, e.g., (11.0) for an X-cut, or inclined, e.g., z(01.1)∧AT-cut=2°58'. A triangular wafer cut from the crystal mounted on the diamond blade saw was marked with a vertical arrow to preserve the azimuthal orientation and pressed by a spring against the reference surface of the X-ray crystal holder. The holder was attached to the goniometer arm and manually rotated by a dial until the peak of the reflection was located. The goniometer was calibrated to directly give the correction angles and both the horizontal and vertical planes were measured. The saw bed was corrected the same amounts to bring the crystal to the required orientation.

The detectors presented some difficult problems in those early days of their development. Geiger counters had a too long dead time to be used in this application. We were fortunate that Dr. Herbert Friedman of the Naval Research Laboratory developed a proportional counter in which the current in the tube was read on a small milliammeter, and the voltage could be changed to vary the deflection from planes of different intensities. High stability was not a problem since intensities were not being compared. The visual readings of the counter tube milliammeter were crucial. We used a small time constant to avoid angular errors and the intensities were made sufficiently high to eliminate needle wiggle. Too high intensities could have caused detector saturation. Since the readings were made all day long, considerable effort was made in designing the instrument to reduce operator fatigue.

The air-cooled X-ray tube was operated at low power (=125w) and narrow slits in the incident beam limited the intensity and peak width. Too narrow slits were avoided because of the possibility of missing the peak in the rapid scans. A wide antiscatter slit was used in front of the detector.

Philips Metaliz Corporation (now Philips Electronic Instruments) and General Electric X-Ray Corporation manufactured hundreds of the instruments which were used in over a hundred plants.

"X-rays are used to make hundreds of thousands of measurements accurate to a few minutes every day by untrained help, taking 10 to 15 seconds for each measurement. In fact, this marks the first application of X-ray diffraction as an integral part of mass production manufacturing." (Parrish and Gordon, 1945b.)

References

Bond, W. L. (1943). Method for Specifying Quartz Crystal Orientation and Their Determination by Optical Means, Bell Sys. Tech. Jour. **22**, **224-262**.

Booth, C. F. (1941). The Application and Use of Quartz Crystals in Telecommunication, J. Inst. Elect. Eng. **88**, 97-144.

Frondel, C. (1945). History of the Quartz Oscillator-Plate Industry, 1941-1944, Am. Mineral. **30**, 205-213.

Gordon, S. G. and Parrish, W. (1945). Cutting Schemes for Quartz Crystals, Am. Mineral. **30**, 347-370.

Parrish, W. (1950). The Manufacture of Quartz Oscillator-Plates, Philips Tech. Rev. **11**, 323-332, 351-360; **12**, 166-176.

Parrish, W. and Gordon S. G. (1945). Orientation Techniques for the Manufacture of Quartz Oscillator-Plates, Am. Mineral. **30**, 296-325; Precise Angular Control of Quartz-Cutting by X-Rays, Ibid., 326-346.

Parrish, W. and Hamacher, E. A. (1952). Accurate Orientation of Quartz Oscillators with X-Rays, Trans. Inst. Meas. Conf. Stockholm, pp. 106-112.

Parrish, W. and Hamacher, E. A. (1945). Symposium on Quartz Oscillator-Plates, Am. Mineral. **30**, 205-468.

CHAPTER 14

BONE AND TOOTH MINERAL

Aaron S. Posner
Cornell University Medical College, New York, NY 10021

In 1771 Scheele showed that calcium phosphate was a major constituent of bone. Since then the list of scientists who studied bone reads like a "Who's Who of Chemistry." Berzelius, Gabriel, Hoppe-Seyler and Werner are but a few of the great classical chemists who sought to define the nature of bone salt. It wasn't until 1926 that de Jong (1926) showed by X-ray powder diffraction that bone-and tooth-mineral are poorly-crystallized analogues of the mineral hydroxyapatite, $Ca_{10}(PO_4)_6(OH)_2$. Structural studies on the apatite family of minerals began to appear in the 1930's [e.g., hydroxyapatite(Mehmel, 1930), fluorapatite (Naray-Szabo, 1930; Beevers and McIntyre, 1946) and lead apatite (Klement, 1938)]. In fact, these early structure determinations helped explain the later biological observations that certain ions (e.g., F^{-1}, Pb^{+2}) were bone seekers when ingested. In this regard, one of the major dangers of above-ground H-bomb testing was the preferential uptake and long-time retention in our bones of radioactive Sr^{90} from fall-out, increasing the possibility of cancer in our population.

The recent application of neutron diffraction by Young and his associates at Georgia Institute of Technology has contributed substantially to our understanding of the well-crystallized apatites. Fluorapatite, $Ca_{10}(PO_4)_6F_2$, occurs in the hexagonal crystal structure form with space group symmetry $P6_3/m$. However, defect-free hydroxyapatite and chlorapatite, $Ca_{10}(PO_4)_6(Cl)_2$, are found to be monoclinic with space group $P2_1/b$. This change in symmetry results from the fact that the OH and Cl are situated above or below the mirror plane (on what was the 6-fold screw axis in hydroxyapatite) in an ordered fashion causing a degeneration of symmetry from hexagonal to monoclinic in near perfect crystals of OH- and Cl-apatite. In F-apatite the F is situated in the mirror plane and is closer to the neighboring three Ca ions in this plane than OH and Cl, explaining why F will substitute for OH and Cl in biological apatites. In most hydroxyapatites, mineral and biological, the OH ions are randomly distributed above and below the mirror planes producing an apparent $P6_3/m$ space group. A thorough discussion of this effect is given by Young (1980).

Structural studies on the well-crystallized apatites have not been as useful for determining the structure of biological apatites as have the studies on poorly-crystallized, chemically precipitated hydroxyapatites. The major characteristics of bone and tooth apatites are: (a) the submicroscopic crystal size yielding a high specific surface, (b) a substantial carbonate content; (c) the lack of stoichiometry usually manifested as a Ca-deficiency, and (d) the high internal disorder of these crystals. Since the mid 1950's the major push for the study of hard tissue mineral came with the growth of the federal health science support from the National Institutes of Health. At the same time the Atomic Energy Commission was interested in funding calcium phosphate research not only because of the danger of fall-out to bone, but also because uranium is found in calcium phosphate mineral deposits. Uranium, again, is the reason why the US Geological Survey has a large internal research program on the calcium phosphate minerals. The value of many of these calcium phosphate deposits as fertilizer was also a reason for the work at the U.S. Department of Agriculture on these materials. For many years the National Aeronautics and Space Administration has backed intramural and extramural research on bone structure because of the apparent bone loss which afflicts the astronauts in space flight. Finally, there are many private funding agencies which have shown interest in this field.

The crystal size and orientation in bone and teeth have been investigated since transmission electron microscopy and X-ray diffraction have been available. The early groups using the electron microscope to study bone were Robinson and Watson (1952), Ascenzi and Bonucci (1966) and Engstrom and his co-workers (Engstrom and Zetterstrom, 1951; Fernandez-Moran and Engstrom, 1957). The major electron microscope studies on tooth enamel and dentine were carried out at the National Institute of Dental Research by Wyckoff, Scott, and Nylen. The discovery of the caries-prevention action of fluoride stimulated this group and many others to begin electron microscope studies of teeth from the early 1950's to this date. At the same time X-ray diffraction work on bone, and tooth crystal size was begun by Engstrom's group in Sweden and by Posner in Washington, D.C.

In general, it was found that bone and dentine mineral crystals are similar in size to apatites precipitated under body conditions, i.e., in the order of 400 x 250 x 45 Å in dimensions. Beebe and his group showed bone apatite crystals had both a high specific surface (100-200 m^2/g) and a highly reactive surface (Posner and Beebe, 1975). Enamel crystals although an order of magnitude larger than bone and dentine crystals, are still much smaller in crystal size than mineral apatites. The c-axis of bone apatite crystals were always parallel to the fiber axis of the bone collagen matrix. Enamel crystals were also shown to display orientations related to the organic matrix. Later work showed bone crystal size to be related to age, disease and diet. Fluoride ingestion which stabilized biological hard tissue was shown to increase the average human and animal bone and enamel crystal size as well as to substitute F for OH in the bone and tooth apatite (Posner, 1969).

Biological and mineral apatites contain in the order of 3-5% (w/w) of carbonate ion. Most mineralogists assumed that carbonate occupied a structural position in apatite but others felt it was adsorbed on the crystal surface (especially in the finely-divided apatites), and, it was even suggested by Hendricks and Hill (1950) that carbonate was coated on internal surfaces of the mosaic crystal of mineral francolite (fluor-carbonate apatite). It has become apparent that both views are correct insofar as bone and tooth mineral are concerned. Infrared spectroscopy (Montel, 1968) has shown that CO_3^{-2} can substitute (in some yet unexplained fashion) for PO_4^{-3} in apatites formed at body temperature. To date, no diffraction studies have been able to clarify this "non-space group" substitution. Isotope exchange studies (Neuman and Mulryan, 1967) have shown that about half of the carbonate in bone is situated on crystal surfaces; in particles 100-200 m^2/g in specific surface one would expect half of the crystal to be exchangeable. It has been suggested that bone surface carbonate is useful in the body function to supply HCO_3^{-1} ions to counteract acidosis in body fluids (Neuman and Neuman, 1958).

One of the earliest problems in bone and tooth structure was the explanation for the apparent Ca-deficiency or lack of stoichiometry of biological apatites (Posner *et al.*, 1954). Neuman and Neuman (1958) showed by surface exchange studies that the low Ca/P ratio hydroxyapatites did not have excess PO_4 adsorbed on the surface. A powder diffraction study of Pb-deficient lead apatites (Posner and Perloff, 1957) established the principle that divalent cations could be missing from apatite structures in the order of ten percent of the total. Chemists (Gee and Dietz, 1955; Berry, 1967) showed that heating synthetic Ca-deficient hydroxyapatites to about 600°C produced condensation to pyrophosphate ions. It was postulated that the missing Ca ions were balanced electrostatically by some combination of missing OH ions and H-bonds (between neighboring PO_4 groups). In the carbonate-free case Winand (1965) suggested this formula: $Ca_{10-x}H_x(PO_4)_6(OH_{2-x}$. Infrared studies (Posner *et al.*, 1960) showed the presence of hydrogen bonds between orthophosphate groups and a later titration study (Meyer, 1979) showed missing OH ions. Pyrolysis studies on bone produced pyrophosphate as predicted from the Ca/P ratios and the Winand formula (Dallemagne and Richelle, 1973).

Brown and his co-workers (Brown, 1965; Brown, 1966) suggested an alternate model to explain the low Ca/P ratio apatites. They suggest an interlayering of octacalcium phosphate (Ca/P = 8/6) and hydroxyapatite (Ca/P = 10/6) in the ratio needed to produce the, roughly, 9/6 ratio seen in bone and synthetic apatites. Since the octacalcium phosphate and hydroxyapatite X-ray patterns are similar they reasoned that X-ray diffraction could not differentiate between their interlayer model and the Winand-Berry, Ca-deficient model. There is little doubt that Brown made large crystals of octacalcium phosphate (OCP) with an epitaxial layer of hydroxyapatite. It does not seem reasonable that bone crystals with longest dimensions about 200Å could consist of interlayers of two compounds, one of which (OCP) has a b axis in the order of 18Å. In all events, this model of bone apatite, though not disproved, remains to be proved.

When hydroxyapatite is precipitated from highly supersaturated calcium phosphate solutions, similar to body fluids, an unstable amorphous calcium phosphate precursor, is seen (Eanes *et al.*, 1965). X-ray radial distribution function studies showed this precursor, which appears in the electron microscope as 300-1000Å spheres, to consist of randomly-packed $Ca_9(PO_4)_6$ ion clusters (Betts and Posner, 1974). The presence of a number of chemical species have been shown to stabilize or moderate the solution-mediated conversion of the amorphous precursor to hydroxyapatite. Earlier workers postulated that mature bone mineral consisted of about 35% amorphous calcium phosphate and 65% Ca-deficient, CO_3-containing hydroxyapatite (Termine and Posner, 1967). Later, radial distribution function analysis of bone ruled out the possibility of anything but a small amount of the amorphous precursor phase in mature bone (Posner and Betts, 1975). It is interesting to note that an amorphous calcium phosphate (stabilized by a combination of Mg and adenosine triphosphate) is seen in the mitochondria of cells involved in tissue mineralization (Betts *et al.*, 1975). An X-ray radial distribution study showed that this intracellular amorphous calcium phosphate resembled the structure of the synthetic amorphous compound. Other people have seen amorphous calcium phosphate in developing bone (Miller and Schraer, 1975; Gay, 1977; Gay *et al.*, 1978). Recently, some workers have observed brushite, Ca $HPO_4.2H_2O$, as a constituent of early chick-embryo bone (Roufosse *et al.*, 1979) and bovine embryo (Betts, 1980) bone. It is not clear at this point whether this acid phosphate phase (which can slowly convert to hydroxyapatite at body pH) is present as a precursor or as a result of some acidic microvolume in the ossifying tissue.

X-ray radial distribution function analysis has been used to formulate synthetic models of mature bone apatite. It was shown using this method (Blumenthal *et al.*, 1975) and concomitantly by infrared spectroscopy (LeGeros *et al.*, 1968) that the entry of carbonate into the structure of precipitated hydroxyapatite produces structural distortion. In addition, a recent study (Blumenthal *et al.*, 1981) shows that Ca-deficiencies produce specific changes in the radial distribution function of stoichiometric hydroxyapatite. It was then shown that the radial distribution function of mature bone could be approximated by a synthetic Ca-deficient hydroxyapatite containing about 4% CO_3^{-2} (w/w). No radial distribution work has been performed on dental enamel or dentine. In recent years Young and his co-workers have been studying enamel by neutron diffraction. To date no definitive structural decisions have been reached but it is hoped that this powerful method will help place the CO_3^{-2} groups, H-bonds and OH-deficiencies in these materials. There have been other attempts to study the atomic misalignment in precipitated and bone apatites. An analysis of the X-ray line broadening by bone apatite was done as a part of the study on the improvement of crystallinity with fluoride ingestion (Posner *et al*., 1963). Later, Lundy and Eanes (1973) performed a detailed analysis of the broadening of the apatite crystals deposited in the ossification of turkey leg tendon. A recent X-ray study (Wheeler and Lewis, 1977) showed that the structural disorder in bone mineral fits the Type II paracrystalline disorder model of Hosemann. Much work remains to be done on the nature and extent of crystalline disorder in bone apatite.

This has been a brief but by no means a complete review of the crystallographic studies of hard tissues and related synthetic and mineral apatites. I have tried to make this essay readable and in doing so I may have failed to refer to many excellent papers in this polyglot field. For this I apologize with the hope that my selective editing of the history of the field has given the outsider a clear picture of our slow progress toward an understanding of the structure of the biological apatites.

References

Ascenzi, A. and Bonucci E. (1966). The osteon calcification as revealed by the electron microscope in "Calcified Tissues" edited by H. Fleisch, J. J. Blackwood and M. Owen. New York: Springer.

Beevers, C. A. and McIntyre, D. B. (1946). Miner. Mag. **27**, 254-257.

Betts, F. and Posner, A. S. (1974). Trans. Am. Crystallogr. Assoc. **10**, 73-84.

Betts, F., Blumenthal, N. C., Posner, A. S., Becker, G. L. and Lehninger, A. L. (1975). Proc. Natl. Acad. Sci. U.S.A. **72**, 2088-2090.

Betts, F. (1980). Unpublished data.

Berry, E. E. (1967). J. Inorg. Nucl. Chem. **29**, 317-327.

Blumenthal, N. C., Betts, F. and Posner, A. S. (1975). Calc. Tiss. Int. **18**, 81-90.

Blumenthal, N. C., Betts, F. and Posner, A. S. (1981). Calc. Tiss. Int., **33**, 111-117.

Brown, W. E. (1965). A mechanism for growth of apatitic crystals. Tooth Enamel, M. V. Stack and R. W. Fearnhead. Bristol, England: John Wright and sons.

Brown, W. E. (1966). Clin. Orthoped. **44**, 205-220.

Dallemagne, M. J. and Richelle, L. J. (1973). Inorganic chemistry of bone. Biological Mineralization, I. Zipkin, John Wiley & Sons, New York: John Wiley and Sons.

de Jong, W. F. (1926). Rec. Trav. Chim. Pays-Bas. **45**, 445-448.

Eanes, E. D., Gillessen, I. H. and Posner, A. S. (1965). Nature (London) **208**, 365-367.

Engstrom, A. and Zetterstrom, R. (1951). Exptl. Cell Res. **2**, 268-274.

Fernandez-Moran, J. and Engstrom, A. (1957). Biochim. Biophys. Acta **23**, 260-264.

Gay, C. V. (1977). Calc. Tiss. Res. **23**, 213-215.

Gay, C. V., Schraer, H. and Hargest, T. E. (1978). Metab. Bone Dis. Rel. Res. **1**, 105-108.

Gee, A. and Dietz, V. R. (1955). Anal. Chem. **77**, 2961-2965.

Hendricks, S. B. and Hill, W. L. (1950). Proc. Natl. Acad. Sci. U.S.A. **36**, 731-737.

Klement, R. (1938). Z. Anorg. Allg. Chem. **237**, 161-171.

LeGeros, R. Z., Trautz, O. R., LeGeros, J. P. and Klein, E. (1968). Bull. Soc. Chim. Fr. (No. special) 2^{e} trimestre, 1712-1718.

Lundy, D. R. and Eanes, E. D. (1973). Archs. Oral Biol. **18**, 813-826.

Mehmel, M. (1930). Z. Kristallogr. **75**, 323-331.

Meyer, J. L. (1979). Calc. Tiss. Int. **27**, 153-160.

Miller, A. L. and Schraer, H. (1975). Calc. Tiss. Res. **18**, 311-324.

Montel, G. (1968). Bull. Soc. Chim., Special No., pp. 1693-1700.

Naray-Szabo, S. (1930). Z. Kristallogr. **75**, 387-398.

Neuman, W. F. and Neuman, M. W. (1958). The Chemical Dynamics of Bone Mineral, Chicago: Chicago University Press.

Neuman, W. F. and Mulryan, B. J. (1967). Calc. Tiss. Res. **1**, 94-104.

Posner, A. S., Fabry, C. and Dallemagne, M. J. (1954). Biochim. Biophys. Acta **15**, 304-305.

Posner, A. S. and Perloff, A. (1957). J. Res. Nat. Bur. Std. **A58**, 279-286.

Posner, A. S., Stutman, J. M. and Lippincott, E. R. (1960). Nature (London) **188**, 485.

Posner, A. S., Eanes, E. D., Harper, R. A. and Zipkin, I. (1963). Archs. Oral Biol. **8**, 549-570.

Posner, A. S. and Beebe, R. A. (1975). Seminars in Arthritis and Rheum. **4**, 445-448.

Posner, A. S. and Betts, F. (1975). Acc. Chem. Res. **8**, 273-281.

Posner, A. S. (1969). Physiol. Revs. **49**, 760-792.

Robinson, R. A. and Watson, M. L. (1952). Anat. Record **114**, 383-410.

Roufosse, A. H., Landis, W. J., Sabine, W. K. and Glimcher, M. J. (1979). J. Ultrastruct. Res. **68**, 235-255.

Termine, J. D. and Posner, A. S. (1967). Calc. Tiss. Res. **1**, 8-23.

Wheeler, E. J. and Lewis, D. (1977). Calc. Tiss. Res. **24**, 243-248.

Winand, L. (1965). Physico-chemical study of some apatitic calcium phosphates. "Tooth Enamel," edited by M. V. Stack and R. S. Fearnhead. Bristol, England: John Wright and Sons.

Young, R. A. (1980). Proc. 2nd Congr. P. Compds., 73-88.

CHAPTER 15

STONE DISEASES

I. N. Rabinowitz, Santa Barbara, CA 93102 and
D. June Sutor, University College, London WC1H 0AJ

The somewhat vernacular term "stone disease" refers to those conditions where a concretion of mineral and/or organic material is formed, most commonly in the kidney, urinary bladder, ureter, or gall bladder. As the chemical composition of the concretions observed in gout (called tophi) are similar to one of the classes of compounds found in kidney and urinary bladder stones, they may perhaps be conveniently reviewed in this chapter rather than in the chapter on biological calcification and bone structure.

The minerals and organic compounds most often found in stone formations are (1) calcium phosphate minerals such as apatites and brushite, (2) magnesium phosphates, such as struvite, (3) calcium oxalates, either the monohydrate form (whewellite) or the dihydrate (weddelite). Considerable doubt remains about reports of a naturally occurring trihydrate form. (4) Uric acid and urate salts, (5) cystine, (7) protein or protein degradation products, (8) miscellaneous compounds, including drugs and drug metabolites, silicas and silicates, et al.

The chief aim of this chapter is to chronicle the contributions of North American crystallographers to the field of stone disease research. Those contributions have of course not been of equal significance in each of the subdivisions of this research. Avoiding a parochial "North American" view of structural research in stone disease is best accomplished by a thorough reading of this chapter's bibliography. It is not the purpose of this chapter to provide an exhaustive review of the subject.

The modern era of structural studies of stone, as well as of bone and tooth begins with W. F. DeJong (1926) who described the use of the Debije-Scherrer powder diffraction method (Debije and Scherrer, 1915, 1916) as a means toward determining the chemical stoichiometry and physical structure of bone material. Samples studied were both "fresh" bone and museum fossil fragments. Both the choice of samples and the judicious discussion of results by DeJong continue to serve as an excellent illustration of the continuing problems of biological mineral structure determination. These problems include the appropriateness or representative nature of a sample, be it bone, tooth, stone or other crystalline material in situ, and the continuing variations in reported structure determinations for seemingly identical materials.

As found by DeJong, apatite is the major bone structure. It is also among the most common constituents of urethral, renal and bladder stones, as well as being frequently present in amorphous or crystalline form in the urine. The subsequent unraveling of the subtleties of "apatitic" bone structures, with particular attention to what we understand today of solid solution structure, has been reviewed by Brown and Chow (1976) and in this volume by Posner (Chapter 14, p. 396.)

The problems of proper sampling in crystal structure analyses of stones and of crystals found free in urine, tissue or exudate have marked the history of crystallographic stone research in much the same manner as it has in bone studies, and most particularly in the description of structure as a function of domains in the stone.

After DeJong, Saupe (1931) was the first to use the powder diffraction method for the structure analysis of urinary calculi. Before and after DeJong and Saupe, optical and polarized light microscopy studies were made of urinary crystals and kidney and bladder stone structures. Thus Keyser (1923), as part of his M.S. thesis in pathology at the University of Minnesota, observed calcium oxalate crystals in the urine of dogs and rabbits fed oxalate and oxamide. He described both calcium oxalate monohydrate and dihydrate habits but did not attempt a correlation of structure with habit. Hammersten (1929) was the first to achieve this correlation, with synthetically grown crystals, using analytical chemistry procedures.

The early history of structural stone studies has been well reviewed by Prien and Frondel (1947). Professor Frondel, then a research associate at Harvard University, collaborated with Dr. Prien of the Newton-Wellesley Hospital, Boston, in the first comprehensive crystallographic study of stones performed in the U.S.A. Prien and Frondel assigned worldwide priority in the field to A. T. Jensen of Sweden (Jensen and Thygesen, 1938). Prien and Frondel (1941) over Jensen's description of and of Winchell in mineralogy and optical mineralogy. The earliest investigators of biological stone and crystal, in Britain and Germany (cf: Keyser, 1923) also referred to the extant mineralogical handbooks and literature. Hence the reasonable adoption of mineralogical terminology for biological samples.

Professor Pepinsky, a research associate at the University of Chicago, shared disagreement (Phemister, Aronsohn and Pepinsky, 1939; Pepinsky, 1941) with Prien and Frondel (1941) over Jensen's description of the apatite in bladder and kidney stones as being amorphous or colloidal in nature. Both investigators reported distinct X-ray diffraction patterns. Pepinsky worked with D. B. Phemister, M.D. of the Dept. of Surgery, University of Chicago, who had a long standing interest in bone diseases. Collaboration of Ph.D. and M.D.s in this country in bone and stone research has been the rule.

The pioneering work of Prien and Frondel was in part instigated by a contemporary major advance in medicine, the introduction of sulfonamide drugs. Use of these drugs resulted in some cases in the appearance of sulfonamide crystals in the urine as well as inclusions in kidney stones. "The use of sulfonamide drugs has awakened interest in the study of crystals in the urinary sediments." (Prien and Frondel, 1941). After this work, major crystallographic stone studies (optical, and/or diffraction) were performed in this country by Herring (1962) and Elliott (1973), in England by Lonsdale (1968a) and Sutor (Sutor and Scheidt, 1968; Sutor 1968; Sutor and Wooley, 1972), in Sweden by Lagergren (1956) and in Germany by Schneider et al., (1973). This last comparative study in Germany reaches the conclusion that crystallographic techniques represent the most cost effective analytical method for the clinical reporting of stone composition.

Just as DeJong utilized some fossil sources for bone samples, Lonsdale (1968a) in a fascinating review described the ever-changing worldwide patterns in stone composition making use, inter alia, of museum and mummy sources. Thanks mainly to sulfonamide and antibiotic drugs, phosphatic stones which can be one of the sequela of chronic kidney infection no longer constitute the major stone problem in North America, where instead the more important problem today is calcium oxalate stone (Prien and Frondel, 1941; Herring, 1962; Elliot, 1973; Lonsdale, 1968a; Schneider et al., 1973; Prien, 1975; Prien and Prien, 1968). Pure calcium oxlate stones constitute the largest single class of stones in the U.S.A. today, and mixed apatitic and oxalate stones occur in comparable numbers (Prien and Prien, 1968).

The formidable modern challenge of the understanding of calcium oxalate stone formation (Prien, 1975; Prien and Prien, 1968; Hautmann, Lehmann and Komor, 1980; Butz and Kohlbecker, 1980) should not however detract attention from the complexities of the less common urate (Howell, Eanes and Seegmiller, 1963; Sutor, 1976) and cholesterol stones (Redinger and Small, 1972).

Redinger and Small in North America (Redinger and Small, 1972) and Small, collaborating with Bourges and Dervichian in France (Bourges, Small and Dervichian, 1967), have made use of low angle X-ray diffraction techniques in the elucidation of the several paracrystalline phases of cholesterol in complex environments, and their possible role in growth of cholesterol stones. A very well known figure in American crystallography and a past president of the American Crystallogrpahic Association, Isidor Fankuchen worked with J. D. Bernal and Dorothy Crowfoot in a survey of a variety of sterol structures (Bernal, Crowfoot and Fankuchen, 1940) while in the process of earning his second Ph.D. in physics at Cambridge University. The crystal structure of cholesteryl iodide was subsequently solved by Carlisle and Crowfoot (1945) in one of the first triumphs of the heavy atom technique. The crystal structure of cholesterol monohydrate itself was solved recently by Professor B. M. Craven of the University of Pittsburgh (Craven, 1976). It is the monohydrate form which appears in gallstones and atherosclerotic lesions.

C. W. Vermeulen, M.D. of the University of Chicago Medical School and associates, presented some of the most compelling experimental evidence for a crystal growth theory of the origin of kidney stones. The theory, broadly, required first the growth of crystal nuclei to macroscopic size, followed by continued growth to a pure or mixed compound stone. Vermeulen and associates also emphasized the importance of a compound's crystal habit in growth mechanisms, most particularly with regard to calcium oxalate stones (Borden and Vermeulen, 1966; Vermeulen, Lyon, Ellis and Borden, 1967).

After the chemical and optical investigations of Hammersten (1929) on synthetic calcium oxalate crystal habits, there elapsed a period of 41 years before Catalina and Cifuentes Delatte (1970) identified the mono-hydrate and dihydrate oxalate crystal habits in urine by electron diffraction. This was further confirmed by Elliott and Rabinowitz (1977) using X-ray powder diffraction methods. It should also be mentioned that calcium oxalate crystals occurring elsewhere in plant and animal sources apparently may not have precisely identical structures (Arnott, Pautard and Steinfink, 1965; Ohnishi, Takahashi, Sonobe and Hayashi, 1968; Teigler and Arnott, 1972).

The crystal structure of calcium oxalate dihydrate (weddelite) was solved by Sterling (1965) at the University of California, Davis, using a crystal chipped from a recovered kidney stone and the structure of calcium oxalate monohydrate (whewellite) was solved by Cocco (1961) and Cocco and Sabelli (1962). The whewellite crystal was from an Alsatian mineral sample at the University of Florence Mineralogical Museum. Weddelite crystals have "zeolitic" water channels which may have some relation to the well known cation adsorption exchange and varying water content properties displayed by calcium oxalate crystals (Kolthoff and Sandell, 1937; Sandell and Kolthoff, 1932). The monohydrate structure is not yet refined enough to say whether it also has zeolite characteristics.

With the need to understand the serious medical problem of stone formation, structural studies of late have moved to the examination of crystal surfaces, to small (approximately 200Å^2) nucleation sites of stones, and nucleation areas in tissue and fluids. Structural techniques used have included electron microprobe (Chambers, Hodgkinson and Hornung, 1972), X-ray microdiffraction (Lagergren, 1956), and Auger spectroscopy (Rabinowitz and Elliott, 1976). Valuable symposium proceedings which treat of recent structural studies pertinent to mechanisms of stone nucleation and growth depending on whether one posits a "free" or "fixed" particle genesis (Finlayson and Reid, 1978) are references (Finlayson, Hench and Smith, 1972; Hodgkinson and Nordin, 1969; Cifuentes Delatte, Rapado and Hodgkinson, 1973; Finlayson and Thomas, 1976; Fleisch, Robertson, Smith and Vahlensieck, 1976).

In the same volume of the Comptes-Rendus in which Hammersten (1929) published, one will find reports by K. Linderstrøm-Lang, a leader in the modern development of protein chemistry and structural principles. It should be recalled that in 1929 there were very few subscribers to the notion of macromolecular structure. The suggestion that a protein surface could serve as the nucleation or growth site for stones was put forward (Boyce, 1968) as was the suggestion that one crystal surface could serve as a site for epitaxial growth of a second different crystal (Londsdale, 1968b; Modlin, 1967). Modlin acknowledged earlier suggestions of Neuman and Neuman (1958) at the University of Rochester, that epitaxial mechanisms might be involved in the growth of bone apatite, thus illustrating further the cross fertilization of ideas in these two areas of structural investigation.

In the older literature (cf: Keyser 1923) extending back into the late nineteenth century, the possible role of organic "colloids" and intact tissue slices in stone formation was studied. This particular theory of stone nucleation and growth therefore bares considerable science history and some of the most stimulating new work to arise from the parallel development of protein and crystal surface chemistry has been the suggestion that crystals may act as initiators of certain disease conditions such as gout, silicosis and arthritis (Kozin and McCarty, 1976; Mandel, 1976; Brown and Gregory, 1976). This is distinct from their role in disease arising from crystal growth into a "stone". Craven (1976) has similarly pointed out that cholesterol and hydroxyapatite could support epitaxial growth of one upon the other in the growth of an atherosclerotic lesion. Professor Chandross of the University of North Carolina has continued work on the solution of the crystal structures of cholesterol derivatives (Chandross and Bordner, 1978) since crystallization modes of these derivatives may help to explain phenomena such as growth of atherosclerotic lesions and gallstones.

References

Arnott, H. J., Pautard, F. G. E. and Steinfink, H. (1965). Nature **208**, 1197-1198.

Bernal, J. D., Crowfoot, D. and Fankuchen, I. (1940). Phil. Trans. **A 239**, 135-182.

Borden, T. A. and Vermeulen, C. W. (1966). Invest. Urol. **4**, 125-132.

Boyce, W. H. (1968). Amer. J. Med. **45**, 673-683.

Bourges, M., Small, D. M. and Dervichian, D. G. (1967). Biochim. Biophys. Acta. **114**, 189-201.

Brown, W. E. and Chow, L. C. (1976). Ann. Rev. Mat. Sci. **6**, 213-236, Ed. Hugginns, R. A., Bube, R. H. and Roberts, R. W., Palo Alto, Ann. Reviews Inc.

Brown, W. E. and Gregory, T. M. (1976). Arch. Rheum. **19**, 446-463.

Butz, M. and Kohlbecker, G. (1980). Urol. Int. **35**, 303-308.

Carlisle, C. H. and Crowfoot, D. (1945). Proc. Roy. Soc. **A 184**, 64-82.

Cataline, F. and Cifuentes Delatte, L. (1970). Science **169**, 183-184.

Chambers, A., Hodgkinson, A. and Hornung, G. (1972). Invest. Urol. **9**, 376-384.

Chandross, R. J. and Bordner, J. (1978). Acta Cryst. **B 34**, 2872-2875.

Cifuentes Delatte, L., Rapado, A. and Hodgkinson, A., Ed., Urinary Calculi, Karger, Basel 1973.

Cocco, G. (1961). Rend. Acc. Naz. dei Lincei Ser. **8**, **31**, 292-298.

Cocco, G. and Sabelli, C. (1962). Atti Soc. Toscana Di Scienze Naturali, Ser A, **69**, 289-298.

Craven, B. M. (1976). Nature **260**, 727-729.

Debije, P. and Scherrer, P. (1915, 1916). Nachr. Kgl. Ges. Wiss. Gottengen.

DeJong, W. F. (1926). Rec. Trav. Chim. Pays-Bas. **45**, 445-448.

Elliot, J. S. (1973). J. Urol. **109**, 82-83.

Elliot, J. S. and Rabinowitz, I. N. (1977). Urol. Digest, Dec. 13-32.

Finlayson, B., Hench, L. L. and Smith L. H., Ed.. Urolithiasis, Physical Aspects,Nat. Acad. Sci. U.S.A., Wash., D.C. 1972.

Colloquim on Renal Lithiasis, Ed. Finlayson, B. and Thomas, Lithiasis. Univ. Fla. Press, Gainesville, 1976.

Finlayson, B. and Thomas, W. C., Jr., Ed., Colloquium on Renal

Fleisch, H., Robertson, W. G., Smith, L. H. and Vahlensieck, W., Ed., Urolithiasis Research. Plenum Press, New York 1976.

Frondel, C. and Prien, E. L. (1942). Science **95**, 431-433.

Hammersten, G. (1929). Comptes-Rendus Trav. Lab. Carlsberg **17**, (11).

Hautmann, R., Lehmann, A., and Komor, S. (1980). J. Urol. **123**, 317-319.

Herring, L. C. (1962). J. Urol. **88**, 545-562.

Hodgkinson, A. and Nordin, B. E. C., Ed., Proc. Renal Stone Research Symposium. J. & A. Churchill Ltd. London 1969.

Howell, R. R., Eanes, E. D. and Seegmiller, J. E. (1963). Arth. Rheum. **6**, 97-103.

Jensen, A. T. and Thygesen, J. E. (1938). Z. Urol. **32**, 659-666.

Keyser, L. (1923). Arch. Surg. **6**, 525-553.

Kolthoff, I. M. and Sandell, E. B. (1937). J. Amer. Chem. Soc. **59**, 1643-1648.

Kozin, F. and McCarty, D. J. (1976). Arch. Rheum. **19**, 433-438.

Lagergren, C. (1956). Acta Rad. Suppl. **133**.

Lonsdale, K. (1968a). Nature **217**, 56-58.

Lonsdale, K. (1968b). Science **159**, 1199-1207.

Mandel, N. S. (1976). Arth. Rheum. **19**, 439-445.

Modlin, M. (1967). Ann. Roy. Coll. Surg. **40**, 155-178.

Neuman, W. F. and Neuman, N. W. (1958). The Chemical Dynamics of Bone Mineral, Univ. Chicago Press, Chicago.

Ohnishi, E., Takahashi, S. Y., Sonabe, H. and Hayashi, T. (1968). Science **160**, 783-784.

Pepinsky, R. (1941). Phys. Rev. **60**, 168.

Phemister, D. B., Aronsohn, H. G. and Pepinsky, R. (1939). Ann. Surg. **109**, 161-186.

Prien, E. L. (1975). Ann. Rev. Med. **26**, 173-179.

Prien, E. L. and Frondel, C. (1941). J. Urol. **46**, 748-758.

Prien, E. L. and Frondel, C. (1947). J. Urol. **57**, 949-991.

Prien, E. L. and Prien, E. L., Jr. (1968). Amer. J. Med. **45**, 654-672.

Rabinowitz, I. N. and Elliot, J. S. (1976). Coll. Renal Lithiasis, Ed. Finlayson, B. and Thomas, W. C., Jr., Univ. Fla. Press.

Redinger, R. N. and Small, D. M. (1972). Arch. Intern. Med. **130**, 618-630.

Sandell, E. B. and Kolthoff, I. M. (1932). J. Phys. Chem. **37**, 153-170.

Saupe, E. (1931). Fort. Gebiete Röntgenstrahlen **44**, 204-211.

Schneider, H. J., Berenyi, M., Hesse, A. and Tscharnke, J. (1973). Intl. Urol. Nephrol. **5**, 9-17.

Sterling, C. (1965). Acta Cryst. **18**, 917-921.

Sutor, D. J. (1968). Brit. J. Urol. **40**, 29-32.

Sutor, D. J. (1976). Crystallographic analysis of urinary calculi. In: Scientific Foundations of Urology. Williams, D. I. and Chisholm, G. D. (eds.), Vol. I, 244-254, Heinemann, London.

Sutor, D. J. and Scheidt, S. (1968). Brit. J. Urol. **40**, 22-28.

Sutor, D. J. and Wooley, S. E. (1972). Brit. J. Urol. **44**, 532-536.

Teigler, D. J. and Arnott, H. J. (1972). Nature **235**, 166-167.

Vermeulen, C. W., Lyon, E. S., Ellis, J. E. and Borden, T. A. (1967). J. Urol. **07**, 573-582.

CHAPTER 16

EARLY STUDIES OF AMINO ACIDS AND PEPTIDES

Richard E. Marsh
Arthur Amos Noyes Laboratory of Chemical Physics
California Institute of Technology, Pasadena, California 91125

X-Ray diffraction studies of amino acids and peptides date back to 50 years ago, when J. D. Bernal (1931) used oscillation photographs to determine the unit-cell dimensions and space groups of 15 such compounds. F. V. Lenel (1931) used rotation and Weissenberg photographs to confirm Bernal's values for α-glycylglycine and to characterize alanylglycylglycine and two modifications of glycylglycylglycine. In the same year J. Hengstenberg and Lenel (1931) made spectrometric measurements of the intensities of *33hk*0 and 0*kl* reflections from crystals of α-glycine and proposed a detailed structure, which turned out to be incorrect. Five years later, A. I. Kitaigorodskii (1936) proposed a different structure for glycine, based on the data of Hengstenberg and Lenel supplemented by a few visually-estimated intensities from rotation and oscillation photographs; this structure was also incorrect. The correct structure of α-glycine was finally derived by G. Albrecht and R. B. Corey (1939).

Essentially all the early successes in crystal-structure studies of amino acids and peptides resulted from a program initiated by L. Pauling at the California Institute of Technology, and implemented by R. B. Corey, E. W. Hughes, and co-workers. The first such structure reported was that of diketopiperazine, the simplest cyclic dipeptide (Corey, 1938). The structure was deduced by trial-and-error methods based on the intensities of 60 *hk*0 reflections recorded on Weissenberg photographs. It is interesting to note that the first model tested was based on a "puckered" six-membered ring as had been found from X-ray and electron diffraction studies of cyclohexane, dioxane, and related compounds; only after this model failed to provide satisfactory agreement between *F*(obs) and *F*(cal) values was a planar model considered. When an orientation of this planar molecule was found which gave relatively good agreement, an electron density projection $\rho(x,y)$ was calculated by hand, from which the final values of the x and y coordinates were obtained. The z coordinates were determined by similar trial-and-error methods followed by a Fourier summation based on 25 *h*0*l* reflections. The final coordinates obtained in this way agree, on the average, within 0.02 Å for x and y and within 0.06 Å for z with values later obtained from least-squares refinements based on 1144 three-dimensional data Degeilh and Marsh, 1959).

The structure determination of glycine (Albrecht and Corey, 1939) was based on more extensive data: zonal reflections about all three axes (72 0*kl*'s., 32*h*0*l*'s, 68 *hk*0's) plus an additional 124 general reflections. obtained from a combination of oscillation and Weissenberg photographs. The structure was derived from various Patterson summations, including Harker sections and lines, and was refined by a juggling technique aided by structure-factor maps of many of the zonal planes. Apparently no electron density maps were calculated.

Other pre-war structure determinations included D,L-alanine (Levy and Corey, 1941) and the dipeptide β-glycylglycine (Hughes and Moore, 1942). The structure of D,L-alanine presented extra difficulties because it contains no center of symmetry (space group, $Pna2_1$); as a lagniappe, the authors derived coordinates for the hydrogen atoms and included them, as well as an extinction coefficient, in the structure factor calculations. Some of the Fourier summations for β-glycylglycine were carried out on a punched-card tabulator.

The end of the war brought rapid development of punched-card techniques, and soon three-dimensional Fourier syntheses could be calculated with relative ease (perhaps 20-30 man-hours for a typical summation). Patterson techniques such as superposition maps allowed more formidable structures to be tackled, and by 1955 three-dimensional analyses had been completed for the amino acids threonine (Shoemaker, Donohue, Schomaker and Corey, 1950) serine (Shoemaker, Barieau, Donohue and Lu, 1953), hydroxyproline (Donohue and Trueblood, 1952), and N-acetylglycine (Carpenter and Donohue, 1950), and for the peptides glycylasparagine (Pasternak, Katz and Corey, 1954) and N, N-diglycylcystine (Yakel and Hughes, 1954). All of these early three-dimensional analyses were carried out at Caltech.

By this time, the basic structural features of a polypeptide chain had become evident: the planarity of the peptide linkage, the strong tendency for the formation of N-H. . .O hydrogen bonds, and the approximate values of bond lengths and angles. Of these three features, the concept of a *trans*-planar peptide linkage was undoubtedly the most important key to Pauling's derivation of the alpha-helix as the basic onformation of the polypeptide chain in many proteins (see, for example, Pauling and Corey, 1952; Corey and Pauling, 1952).

The decade 1955-1965 saw three-dimensional studies for five additional amino acids, two dipeptides, three tripeptides, and the cyclic hexapeptide cyclo (glycylglycylglycylglycylglycylglycyl); a summary of the work is given by Marsh and Donohue (1967). Of special significance was the study of cyclo(glyglyglyglyglygly) by I. L. Karle and J. Karle (1963). The triclinic unit cell contains eight molecules, but there is a prominent subcell containing two molecules in space group $P1$ The sub-structure was solved by direct methods, using symbolic signs for four reflections; this was only the second application of the "symbolic addition" procedure (originated by W. H. Zachariasen (1952), which, used in conjunction with the probability methods developed by J. Karle and H. Hauptman, released a flood of structural studies of organic and biologic molecules that had heretofore been too complicated for diffraction analysis.

References

Albrecht, G. and Corey, R. B. (1939). J. Am. Chem. Soc. **61**, 1087-1103.
Bernal, J. D. (1931). Z. Kristallogr. **78**, 363-368.
Carpenter, G. B. and Donohue, J. (1950). J. Am. Chem. Soc. **72**, 2315-2328.
Corey, R. B. (1938). J. Am. Chem. Soc. **60**, 1598-1604.
Corey, R. B. and Pauling, L. (1952). Proc. Roy. Soc. (London) **B141**, 10-20.
Degeilh, R. and Marsh, R. E. (1959). Acta. Cryst. **12**, 1007-1014.
Donohue, J. and Trueblood, K. N. (1952). Acta Cryst. **5**, 414-431.
Hengstenberg, J. and Lenel, F. V. (1931). Z. Kristallogr. **77**, 424-436.
Hughes, E. W. and Moore, W. J. (1942). J. Am. Chem. Soc. **64**, 2236.
Karle, I. L. and Karle, J. (1963). Acta Cryst. **16**, 969-975.
Kitaigorodskii, A. I. (1936). Acta Physiochim. URSS **5**, 749-755.
Lenel, F. V. (1931). Z. Kristallogr. **81**, 224-229.
Levy, H. A. and Corey, R. B. (1941). J. Am. Chem. Soc. **63**, 2095-2108.
Marsh, R. E. and Donohue, J. (1967). Adv. Prot. Chem. **22**, 235-256.
Pasternak, R. A., Katz, L. and Corey, R. B. (1954). Acta Cryst. **7**, 225-236.
Pauling, L. and Corey, R. B. (1952). Proc. Roy. Soc. (London) **B141**, 21-33.
Shoemaker, D. P., Barieau, R. E., Donohue, J. and Lu, C. S. (1953). Acta Cryst. **6**, 241-256.
Shoemaker, D. P., Donohue, J., Schomaker, V. and Corey, R. B. (1950). J. Am. Chem. Soc. **72**, 2328-2349.
Yakel, H. L. and Hughes, E. W. (1954). Acta Cryst. **7**, 291-297.
Zachariasen, W. H. (1952). Acta Cryst. **5**, 68-73.

CHAPTER 17

SOME ASPECTS OF CRYSTALLOGRAPHY IN CANCER RESEARCH IN NORTH AMERICA

Jenny P. Glusker
The Institute for Cancer Research, The Fox Chase Cancer Center
Philadelphia, PA 19111

The observation that chimney sweeps were particularly prone to scrotal cancer was first made by Sir Percivall Pott, a noted London surgeon, in 1775 (Pott, 1775). He attributed this occupational hazard to the soot that covered the small boys who had to crawl up inside the chimneys of the time in order to clean them. This conclusion was verified in 1915 by Yamagiwa and Ichikawa when they induced malignant tumors on the ears of rabbits by painting them with undiluted coal tar. These experiments were repeated by Kennaway in 1930 using the polycyclic aromatic hydrocarbon, dibenz[*a,h*] anthracene on mouse skin (Kennaway, 1930). One of the carcinogenic components of coal tar, extracted by Cook (Cook, Hewett and Hieger, 1933), was shown to be benzo[*a*]pyrene, BP (**I**). Thus a direct relationship was established between certain polycyclic aromatic hydrocarbons and cancer. Another potent carcinogen is the methylated compound 7,12-dimethylbenz[*a*]anthracene, DMBA (**II**).

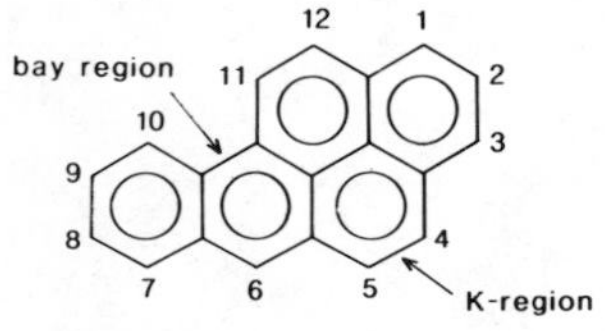

bay region 12 1 11 2 10 3 9 8 4 7 6 5 K-region

bay region 2 CH_3 1 3 11 12 4 10 9 5 8 7 6 K-region CH_3

(I) 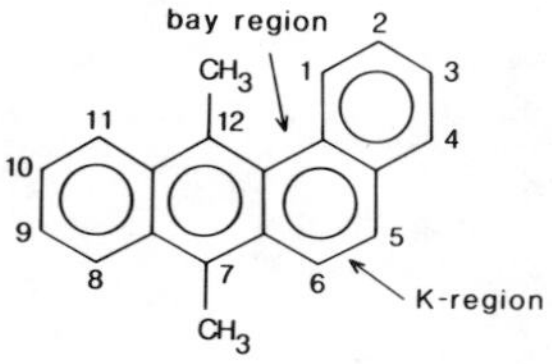 *(II)*

These were the beginning of studies of chemical carcinogenesis. Other epidemiological studies have attributed excessive use of tobacco snuff to nasal cancer, certain aromatic amines such as are used in "aniline" dye factories to bladder cancer, and tobacco smoke to lung cancer.

On contact with the skin some of polycyclic aromatic hydrocarbons may penetrate the body. However, the body has mechanisms for eliminating such foreign substances. Hydroxylating and epoxidizing enzymes, known as microsomal "mixed function oxidases," detoxify polycyclic aromatic hydrocarbons by adding hydroxyl and epoxide groups to them, thereby making the hydrophobic aromatic hydrocarbons more soluble so that they can be excreted. In some cases, however, the result of such enzyme action is a compound that is more cytotoxic and carcinogenic than the parent compound (Gelboin, Wiebel and Kinoshita, 1972), in which case the process is known as "activation." The important activated intermediates of the potent carcinogenic polycyclic aromatic hydrocarbons (**I** and **II**) listed above have been shown to be diol epoxides (shown as formulae **III** and **IV** respectively, below) (Sims and Grover, 1974).

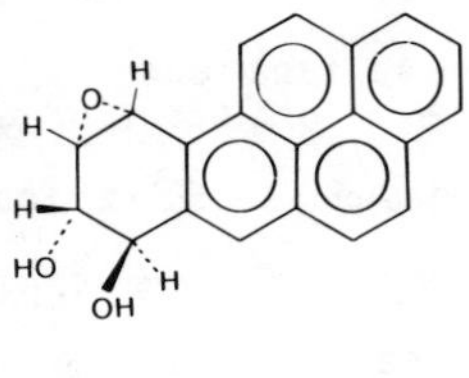

H O H H HO OH H

CH_3 H O H H OH OH H CH_3

(III) 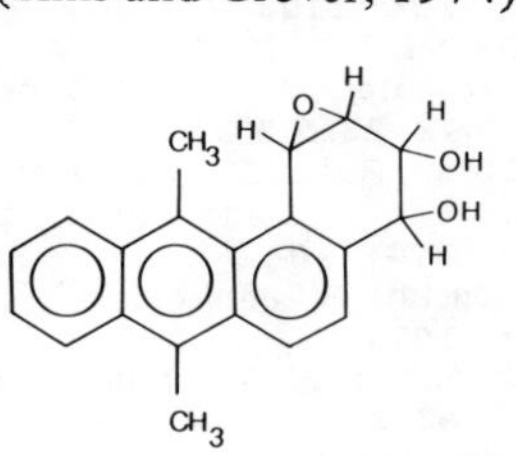 *(IV)*

These are alkylating agents by virtue of the epoxide ring, which can open and interact with another molecule such as a biological macromolecule. Since, once a cell becomes cancerous, all its progeny inherit its carcinogenic capacity, the target for such an alkylating action is believed to be an informational macromolecule such as DNA or the proteins, such as repressors, histones, polymerases, etc. that are associated with the DNA. Thus the interaction of the diol epoxide of a carcinogenic polycyclic aromatic hydrocarbon with DNA or protein is believed to be an important step in the development of a malignancy. In this chapter the role of crystallography in the study of this process and its prevention will be described. Since space is short, only a few selected examples have been chosen for description.

Crystallography and cancer research were interwoven at an early stage. The crystal structures of several polycyclic aromatic hydrocarbons were studied by J. Monteath Robertson at the University of Glasgow in Scotland. His student, John Iball, continued such studies at the University of Dundee in Scotland, but extended them to carcinogenic polycyclic aromatic hydrocarbons. He first studied chrysene and 1:2-cyclopentenophenanthrene (Iball, 1935) because he was interested in the similarities of the formulae of carcinogenic polycyclic aromatic hydrocarbons, steroids, bile acids, and toad poisons and many other biologically active molecules. Then he continued with studies of compounds having more carcinogenicity. John Iball proposed the measure of carcinogenicity of a compound, expressing it in terms of the latent period and the number of tumors per standard dose (Iball, 1939). This is referred to as the "Iball index" and is still in use.

Early Studies in America

The first structure of a carcinogenic polycyclic aromatic hydrocarbon that was studied in America was

7,12-dimethylbenz[*a*]anthracene, determined by Sayre and Friedlander in the 1950s. David Sayre had started to work at the Johnson Foundation in the University of Pennsylvania in Philadelphia and wanted to start a crystal structure analysis program. He considered that a study of carcinogenic molecules would provide a good combination of structural and biological interest. David chose to work on benz[*a*]anthracene (BA) and 7,12-dimethylbenz[*a*]anthracene (DMBA) because these are far apart in the activity spectrum, the former almost inactive and the latter very active as a chemical carcinogen. He was aided in the decision to study these two compounds by Hugh J. Creech, at the Institute for Cancer Research in Philadelphia, who supplied the material. Peter Friedlander, who had worked for J. Monteath Robertson, came to work for David on this problem.

After roughly solving the structures from poor crystals of benz[*a*]anthracene and good crystals of 7,12-dimethylbenz[*a*]anthracene, David wanted to refine the structures. In order to do this he went to IBM in New York (as a customer) and wrote a least squares refinement program for the IBM 701. This was one of the first refinement programs written in the U.S. for an electronic computer (approximately contemporaneous with such programs written for the National Bureau of Standards computer SWAC at UCLA, and for the computer in Manchester, England). The structures of BA and DMBA were, as a result, refined and published (Sayre and Friedlander, 1956, 1960). This programming venture led to a job offer for David at IBM, which he accepted. Surprisingly several thought that this structure reported for DMBA was wrong because it was not planar. Of course it was a correct structure, a fact that should have been expected by anyone who is used to the concept that atoms, even hydrogen atoms, have finite sizes and therefore cannot lie too near to other atoms. If the DMBA molecule were planar, two hydrogen atoms (on C(1) and the methyl group on C(12)) would lie almost on top of each other. As a result of out-of-plane distortion in DMBA, their distance apart is increased to a reasonable value. Refinements of the original data on DMBA and a low temperature redetermination have since been made (Iball, 1964; Zacharias and Glusker, 1981) and have verified the buckled nature of the molecule.

Such a distortion from planarity is not essential for carcinogenicity, although it appears to enhance this property. The carcinogen, benzo[*a*]pyrene (BP), is almost planar (Iball and Young, 1956; Iball, Scrimgeour and Young, 1976), but when an additional methyl group is introduced in the 11-position the molecule becomes much more carcinogenic. Similarly the buckled molecule 5-methyl chrysene is the only monomethylchrysene that is carcinogenic, while chrysene itself is not carcinogenic. The amount of buckling is represented by the maximum values of torsion angles that would be expected to be 0° if the molecule were completely flat. Values for such maximum torsion angles are 2° for BP and 23° for DMBA. To date the structures of many carcinogenic polycyclic aromatic hydrocarbons have been determined. When diol epoxides are formed from such molecules they are more buckled because a more saturated ring is formed.

Carcinogenic polycyclic aromatic hydrocarbons and their metabolites

What are the structural characteristics of a carcinogenic polycyclic aromatic hydrocarbon? First, it seems important that it have a characteristic size - one that approximates those of base pairs in nucleic acids, and certain steroid hormones, such as estradiol. Secondly, most have a phenanthrene-like group, which means that they have a reactive double bond (the K-region) and a "bay-area" opposite it. Buckling due to methyl substitution in the "bay area" apparently enhances carcinogenicity. Originally it was thought that the K-region is the important area for carcinogenesis, and several K-region epoxides were studied. However, the active intermediate in chemical carcinogenesis was shown to be a diol epoxide (formulae **III** for BP, **IV** for DMBA).

There are two diastereomers of each diol epoxide, defined with respect to the oxygen of the epoxide ring and the more distant hydroxyl group. The *anti*-diol epoxide has these on opposite sides of the ring and the compound derived from BP was studied by Neidle and co-workers (Neidle, Subbiah, Cooper and Ribeiro, 1980). A *syn*-monol epoxide (i.e., with only one hydroxyl group) derived from naphthalene has been studied by Glusker, Zacharias and co-workers (Glusker, Zacharias, Whalen, Friedman and Pohl, 1981) and shows that when the epoxide group and opposite hydroxyl group are on the same side of the ring an internal hydrogen bond is formed. Several other epoxides and diols have been the subject of X-ray and nuclear magnetic resonance investigations and have shown that, in hindered positions the hydroxyl group will generally lie in an axial position, while in unhindered positions an equilibrium between axial and equatorial conformers exist in solution (Zacharias, Glusker, Fu and Harvey, 1979).

There are, of course, many other types of chemical carcinogens, but few potent carcinogens have been studied. The carcinogenic component of moldy peanuts and grain is aflatoxin B1, one of the most powerful carcinogens known. It also exerts its action as a result of the intermediate formation of an epoxide. The structure has been determined in Holland (van Soest and Peerdeman, 1970).

Other areas of investigation include azo dyes, aromatic amines, acetylaminofluorene and nitroso compounds. The structure of the pyrrolizidine alkaloids fulvine and heliotrine were studied by Wodak and Sussman (Sussman and Wodak, 1973; Wodak, 1975). These contain two five membered rings fused at a common C - N bond, and a possible mechanism of toxicity of these compounds has been suggested.

The effect of ultraviolet and γ-radiation on nucleic acids is to cause polymerization or rearrangement to cyclobutyl dimers and thymine-thymine adducts. The molecular formulae of several such photoproducts have been confirmed and the configurations established by X-ray structure analyses (Camerman, Nyburg and Weinblum, 1967; Camerman and Camerman, 1968; Karle, Wang and Varghese, 1969; Camerman, Weinblum and Nyburg, 1969; Flippen, Karle and Wang, 1970). Such photodimers are probably responsible for the majority of photobiological effects of irradiation on DNA, perhaps even carcinogenesis.

Interactions of carcinogens and mutagens with DNA

Once the carcinogen has been "activated" it reacts with intracellular macromolecules in the body. As mentioned earlier these macromolecules can be either DNA or the proteins associated with it. Presumably there is a specific location on the "critical" target molecule, such as a promoter sequence on DNA, that is alkylated during the progress of the carcinogenic process. There are almost no studies of carcinogen- protein interactions. The structure of a model, a tripeptide alkylated by a carcinogen, has been determined (Glusker, Carrell, Berman, Gallen and Peck, 1977). This showed that the introduction of a large hydrophobic residue (the carcinogen) greatly altered the peptide conformation.

However, most studies have been done on nucleic acid interactions with polycyclic molecules. Lerman (Lerman, 1961) was the first to propose, from X-ray studies, that one of the modes of interaction of acridines with DNA is by "intercalation." In this process the acridine becomes inserted between the bases of DNA as a result of extension of the phosphodiester backbone of DNA. When the distance between base pairs is extended from 3.4 to 6.8 Å the flat molecule can become completely enveloped within the hydrophobic area of the nucleic acid.

Many such intercalated complexes have been studied, of which the first was the model for the interaction of actinomycin with a nucleic acid (Sobell, Jain, Sakore and Nordman, 1971) from crystal structure data on an actinomycin-guanine complex. Many other complexes have been studied including many from the laboratories of Sobell, Neidle, Berman, Rich and Sundaralingam (Tsai, Jain and Sobell, 1975; Neidle, Achari, Taylor, Berman, Carrell, Glusker and Stallings, 1977; Wang, Nathans, van der Marel, van Boom and Rich, 1978; Shieh, Berman, Dabrow and Neidle, 1980; Westhof and Sundaralingam, 1980). These complexes generally involve self-complementary dinucleoside phosphates that can dimerize by hydrogen bond formation to give a portion (two Watson-Crick types of base pairs) of a nucleic acid. These structures demonstrate the types of conformational changes that the phosphodiester backbone has undergone, the insertion of the flat molecule between bases, and any hydrogen bonding that the inserted molecule can make such as to the phosphate groups in the case of proflavine. Each base pair is found to be twisted in a propeller-like fashion. The detailed structures of a dodecamer with a right-handed helical twist as in normal B-DNA (Wing, Drew, Takano, Broka, Tanaka, Itakura and Dickerson, 1980) and of a hexamer with a left-handed helical twist, representative of a high salt form (Wang, Quigley, Kolpak, Crawford, van Boom, van der Marel and Rich, 1979; Wang, Quigley, Kolpak, van der Marel, van Boom and Rich, 1981) have been determined. These may be used for comparison with the perturbed structures. A complex of daunomycin with a hexanucleotide has been determined and shows some of the distortions caused by intercalation and some of the interactions that can occur in such complexes (Quigley *et al.*, 1981). A complex of ethidium bromide with the macromolecule tRNA (transfer RNA) has been studied in Sundaralingam's laboratory (Liebman, Rubin and Sundaralingam, 1977), but this is not an intercalation complex.

There are three major ways in which a flat polycyclic molecule can interact with DNA. It can intercalate between the bases, it can hydrogen bond (if the appropriate groups are available on the flat molecule) to the phosphate groups and lie outside the helix, or it can hydrogen bond to the bases and lie in a groove of DNA. If the molecule to be inserted is buckled but is hydrophobic, such as an activated derivative a non planar carcinogenic polycyclic aromatic hydrocarbon, the last situation will probably occur when the molecule interacts with DNA. The exposed bases would be a likely target for alkylation, and then the carcinogen could lie in a groove of DNA, with little perturbation of its structure, or between the bases causing considerable distortion of helical structure because of the buckled nature of the activated carcinogen (Glusker, Zacharias and Carrell, 1976). A model for the former is found in the product of alkylation of deoxyadenosine by a chloromethylbenz[*a*]anthracene. This structure shows the base and carcinogen lying nearly perpendicular to each other (Carrell, Glusker, Moschel, Hudgins and Dipple, 1981).

Activating enzymes

Not many studies have been done on the metabolizing enzymes. They are difficult to purify. The cytochrome P-450 enzymes are members of the monooxygenase class of enzymes and catalyze the incorporation of one atom of oxygen (from dioxygen) into a substrate. Crystals of cytochrome P-450 from *Pseudomonas putida* are presently being studied by Thomas Poulos at the University of California at San Diego, La Jolla.

Several porphyrin model compounds have been studied (Tang, Koch, Papaefthymiou, Foner, Frankel, Ibers and Holm, 1976).

Modifiers of Chemical Carcinogenesis

There are two ways in which chemical carcinogenesis can be modified. Addition of certain chemicals may make a carcinogen less potent, either because the activating enzyme is inhibited or because the detoxification pathways are made more likely. For example, 7,8-benzoflavone can, in many cases, inhibit carcinogenesis by DMBA. The structure is non-planar, unlike the isomer, the 5,6-benzoflavone, which is an inducer of the activating enzyme and is flat (Rossi, Cantrell, Farber, Dyott, Carrell and Glusker, 1980). Presumably such flavones interact in some way with the iron porphyrin of cytochrome P-450, inhibiting this activating enzyme. Vitamin A derivatives, studied by Caroline McGillavry in Holland, also have a chemopreventive action.

The other types of chemicals that affect carcinogens are the tumor promoters, which enhance the carcinogenic action of a chemical. Examples include phorbol esters and also, probably, tobacco smoke. Some structures of phorbol derivatives have been determined in Europe.

Antitumor agents

Much more crystallographic work has been done on the structures of antitumor agents than on carcinogens. There are several reasons for this (as well as the safety factor). In the first place the structures of many naturally occurring antitumor agents can be established only by X-ray work. It is also clearer what the target molecule of such antitumor agents is than is the case for carcinogens. For example, many interact with enzymes (such as dihydrofolate reductase) and others (such as daunomycin) interact with nucleic acids. Thus X-ray work gives information on the conformations of the molecules and the nature of the intermolecular interactions they undergo.

The "unknown" chemical formulae of many naturally-occurring antitumor agents have been determined by X-ray crystallographic techniques. One such example is maytansine (Bryan, Gilmore and Haltiwanger, 1973), a powerful anti-leukemic agent which acts as a mitotic spindle inhibitor. This work was done in collaboration with Morris Kupchan, who extracted active components from plants from around the world supplied to him after preliminary screening at NIH. Since the amount of material extracted was extremely small, the use of X-ray techniques was most advantageous. Maytansine was studied as the 3-bromopropyl ether ($C_{37}H_{51}Br$ $Cl\ N_3O_{10}$) and, as a result of the presence of bromine, its absolute configuration could be determined (there are 8 asymmetric carbon atoms in the structure). It was shown to contain a 19-membered ring and an epoxide group. Some other naturally occurring antitumor agents studied in this way were shown to be diterpenoids, and include the unusual triepoxides triptolide and tripdiolide (Gilmore and Bryan, 1973), gnidimacrin and its 20-palmitate (Kupchan, Shizuri, Murae, Sweeney, Haynes, Shen, Barrick, Bryan, van der Helm and Wu, 1976), and jatrophone (Kupchan, Sigel, Matz, Gilmore and Bryan, 1976). All these five compounds are anti-leukemic agents. The conformation of the bisbenzylisoquinoline alkaloid *dl*-tetrandrine, active against carcinosarcomas, also was determined (Gilmore, Bryan and Kupchan, 1976). In this case the formula and absolute configuration were confirmed by the X-ray analysis, but the different reactivities of the two chemically equivalent tertiary nitrogen atoms could be explained in terms of their observed conformations. In a similar manner Dick van der Helm has studied several antitumor agents isolated from marine organisms. For example, the chemical formulae and absolute configurations of two compounds, sinularin and dihydrosinularin have been established by X-ray work (Hossain, van der Helm, Matson and Weinheimer, 1979). These compounds are also diterpenoids and were isolated from soft coral collected from the Great Barrier Reef of Australia. The molecular structure of the anti-leukemic agent, vincristine, isolated from a periwinkle species, was determined by Moncrief and Lipscomb (Moncrief and Lipscomb, 1965, 1966). Van der Helm and his co-workers have also determined the formulae of similar compounds extracted from marine organisms in the Caribbean. These compounds include jeunicin (van der Helm, Enwall, Weinheimer, Karns and Ciereszko, 1976), crassin (Hossain and van der Helm, 1969) and acanthifolicin (Schmitz, Prasad, Gopichand, Hossain, van der Helm and Schmidt, 1981).

One of the most widely used antitumor agents is the drug cyclophosphamide. This contains phosphorus, nitrogen, oxygen and three carbon atoms in a six-membered ring. It has been shown that the more active (+)-enantiomer has the *R* absolute configuration (Karle, Karle, Egan, Zon and Brandt, 1977). Cyclophosphamide itself is inactive as a cytotoxic agent; it needs to be activated by C4-hydroxylation by the liver mixed function oxidase system.

cyclophosphamide (CPA) → 4-hydroxy CPA ⇌ aldophosphamide → phosphoramide mustard + acrolein ($CH_2{=}CH{-}CHO$)

($R = N(CH_2CH_2Cl)_2$)

The nature of the real anticancer alkylating agent is uncertain at this time. It has variously been held to be either 4-hydroxycyclophosphamide or phosphoramide mustard. Cyclophosphamide can be detoxified enzymatically to 4-ketocyclophosphamide or carboxyphosphamide which are inactive metabolites. The most significant result of X-ray structural studies of cyclophosphamide derivatives by Arthur and Norman Camerman and co-workers is that in all the "pre-activated" 4-hydroxy compounds the hydroxyl group is axial to the ring system (Camerman, Smith and Camerman, 1975). Another effective antitumor agent is 1-(2-chloroethyl)-3-(*trans*-4-methycyclohexyl)-1-nitrosourea (MeCCNU). The structure has been determined (Smith, Camerman and Camerman, 1978) and consists of a cyclohexyl ring in the chair conformation with the nitrosourea group twisted almost perpendicular to the ring.

One particularly potent antitumor agent, discovered by Rosenberg, is *cis* diamminoplatinum dichloride, and it is believed to interact with DNA. Much work by Bau and his co-workers on platinum-nucleic acid base complexes (Gellert and Bau, 1975; Wu and Bau, 1979) suggest that the platinum is complexed with N(7) of guanine (the nitrogen in the 5-ring, exposed on the DNA surface) with subsequent cross-links *via* hydrogen bonds between DNA molecules. Other models, such as one involving an exocyclic oxygen atom of a base, have been suggested from X-ray studies (Barton, Szalda, Rabinowitz, Waszczak and Lippard, S. J., 1979). A complex with a polynucleotide has been studied in the laboratory of Dickerson at Cal Tech. Another interesting class of antitumor agents includes daunomycin and adriamycin which are believed to act by intercalating in DNA. Crystals of complex of a hexanucleotide with daunomycin have been studied by Gary Quigley and co-workers at MIT (Quigley et al., 1981). The mode of interaction of the antitumor antibiotic anthramycin with DNA has been deduced by Arora (Arora, 1979) from an X-ray structure determination of the antibiotic, plus model building. It was shown that it is likely that anthramycin binds in the wide groove of DNA.

One of the most extensively studied enzyme-inhibitor systems is that involving the protein dihydrofolate reductase. The enzyme and its complexes with both substrates and inhibitors have been analyzed crystallographically. This enzyme catalyzes the reduction of 7,8-dihydrofolate to 5,6,7,8-tetrahydrofolate which is an essential coenzyme used in the synthesis of thymidylate, inosinate and methionine. Inhibition of this enzyme by antitumor and antibiotic agents such as methotrexate and trimethoprim respectively lead to a cellular deficiency of thymidylate ("thymine-less death"). Crystallographic studies of folic acid (Mastropaolo, Camerman and Camerman, 1980), trimethoprim (Koetzle and Williams, 1976), and several other antifolates (Camerman, Smith and Camerman, 1979; Hunt, Schwalke, Bird and Mallinson, 1980) have been done.

The three-dimensional structure of dihydrofolate reductase is being studied from two bacterial sources – *Escherichia coli* (Matthews, Alden, Bolin, Freer, Hamlin, Xuong, Kraut, Poe, Williams and Hoogsteen, 1977), and *Lactobacillus casei* (Matthews, Alden, Bolin, Filman, Freer, Hamlin, Hol, Kisliuk, Pastore, Plante, Xuong and Kraut, 1978; Matthews, Alden, Freer, Xuong and Kraut, 1979). The structures of these bacterial enzymes have been refined to R $\sim$ 0.15 at 1.8 Å resolution. The structure of the chicken liver enzyme has also been solved by the same group. One of their aims is to obtain basic rules for defining the architecture of dihydrofolate reductase from various organisms. Then if the amino acid sequence of the enzyme from another organism is known, the three-dimensional structure can hopefully be deduced. Both the *E. Coli* and *L. casei* enzymes have been studied as complexes with methotrexate bound at the active site and, in the case of the *L. casei* enzyme, the cofactor, NADPH, was also present. In the *E. coli* enzyme complex methotrexate is bound in a cavity, 15 Å deep, in the enzyme. The pteridine ring of methotrexate lies nearly perpendicular to the aromatic ring of the *p*-amino benzoyl glutamate group. The entire inhibitor molecule is bound by at least 13 amino acid residues to the enzyme by a system of hydrogen bond and hydrophobic interactions. A similar binding is found for the *L. casei* enzyme complex. Matthews and co-workers (Matthews *et al.*, 1978) pointed out, however,

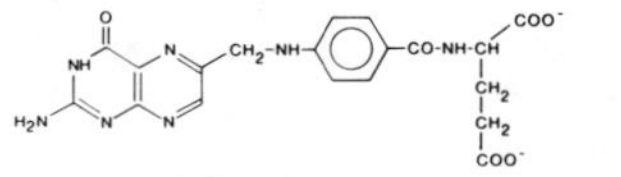

folate

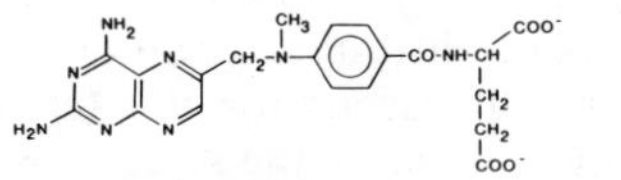

methotrexate

that the pteridine ring could be rotated about the C(6)-C(9) and C(9)-N(10) bonds and still bind well in the crevice. The absolute configuration of tetrahydrofolate has been established (Fontecilla-Camps, Bugg, Temple, Rose, Montgomery and Kisliuk, 1979; Armarego, Waring and Williams, 1980) to be *S* at C(6), but if dihydrofolate is bound in the same orientation as methotrexate, the configuration should be *R*. Thus dihydrofolate and methotrexate must bind to the enzyme with different orientations in accord with the earlier suggestions by Matthews *et al.* A crystallographic study of folate bound to chicken liver dihydrofolate reductase is now underway in the laboratory of David Matthews and the results should reveal the detailed differences in folate and methotrexate binding.

Acknowledgment

Author supported by grants BC-242 from the American Cancer Society, CA-10925, CA-06927, and RR-05539 from the National Institutes of Health, U.S. Public Health Service, and by an appropriation from the Commonwealth of Pennsylvania.

References

Armarego, W. L. F., Waring, P. and Williams, J. W. (1980). J. Chem. Soc. Chem. Commun.; p. 334.

Arora, S. K. (1979). Acta Cryst. **B35**, 2945.

Barton, J. K., Szalda, D. J., Rabinowitz, H. N., Waszczak, J. V. and Lippard, S. J. (1979). J. Am. Chem. Soc. **101**, 1434.

Bryan, R. F., Gilmore, C. J. and Haltiwanger, R. C. (1973). J. Chem. Soc. Perkin Trans. **2**, p. 897.

Camerman, N. and Camerman, A. (1968). Science **160**, 1451.

Camerman, N. and Camerman, A. (1973). J. Am. Chem. Soc. **95**, 5083.

Camerman, N., Nyburg, S. C. and Weinblum, D. (1967). Tetrahedron Lett p. 4127.

Camerman, A., Smith, H. W. and Camerman, N. (1975). Biochem. Biophys. Res. Commun. **65**, 828.

Camerman, A., Smith, H. W. and Camerman, N. (1979). Acta Cryst. **B35**, 2113.

Camerman, N., Weinblum, D. and Nyburg, S. C. (1969). J. Am. Chem. Soc. **91**, 982.

Carrell, H. L., Glusker, J. P., Moschel, R., Hudgins, W. R. and Dipple, A. (1981). Cancer Research **41**, 2230.

Cook, J., Hewett, C. L. and Hieger, I. (1933). J. Chem. Soc., p. 396.

Flippen, J. L., Karle, I. L. and Wang, S. Y. (1970). Science **169**, 1084.

Fontecilla-Camps, J. C., Bugg, C. E., Temple, C. Jr., Rose, J. D., Montgomery, J. A. and Kisliuk, R. L. (1979). J. Am. Chem. Soc. **101**, 6114.

Gelboin, H. V., Wiebel, F. J. and Kinoshita, N. (1972). In Biological Hydroxylation Mechanisms (Edited by G. S. Boyd and R. M. S. Smellie. p. 103. New York: Academic Press.

Gellert, R. W. and Bau, R. (1975). J. Am. Chem. Soc. **97**, 7379.

Gilmore, C. J. and Bryan, R. F. (1973). J. Chem. Soc. Perkin Trans. **2**, p. 816.

Gilmore, C. J., Bryan, R. F. and Kupchan, S. M. (1976). J. Am. Chem. Soc. **98**, 1947.

Glusker, J. P., Carrell, H. L., Berman, H. M., Gallen, B. and Peck, R. M. (1977). J. Am. Chem. Soc. **99**, 595.

Glusker, J. P., Zacharias, D. E. and Carrell, H. L. (1976). Cancer Research **36**, 2428.

Glusker, J. P., Zacharias, D. E., Whalen, D. L., Friedman, S. and Pohl, T. M. (1982). Science **215**, 695.

Hossain, M. B. and van der Helm, D. (1969). Rec. Trav. Chim. Pays-Bas, **88**, 1413.

Hossain, M. B., van der Helm, D., Matson, J. A. and Weinheimer, A. J. (1979). Acta Cryst. **B35**, 660.

Hunt, W. E., Schwalke, C. H., Bird, K. and Mallinson, P. D. (1980). Biochem. J., **187**, 533.

Iball, J. (1935). Z. Kristallogr. **92**, 293.

Iball, J. (1939). Amer. J. Cancer (London) **35**, 188.

Iball, J. (1964). Nature (London) **201**, 916.

Iball, J., Scrimgeour, S. N. and Young, D. W. (1976). Acta Cryst. **B32**, 328.

Iball, J. and Young, D. W. (1956). Nature (London) **177**, 985.

Karle, I. L., Karle, J. M., Egan, W., Zon, G. and Brandt, J. A. (1977). J. Am. Chem. Soc. **99**, 4803.

Karle, I. L., Wang, S. Y. and Varghese, A. J. (1969). Science **164**, 183.

Kennaway, E. L. (1930). Biochem. J. **24**, 497.

Koetzle, T. F. and Williams, G. J. B. (1976). J. Am. Chem. Soc. **98**, 2074.

Kupchan, S. M., Shizuri, Y., Murae, T., Sweeney, J. G., Haynes, H. R., Shen, M-S., Barrick, J. C., Bryan, R. F., van der Helm, D. and Wu, K. K. (1976). J. Am. Chem. Soc. **98**, 5719.

Kupchan, S. M., Sigel, C. W., Matz, M. J., Gilmore, C. J. and Bryan, R. F. (1976). J. Am. Chem. Soc. **98**, 2295.

Lerman, L. S. (1961). J. Mol. Biol. **3**, 18.

Liebman, M. N., Rubin, J. and Sundaralingam, M. (1977). Proc. Natl. Acad. Sci. U.S.A. **74**, 4821.

Mastropaolo, D., Camerman, A. and Camerman, N. (1980). Science **210**, 334.

Matthews, D. A., Alden, R. A., Bolin, J. T., Filman, D. J., Freer, S. T., Hamlin, R., Hol, W. G. J., Kisliuk, R. L., Pastore, E. J., Plante, L. T., Xuong, N-h. and Kraut, J. (1978). J. Biol. Chem. **253**, 6946.

Matthews, D. A., Alden, R. A., Bolin, J. T., Freer, S. T., Hamlin, R., Xuong, N-h., Kraut, J., Poe, M. Williams, M. and Hoogsteen, K. (1977) . Science **197**, 452.

Matthews, D. A., Alden, R. A., Freer, S. T., Xuong, N-h. and Kraut, J. (1979). J. Biol. Chem. **254**, 4144.

Moncrief, J. W. and Lipscomb, W. N. (1965). J. Am. Chem. Soc. **87**, 4963.

Moncrief, J. W. and Lipscomb, W. N. (1966). Acta Cryst. **21**, 322.

Neidle, S., Achari, A., Taylor, G. L., Berman, H. M., Carrell, H. L., Glusker, J. P. and Stallings, W. C. (1977). Nature (London) **269**, 304.

Neidle, S., Subbiah, A., Cooper, C. S. and Ribeiro, O. (1980). Carcinogenesis **1**, 249.

Pott, P. (1775). "Chirurgical Observations Relative to the Cataract, the Polypus of the Nose, the Cancer of the Scrotum, the Different Kinds of Ruptures, and the Mortification of the Toes and Feet." p. 63. London: L. Harves, W. Clark and R. Collins.

Quigley, G. J., Wang, A. H.-J., Ughetto, G., van der Marel, G., van Boom, J. H. and Rich, A. (1980). Proc. Natl. Acad. Sci. U.S.A. **77**, 7204.

Rossi, M., Cantrell, J. S., Farber, A. J., Dyott, T., Carrell, H. L. and Glusker, J. P. (1980). Cancer Research **40**, 2774.

Sayre, D. and Friedlander, P. H. (1956). Nature (London) **178**, 999.

Sayre, D. and Friedlander, P. H. (1960). Nature (London) **187**, 139.

Schmitz, F. J., Prasad, R. S., Gopichand, Y., Hossain, M. B., van der Helm, D. and Schmidt, P. (1981). J. Am. Chem. Soc. **103**, 2467.

Shieh, H-S., Berman, H. M., Dabrow, M. and Neidle, S. (1980). Nucl. Acids Res. **8**, 85.

Sims, P. and Grover, P. L. (1974). Nature (London) **252**, 326.

Smith, H. W., Camerman, A. and Camerman, N. (1978). J. Med. Chem. **21**, 468.

Sobell, H. M., Jain, S. C., Sakore, T. D. and Nordman, C. E. (1971). Nature New Biol. (London) **231**, 200.

Sussman, J. L. and Wodak, S. J. (1973). Acta Cryst. **B29**, 2918.

Tang, S. C., Koch, S., Papaefthymiou, G. C., Foner, S., Frankel, R. B., Ibers, J. A. and Holm, R. H. (1976). J. Am. Chem. Soc. **98**, 2414.

Tsai, C. C., Jain, S. C. and Sobell, H. M. (1975). Proc. Natl. Acad. Sci. U.S.A. **72**, 628.

van der Helm, D., Enwall, E. L., Weinheimer, A. J., Karns, T. K. B. and Ciereszko, L. S. (1976). Acta Cryst. **B32**, 1558.

van Soest, T. C. and Peerdeman, A. F. (1970). Acta Cryst. **B26**, 1940, 1947 and 1956.

Wang, A. H. J., Nathans, J., van der Marel, G., van Boom, J. H. and Rich, A. (1978). Nature (London) **276**, 471.

Wang, A. H.-J., Quigley, G. J., Kolpak, F. J., Crawford, J. L., van Boom, J. H., van der Marel, G. and Rich, A. (1979). Nature (London) **282**, 680.

Wang, A. H.-J., Quigley, G. J., Kolpak, F. J., van der Marel, G., van Boom, J. H. and Rich, A. (1981). Science **211**, 171.

Westhof, E. and Sundaralingam, M. (1980). Proc. Natl. Acad. Sci. U.S.A. **77**, 1852.

Wing, R., Drew, H., Takano, T., Broka, C., Tanaka, S., Itakura, K. and Dickerson, R. E. (1980). Nature (London) **287**, 755.

Wodak, S. J. (1975). Acta Cryst. **B31**, 569.

Wu, S-M. and Bau, R. (1979). Biochem. Biophys. Res. Commun. **88**, 1435.

Zacharias, D. E. and Glusker, J. P. (1981). Manuscript in preparation.

Zacharias, D. E., Glusker, J. P., Fu, P. P. and Harvey, R. G. (1979). J. Am. Chem. Soc. **101**, 4043.

CHAPTER 18

VIRUS STRUCTURE AND CRYSTALLOGRAPHY: AN HISTORICAL PERSPECTIVE

John E. Johnson
Department of Biological Sciences
Purdue University
West Lafayette, Indiana 47907

Simple spherical and helical viruses represent an extreme example of oligomeric biological molecules. These obligate parasites are composed of protein subunits ranging in size from 17,000 to 45,000 daltons* which are distributed on a helical or icosahedral surface lattice, thus providing protection for a nucleic acid core. Such nucleoprotein complexes have been aptly termed the "threshold of life" (Fraenkel-Conrat, 1962). When isolated from the host system, viruses can be purified and studied using standard methods of biochemistry. In the highly purified state simple spherical viruses can be crystallized using well established procedures for biological molecule crystallization (McPherson, 1976). The first report of X-ray diffraction studies of viruses (Bernal and Fankuchen, 1939) occurred shortly after the first reports of protein diffraction studies. Thus, in a sense, virus crystallography has a rather long history. In this review the development of the principles governing virus quaternary structure as elucidated by electron microscopy, theoretical considerations of shell formation by identical subunits and biochemistry will be discussed as well as X-ray diffraction. Without these principles it would have been impossible to determine the virus tertiary structure using single crystal X-ray diffraction.

When addressing the topic of virus crystallography the first striking feature is the sheer size of the problem to be solved. With most simple viruses having a diameter of 300Å or greater, the most favorable packing of particles leads to unit cell dimensions of this order. In some cases axial lengths in excess of 1000Å have been reported for virus crystals (White and Johnson, 1980). The molecular weights of simple spherical viruses range from 2 to 9 million daltons leading to typical values of 600,000 to a million daltons in the crystallographic asymmetric unit when the virus particle is situated at points of lattice symmetry. Crystallographic asymmetric units containing 2 to 3 million daltons are not uncommon. Unit cells of this size and content allow measurement of 300,000 to 500,000 unique structure factors at high resolution. In a sense, virus diffraction studies are extending the method of X-ray crystallography to its apparent limit.

The second striking feature of crystalline viruses is the frequent display of extremely high order. Many virus crystals diffract X-rays to 3.0Å resolution or higher. From the earliest days of virus crystallography the goal of achieving a structure in which the polypeptide chain could be traced, nuances of protein-protein and protein-nucleic acid interactions observed, and the mechanisms of self-assembly ascertained has appeared intrinsically possible. Theoretical, experimental, and computational advances in macromolecular crystallography during the last 25 years have allowed this goal to be achieved. The structure determinations of tomato bushy stunt virus (TBSV) at 2.9Å resolution (Harrison, Olsen, Schutt, Winkler and Bricogne, 1978), southern bean mosaic virus (SBMV) at 2.8Å resolution (Abad-Zapatero, Abdel-Meguid, Johnson, Leslie, Rayment, Rossmann, Suck and Tsukihara, 1980) and satellite tobacco necrosis virus (STNV) at 4.0Å resolution (Unge, Liljas, Strandberg, Vaara, Kannan, Fridborg, Nordman and Lentz, 1980), have provided a structural basis for the chemistry, stability and self assembly properties displayed by spherical viruses (Kaper, 1975). Likewise, the high resolution single crystal study of the disk protein of tobacco mosaic virus (TMV) (Bloomer, Champness, Bricogne, Staden and Klug, 1978) and the fibre diffraction studies of TMV (Stubbs, Warren and Holmes, 1977), when combined with the comprehensive physical chemical and chemical studies of the virus (Butler and Durham, 1977) provide a picture of unprecedented detail regarding structure and assembly of a helical virus.

The earliest X-ray investigations of spherical viruses (Bernal and Fankuchen, 1939; Crowfoot and Schmidt, 1945; Carlisle and Dornberger, 1948) concentrated on the determination of unit cell parameters and crystal symmetry based on "still" and oscillation photographs of wet and dried crystals. At the time of these investigations the true nature of the virus as a nucleoprotein chemical entity was just being revealed (Stanley and Laufer, 1939); thus it was of substantial significance that Bernal and Fankuchen, (1941) could provide evidence from X-ray diffraction studies of TMV that it was composed of a regular substructure. Systematic biophysical and crystallographic investigations of the spherical turnip yellow mosaic virus (TYMV) were initiated in the late 1940s. In 1948 Bernal and Carlisle crystallized TYMV in a cubic space group thus prompting the suggestion by Hodgkin (1949) that the virus is situated at a lattice site in the crystal requiring a minimum of 12n identical subunits in the particle distributed with tetrahedral symmetry. Subsequently it was shown (Klug, Finch and Franklin, 1957) that the particle did utilize the tetrahedral symmetry of the cell. In the biophysical investigations of TYMV, Markham

*The dalton is a unit of molecular weight, the mass of a hydrogen atom.

(1951), reported a low density form of the particle with the same size and shape as the virus, but containing no RNA. He proposed from these studies that the protein was in the form of a spherical shell enclosing the nucleic acid. This model was confirmed by solution X-ray scattering studies of a number of different plant viruses (Schmidt, Kaesberg and Beeman, 1954). Crick and Watson (1956) addressed virus structure in terms of the limited genetic information available in the encapsulated nucleic acid and suggested that multiple copies of identical subunits would provide a genetically efficient way of generating a protein shell such as that proposed by Markham. To be consistent with the spherical appearance of the virus and the isometric symmetry apparently imposed by the lattice in cubic crystals of both TBSV and TYMV, Crick and Watson suggested that the capsid must have the rotational symmetry of the tetrahedron (23), octahedron (432) or icosahedron (532). These symmetries would require 12, 24 or 60 subunits, respectively, each in identical environments, a chemically satisfying result, since in the process of self assembly there are presumably certain minimal free energy interactions between subunits that repeat, leading to the closed symmetric shell. With sixty identical subunits, it is clear that an icosahedral distribution of protein would provide the largest shell for a given sized subunit and was proposed to be the likely symmetry for spherical viruses. This prediction was immediately verified as precession photographs of TBSV crystals (Caspar, 1956) provided conclusive evidence that the virus possessed icosahedral symmetry. Likewise TYMV diffraction patterns were shown to be consistent with this symmetry (Klug, Finch and Franklin, 1957).

In the early 1960s a second major advance in the understanding of virus quaternary structure occurred. High resolution electron microscopy of TYMV particles (Huxley and Zubay, 1960; Nixon and Gibbs, 1960) and chemical studies of the capsid composition (Harris and Hindley, 1961) revealed that, while the virions possessed icosahedral symmetry, there were probably 180 subunits rather than the sixty expected. The electron microscopy indicated that the subunits were grouped in pentamers at the five-fold axes and hexamers at the icosahedral three-fold axes. Such an arrangement was suggested by both Huxley and Zubay (1960) and Klug and Finch (1960). The presence of 180 subunits requires that three monomers occupy an icosahedral asymmetric unit and releases the restriction that all particles have identical environments, a suggestion that was initially dissatisfying as it seemed to violate the concept of repeated interactions leading to a minimal free energy shell. In 1962, however, the theory of quasi equivalence (Caspar and Klug, 1962) revealed that the concept used to explain the structure of TYMV was a special application of a generalized theory that allowed for shells of even larger size to be constructed of identical subunits. These shells, like that of TYMV, would be composed of hexamer and pentamer clusters of subunits which display icosahedral symmetry. The environment of subunits arranged in pentamers is quasi equivalent to that of those arranged in hexamers. It is assumed that the same chemical interaction will stabilize the pentamers and hexamers which, given the observed flexibility of biological molecules, is a reasonable assumption. The possible shells that can be constructed are determined by selection rules. Conceptually the capsid can initially be viewed as a sheet of hexamers which extend indefinitely. Selection rules determine positions at which pentamers are inserted in place of hexamers, in this sheet, thus adding a curvature to the surface. When all pentamers are appropriately inserted the shell is closed and displays exact icosahedral symmetry. The selection rules are summarized by the T number which indicates the total number of subunits required to make up the shell as 60T. Thus, TYMV, and in fact most other spherical plant viruses, are in the T=3 class. Shells with higher T number (≥ 7), possess a hand, a result which has been observed in human wart virus (Klug and Finch, 1965) and other animal viruses. To date, the theory of quasi-equivalence has adequately accounted for all spherical virus structures observed. It is interesting to note that geodesic domes designed by Buckminster-Fuller are also based on these principles (Marks, 1960). The development of this theory with contributions from biochemistry, electron microscopy and X-ray diffraction set the stage for more quantitative X-ray diffraction studies.

The first quantitative single crystal diffraction studies of a virus were performed using the cubic form of TYMV (Klug, Longley and Lieberman, 1966). Using data measured from the diffraction patterns of the F23 crystals (a=700Å) comparisons were made with calculated structure factors of various model distributions of protein leading to a detailed description of the low resolution TYMV structure. Other virus crystals that received study during this period were polio virus (Finch and Klug, 1959); southern bean mosaic virus (Magdoff, 1960); broad bean mottle virus (Finch, Lieberman and Berger, 1967); satellite tobacco necrosis virus (Fridborg, Hjerten, Hogland, Liljas, Lundberg, Oxelfelt, Phillipson and Strandberg, 1965); and tomato bushy stunt virus (Harrison, 1969). The authors of the latter two papers expressed intentions of carrying the structural studies to high resolution thus leading virus crystallography to a new and very challenging stage.

The high resolution structure solution of any crystalline biological molecule involves a number of stages including data collection, preparation of derivatives for isomorphous replacement, heavy atom position determination for the derivatives and phasing by isomorphous replacement.

A primary frustration in the early work on virus structure was resolving the individual diffraction maxima that are obtained from crystals with unit cell dimensions in excess of 300Å. Low resolution data could be effectively collected, but collection of high resolution data was impossible. Three major factors have lead to the current state of high resolution data collection from virus crystals. In 1968 Harrison reported the adaptation of Frank's focusing mirrors (Franks, 1955) for use in single crystal diffraction studies, theoretically permitting the resolution of diffraction maxima from a 1000Å lattice. This experimental design is now widely employed in virus crystallography. The second factor improving the data collection process is the wide availability of high intensity micro-focus rotating anode X-ray generators. While the rotating anode concept had been utilized for decades, the commercial production and general availability of such instruments is a more recent phenomenon. The last major event allowing the high resolution data collection from virus crystals was the rejuvenation of the screenless oscillation method of photography employing flat films (Arndt, 1969). The development of such camers was rapid, leading quickly to the availability and utilization of commercial models (Arndt, Champness, Phizackerley and Wonacott, 1973). Processing of these films involves sophisticated procedures due to the complicated pattern produced by the screenless technique. Programs for processing such films developed rapidly with many laboratories making unique contributions to the philosophy and procedures of processing. This, as well as many other aspects of oscillation photography have been compiled in the book *"The Rotation Method in Crystallography,"* (Arndt and Wonacott, 1977). Unique approaches to processing virus diffraction data have recently been reported (Rossmann, 1979; Winkler, Schutt and Harrison, 1979; Rossmann, Leslie, Abdel-Meguid and Tsukihara, 1979). All of the procedures employed for processing such films require high speed accurate film scanners and the availability of high speed computers. The technological advances in both of these areas have played a pivotal role in film processing.

The preparation of heavy atom derivatives of virus crystals has followed procedures used for derivatizing other protein crystals, usually soaking the crystals in solutions of the appropriate heavy atom derivatizing agent (Blundell and Johnson, 1976). The structural study of STNV has employed a covalently linked iodine derivative which has aided the chain tracing since the iodinated tyrosine and histidine positions were known from the sequence (Unge and Strandberg, 1979).

The determination of heavy atom positions from difference Patterson maps of virus crystals would be an enormous, if not an impossible task, without utilizing the icosahedral symmetry of the virus particle. Even a singly substituted protein site in a T=3 virus with only the five fold axis being non-crystallographic, would yield 210 heavy atom vectors. Rossmann and Blow (1962) first described a procedure useful for establishing the direction of non-crystallographic symmetry axes in biological molecules using only intensity information. Assuming the availability of the virus orientation, Argos and Rossmann (1976) developed a procedure for systematically investigating the difference Patterson of a virus derivative by verifying the heavy atom vector relationships generated by non-crystallographic symmetry. The procedure has been used in solving heavy atom derivatives of STNV (Lentz, Strandberg, Unge, Vaara, Borell, Fridborg and Petef, 1976) and SBMV (Rayment, Johnson, Suck, Akimoto and Rossmann, 1978). The heavy atom positions for TBSV were found using phases of a low resolution spherical model (Harrison, 1971). The interpretation was aided by the fact that the $PtCl_6^{2-}$ ions were clustered near the quasi-six fold axes; thus producing a large peak in the spherically phased low resolution difference Fourier. With the determination of one derivative site, other derivatives can be solved using the single-isomorphous replacement phasing (Rossmann and Blow, 1961). For SBMV other derivatives were solved at 10Å resolution (Suck, Rayment, Johnson and Rossmann, 1978) using this procedure. Following the heavy atom position determination, refinement of these positions for spherical viruses has been performed using the non-crystallographic icosahedral symmetry to constrain the atom shifts (Rossman, 1976; Harrison and Jack, 1975; Lentz, Strandberg, Unge, Vaara, Borell, Fridborg and Petef, 1976). With heavy atom positions determined and refined for multiple isomorphous derivatives, phases can be calculated and weighted using standard procedures (Blow and Crick, 1959).

Since 1962 (Rossmann and Blow, 1962) it has been recognized that the symmetry of an oligomeric arrangement of protein subunits which is not incorporated in the lattice, provides constraints on the phases of the reflections. Procedures for applying these constraints in both reciprocal space (Rossmann and Blow, 1963, 1964; Crowther, 1967, 1969) and electron density space (Main and Rossmann, 1966) were essentially *ab initio* in the early stages and were not particularly successful, although Crowther's approach was used at low resolution with both the TMV disk protein (Jack, 1973) and TBSV (Harrison and Jack, 1975). In 1974 the power of the non-crystallographic symmetry was shown to be more applicable to phase refinement than *ab initio* phasing. A single isomorphous derivative electron density map of glyceraldehyde-3-phosphate dehydrogenase was greatly improved by averaging the electron density over the 222 non-crystallographic symmetry of the oligomer (Buehner, Ford, Moras, Olsen and Rossman, 1974), allowing a chain tracing of a previously uninterpretable map. The process of electron density averaging, followed by Fourier transformation of the averaged map to yield improved phases has been shown to be equivalent to procedures employed in reciprocal space (Bricogne, 1974). This strategy for phase improvement has been employed in the high resolution structure determination of all the spherical viruses and the TMV disk protein. Computational methods for carrying out the averaging have been described by Bricogne (1976), Johnson (1978) and Nordman (1980). The first two procedures minimize the amount of computer memory required while Nordman's approach is substantially faster, but stores the entire electron density map in core. The application of molecular replacement procedures to proteins and viruses has recently been reviewed by Argos and Rossmann (1980).

The computational effort in isomorphous replacement refinement and phasing, molecular replacement averaging and Fourier transformations is substantial. The structure determination of SBMV required phasing of 276,000 unique structure factors and Fourier transformations of electron density maps containing 2.7 x 10^7 grid points in the asymmetric unit. Fourier computations of this magnitude were efficiently carried out using the fast Fourier transform algorithm (Ten Eyck, 1973). In SBMV phase refinement was achieved by averaging electron density over the ten icosahedral asymmetric units in the crystallographic asymmetric unit then calculating structure factors from the symmetrized map. The calculated phases were combined with the observed amplitudes, appropriately weighted (Sim, 1959) and a new map computed. This led to virtual convergence after two cycles at 2.8Å resolution.

It is impossible to discuss in detail the significance of the structures in a review of this nature. Harrison (1980) has summarized the recent results. In short, the structures have revealed that assembly of the T=3 viruses probably occurs through a T=1 initial framework for both TBSV and SBMV. Both viruses have very similar tertiary folds in their shell domains which are composed primarily of antiparallel β-pleated sheet. There is some evidence that STNV and possibly other plant viruses may also have the same "β barrel" fold (Argos, 1981). It is clear from the structure that the N-terminal region of the virus interacts with the RNA, but not in the "lock and key" manner in which nucleotide cofactors interact with enzymes. It appears rather that the protein conforms to the variation in RNA structure. There is substantial disorder of the nucleic acid due to the high symmetry of the capsid rendering the RNA and the last fifty residues of the capsid protein, which interacts with the RNA, invisible in both the SBMV and TBSV electron density. The combined structures of the disk protein and intact virion of tobacco mosaic virus, reveal that the interaction with nucleic acid is also through a flexible region of protein, but this does form a specific interaction with phosphates and bases of the RNA.

The future of virus crystallography will include studies of plant viruses with properties different from those whose structures have already been determined. Recent reports of viruses at early stages of crystallographic study include cowpea mosaic virus, a T=1 virus with two types of protein subunits in the capsid, (White and Johnson, 1980) belladonna mottle virus (Heuss, Rao and Argos, 1980) and Erasmus latent virus (Colman, Tulloch, Shukla and Gough, 1980) both tymo viruses and cowpea chlorotic mottle virus (Heuss, Rao and Argos, 1980) a bromo virus. Animal viruses such as poliovirus and rhino virus, both of which have been crystallized, will undoubtedly be studied in the near future. Low resolution structural studies of the oncogenic viruses polyoma (Adolph, Caspar, Hollingshead, Lattman, Phillips and Murakami, 1979) and simian virus 40 (Lattman, 1980) are currently under way. In addition, the constituent proteins of more complex viruses have been isolated and crystallized. Recently the structure of the surface glycoprotein of influenza virus was determined using this procedure (Wilson, Skehel and Wiley, 1980).

In the last two years the high resolution structure determination of viruses has provided great insight into the process of assembly, capsid stability and the structural relationship between viruses. With more virus structures becoming available in the near future, it is a certainty that molecular virology will have a firm structural foundation.

References

Abad-Zapatero, C., Abdel-Meguid, S., Johnson, J. E., Leslie, A., Rayment, I., Rossmann, M. G., Suck, D. and Tsukihara, T. (1980). Nature (London) **286**, 33-39.

Adolph, K. W., Caspar, D. L. D., Hollingshead, C. J., Lattman, E. E., Phillips, W. C. and Murakami, W. T. (1979). Science **203**, 1117-1119.

Argos, P. (1980). Virology, **110**, 55-62.

Argos, P. and Rossmann, M. G. (1980). In "Theory and Practice of Direct Methods in Crystallography," edited by M. F. C. Ladd and R. A. Palmer. New York: Plenum Publishing Corporation.

Argos, P. and Rossmann, M. G. (1976). Acta Cryst. **A32**, 2975-2979.

Arndt, U. W. (1969). Acta Cryst. **B24**, 1355-1357.

Arndt, U.W., Champness, J. N., Phizackerley, R. P. and Wonacott A. J. (1973). J. Appl. Cryst. **6**, 457-463.

Arndt, U. W. and Wonacott, A. J. (1977). The Rotation Method in Crystallography, Amsterdam: North Holland Publishing Co.

Bernal, J. D. and Carlisle, C. H. (1948). Nature (London) **162**, 139-140.

Bernal, J. and Fankuchen, I. (1939). Nature (London) **139**, 923-924.

Bernal, J. D. and Fankuchen, I. (1941). J. Gen. Physiol. **25**, 147-165.

Bloomer, A. C., Champness, J. N., Bricogne, G., Staden, R. and Klug, A. (1978). Nature (London) **276**, 362-368.

Blow, D. M. and Crick, F. (1959). Acta Cryst. **12**, 794-802.

Blundell, T. L. and Johnson, L. N. (1976). Protein Crystallography. London: Academic Press.

Bricogne, G. (1974). Acta Cryst. **A30**, 395-405.

Bricogne, G. (1976). Acta Cryst. **A32**, 832-847.

Buehner, M., Ford, G. C., Olsen, K. W., Moras, D. and Rossmann, M. G. (1974). J. Mol. Biol. **82**, 563-585.

Butler, P. J. G. and Durham, A. C. H. (1977). Advances in Protein Chemistry **31**, 187-251.

Carlisle, C. H. and Dornberger, K. (1948). Acta Cryst. **1**, 194-196.

Caspar, D. L. D. (1956). Nature (London) **177**, 475-476.

Caspar, D. L. D. and Klug, A. (1962). Cold Spring Harbor Symp. Quant. Biol. **27**, 1-24.

Colman, P. M., Tulloch, P. A., Shukla, D. D. and Gough, K. H. (1980). J. Mol. Biol. **142**, 263-268.
Crick, F. H. C. and Watson, J. D. (1956). Nature (London) **177**, 473-475.
Crowfoot, D. and Schmidt, G. M. J. (1945). Nature (London) **155**, 504-505.
Crowther, R. A. (1967). Acta Cryst. **22**, 758-764.
Crowther, R. A. (1969). Acta Cryst. **B25**, 2571-2580.
Fraenkel-Conrat, H. (1962). Design and Function at the Threshold of Life: The Viruses. New York: Academic Press.
Finch, J. T. and Klug, A. (1959). Nature (London) **183**, 1709-1714.
Finch, J. T., Lieberman, R. and Berger, J. E. (1967). J. Mol. Biol. **27**, 17-24.
Franks, A. (1955). Proc. Phys. Soc. (London) **B68**, 1054-1064.
Fridborg, K., Hjerten, S., Hogland, S., Liljas, A., Lundberg, B. K. S., Oxelfelt, P., Phillipson, L. and Strandberg, B. (1965). Proc. Natl. Acad. Sci., U.S.A. **54**, 513-521.
Harris, J. and Hindley, J. (1961). J. Mol. Biol. **3**, 117-120.
Harrison, S. C. (1968). J. Appl. Cryst. **1**, 84-90.
Harrison, S. C. (1969). J. Mol. Biol. **42**, 457-483.
Harrison, S. C. (1971). Cold Spring Harbor Symp. Quant. Biol. **36**, 495-501.
Harrison, S. C. (1980). Nature (London). **286**, 558-559.
Harrison, S. C., Olsen, A. J., Schutt, C. E., Winkler, F. K. and Bricogne, G. (1978). Nature (London) **276**, 368-373.
Harrison, S. C. and Jack, A. (1975). J. Mol. Biol. **97**, 173-191.
Heuss, K., Mohana Rao, J. K. and Argos, P. (1980). J. Mol. Biol., in press.
Heuss, K., Mohana Rao, J. K. and Argos, P. (1980). J. Mol. Biol., submitted.
Hodgkin, D. C. (1949). Cold Spring Harbor Symp. Quant. Biol. **14**, 65-78.
Huxley, H. and Zubay, G. (1960). J. Mol. Biol. **2**, 189-196.
Jack, A. (1973). Acta Cryst. **A29**, 545-554.
Johnson, J. E. (1978). Acta Cryst. **B34**, 576-577. (Appendix II).
Kaper, J. M. (1975). The Chemical Basis of Virus Structure, Dissociation and Reassembly. Amsterdam: North Holland Publishing Company.
Klug, A., Finch, J. and Franklin, R. (1957). Biochim. Biophys. Acta **25**, 242-252.
Klug, A. and Finch, J. (1960). J. Mol. Biol. **2**, 201-215.
Klug, A. and Finch, J. (1965). J. Mol. Biol. **11**, 403-423.
Klug, A., Longley, W. and Lieberman, R. (1966). J. Mol. Biol. **15**, 315-343.
Lattman, E. E. (1980). Science **208**, 1048-1050.
Lentz, P. J., Strandberg, B., Unge, T., Vaara, J., Borell, A., Fridborg, K. and Petef, G. (1976). Acta Cryst. **B32**, 2979-2983.
Magdoff, B. (1960). Nature (London) **185**, 673-674.
Main, P. and Rossmann, M. G. (1966). Acta Cryst. **21**, 67-72.
Markham, R. (1951). Discuss. Faraday Soc. **11**, 221-227.
Markham, R. and Smith, K. M. (1949). Parasitology **39**, 330-342.
McPherson, A., Jr. (1976). In "Methods of Biochemical Analysis" edited by D. Glick **23**, 249-345.
Nixon, H. and Gibbs, A. (1960). J. Mol. Biol. **2**, 197-200.
Nordman, C. E. (1980). Acta Cryst. **A36**, 747-754.
Rayment, I., Johnson, J. E., Suck, D., Akimoto, T. and Rossmann, M. G. (1978). Acta Cryst. **B34**, 567-578.
Rossmann, M. G. (1976). Acta Cryst. **A32**, 774-777.
Rossmann, M. G. (1979). J. Appl. Cryst. **12**, 225-238.
Rossmann, M. G. and Blow, D. M. (1961). Acta Cryst. **14**, 1195-1202.
Rossmann, M. G. and Blow, D. M. (1962). Acta Cryst. **15**, 24-31.
Rossmann, M. G. and Blow, D. (1963). Acta Cryst. **16**, 39-45.
Rossmann, M. G. and Blow, D. (1964). Acta Cryst. **17**, 1474-1475.
Rossmann, M. G., Leslie, A. W., Abdel-Meguid, S. S. and Tsukihara, T. (1979). J. Appl. Cryst. **12**, 570-581.
Schmidt, P., Kaesberg, P. and Beeman, W. W. (1954). Biochim. Biophys. Acta **14**, 1-11.
Sim, G. A. (1959). Acta Cryst. **12**, 813-815.
Stanley, W. M. and Laufer, M. A. (1939). Science **89**, 345-347.
Stubbs, G., Warren, S. and Holmes, K. (1977). Nature (London) **267**, 216-221.
Suck, D., Rayment, I., Johnson, J. E. and Rossmann, M. G. (1978). Virology **85**, 187-197.
Ten-Eyck, L. F. (1973). Acta Cryst. **A29**, 183-191.
Unge, T. and Strandberg, B. (1979). Virology **96**, 80-87.
Unge, T., Liljas, L., Strandberg, B., Vaara, I., Kannan, K. K., Fridborg, K., Nordman, C. E. and Lentz, P. J., Jr. (1980). Nature (London) **285**, 373-377.
White, J. and Johnson, J. E. (1980). Virology **101**, 319-324.
Wilson, I. A., Skehel, J. J. and Wiley, D. C. (1980). In "Structure and Variation in Influenza Virus. Edited by T. Laver and G. Air. New York: Elsevier, North Holland Publishing Co.
Winkler, F. K., Schutt, C. E. and Harrison, S. C. (1979). Acta Cryst. **A35**, 901-911.

INDEXES

These indexes were prepared at the Institute for Cancer Research, Philadelphia, PA with the assistance of H. Steinfink, E. Pytko, M. H. Carrell, M. J. Glusker, J. Hurley and C. Zobel.

Subject Index

SUBJECT INDEX

SUBJECT INDEX

SUBJECT INDEX

SUBJECT INDEX

Name Index

NAME INDEX

NAME INDEX

NAME INDEX

Name Index

NAME INDEX

Name Index

NAME INDEX